A Practical Guide
To Intensity-Modulated Radiation Therapy

A Practical Guide
To Intensity-Modulated Radiation Therapy

Published in cooperation with members of the staff
of Memorial Sloan-Kettering Cancer Center

With a Foreword by

Samuel Hellman, M.D.
University of Chicago

Medical Physics Publishing
Madison, Wisconsin

Printed in the United States of America

09 07 05 03   4 3 2 1

Library of Congress Catalog Card Number: 2003102324

ISBN 1-930524-13-7

Medical Physics Publishing
4513 Vernon Blvd.
Madison, WI 53705-4964
1-800-442-5778
1-608-262-4021
Fax: 1-608-265-2121
Web: www.medicalphysics.org
E-mail: mpp@medicalphysics.org

Contents

Preface

According to a Collaborative Working Group sponsored by the National Cancer Institute, "Intensity modulated radiotherapy (IMRT) . . . represents one of the most important technical advances in radiation therapy since the advent of the medical linear accelerators." In the last ten years, the clinicians and scientists at Memorial Sloan-Kettering Cancer Center have been fortunate to be involved in the development and implementation of this advanced form of radiotherapy. Indeed, it was in part serendipity that several key components of the IMRT process became available at MSKCC that permitted us to implement IMRT treatment with the use of dynamic multileaf collimation (DMLC) in 1995. Since then, we have amassed a body of technical and clinical experience in the use of this modality.

This book is an attempt to provide an account of our perspective, methodology, and experience in the physical and medical aspects of IMRT. For the medical physicists, there are detailed discussions on the inverse method, the commissioning and acceptance testing of DMLC, dose calculation and independent monitor unit (MU) check. For the radiation oncologists and clinical physicists, the relevant material includes treatment planning, quality assurance protocols, disease-specific treatment procedures, and emerging clinical outcome data. In addition, advanced topics relevant to IMRT are addressed such as the estimation of tumor control and normal tissue complication probabilities, the quantification and minimization of treatment uncertainties, the use of respiration-controlled techniques, and the emerging importance of biological imaging. Given the enthusiasm for the potential benefits of this modality, we hope this book will be useful for our colleagues in the radiation oncology community interested in applying IMRT.

Lastly, we wish to acknowledge colleagues who in the past have contributed to our IMRT program. These include John Laughlin, Jerry Kutcher, Radhe Mohan, Thomas Bortfeld, among others.

Zvi Fuks
Steven A. Leibel
C Clifton Ling
New York, New York
January 2003

List of Authors

C Clifton Ling, Ph.D.[1]
Steven A. Leibel, M.D.[2]
Zvi Fuks, M.D.[2]
John L. Humm, Ph.D.[1,3]
Jason A. Koutcher, M.D., Ph.D.[1,3]
Hedvik Hricak, M.D., Ph.D.[3]
Howard I. Amols, Ph.D.[1]
Spiridon V. Spirou, Ph.D.[1]
Chen-Shou Chui, Ph.D.[1]
Thomas J. LoSasso, Ph.D.[1]
Asa Palm, Ph.D.[1]
Margie A. Hunt, M.S.[1]
Chandra M. Burman, Ph.D.[1]
Michael J. Zelefsky, M.D.[2]
Lanceford M. Chong, M.D.[2]
Kamil Yenice, Ph.D.[1]
Suzanne Wolden, M.D.[2]
Linda X. Hong, Ph.D.[1]
Beryl McCormick, M.D.[2]
Kenneth Rosenzweig, M.D.[2]
Kaled Alektier, M.D.[2]
Jie Yang, Ph.D.[1]
Andrew Jackson, Ph.D.[1]
Ellen D. Yorke, Ph.D.[1]
Gikas S. Mageras, Ph.D.[1]
Jenghwa Chang, Ph.D.[1]
D. Michael Lovelock, Ph.D.[1]
Jean St. Germain, Ph.D.[1]
James Mechalakos, Ph.D.[1]

Departments of Medical Physics[1], Radiation Oncology[2] and Radiology[3]
Memorial Sloan Kettering Cancer Center
New York, New York

Foreword

The adoption of intensity-modulated radiation therapy (IMRT) is a major shift in the practice of radiation oncology, perhaps the most important one since the development of gantry-mounted linear accelerators. Supervoltage energy combined with gantry mounting providing the ability to use both fixed and moving radiation sources was a quantum leap over the previously available fixed source to skin treatment from limited directions. These advances were due to both the more penetrating nature of supervoltage radiation and its skin sparing as compared to fixed mounted orthovoltage machines. This technology made rotational and arc therapy available and allowed the patient to be treated from many directions while in a single position. It was these improvements that ushered in current radiotherapy practices.

A longstanding goal of radiation therapy has been to make the radiation dose conform as closely as possible to the target volume while minimizing the dose to transited normal tissues. IMRT is the next major saltation in the quest for such a treatment. This method has come about because of advances in radiation physics, collimators, computers, software, and the armamentaria available for delivering radiation treatment. Perhaps not so clearly realized is the expanding revolution in medical imaging. It is cross-sectional tomographic and three-dimensional imaging using X-ray and magnetic resonance imaging (MRI) that allow much improved delineation of the tumor and the dose limiting normal tissues. Positron emission tomography (PET) and magnetic resonance spectroscopy (MRS) provide additional information important to tumor identification, localization, and, most intriguingly, the ability to anatomically evaluate tumor physiology. The increased diagnostic imaging ability coupled with IMRT has allowed conforming of radiation dose to the target volume to become a reality. Further, rather than continue pursuing homogeneity of dose within the target volume, radiation dose can now be made to conform to localized tumor using physiology as a guide; for example, raising the dose administered to regions of hypoxia or varying it consistent with tumor cell concentration. It can be modified during a course of multifraction radiation therapy in response of these variables to treatment.

There is an old adage cautioning one to be careful of what one wishes for, because the wish might be granted. Such may be the case with IMRT. We now are able to conform the radiation to the target volume: we are even able to sculpt the dose within the target volume consistent with our desires while minimizing unwanted radiation—or at least—placing it in the least damaging anatomic locations. But in order to do so we must be much more confident of the tumor location. The target definition needs to be accurately arrived at since we are now able to reduce dose at the edges of the tumor quite abruptly with this technique. Since it is characteristic of cancer to have subclinical disease beyond the gross tumor margins, we must either improve our diagnostic imaging techniques to visualize such subclinical disease or at least give us an estimate of tumor cell concentration and the rate at which this diminishes at the tumor periphery. With such information the shape of the dose distribution surrounding the high dose volume can be appropriately configured. The optimal shape of the dose fall-off could then be made to conform to the tumor cell concentration. As with previous methods, it is also determined by the proximity of dose-limiting normal tissues as well as by knowledge of the natural history of the tumor.

Radiation dose delivered to transited normal tissues is inherent in all radiation therapy for internal cancers. This unwanted radiation must be deposited consistent with the nature of the toxicity associated with irradiating the specific tissue. This is dependent on the dose, dose rate, the beam arrangement, and the resulting volume of the specific tissue irradiated to high doses. For the same beam arrangement, the integral dose of IMRT is no different from that of non-intensity-modulated 3DCRT plans, or that of simpler plans with similar radiation path-lengths. However, the ability to place and shape the unwanted radiation is far greater with IMRT and may result in dose deposition in normal tissues different from that of past practices. Addressing unfamiliar dose distributions encountered in IMRT may become important in treatment planning, and organ response to large volume moderate dose radiation needs to be carefully quantified.

In order to use IMRT properly not only must we have confidence as to the tumor location but we must also be confident of the position of the patient as well as the target volume during treatment. Immobilization of the patient, the use of fiducial-markers, real-time monitoring of the tumor location as well as gating of the treatment to breathing or other physiologic functions such a bowel gas movement are important if we are to successfully use intensity-modulated radiation therapy. There is far less margin for error in patient setup and tumor or organ motion with this method. Ideally, the technique should have imaging feedback at regular intervals during the treatment with the hope that eventually one might have real-time feedback and control of patient and tumor position, i.e., cybernetic radiation therapy.

Dose sculpting within the target volume requires that PET scanning and MRS be anatomically fused with computed tomography (CT) and MRI. Despite data being presented and evaluated as two-dimensional images, treatment is three-dimensional, requiring that the target volume be visualized this way. Once a three-dimensional presentation of the target volume and adjacent normal tissues is realized, then a vast increase in therapeutic alternatives might be achieved with non-coplanar field arrangements.

IMRT offers the opportunity to reconsider radiation fractionation since individual fraction sizes can be greatly increased using this technique. It may well be true that different types of tumors require very different fractionation schemes to achieve optimal benefit. The devices for delivering the radiation dose also need to be reevaluated. Gantry-mounted linear accelerators with dynamic multileaf collimators currently are the most commonly used technique, but there are others, either presently in use or in development. It will be important to determine what the appropriate armamentaria should be for an effective radiation oncology practice. These considerations will depend upon opportunities provided for off-axis treatment, reduced treatment time, increased flexibility, and the facility to use images in both treatment planning and treatment monitoring in real or close to real time.

Considerations of radiation dose have conventionally been arrived at more often by the tolerance of the normal tissues than by determining a dose consistent with the high probability of curing the tumor. The proper dose to maximize tumor control is now being reconsidered. Dose escalation is now possible with IMRT. While this offers opportunities for improved tumor cure, it also increases the considerable risks of damage to the matrix normal tissue in which the tumor resides. Damage to these normal tissues rather than those outside the target volume becomes the major limitation on the dose delivered.

Even the current cancer paradigms must be revisited. It is the accepted premise that it is not possible to visualize all of the cancer. Should this concept continue to underlie therapeutic planning as we greatly reduce the uncertainties of tumor extension with evolving imaging methods? Perhaps most important is reconsideration of the notion that metastases are always in great number and widely scattered throughout the body. With early diagnosis and current imaging techniques, it may well be that there are circumstances where the number of metastases are relatively limited and such oligometastases are restricted to one of the major sites of metastases: the brain, bone, lung, or liver. If we can identify such patients then perhaps these oligometastases can be individually treated with the goal of cure, or at least, long-term control. With the increasing effectiveness of systemic treatment, the oligometastatic state may also be achieved by the success of systemic treatment destroying small tumor volumes, leaving only a limited number of anatomically definable larger metastases. These residual tumors could be eradicated by targeted radiation delivery, perhaps delivered with carefully limited IMRT using hypofractionation with very precise immobilization and tumor conformality.

So what we have been wishing for is becoming a reality: one may be able to offer real improvement in cancer care if precise and accurate radiation therapy is combined with improved knowledge of tumor location and extent or conversely such treatment may result in increased marginal recurrences because of improper assessment of the target volume. In order for this approach to be successful we must extract all we can from diagnostic imaging methods both before and during a course of radiation therapy. Despite the clear decrease in the acute toxicity of IMRT because of the improved sparing of transited normal tissues it is not yet clear that there may not be increased toxicity due to increased dose administered to normal structures within the target volume. Nor is it clear what the effects of different dose distributions on normal tissue function will be or whether the increase in normal tissue volume receiving moderate radiation dose due to dose-escalation will increase the development of second tumors. These opportunities and uncertainties require that IMRT be used with care and based on a carefully documented experience.

Memorial Sloan-Kettering Cancer Center has been one of the pioneers of conformal radiation therapy and intensity-modulated radiation therapy. This book offers the reader an opportunity to consider how IMRT is currently used in clinical practice at that institution. While the IMRT technique is evolving, it is important for radiation oncologists to learn from the experience of these investigators so as to be most effective in bringing this new technique to practice. Consider this "A User's Guide to Intensity-Modulated Radiation Therapy."

Samuel Hellman, M.D.
University of Chicago
Chicago, Illinois

January 2003

1

Imaging For IMRT

C. Clifton Ling
Steven Leibel
Zvi Fuks
John Humm
Jason Koutcher
Hedvig Hricak

Introduction

A century has passed from the discovery of X-rays by Roentgen, and of radium by Curie, to the clinical implementation of intensity-modulated radiation therapy (IMRT). Over this time, the use of radiation for disease diagnosis and treatment has steadily matured and become extremely effective. From the very beginning radiotherapy has been image based, and, in fact, initially the same x-ray equipment was used for both diagnosis and therapy. Subsequent advances in imaging technology have resulted in improvement in radiotherapy planning. For example, the introduction of fluoroscopy led to its use in radiotherapy simulators and the invention of computed tomography (CT) ushered in the era of three-dimensional (3-D) anatomy-based radiotherapy. The development of image-based computer treatment planning systems during the 1980's allowed CT image data to be incorporated into radiotherapy treatment plans, heralding 3-D conformal radiation therapy (3-D CRT) (Ling and Fuks 1995). With the recent advances in biological imaging, the nature of the "image" is evolving and there are expectations that its application will lead to further improvement in cancer radiotherapy.

During the last twenty years, with the emergence of commercial 3-D treatment planning systems and CT simulators, 3-D CRT has become the standard of radiotherapy. In 3-D CRT, the target and non-target structures are delineated from patient-specific 3-D image data sets [primarily CT, more recently often supplemented with magnetic resonance imaging (MRI), positron emission tomography (PET), and ultrasound]. The treatment portals are determined within beam's eye view (BEV). From these inputs, the planned dose distributions are then calculated and displayed by powerful workstations (Goitein et al. 1983; McShan, Fraass, and Lichter 1990). Increasingly, the evaluation of radiotherapy treatment plans includes the analysis of structure-specific dose-volume data [dose volume histogram (DVH)] and consideration of biological indices, e.g., tumor control probability (TCP) and normal tissue complication probability (NTCP). Biological indices are surrogate estimations of treatment outcome: local TCP and NTCP, respectively. In the delivery of

3-D CRT, computer-controlled radiation machines, multileaf collimators (MLCs), and treatment verification with electronic portal images, are increasingly used (Ling and Fuks 1995).

Within the past decade, IMRT has emerged, leading to further improvement in dose distribution conformality (Brahme 1988; Carol et al. 1996; Spirou and Chui 1998; Ling et al. 1996). In addition, IMRT has the potential of reducing the manual iterative and time-consuming steps in the treatment design phase of 3-D CRT, particularly when the treatment design is complex. In brief, IMRT differs from 3-D CRT in two key elements: the use of intensity-modulated radiation beams (Burman et al. 1997), and computerized iterative plan optimization. In terms of delivery methods, intensity modulated radiation fields at fixed gantry angles can be delivered with MLCs, in either the static or dynamic mode (Spirou and Chui 1994; LoSasso, Chui, and Ling 1998), or using the tomotherapy approach with beams directed from 360° and modulated either with a slit MLC or several MLC-shaped fields (NCI 2001; Yu 1995).

The improvement in the dose distribution conformality of IMRT brings new challenges in the planning and verification of radiation delivery, and offers opportunities for exploiting biological imaging. The following sections discuss some aspects of this ongoing synthesis of advances in what some would say is "physics meets biology."

Anatomical Imaging with CT and MRI

CT has been used in radiotherapy treatment planning for more than twenty years. However, some recent developments are important for IMRT. To produce high-resolution sagittal or coronal images, and digitally-reconstructed radiographs (DRRs), many contiguous CT slices with an inter-slice spacing of 3 to 5 millimeters (mm) are needed. In this regard, the introduction of the so-called *spiral*, or (more appropriately) *helical*, scanners is important. This advance was made possible by the slip ring technology, with continuous electrical connections between the rotating and the fixed components of the scanner. Volumetric scan data is acquired by having the table move continuously relative to the gantry, concomitant with the continuous rotation of the x-ray tube and detector array around the patient. Recently, multislice helical CT has become available, providing high image quality at a fraction of the previously required time. This permits acquiring a complete data set for the thorax in a few seconds, allowing the lungs to be scanned during one breath hold. This results in high-resolution images devoid of motion artifact.

The advent of high-speed helical scanners has led to the so-called *CT simulation*, which, by combining the simulation and CT imaging sessions into one, can lead to greater operation efficiency (Sherouse et al. 1990). The CT-simulator combines the capabilities of spiral (or helical) scanners for volumetric data acquisition with the computer software required to assist in the process of treatment planning. This permits the so-called *virtual simulation*, carried out on a computer workstation using the 3-D CT data set as the virtual patient. The CT data, acquired with the patient in the treatment position, can be reconstructed and displayed in BEV and *simulation* fluoroscopic and radiographic images digitally generated for viewing, decision-making, and documentation. Aside from improved efficiency, CT simulation eliminates or minimizes systematic uncertainties in the registration of simulation films to CT data sets, and in set-up errors when transferring the patient from one mechanical co-ordinate system to another. Taking these factors together, CT simulation is considered an important ingredient of IMRT.

The improved soft tissue contrast, relative to CT, underlies the usefulness of MRI as a vital modality for tumor delineating for radiotherapy planning. MRI by itself may not be sufficient for treatment planning purposes since it does not provide electronic density data needed for dose calculations. However, by spatially registering the images from the two modalities, the combination of MRI image detail and x-ray attenuation data from CT provides a powerful planning tool. Combined MRI/CT images are increasingly and effectively used in the treatment planning of brain, head and neck, and pelvic diseases.

Biological Imaging

Thus far, primarily anatomical or structural data from CT and MRI images are used in radiotherapy treatment planning. However, recent advances based on increased understanding of cellular and molecular processes and novel approaches in imaging are providing new types of images. These images yield biological and mechanistic data, for example, metabolic information from PET scanning with FDG (fluorodeoxyglucose radiolabeled with ^{18}F), functional/metabolic data from magnetic resonance spectroscopy (MRS), and gene expression using molecular imaging (Blasberg and Gelovani 2002; Weissleder et al. 2000; Urbain 1999; Ling et al. 2000). Indeed, there is a wide spectrum of information that the *biological* (in contrast to anatomical) imaging techniques can unfold, including metabolic, biochemical, physiological, functional, and molecular (genotypic and phenotypic). For radiation therapy, images that give information about factors (e.g., tumor hypoxia, tumor burden) that influence radiosensitivity and treatment outcome can be regarded as radiobiological images (Ling et al. 2000).

Biological imaging encompasses a wide range of topics and is under rapid development, Therefore, a detailed account on this topic would be difficult and rapidly out of date. Instead, in the following sections we shall describe several topics of particular interest for application to IMRT. In addition, we shall briefly summarize other areas of ongoing investigation of relevance to radiotherapy.

Incorporation of FDG-PET in Radiotherapy Treatment Planning

The increased glucose metabolism of cancer cells, as compared to normal tissues, underlies the enhanced uptake of FDG in malignant growth. Clinical studies of many disease sites, including brain, breast, head and neck, lung, colorectal, and ovarian, have shown that PET imaging with FDG has the potential to improve the detection, staging, treatment design and evaluation (Scheidhauer et al. 1996; Rigo et al. 1996; Utech, Young, and Winter 1996; Avril et al. 1997; Brock, Meikle, and Price 1997). The efficacy of FDG-PET in diagnosing and staging thoracic lesions is now well established by a large number of clinical studies (MacManus et al. 2001).

For radiotherapy treatment planning for lung cancers, there are now numerous reports indicating that CT is inadequate in defining the gross target volume (GTV), and that PET is needed to more fully delineate the entire tumor. In a retrospective study, Munley et al. (1999) reported that in 12 of 35 patients PET resulted in enlarging portions of the beam aperture up to 15 mm. Similarly, in the retrospective analysis of Nestle et al. (1999) 12 of 34 patients would have altered treatment portals due to PET. Vanuytsel et al. (2000) compared CT data, and CT + PET data, with the surgical assessment of 988 lymph nodes in 73 patients. Adequate tumor coverage occurred 75% when CT was used alone, 89% when PET was added. In 45 patients (62%), PET altered the treatment volume. In a prospective study, Erdi et al. (2002) combined PET images with CT data in 11 non small cell lung cancer (NSCLC) patients for treatment planning purposes and reported increases in the planning target volume (PTV) of 7 patients, with an average increase of 19%. For the other 4 patients, their PTVs decreased by an average of 18%. Finally, in a detailed study by Mah et al. (2002) the treatment volumes of 30 patients were outlined by three radiation oncologists using combined CT and PET data. As a result of the addition of PET data, the treatments in 7 of the 30 patients were changed to palliative. In 5 of the remaining 23 patients, PET-positive nodes were discerned within 5 centimeters (cm) of primary as defined by CT. CT-derived PTV would have led to 17 to 29% marginal misses based on combined data. Both PTV reduction (24–70%) and increase (30–76%) were observed, in spite of interobserver differences.

An important recent development is the commercial availability of combination PET/CT units, in which both instruments are housed in a common gantry, thus allowing patients to be scanned in the same position and imaging session (Townsend and Cherry 2001). This will permit accurate anatomical localization of functional PET tracers, which is a prerequisite of modern radiotherapy treatment planning. The CT serves two functions in these imaging devices. First, the CT is used to perform attenuation correction of the PET emission data. This substantially speeds up the duration of a PET scan by approximately 30%, relative to dedicated PET scanners utilizing transmission rod sources. Second, it allows the patient data sets to be automatically fused by the application of a simple coordinate transformation given by the displacement between the two isocenters of the respective imaging units.

Just as the CT scanner became adapted to the needs of the radiation oncologist and led to the CT simulator, efforts are ongoing to create the PET/CT simulator. One of the drawbacks of the PET/CT is the duration of the PET exam, (currently about 30 to 40 minutes whole body) and the width of the gantry aperture, which places constraints on the size of the patient who can be scanned. However, improvements in scanner and detector design, leading to greater sensitivity, performance, and throughput, hold the promise of reducing PET scan times down to <10 minutes.

MRSI and Functional Imaging for IMRT Planning

Magnetic resonance spectroscopy/imaging (MRSI) is a powerful approach to provide biological information associated with different biomolecules. Proton (^{1}H) spectroscopy is particularly attractive in terms of sensitivity, spatial resolution, signal to noise, and acquisition time. Molecules that can be studied with ^{1}H spectroscopy include water, lipids (which are suppressed in most studies), choline, citrate, lactate, and creatine. In the prostate, an elevated choline level relative to that of citrate (choline/citrate ratio of 2.1±1.3) has been hypothesized as an indicator of active tumor (Kurhanewicz et al. 1995). In a study comparing 3-D MRSI data with step-section histopathologic examination after radical prostatectomy, Scheidler et al. (1999) concluded that "high specificity (up to 91%) was obtained when combined MR imaging and 3-D MR spectroscopic imaging indicated cancer." Based on this information, the investigators at University of California San Francisco (UCSF) have suggested the use of patient-specific MRSI parametric images in IMRT planning to deliver a higher dose to those regions with a higher than normal choline/citrate ratio (Pickett et al. 1999). A more recent paper by Xia et al. (2001) of UCSF extended this study to show feasibility of IMRT to deliver a dose of 90 gray (Gy) to two dominant intraprostatic lesions, while treating the entire prostate to 73.8 Gy and keeping the dose to the rectum and bladder below tolerance.

MRSI studies of choline/citrate in prostatic cancers are being conducted at Memorial Sloan-Kettering Cancer Center (MSKCC). In one important study, abnormal choline/citrate levels identified by MRSI are being correlated with results from whole-mount pathological examinations to evaluate MRSI as a measure of tumor characteristics (e.g., Gleason score). In addition, the tumor extent as assessed by MRSI is compared to that determined pathologically. Preliminary results from these studies indicate that MRSI data can identify and localize regions of high Gleason scores (≥8). Given our recent findings that high-risk patients may require >90 Gy to optimize local control (Levegrun et al. 2002) and that high Gleason is one characteristic of high-risk patients, we are evaluating the possibility of MRSI-based dose painting using IMRT. Our goal would be to deliver 90 Gy to regions of high Gleason score (GTV) as identified by MRSI, while covering the entire PTV with 81 Gy.

Our dose painting study of prostate cancer, however, will only be initiated after the development of certain technical capability. First, since the radiofrequency (RF) applicator inserted into the rectum for MRSI data acquisition substantially distorts the anatomy, image correlation for treatment planning must use deformable object registration to take into account the differences in the shapes of the prostate between the MRSI and CT images (figure 1–1). Second, treatment uncertainty is extremely important for dose painting of small volumes, and we must assess the effect of treatment uncertainty on the dose distribution achieved by dose painting.

Another clinical application of nuclear magnetic resonance (NMR) technology to IMRT planning involves the mapping of brain function [functional MRI (fMRI)]. One type of fMRI is based

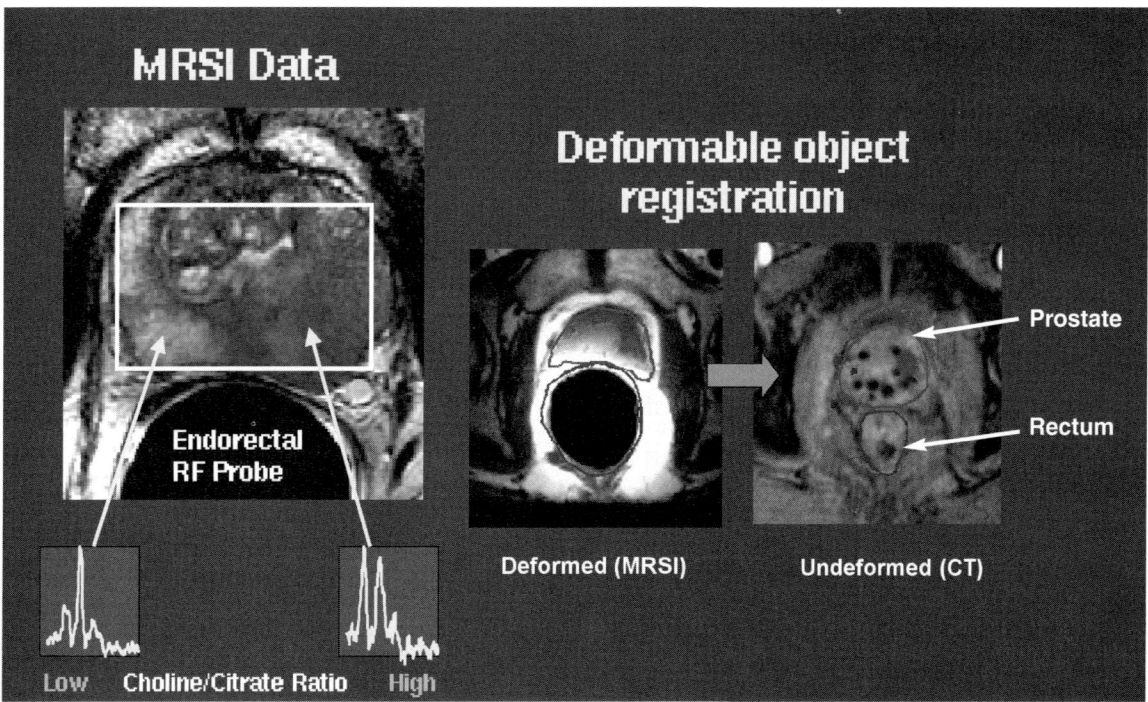

FIGURE 1–1. The left panel shows magnetic resonance spectroscopy and imaging data; the color coding is based on the respective levels of choline and citrate (based on the NMR spectra below), with red indicating a high level of the former and high probability of aggressive tumor. Note the distention of the rectum due to the endorectal RF probe. The right panels illustrate that deformable object registration is needed to map the region of suspected tumor from the MRSI data to the CT scans; note the shape of the distended prostate (red contours) and of the rectum (blue contours) in the two images.See COLOR PLATE 1.

on the paramagnetic property of deoxygenated hemoglobin that induces local magnetic inhomogeneities, enhances the relaxation of the adjacent water molecules, and thereby decreases the T2 signal adjacent to the blood vessels. Task activation stimulates the brain and increases blood flow and oxygen to the activated region, resulting in a higher oxyhemoglobin level and increased T2 signal. Such *blood oxygenation level-dependent* (*BOLD*) changes have been observed during sensory stimulation, manual tasks, and a wide variety of other forms of activation (Kim, K., et al. 1997; Kim, S., et al. 1993; Belliveau et al. 1991). An important application of this noninvasive procedure for surgical and radiation oncology is in conformal avoidance of critical areas of the brain to minimize the possibility of loss of critical functions (Fried et al. 1995; Hamilton et al. 1997). The potential use of fMRI data for radiotherapy treatment planning with function-specific dose-volume histograms has been described by Hamilton et al. (1997).

Imaging of Tumor Hypoxia

Results of recent clinical trials have clearly demonstrated correlations between hypoxia and radiocurability in cervical cancer and in both metastasis containing lymph nodes in head and neck cancer (Hockel et al. 1996a,b; Nordsmark, Overgaard, and Overgaard 1996). Even among the patients who underwent surgery for cancer of the cervix, survival and relapse free survival were poorer for those with hypoxic tumors (Hockel et al. 1996b). And, in high-grade soft tissue sarcomas, there was an association between tumor hypoxia and the development of metastases (Brizel et al. 1996). In addition, hypoxia-inducible factor 1α (HIF-1α), a transcription factor that regulates genes involved in adaptation to hypoxia leading to tumor angiogenesis and progression, has been shown to be substantially up-regulated in hypoxic human prostate cancer cells (Zhong et al. 1998). Thus, in addition to radioresistance, hypoxia may be associated with a more aggressive tumor phenotype that is more likely to metastasize. Accordingly, it appears desirable to identify and localize hypoxic cells in tumors so that they may be specifically targeted with additional therapy.

One promising approach to identify and localize hypoxic tumor cells with spatial precision comparable to that of therapeutic irradiation achievable with IMRT is to combine high-resolution PET and hypoxia-specific positron-emitting radiotracers. In this regard, a number of radiotracers for viable hypoxic cells in solid tumors have been developed, and several of these are now being evaluated clinically (Chapman et al. 2001). The initial imaging studies of tumor hypoxia used the 2-nitroimidazoles, iodine-123-labeled iodoazomycin arabinoside (IAZA) with single photon emission computed tomography (SPECT) (Chapman et al. 2001) and fluorine-18-labeled fluoro-misonidazole (FMISO) with PET (Rasey et al. 1996). These two agents exhibited similar hypoxia-localizing ability in cells *in vitro* and in tumor sections *in vivo*, but in clinical studies, the percent of tumors exhibiting hypoxia and the associated HF varied depending on the type of cancer. It is possible that these differences are due to variation in cancer types, the difference in oxygen-dependence of uptake of the tracers, or to differences in imaging modality.

Recently, Lehtiö et al. (2001) studied, in eight head-and-neck cancer patients, tumor hypoxia using F18-fluoro-erythronitroimidazole (FETNIM), metabolically active volume with FDG, blood flow with ^{15}O-H$_2$O and blood volume using ^{15}O-CO. FETNIM levels were higher in tumors than in muscle and, in four of eight patients, higher than in blood, and appeared to be governed by blood flow, at least in the early phase of accumulation. Their results compare favorably with those reported previously for FMISO, but without significant advantage. Another 2-nitroimidazole, 2-(2-nitro-1H-imidazol-1-yl)-N-(2,2,3,3,3-pentafluoropropyl acetamide (EF5), shows promise in hypoxia imaging studies (Evans et al. 2000). More recently, the 2-nitronidazole fluoro-azomycin arabinoside (FAZA), labeled with fluorine-18, has been studied in tumor-bearing mice using PET (Piert et al. 2002). Compared to FMISO, FAZA showed similar tumor uptake, but lower liver uptake and faster clearance, and thus appears to be a promising PET tracer for the visualization of tumor hypoxia.

Cu(II)-diacetyl-bis(N4-methylthiosemicarbazone) (Cu-ATSM), stably labeled with the positron-emitters copper-62 (T$_{1/2}$: 9.74 min) and -64 (T$_{1/2}$: 12.7 hr), is another promising hypoxia imaging agent under evaluation. Based on *in vitro* results and *in vivo* studies indicating the targeting of hypoxic cells in rodent tumors (Lewis et al. 1999), clinical studies are currently underway. In fact, Chao et al. (2001) has reported the first clinical study on planning radiotherapy treatment based on hypoxic imaging. Thus, the enthusiasm for hypoxia imaging, is that it provides the opportunity to visualize and localize radioresistant tumor cell populations, and thereby to selectively target these regions by intensity modulated radiotherapy.

Beside the nuclear medicine approach, the use of NMR techniques can also provide information relative to tumor hypoxia. For example, tissue perfusion (related to blood flow and tumor hypoxia) with sub-millimeter resolution can be estimated from the increase in the T1 signal after the bolus administration of a contrast agent (e.g., gadolinium diethylenetriaminepenta-acetic acid, or Gd-DTPA). With faster pulse sequences such as the echo-planar dynamic imaging (EPI) method, serial images at <1 second per frame can be obtained to track the uptake of Gd-DTPA. The derived parametric image can yield pixel-by-pixel information on blood volume, blood brain barrier permeability, blood perfusion, diffusion, extravascular space, etc.

Recently, we have provided evidence that I-124 labeled iodo-azomycin-galactoside (I124-IAZG) is effective in identifying and localizing tumor hypoxia in animal models (Zanzonico et al. 2002). To validate the serial microPET imaging data, we also performed direct measurement of tumor oxygenation status with the Oxylite probe system. In addition, we are comparing the use of several hypoxic cell markers, I124-IAZG, F18-FMISO, and Cu-ATSM. Preliminary data show that the relative long 4.2-d half-life may be an important advantage for IAZG labeled compounds for imaging tumor hypoxia with potential for clinical application (see figure 1–2). In parallel with the PET studies, we are also using NMR techniques to measure perfusion and tumor lactate levels as surrogates of hypoxia. We believe validation and comparative studies using different approaches are important to optimizing the methods for assessing tumor hypoxia.

Research in Biological Imaging

As discussed previously, MRS has the potential to provide cancer cell signatures due to their different chemical moieties. A current disadvantage is the large voxel size required to obtain reasonable signal-to-noise ratio, e.g., 0.25 cm^3 for ^{1}H (proton) MRS and 2 cm^3 for in vivo ^{31}P MRS, although these should be reduced significantly with further technological advances, in particular the availability of MRI units with greater magnetic field strengths. Compounds that have been detected using ^{1}H NMR spectroscopy include choline, lactate, creatine, alanine, glutamate (a major excitatory neurotransmitter and energy source), myo-inositol (involved in intracellular

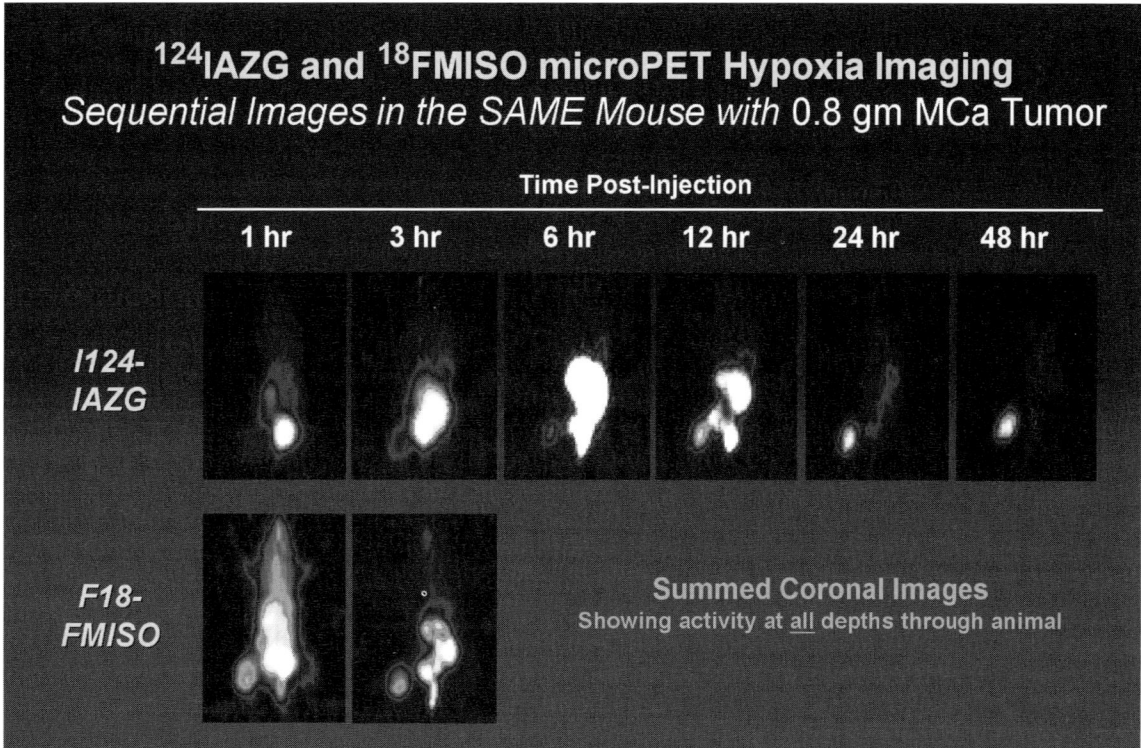

FIGURE 1–2. Positron emission tomographs of an 0.8 gm MCa breast tumor xenograft on the right thigh of a nude mouse, obtained with a microPET scanner. The lower panels were imaged at 1 and 3 hours after the injection of 200 microCi of F-18 labeled fluoromisonidazole (FMISO). The upper panels were imaged on the subsequent day on the same animal, at 1 to 48 hours after the injection of 400 microCi of I-124 labeled iodo-azomycin-galactoside (IAZG). The FMISO images visualize the hypoxic tumor at 1 to 3 hours post-injection, but not at longer time points due to the 1.9 h half-life of F-18. In contrast, the IAZG images of the tumor become distinct at longer times and exhibit much high signal-to-background ratio than the FMISO images.

signaling pathways related to growth), gamma aminobutyric acid (GABA) (a major inhibitory neu-rotransmitter), and taurine. Lactate is of particular interest since it increases in direct relation to anaerobic glycolysis caused by tumor hypoxia (Behar et al. 1983). ^{1}H NMR has been applied to non-invasive grading and classification of brain tumors, to attempting to characterize metastatic brain tumors from different primary sites, and to monitoring response after treatment with radi-ation and other modalities. At MSKCC, there is an ongoing study following patients with primary central nervous system (CNS) lymphoma prior to, during, and after treatment. The feasibility of discriminating between tumor regrowth and radiation necrosis based on higher choline in tumors and reduction of creatine, choline and NAA in necrosis has been proposed but requires further study (Nelson et al. 2001). In the NMR spectra of ^{31}P we can detect nucleoside triphosphates (NTP), phosphocreatine (PCR), inorganic phosphates (P_i), phosphocholine (PC), phospho-ethanolamine (PE), and can therefore provide information about tumor energy status and phos-pholipid metabolism (Koutcher et al. 1990). ^{31}P NMR, in providing data on energy status, may have prognostic value since hypoxic tumors have a lower energy status compared to well-oxy-genated tumors (Koutcher et al. 1992). Also, changes in ^{31}P spectra have been related to changes in tumor growth fraction, response to therapy, and percent necrosis at surgery. With further research in MRSI, and with the clinical deployment of magnets of greater field strengths, other specific chemical spectroscopic markers may be identified to provide a biological rationale for dose painting/sculpting in IMRT.

Despite the usefulness of FDG in PET imaging, FDG has limitations in distinguishing tumors from inflammation, and its usefulness in detecting sites of disease close to the bladder has been questioned. For this reason, alternative tracers with different attributes are being developed for cancer detection and characterization. For example, ^{11}C-methionine is a tracer that can differen-tiate tumor from normal tissue based on elevated levels of protein synthesis. Although ^{11}C has a half-life of 20 min (even shorter than ^{18}F), the uptake of ^{11}C-methionine is rapid, reaching a plateau 10 minutes post injection. This permits whole body imaging with ^{11}C-methionine with decay cor-rection. ^{11}C-methionine may be superior to FDG for imaging prostate cancers, because of the minimal uptake in the bladder, as seen in figure 1–3 (Macapinlac et al. 1999). However, the high metabolism of methionine in the liver would render this radiotracer sub-optimal for the detection of liver metastases. Therefore, the optimum choice of radiotracer for tumor diagnosis depends on the organ site. Another important discovery of multiple tracer studies, on the same patient, is that it frequently displays the heterogeneity of tumor biology. Patients who receive ^{11}C-methionine and FDG scans on the same day can display metastases, which are positive by both tracers, or which are positive by ^{11}C-methionine or FDG only. Such features indicate differences in tumor biology as it adapts to each sanctuary site.

Complementary to choline studies in MRSI, there are numerous ongoing studies with ^{11}C-choline, based on the premise that increased membrane synthesis, which occurs in prostate and possibly other cancers, results in a greater uptake and retention of the tracer there (De Jong et al. 2002).

There are emerging imaging techniques that provide information about molecular processes that underlie various biological functions at the molecular, cellular, and organ levels in contra-distinction to classical diagnostic imaging with its focus on anatomical and structural abnor-malities. These methods and the associated probes are often based on molecular biological processes they are designed to assess and, therefore, are potentially valuable tools to understand and treat disease processes at the molecular level.

Molecular imaging, in the context of cancer diagnosis and treatment, promises to provide spatial and quantitative information concerning DNA, RNA, and proteins that are involved in the disease process. Thus, the imaging methods could be for genotype, gene transcription, post-transcrip-tional modulation/stabilization of mRNA, protein-protein interactions in signal transduction path-ways, and other molecular information that influences tumor phenotype. Examples include the

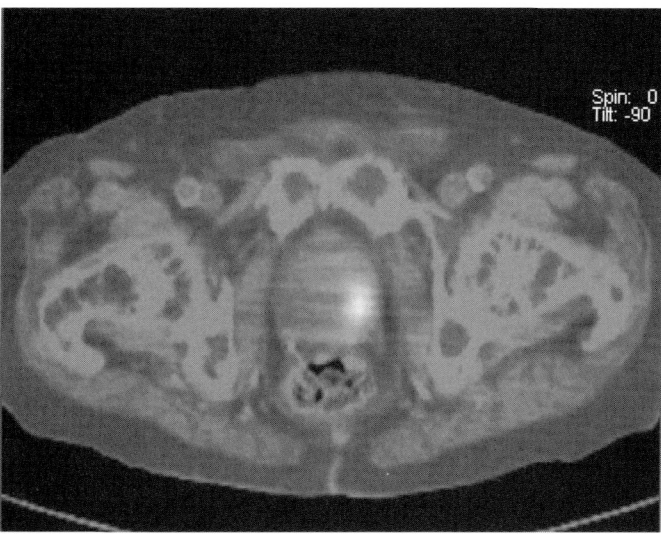

FIGURE 1–3. A PET image of a tumor of the prostate obtained using C-11 labeled methionine, a substrate for protein synthesis.

ability to image transcriptional regulation of p53 using a PET (and optical) reported gene or to image gene transfection and expression (Tjuvajev et al. 1996; Doubrovin et al. 2001). Clearly, such methodologies, once validated and approved, will have a major impact in disease evaluation, treatment planning, and monitoring outcome, so as to improve the clinical management of individual patients.

GTV, CTV, PTV, BTV, and Dose Painting/Sculpting

A common nomenclature for describing the patient's disease relative to radiotherapy treatment is useful, and International Commission on Radiation Units and Measurements (ICRU) Reports 50 and 62 have defined the terms of gross, clinical, and planning target volumes (GTV, CTV and PTV), and internal and set-up margins (IM and SM) (ICRU 1993, 1999). The PTV is essentially derived from the CTV, with appropriate inclusion of the internal and set-up margins. In general, cross-sectional images (CT and MRI) are used to delineate the GTV and CTV, and knowledge on physiological effects and set-up uncertainty incorporated to delineate the PTV. Then, radiation treatment portals are designed to cover the PTV entirely.

In conventional radiotherapy a uniform dose distribution within the PTV has been the *modus operandi* of treatment planning. This is largely because there was previously no basis to suggest any alternative strategy. With the ability of IMRT to deliver non-uniform dose patterns by design, there are investigations to evaluate the efficacy of "dose painting" or "dose sculpting" based on information from biological images. An example was given previously on the use of IMRT to deliver a non-uniform dose distribution within the PTV, with a higher dose given to specific regions identified by MRSI. Another example is the dose painting in head and neck cancers based on tumor hypoxia image obtained with PET and the radioactive tracer Cu-ATSM (Chao et al. 2001).

The above discussion advances the concept of IMRT dose painting based on "biological target volume" (BTV) derived from the new imaging methods (figure 1–4). Whereas up to the present radiological images are largely anatomical, the new imaging techniques can provide biological and mechanistic data, metabolic information from PET-FDG, functional/metabolic data from MRI/MRS, tumor hypoxia distribution from PET, and developing approaches to characterize tumor genotype and phenotype. Thus, such *biological images* may yield insights for defining the BTV. For relevance to radiation therapy, characteristics of the BTV should relate to the tumor extent and burden, growth kinetics, and factors that influence radiosensitivity and treatment outcome.

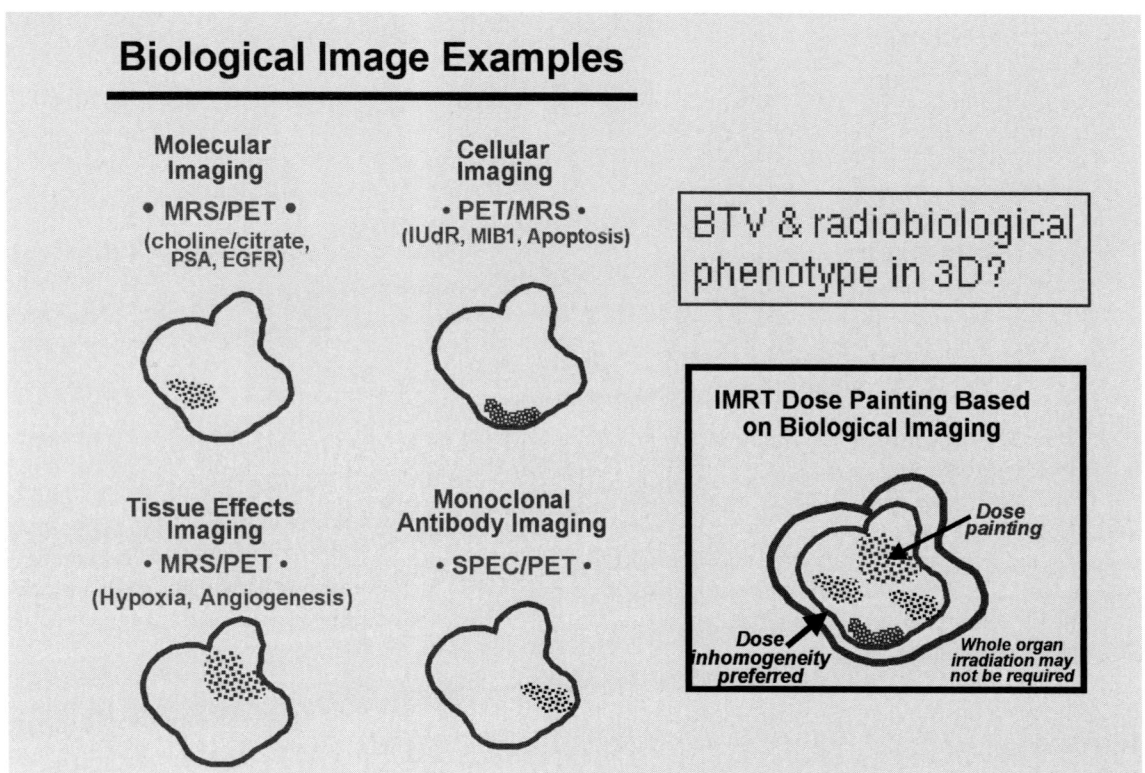

FIGURE 1–4. Examples of biological images that can yield molecular, cellular, physiological, and other types of information. Of interest to radiation therapy are factors that govern radioresponse and their 3-D presentation. Biological target information may be exploited for IMRT dose painting.

References

Avril, N., S. Bense, S. I. Ziegler, J. Dose, W. Weber, C. Laubenbacher, W. Romer, F. Janicke, and M. Schwaiger. (1997). "Breast imaging with fluorine-18-FDG PET: Quantitative image analysis." *J. Nucl. Med.* 38:1186–1191.

Behar, K. L., J. A. den Hollander, M. E. Stromski, T. Ogino, R. G. Shulman, O. A. Petroff, and J. W. Prichard. (1983). "High-resolution 1H nuclear magnetic resonance study of cerebral hypoxia in vivo." *Proc. Natl. Acad. Sci. USA* 80:4945–4948.

Belliveau, J. W., D. N. Kennedy, Jr., R. C. McKinstry, B. R. Buchbinder, R. M. Weisskoff, M. S. Cohen, J. M. Vevea, T. J. Brady, and B. R. Rosen. (1991). "Functional mapping of the human visual cortex by magnetic resonance imaging." *Science* 254:716–719.

Blasberg, R. G., and J. Gelovani. (2002). "Molecular-genetic imaging: A nuclear medicine based perspective." *Molecular Imaging* 1:160–180.

Brahme, A. (1988). "Optimization of stationary and moving beam radiation therapy techniques." *Radiother. Oncol.* 12: 129–140.

Brizel, D. M., S. P. Scully, J. M. Harrelson, L. J. Layfield, J. M. Bean, L. R. Prosnitz, and M. W. Dewhirst. (1996). "Tumor oxygenation predicts for the likelihood of distant metastases in human soft tissue sarcoma." *Cancer Res.* 56:941–943.

Brock, C. S., S. R. Meikle, and P. Price. (1997). "Does fluorine-18 fluorodeoxyglucose metabolic imaging of tumours benefit oncology? [see comments]." *Eur. J. Nucl. Med.* 24:691–705.

Burman, C., C. S. Chui, G. Kutcher, S. Leibel, M. Zelefsky, T. LoSasso, S. Spirou, Q. Wu, J. Yang, J. Stein, R. Mohan, Z. Fuks, and C. C. Ling. (1997). "Planning, delivery, and quality assurance of intensity-modulated radiotherapy using dynamic multileaf collimator: A strategy for large-scale implementation for the treatment of carcinoma of the prostate." *Int. J. Radiat. Oncol. Biol. Phys.* 39:863–873.

Carol, M., W. H. Grant 3rd, D. Pavord, P. Eddy, H. S. Targovnik, B. Butler, S. Woo. J. Figura, V. Onufrey, R. Grossman, and R. Selkar. (1996). "Initial clinical experience with the Peacock intensity modulation of a 3-D conformal radiation therapy system." *Stereotact. Funct. Neurosurg.* 66:30–34.

Chao, K. S., W. R. Bosch, S. Mutic, J. S. Lewis, F. Dehdashti, M. A. Mintun, J. F. Dempsey, C. A. Perez, J. A. Purdy, and M. J. Welch. (2001). "A novel approach to overcome hypoxic tumor resistance: Cu-ATSM-guided intensity-modulated radiation therapy." *Int. J. Radiat. Oncol. Biol. Phys.* 49:1171–1182.

Chapman, J. D., R. F. Schneider, J. L. Urbain, and G. E. Hanks. (2001). "Single-photon emission computed tomography and positron-emission tomography assays for tissue oxygenation." *Semin. Radiat. Oncol.* 11: 47–57.

De Jong, I. J., J. Pruim, P. H. Elsinga, W. Vaalburg, and H. J. Mensink. (2002). "Visualization of prostate cancer with 11C-choline positron emission tomography." *Eur. Urol.* 42:18–23.

Doubrovin, M., V. Ponomarev, T. Beresten, J. Balatoni, W. Bornmann, R. Finn, J. Humm, S. Larson, M. Sadelain, R. Blasberg, and J. Gelovani Tjuvajev. (2001). "Imaging transcriptional regulation of p53 dependent genes with positron emission tomography in vivo." *Proc. Natl. Acad. Sci. USA* 98:9300–9305.

Erdi, Y. E., E. D. Yorke, A. K. Erdi, Y. C. Hu, L. E. Braban, H. M. Macapinlac, J. L. Humm, O. D. Squire, C.-S. Chui, S. M. Larson, and K. Rosenzweig. (2002). "Radiotherapy treatment planning for patients with non-small cell lung cancer using positron emission tomography." *Radiother. Oncol.* 62:51–60.

Evans, S. M., A. V. Kachur, C. Y. Shiue, R. Hustinx, W. T. Jenkins, G. G. Shive, J. S. Karp, A. Alavi, E. M. Lord, W. R. Dolbier Jr., C. J. Koch. (2000). "Noninvasive detection of tumor hypoxia using the 2-nitroimidazole [18F]EF1." *J. Nucl. Med.* 41:327–336.

Fried, I., V. I. Nenov, S. G. Ojemann, and R. P. Woods. (1995). "Functional MR and PET imaging of rolandic and visual cortices for neurosurgical planning." *J. Neurosurg.* 83:854–861.

Goitein, M., M. Abrams, D. Rowell, H. Pollari, and J. Wiles. (1983). "Multidimensional treatment planning. II: Beam's eye view, back projection, and projection through CT sections." *Int. J. Radiat. Oncol. Biol. Phys.* 9:789–797.

Hamilton, R. J., P. J. Sweeney, C. A. Pelizzari, F. Z. Yetkin, B. L. Holman, B. Garada, R. R. Weichselbaum, and G. T. Chen. (1997). "Functional imaging in treatment planning of brain lesions." *Int. J. Radiat. Oncol. Biol. Phys.* 37:181–188.

Hockel, M., K. Schlenger, B. Aral, M. Mitze, U. Schaffer, and P. Vaupel. (1996a). "Association between tumor hypoxia and malignant progression in advanced cancer of the uterine cervix." *Cancer Res.* 56:4509–4515.

Hockel, M., K. Schlenger, M. Mitze, U. Schaffer, and P. Vaupel. (1996b). "Hypoxia and radiation response in human tumors." *Semin. Radiat. Oncol.* 6:3–9.

ICRU Report 50. Prescribing, Recording and Reporting Photon Beam Therapy. Washington, DC: International Commission on Radiation Units and Measurements, 1993.

ICRU Report 62. Prescribing, Recording and Reporting Photon Beam Therapy. Supplement to ICRU Report 50. Washington, DC: International Commission on Radiation Units and Measurements, 1999.

Kim, K. H., N. R. Relkin, K. M. Lee, and J. Hirsch. (1997). "Distinct cortical areas associated with native and second languages." *Nature* 388:171–174.

Kim, S. G., J. Ashe, K. Hendrich, J. M. Ellermann, H. Merkle, K. Ugurbil, and A. P. Georgopoulos. (1993). "Functional magnetic resonance imaging of motor cortex: Hemispheric asymmetry and handedness." *Science* 261:615–617.

Koutcher, J. A., D. Ballon, M. Graham, J. H. Healey, E. S. Casper, R. Heelan, and L. E. Gerweck. (1990). "31P NMR spectra of extremity sarcomas: diversity of metabolic profiles and changes in response to chemotherapy." *Magn. Reson. Med.* 16:19–34.

Koutcher, J. A., A. A. Alfieri, M. L. Devitt, J. G. Rhee, A. B. Kornblith, U. Mahmood, T. E. Merchant, and D. Cowburn. (1992). "Quantitative changes in tumor metabolism, partial pressure of oxygen, and radiobiological oxygenation status postradiation." *Cancer Res.* 52:4620–4627.

Kurhanewicz, J., D. B. Vigneron, S. J. Nelson, H. Hricak, J. M. MacDonald, B. Konety, and P. Narayan. (1995). "Citrate as an in vivo marker to discriminate prostate cancer from benign prostatic hyperplasia and normal prostate peripheral zone: detection via localized proton spectroscopy." *Urology* 45:459–466.

Lehtiö, K., V. Oikonen, T. Gronroos, O. Eskola, K. Kalliokoski, J. Bergman, O. Solin, R. Grenman, P. Nuutila, and H. Minn. (2001). "Imaging of blood flow and hypoxia in head and neck cancer: Initial evaluation with [(15)O]H(2)O and [(18)F]fluoroerythronitroimidazole PET." *J. Nucl. Med.* 42:1643–1652.

Levegrun, S., A. Jackson, M. J. Zelefsky, E. S. Venkatraman, M. W. Skwarchuk, W. Schlegel, Z. Fuks, S. A. Leibel, and C. C. Ling. (2002). "Risk group dependence of dose response for biopsy outcome after three-dimensional conformal radiation therapy of prostate cancer." *Radiother. Oncol.* 63:11–26.

Lewis, J. S., D. W. McCarthy, T. J. McCarthy, Y. Fujibayashi, and M. J. Welch. (1999). "Evaluation of ^{64}Cu-ATSM in vitro and in vivo in a hypoxic tumor model." *J. Nucl. Med.* 40:177–183.

Ling, C. C., and Z. Fuks. (1995). "Conformal radiation treatment: A critical appraisal." *Eur. J. Cancer* 5:799–803.

Ling, C. C., C. Burman, C. S. Chui, G. J. Kutcher, S. A. Leibel, T. LoSasso, R. Mohan, T. Bortfeld, L. Reinstein, S. Spirou, X. H. Wang, Q. Wu, M. Zelefsky, and Z. Fuks. (1996). "Conformal radiation treatment of prostate cancer using inversely-planned intensity-modulated photon beams produced with dynamic multileaf collimation [see comments]." *Int. J. Radiat. Oncol. Biol. Phys.* 35:721–730.

Ling, C. C., J. L. Humm, S. M. Larson, H. Amols, Z. Fuks, S. Leibel, and J. A. Koutcher. (2000). "Towards multidimensional radiotherapy (MD-CRT): Biological imaging and biological conformality." *Int. J. Radiat. Oncol. Biol. Phys.* 47:551–560.

LoSasso, T., C. S. Chui, and C. C. Ling. (1998). "Physical and dosimetric aspects of a multileaf collimation system used in the dynamic mode for implementing intensity modulated radiotherapy." *Med. Phys.* 25:1919–1927.

Macapinlac, H. A., J. L. Humm, T. Akhurst, I. Osman, K. Pentlow, S. Cai, W. D. Yeung, O. Squire, R. D. Finn, H. I. Scher, and S. M. Larson. (1999). "Differential metabolism and pharmacokinetics of C-11 methionine and F-18 fluorodeoxyglucose (FDG) in androgen independent prostate cancer." *Clin. Positron Imaging* 2:173–181.

MacManus, M. P., R. J. Hicks, D. L. Ball, V. Kalff, J. P. Matthews, E. Salminen, P. Khaw, A. Wirth, D. Rischin, and A. McKenzie. (2001). "F-18 fluorodeoxyglucose positron emission tomography staging in radical radiotherapy candidates with non-small cell lung carcinoma." *Cancer* 92:886–895.

Mah, K., C. B. Caldwell, Y. C. Ung, C. E. Danjoux, J. M. Balogh, S. N. Ganguli, L. E. Ehrlich, and R. Tirona. (2002). "The impact of 18FDG-PET on target and critical organs in CT-based treatment planning of patients with poorly defined non-small-cell lung carcinoma: A prospective study." *Int. J. Radiat. Oncol. Biol. Phys.* 52:339–350.

McShan, D. L., B. A. Fraass, and A. S. Lichter. (1990). "Full integration of the beam's eye view concept into computerized treatment planning." *Int. J. Radiat. Oncol. Biol. Phys.* 18:1485–1494.

Munley, M. T., L. B. Marks, C. Scarfone, G. S. Sibley, E. F. Patz Jr., T. G. Turkington, R. J. Jaszczak, D. R. Gilland, M. S. Anscher, and R. E. Coleman. (1999). "Multimodality nuclear medicine imaging in three-dimensional radiation treatment planning for lung cancer: Challenges and prospects." *Lung Cancer* 23:105–114.

NCI (National Cancer Institute). (2001)."NCI intensity-modulated radiotherapy: Current status and issues of interest." *Int. J. Radiat. Oncol. Biol. Phys.* 51:880–914.

Nelson, S. J. (2001). "Analysis of volume MRI and MR spectroscopic imaging data for the evaluation of patients with brain tumors." *Magn. Reson. Med.* 46:228–239.

Nestle, U., K. Walter, S. Schmidt, N. Licht, C. Nieder, B. Motaref, D. Hellwig, M. Niewald, D. Ukena, C. M. Kirsch, G. W. Sybrecht, and K. Schnabel. (1999). "18F-deoxyglucose positron emission tomography (FDG-PET) for the planning of radiotherapy in lung cancer: High impact in patients with atelectasis." *Int. J. Radiat. Oncol. Biol. Phys.* 44:593–597.

Nordsmark, M., M. Overgaard, and J. Overgaard. (1996). "Pretreatment oxygenation predicts radiation response in advanced squamous cell carcinoma of the head and neck." *Radiother. Oncol.* 41:31–39.

Pickett, B., E. Vigneault, J. Kurhanewicz, L. Verhey, and M. Roach. (1999). "Static field intensity modulations to treat a dominant intra-prostatic lesion to 90 Gy compared to seven field 3-dimensional radiotherapy." *Int. J. Radiat. Oncol. Biol. Phys.* 43:921–929.

Piert, M., et al. (2002). "18F labeled fluoroazomycin arabinoside (FAZA): Imaging tumor hypoxia with improved biokinetics." *J. Nucl. Med.* 43:278.

Rasey, J. S., W. J. Koh, M. L. Evans, L. M. Peterson, T. K. Lewellen, M. M. Graham, and K. A. Krohn. (1996). "Quantifying regional hypoxia in human tumors with positron emission tomography of [18F]fluoromisonidazole: A pretherapy study of 37 patients." *Int. J. Radiat. Oncol. Biol. Phys.* 36:417–428.

Rigo, P., P. Paulus, B. J. Kaschten, R. Hustinx, T. Bury, G. Jerusalem, T. Benoit, and J. Foidart-Willems. (1996). "Oncological applications of positron emission tomography with fluorine-18 fluorodeoxyglucose." *Eur. J. Nucl. Med.* 23:1641–1674.

Scheidhauer, K., A. Scharl, U. Pietrzyk, R. Wagner, U. J. Gohring, K Schomacker, and H. Schicha. (1996). "Qualitative [18F]FDG positron emission tomography in primary breast cancer: Clinical relevance and practicability." *Eur. J. Nucl. Med.* 23:618–623.

Scheidler, J., H. Hricak, D. B. Vigneron, K. Yu, L. Dahila, L. R. Hunag, C. J. Zaloudek, S. Nelson, P. Carroll, and J. Kurhanewicz. (1999). "Prostate cancer: Localization with 3D proton MR spectroscopic imaging—clinicopathological study." *Radiol.* 213:474–480.

Sherouse, G. W., J. D. Bourland, K. Reynolds, H. L. McMurry, T. P. Mitchell, and E. L. Chaney. (1990). "Virtual simulation in the clinical setting: some practical considerations." *Int. J. Radiat. Oncol. Biol. Phys.* 19:1059–1065.

Spirou, S. V., and C. S. Chui. (1994). "Generation of arbitrary intensity profiles by dynamic jaws or multileaf collimators." *Med. Phys.* 21:1031.

Spirou, S. V., and C. S. Chui. (1998). "A gradient inverse planning algorithm with dose-volume constraints." *Med. Phys.* 25:321–333.

Tjuvajev, J., R. Finn, K. Watanabe, R. Joshi, T. Oku, J. Kennedy, B. Beattie, J. Koutcher, S. Larson, and R. Blasberg. (1996). "Noninvasive imaging of herpes virus thymidine kinase gene transfer and expression: A potential method for monitoring clinical gene therapy." *Cancer Res.* 56:4087–4095.

Townsend, D. W., and S. R. Cherry. (2001). "Combining anatomy and function." *Eur. Radiol.* 11:1968–1974.

Urbain, J. L. (1999). "Oncogenes, cancer and imaging." *J. Nucl. Med.* 40: 498–504.

Utech, C. I., C. S. Young, and P. F. Winter. (1996). "Prospective evaluation of fluorine-18 fluorodeoxyclucose positron emission tomography in breast cancer for staging of the axilla related to surgery and immunocytochemistry." *Eur. J. Nucl. Med.* 23:1588–1593.

Vanuytsel, L. J., J. F. Vansteenkiste, S. G. Stroobants, P. R. De Leyn, W. De Wever, E. K. Verbeken, G. G. Gatti, D. P. Huyskens, and G. J. Kutcher. (2000). "The impact of 18F- fluro-2-deoxy-D-glucose positron emission tomography (FDG-PET) lymph node staging on the radiation treatment volumes in patients with non-small cell lung cancer." *Radiother. Oncol.* 55:317–324.

Weissleder, R., A. Moore, U. Mahmood, R. Bhorade, H. Benveniste, E. A. Chiocca, and P. Basilion. (2000). "In vivo magnetic resonance imaging of transgene expression." *Nature Med.* 6:351–354.

Xia, P., B. Pickett, E. Vigneault, L. Verhey, and M. Roach. (2001). "Forward or inversely planned segmental multileaf collimator IMRT and sequential tomotherapy to treat multiple dominant intraprostatic lesions of prostate cancer to 90 Gy." *Int. J Radiat. Oncol. Biol. Phys.* 51:244–254.

Yu, C. X. (1995). "Intensity-modulated arc therapy with dynamic multileaf collimation: An alternative to tomotherapy." *Phys. Med. Biol.* 40:1435–1449.

Zanzonico, P., R. Finn, R. Schneider, J. D. Chapman, S. Cai, Y. Chen, S. Ruan, M. Urano, B. Beattie, K. S. Pentlow, J. Humm, S. M. Larson, and C. Ling. (2002). "MicroPET imaging of tumor hypoxia: Direct comparison in tumor-bearing mice of F18-FDG, F18-Fluoromisonidazole and I124-iodo-azomycin galactoside." *J. Nucl. Med.* 43(suppl):278.

Zhong, H., F. Agani, A. A. Baccala, E. Laughner, N. Rioseco-Camacho, W. B. Isaacs, J. W. Simons, and G. L. Semenza. (1998). "Increased expression of hypoxia inducible factor-1 alpha in rat and human prostate cancer." *Cancer Res.* 58:5280–5284.

2

Overview Of The IMRT Process

Howard I. Amols
C. Clifton Ling
Steven A. Leibel

Introduction

It is well accepted that local tumor control and normal tissue complications have sigmoidally shaped dose-response curves. For normal tissue complications, radiation response also depends on the volume of tissue irradiated, with some tissues (such as lung, liver, bowel, and kidney) having greater volume dependence than others (such as spinal cord and brain). The success of radiotherapy is therefore highly dependent on the radiosensitivity of the particular tumor being treated relative to that of the surrounding normal tissues. For tumor sites where the tumor control curve is less steep than the normal tissue complication curve, the high doses required for tumor cure may cause unacceptable normal tissue complications. The goal in radiation therapy, therefore, is to sufficiently separate the dose-response curves of local tumor control and normal tissue complications. Basically, the only physical or dosimetric method for achieving this involves configuring the radiation portals to reduce the dose delivered to the normal tissues, and also the total volume of normal tissue irradiated.

During the past two decades, advances in radiological imaging and computer technology have significantly enhanced our ability to achieve this goal through the development of three-dimensional image-based conformal radiotherapy (3DCRT). Intensity-modulated radiation therapy (IMRT) is an especially advanced method of 3DCRT (super conformal therapy, if you will) that utilizes sophisticated computer controlled radiation beam delivery to improve the conformality of the dose distribution to the shape of the tumor. This is achieved by varying beam intensities within each beam portal, as opposed to uniform beam intensities as in conventional 3DCRT. IMRT usually (but not always) also incorporates computerized inverse treatment plan (ITP) optimization as opposed to the manual optimization techniques of conventional 3DCRT. Both 3DCRT and IMRT utilize sophisticated strategies for patient immobilization and positioning, image-guided treatment planning, and (often) computer-enhanced treatment verification. At the heart of these techniques is advanced computer technology, and in particular, 3-D patient imaging with computed tomography (CT) [more recently augmented with magnetic resonance (MR) and/or positron emission tomography (PET) studies].

This book is designed to guide the medical physicist and radiation oncologist in implementing IMRT in their clinical practice. Emphasis is placed on the practical aspects of IMRT. It is assumed that your clinic is already experienced in CT-based 3-D treatment planning, designing fields with beam's-eye view (BEV), and using dose-volume histograms (DVHs) as part of treatment plan evaluation. Indeed, it is difficult, and not recommended, to attempt a transition from non-CT, 2-D treatment planning directly to IMRT. One must walk before one can run!

There are two principal motivations for implementing IMRT. First, for treatment sites where radiation therapy is currently not very successful, IMRT may permit the delivery of higher radiation doses with less increase in normal tissue toxicity than conventional radiation therapy (RT). For such sites the hypothesis is that dose escalation may result in improved local control. Prostate cancer is a good example where IMRT permits substantial dose escalation, with proven increase in local control, while concurrently maintaining acceptable rectal, urethral, and bladder toxicity. In 1995 this was the first disease site chosen at MSKCC for IMRT. We now routinely treat prostate patients to 86 Gy with IMRT (Zelefsky et al. 1999, 2002). Other disease sites including nasopharynx (Hunt et al. 2001), recurrent head and neck tumors (Hsiung et al. 2002), lung, and paraspinal tumors have been added over the years.

A second category where IMRT is advantageous includes treatment sites such as pediatric cancers, breast, thyroid, and whole abdomen radiation (WAR), where current techniques already provide good tumor response and local control, but where additional normal tissue sparing and/or dose homogeneity within the planning target volume (PTV) may reduce treatment morbidity. In breast, for example, IMRT provides 3-D tissue compensation and more homogeneous PTV doses as compared to conventional 2-D wedged treatments, and also reduces dose to the heart and lung (Hong et al. 1999). For pediatric tumors, IMRT can significantly reduce normal tissue doses which, we hypothesize, may improve the quality of life for therapy survivors. For WAR, IMRT permits reduced dose to bone marrow, which is particularly important for patients receiving chemotherapy (Hong et al. 2002).

Clinical implementation of an IMRT treatment planning and beam delivery system entails a significant change in the *modis avendi* of treatment planning strategy. Clinicians and treatment planners must learn to think about treatment design and optimization in a totally new way, particularly the concept of ITP rather than forward treatment planning. Considerable physics effort is required to acceptance test and commission the hardware and software required for IMRT, and significant changes are required in schema for treatment plan verification and continuing quality assurance (QA). Weeks, or even months of work, are involved, in addition to which you will also have to establish a reliable and practical system of continuing dosimetry and QA tests for your IMRT system.

In this chapter, we briefly introduce the rationale and processes for IMRT, including patient imaging and simulation, treatment planning, dose calculation algorithms, plan evaluation, treatment delivery, and QA issues. In the discussion, we will identify the similarities and differences between 3DCRT and IMRT, and emphasize those features and benefits unique to IMRT, with details being presented in subsequent chapters.

IMRT Treatment Planning Process

3DCRT and IMRT entail more sophisticated shaping of the dose distribution than conventional radiation therapy with the shaping of fields and selection of beam directions based on 3-D images of the patient (usually CT, but often augmented with MR, PET, and/or other functional imaging studies) (Ling et al. 2000). These images, displayed in so-called *beam's eye view* (*BEV*) format, permit better avoidance of normal tissues. IMRT goes one step beyond 3DCRT by, in addition, enabling variations of the *radiation intensity* within each beam. This intensity modulation can be achieved via several different approaches including fabrication of complex physical compensators (often constructed with computer-controlled milling machines) to be placed in the beam between the radiation source and the patient, but more commonly via the use of a multileaf collimator (MLC) capable of dynamic beam delivery or multiple static beams sequentially altered in shape by the MLC.

Implementation of IMRT entails a fundamental change in treatment planning strategy and concept. Most importantly, IMRT treatment planning is usually performed using *inverse treatment planning* (*ITP*) algorithms as opposed to *forward planning*, meaning that the treatment planning team (radiation oncologist, physicist, and dosimetrist) ***does not*** specify beam weights, wedge angles, beam modifiers, or the like, nor do they have *direct* control over these parameters at any point in the treatment planning process. The number, energy, and direction of the beams are still chosen by the planner in the conventional manner but, once beams have been specified, the computer usurps complete control over all other beam properties such as monitor units or weight, wedge angles, beam modifiers, etc., and designs a customized intensity pattern to best meet the specified dose-volume constraints for the PTV and normal tissues which have been specified *in advance.* In other words, the user must define the desired dose distribution, and the computer then calculates the beam weights, intensity modulation, and dynamic multileaf collimator (DMLC) motions that come closest to producing the requested dose distribution. Obviously, a key to the entire process is that the treatment planner specifies dose-volume constraints that are physically realistic and achievable.

Let us first briefly describe the overall process of IMRT, after which we discuss the individual steps. The treatment planning steps for IMRT are similar to 3DCRT during the initial and final stages, but diverge in the middle. As shown in figure 2–1, the treatment planning process begins with *treatment simulation*, much the same as 3DCRT, during which the patient is set up on the CT unit in the treatment position. More and more frequently, patient CT imaging and simulation is being augmented with magnetic resonance imaging (MRI), PET, and other functional imaging studies to better define the tumor and target volume.

Simulation usually first entails fabrication of custom-designed body molds to facilitate accurate reproduction of patient position during simulation and planning, as well as during multi-fraction treatment delivery. After 3-D images are acquired the target and non-target structures are delineated, usually directly on a computer display of transverse CT (or MR, PET, etc.) images using standard computer graphics options such as mouse, track ball, light pen, etc.

Similarly, the design of treatment portal shapes also takes place on the computer using BEV. Using such displays one can adjust the beam directions and shapes so as to minimize the volume of normal tissues included in each radiation portal. Figure 2–2 depicts a BEV image for a brainstem lesion showing the tissues involved in a so-called *color wash display*. Also superimposed on the BEV are MLC leaf settings, such as one might design for a conventional 3DCRT treatment plan. When doing an ITP, however, the treatment planner would NOT specify the MLC settings, as these would be calculated by the computer. Once all relevant tissues have been

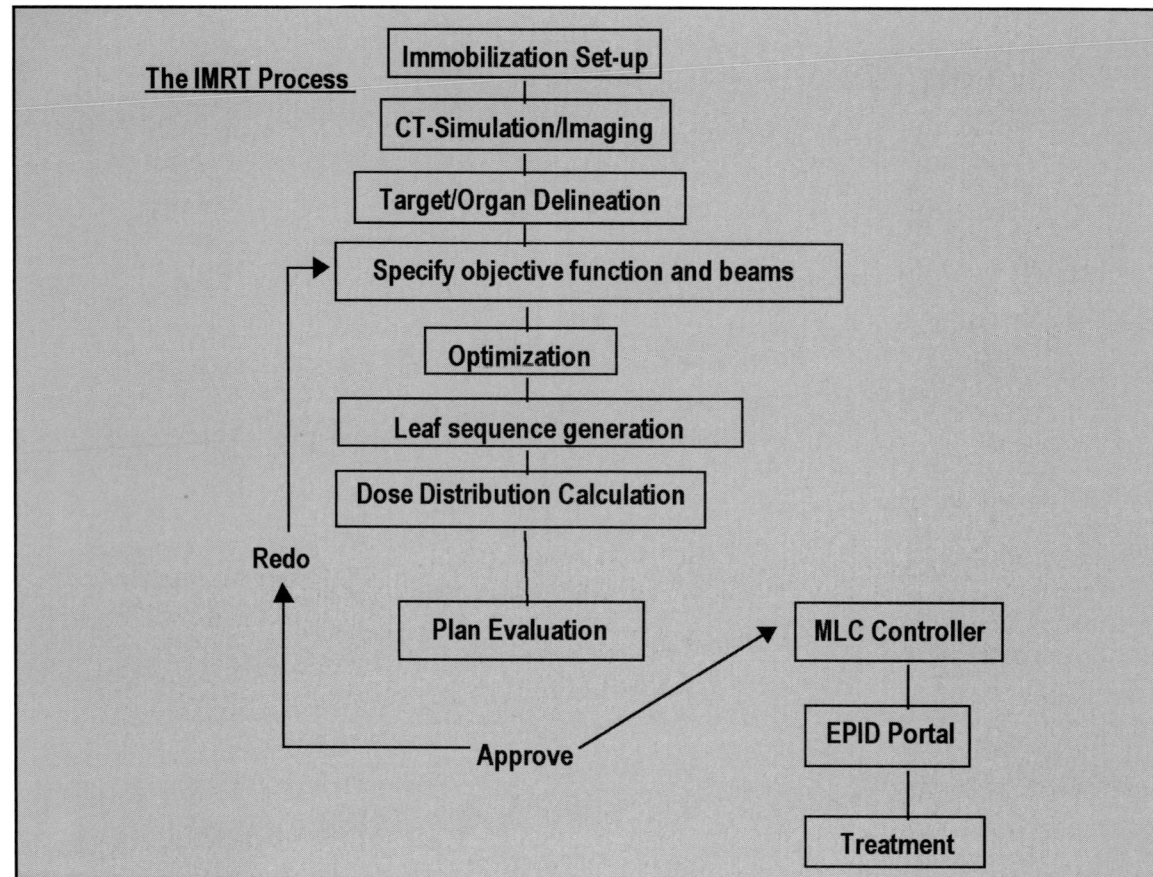

FIGURE 2–1. Block diagram showing sequence for the IMRT-ITP process.

delineated, and beam directions specified, the radiation oncologist specifies the desired doses to tumor and normal tissues. From these specifications the ITP algorithm adjusts beam shapes and intensities so as to best meet these dose criteria.

IMRT incorporates (usually) computerized iteration of radiation beams as opposed to the manual optimization procedures used in conventional treatment planning. With ITP, the computer performs hundreds of iterations in less time than a human could perform a few iterations, with each iteration entailing the adjustment of thousands of individual pencil beams. In this way the computer can modulate the dose intensity within each radiation portal. This procedure is different from conventional 3DCRT plan optimization, designated as *forward planning*, where the treatment planner manually iterates beam weights, directions, and shapes. With manual forward planning flexibility is constrained by time, by the fact that x-ray beam intensities within each individual beam are uniform or smoothly varying, and by the experience, skill, and imagination of the treatment planner.

IMRT plan evaluation, as for 3DCRT, relies largely on analysis of dose distributions and DVHs. When the treatment plan has been accepted, all planning data on beam configurations and intensities are transferred to the linear accelerator (usually via a computer network, electronic chart or "record and verify" system) and patient treatment proceeds. For IMRT, the data transferred includes information on DMLC motion files required to deliver the desired X-ray intensity profiles. Finally, the physicist must perform dosimetric and QA tasks to verify that all equipment

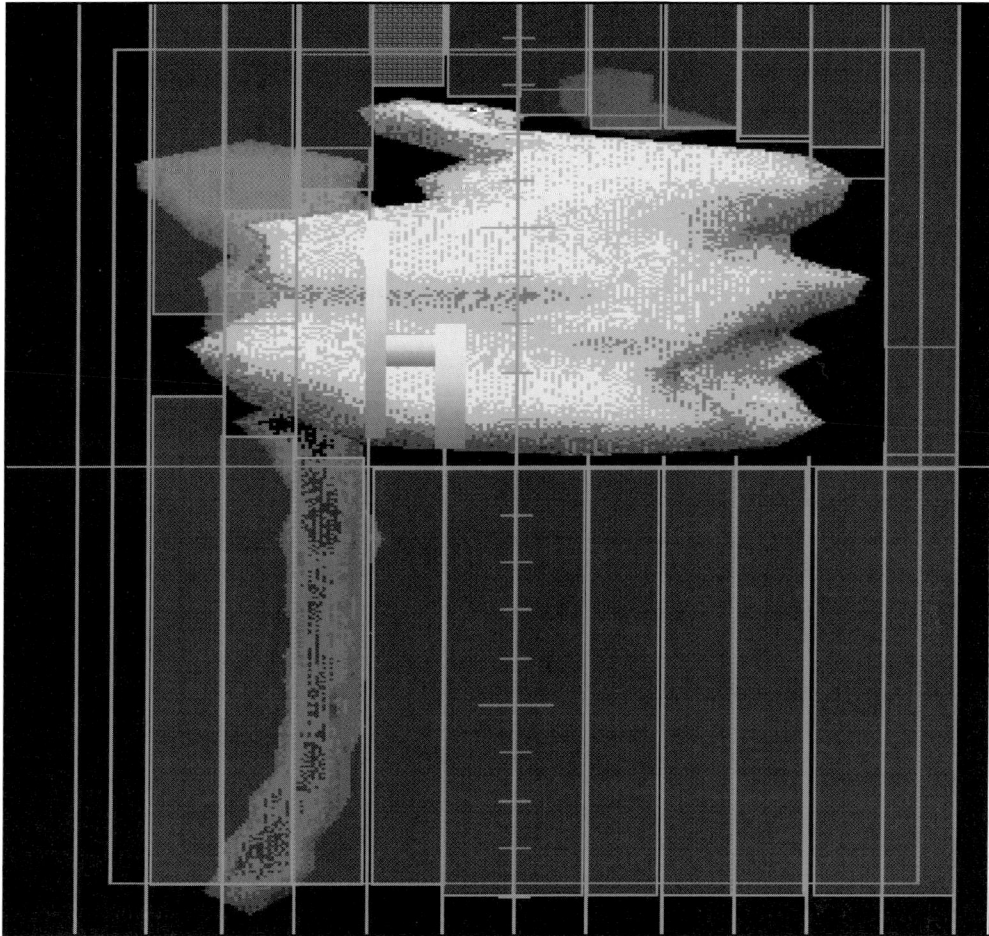

FIGURE 2–2. The BEV of a right anterior oblique field for a head-neck patient in a supine position. PTV is depicted in yellow, brainstem in green, cord in purple, and pituitary gland in red. For conventional 3DCRT treatment planning the beam shape would be defined by the MLC leaves as shown. For ITP, the MLC shape would be designed automatically by the optimization algorithms. See COLOR PLATE 2.

is functioning properly, and that the specifics of the dose prescription and treatment plan are accurately delivered to the patient on a daily basis.

In the following sections we briefly discuss each of these treatment planning steps in slightly more detail, with full discussions provided in subsequent chapters.

Patient Setup, Immobilization, Image Acquisition

Minimization of patient setup uncertainty is particularly critical for IMRT because field margins are often intentionally minimized and dose gradients can be exceptionally large. Thus, small errors in patient position can result in larger treatment errors than for conventional therapy. It is important that the same techniques of patient immobilization be used throughout the entire treatment process including image acquisition, simulation, and treatment. Many different techniques and devices are in use to achieve this, including conventional devices such as vacuum cradles, plaster casts, face masks, etc. More elaborate devices are also being introduced such as stereotactic body frames (SBF) that can achieve setup accuracy approaching that of stereotactic radiosurgery, as shown in figure 2–3.

The utility of conventional radiation therapy simulators is greatly decreased with IMRT as treatment planning and "virtual simulation" are based almost entirely on 3-D CT images instead of conventional 2-D simulation films. A conventional simulator may still be used for an initial (pre-CT) simulation in conjunction with the fabrication of patient immobilization devices, establishment of fiduciary skin marks, identification of a tentative treatment isocenter, etc. Some centers also continue to utilize a second verification simulation to position the patient at the established isocenter and to acquire reference radiographs of each field that can later be compared to digitally-reconstructed radiographs (DRRs) and treatment portal images. However, conventional simulation is becoming something of an anachronism with IMRT and 3DCRT.

Indeed, DRRs provide more flexibility than conventional simulation films as they can easily be manipulated on the computer to enable contrast and image enhancement, plus automatic comparison with digital port films obtained during patient treatments with electronic portal imaging systems (EPIDs). Figure 2–4 shows a typical posterior-anterior DRR for a prostate patient. One current limitation of DRRs, however, is their degraded image quality as compared to conventional simulation films because of the relatively large size of the CT pixels (typically 0.6×0.6 mm^2), and more importantly the slice thickness (typically 3 to 5 mm). Newer CT scanners, however, can produce slice thicknesses of 1.5 mm or less, which greatly improves the quality of DRRs but at a cost of greatly increased file sizes.

In any case, the entire sequence of initial simulation, CT simulation, and confirmation simulation can often be completely replaced by a single CT-based *virtual simulation* that is more efficient and also permits electronic transfer of all required treatment information such as target volume and normal tissues contours, isocenter coordinates, etc., directly to the treatment planning

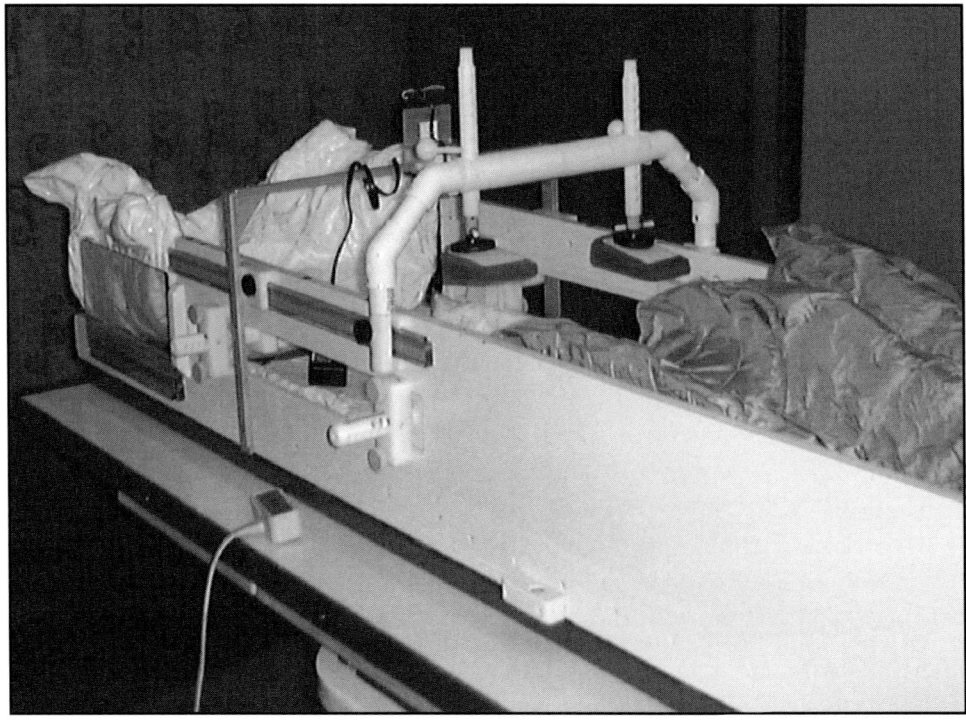

FIGURE 2–3. Photograph of a whole body stereotactic patient positioning system.

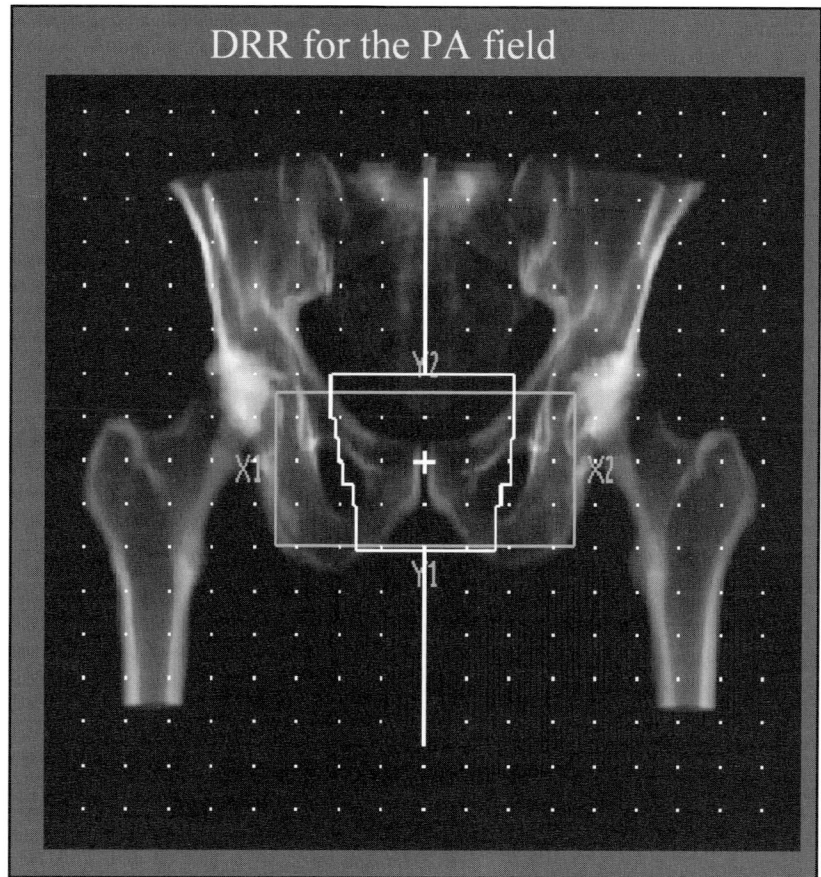

FIGURE 2–4. A posterior-anterior DRR for a prostate field showing MLC settings superimposed on 2-D patient anatomy.

system. In theory, data transfer between CT computers, treatment planning computers, record and verify systems, and EPID computers occurs seamlessly using programs all conforming to a single internationally accepted format known as DICOM (Digital Imaging and Communications in Medicine). In practice, however, it is not uncommon to find small differences in file formats between different manufacturers, all claiming to be DICOM compatible, that will require intervention by a knowledgeable computer system expert.

Delineation Of Treatment Volume And Critical Organs

Treatment volumes are defined on CT images according to the International Commission on Radiation Units and Measurements (ICRU) Report 50 (ICRU 50 1993) nomenclature, with the gross target volume (GTV) defined as the visible tumor as seen on CT or other imaging studies; the clinical target volume (CTV) being the visualized tumor plus regions at risk such as microscopic extension of disease, nodal chains, etc.; and planning target volume (PTV) being an expanded CTV that includes setup errors, patient motion, linear accelerator alignment errors, and other uncertainties. The manual delineation of these volumes, plus adjacent normal tissues, is time consuming and sometimes subjective. Many virtual simulation and treatment planning systems incorporate various software tools for performing auto-contouring of tissue structures, but current algorithms are of limited utility, and usually only accurate for outlining external body contours and internal contours of very high contrast such as bone, lung, and air cavities. Human intervention will likely remain a necessary and important component for tissue contouring for the foreseeable future.

With IMRT, accurate delineation of the PTV and normal tissues is critical because these contours are used directly by the ITP optimization algorithm in the assessment of whether or not a resulting dose distribution conforms to the prescribed dose constraints. For similar reasons, it is often necessary with IMRT and ITP to contour many normal tissues in the vicinity of the PTV that would not normally be contoured for 3DCRT treatment planning. This is necessary in order to *advise* the optimization algorithm as to whether or not it is permissible to deposit high doses in a particular region. For example, in a typical head and neck case, in addition to the commonly contoured structures such as spinal cord and brain stem, it may also be necessary to contour parotid glands, optical structures, tongue, mandible, ears, etc., lest the optimization algorithms *unwittingly* deposit large doses to these volumes in order to meet dose constraints placed on other structures.

Selection Of Treatment Beams

Various display schema have been developed to represent internal patient anatomy in 3-D perspective on a 2-D computer display monitor. The main objective being to facilitate the selection of beam angles and field shapes based on the relative positions of these various anatomical structures. These structures are often displayed as wire frames or solid or translucent structures, often using color or intensity to indicate depth below the surface. An example of such a display in BEV perspective was shown previously in figure 2–2. From this BEV, beam orientations are chosen by observing the patient from various orientations and selecting those directions for which the PTV appears to be best separated from the normal tissues, and from which the volume of irradiated normal tissue can be minimized. Once beam orientations have been chosen, their shapes (MLC apertures) can be determined, as shown in figure 2–2, which depicts a right anterior oblique head and neck field with the beam shape defined by a MLC. Note in figure 2–2 that the brainstem (green), spinal cord (purple), and pituitary gland (red) are all partially or totally shielded by the MLC whereas the PTV is completely covered by the radiation field. We should note that the MLC field shape shown in figure 2–2 is one that would be set only when doing conventional 3DCRT treatment planning. With ITP the treatment planner only defines the maximum field shape while the computer optimization algorithm determines the *best* IMRT field shape and dose intensity pattern.

Selection of beams is based on a combination of experience, (sometimes) standard protocols, and patient specific anatomy. For some sites, such as prostate, the relatively small variations between patients in anatomic details permit the use of standard templates for initial parameters such as beam directions and dose-volume constraints. For other sites, such as brain and head and neck, beam selection is highly dependent on the patient specific anatomy and the experience of the individual planner.

Selecting optimum beam directions will often have an effect on the required degree of intensity modulation within each field, which one would like to minimize. IMRT fields with large intensity fluctuations require higher velocities and accelerations of MLC leaves (when using dynamic, or sliding window IMRT) during beam delivery, which can be impractical to deliver. Similarly, for delivery of static IMRT fields (or step and shoot) large dose fluctuations require more beam segments, which will likewise increase delivery time. Also, as the complexity of an intensity profile increases, so does the total number of monitor units, and hence the contribution of scatter and leakage dose, which at best can only be accounted for approximately by most dose calculation algorithms. Selection of optimum beam directions is particularly important when treating concave shaped PTVs or where there is minimal separation between the PTV and a critical normal structure. Unfortunately, none of the available IMRT treatment planning systems can effectively determine optimum beam directions, and this must still be done based on the treatment planner's experience and intuition.

Plan Optimization

In forward 3DCRT planning, the starting point for dose calculations (by the computer) is the number of beams, their directions and shapes, inclusion or exclusion of hard or dynamic wedges, and the *static* beam intensities *selected by the planner*. In forward planning the computer merely calculates the resulting dose distribution from the beam intensities and shapes selected by the human treatment planner. This is fundamentally different from ITP where the human treatment planner specifies the desired dose distribution and the computer calculates the required beam intensities and shapes to best meet the specified dose distribution.

In either case, after the computer calculates the resultant dose distribution, the planner may choose to make adjustments in order to improve the plan. In forward planning the user adjusts the beam intensities (or weights), wedges, field shapes and/or directions, and then asks the computer to recalculate the dose distribution. This process, called *manual optimization,* is repeated until an *optimal* dose distribution is obtained. The experience of the individual planner is critical in manual planning, but even for a skilled planner, the quality of the plan is limited by the restricted number of degrees of freedom one has, particularly the constraint that intensity within each field must be uniform. In addition, each iteration is time consuming and the ability to truly optimize a dose distribution with 3DCRT techniques is often quite limited.

With inverse planning the philosophy behind optimization is completely reversed as the user does not directly optimize or re-adjust beam intensities. Instead, after defining the orientation and energies of all beams (but *not* their intensities) the planner specifies the *desired* dose limits for the PTV and all tissues of interest, from which the *computer* optimizes the dose intensity pattern within each. This combination of different intensity sub-beams plus hundreds (or even thousands) of iterations often enables significantly improved dose distributions with IMRT as compared to 3DCRT. If, however, the resulting dose distribution is not acceptable, the only recourse the treatment planner has is to modify the dose-volume constraints and restart the computer optimization algorithm, although the planner can also add or subtract fields, change field directions, and modify contours used for optimization. This process of inverse treatment planning was first proposed by Brahme and Bortfeld (Brahme 1988, 1999; Bortfeld 1999).

Central to the success of any optimization schema is the specification of an objective or cost function. The cost function is a mathematical definition of the *goodness* of a treatment plan, and the computer optimization algorithm attempts to minimize the cost function as it adjusts the beam intensities from one iteration to the next. Objective functions can be based on biological criteria, or more often dose-based criteria. Biological-based objective functions use a calculated radiobiological response as a measure of the merit of a plan (with calculations based on some model that relates radiation dose plus volume of irradiated tissue to predicted biological response). The use of biologically weighted objective functions is in principle more relevant because treatment outcome is determined by the biological response. However, a universally accepted biological model that predicts treatment outcome is yet to be developed (Niemierko, Urie, and Goitein 1992; Lyman 1985). Hence, most inverse planning algorithms currently rely on dose-based cost functions.

The numerical value of a dose-based objective function is calculated from a weighted average of the differences between delivered and prescribed doses for every voxel in every tissue defined in the treatment plan (i.e., the PTV plus all normal tissues for which a dose constraint has been specified). The prescription dose to the target, specified tolerance doses to normal tissues, and weighting factors for each tissue are designated as the *constraints*. For the PTV, one (of many) possible Objective Functions (OF) would be:

$$OF = \sum_i w_i * \left(d_{cal} - d_{pres}\right)_i^2 \qquad (1)$$

where

$\sum_i$ = a summation over all voxels in the PTV,

w_i = the weighting (or penalty) factor for the Ith voxel (usually within a PTV all voxels would have equal weighting factors, but this does not have to be true),

d_{cal} = the calculated dose (using the current beam parameters) for the ith voxel,

d_{pres} = the prescribed dose for the ith voxel.

The exact formulation of the objective however could take many other algebraic forms. For example, the absolute value of $(d_{cal} - d_{pres})_i$ could be used rather than its square. Or, one could include $(d_{cal} - d_{pres})_i$ in the summation only if $(d_{cal} - d_{pres})_i$ is negative; that is assign a penalty only if the calculated dose is less than the prescribed dose. Conversely, when calculating the contribution to the objective function for a normal tissue, one would usually assign a penalty only if $d_{cal} > d_{pres}$. The total objective function is then the sum of the objective functions for each tissue, weighted by the w_i values, which will likely differ for each tissue. For example, one would often assign a higher penalty or weighting factor to the spinal cord than to rectum or bladder, as the former is a more radiosensitive structure with more serious consequences if overdosed.

For many tissues, however, acceptable doses cannot be specified by a single dose value, and *dose-volume* effects are often incorporated into the objective function. A typical dose-volume constraint may be stated as "no more than q% of the particular organ may receive a dose greater than d." This is equivalent to specifying a single point on a DVH with the constraint that the value of the DVH at a dose value of d must be less than q%. Most inverse treatment planning algorithms allow the planner to define multiple such points for each tissue, with different penalties assigned to each point if desired. DVH type constraints, rather than a single dose limit, can also be assigned to the PTV if desired.

Several objective functions are illustrated graphically in figure 2– 5. For the PTV or target, the graph illustrates the concept of the *allowable inhomogeneity*. That is, if the dose is between a lower limit P_1 and an upper limit P_u, no penalty is assessed. Also, a larger weight can be assigned to penalize underdose as opposed to overdose (or vice versa). For the normal tissue, a penalty can be applied if dose exceeds a certain critical value (Dc) or based on dose-volume considerations, as represented schematically in the rightmost panel of figure 2–5.

Thus dose constraints are the prescribed dose and allowable dose inhomogeneities to the PTV plus the dose limits to various sensitive structures, specified by the treatment planner to the inverse planning computer system. The resultant inverse plan is highly dependent on the specification of these constraints. Overly lax or overly restrictive constraints may lead computer optimization to produce an inferior plan. In particular, if one specifies dose constraints that are physically impossible (for example, prescribing 100% dose to a paraspinal tumor with large penalty for underdose, plus 5% dose to an adjacent section of spinal cord with large penalty for overdose) the resultant inverse plan may be worse than a conventional plan. The reason for this seemingly incongruous result is that the optimization algorithm may attempt to adjust beamlet weights to extreme values in a futile attempt to meet dose constraints that it does not *know* are physically impossible.

Sometimes, defining subsections within certain sensitive normal tissues, such as the anterior rectal wall when treating the prostate, and assigning to that substructure less stringent dose constraints than have been assigned to the rest of the rectum is a more reasonable physical set of constraints, and may permit the optimization algorithm to find a better plan, rather than futilely searching for a physically impossible plan. Many treatment planning systems allow the use of Boolean logic for tissue contouring. This enables you, for example, to contour the prostate, the

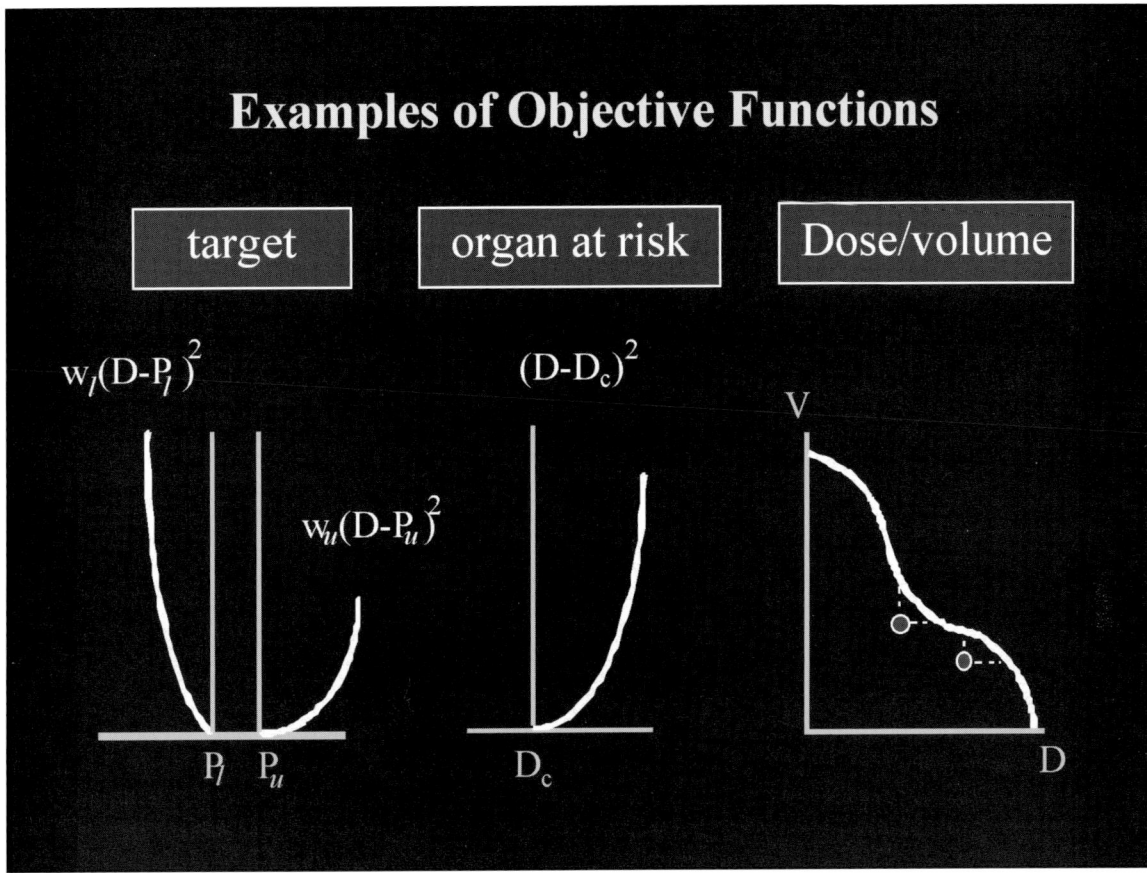

Examples of Objective Functions

| target | organ at risk | Dose/volume |

$$w_l(D\text{-}P_l)^2$$

$$w_u(D\text{-}P_u)^2$$

$$(D\text{-}D_c)^2$$

FIGURE 2–5. Examples of possible objective functions, illustrated graphically. For the PTV (leftmost graph) doses within the "allowable inhomogeneity" are permitted, with different penalties assessed for underdose as opposed to overdose. For the normal organ, penalty can be applied if dose exceeds a certain critical value (D_c) (middle graph), or based on dose-volume considerations (rightmost graph).

rectum, and then to define the overlap of the two contours as a separate structure, which can be assigned separate dose-volume constraints.

If the computer cannot find a set of beam intensity profiles to *exactly* match the desired dose constraints (as will inevitably be the case), then the software must select the *best* suboptimal plan; i.e., the plan that comes closest to meeting the desired constraints. In order to do this, the treatment planner must define for the computer a set of penalty values, or relative weights quantifying the relative importance of meeting the dose constraints set for each tissue. Normal tissues for which overdose must be critically avoided, such as spinal cord, would typically be assigned a high weight, indicating to the computer that the dose-volume constraints designated must be scrupulously adhered to, even if such restrictions result in underdosing parts of the PTV. Clearly, the choice of tissue weighting factors is a subjective decision on the part of the treatment planner.

Thus, ITP still requires a significant level of human input and control, but the human control aspect takes on a totally different form than for conventional treatment planning. To specify realistic dose constraints, users need to learn the relationship between the dose constraints and the resulting dose distributions for the specific inverse planning system they are using, plus have some familiarity with basic radiation dosimetry concepts. This learning process is one of trial and error, and is less intuitive than the trial and error planning process involved in learning 3DCRT planning. Once the treatment planner has gained this experience, it becomes possible to develop disease-specific templates which can be used as starting point dose constraints for commonly treated tumor sites such as prostate, nasopharyngeal, and oropharyngeal cancers. Patient-to-patient anatomic variations obviously limit the use of standard dose constraints, but they can be good starting points.

The bottom line, however, is that the planner must learn to specify dose constraints that are reasonable and physically achievable, otherwise ITP may turn out to be worse than conventional plans. These fundamental differences between ITP and conventional forward planning render most of our previous experience on designing, modifying, and improving treatment plans virtually useless. Thus, when you start IMRT, you also start at ground zero on a new learning curve about how to do treatment plans! With ITP, modifications of treatment plans and adjustment of isodose distributions can only be accomplished by changing dose-volume constraints and penalties; and not by user modification of beam weights or beam modifiers. Because different computer systems have different penalty functions and different scoring systems, dose constraints that work on one system may not work on another.

Plan Evaluation

The tools used for plan evaluation are similar for 3DCRT and IMRT, although the schema for adjusting plans based on this evaluation are quite different, as described above. Evaluation tools include 2-D dose distributions superimposed on CT images, 3-D volumetric and/or color wash rendering of dose distributions overlaid on the PTV and critical organs, structure-specific DVHs, and biological indices. But there is now more reliance on DVHs for plan evaluation. Additionally, inspection of beam intensity profiles (as shown in figure 2–6) is often useful for evaluating an individual beam's contribution to the dose distribution.

For reasons previously discussed, it is often difficult and unintuitive to adjust or improve an IMRT treatment plan by adjustment of dose-volume constraints, although this is what is commonly done. Sometimes the treatment planner may find it necessary to resort to artificial tricks to adjust an IMRT dose distribution. For example, if a resultant IMRT treatment plan has an undesirable hot spot that cannot be eliminated by adjusting penalty weights or dose-volume constraints, one could deceive the computer by contouring the overdosed tissues separately, defining them as new structures, and assigning different and separate dose constraints. An example of this is shown in figure 2–7 where we have added an axillary contour around the PTV (middle and outermost rings) to prevent the occurrence of a hotspot near the intersection of beams 1 and 2 outside the PTV (shaded zone), that would unwittingly result if the ITP optimization algorithm was not given these tissues as a dose-limiting structure. Resorting to such tricks is a recognition of the fact that DVH-based objective functions such as Equation (1) take no account of the spatial distribution of *hot* or *cold* dose regions. Nor, according to Equation (1), is a hot dose region of any importance whatsoever unless it occurs in a tissue having a dose-volume constraint. In reality, of course this is not always clinically realistic, hence the need sometimes to define substructures.

Generation Of Leaf Motion Files

After the computer calculates the required x-ray fluence for each field, a second calculation must be performed to determine the sequence of MLC leaf motion that will deliver this x-ray fluence. This is called the *leaf sequence file*, or *leaf motion file,* and is transferred from the treatment planning system to the linear accelerator either by floppy disk or via a networked data transfer system.

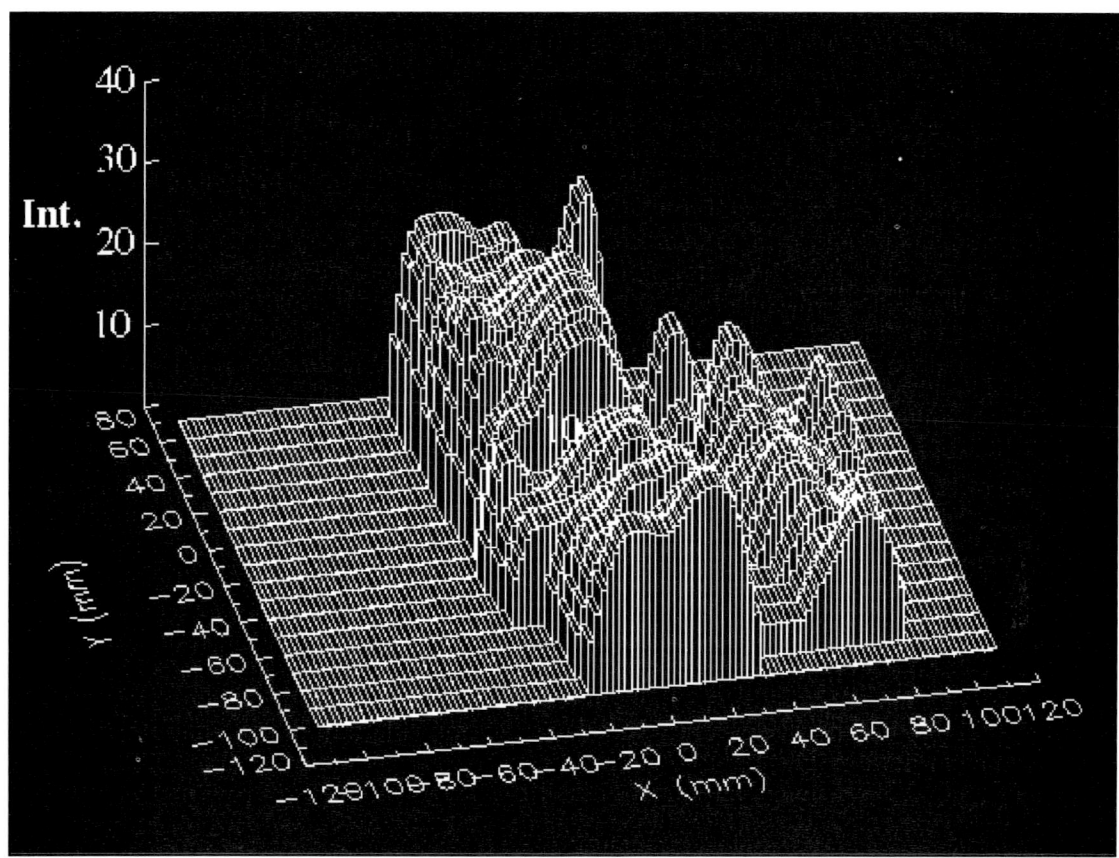

FIGURE 2–6. Intensity profile for a typical left posterior oblique DMLC field for treatment of nasopharynx. Note that for such fields it is not uncommon to have very steep dose intensity gradients.

When treating with sliding window IMRT, there are many possible sequences of dynamic leaf motions capable of producing a desired intensity pattern (or equivalently, combinations of multiple static field segments if you are using step-and-shoot IMRT). For a given intensity pattern, the optimum leaf sequence pattern is a compromise between efficient delivery, minimization of the number of segments, total monitor units (MUs), leaf travel distance, total delivery time, etc. (Spirou et al. 2001). Here, we describe two basic leaf sequencing methods using first dynamic MLC (sliding window), and second, static MLC (step-and-shoot) delivery.

One can consider a dynamic MLC intensity pattern to be a series of many 1-D strips of intensity profiles, with each strip or profile delivered by one pair of leaves. A photograph of an MLC is given in figure 2–8, and one can envision each pair of leaves sliding across the field at their own pre-programmed velocities to deliver a specific intensity pattern. This is illustrated in figure 2–9 where we show a schematic representation of radiation delivery using the sliding window method. The dotted lines represent the positions of a leaf pair (x-axis) as a function of beam-on time (y-axis). Both leaves start at the extreme left edge of the intended treatment field. As the beam is turned on (point a), both leaves move, with different speeds, from left to right (initially the right leaf moves more rapidly than the left leaf). The point P begins to receive radiation when the right leaf edge moves past it (point b), and continues to receive radiation until the left leaf blocks the beam (point c). By controlling the movement of the leaves and therefore the *beam-on-time* duration between b and c, one can deliver any desired intensity to point P, or

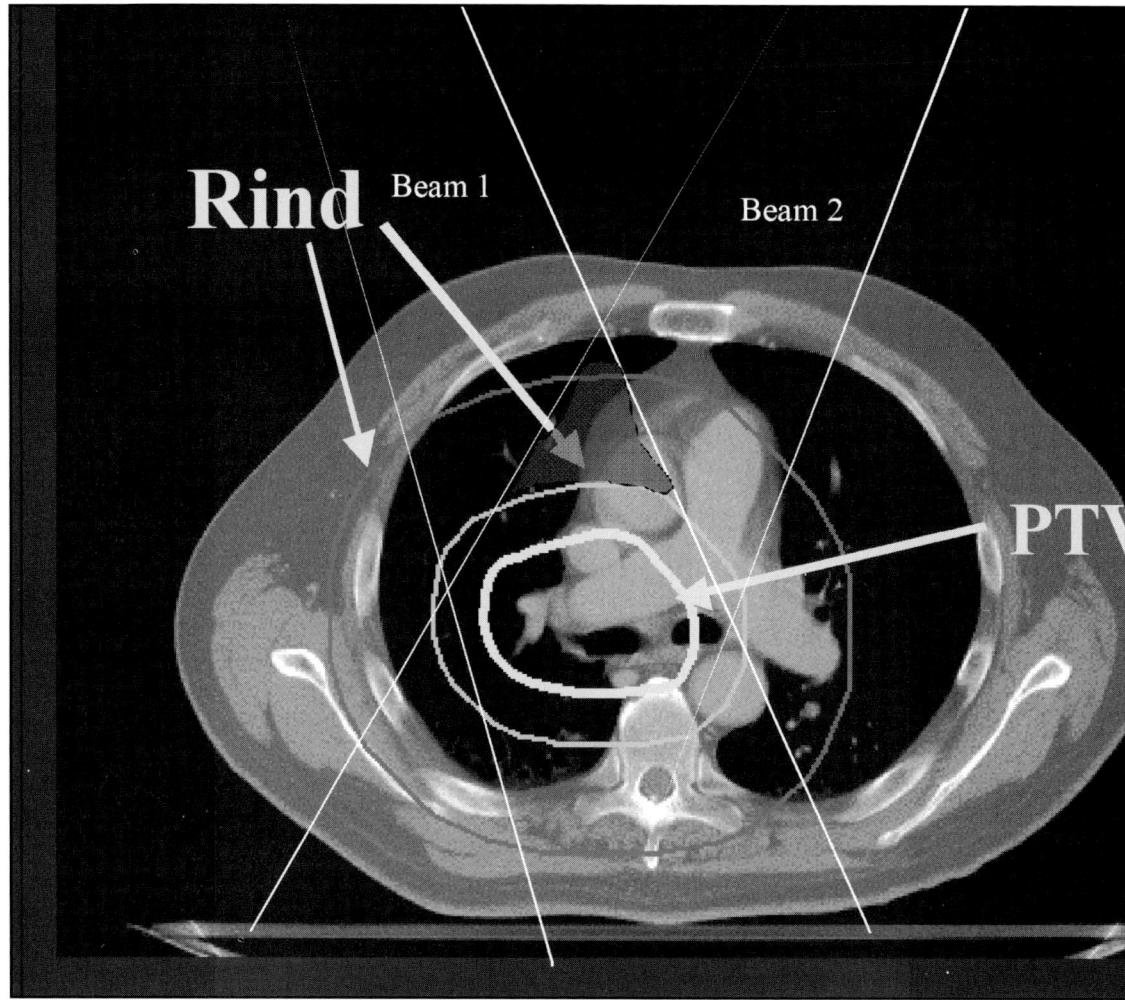

FIGURE 2–7. Example of contouring "false structures," as designated by the rind defined by the middle and outermost contours (PTV contoured in center, thicker line) to avoid a possible hot spot in the region of beam overlap (shading). This is often required for ITP since the optimization algorithms will not otherwise avoid overdosing tissues without designated dose-volume constraints. See COLOR PLATE 37 (Figure 13–5).

any other point under this leaf pair. The solid line depicts the total integrated beam intensity to all points underneath the strip of tissue being treated by this leaf pair.

By extending this concept to multiple pairs of leaves, any desired intensity-modulation pattern can be produced with designed sequences of leaf positions. Since each leaf pair must deliver a different intensity profile, the speed of each leaf, and its position as a function of MU delivered must be individually controlled. For maximum efficiency in beam delivery (i.e., shortest possible treatment times) it is necessary to be able to move all leaves simultaneously. This requires both leaf speed modulation and dose rate modulation. Ideally, one would always be treating at the maximum dose rate of the linear accelerator, but when a leaf is required to move a large distance, requiring a speed exceeding the maximum mechanical leaf speed, the dose rate must be reduced. The shape of the field at any point in time can be monitored (if desired) by an EPID (Chang et al. 2000; Curtin-Savard and Podgorsak 1999), as shown in figure 2–10.

FIGURE 2–8. Photograph of MLC located in "head" of linear accelerator. Note multiple leaf pairs and individual control motors for each leaf (courtesy Varian Medical Systems, Palo Alto, CA).

In practice, a separate computer program (independent of the inverse planning algorithm), sometimes called the *leaf-sequencer,* is used to translate the intensity profiles of the IMRT beam into the so-called DMLC file that contains the data of the leaf positions as a function of monitor units. In practice, not all desired intensity profiles are (exactly) achievable because of the constraints on leaf motion imposed by the design of the MLC and the clinical dose rate of the machine. The MLC manufactured by Varian Medical Systems (Palo Alto, CA), for example, is constrained by the maximum leaf speed of 2.5 cm/sec and a maximum DMLC field width of approximately 14.5 cm without carriage movement.

For static, or step-and-shoot IMRT, the continuous 2-D intensity distribution is approximated by dividing it into a discrete matrix with a user-selected number of intensity levels. The greater the number of intensity levels, the larger the number of segments required for delivery, and the closer the step-and-shoot method comes to sliding window technique. As an example, we show in figure 2–11 a discretized intensity pattern having eight levels of dose intensity (figure 2–11a). There are many ways to deliver the intensity pattern of figure 2–11a, one such method being to first determine the maximum contiguous field shape that can be exposed to an intensity equal to one half of the maximum intensity (i.e., 4 units). This area would be the first field shape to be treated, as shown in figure 2–11b (white boxes treated in the first step, with the light and dark gray boxes being blocked by the right and left bank of MLC leaves respectively). Subsequent field shapes are shown in panels 11c–11f, and have intensity levels of 2, 2, 1, and 1 respectively. Each segment or step in the treatment, i.e., panels 11b–11f, can be considered individual con-

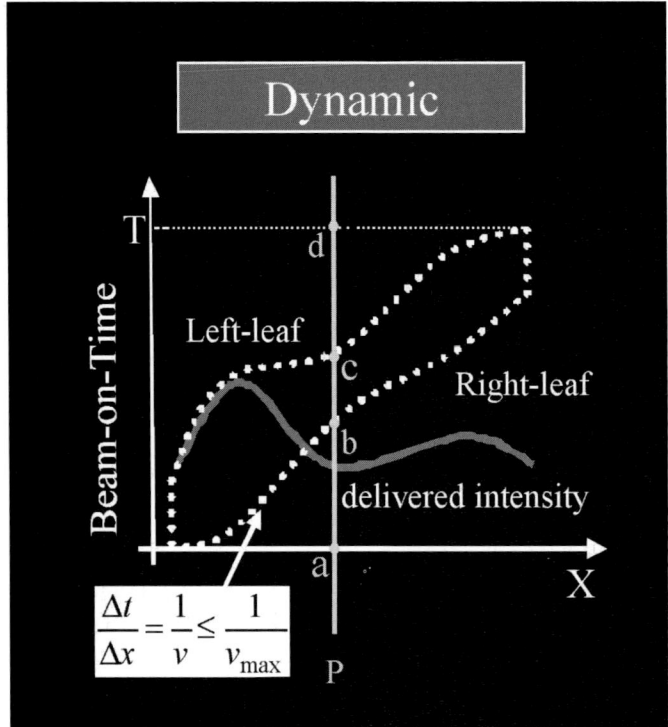

FIGURE 2–9. Schematic representation of radiation delivery using the sliding window method. The dotted lines represent the positions of a leaf pair (x-axis) as a function of beam-on time (y-axis), and the solid line depicts the total integrated beam intensity to all points underneath the strip of tissue being treated by this leaf pair. See text for details.

ventional fields, but the summation of b–f equals the desired intensity level distribution of panel 11a.

Some clinicians feel more comfortable with step-and-shoot than with sliding window, because it is more similar to conventional therapy, as nothing moves while the beam is on. But one pays a price in that step-and-shoot cannot deliver infinitely small gradations in intensity variations as can sliding window technique.

Delivery Of Intensity-Modulated Treatment

Several different techniques have been developed for delivery of IMRT. The first method is the relatively "low tech" fabrication of custom-designed 3-D physical compensators. These can be fabricated using a computer-controlled milling machine to cut either the compensator itself or a negative mold of the compensator to be filled with Cerrobend™ or similar type material. This approach can be implemented without a significant capital investment as it does not require an MLC. There is, however, additional personnel time required for fabricating the filters and for inserting them unto each field during treatment. Physical compensators also harden the beam, generate scattered radiation, and increase skin doses and doses outside the field. Nonetheless, the intuitive and low-tech aspects of this technique make it appealing at some centers, and with the exception of increased scatter and skin dose, it is technically capable of delivering dose distributions comparable to more technologically complex MLC-based beam delivery systems.

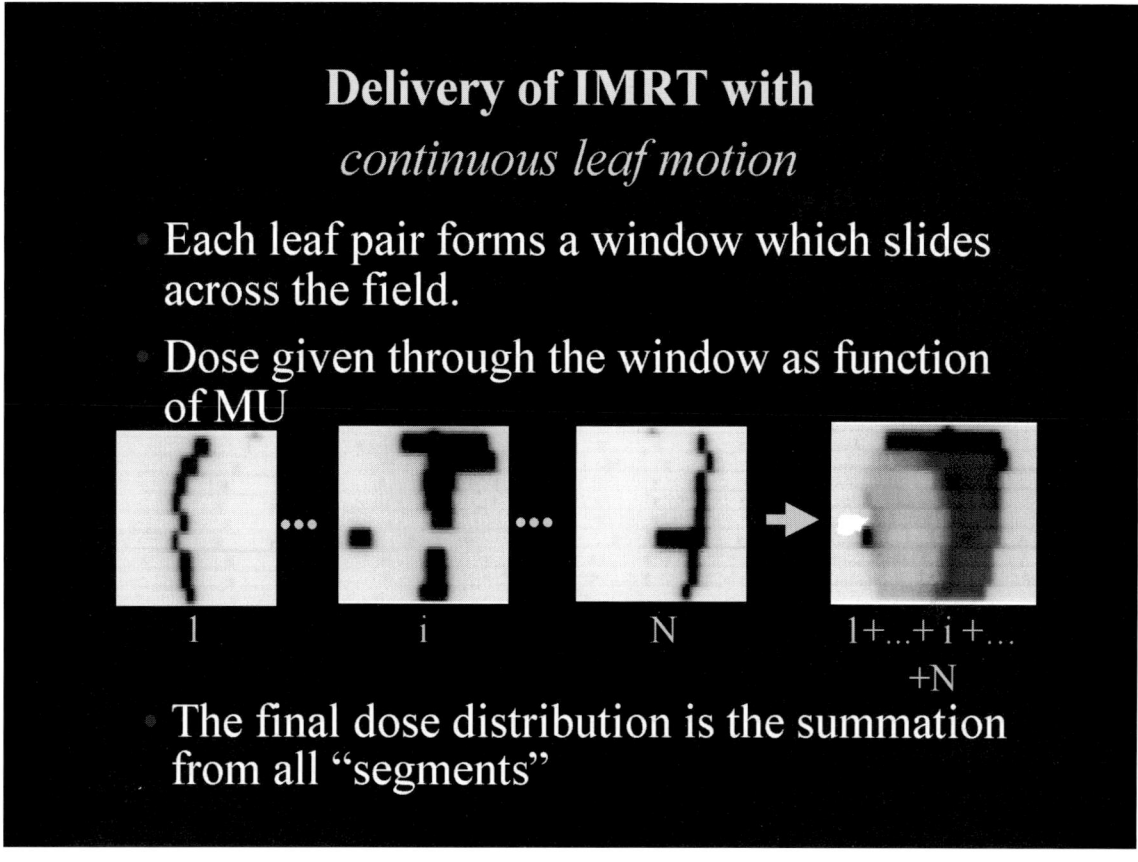

FIGURE 2–10. Use of an EPID system to monitor the delivery of an IMRT treatment. At any instant in time the IMRT field is an irregular slit (panels 1–3 on the left), as recorded by a transmission EPID image. The total dose from this beam is the summation of all segments, as indicated in gray scale (white = low dose, black = high dose) in the rightmost panel.

Much more commonly, however, IMRT is delivered using a computer-controlled MLC, such as shown in figure 2–8. MLC-based IMRT can, as discussed above, follow one of two basic approaches; step and shoot technique, or sliding window. Both techniques require a computer-controlled MLC, but differ in that the latter approach utilizes continuously moving MLC leaves (while the radiation beam is on), whereas the former approach utilizes a sequence of multiple, fixed field shapes, each with its own predetermined incremental radiation dose. In the step-and-shoot technique, after the dose from the first subfield is delivered, the beam is turned off, the field shape reset by the MLC, the second incremental dose delivered, etc. The number of steps or field shapes in the sequence can be as small as a few, or as large as one hundred or more.

A technologically different approach to either of these techniques is tomotherapy, wherein a rotating fan-beam (broad in the transverse direction, but narrow in the superior-inferior direction) is intensity modulated with a bimodal MLC. The bimodal MLC also has multiple leaf pairs, but each leaf pair has only two positions—completely open or completely closed. Two basic tomotherapy systems have been proposed. A group at the University of Wisconsin (Mackie et al. 1999) has designed a dedicated tomotherapy machine that is a hybrid between a spiral CT scanner and a linear accelerator. The second approach, which is commercially available (NOMOS Corp, Sewickley, PA), utilizes a custom-designed bimodal MLC that is attached to a conventional

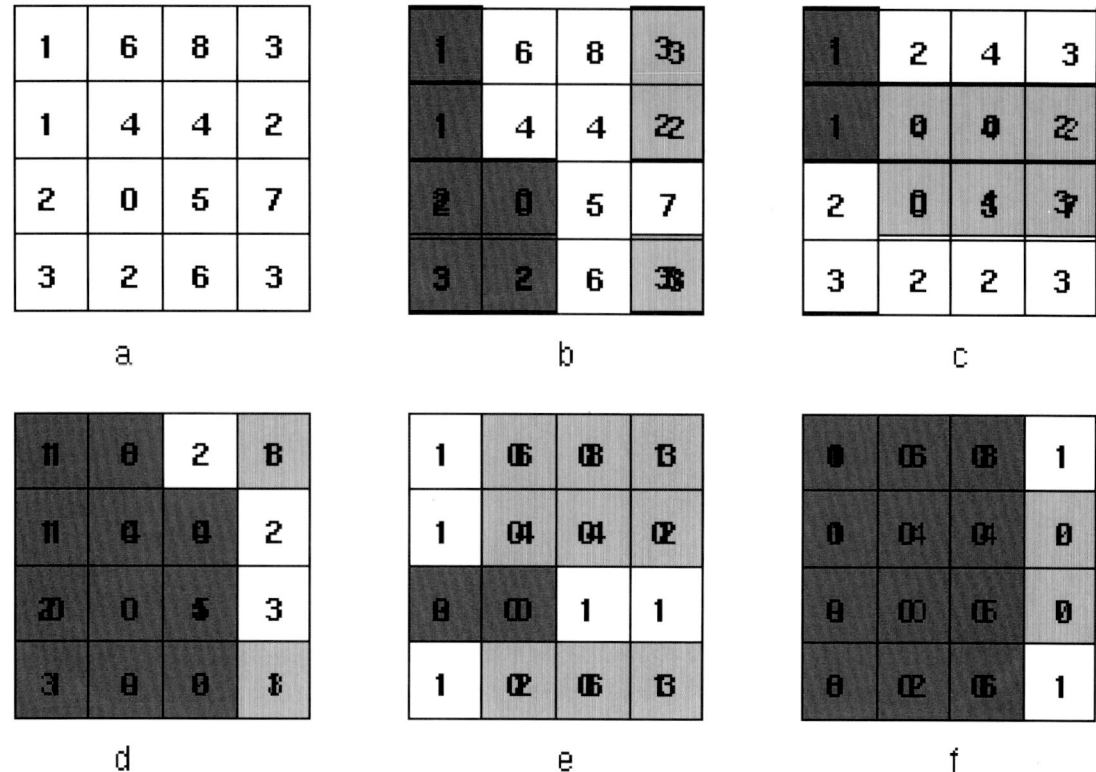

FIGURE 2–11. An example of an IMRT treatment using five step-and-shoot beam segments. The desired final intensity map is shown in panel a. Panels b–f depict five separate MLC-shaped fields (open portion of field in white, blocked portions of field in dark and light gray) with beam intensities of 4, 2, 2, 1, and 1 respectively used to deliver this total intensity pattern. See text for details.

linear accelerator that is programmed to deliver arc type beam delivery. In this system, a spiral CT-like beam delivery is approximated by precise stepping of the linac treatment couch to deliver abutting slices of RT.

Characteristics Of MLCs

To better understand some of the technical details of MLC-based IMRT delivery, we describe below the basic physical and geometric characteristics of MLC design, as implemented by three major manufacturing companies (Elekta, Norcross, GA; Siemens Medical Systems, Concord, CA; and Varian Medical Systems, Palo Alto, CA). The three designs differ in many ways including the location of the MLC relative to the conventional jaws, whether MLC leaves are single focused or double focused, the physical characteristics or shapes of leaves, leaf movement restriction, and maximum achievable field sizes. These design features all have an impact on IMRT beam delivery and dosimetry characteristics.

The MLC can be installed within the accelerator gantry head either above the upper set of secondary collimator jaws or below the lower set of secondary collimator jaws as a tertiary collimator. In the Elekta linear accelerator, for example, the MLC replaces the upper (Y) jaws although it is augmented with an additional 3-cm thick Y jaw to reduce the radiation leakage. The Siemens MLC completely replaces the lower (X) jaws of the LINAC. The Varian MLC is a tertiary collimator located below and in addition to the conventional X and Y secondary collimators. Placing the MLC below both sets of secondary collimators (as in the Varian design) places it closer to the patient, which in principle improves the geometric penumbra and reduces leakage radiation, but at the expense of a physically larger device with less clearance between the MLC and the patient. Placing the MLC above the secondary collimators (as in the Elekta design) produces a poorer geometric penumbra, but results in a smaller and lighter MLC design with increased physical clearance between the accelerator head and the patient. The Siemens design obviously takes a middle course in this unavoidable tradeoff of geometric penumbra versus MLC size.

Adjacent MLC leaves must slide smoothly across each other with minimal gaps between leaves to reduce leakage and transmission radiation. This is accomplished by machining tongue-and-groove patterns into the sides of the leaves as shown schematically in figure 2–12. The side of one leaf has an extended portion called the *tongue*, while the abutting side of the adjacent leaf has an indented portion called the *groove*. Two adjacent leaves are coupled together as the tongue of one leaf slides within the groove of the adjacent leaf. Each leaf has a tongue on one side, and a groove on the opposite side. The manufacture of precise tongues and grooves into extremely small individual leaves is an exacting process that places a lower limit on the size of individual MLC leaves.

While this tongue-and-groove design reduces radiation leakage, it also complicates treatment planning dose calculations because the transmission through any leaf depends on whether the beam passes through the tongue, the center, or the groove portion of the leaf. If adjacent leaves are set to the same position, this is not an issue as transmission through the tongue of one leaf plus the groove of the adjacent leaf is very nearly equal to the transmission through the center of a leaf. But when two adjacent leaves have different degrees of extension, the tongue side of the

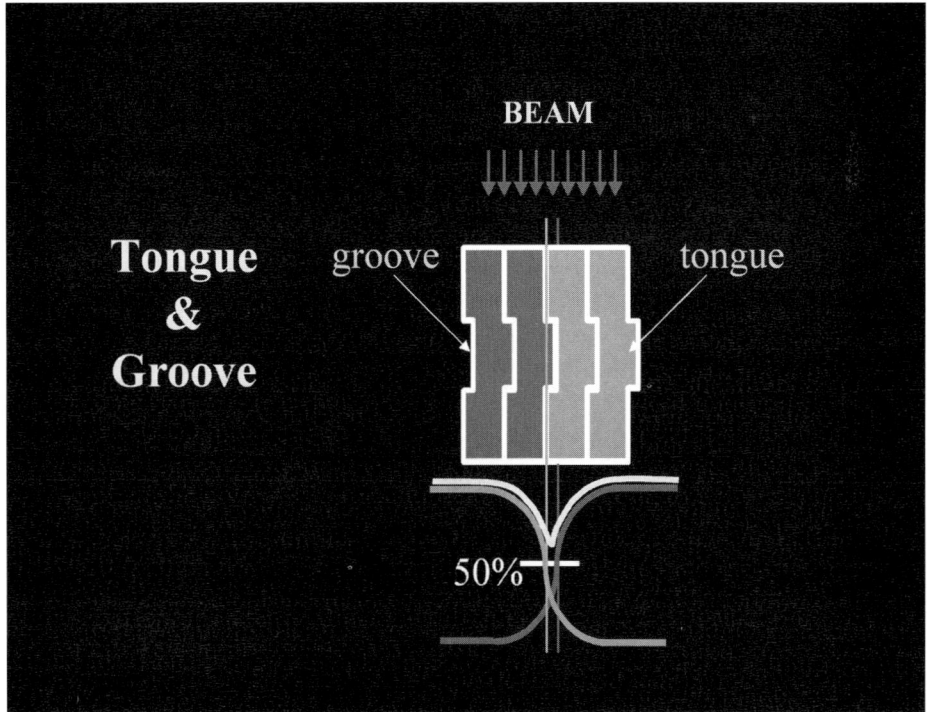

FIGURE 2–12. Schematic "end-on" view of MLC showing tongue-and-groove design and approximate resulting beam transmission.

more extended leaf produces an underdose region near the leaf edge. This phenomenon is called the *tongue-and-groove effect* (Sykes and Williams 1998, van Santvoort and Heijmen 1996).

Another design consideration is how to focus the ends of MLC leaves to match radiation beam divergence. This is of course a 2-D problem as the beam diverges both along the direction of leaf motion and perpendicular to it (designated as the X and Y directions respectively). Focusing MLC leaves in the Y direction can be accomplished in the same manner as in a conventional secondary collimator; namely, the width of each MLC leaf is slightly narrower at the top of the leaf (nearer the radiation source) than at the bottom of the leaf (nearer the patient). MLCs focused only in the Y direction are called *single focused.*

Double-focused MLCs are also focused in the X direction, parallel to the leaf motion. This is a more difficult problem because in the X direction beam divergence is a function of leaf position, being equal to the arctangent of leaf position (i.e., field size) divided by the distance of the MLC from the source. Of the three manufacturer designs discussed here, only the Siemens MLC is double-focused. "Pseudo" 2-D focusing can be achieved by making the ends of the leaves slightly convex as shown in figure 2–13. This provides a reasonably faithful geometric penumbra over

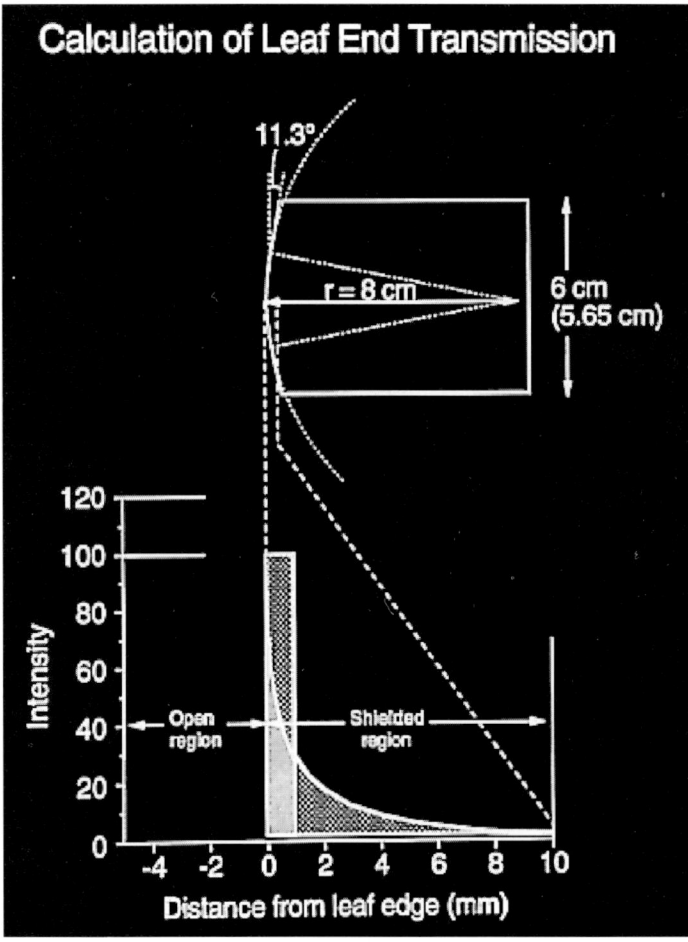

FIGURE 2–13. Schematic of a typical rounded leaf end design of an MLC. Note that the variable beam transmission affects the shape of the beam dose penumbra, and that both penumbra and coincidence of light and radiation fields are functions of leaf position.

all field sizes, but compromises beam transmission through the leaf ends. Coincidence between radiation and light fields is also dependent on leaf position when leaves are not double focused.

The number of leaves, leaf thickness, and width also differs between manufacturers. The Elekta MLC consists of 40 pairs of tungsten alloy leaves with 7.5 cm thickness with each leaf projecting as 1.0 cm wide and 32.5 cm in length at the isocenter (the physical dimensions of the leaves are of course less than half this size). The Siemens MLC consists of 29 pairs of tungsten alloy leaves, also with 7.5 cm thickness. The inner 27 pairs of leaves project a 1.0 cm leaf width and 31 cm leaf length at the isocenter, while each outer pair of leaves (called *mage leaves*) project a 6.5 cm leaf width at the isocenter. Several different Varian MLC designs exist incorporating 26, 40, or 60 pairs of tungsten alloy leaves with 5 cm thickness. In the 26 and 40 pair designs, all leaves project to 1 cm width at isocenter, but in the 60 pair design the central 40 leaves project to 0.5 cm width at isocenter, and the outer 20 pairs project to 1 cm width.

Leaf motion restrictions also affect the delivery, efficiency, and dose accuracy of IMRT. Figure 2–14 shows three such leaf motion restrictions. Figure 2–14a shows a design incorporating a minimum gap between opposing leaves, usually incorporated in the design to prevent leaves from hitting each other (left bank leaves are denoted in gray, and right bank leaves in black). Figure 2–14b shows a design that prohibits interdigitation, which occurs when one (black) leaf end extends beyond the tips of its two neighboring opposing (gray) leaves. Figure 2–14c shows a design restricting leaf over-travel, either beyond the field center, or beyond the back ends of its neighboring

Leaf Motion Constraints

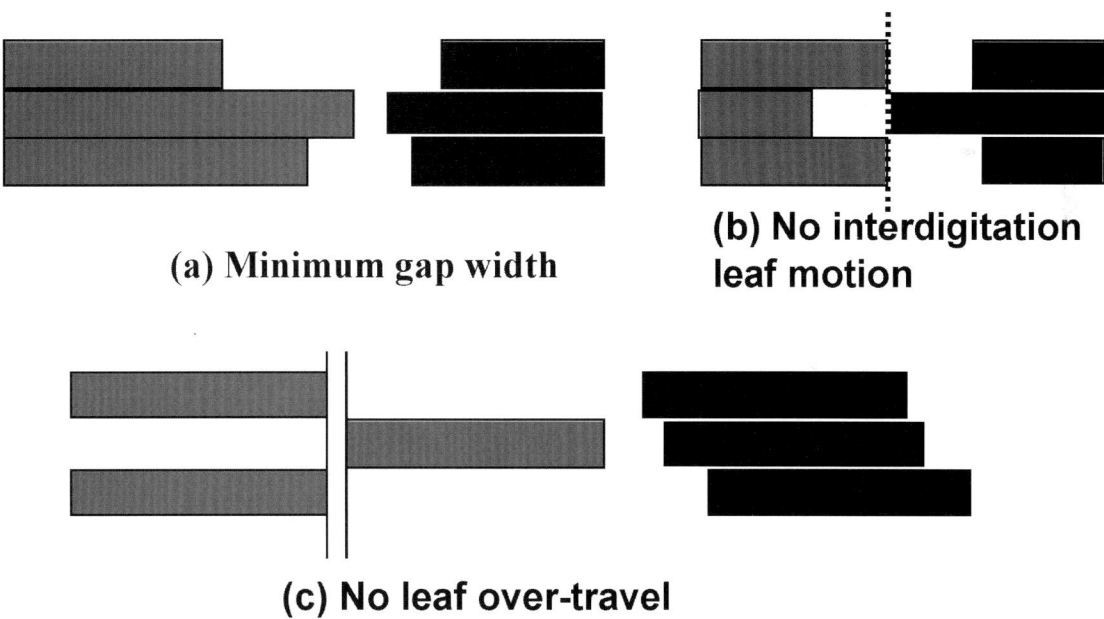

(a) Minimum gap width

(b) No interdigitation leaf motion

(c) No leaf over-travel

FIGURE 2–14. Three types of leaf motion constraints. (a) No leaf end can come closer than a fixed distance (typically 0.5–5.0 mm) to the tip of its directly opposing leaf or to the tips of its two neighboring opposing leaves. (b) A design that prohibits interdigitation. (c) Leaves are not allowed to extend either a fixed distance beyond field center (typically 15 cm), or beyond the back end of the carriage (thus leaving an uncollimated rectangle of beam at the trailing edge of the leaf).

leaves. There are also restrictions on the distance any leaf can travel beyond field midline, called the *over travel distance.*

These restrictions affect the maximum IMRT field size, which is always smaller than the maximum static field size, and usually smaller than the maximum static MLC field size (i.e., when the MLC is used simply to replace a conventional Cerrobend block). These over-travel restrictions are required to prevent the occurrence of a hole in the field at the "tail end" of a fully extended leaf. Full over-travel freedom would only be possible if each leaf were as long as the maximum field width (usually 40 cm), but then, the physical size of the MLC would be inconveniently large.

The maximum field size in IMRT delivery can be increased if the secondary jaws can be moved to cover the region behind the most extended MLC leaves, such as the case for Varian's MLC. Thus, a combination of carriage moves plus secondary jaw moves can be used to increase the maximum field size for IMRT, although this complicates treatment planning in that each translation of either the carriage or the jaws may need to be considered as a separate field. (Even with this, the maximum IMRT field size is still less than 40 cm due to over-travel limits on the secondary jaw and MLC leaves.) With the Varian system, for example, fields wider than approximately 14.5 cm must be split into two or three fields. Finally, there are limits on maximum leaf speed that are on the order of 2 cm/second at the isocenter for all major MLC designs.

Dose Calculation Algorithms

Dose calculation models for IMRT are not fundamentally different from those used for conventional forward planning. This is partly because the dose calculations and the actual optimization processes are decoupled. That is, first the computer divides each treatment field into multiple pencil beams and calculates the dose distribution for each individual pencil beam. The pencil beams, or beamlets, are on the order of several square millimeters. These dose distributions are stored in memory and need only be calculated once. The optimization algorithm then iteratively adjusts the weight of each pencil beam until the composite 3-D dose distribution best conforms to the specified dose objectives (Spirou and Chui 1998; Chui, LoSasso, and Spirou 1994). But the initial pencil beam dose calculation algorithm need not be fundamentally different from calculation algorithms used for conventional treatment planning.

Having said that, we must note that there are still some complications associated with MLC IMRT dose calculations. First, when delivering IMRT dose, only a portion of the entire field is exposed at any given time. As a result, the total beam-on time (or monitor units) required can be 2 to 4 times longer than for conventional treatments. In general, the more complex the intensity modulation, the more monitor units required.

This means that factors such as variable beam transmission through rounded leaf ends or through tongues and grooves, which are ignorable in conventional dose calculations must be more accurately accounted for with IMRT. Radiation leakage through MLC leaves can also amount to several percent of total IMRT dose as opposed to less than 1% for conventional treatments. Extremely variable effective field sizes during IMRT beam delivery also complicate calculation of output factor, as the relationship between Collimator Scatter Factor (CSF) and Phantom Scatter Factor (PSF) becomes more complicated than for conventional dose calculations (LoSasso et al. 1993; LoSasso, Chui, and Ling 1998).

Another problem is the requirement for independent, or second check, monitor unit calculations. For dynamic IMRT treatments in particular, these are too difficult to perform manually, and must be made with a second, completely independent computer program. Many IMRT treatments require as many as 800 or 900 MUs per fraction; the clinical delivery of which can be particularly unnerving without the "security blanket" of old-fashioned central axis dose calculation checks. But independent second check dose calculation programs are only now becoming commercially available. Some institutions have developed "in-house" software for this purpose.

Without such software, the only method for independently verifying IMRT dose calculations is tedious physical measurement.

When IMRT treatments were started at MSKCC in 1995 we did not have independent calculational methods for verification of IMRT doses or MU calculations, and we relied instead on *in vivo* and phantom dosimetry measurements for *every* IMRT patient. These phantom dosimetry tests were performed by having the treatment planning system recalculate the dose distributions that would ensue in a regular geometric phantom (usually rectangular or cylindrical in shape, constructed of either water or plastic) using the same intensity-modulated beams as were calculated for the patient treatment. Film or ion chamber dosimetry can then be performed in the phantom, and if agreement between measurements and calculations *in the phantom* are acceptable, one assumes that the computer-calculated dose distributions in the patient are also correct. One type of possible dosimetry phantom for these measurements is shown in figure 2–15. In figure 2–16 we show sample results comparing a calculated "treatment plan" in a phantom to the results of film dosimetry (LoSasso et al. 1993; LoSasso, Chui, and Ling 1998, 2001; Wang et al. 1996).

In addition to phantom studies, *in vivo* dosimetry measurements can be performed in the patient using diodes, thermoluminescent dosimeters, etc. The techniques for *in vivo* dosimetry for IMRT are similar to measurements of conventional radiation fields, except that for IMRT the dose gradients within a single field can be very large (10% to 20% per millimeter or more), and accurate placement of dosimeters can be critical. Small errors in dosimeter placement could result in what appear to be erroneous dose measurements, and vice versa. The difficulties of experimentally

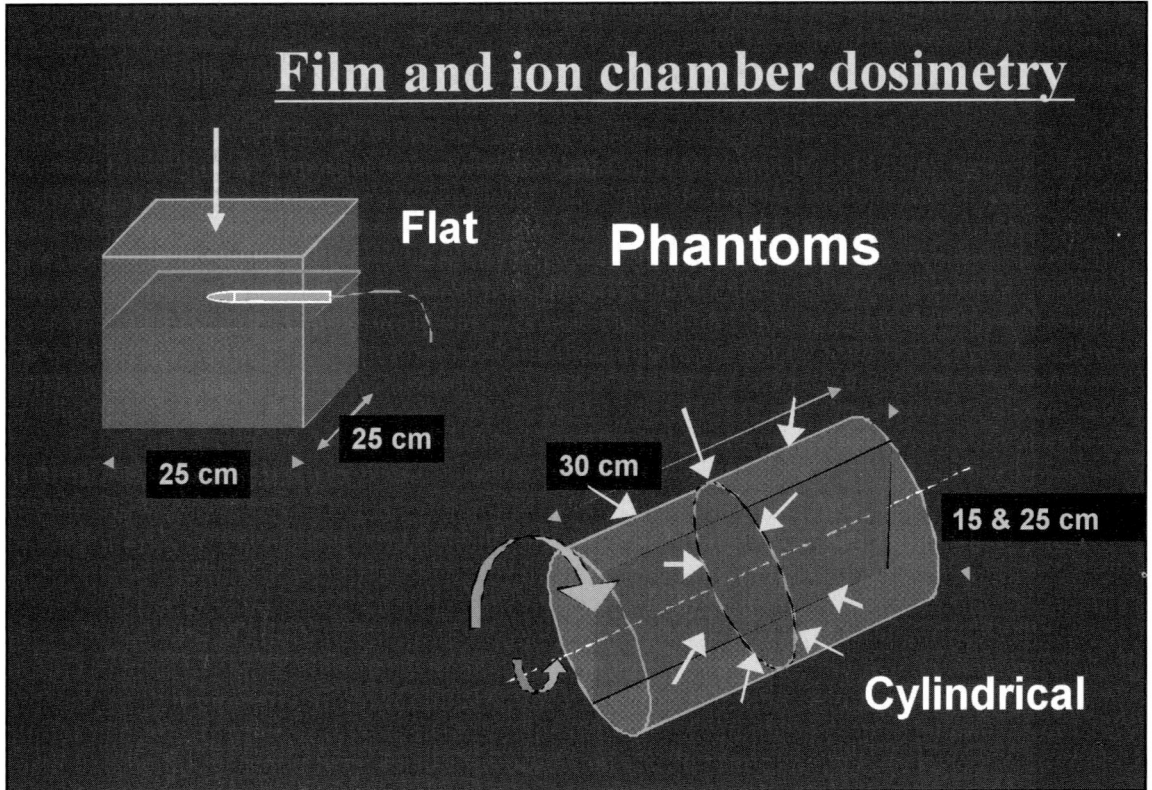

FIGURE 2–15. Schematic of rectangular and cylindrical plastic phantoms for ion chamber or film dosimetry.

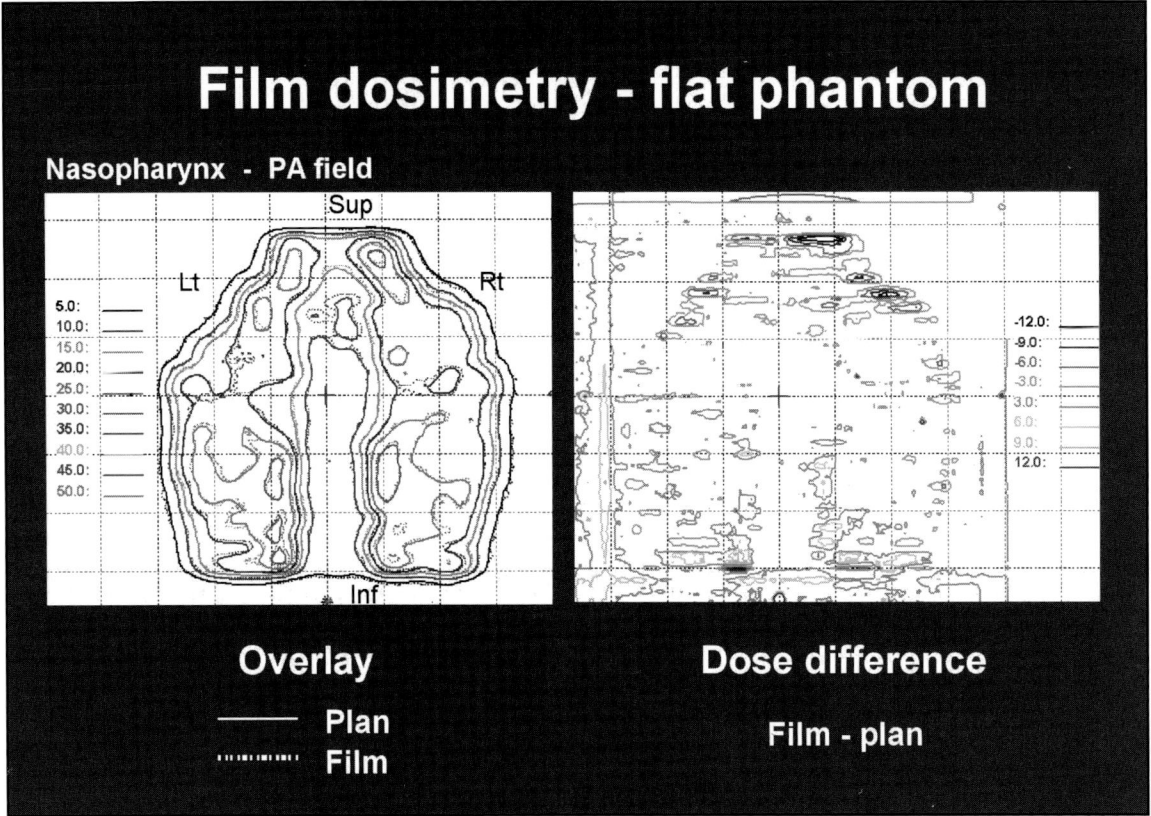

FIGURE 2–16. Using film dosimetry to compare treatment planning dose calculations with radiographic film measurements. Left panel: Overlay of calculated and film dosimetry isodose contours as measured in a flat phantom for a typical posterior prostate IMRT field. Note low-dose region in center of field designed to spare rectal dose. Right panel: Difference plot of left panel, shows more clearly differences in calculated and measured dose. See COLOR PLATE 14 (Figure 8–12).

verifying the dose accuracy of IMRT fields with high dose gradients, coupled with the complete inability to perform human checks of dose calculations and monitor units can be a particularly unnerving introduction to IMRT!

Quality Assurance For IMRT Delivery

IMRT requires a more extensive physics quality assurance (QA) program than does conventional therapy. Items of concern include mechanical tests of the MLC, dosimetric measurements required for acceptance testing and commissioning, plus tests specific to each individual patient's treatment plan.

Mechanical Tests

The geometric accuracy of leaf positions is critically important for both static and dynamic IMRT, much more so than collimator settings for conventional radiotherapy. For dynamic IMRT the dose delivered to each strip of tissue is directly related to the gap width of the leaf pair passing over it, and small errors in average gap width can result in significant dose errors. Consider for example an MLC field for which the average gap width is intended to be 1 cm, but for which one of the leaves has a systematic positional error of 1 mm. The resulting gap, and consequently the delivered dose to that strip, would be in error by 10% (i.e., 1 mm/1 cm). Similarly, for static IMRT such a systematic error in leaf position would affect the dose not just at a single field edge (as for conventional therapy), but at the edge of every subfield, or segment in that particular treatment beam.

Mechanical calibration of MLC leaf positions can be tested using procedures and software supplied by the manufacturers, although independent physical verification methods such as caliper/micrometer measurements of leaf positions are also recommended. Several MLC test patterns have been designed for periodic checking of MLC accuracy. One such test pattern (see figure 2–17) can be used to monitor MLC performance via radiographic film measurements. A leaf motion file designed expressly for this test instructs the MLC to deliver 5 narrow vertical bands of radiation. The leftmost panel shows a film exposure in which the MLC functioned properly, as indicated by the fact that all five strips are of equal intensity and thickness (lighter horizontal bands and streaks are the result of normal MLC leakage). In the rightmost panel, however, several individual leaves have been intentionally miscalibrated by –0.5, –0.2, +0.2, and +0.5 mm (top to bottom of film, respectively). Note the narrow spots in all five vertical bands for the –0.5 and –0.2mm errors in calibration, and the broad spots for the +0.5 and +0.2 mm errors. This

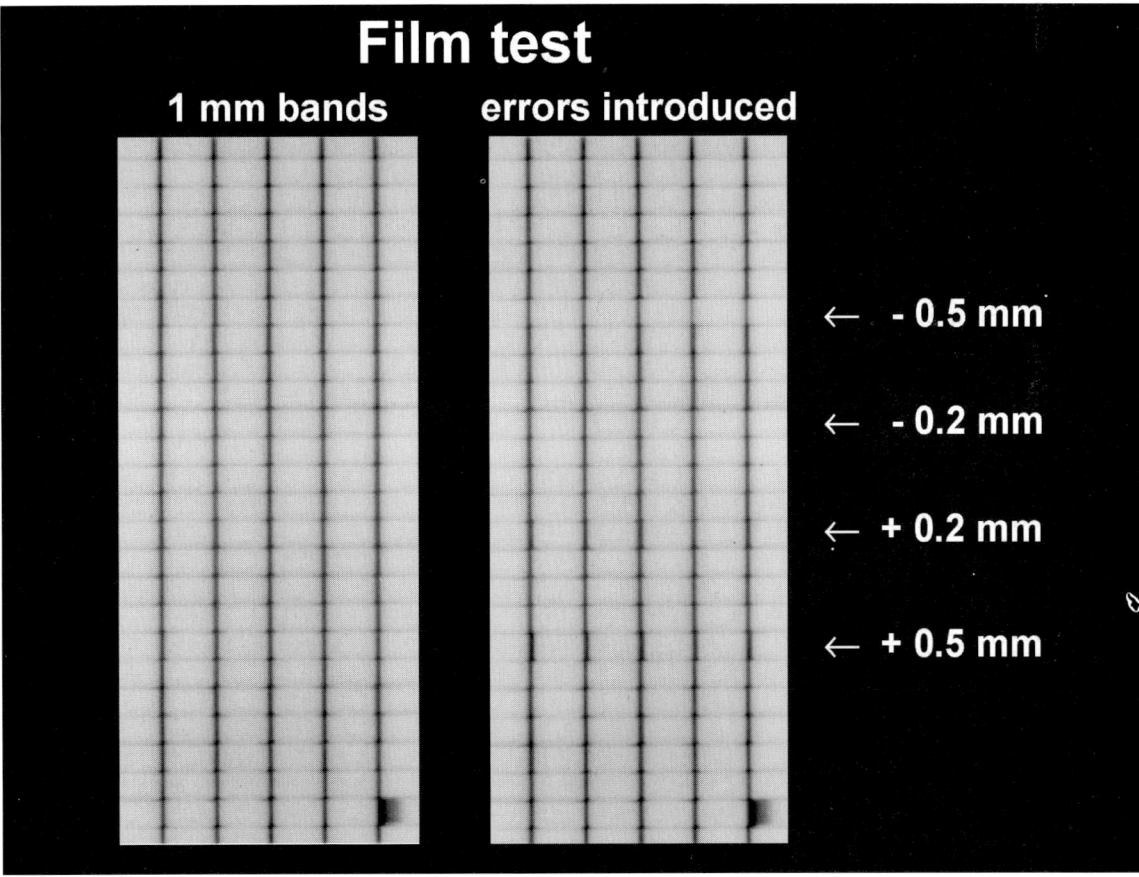

FIGURE 2–17. MLC film test pattern in which leaves are programmed to deliver five separate 5 mm wide dose strips. Note that in the rightmost film certain leaf pairs were intentionally programmed to improper gap widths of ±0.2 and ±0.5 mm, respectively, which is clearly visible on the films.

demonstrates that a simple daily film QA test can detect MLC positioning errors of a fraction of a millimeter.

One may also wish to monitor DMLC performance during treatment. We rely on two independent methods for such verification. First, the MLC control computer checks all leaf positions every 55 msec, compares them to the planned leaf positions in the DMLC file, and records them in a log file. If any leaf deviates from its planned position beyond a preset tolerance, the control computer invokes a *beam hold-off,* and radiation delivery is withheld until all leaves are again within tolerance. Inspection of these log files provides reassurance that treatment was completed as planned.

Of course, using the MLC control computer to monitor its own performance is akin to asking the fox to guard the chicken coop. An EPID, shown schematically in figure 2–18, can also be used to take snapshots of IMRT leaf positions during treatment (see figure 2–10) to provide truly independent confirmation of MLC performance (Chang et al. 2000; Curtin-Savard and Podgorsak 1999; Pasma et al. 1999).

Dosimetric Tests

Dosimetric characterization of the MLC during acceptance testing and commissioning is very time consuming. Tests using film, ion chambers, and other dosimeters are commonly performed.

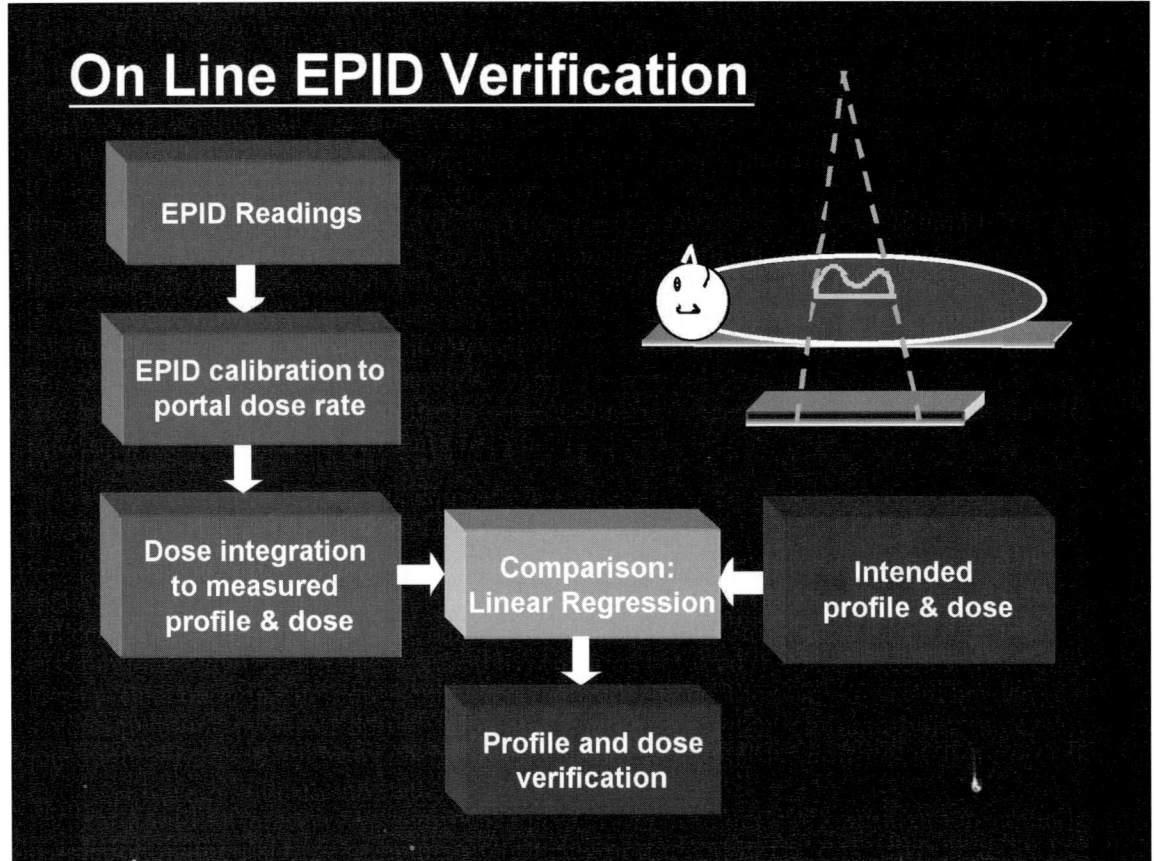

FIGURE 2–18. Schematic showing the use of EPID as a transit dosimeter to monitor the accuracy of IMRT beam delivery.

This includes measurements of radiation transmission through the leaves and their rounded ends, and the determination of head scatter. The dosimetric contribution of these factors can amount to as much as 15% for large, highly modulated fields. Even after an IMRT system has been commissioned, periodic dosimetric and geometric verification of the MLC and IMRT fields will be required as part of an ongoing physics QA program.

Patient-Specific Issues

Finally, specific tests should be performed to ensure the accuracy of each patient IMRT treatment field. This includes verification that all field and file names are correct, and that the proper DMLC leaf motion files have been downloaded to the linear accelerator. Once an acceptable treatment plan has been calculated, there is an enormous amount of data that must be transferred from the treatment planning system to the linear accelerator and MLC computers—much more data than can be transferred manually, and much more data than can even be checked manually. Again, one is confronted with having to trust one computer to check another computer.

For dynamic IMRT, both the outer boundary of the treatment field and the intensity pattern should be verified prior to the first patient treatment. Such tests can be made using radiographic film or EPIDs. As discussed earlier, verification of absolute dose to specific points within each IMRT field must also be performed, either with an independent computer program or via in-phantom dosimetry measurements. The latter technique entails recalculating the dose in a phantom plus time-consuming physical measurements. For film measurements of IMRT fields, one should be particularly cognizant of possible film calibration errors due to the large amounts of low-energy scatter radiation that are present in large, highly modulated IMRT fields (radiographic film overresponds to low-energy X-rays because of its high silver content). *In vivo* TLD measurements can also be performed on the patient.

Clinical Experience

IMRT should be reserved for tumor sites where either dose escalation or significantly improved normal tissue dose sparing is deemed particularly beneficial. At MSKCC, IMRT has been most commonly applied for the treatment of prostate, head and neck, and breast cancers, although other sites are also treated. We present below a brief description of the dosimetric and treatment planning aspects of IMRT for some of these sites.

Prostate Cancer

Since 1986, well over one thousand patients with cancers of the prostate have been treated using IMRT at MSKCC, with the standard treatment being a five-field technique using accelerator gantry angles of 0°, 75°, 135°, 225°, and 285° and 15 MV photons to 81.0 Gy (see figure 2–19). Patients are immobilized in the prone position with a thermoplastic mold for both CT simulation and treatment. ITP optimization criteria include dose or dose-volume constraints for the PTV, rectal wall, and bladder. The overlap region between the PTV and the rectal wall is defined as a separate structure with an independent dose constraint to permit greater control of the dose gradient between the PTV and the rectum, and to avoid falling into the previously alluded to trap of obtaining a poor quality plan because one specified physically impossible dose constraints (such as specifying 81 Gy to the PTV, but less than 75 Gy to the *entire* rectal wall).

Figures 2–19 and 2–20 show typical dose distributions for such a plan. The PTV and rectum are outlined. Note how the dose distribution conforms to the PTV and restricts the rectal volume irradiated to high doses (i.e., the 100% isodose contour scallops around the rectum).

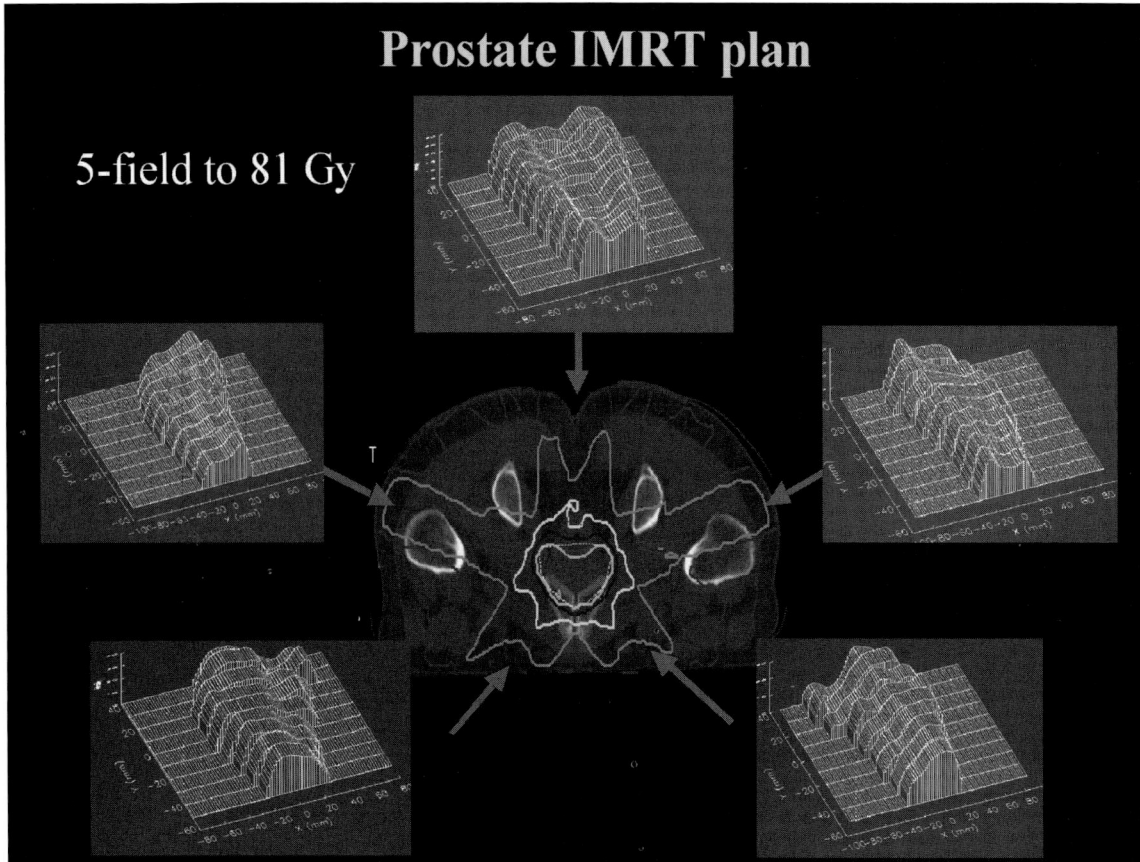

FIGURE 2–19. Five IMRT photon beams for treatment of prostate showing intensity profiles in relief map format. Note the decrease in beam intensity in the center of the posterior field designed to reduce dose to the rectum, and the resulting scalloping of the high isodose contours in the posterior portions of the prostate closest to the rectal wall.

Head And Neck Cancer

IMRT treatment of nasopharyngeal tumors often results in significant clinical advantage, particularly when *concave*-shaped dose distributions are required—as is often the situation since these tumors are in close proximity to the spinal cord and brain stem. At MSKCC a particularly useful, although somewhat unintuitive, beam geometry for creating such concave distributions is used, consisting of seven coplanar equi-spaced IMRT beams directed from the *posterior* side of the patient *directly through the brain stem* (figure 2–21). With intensity modulation, however, even beams aimed directly at the brain stem can be designed with lower dose intensity to the brain stem, while still delivering high dose to the necessary portions of the PTV. Thus, such a treatment geometry requires complicated dose intensity patterns as shown previously in figure 2–6 (for a left posterior oblique IMRT beam).

Optimization parameters for these plans include the dose or dose-volume constraints for the PTV, spinal cord, parotid glands, cochlea, and brain stem. Comparisons between conventional, 3DCRT, and IMRT treatment plans in the axial plane for a typical patient are shown in figure 2–22, and demonstrate that target dose is improved and doses to parotids, mandible, temporal lobes, spinal cord, and brain stem are all improved with IMRT. DVHs for the PTV and brainstem are

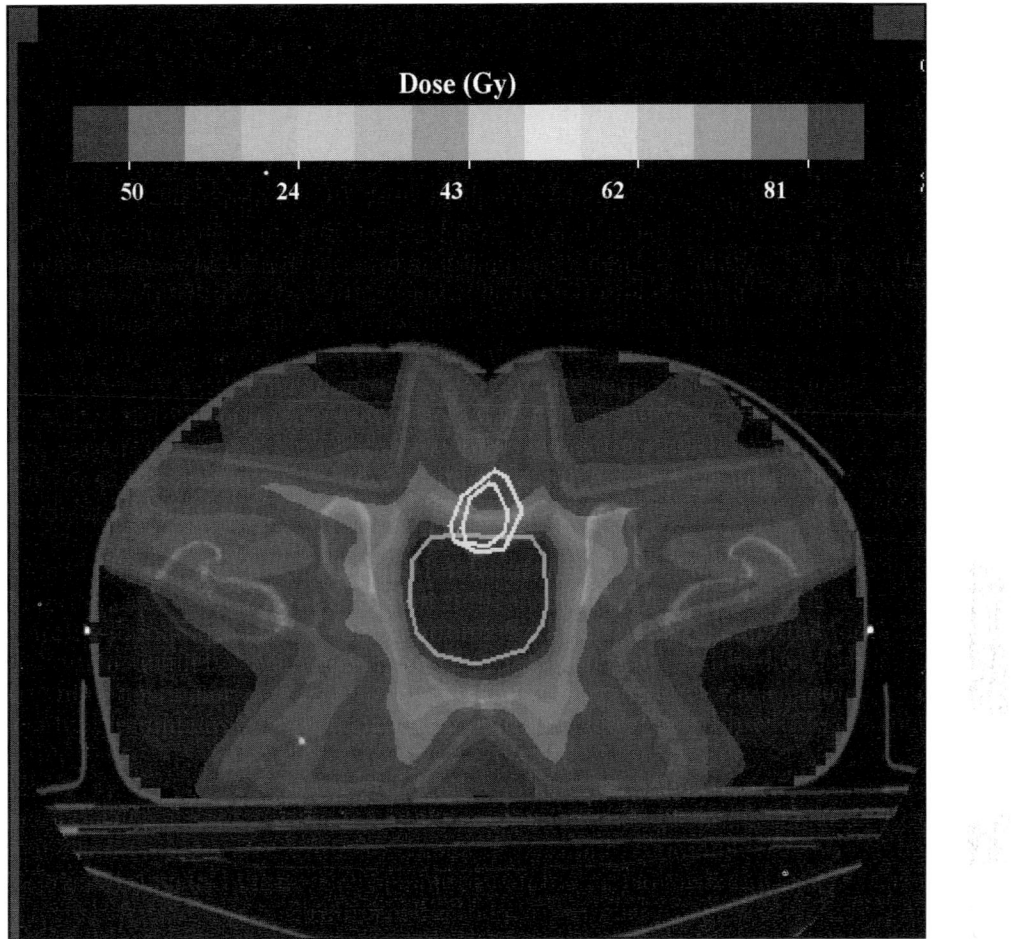

FIGURE 2–20. Color wash style display isodose distribution for an 81 Gy IMRT prostate plan with beam configuration as in figure 2–19. The PTV is shown in green, and the rectal wall in yellow. See COLOR PLATE 3.

compared for these three plans in figure 2–23. The improved PTV coverage can also be seen using a so-called region of regret plot shown in figure 2–24. In this type of plot, the prescription isodose contour is plotted in white, and regions of the PTV receiving less than the prescription dose are plotted in red.

Other Treatment Sites

IMRT can also be highly beneficial in the treatment of recurrent paraspinal lesions where patients have had previous surgical resections and/or previous RT. Like nasopharynx, such treatments require a very sharp dose gradient between a concave-shaped tumor and the adjacent spinal cord. An example of a highly conformal seven-field IMRT plan is shown in figure 2–25.

IMRT for irradiation of breast has several goals, but dose escalation is *not* one of them. Instead, we attempt to achieve a more uniform dose to the breast itself (via 3-D dose compensation rather that the 2-D compensation possible with a conventional wedge), plus reduced dose to the contralateral breast, and to the heart, and lungs. Comparison plans are shown in figure 2–26, and clearly show the superiority of the IMRT plan.

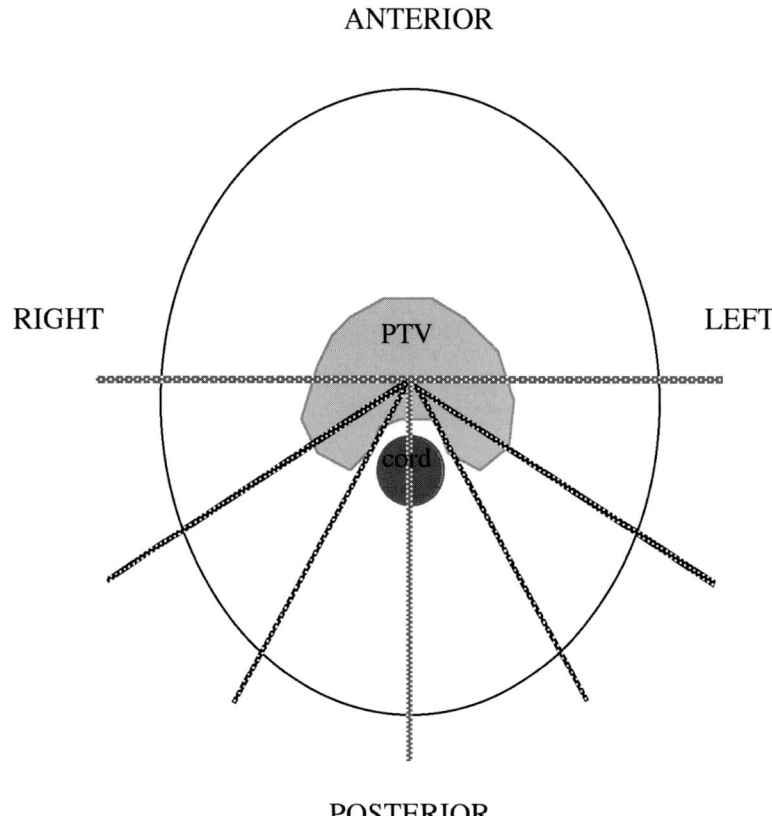

FIGURE 2–21. The seven-field beam arrangement used for IMRT treatment of a concave-shaped nasopharyngeal tumor in close proximity to the brain stem and spinal cord.

Treatment errors induced by respiratory motion, always a concern in RT of the breast is perhaps more of a concern with IMRT. In particular, if by some unfortuitous circumstance the patient's breathing is in synchrony with the motion of the leaves, it is possible for a strip of tissue to follow the leaf motion and thus receive a higher dose than intended. Monte Carlo simulations of such treatments, however, demonstrated that synchrony between tissue motion and leaf motion is extremely improbable, as the start of each treatment beam is random with respect to the patient's respiratory cycle. Any such dose errors will therefore be averaged out over a multiple fraction course of treatment.

For other sites, however, such as lung, liver, and others, respiratory gated beam delivery may be of major importance (Kubo and Wang 2000, Mageras et al. 2000). Respiratory gated IMRT is indeed feasible using a respiratory monitoring system such as the one shown in figure 2–27.

Miscellaneous Issues And Conclusions

Other issues germane to the implementation of an IMRT program include such factors as the selection of the optimum beam energy for IMRT, special room shielding requirements, selection of appropriate treatment sites, manpower requirements, and just simply how to get started. Let us look now at some of these issues.

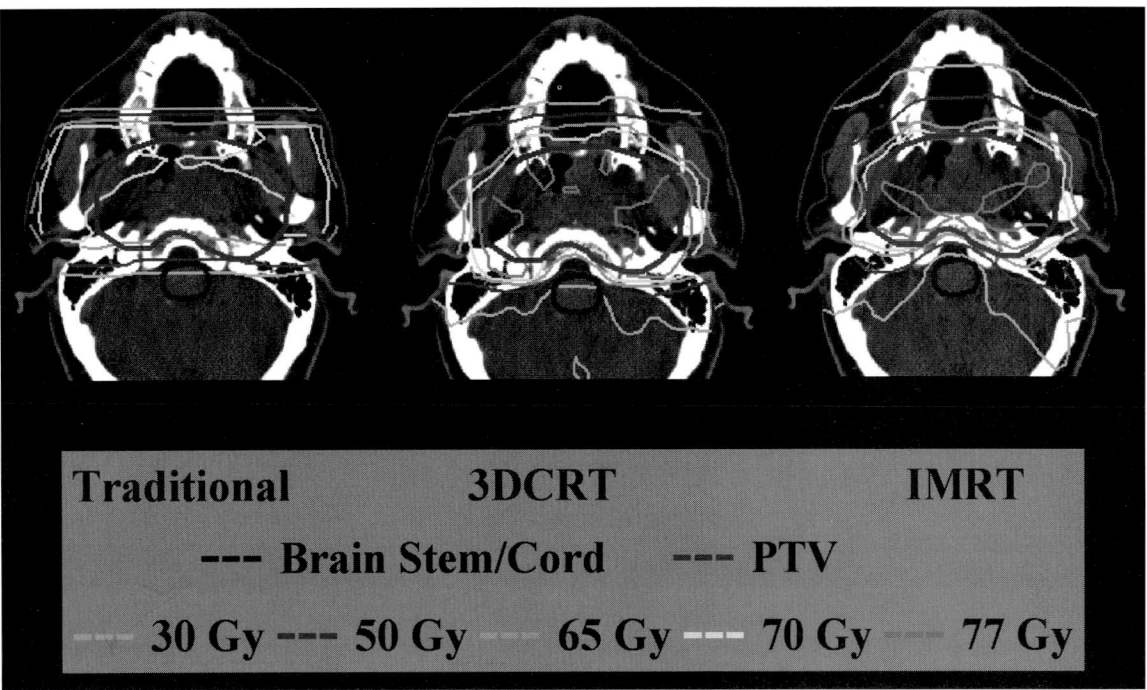

FIGURE 2–22. Axial dose distributions through the nasopharynx for the IMRT, 3DCRT, and traditional treatment plans. Note the relatively poor coverage of the skull base using the Traditional plan and the improved conformality of the IMRT dose. See COLOR PLATE 4.

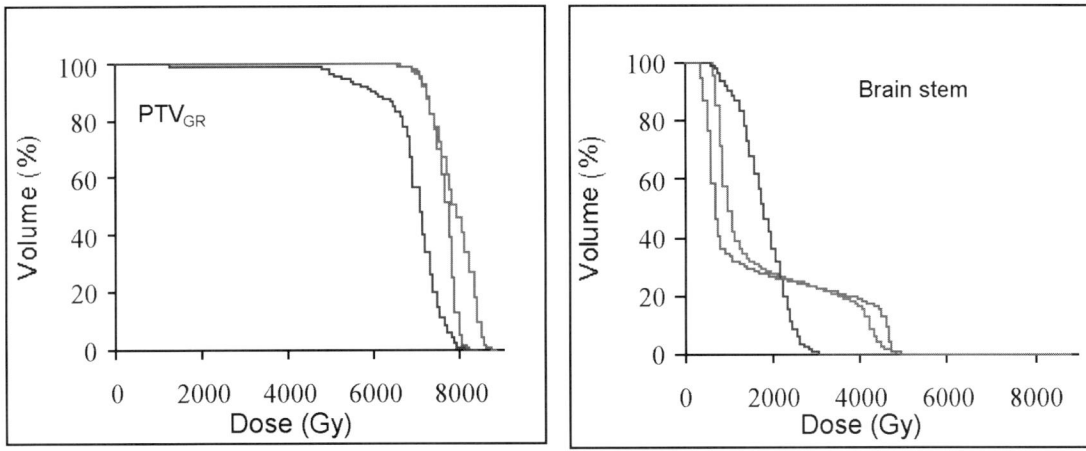

FIGURE 2–23. Dose-volume histograms comparing the IMRT, 3DCRT, and traditional plans for the treatment plans in figure 2–22. See COLOR PLATE 5.

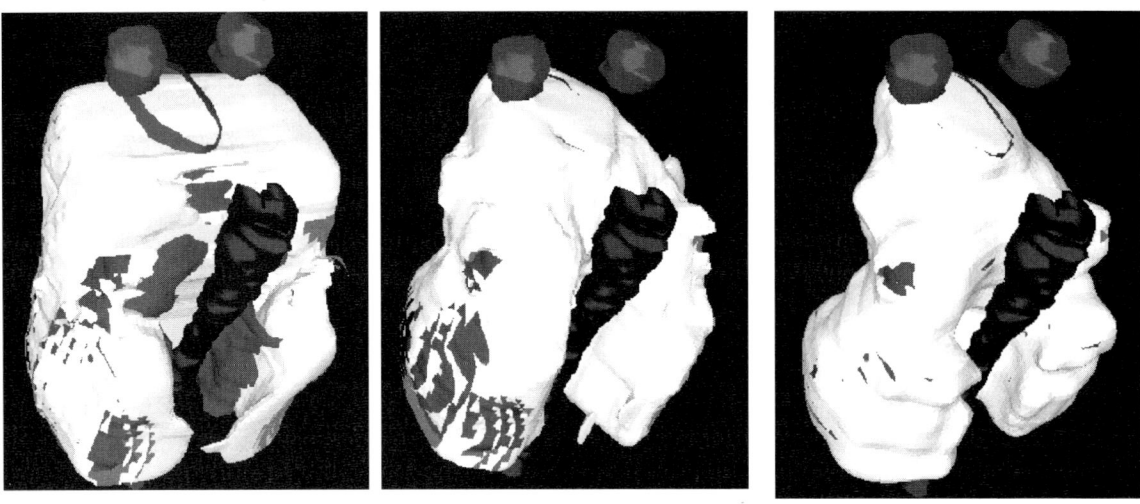

Traditional   **3DCRT**   **IMRT**

FIGURE 2–24. Region of regret plot showing the 65 Gy prescription isodose surface (white), and regions of the PTV receiving less than the prescription dose (red) for traditional, 3DCRT, and IMRT nasopharyngeal treatment plans. The brainstem and cord are shown in blue, and the eyes in purple. See COLOR PLATE 6.

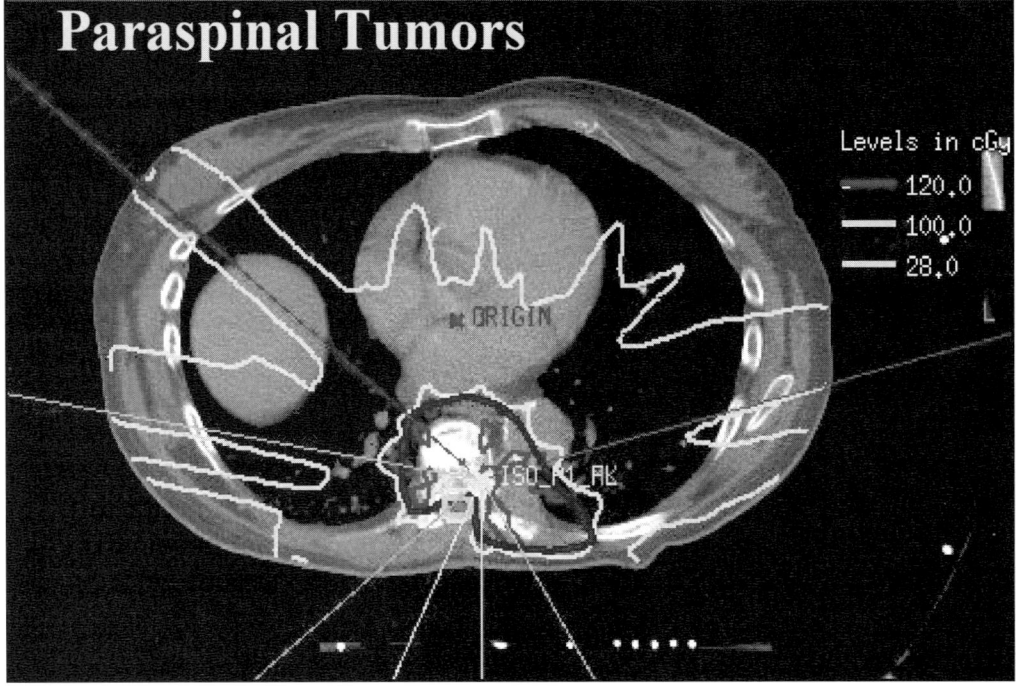

FIGURE 2–25. Seven-field IMRT treatment plan for a paraspinal tumor in close proximity to the spinal cord. These patients would typically be treated in a stereotactic whole body frame such as shown in figure 2–3.

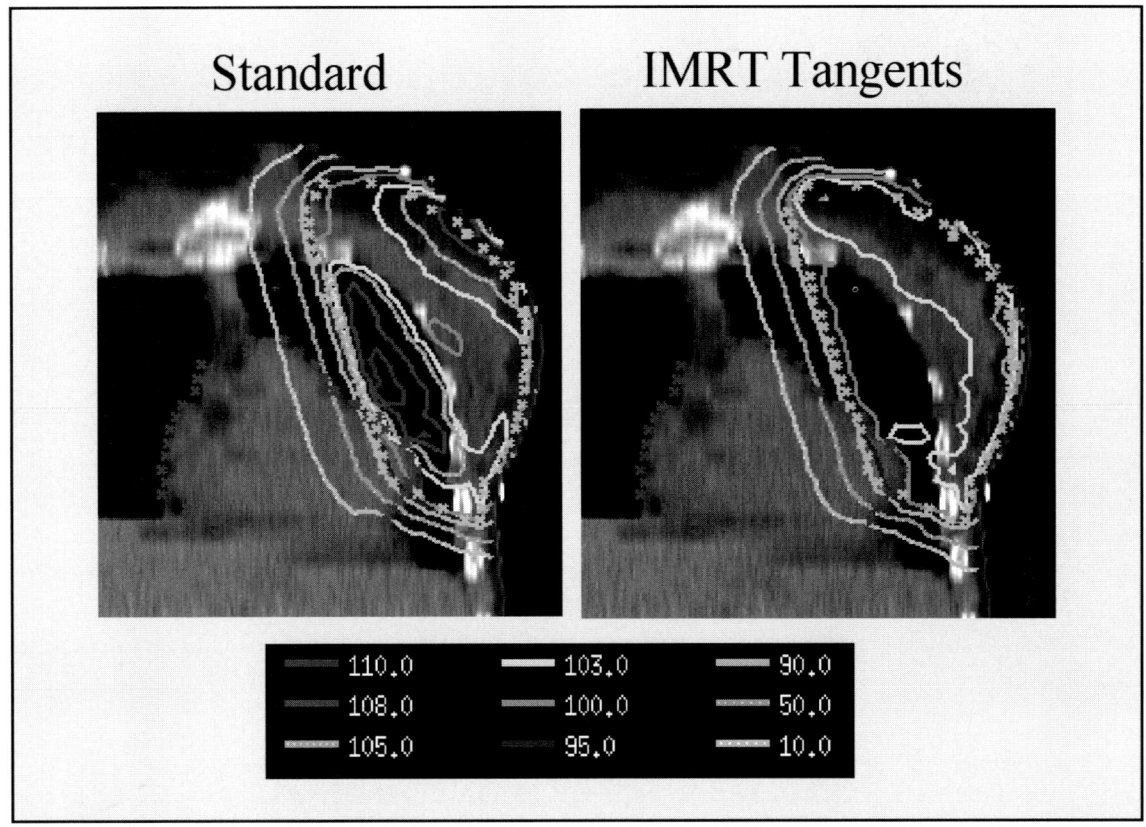

FIGURE 2–26. Comparison of standard wedges versus IMRT treatment plan in the coronal plane for breast. Not improved dose uniformity and reduced lung and heart dose for the IMRT plan. See COLOR PLATE 34 (Figure 12–1).

Because of the exquisite dose shaping potential of IMRT there is little need for beam energies greater than 15 MV, even for treatments in the abdomen and pelvis. In fact, 6 MV is optimum for most head and neck, breast, and mediastinal treatments. In addition, because of the excess number of monitor units required for IMRT, neutron contamination and radiation safety become of somewhat more concern when beam energy exceeds 15 MV. Fifteen MV photons also have a slightly better penumbra than 18 MV photons.

Radiation safety workloads for the purposes of room shielding design may have to be revisited, particularly for secondary barriers (i.e., scatter only), since many IMRT treatments require a significantly increased number of monitor units compared to conventional treatments.

If our experience is in any way typical, you will most likely introduce IMRT first for the most commonly treated sites such as prostate and head and neck. There will be, however, a significant learning curve, as both physicians and physicists become comfortable with the philosophical differences between forward and inverse planning. In addition, both physicians and physicists will spend a great deal of time relearning treatment planning strategies as they become familiar with the fundamentally different mode of thinking required for ITP optimization as compared to conventional treatment planning. But the good news is, once you get started, the time required for treatment planning and QA is not significantly different than for conventional 3DCRT. Also, the time required for actual patient treatments is also not very different from conventional therapy. However, when you start treating a new site, it is going to take a lot of time to work out optimum treatment planning parameters, dose-volume constraints, penalty weights, beam directions, etc.

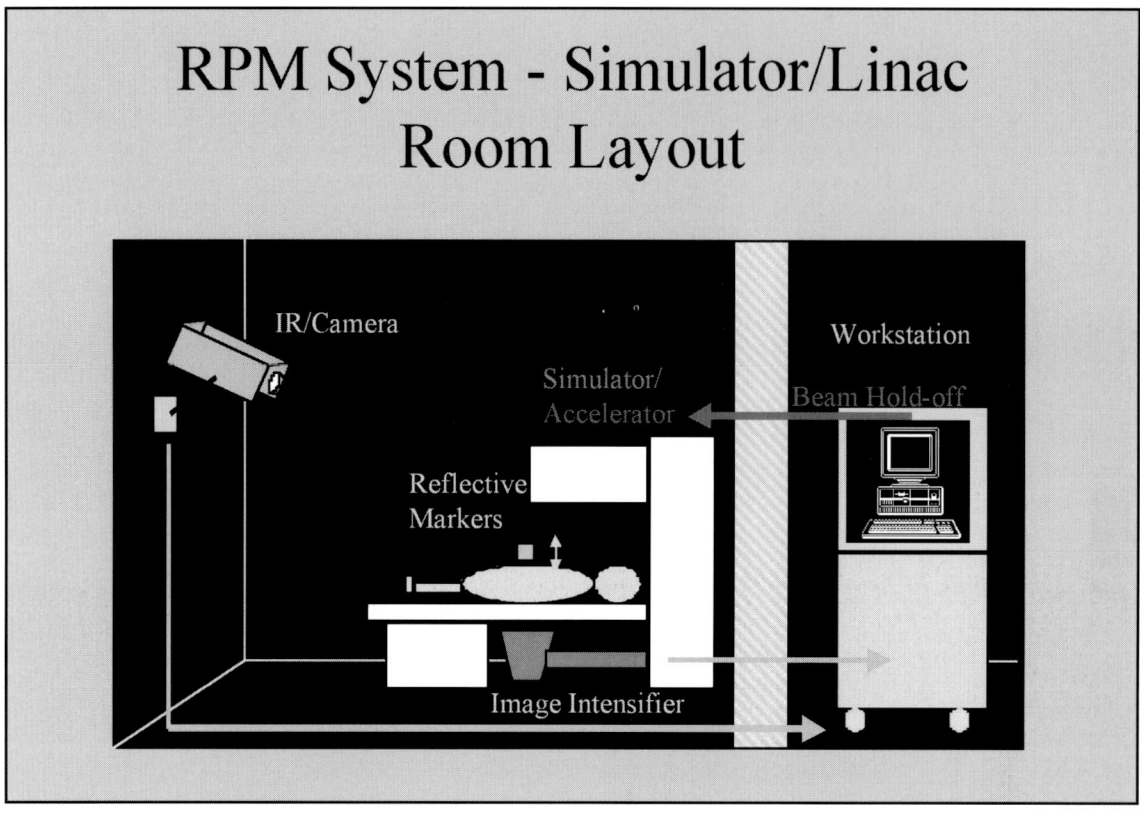

FIGURE 2–27. Schematic of respiratory gating system utilizing an infrared camera to track the motion of a reflective marker placed on the patient's chest or abdomen.

As you introduce IMRT to other disease sites, each new application will represent a new challenge, requiring adaptation and/or new custom features, plus a better understanding of the use of optimization criteria for inverse planning. Expect a lot of "trial and error," as optimum treatment strategies are disease site specific as well as functions of the specific treatment planning system being used.

The question of whether to rely solely on dose volume constraint optimization or whether to incorporate the use of biological models of tumor tissue response may also be raised. Usually biology-based score functions rely on some weighted combination of biological indices such as normal tissue complication probability (NTCP) and tumor control probability (TCP). These indices condense structure specific dose-distribution data to yield relative figures of merit for the respective objects of interest. However, such models are based on rather simplistic assumptions with little supportive data. Nonetheless, we have found some of these models to be useful for some tissues, particularly lung. Also, 3DCRT is now more than a decade old, and relevant clinical data relating treatment outcome with DVHs are now available, and it is likely that these models will improve in the future. Treatment verification of IMRT is also problematic, or at least labor intensive. The maturation of EPID technology, however, offers the potential for greatly simplifying this in the near future.

The ability of IMRT to deliver non-uniform dose patterns brings to the fore the possibility of "dose painting" or "sculpting," which becomes more intriguing as biological and functioning imaging studies such as magnetic resonance spectroscopy (MRS) and PET provide us with more information on tumor kinetics and hypoxia.

In conclusion, IMRT is a powerful technique that provides extra degrees of freedom in customizing the dose distribution for photon radiotherapy. With the development of the computer-controlled treatment machines equipped with MLCs, it is now possible to deliver these treatments reliably. The clinical implementation of inverse planning and treatment delivery is complex and involves a substantial developmental effort. However, once accomplished, the process is efficient and capable of providing the dual benefits of improved dose distribution and cost savings.

References

Bortfeld, T. (1999). "Optimized planning using physical objective and constraints." *Semin. Radiat. Oncol.* 9(1):20–34.

Brahme, A. (1988). "Optimization of stationary and moving beam radiation therapy techniques." *Radiother. Oncol.* 12:129–140.

Brahme, A. (1999). "Optimized radiation therapy based on radiobiological objectives." *Semin. Radiat. Oncol.* 9(1):35–47.

Chang, J., G. Mageras, C. S. Chui, C. C. Ling, and W. Lutz. (2000). "Relative profile and dose verification of intensity-modulated radiation therapy." *Int. J. Radiat. Oncol. Biol. Phys.* 47:231–240.

Chui, C. S., T. LoSasso, and S. Spirou. (1994). "Dose calculation for photon beams with intensity modulation generated by dynamic jaw or multileaf collimations." *Med. Phys.* 21(8):1237–1244.

Curtin-Savard, A. J., and E. B. Podgorsak. (1999). "Verification of segmented beam delivery using a commercial electronic portal imaging device." *Med. Phys.* 26:737–742.

Hong, L., M. Hunt, C. Chui, S. Spirou, K. Forster, H. Lee, J. Yahalom, G. J. Kutcher, and B. McCormick. (1999). "Intensity-modulated tangential beam irradiation of the intact breast." *Int. J. Radiat. Oncol. Biol. Phys.* 44(5):1155–1164.

Hong, L., K. Alektiar, C. Chui, T. LoSasso, M. Hunt, S. Spirou, J. Yang, H. I. Amols, C. C. Ling, Z. Fuks, and S. Leibel. (2002). "IMRT of large fields: Whole abdomen irradiation." *Int. J. Radiat. Oncol. Biol. Phys.* 54:278–289.

Hsiung, C. Y., E. D. Yorke, C. S. Chui, J. Hu, J. P. Xiong, M. A. Hunt, C. C. Ling, E. Y. Huang, C. C. Sung, Y. J. Huang, C. J. Wang, H. C. Chen, S. A. Yeh, H. C. Hsu, and H. I. Amols. (2002). "Intensity modulated radiation therapy versus conventional three-dimensional conformal radiotherapy for the boost or salvage treatment of nasopharyngeal carcinoma." *Int. J. Radiat. Oncol. Biol. Phys.* 53(3):638–647.

Hunt, M. A., M. Zelefsky, S. Wolden, C. Chui, T. LoSasso, K. Rosenzweig, L. Chong, S. Spirou, L. Fromme, M. Lumley, H. I Amols, C. C. Ling, and S. Leibel. (2001). "Treatment planning and delivery of intensity modulated radiation therapy for primary nasopharynx cancer." *Int. J. Radiat. Oncol. Biol. Phys.* 49:623–632.

International Commission on Radiation Units and Measurements (ICRU) Report 50. Prescribing, Recording and Reporting Photon Beam Therapy. Washington, D. C.:ICRU, 1993.

Kubo, H. D., and L. Wang. (2000). "Compatibility of Varian 2100C gated operations with enhanced dynamic wedge and IMRT dose delivery." *Med. Phys.* 27:1732–1738.

Ling, C. C., J. Humm, S. Larson, H. Amols, Z. Fuks, S. Leibel, and J. A. Koutcher. (2000). "Towards multi-dimensional radiotherapy (MD-CRT): Biological imaging and biological conformality." *Int. J. Radiat. Oncol. Biol. Phys.* 47:551–560.

LoSasso, T., C. S. Chui, C. C. Ling. (1998). "Physical and dosimetric aspects of a multileaf collimation system used in the dynamic mode for implementing intensity modulated radiotherapy." *Med. Phys.* 25; 1919–1927.

LoSasso, T., C. S. Chui, and C. C. Ling. (2001). "Comprehensive quality assurance for the delivery of intensity modulated radiotherapy with a multileaf collimator used in the dynamic mode." *Med. Phys.* 28(11):2209–2219.

LoSasso, T., C. S. Chui, G. J. Kutcher, S. A. Leibel, Z. Fuks, and C. C. Ling. (1993). "The use of a multi-leaf collimator for conformal radiotherapy of carcinomas of the prostate and nasopharynx." *Int. J. Radiat. Oncol. Biol. Phys.* 25:161–170.

Lyman, J. T. (1985). "Complication probability as assessed from dose volume histograms." *Radiat. Res.* 8:104–113.

Mackie, T. R., J. Balog, K. Ruchala, D. Shepard, S. Aldridge, E. Fitchard, P. Reckwerdt, G. Olivera, T. McNutt, and M. Mehta. (1999). "Tomotherapy." *Semin. Radiat. Oncol.* 9(1):108–117.

Mageras, G. S., E. Yorke, K. Rosenzweig, F. Fontenla, E. Keatley, and C. Ling. (2000). "Initial clinical evaluation of a respiratory gated radiotherapy system." *Med. Phys.* 27(6):1419–14xx.

Niemierko, A., M. Urie, and M. Goitein. (1992). "Optimization of 3D radiation therapy with both physical and biological end points and constraints." *Int. J. Radiat. Oncol. Biol. Phys.* 23(1):99–108.

Pasma, K. L., M. L. Dirkx, M. Kroonwijk, A. G. Visser, and B. J. Heijmen. (1999). "Dosimetric verification of intensity modulated beams produced with dynamic multileaf collimation using an electronic portal imaging device." *Med. Phys.* 26:2373–2378.

Spirou, S. V., N. Fournier-Bidoz, J. Yang, C. S. Chui, and C. C. Ling. (2001). "Smoothing intensity-modulated beam profiles to improve the efficiency of delivery." *Med. Phys.* 28(10):2105–2112.

Spirou, S. V., and C. S. Chui. (1998). "A gradient inverse planning algorithm with dose-volume constraints." *Med. Phys.* 25(3):321–333.

Sykes, J. R., and P. C. Williams. (1998). "An experimental investigation of the tongue and groove effect for the Philips multileaf collimator." *Phys. Med. Biol.* 43(10):3157–3165.

van Santvoort, J. P., and B. J. Heijmen. (1996). "Dynamic multileaf collimation without 'tongue-and-groove' underdosage effects." *Phys. Med. Biol.* 41(10):2091–2105.

Wang, X., S. Spirou, T. LoSasso, J. Stein, C. Chui, and R. Mohan. (1996). "Dosimetric verification of intensity-modulated fields." *Med. Phys.* 23(3):317–327.

Zelefsky, M. J., D. Cowen, Z. Fuks, M. Shike, C. Burman, A. Jackson, E. S. Venkatramen, and S. A. Leibel. (1999). "Long term tolerance of high dose three-dimensional conformal radiotherapy in patients with localized prostate carcinoma." *Cancer* 85(11):2460–2468.

Zelefsky, M., Z. Fuks, M. Hunt, Y. Yamada, C. Marion, C. Ling, H. Amols, E. S. Venkatramen, and S. A. Leibel. (2002). "High-dose intensity modulated radiation therapy for prostate cancer: Early toxicity and biochemical outcome in 772 patients." *Int. J. Radiat. Oncol. Biol. Phys.* 53(5):1111–1116.

Glossary: Basic Term and Definitions

autosequencing:
Accelerator control software which automatically sets up gantry, collimators, etc., for multi-field treatments. Particularly useful for step-and-shoot IMRT.

Beam's Eye View (BEV):
Computer rendition of tumor and patient anatomy from the direction of a proposed radiation treatment portal.

Conformal Therapy:
Treatment that creates a high dose volume that closely conforms to the shape of the PTV(s) in 3-D, while also minimizing dose to normal structures.

Digitally Reconstructed Radiograph (DRR):
Computer reconstruction (from CT data) of a 2-D x-ray projection image.

DMLC:
Dynamic multileaf collimation. A technique used to deliver intensity-modulated beams using a multileaf collimator, with the leaves in motion during radiation delivery.

Forward Planning:
Treatment planning in which the planner defines beam directions, beam weights, wedges, blocks, margins, etc., followed by dose calculation and then display and evaluation of the dose distribution. Iteration through the process is performed manually in order to reach an optimal (or at least acceptable) plan.

Intensity-Modulated Radiation Therapy (IMRT):
An advanced form of image-guided 3DCRT that utilizes variable beam intensities across the irradiated volume that have been determined using computer optimization techniques.

Inverse Planning:
Treatment planning in which the clinical objectives (usually dose-volume constraints) are specified mathematically and computer algorithms iteratively optimize beam parameters (mainly beamlet weights) to best meet the desired dose constraints.

leaf motion file, or leaf sequence file (DVA):
Data file generated by treatment planning system which defines position of each MLC leaf as a function of the fraction of total monitor units delivered. Down loaded to MLC control computer on linear accelerator for patient treatment.

Objective Function (cost function):
A mathematical description of criteria for treatment plan optimization (i.e., clinical objectives). Optimization criteria may be specified in terms of dose-limits, dose-volume limits, and/or in terms of dose-response functions (TCP, NTCP, etc.)

Score:
The numerical value of the objective function that represents a figure of merit indicating the quality of the treatment plan. The best plan corresponds to the extremum score. The extremum may be a minimum or a maximum depending upon the way the objective function is defined.

sliding window technique:
A form of IMRT in which the gap, or window, formed by each opposing pair of leaves traverses across the tumor volume while the beam is on.

step and shoot:
A form of IMRT in which variable fluence across a field is created by multiple static fields. For each "step," the MLC leaves are set, and a pre-calculated number of MUs are delivered (shoot). The beam is then halted, the MLC shape changed, and the beam comes on again. Repeated multiple times. The more steps, the finer the gradation in fluence changes.

virtual simulation:
Designing treatment beams from 3-D image data only, without use of conventional x-ray films.

3

IMRT Plan Optimization

Spiridon V. Spirou
Chen-Shou Chui

Introduction—Objectives

This chapter shall discuss the process of Inverse Planning, or Optimization. This is the process by which the intensity distribution of each beam employed in a plan is determined, such that the resultant dose distribution can best meet the criteria specified by the planner. These criteria are typically specified in terms of dose and dose-volume requirements, or biological indices such as tumor control probability (TCP) and normal tissue complication probability (NTCP).

The concept of inverse planning was first suggested by Brahme (Brahme 1988). Since then, a variety of optimization algorithms have been proposed, based on either dose and dose-volume considerations (Bortfeld et al. 1990; Cho et al. 1998; Gustafsson, Lind, and Brahme 1994; Holmes, Mackie, and Reckwerdt 1995; Llacer 1997; Olivera et al. 1998; Sauer, Shepard, and Mackie 1999; Spirou and Chou 1998; Webb 1992), or on biological indices (Brahme 1999a,b; Kaver et al. 1999; Wang et al. 1995; Wu et al. 2002). A number of methods based on dose and dose-volume considerations have already been used to treat a variety of diseases in recent years (Burman et al. 1997; Butler et al. 1999; Carol 1995; Carol et al. 1996; Chao et al. 2000, 2001; De Neve et al. 1999; Eisbruch et al. 1999, 2001; Grant and Cain 1998; Hong et al. 1999; Huang et al. 2002; Hunt et al. 2001; Kuppersmith et al. 1999, 2000; Ling et al. 1996; Mundt et al. 2001; Nutting, Dearnaley, and Webb 2000; Sultanem et al. 2000; Verellen et al. 1997; Zelefsky et al. 2000). Methods based on biological indices have not been widely implemented in the clinic (De Neve et al. 1999), primarily because the radiation-biological models are not well established yet, even though TCP and NTCP values are sometimes calculated for plan evaluation.

An optimization algorithm can generally be said to consist of two parts: the objective function that encapsulates the clinical objectives of planning and assigns a numerical score to each plan, and a method to minimize (or maximize) the objective function. In the following sections, we shall present and compare several forms of objective function as well as methods for minimizing them. We shall also present other issues that relate to optimization, namely: local vs. global minima, target dose homogeneity, more accurate inclusion of scattered dose, smoothing, skin flash, and the existence of unexpected hot spots. Finally, we shall present a simplified form of optimization for the breast.

Objective Functions

For dose-based algorithms the objective function should include the clinical criteria typically used in routine planning. These include: (a) target prescription dose, (b) target dose homogeneity, (c) critical organ maximum dose, and (d) critical organ dose-volume constraints. Dose-volume constraints are generally stated as "no more than $q\%$ of the organ may exceed a dose D_{dv}." In practice, one is usually dealing with uniformly placed dose calculation points rather than volumes; hence, the dose-volume constraint is re-stated as "no more than N_q points of the organ may exceed a dose D_{dv}."

The most commonly used form of objective function is the quadratic. For the target it is given by:

$$
F_{target} = \frac{1}{N_t}
\begin{bmatrix}
\sum_{i=1}^{N_t} (D_i - D_{presc})^2 \\
+ w_{t,min} \cdot \sum_{i=1}^{N_t} (D_i - D_{min})^2 \cdot \Theta(D_{min} - D_i) \\
+ w_{t,max} \cdot \sum_{i=1}^{N_t} (D_i - D_{max})^2 \cdot \Theta(D_i - D_{max})
\end{bmatrix}
\tag{1}
$$

where N_t is the number of points in the target, D_i is the dose to point i, and D_{presc} is the prescription dose. The second and third terms inside the brackets implement the target homogeneity criterion: D_{min} and D_{max} are the minimum and maximum tolerance doses, and $w_{t,min}$ and $w_{t,max}$ are the penalties associated with under- and overdosing. $\Theta(x)$ is the Heaviside function, defined as:

$$
\Theta(x) = \begin{cases} 1 & x \geq 0 \\ 0 & x < 0 \end{cases}
\tag{2}
$$

Similarly, the objective function term for an organ-at-risk (OAR) is given by:

$$
F_{OAR} = \frac{1}{N_{OAR}}
\begin{bmatrix}
w_{OAR,max} \cdot \sum_{i=1}^{N_{OAR}} (D_i - D_{max})^2 \cdot \Theta(D_i - D_{max}) \\
+ w_{OAR,dv} \cdot \sum_{i=1}^{N_{dv}} (D_i - D_{dv})^2 \cdot \Theta(D_i - D_{dv})
\end{bmatrix}
\tag{3}
$$

where the first term inside the brackets implements a maximum dose constraint on the OAR and the second a dose-volume constraint. The relative penalty weights are given by $w_{OAR,max}$ and $w_{OAR,dv}$ respectively. N_{OAR} is the number of points in the OAR and N_{dv} the number of points whose dose must be below the dose-volume constraint dose D_{dv}. In evaluating the second term, we can assume without loss of generality that the points are always ordered in ascending dose (computationally this re-ordering operation can always be carried out before evaluating the objective function, its gradient, etc.).

The following example illustrates the dose-volume term. Suppose that the OAR has 10 points. The dose-volume constraint is defined as "no more than 4 points may exceed a dose of 70%." That means that $D_{dv} = 70$, and $N_{dv} = 6$. Suppose that at a particular iteration the doses to the points in the OAR are {42, 74, 63, 24, 58, 79, 81, 93, 87, 35}. Five points, namely 2, 6, 7, 8, and 9, have doses higher than D_{dv}, so that the constraint is violated. Since the constraint allows 4 points to have doses higher than D_{dv}, it is necessary to modify the dose distribution to reduce the dose to just one of the five points. The question is "which of the five points?"

Critical organ protection competes with target coverage; therefore, any modification of the dose distribution to reduce the dose to a critical organ will degrade the distribution in the target. Hence, the modification of the dose distribution should be minimal. This is achieved by selecting the point whose dose is closest to D_{dv}, namely point 2.

The objective function is calculated by first sorting the points in the OAR into order of ascending dose: {24, 35, 42, 58, 63, 74, 79, 81, 87, 93}. The last four points in the sequence are ignored and the first six points are examined. Of those, the first five have doses less than D_{dv}, so that the Θ term is zero and these points contribute nothing to the objective function. The next point, whose dose is 74, has dose higher than D_{dv} and does contribute to the objective function.

For clarity, only one constraint of each type has been included in the above equations. It is straightforward to include more constraints, with different constraint doses and varying penalty weights.

The quadratic form is used primarily for mathematical convenience. It is also reasonable as the objective function increases when the actual dose deviates from the desired dose in the target or exceeds the tolerance in critical organs. But it should be noted that the quadratic form has no fundamental physical meaning. The objective function can take any other form as long as it is qualitatively consistent with the clinical goals. For example, Vineberg et al. (2002) use higher powers to make the cost of violating a constraint rise more rapidly.

As mentioned in the **Introduction**, only dose and dose–volume-based optimization methods are presently in clinical use. The main reason biological indices-based methods have not been used is that the models are not well established yet. For example, current TCP models predict that a cold spot in the target would seriously degrade the probability of tumor control. In reality, however, the treatment outcome very much depends on the location of the cold spot, i.e., whether it is in the periphery or in the middle of the target. This effect is not accounted for in the current TCP models. It is hoped that when more clinical data become available in the future, biological models will one day be used in the clinic.

Minimization Methods

Once the objective function has been defined, it must be minimized. In general, there exist two types of methods for minimizing a mathematical function: deterministic and stochastic. Deterministic methods are completely determined by the setup of the problem, i.e., patient geometry and anatomy, beam setup, optimization parameters, etc., and the initial conditions, i.e., the starting value of the solution. Given the same setup and the same initial conditions, the same solution is always obtained. Stochastic methods, on the other hand, involve an element of randomness, so that repeating the minimization process with the same setup and initial conditions will not necessarily yield the same results. This element of randomness in principle allows stochastic methods to escape from local minima, whereas deterministic methods converge to the closest minimum, whether local or global. However, for stochastic methods to succeed in escaping from local minima, the changes at each step must be small, thus making stochastic methods slow.

The most common class of deterministic methods is gradient techniques. These techniques use a vector derived from the gradient of the objective function as the direction of minimization.

Then a step is taken along the direction of minimization. Usually, the step is such that the objective function is minimized along the direction of minimization.

The dose to any point D_i can be written in vector form as:

$$D_i = \vec{a}_i \cdot \vec{x} \text{ with } \vec{a}_i = \left\{a_{ij}\right\} \text{ and } \vec{x} = \left\{x_j\right\}, j = 1,...,N_r \tag{4}$$

where N_r is the number of rays or beam elements, x_j is the intensity of the j-th ray, a_{ij} the dose deposited to the i-th point from the j-th ray, and the dot product is summed over all j. The dose-deposition coefficients a_{ij} depend on the beam characteristics (energy, modality, etc.) and the anatomical geometry between the ray and the point. For IMRT with fixed beam directions, the dose deposition coefficients are constant and minimization of the objective function takes place over the beam elements $\{x_j\}$

Using Equations (1) and (4), the gradient for a target is given by:

$$\nabla F_{\text{target}} = \frac{2}{N_t} \begin{bmatrix} \sum_{i=1}^{N_t} (D_i - D_{presc}) \cdot \vec{a}_i \\ + w_{t,\min} \cdot \sum_{i=1}^{N_t} (D_i - D_{\min}) \cdot \Theta(D_{\min} - D_i) \cdot \vec{a}_i \\ + w_{t,\max} \cdot \sum_{i=1}^{N_t} (D_i - D_{\max}) \cdot \Theta(D_i - D_{\max}) \cdot \vec{a}_i \end{bmatrix} \tag{5}$$

and for an OAR by:

$$\nabla F_{OAR} = \frac{2}{N_{OAR}} \begin{bmatrix} w_{OAR,\max} \cdot \sum_{i=1}^{N_{OAR}} (D_i - D_{\max}) \cdot \Theta(D_i - D_{\max}) \cdot \vec{a}_i \\ + w_{OAR,dv} \cdot \sum_{i=1}^{N_{dv}} (D_i - D_{dv}) \cdot \Theta(D_i - D_{dv}) \cdot \vec{a}_i \end{bmatrix} \tag{6}$$

The simplest implementation of the gradient technique is called *steepest descent*, and it uses the gradient itself as the direction of minimization:

$$\vec{x}^{k+1} = \vec{x}^k + s\vec{G}(\vec{x}^k) \tag{7}$$

where the superscript k indicates the iteration, $\vec{G}(\vec{x}^k)$ is the gradient at the k-th iteration, and s the size of the step.

The steepest descent is not the most efficient implementation of the gradient technique, because the gradient at the minimum of the current iteration, which is going to be the next direction of minimization, is perpendicular to the current direction of minimization. Therefore, minimization at the following iteration may partially spoil the minimization at the current iteration. For example, if the minimum lies in a long narrow valley, the steepest descent will require many small steps to reach the minimum. A more efficient approach is to use the conjugate gradient method. This method constructs a series of non-interfering directions of minimization, based upon the gradients and directions of minimization at the previous iterations, in an attempt to reduce the spoiling associated with the steepest descent (Press et al 1992; Spirou and Chui 1998).

Another deterministic method is based on the ratio of the current and prescription or constraint doses. It has the same appearance as the maximum likelihood method used in nuclear medicine for emission image reconstruction. The details of this method can be found in Llacer (1997).

A well-known stochastic method is simulated annealing (Webb 1992). Simulated annealing mimics the way a thermalized system reaches its ground state as the temperature slowly decreases. At each iteration, a small change, either positive or negative and of varying magnitude, is attempted in the ray weights. If the score decreases, then the change is accepted. If the score increases, the change is not automatically rejected, but accepted with a probability of $e^{-\Delta V/kT}$, where ΔV is the change in score, k is Boltzmann's constant and T the "temperature" at this stage. By accepting changes that actually worsen the dose distribution, the method has the potential to avoid getting trapped in local minima. In the early stages of the optimization, the temperature is relatively high so as to provide an opportunity to search the entire solution space. As the process progresses, the temperature slowly drops in order to reduce the search space. The main advantage of this method is that it is easy to implement and in principle it has the ability to escape from local minima. However, it is relatively inefficient compared to deterministic methods.

Inverse planning with intensity-modulated beam profiles was preceded by inverse planning with fixed open or wedged beams. In this case, the algorithm would not modulate the intensity within the beam but would determine the optimal weight of each beam. In addition to the methods described above, a number of other methods, such as linear programming and genetic algorithms (Langer et al. 1990, 1996; Langer and Leong 1987; Morill et al. 1991; Rosen et al. 1991), were investigated for beam-weight optimization. Though the results obtained were promising, the methods were computationally restricted to, at most, a few hundred dose-calculation points. Thus the applicability of these methods for intensity-modulated radiation therapy (IMRT) was very limited.

Local vs. Global Minima

A question frequently asked in optimization is the possible existence of local minimal (Deasy 1997), and if so, how one finds the global minimum among the local ones. It is not difficult to set up situations where local minima do exist and can be identified (Chui and Spirou 2001). The question is how to deal with this problem.

In principle, one could use a stochastic minimization method, which has the ability to escape from a local minimum. In practice, however, it is difficult to determine whether the current solution is at a global minimum or a local minimum, regardless of which type of method is used. Therefore, the theoretical advantage of the stochastic methods over the deterministic ones may not be realized in practice. But this question may not be important in a clinical application. If a solution already meets all the requirements specified by the planner, then it is acceptable, even though it may not be the best possible solution, that is, at the global minimum.

On the other hand, if the current solution is not adequate, then either (a) no acceptable solution exists, i.e., the clinical requirements cannot be met so they must be relaxed, or (b) an acceptable solution does exist and may be found with further effort. For the simulated annealing method, for example, a larger number of iterations may be needed, perhaps in conjunction with a different annealing scheme, such as a different rate of cooling. For the deterministic methods, a different initial guess may be used to lead to a different solution.

Target Dose Uniformity

Most, if not all, investigators have reported that IMRT plans are more conformal than those achieved with traditional three-dimensional conformal radiation therapy (3DCRT) methods. At the same time, these IMRT plans sometimes feature reduced target dose homogeneity compared with traditional 3DCRT plans. This has lead to speculation as to whether the loss of target dose homogeneity is the price to pay for the higher conformality of IMRT (Vineberg et al. 2002).

Before the issue of target dose homogeneity is addressed, one needs to ask whether this question is clinically important. If dose uniformity in the target is not important (at least to some physicians for some disease sites), then there is no cause for concern. If, however, it is important, then an IMRT plan should, in principle, do no worse than a conventional plan, for the former has more degrees of freedom. In fact, the conventional plan can be seen as a special case of an IMRT plan, where all the rays are either equal (open beam), or smoothly varying according to the wedge angle (wedged beam).

It is possible that the reported lack of target dose homogeneity is not a result of IMRT itself but rather of the physics of radiation delivery. Since IMRT is a relatively new technique and experience with it is limited, it may be that unrealistic demands are made of it. In other words, if the constraints on the organs-at-risk are too stringent, then it may simply be impossible to get good target dose homogeneity, regardless of whether one uses IMRT or traditional 3DCRT. Conversely, if a given level of target dose homogeneity is to be attained, then there may be a limit to the protection that can be given to organs-at-risk.

The improved dose uniformity achieved with IMRT is illustrated using the example shown in figures 3–1a and 3–1b, in which two perpendicular beams are used to treat a square target. The conventional plan uses a pair of 45° wedged fields whereas the IMRT plan uses a pair of intensity-modulated fields, with the same beam directions. Both plans were normalized so that the 100% isodose covers the target. It can be seen that the IMRT plan has less target volume receiving the 105% isodose and less normal tissue (the rest of the phantom outside the target) covered by the 70% isodose than the conventional plan. This comparison is evident in the dose-volume histograms (DVHs) in figure 3–1c in which the target DVH is shown as differential distribution whereas the normal tissue DVH is shown as cumulative distribution. For the target, the DVH for the IMRT plan (shown in solid lines) has a sharper peak closer to the prescribed dose 100% than the conventional plan (shown in dotted lines), indicating a more homogeneous dose distribution. For the normal tissues, the DVH for the IMRT plan is consistently lower than that for the conventional plan, indicating better sparing. This simple example shows that the IMRT plan can achieve more uniform dose in the target while at the same time providing better protection to normal tissues than the conventional plan.

More Accurate Inclusion Of Scattered Dose

The process of optimization is computationally intensive and time consuming. One way to make it more efficient is to account only partially for scattered dose while optimizing. This creates a discrepancy with the final dose calculation that includes full scatter, often degrading the dose distributions achieved. In other words, the correlation between the parameters used in optimization and the final dose distributions achieved is reduced. Hence, the planner would have to go through several planning cycles of optimization and final dose calculation, modifying the optimization parameters in each cycle, in order to determine whether the *optimal* results produced by optimization are, indeed, optimal.

A solution that preserves a significant part of the efficiency of optimization is the iterative correction method shown in figure 3–2. During optimization the total dose D_i to each point is defined to be the dose P_i from the partial-scatter calculation plus a correction factor C_i. After optimization has completed, the full-scatter dose, F_i, is calculated and compared with D_i. If the difference is small, i.e., below a user-selected tolerance, the process terminates. Otherwise, the correction factor C_i is updated and another cycle is performed. In practice, for most sites this process terminates after 2 to 3 cycles. The DVHs for a head-neck case calculated for the same beam

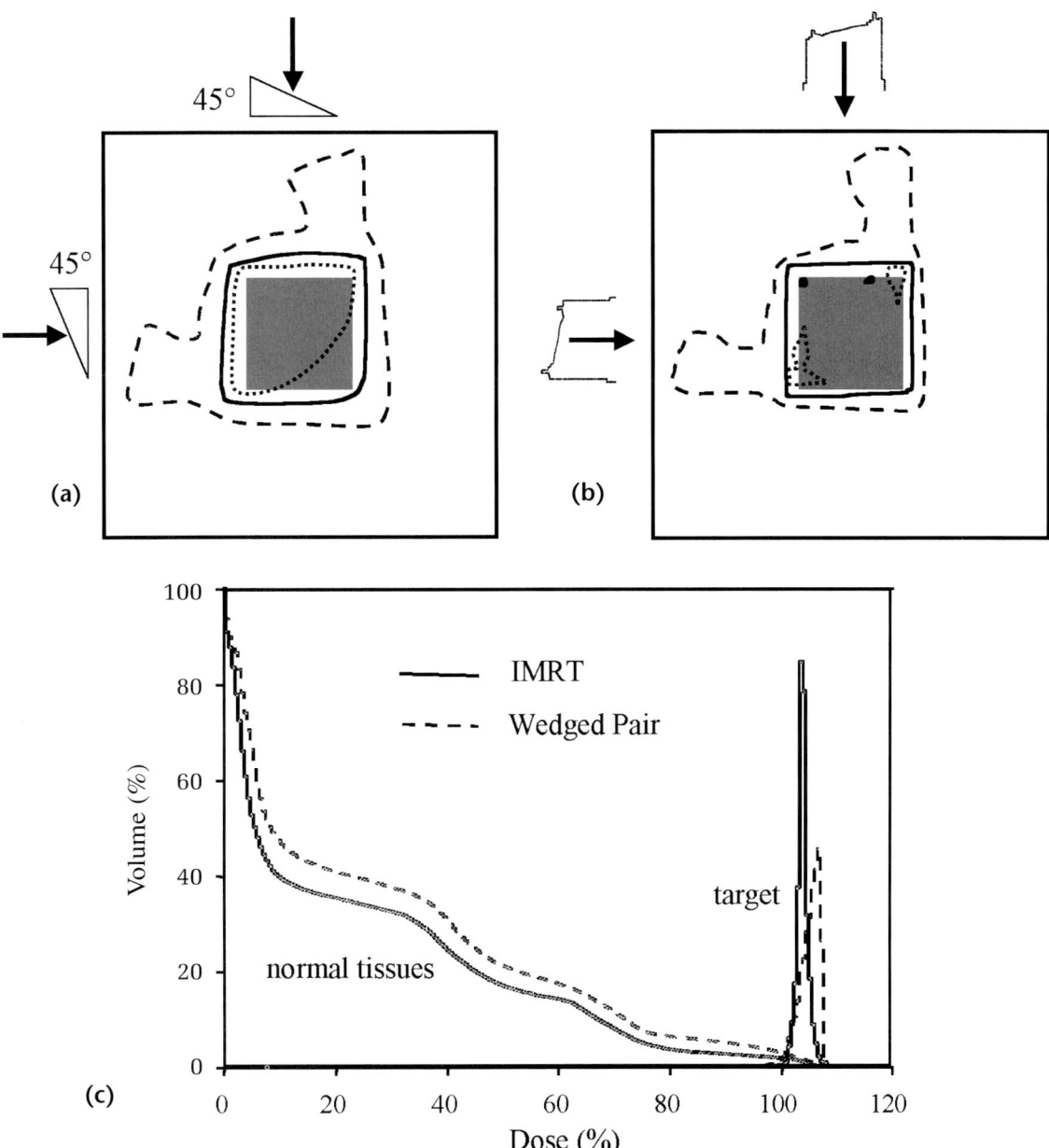

FIGURE 3–1. Isodose distributions for: (a) a conventional and (b) an IMRT plan for a square target in a square phantom. Isodoses shown are the 105% (dotted), 100% (solid), and 70% (dashed). (c) Differential DVHs for the two plans. Both plans were normalized so that the target is covered by the 100% isodose.

profiles using partial and full scatter with and without the correction process are shown in figure 3–3.

The advantage of the correction process from the planner's perspective is illustrated in figure 3–4. Without the correction process the planner would get the dose distribution shown on the left. The dose distribution shown on the right is the one obtained when the correction process is employed. Notice the improved sparing of the brainstem.

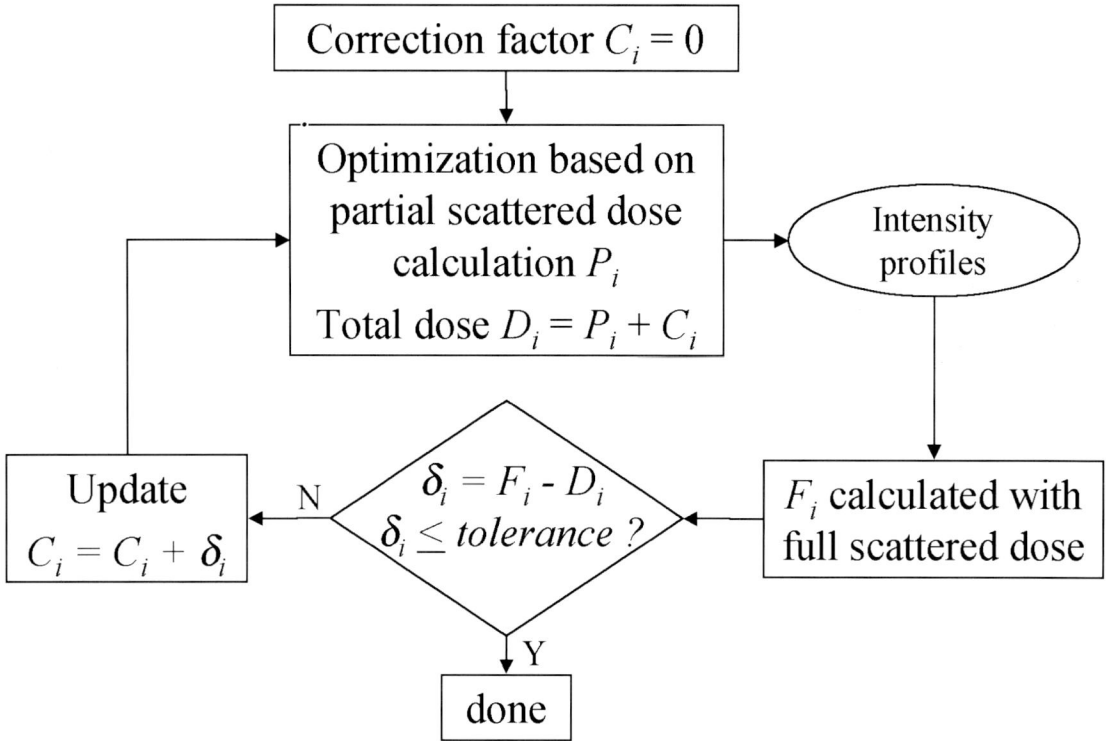

FIGURE 3–2. The iterative process that corrects for the fact that partial rather than full scatter is used during optimization. F_i is the dose to the i-th point with full scatter included, P_i is the dose to the i-th point with only partial scatter included, C_i is a correction factor, and D_i is the total dose to the i-th point during optimization.

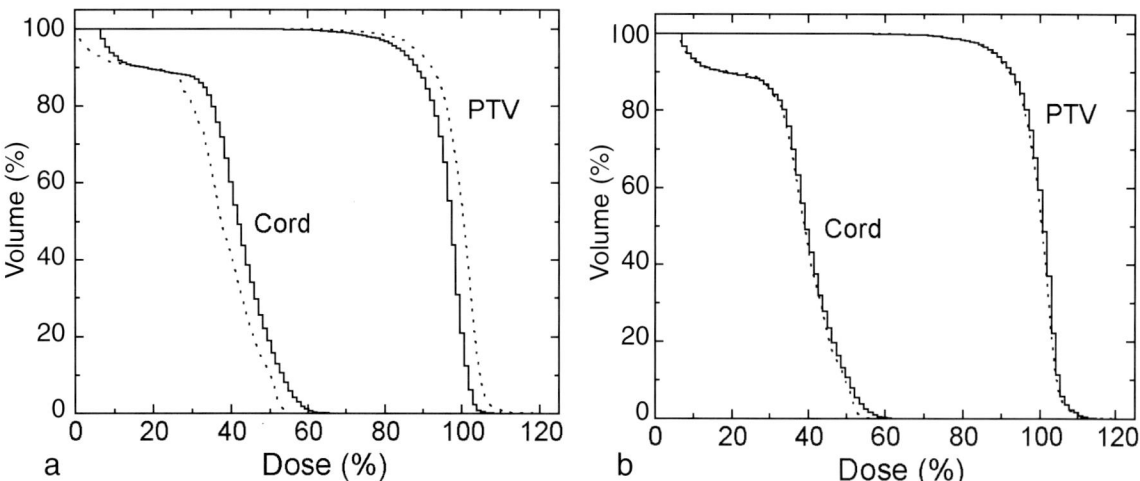

FIGURE 3–3. DVHs for a head-and-neck case calculated for the same beam profiles with partial (dotted) and full scatter (solid), without the iterative correction process (left) and with the correction process (right).

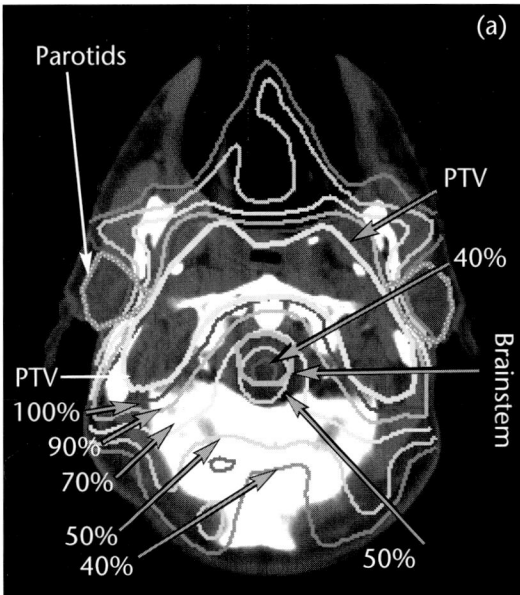 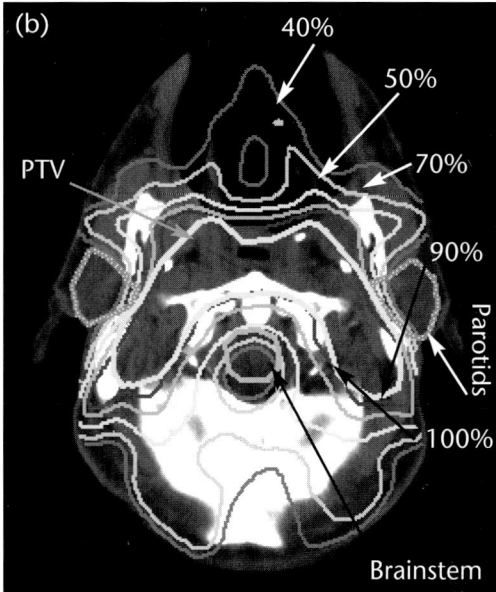

FIGURE 3–4. The isodose distributions obtained by optimizing without (left) and with (right) the correction process. Structures shown are the PTV, the brainstem and the parotids. Isodose levels shown are 100%, 90%, 70%, 50%, and 40%. The same optimization parameters were used for both plans. Notice the improved sparing of the brainstem.

Smoothing of the Intensity Profile

Regardless of which optimization method is used, the resulting intensity distributions will likely have local noise (fluctuation) due to numerical artifacts. This noise is undesirable as it increases the delivery time and makes the delivery more susceptible to treatment uncertainites. Therefore, it is useful to apply smoothing to the intensity distribution. Smoothing can be applied either after the intensity distributions have been calculated (e.g., after each iteration) or can be included in the objective function as a criterion of optimization (Alber and Nusslin 2000; Spirou et al. 2001). The advantage of the former approach is that any smoothing algorithm can be employed. On the other hand, since smoothing is a separate process, its dosimetric effect cannot be calculated at the same time. Hence, it is impossible to tell whether a particular fluctuation or gradient is dosimetrically desirable or not. In constrast, when smoothing is included within the objective function, its dosimetric effect can be taken into account and the method can, in principle, distinguish between necessary and unnecessary fluctuations. The disadvantage of the method is that it is more restrictive, since the smoothing terms must conform to the objective function.

One smoothing method is the *median window filter* (Webb, Convery, and Evans 1998). This method is illustrated on the right side of the curve shown in figure 3–5. The value of each beam element in the intensity profile is replaced by the median value of a set that consists of the particular beam element and its nearest neighbors. This method performs smoothing after each iteration has completed.

Another method is the "Savitzky-Golay" method (left side of figure 3–5), in which a polynomial (typically quadratic) is least-squares fit through these points and the current point is changed to the value on the fitted polynomial (Spirou et al. 2001). Since the value of the point after smoothing can be algebraically expressed in terms of the values **before** smoothing, it is possible to make

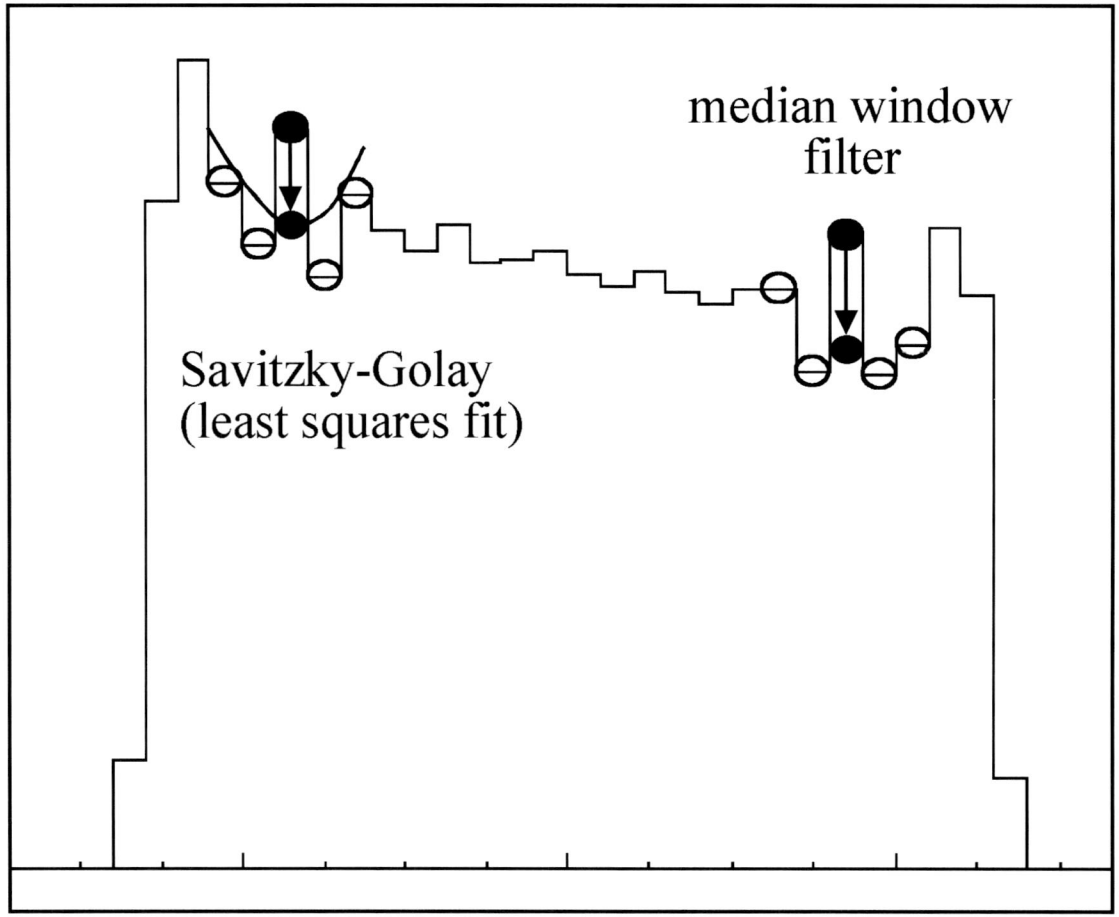

median window
filter

Savitzky-Golay
(least squares fit)

FIGURE 3–5. The "median window" and "Savitzky-Golay" filters.

the smoothness of the profiles a criterion of optimization by adding a third term in the objective function:

$$F_{obj} = \sum_{i \in \text{target}} F_{\text{target},i} + \sum_{i \in \text{critical organ}} F_{\text{organ},i} + \sum_{j \in \text{ray}} \left(x'_j - x_j\right)^2 \tag{8}$$

where x_j and x'_j are the ray weights before and after smoothing, respectively. The advantage of this approach is that the intensity profile only gets smoothed where it does not adversely affect the desired dose and, therefore, a sharp gradient can be maintained near the boundary of critical organs.

The behavior of the two methods is illustrated in figure 3–6. Three identical rectangular structures, an organ-at-risk (OAR) sandwiched between two planning target volumes (PTVs), are in a flat phantom and a single beam is incident on the phantom perpendicularly (figure 3–6a). The prescription dose to the PTVs is 100% and the constraint dose on the OAR is 5%. When smoothing is applied at the end of each iteration, the gradient in the profile at the interface between the OAR and the PTVs is smooth and gentle (figure 3–6b). In contrast, when smoothing is included within the objective function (figure 3–6c), the gradient is much sharper, since the

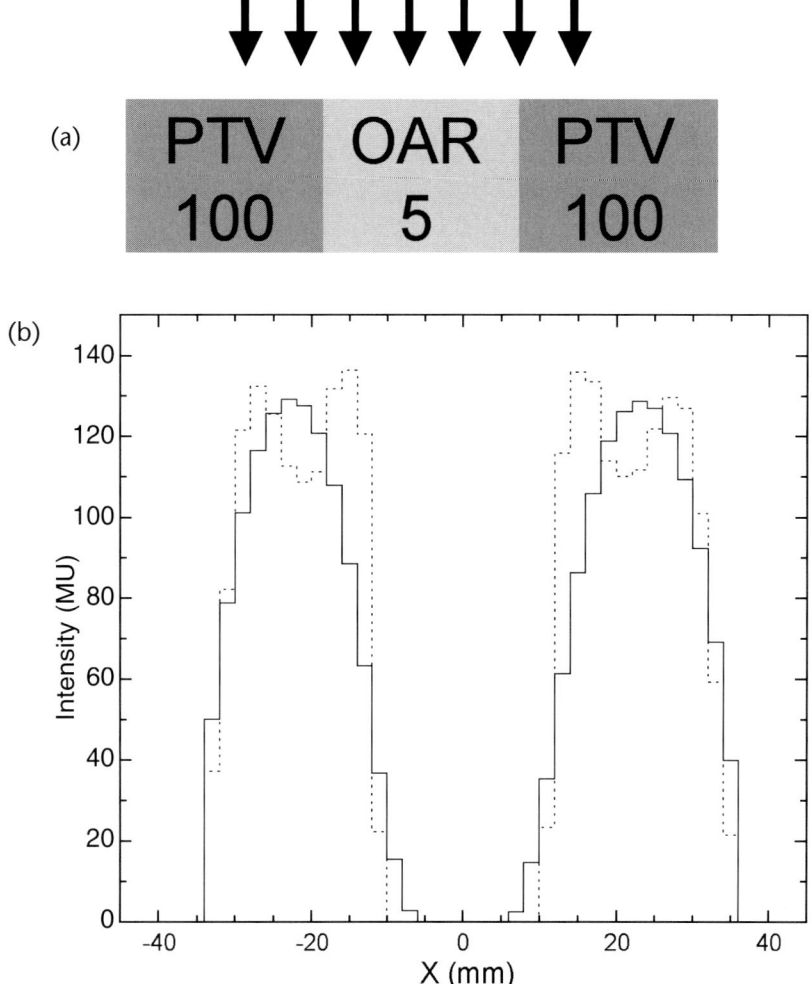

FIGURE 3–6. (a) The setup of the PTVs and the organ-at-risk; (b) the beam profiles resulting from smoothing at the end of each iteration (solid) and from including smoothing the objective function (dotted).

algorithm does realize that a sharp gradient is dosimetrically useful. The drawback of maintaining sharper gradients and more modulation on the beam profiles is that they put more strain on the multileaf collimator (MLC) and, if the delivery technique is dynamic MLC (DMLC), require more monitor units (MUs). The application of the two methods to a real clinical case, a paraspinal patient, shall be presented in chapter 6.

Skin Flash

Treatment uncertainty is an unavoidable factor that must be taken into account during planning. The standard practice is to include its effects in the delineation of the PTV. However, when the target is sufficiently close to the skin, then a *skin flash* is applied to extend the field edge into the air artificially. This is routinely done for conventional plans of the breast and head and neck regions. For IMRT, however, the intensity distribution as well as the field edge is determined by the optimization process, and therefore no skin flash is automatically included. Ideally, the optimization algorithm should account for treatment uncertainties, but this is not routinely done. An alternative method is to take the intensity distribution as determined by optimization, and extend the distribution into the air using the intensity level at the skin. One such example is shown in figure 3–7, in which a skin flash is added to the intensity distribution of a medial field used in a breast treatment. The general question of treatment uncertainties has not been adequately addressed in IMRT optimization and should receive more attention.

Unexpected Hotspots

In conventional treatments the beam profiles are usually uniform (open beams) or smoothly varying (wedged beams). Therefore, hotspots are likely to occur within the PTV or near it, since that is where the dose is concentrated. Intensity-modulated beams, on the other hand, sometimes have sharp peaks at the field edge in order to improve conformality at the periphery of the PTV and to sharpen the dose gradient. It is possible then to have hotspots where the beam edges meet (figure 3–8). These hotspots may be located at a distance to the PTV, outside any of the organs used for plan evaluation, and may not be visible on any of the standard planes routinely used for plan evaluation (transverse, coronal, sagittal through the isocenter). An example, from a lung case, is illustrated in figure 3–9.

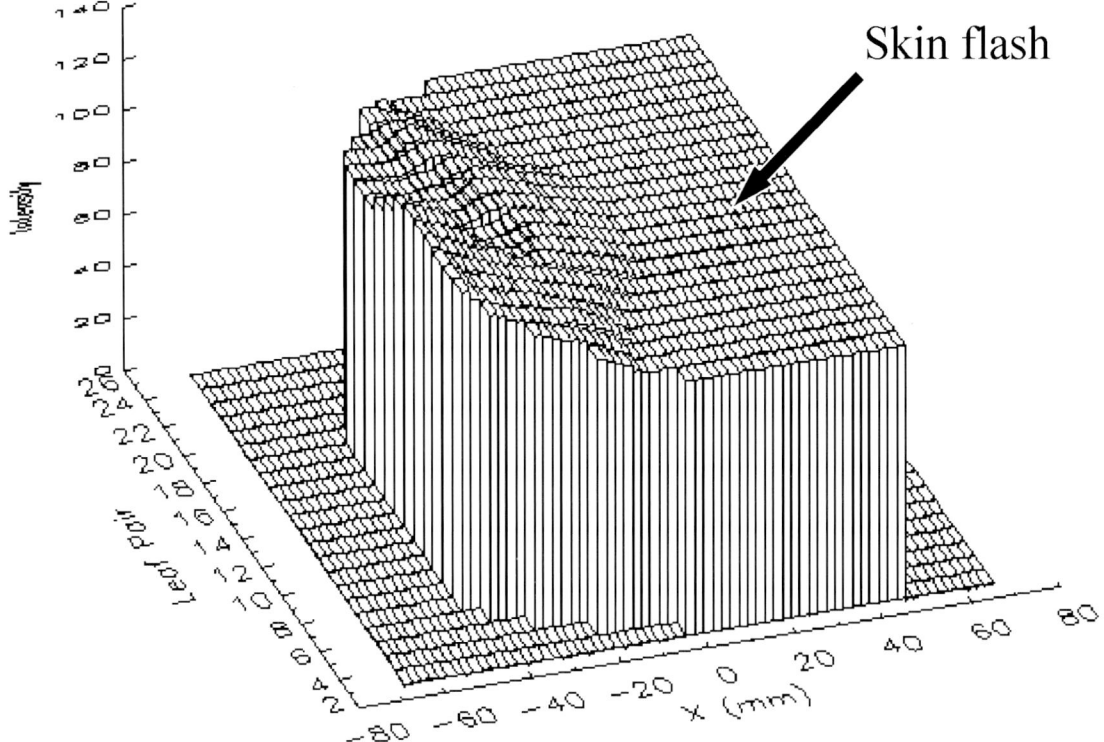

FIGURE 3–7. A medial field used in a breast treatment. The flat portion of the profile for $x \geq 0$ is the skin flash extension.

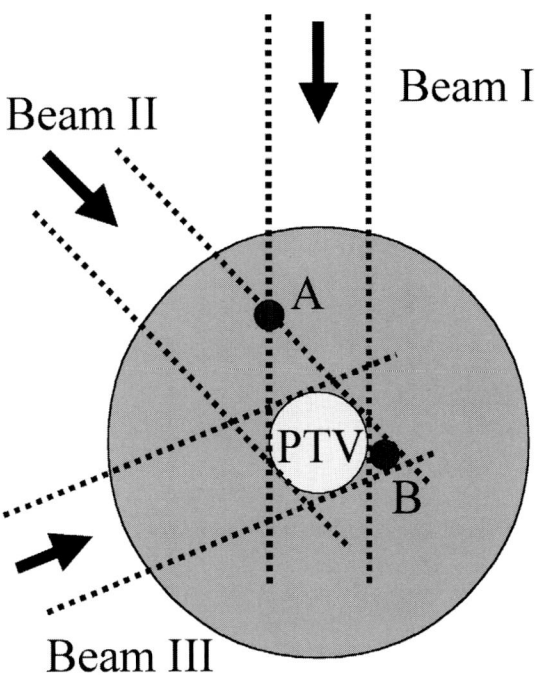

FIGURE 3–8. IMRT beams sometimes have high intensities at the ends in order to achieve high gradients at the periphery of the PTV. Where the beam edges meet (point A), it is possible to have hotspots.

Such hotspots can only be found by doing a dose calculation over the entire part of the body that is irradiated. This calculation is very time-consuming due to the size of the dose calculation matrix. Hence, it can be performed occasionally, whenever the plan is evaluated, but cannot be done continuously, as required during optimization. One possible solution is to reduce the resolution of the dose calculation matrix to, say, 5 mm or 10 mm. The drawback of this solution is that the hotspots may be missed due to the reduced resolution. An alternative solution is to create auxiliary *rind* structures on which maximum-dose constraints are placed. Usually a single rind around the PTV is sufficient. See chapter 13 on IMRT of cancer of the lung.

A Simplified IMRT Method for the Breast

The characteristic geometry of the breast and standardized tangential beam arrangement has prompted the investigation of simpler IMRT techniques (Carruthers, Redpath, and Kunkler 1999; Evans et al. 2000; Lo et al. 2000; van Asselen et al. 2001; Zackrisson, Arevarn, and Karlsson 2000). Chui et al. (2002) proposed that the weight of each beam element be such so as to deliver one-half of the prescribed dose to the mid-point of the pencil beam segment that intercepts the treatment volume. The corresponding beam elements from the opposed tangential beams would together deliver the prescription dose to the mid-point. The optimum intensity of each pencil beam is then simply the inverse of the dose to the mid-point due to the open beam. This method has the advantage of producing dose distributions that are similar to those achieved with full-volume IMRT, yet is much simpler to implement and to use, and less time-consuming. Specifically, the technique does not require most of the contours required for full-volume IMRT. It only requires the external contour, which can be traced automatically. Moreover, the optimal plan is obtained in a single operation without adjusting any parameters and other trial-and-error

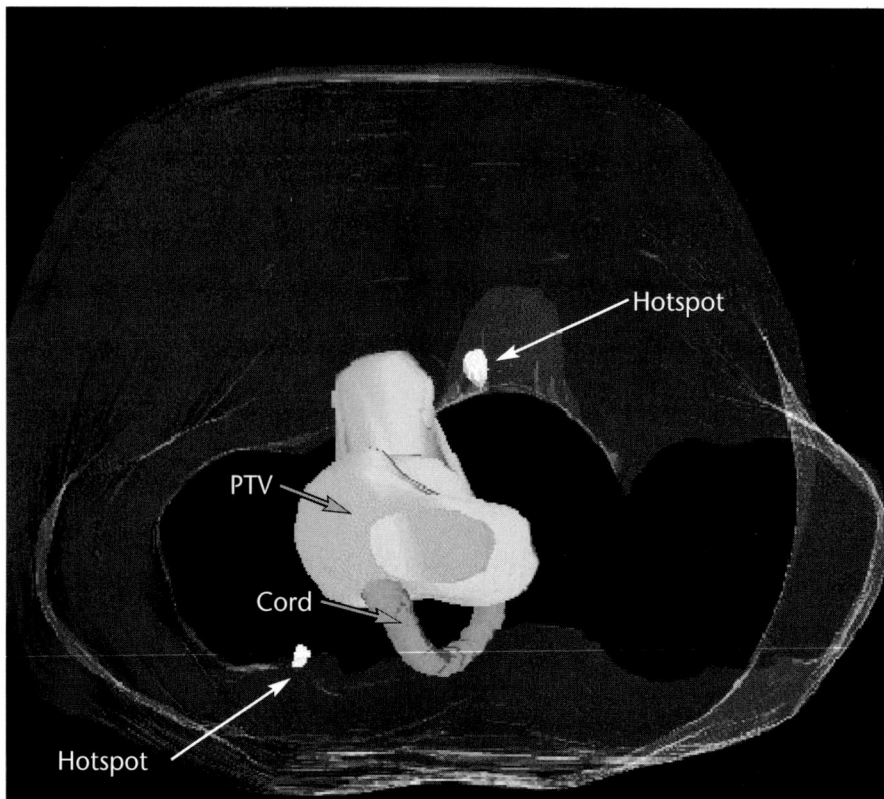

FIGURE 3–9. A lung case showing the existence of unexpected hotspots (white). These hotspots are located more than 3 cm away from the PTV and exceed 120% in dose. Structures shown are the PTV and cord.

iterations. The beam profiles require about the same number of MUs to be delivered using DMLC as conventional wedged plans, and fewer MUs than the profiles obtained from full-volume IMRT.

Summary

IMRT is a new treatment technique that promises to produce superior dose distributions than conventional techniques, in terms of both target coverage and normal tissue sparing. An important step in IMRT is inverse planning, or optimization. This is the process by which the optimum intensity profile of each beam is determined. The goodness of a solution is usually measured by an objective function, which at present is dose or dose/volume based. There are two types of search algorithms, stochastic and deterministic. The stochastic methods in principle can find the global solution whereas the deterministic methods may get trapped in a local solution. In practice, however, it is difficult to determine whether a current solution is a global or a local one, regardless of which method is used. For clinical applications, there are other issues that need to be considered, such as smoothing of the intensity profile and skin flash.

The IMRT technique today is still in its infancy, and more research and improvements are needed. For example, the effects of treatment uncertainties on the planning and delivery of IMRT need more attention. Dose response or biological indices-based optimization needs to be established and validated. As with any new technology, IMRT should be used with great caution. Comprehensive quality assurance is essential to ensure accurate and safe delivery of IMRT.

References

Alber, M., and F. Nusslin. (2000). "Intensity modulated photon beams subject to a minimal surface smoothing constraint." *Phys. Med. Biol.* 45:N49–52.

Bortfeld, T., J. Burkelbach, R. Boesecke, and W. Schlegel. (1990). "Methods of image reconstruction from projections applied to conformation radiotherapy." *Phys. Med. Biol.* 35:1423–1434.

Brahme, A. (1988). "Optimization of stationary and moving beam radiation therapy techniques." *Radiother. Oncol.* 12:129–140.

Brahme, A. (1999a). "Biologically based treatment planning." *Acta Oncol.* 38:61–68.

Brahme, A. (1999b). "Optimized radiation therapy based on radiobiological objectives." *Semin. Radiat. Oncol.* 9:35–47.

Burman, C., C. S. Chui, G. Kutcher, S. Leibel, M. Zelefsky, T. LoSaso, S. Spirou, Q. Wu, J. Yang, J. Stein, R. Mohan, Z. Fuks, and C. C. Ling. (1997). "Planning, delivery, and quality assurance of intensity-modulated radiotherapy using dynamic multileaf collimator: A strategy for large-scale implementation for the treatment of carcinoma of the prostate." *Int. J. Radiat. Oncol. Biol. Phys.* 39:863–73.

Butler, E. B., B. A. Teh, W. H. Grant III, B. M. Uhl, R. B. Kuppersmith, J. K. Chiu, D. T. Donovan, and S. Y. Woo. (1999). "Smart (simultaneous modulated accelerated radiation therapy) boost: A new accelerated fractionation schedule for the treatment of head and neck cancer with intensity modulated radiotherapy." *Int. J. Radiat. Oncol. Biol. Phys.* 45:21–32.

Carol, M. P. (1995). "Peacock: A system for planning and rotational delivery of intensity-modulated fields." *Int. J. Imaging. Sys. Tech.* 6:56–61.

Carol, M., W. H. Grant III, D. Pavord, P. Eddy, H. S. Targovnik, B. Butler, S. Woo, J. Figura, V. Onufrey, R. Grossman, and R. Selkar. (1996). "Initial clinical experience with the Peacock intensity modulation of a 3-D conformal radiation therapy system." *Stereotact. Funct. Neurosurg.* 66:30–34.

Carruthers, L. J., A. T. Redpath, and I. H. Kunkler. (1999). "The use of compensators to optimise the three dimensional dose distribution in radiotherapy of the intact breast." *Radiother. Oncol.* 50:291–300.

Chao, K. S., D. A. Low, C. A. Perez, and J. A. Purdy. (2000). "Intensity-modulated radiation therapy in head and neck cancers: The Mallinckrodt experience." *Int. J. Cancer* 90:92–103.

Chao, K. S., N. Majhail, C. J. Huang, J. R. Simpson, C. A. Perez, B. Haughey, and G. Spector. (2001). "Intensity-modulated radiation therapy reduces late salivary toxicity without compromising tumor control in patients with oropharyngeal carcinoma: A comparison with conventional techniques." *Radiother. Oncol.* 61:275–280.

Cho, P. S., S. Lee, R. J. Marks 2nd, S. Oh, S. G. Sutlief, and M. H. Phillips. (1998). "Optimization of intensity modulated beams with volume constraints using two methods: Cost function minimization and projections onto convex sets." *Med. Phys.* 25:435–443.

Chui, C. S., L. Hong, M. Hunt, and B. McCormick. (2002). "A simplified intensity modulated radiation therapy technique for the breast." *Med. Phys.* 29:522–529.

Chui, C. S., and S. V. Spirou. (2001). "Inverse planning algorithms for external beam radiation therapy." *Med. Dosim.* 26:189–197.

Deasy, J. O. (1997). "Multiple local minima in radiotherapy optimization problems with dose-volume constraints." *Med. Phys.* 24:1157–1161.

De Neve, W., W. De Gersem, S. Derycke, G. De Meerleer, M. Moerman, M. T. Bate, B. Van Duyse, L. Vakaet, Y. De Deene, B. Mersseman, C. De Wagter, and C. De Waeter. (1999). "Clinical delivery of intensity modulated conformal radiotherapy for relapsed or second-primary head and neck cancer using a multileaf collimator with dynamic control." *Radiother. Oncol.* 50:301–314.

Eisbruch, A., L. A. Dawson, H. M. Kim, C. R. Bradford, J. E. Terrell, D. B. Chepeha, T. N. Teknos, Y. Anzai, L. H. Marsh, M. K. Martel, R. K. Ten Haken, G. T. Wolf, and J. A. Ship. (1999). "Conformal and intensity modulated irradiation of head and neck cancer: The potential for improved target irradiation, salivary gland function, and quality of life." *Acta Otorhinolaryngol. Belg.* 53:271–275.

Eisbruch, A., H. M. Kim, J. E. Terrell, L. H. Marsh, L. A. Dawson, and J. A. Ship. (2001). "Xerostomia and its predictors following parotid-sparing irradiation of head-and-neck cancer." *Int. J. Radiat. Oncol. Biol. Phys.* 50:695–704.

Evans, P. M., E. M. Donovan, M. Partridge, P. J. Childs, D. J. Convery, S. Eagle, V. N. Hansen, B. L. Suter, and J. R. Yarnold. (2000). "The delivery of intensity modulated radiotherapy to the breast using multiple static fields." *Radiother. Oncol.* 57:79–89.

Grant, W., III, and R. B. Cain. (1998). "Intensity modulated conformal therapy for intracranial lesions." *Med. Dosim.* 23:237–241.

Gustafsson, A., B. K. Lind, and A. Brahme. (1994). "A generalized pencil beam algorithm for optimization of radiation therapy." *Med. Phys.* 21:343–356.

Holmes, T. W., T. R. Mackie, and P. Reckwerdt. (1995). "An iterative filtered backprojection inverse treatment planning algorithm for tomotherapy." *Int. J. Radiat. Oncol. Biol. Phys.* 32:1215–1225.

Hong, L., M. Hunt, C. Chui, S. Spirou, K. Forster, H. Lee, J. Yahalom, G. J. Kutcher, and B. McCormick. (1999). "Intensity-modulated tangential beam irradiation of the intact breast." *Int. J. Radiat. Oncol. Biol. Phys.* 44:1155–1164.

Huang, E., B. S. Teh, D. R. Strother, Q. G. Davis, J. K. Chiu, H. H. Lu, L. S. Carpenter, W. Y. Mai, M. M. Chintagumpala, M. South, W. H. Grant III, E. B. Butler, and S. Y. Woo. (2002). "Intensity-modulated radiation therapy for pediatric medulloblastoma: Early report on the reduction of ototoxicity." *Int. J. Radiat. Oncol. Biol. Phys.* 52:599–605.

Hunt, M. A., M. J. Zelefsky, S. Wolden, C. S. Chui, T. LoSasso, K. Rosenzweig, L. Chong, S. V. Spirou, L. Fromme, M. Lumley, H. A. Amols, C. C. Ling, and S. A. Leibel. (2001). "Treatment planning and delivery of intensity-modulated radiation therapy for primary nasopharynx cancer." *Int. J. Radiat. Oncol. Biol. Phys.* 49:623–632.

Kaver, G., B. K. Lind, J. Lof, A. Liander, and A. Brahme. (1999). "Stochastic optimization of intensity modulated radiotherapy to account for uncertainties in patient sensitivity." *Phys. Med. Biol.* 44:2955–2969.

Kuppersmith, R. B., S. C. Greco, B. S. Teh, D. T. Donovan, W. Grant, J. K. Chiu, R. B. Cain, and E. B. Butler. (1999). "Intensity-modulated radiotherapy: First results with this new technology on neoplasms of the head and neck." *Ear Nose Throat J.* 78:238, 241–246, 248 passim.

Kuppersmith, R. B., B. S. Teh, D. T. Donovan, W. Y. Mai, J. K. Chiu, S. Y. Woo, and E. B. Butler. (2000). "The use of intensity modulated radiotherapy for the treatment of extensive and recurrent juvenile angiofibroma." *Int. J. Pediatr. Otorhinolaryngol.* 52:261–268.

Langer, M., and J. Leong. (1987). "Optimization of beam weights under dose-volume restrictions." *Int. J. Radiat. Oncol. Biol. Phys.* 13:1255–1260.

Langer, M., R. Brown, M. Urie, J. Leong, M. Stracher, and J. Shapiro. (1990). "Large scale optimization of beam weights under dose-volume restrictions." *Int. J. Radiat. Oncol. Biol. Phys.* 18:887–893.

Langer, M., R. Brown, S. Morrill, R. Lane, and O. Lee. (1996). "A generic genetic algorithm for generating beam weights." *Med. Phys.* 23:965–971.

Ling, C. C., C. Burman, C. S. Chui, G. J. Kutcher, S. A. Leibel, T. LoSasso, R. Mohan, T. Bortfeld, L. Reinstein, S. Spirou, X. H. Wang, Q. Wu, M. Zelefsky, and Z. Fuks. (1996). "Conformal radiation treatment of prostate cancer using inversely-planned intensity-modulated photon beams produced with dynamic multileaf collimation." *Int. J. Radiat. Oncol. Biol. Phys.* 35:721–730.

Llacer, J. (1997). "Inverse radiation treatment planning using the Dynamically Penalized Likelihood method." *Med. Phys.* 24:1751–1764.

Lo, Y. C., G. Yasuda, T. J. Fitzgerald, and M. M. Urie. (2000). "Intensity modulation for breast treatment using static multi-leaf collimators." *Int. J. Radiat. Oncol. Biol. Phys.* 46:187–194.

Morill, S. M., R. G. Lane, J. A. Wong, and I. I. Rosen. (1991). "Dose-volume considerations with linear programming optimization." *Med. Phys.* 18:1201–1210.

Mundt, A. J., J. C. Roeske, A. E. Lujan, S. D. Yamada, S. E. Waggoner, C. Fleming, and J. Rotmensch. (2001). "Initial clinical experience with intensity-modulated whole-pelvis radiation therapy in women with gynecologic malignancies." *Gynecol. Oncol.* 82:456–463.

Nutting, C., D. P. Dearnaley, and S. Webb. (2000). "Intensity modulated radiation therapy: A clinical review." *Br. J. Radiol.* 73:459–469.

Olivera, G. H., D. M. Shepard, P. J. Reckwerdt, K. Ruchala, J. Zachman, E. E. Fitchard, and T. R. Mackie. (1998). "Maximum likelihood as a common computational framework in tomotherapy." *Phys. Med. Biol.* 43:3277–3294.

Press, W., S. A. Teukolsky, W. T. Vetterling, and B. P. Flannery. *Numerical Recipes in C: The Art of Scientific Computing.* New York: Cambridge University Press, 1992.

Rosen, I. I., R. G. Lane, S. M. Morrill, and J. A. Belli. (1991). "Treatment plan optimization using linear programming." *Med. Phys.* 18:141–152.

Sauer, O. A., D. M. Shepard, and T. R. Mackie. (1999). "Application of constrained optimization to radiotherapy planning." *Med. Phys.* 26:2359–2366.

Spirou, S. V., and C. S. Chui. (1998). "A gradient inverse planning algorithm with dose-volume constraints." *Med. Phys.* 25:321–333.

Spirou, S. V., N. Fournier-Bidoz, J. Yang, C. S. Chui, and C. C. Ling. (2001). "Smoothing intensity-modulated beam profiles to improve the efficiency of delivery." *Med. Phys.* 28:2105–2112.

Sultanem, K., H. K. Shu, P. Xia, C. Akazawa, J. M. Quivey, L. J. Verhey, and K. K. Fu. (2000). "Three-dimensional intensity-modulated radiotherapy in the treatment of nasopharyngeal carcinoma: The University of California-San Francisco experience." *Int. J. Radiat. Oncol. Biol. Phys.* 48:711–722.

van Asselen, B., C. P. Raaijmakers, P. Hofman, and J. J. Lagendijk. (2001). "An improved breast irradiation technique using three-dimensional geometrical information and intensity modulation." *Radiother. Oncol.* 58:341–347.

Verellen, D., N. Linthout, D. van den Berge, A. Bel, and G. Storme. (1997). "Initial experience with intensity-modulated conformal radiation therapy for treatment of the head and neck region." *Int. J. Radiat. Oncol. Biol. Phys.* 39:99–114.

Vineberg, K. A., A. Eisbruch, M. M. Coselmon, D. L. McShan, M. L. Kessler, and B. A. Fraass. (2002). "Is uniform target dose possible in IMRT plans in the head and neck?" *Int. J. Radiat. Oncol. Biol. Phys.* 52:1159–1172.

Wang, X. H., R. Mohan, A. Jackson, S. A. Leibel, Z. Fuks, and C. C. Ling. (1995). "Optimization of intensity-modulated 3D conformal treatment plans based on biological indices." *Radiother. Oncol.* 37:140–152.

Webb, S. (1992). "Optimization by simulated annealing of three-dimensional, conformal treatment planning for radiation fields defined by a multileaf collimator: II. Inclusion of two-dimensional modulation of the x-ray intensity." *Phys. Med. Biol.* 37:1689–1704.

Webb, S., D. J. Convery, and P. M. Evans. (1998). "Inverse planning with constraints to generate smoothed intensity-modulated beams." *Phys. Med. Biol.* 43:2785–2794.

Wu, Q., R. Mohan, A. Niemierko, and R. Schmidt-Ullrich. (2002). "Optimization of intensity-modulated radiotherapy plans based on the equivalent uniform dose." *Int. J. Radiat. Oncol. Biol. Phys.* 52:224–235.

Zackrisson, B., M. Arevarn, and M. Karlsson. (2000). "Optimized MLC-beam arrangements for tangential breast irradiation." *Radiother. Oncol.* 54:209–212.

Zelefsky, M. J., Z. Fuks, L. Happersett, H. J. Lee, C. C. Ling. C. M. Burman, M. Hunt, T. Wolfe, E. S. Venkatraman, A. Jackson, M. Skwarchuk, and S. A. Leibel. (2000). "Clinical experience with intensity modulated radiation therapy (IMRT) in prostate cancer." *Radiother. Oncol.* 55:241–249.

4

Delivery Of Intensity-Modulated Beam Profiles With A Multileaf Collimator

Spiridon V. Spirou
Chen-Shou Chui

Introduction

Intensity-modulated radiation therapy (IMRT) can be delivered using custom-made devices, such as compensators, or a conventional multileaf collimator (MLC). Custom-made devices have several drawbacks: (a) they are labor-intensive to fabricate; (b) they require re-entry into the treatment room between fields, thus prolonging the treatment time; (c) they modify the beam characteristics, such as the spectrum and flatness (Jiang and Ayyangar 1998); and (d) they generate scattered radiation which, due to the proximity of the compensator to the patient, increases the skin dose and the dose outside the field (Scrimger and Kolitsi 1979; Kase et al. 1983; Fraas, Roberson, and Lichter 1985).

Multileaf collimators, on the other hand, suffer from none of these drawbacks. MLCs can be used either in the segmental mode (SMLC, also known as the *step-and-shoot* method) (Bortfeld et al. 1994; Galvin, Chen, and Smith 1993; Ma et al. 1998; Siochi 1999; Que 1999; Xia and Verhey 1998) or in the dynamic mode (DMLC) (Convery and Rosenbloom 1992; Spirou and Chui 1994; Stein et al. 1994; Svensson, Kallman, and Brahme 1994). In the SMLC method, the entire delivery sequence consists of segments of *step*, in which the leaves move to their respective designated positions while the beam is off; and segments of *shoot,* in which the leaves remain stationary while the beam is turned on. In the DMLC method, the leaves move continuously at variable speeds while the beam is on. The sequence of the alternating step and shoot segments in the SMLC method or the variable speeds of leaf motion in the DMLC method can be properly designed so as to deliver almost any desired intensity profile, within the constraints imposed by the MLC design.

The drawback of delivery with an MLC is that only part of the field is exposed at any given instant. The rest of the field is shielded by the leaves and receives radiation transmitted through the leaves, as well as scattered radiation. Therefore, a higher number of monitor units (MU) is required to deliver the profile with an MLC compared with a compensator. This, in turn, increases the contribution of transmitted and scattered radiation relative to direct radiation.

Along the direction of leaf motion, the amount transmitted through the leaves at a given instant depends on the location of the point of interest relative to the rounded leaf edge (Figure 4–1). Perpendicular to the direction of leaf motion, the amount transmitted depends on whether the point of interest is in the midleaf or the tongue-and-groove region. Therefore, to calculate the total amount of radiation fluence at a particular point, one has to integrate the instantaneous leaf transmission over the entire beam-on time, taking into account the exact geometry between the point and the MLC.

These considerations apply to static fields as well; however, in the static case they affect the periphery of the field and can be accounted for using measurements. For dynamic fields, these factors can have a significant effect at every point in the interior of the field. For example, as described in chapter 7, for a 2 cm leaf opening sweeping the field, transmission through the rounded ends of the leaves will yield a 10% dose error throughout the field, if it is not accounted for.

The Sliding Window Method

As stated in the **Introduction,** in the sliding window technique the leaves move continuously, with varying speeds, while the beam is on. Usually the motion is unidirectional, from one end of the field to the other.

In principle, one can write down the equations describing the total fluence at each point, including all factors, as a function of leaf positions. Solving the equations for the leaf sequences, however, is a non-trivial task as they are coupled integral equations. An alternative approach is to implement

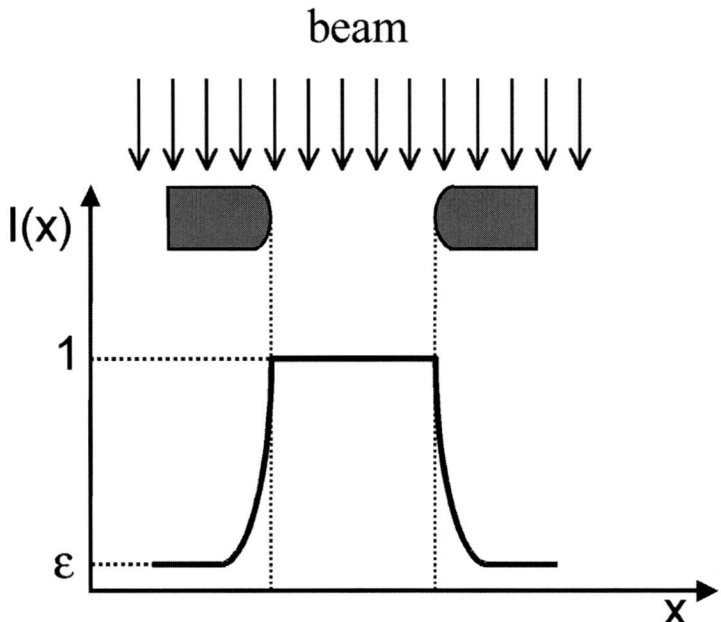

FIGURE 4–1. The intensity under the leaves.

an iterative procedure. First, the paths of the leaves are calculated under a number of simplifications, ignoring all secondary effects. Subsequently, all secondary effects are taken into account during a forward calculation.

Basic Method

If (a) the radiation source is a point source, (b) there is no head scatter, and (c) the leaves are perfect attenuators, i.e. the transmission is 100% in the exposed area and 0% in the shielded area, then a point under the leaves receives radiation only when it is directly exposed. Let $F(x)$ be the profile to be generated. Then:

$$F(x) = M_t(x) - M_l(x) \tag{1}$$

where $M_l(x)$ and $M_t(x)$ are the arrival times of the leading and trailing leaves, respectively, at x. Note that $F(x)$, $M_l(x)$ and $M_t(x)$ are all measured in MU.

Equation (1) can be differentiated to derive the relationship between the speeds of the leaves:

$$\frac{1}{V_t(x)} - \frac{1}{V_l(x)} = \frac{dF(x)}{dx} \tag{2}$$

where $V_l(x)$ and $V_t(x)$ are the speeds of the leading and trailing leaves, respectively. Hence, if the profile is increasing, the leading leaf moves faster than the trailing leaf, whereas if the profile is decreasing, the leading leaf moves slower than the trailing leaf. If the profile is constant, then both leaves move at the same speed.

In practice, $F(x)$ is a discretized, piecewise linear function defined at N points along the positive x-axis. For clarity of presentation, it can be assumed that the points at which $F(x)$ is defined are equispaced, with the distance between them being Δx. Then:

$$F(x_i) = M_t(x_i) - M_l(x_i) \qquad \text{for } i = 1,\dots, N \tag{3}$$

$$\frac{1}{V_t(x_i)} - \frac{1}{V_l(x_i)} = \frac{F(x_i) - F(x_{i-1})}{\Delta x} \qquad \text{for } i = 2,\dots, N \tag{4}$$

Equations (3) and (4) can be used to iteratively determine the leaf paths; i.e., if $M_l(x_i)$ and $M_t(x_i)$ are known, then $M_l(x_{i+1})$ and $M_t(x_{i+1})$ can be calculated. There is an infinite number of solutions to the problem. The most efficient solution, i.e., the one that minimizes the total beam-on time, is obtained by setting the speed of the faster of the two leaves at each segment to be equal to the maximum allowed leaf speed (Spirou and Chui 1994):

Set:
$$M_t(x_1) = F(x_1) \tag{5}$$

and
$$M_l(x_1) = 0$$

For $i = 2,\dots, N$:

$$\text{if } F(x_i) > F(x_{i-1}) \text{ then set } \begin{cases} M_l(x_i) = M_l(x_{i-1}) + \Delta t \\ M_t(x_i) = M_t(x_i) + F(x_i) \end{cases}$$

$$\text{if } F(x_i) < F(x_{i-1}) \text{ then set } \begin{cases} M_t(x_i) = M_t(x_{i-1}) + \Delta t \\ M_l(x_i) = M_t(x_i) - F(x_i) \end{cases}$$

$$\text{if } F(x_i) = F(x_{i-1}) \text{ then set } \begin{cases} M_l(x_i) = M_l(x_{i-1}) + \Delta t \\ M_t(x_i) = M_t(x_{i-1}) + \Delta t \end{cases}$$

where $\Delta t = \dfrac{\Delta x}{V_{max}}$ and V_{max} is the maximum speed of the leaves. V_{max} is measured in cm/MU and is equal to the mechanical speed of the leaves divided by the dose rate.

The total beam-on time (in MU) is then given by:

$$T_{bo} = \frac{x_N - x_1}{V_{max}} + F(x_1) + \sum_{\substack{i=2 \\ F(x_i) > F(x_{i-1})}}^{N} \left[F(x_i) - F(x_{i-1}) \right] \tag{6}$$

Note that the sum is carried out over the increasing parts of the desired profile.

Incorporation Of Secondary Effects

As stated previously, it is a non-trivial task to calculate the leaf paths directly including all factors, such as transmission through different parts of the leaves, head scatter, and the finite source size. An alternative solution is shown in figure 4–2. The leaf trajectories are calculated based on a temporary *working* profile $F_{work}(x)$, using equation (5) and ignoring all secondary effects. After the leaf paths have been thus calculated, the *generated* profile $F_g(x)$ can be calculated, taking all factors into account. If the difference between the desired and generated profiles is small, the procedure terminates; otherwise, the working profile is modified and the procedure repeated.

The details on how each factor is taken into account are described in chapter 5 and are not repeated here.

Correction For Tongue-And-Groove

To reduce interleaf leakage, each leaf has a protruding tongue on one side that fits into the groove of the adjacent leaf. This is illustrated in figure 4–3, which shows a cross section of the MLC, with the leaves moving in and out of the plane of the paper. The dosimetric effect of this design was first described by Galvin, Smith, and Lally (1993). Suppose that half the leaves are retracted and half remain within the field (the light-shaded ones). The intensity under the MLC is shown in figure 4–3a. When the situation is reversed, i.e., the leaves that were previously in the field are retracted and the leaves that were previously retracted are moved into the field, the intensity under the MLC is as shown in figure 4–3b.

However, the two intensity curves do not match spatially, so that if two successive irradiations are performed with the two setups described above, the total intensity under the MLC would exhibit a decrease under the tongue-and-groove region, as shown in figure 4–3c.

A solution to the tongue-and-groove problem in the context of DMLC was first proposed by van Santvoort and Heijmen (1996). They proposed a segment-by-segment modification of the leaf paths in order to make, for any two adjacent leaf pairs, the smaller leaf gap be inside the larger

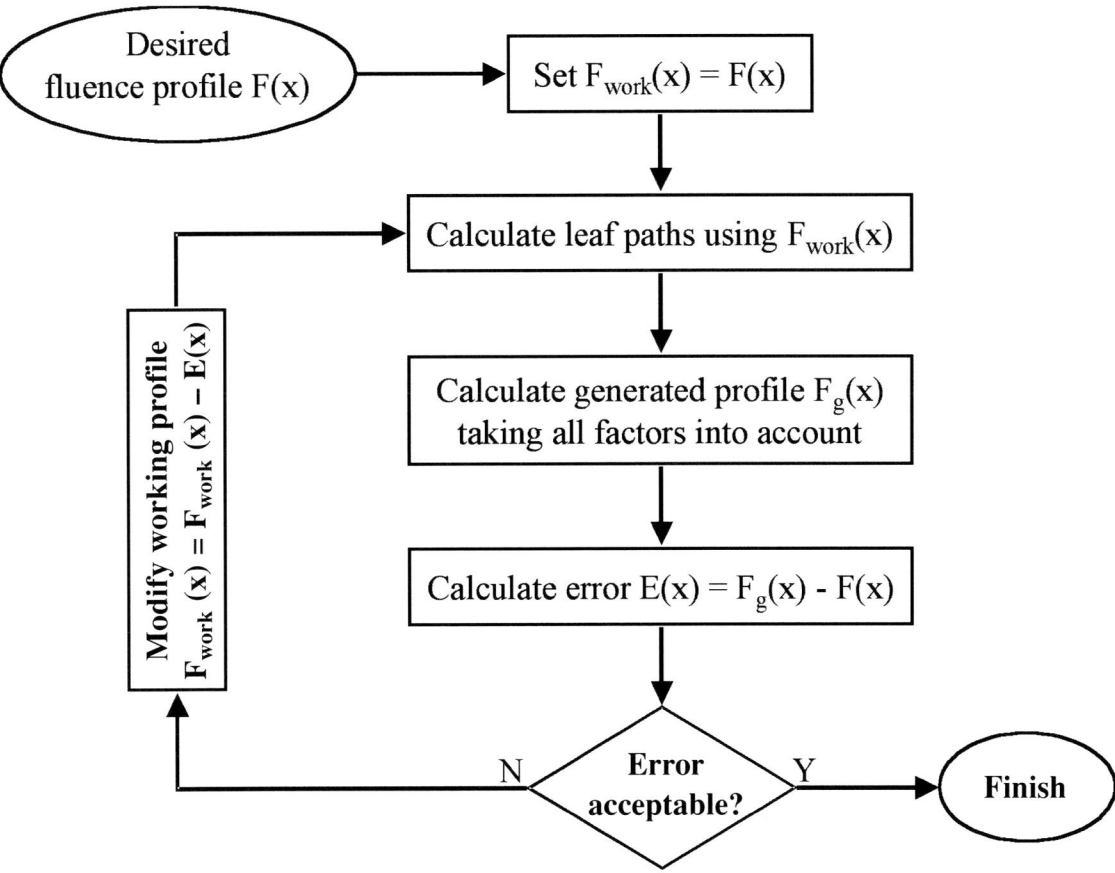

FIGURE 4–2. The iterative process for generating the leaf paths in DMLC.

one. This would make direct exposure under the tongue-and-groove equal to the lower of the direct exposures of the two adjacent leaf pairs.

The solution was partial in that it accounted only for direct exposure and ignored radiation transmitted through the tongue-and-groove region. Transmitted radiation was later incorporated by Webb et al. (1997). With the inclusion of transmitted radiation, the leaf paths can be adjusted so that the intensity under the tongue-and-groove has any value between the intensities of the adjacent leaf pairs.

The procedure by van Santvoort and Heijmen shall be described below for two adjacent leaf pairs. It can easily be extended to more than two leaf pairs.

Consider two adjacent leaf pairs, labeled "A" and "B", as shown in figure 4–4a. Points P_a, P_b, and P_{tg} all have the same coordinate along the direction of leaf travel, x_p, but different coordinates perpendicular to the direction of leaf travel. Points P_a and P_b are under leaf pairs A and B, respectively, in the midleaf region. Under the assumptions used to derive equation (1), the intensity at P_a and P_b is given by:

$$F(P_a) = M_t^a(x_p) - M_l^a(x_p)$$
$$F(P_b) = M_t^b(x_p) - M_l^b(x_p)$$

(7)

Point P_{tg}, however, is in the tongue-and-groove region, so the intensity at P_{tg} depends on the order of arrival of the leaves at x_p. Exposure begins when the second leading leaf reaches x_p and

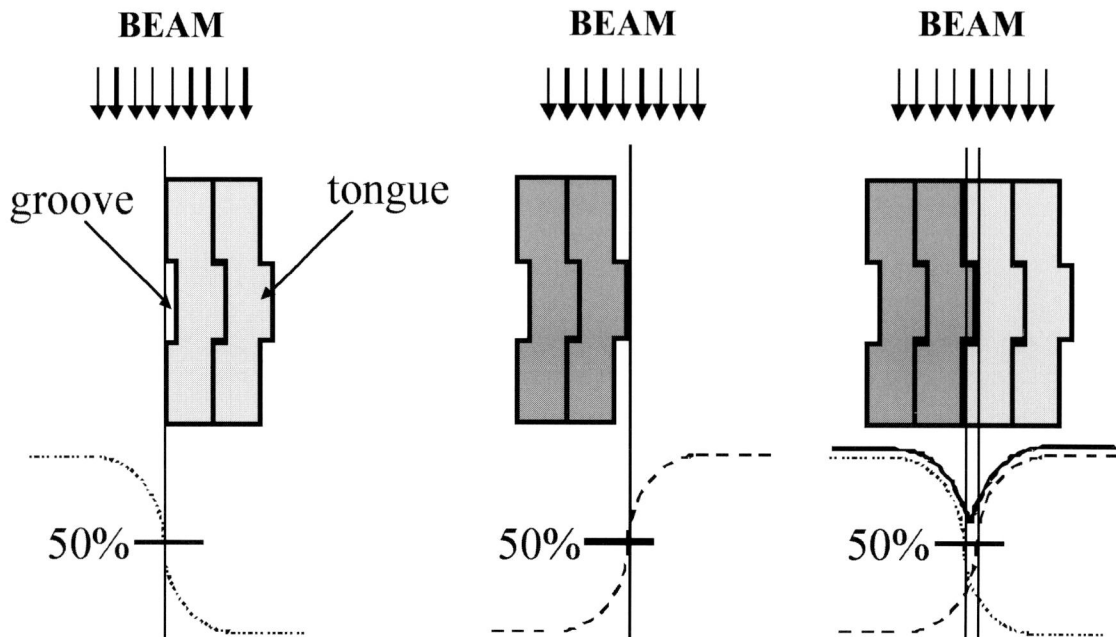

FIGURE 4–3. Cross section of the MLC, illustrating the tongue-and-groove effect (the leaves move in and out of the plane of the paper): (a) The intensity under the MLC when the dark-shaded leaves are retracted and only the light-shaded ones are in the field; (b) The intensity under the MLC when the situation is reversed; (c) The total intensity showing the decrease in the tongue-and-groove region when successive irradiations are performed with the two setups described in (a) and (b).

ends when the first trailing leaf reaches x_p. Let the B-leading leaf be the first one to arrive at x_p. Then there are three possibilities for the order of arrival of the leaves at x_p: {B$_l$, A$_l$, A$_t$, B$_t$}, {B$_l$, A$_l$, B$_t$, A$_t$}, and {B$_l$, B$_t$, A$_l$, A$_t$}. The first case is actually trivial since the intensity at point P_{tg} is equal to the intensity at P_a:

$$F(P_{tg}) = M_t^a(x_p) - M_l^a(x_p) = F(P_a)$$
(8)

The second and third cases are similar, so only the second one shall be described in detail. In the second case, the intensity at point P_{tg} is given by:

$$F(P_{tg}) = M_t^b(x_p) - M_l^a(x_p)$$
(9)

Since the A-leading leaf arrives at x_p after the B-leading leaf, i.e., $M_l^a(x_p) > M_l^b(x_p)$, it follows that $F(P_{tg}) < F(P_b)$. Similarly, $F(P_{tg}) < F(P_a)$.

If $F(P_a) < F(P_b)$, then leaf pair B can be slowed down by an amount $\Delta t = M_t^a(x_p) - M_t^b(x_p)$ so that the two trailing leaves arrive simultaneously at x_p (figure 4–4b and 4–4c). The arrival time then for the B-leading leaf is:

$$\left(M_l^b\right)'(x_p) = M_l^b(x_p) + \Delta t = M_l^b(x_p) + M_t^a(x_p) - M_t^b(x_p)$$
(10)

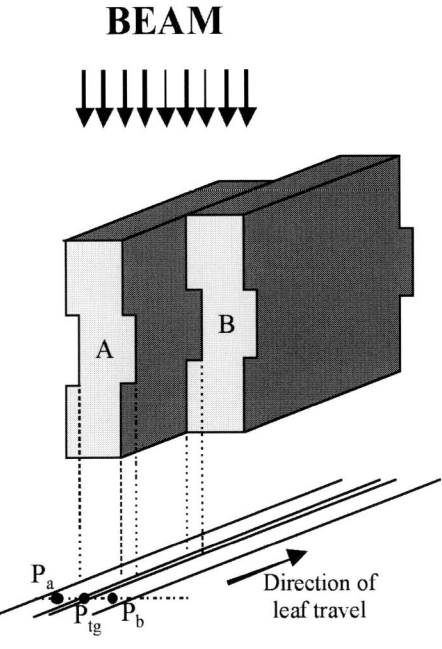

(a)

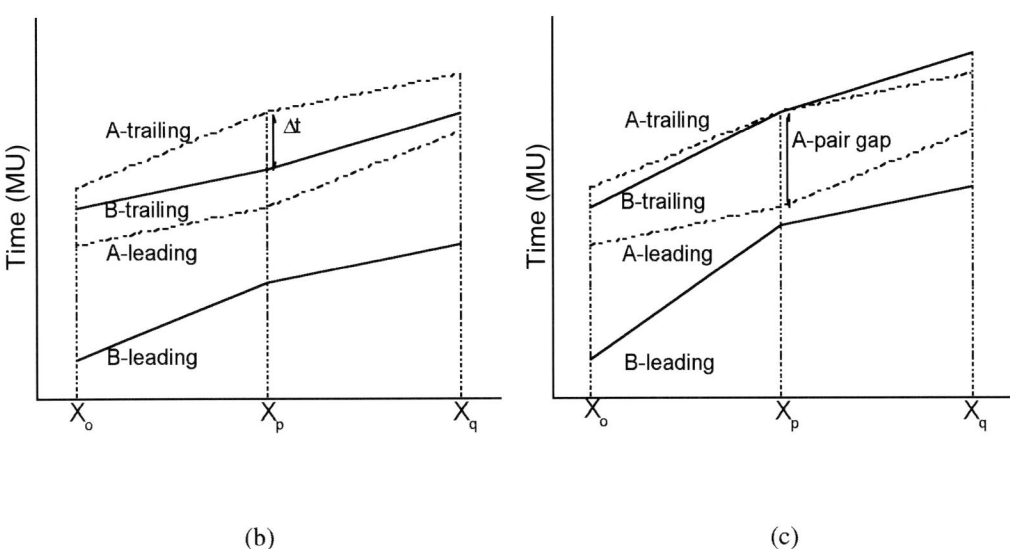

(b)   (c)

FIGURE 4–4. Synchronization of the leaf paths to reduce the tongue-and-groove effect: (a) "A" and "B" are adjacent leaf pairs. Points P_a and P_b are under leaf pairs A and B, respectively, in the mid-leaf region, whereas point P_{tg} is in the tongue-and-groove region under leaf pairs A and B. (b) The leaf paths at point x_p before synchronization and (c) the leaf paths after synchronization at point x_p. After synchronization the gap of leaf pair A is "inside" the gap of leaf pair B.

The intensity at point P_{tg} becomes:

$$F'(P_{tg}) = \left(M_t^b\right)'(x_p) - M_l^a(x_p) = M_t^a(x_p) - M_l^a(x_p) = F(P_a) \tag{11}$$

This procedure is graphically illustrated in figure 4–4b and 4-4c. For simplicity, only three points along the x-axis are shown: x_p, x_o, the point just before x_p, and x_q, the point just after x_p. In both figures the paths for the A-leaves are not modified. The arrival times of the B-leaves at x_o are also not modified since synchronization only takes place at x_p. After synchronization both trailing leaves arrive simultaneously at x_p. The arrival time of the B-leading leaf at x_p, as well as the arrival times of both B-leaves at points subsequent to x_p are also modified.

Similarly, if $F(P_a) > F(P_b)$ then leaf pair B can be slowed down so that $F'(P_{tg}) = F(P_b)$. Therefore, in both cases the intensity at P_{tg} is no less than the intensity at either P_a or P_b.

Figure 4–4b and 4–4c also illustrate the fact that even though the beam-on time at each segment does not increase by slowing down the faster leaf pair, the total beam-on time required to deliver the entire profile does increase. Before synchronization (Figure 4–4b) the A-trailing leaf is the last one to arrive at both points x_p and x_q. After synchronization, however, it is the B-trailing leaf that is the last one to arrive at x_q. Therefore, the number of MU required to generate the beam profile up to x_p is the same before and after synchronization, but the number of MU required to generate the beam profile up to x_q increases.

DMLC vs. SMLC (Step-and-Shoot)

As mentioned in the **Introduction,** an alternative method for delivering intensity-modulated beam profiles with a multileaf collimator is SMLC. The main advantage of DMLC is that the continuous leaf motion enables the delivered intensity profile to closely match the desired one, accurately preserving both the spatial and intensity resolutions. The SMLC method, on the other hand, resembles the conventional multi-segment treatment, and can therefore be verified more easily. Moreover, quality assurance for the latter is less demanding, for only the leaf position needs to be checked, whereas in the DMLC method both the leaf position and the speed must be checked.

On the other hand, the SMLC method requires that the desired profile be approximated by discrete levels of intensity. Inevitably, the conversion from the original continuous profile to the discrete one leads to degradation in resolution. Clearly, the degree of degradation of the SMLC method depends on the number of intensity levels used in the approximation. A large number of levels makes the SMLC method almost equivalent to the DMLC method but requires many segments of short beam-on time, which may present machine stability problems for some linear accelerators. A small number of levels degrades the approximation, and thereby affects the resultant dose distribution.

DMLC and SMLC have been compared in a variety of ways. Keller-Reichenbecher et al. (1999) compared the dose distributions obtained with the two methods for two clinical head-and-neck cases. The resolution along the direction of leaf travel was 5.5 mm. The results of the study suggested that the SMLC method with about five levels was, generally, comparable to the DMLC method. Budgell (1999) looked at the root-mean-square error between the generated and desired profiles and concluded that the SMLC method requires 15 to 25 intensity levels. Chui et al. (2001) examined the dose distributions obtained with the two methods for prostate, head and neck, and breast cases. Using a 2 mm resolution along the direction of leaf travel, they concluded that the SMLC method requires about 10 levels in order to be comparable to the DMLC method. The total delivery time (in MU) is about 20% less for the SMLC method but the total treatment time (in minutes) is about twice as long as the DMLC method.

Splitting A Large IMRT Field

Hardware and/or software limitations in the design of the MLC oftentimes restrict the maximum field size when the MLC is used in the DMLC mode. For example, on the Varian Mark II MLC, the most extended and the most retracted leaves on the same carriage cannot differ by more than 14.5 cm (the maximum MLC span). Since, in many cases the leaves start closed on one side of the field and end up closed on the other side of the field, this restriction becomes the maximum field size of the intensity-modulated beam profiles that can be generated.

The same problem also occurs in conventional treatments, e.g., whole CNS (central nervous system). The solution generally employed is to split the field at a particular location and use a moving junction to account for treatment uncertainties. Although this complicates treatment, it is a situation that is not frequently encountered since the maximum field size for static treatments is about 40 cm.

The same approach can be used for IMRT treatments. In certain cases there may be a natural place at which to split. For example, in posterior head-and-neck fields there is a low-intensity region in the field corresponding to the location of the spinal cord. In general, however, the low-intensity region may not exist or, if it exists, may not be perpendicular to the direction of leaf travel. If it is at an angle, splitting along the low-intensity region may still violate the maximum field size restriction. Furthermore, the much smaller maximum field size for DMLC treatments implies that this situation would be much more frequent, thus making the process impractical. Finally, it is a more complex and labor-intensive solution.

An alternative approach is to split the input beam profile into two or more overlapping sub-profiles that are then generated in sequence. The region of overlap between two adjacent sub-profiles should be chosen so as to avoid discontinuities in the direction perpendicular to leaf motion. In other words, if the input profile contains leaf pairs, say, 10 thru 20 one should avoid a subprofile that contains leaf pairs 10 thru 14 and 16 thru 20 but not leaf pair 15. In addition, feathering applied in the region of overlap accounts for treatment uncertainties (Wu et al. 2000), and eliminates the need for a moving junction, thus simplifying the process (Figure 4–5). The drawback of the splitting with feathering technique is that it increases the beam-on time, i.e.,

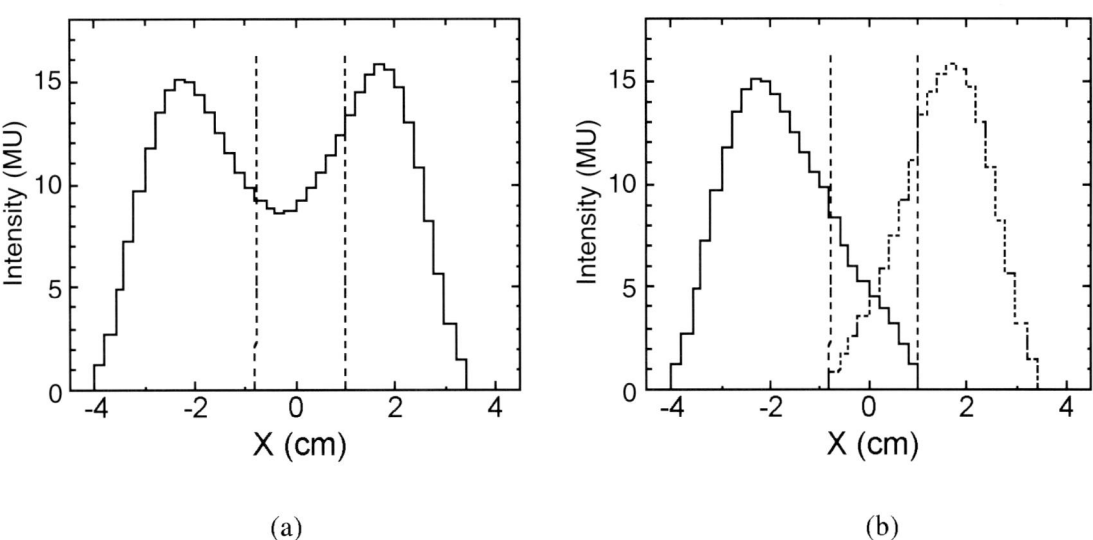

(a)   (b)

FIGURE 4–5. (a) The desired profile to be split and (b) the resulting subprofiles after splitting. The region of overlap is between the dashed lines.

the total number of MU required to deliver the subprofiles is larger than the number of MU required to deliver the original profile.

Summary

Delivery of intensity-modulated beam profiles can be efficiently accomplished with a multileaf collimator, using either the segmental or the dynamic technique. Each technique has its advantages and disadvantages in terms of accuracy of delivery, number of monitor units required, total treatment time, dosimetry, and verification. The leaf trajectories can be designed so as to take into account mid-leaf and interleaf transmission, the rounded leaf ends, head and collimator scatter, to the extent that these can be modeled and/or measured. Finally, the leaf trajectories can be designed in such a way as to overcome mechanical limitations of the MLC, such as the maximum field size.

References

Bortfeld, T. R., D. L. Kahler, T. J. Waldron, and A. L. Boyer. (1994). "X-ray field compensation with multileaf collimators." *Int. J. Radiat. Oncol. Biol. Phys.* 28:723–730.

Budgell, G. J. (1999). "Temporal resolution requirements for intensity modulated radiation therapy delivered by multileaf collimators." *Phys. Med. Biol.* 44:1581–1596.

Chui, C. S., M. F. Chan, E. Yorke, S. Spirou, and C. C. Ling. (2001). "Delivery of intensity-modulated radiation therapy with a conventional multileaf collimator: Comparison of dynamic and segmental methods." *Med. Phys.* 28:2441–2449.

Convery, D. J., and M. E. Rosenbloom. (1992). "The generation of intensity-modulated fields for conformal radiotherapy by dynamic collimation." *Phys. Med. Biol.* 37:1359–1374.

Fraass, B. A., P. L. Roberson, and A. S. Lichter. (1985). "Dose to the contralateral breast due to primary breast irradiation." *Int. J. Radiat. Oncol. Biol. Phys.* 11:485–497.

Galvin, J. M., X. G. Chen, and R. M. Smith. (1993). "Combining multileaf fields to modulate fluence distributions." *Int. J. Radiat. Oncol. Biol. Phys.* 27:697–705.

Galvin, J. M., A. R. Smith, and B. Lally. (1993). "Characterization of a multi-leaf collimator system." *Int. J. Radiat. Oncol. Biol. Phys.* 25:181–192.

Jiang, S. B., and K. M. Ayyangar. (1998). "On compensator design for photon beam intensity-modulated conformal therapy." *Med. Phys.* 25:668–675.

Kase, K. R., G. K. Svensson, A. B. Wolbarst, and M. A. Marks. (1983). "Measurements of dose from secondary radiation outside a treatment field." *Int. J. Radiat. Oncol. Biol. Phys.* 9:1177–1183.

Keller-Reichenbecher, M. A., T. Bortfeld, S. Levegrun, J. Stein, K. Prieser, and W. Schlegel. (1999). "Intensity modulation with the 'step and shoot' technique using a commercial MLC: A planning study. Multileaf collimator." *Int. J. Radiat. Oncol. Biol. Phys.* 45:1315–1324.

Ma, L., A. L. Boyer, L. Xing, and C.-M. Ma. (1998). "An optimized leaf-setting algorithm for beam intensity modulation using dynamic multileaf collimators." *Phys. Med. Biol.* 43:1629–1643.

Que, W. (1999). "Comparison of algorithms for multileaf collimator field segmentation." *Med. Phys.* 26:2390–2396.

Scrimger, J., and Z. Kolitsi. (1979). "Scattered radiation from beam modifiers used with megavoltage therapy units." *Radiol.* 130:233–236.

Siochi, R. A. (1999). "Minimizing static intensity modulation delivery time using an intensity solid paradigm." *Int. J. Radiat. Oncol. Biol. Phys.* 43:671–680.

Spirou, S. V., and C. S. Chui. (1994). "Generation of arbitrary intensity profiles by dynamic jaws or multileaf collimators." *Med. Phys.* 21:1031–1041.

Stein, J., T. Bortfeld, B. Dorschel, and W. Schlegel. (1994). "Dynamic X-ray compensation for conformal radiotherapy by means of multi-leaf collimation." *Radiother. Oncol.* 32:163–173.

Svensson, R., P. Kallman, and A. Brahme. (1994). "An analytical solution for the dynamic control of multileaf collimators." *Phys. Med. Biol.* 39:37–61.

van Santvoort, J. P., and B. J. Heijmen. (1996). "Dynamic multileaf collimation without 'tongue-and-groove' underdosage effects." *Phys. Med. Biol.* 41:2091–2105.

Webb, S., T. Bortfeld, J. Stein, and D. Convery. (1997). "The effect of stair-step leaf transmission on the 'tongue-and-groove problem' in dynamic radiotherapy with a multileaf collimator." *Phys. Med. Biol.* 42:595–602.

Wu, Q., M. Arnfield, S. Tong, Y. Wu, and R. Mohan. (2000). "Dynamic splitting of large intensity-modulated fields." *Phys. Med. Biol.* 45:1731–1740.

Xia, P., and L. J. Verhey. (1998). "Multileaf collimator leaf sequencing algorithm for intensity modulated beams with multiple static segments." *Med. Phys.* 25:1424–1434.

5

Computational Algorithms For Independent Verification Of IMRT

Chen-Shou Chui
Thomas LoSasso
Asa Palm

Objectives

Independent verification is an important component in radiation therapy quality assurance as recommended by the American Association of Physicists in Medicine (AAPM) (Kutcher et al. 1994). In some states, it is a regulatory requirement that each plan must be independently verified for its accuracy. For conventional treatment techniques, independent verification can be performed by hand calculation or by a stand-alone verification program. For intensity-modulated radiation therapy (IMRT) delivered with a multileaf collimator (MLC), hand calculation is impractical due to its complexity. Stand-alone programs have been reported, but at present they are limited to point dose calculations (Xing et al. 2000; Kung, Chen, and Kuchnir 2000; Watanabe 2001), which are inadequate to provide a complete picture of IMRT. Two-dimensional verification has also been reported. However, it was limited to calculating the intensity distribution in-air only (Xing and Li 2000); dose calculation at depth is not currently available.

Alternatively, IMRT plans can be independently verified with measurement. This has been done with chamber, film, thermoluminescent dosimeter (TLD) chips (Low et al. 1998; Tsai et al. 1998; LoSasso, Chui, and Ling 2001); electronic portal imaging devices (Partridge et al. 1998; Chang et al. 2000; James et al. 2000; Van Esch et al. 2001); or special devices such as the water-beam-imaging system (Li, Boyer, and Ma 2001). These measurements indeed provide independent verification to the planning system, but they are time consuming; hence they would not be practical for large-scale implementation of IMRT in a busy clinic.

In this chapter, we describe computational algorithms for independent verification of IMRT plans delivered with an MLC. These algorithms are independent of any planning systems, and are applicable to any linear accelerators (linacs) with any MLC designs, with the leaves moving in either dynamic mode or segmental mode. The general scheme of the calculation consists of two parts. The first part calculates the delivered intensity distribution in-air based on the leaf motion specified in the leaf sequencing file and other relevant data such as the collimator settings and the total beam on time [MU (monitor unit)]. The majority of the delivered intensity comes from direct exposure through the MLC openings. For improved accuracy, however, it is necessary to consider other effects. These include mid-leaf and interleaf transmissions, rounded leaf-end, tongue-and-groove effect, extra-focal source, and scatter from the MLC. Using this calculated intensity distribution, the second part computes the absolute dose (cGy) at depth in a flat phantom, either to individual points, or on two-dimensional dose grids. Dose calculation is primarily based on measured data including the tissue-maximum ratios (TMRs), off-center ratios (OCRs), and output factors. The effect of intensity modulation is accounted for by pencil beam convolution.

Before implementing these algorithms for clinical use, their accuracy must be validated. This is done by comparing the calculation results with chamber and film measurements for a variety of IMRT fields, ranging from simple patterns to clinical fields. These fields represent a wide range of field sizes and varying degrees of intensity modulation. In general, the agreement between calculation and measurement is within 2% to 3% in absolute dose, or 1 mm in distance in high dose gradient regions.

For practical clinical use, the computation must be reasonably fast. With these algorithms, typical computation time per beam on a 512×512 grid is less than 30 seconds on a 266 MHz Alpha station. These algorithms have been incorporated in a stand-alone independent verification program, which has been in routine clinical use in place of measurement since 1999.

In this chapter, we describe in detail these algorithms and present examples comparing the results from calculation and measurement for several IMRT fields.

Delivered Intensity Distribution

We first describe the algorithms for calculating the intensity[1] distribution delivered in-air based on the leaf motion defined in the leaf sequencing file. The calculation is typically performed on a 512×512 grid with a grid spacing of 1 mm at the level of the isocenter. We assume that outside the field defined by the collimating jaws the intensity is zero, hence only those grid points inside the collimator opening are considered. The algorithms described below are applicable to MLC delivery either in dynamic mode or in segmental mode.

Direct Exposure, Mid-Leaf Transmission

The majority of the delivered intensity comes from direct exposure through the MLC opening. Transmission through the leaves constitutes a minor, but not negligible contribution to the total delivered intensity and hence must also be considered. The transmission factor through the mid-leaf, ε, can be easily measured by taking the ratio of doses with all the leaves closed and all leaves retracted in a small field, say 2×2 cm. Its value ranges from approximately 1.1% to 1.5%, depending on the MLC model and photon energies (LoSasso, Chui, and Ling 1998; Arnfield

---

[1] As pointed out by Webb (Webb and Lomax 2001), a more precise term to use is "fluence" rather than "intensity." However, to adhere to the common practice, we continue to use the term "intensity" in this manuscript.

et al. 2000). Here we use the values 1.4 and 1.1 for the Varian Mark II and Millennium™ MLCs, respectively.

For segmental MLC delivery, the entire delivery sequence is divided into a number of segments. For each segment-k, the leaves move to their respective positions when the beam is off, and stay there while the beam is turned on for the duration of the associated beam-on-time T_k. Referring to figure 5–1a, suppose point $P(x,y)$ corresponds to leaf pair-j as determined by its y-coordinate, then the fluence rate, $\phi_k^{midleaf}$ to $P(x,y)$ for segment-k is due to either direct exposure or mid-leaf transmission, depending on the x-coordinate relative to the leaf positions:

$$\phi_k^{midleaf}(x,y) = \begin{cases} 1 & \text{if } L_{j,k} \leq x \leq R_{j,k}, \\ \varepsilon & \text{if } x < L_{j,k} \text{ or } x > R_{j,k}, \end{cases} \tag{1}$$

where $L_{j,k}$ and $R_{j,k}$ are the left and the right leaf positions of pair-j for segment-k, respectively.

For the entire delivery, the intensity to point $P(x,y)$ is the sum of all segments:

$$\phi^{midleaf}(x,y) = \sum_{k=1}^{K} T_k \phi_k^{midleaf}(x,y), \tag{2}$$

where K is the total number of segments.

For dynamic MLC delivery, the entire sequence is defined by a series of control points in beam-on-time. At the k-th control point, the field opening is again defined by the leaf positions $L_{j,k}$ and $R_{j,k}$, and the associated cumulative beam-on-time M_k. Between successive control points k and $k+1$, the left and right leaves move continuously at the appropriate speeds from $L_{j,k}$ to $L_{j,k+1}$ and $R_{j,k}$ to $R_{j,k+1}$, respectively; while the cumulative beam-on-time is increased from M_k to M_{k+1}.

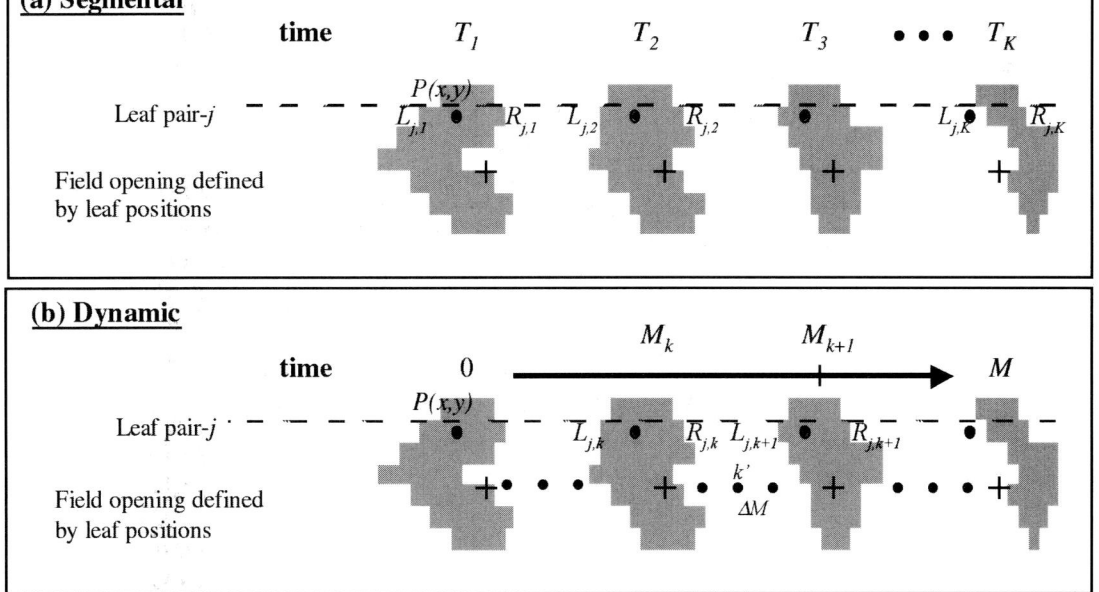

FIGURE 5–1. Field opening defined by leaf positions as a function of beam-on time (MU) for (a) segmental, and (b) dynamic MLC delivery. The isocenter is indicated by the symbol "+". Point $P(x,y)$ is in the path of leaf pair-j, whose left and right leaf positions at segment-k are denoted as $L_{j,k}$ and $R_{j,k}$, respectively.

For computational purposes, we divide the continuous leaf motion into many discrete short segments with beam-on-time increment ΔM, as shown in figure 5–1b. Typically ΔM is set to 0.1 MU in our calculations. For each such short segment, the leaf positions are interpolated from those at the two control points bracketing the short segment. The fluence rate, $\phi^{midleaf}_{k'}(x,y)$, at point $P(x,y)$ for the short discrete segment-k' is given by the same expression (1) as for segmental MLC delivery. For the entire delivery, the intensity to point $P(x,y)$ is given by:

$$\phi^{midleaf}(x,y) = \Delta M \sum_{k'=1}^{K'} \phi^{midleaf}_{k'}(x,y), \qquad (3)$$

where K' is the total number of short discrete segments, and the total beam-on-time is $M = K'\Delta M$.

Note that expressions (2) and (3) for segmental and dynamic delivery, respectively, are equivalent by setting $T_k = \Delta M$.

Tongue-and-Groove, Interleaf Transmission

To reduce the interleaf transmission, most MLCs employ some types of tongue-and-groove design (LoSasso, Chui, and Ling 1998; Arnfield et al. 2000; Deng et al. 2001; Klein and Low 2001). One such example used by the Varian Millennium MLC is illustrated in figure 5–2a. Due to the different thicknesses traversed by the radiation, the transmission factor through the mid-leaf, ε, is different from that through the tongue or the groove, τ. When two neighboring leaves are in the field at the same time, the interleaf transmission factor is λ as shown in figure 5–2b. If the thickness of the tongue or groove is exactly one half of the mid-leaf thickness, then in theory the interleaf transmission $\lambda = \varepsilon = \tau^2$. In practice, however, this is not quite true because of the small

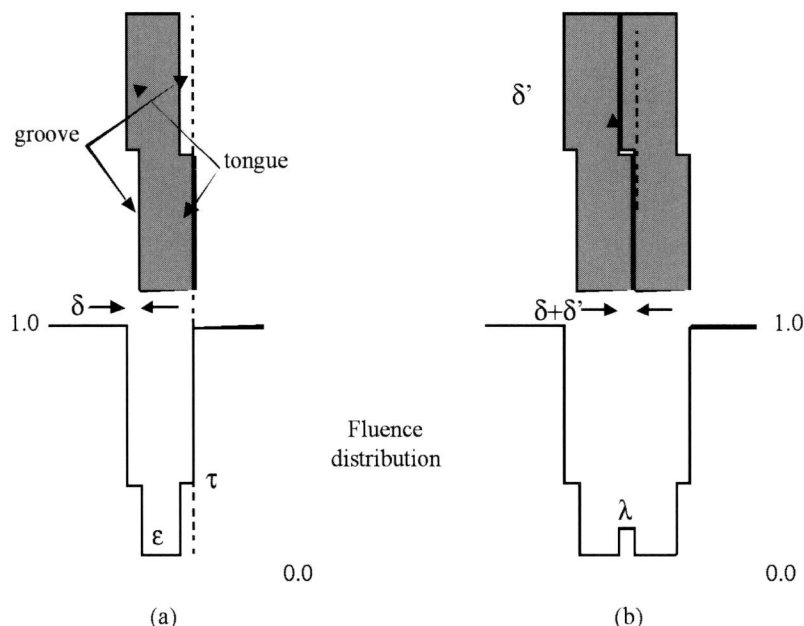

FIGURE 5–2. (a) Diagram illustrating the tongue-and-groove design of the Varian Millennium MLC, where δ is the width of the tongue or the groove, ε is the mid-leaf transmission, and τ is the transmission through the tongue or the groove. (b) When neighboring leaves are in the field at the same time, λ is the interleaf transmission, and δ' is the width of the airspace between the neighboring leaves.

air space δ' between the neighboring leaves. More discussion on the determination of these factors will be given later. When the neighboring leaves are alternately in the field, we have the so-called *tongue-and-groove* underdose effect. In figure 5–3a, leaf A is in the field and leaf B out of the field, the intensity distribution has three distinct levels, λ, τ, and ε, corresponding to the open, tongue or groove, and mid-leaf regions, respectively. Conversely, in figure 5–3b, when leaf A is out of the field and B in the field, we have an identical but reversed distribution with the intensity τ over the same tongue-or-groove region as in figure 5–3a. When the two complementing distributions are added together, instead of getting a uniform intensity distribution, we have a region of reduced intensity of 2τ. Both these effects, interleaf transmission and tongue-and-groove, have been noted before (LoSasso, Chui, and Ling 1998; Arnfield et al. 2000; Deng et al. 2001; Klein and Low 2001).

As discussed previously, if we use the value $\tau = \sqrt{\varepsilon}$ in calculation, the calculated values of interleaf transmission $\lambda = \tau^2$ and the tongue-and-groove underdose 2τ, would not agree with the measurement. This is due to the presence of the narrow airspace, δ', between the leaves. Moreover, the grid size used in calculation (typically at 1 mm) is generally different from the width of the tongue or groove δ. This means that in calculation the entire width of the grid size (e.g., 1 mm) is affected by the value τ, whereas in reality the affected width should have been $\delta + \delta'$. For these reasons, the theoretical value of $\tau = \sqrt{\varepsilon}$ would not give good agreement with measurement. Thus, for computational purposes, λ and τ are inferred independently from measurement. Radiographic film is exposed in phantom to reference fields to generate tongue-and-groove effects (adjacent leaves are present alternately) or interleaf transmission (adjacent leaves are present

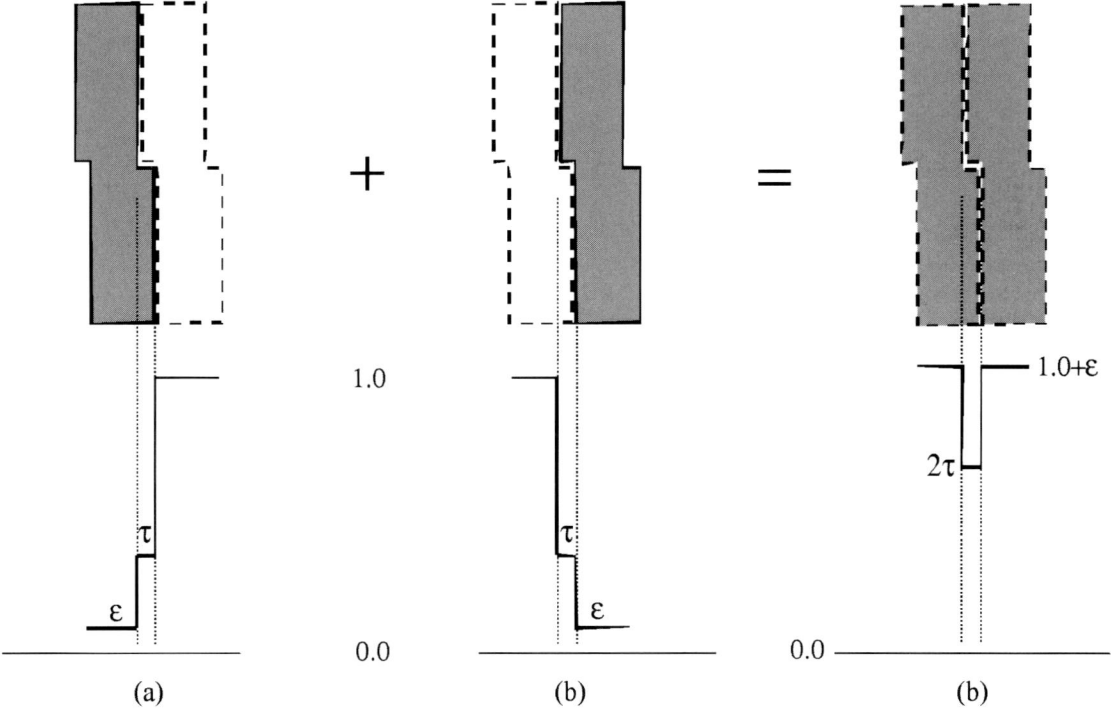

FIGURE 5–3. (a) Leaf A is in the field, leaf B is out of the field, transmission through the tongue of leaf A is τ. (b) Leaf A is out of the field, leaf B is in the field, transmission through the tongue of leaf B is also τ. (c) The combined transmission from (a) and (b) produces the tongue-and-groove underdose effect. The total transmission in the tongue-and-groove region is 2τ.

simultaneously). For the same reference fields, dose distributions are calculated and compared with measured data. Values for λ and τ are adjusted until the calculations agree with the measurements. A summary of various transmission factors for the Varian Mark II and Millennium MLCs are given in table 5–1.

It should be noted that since the values for λ and τ were fit using the treatment planning system, their values depend on the particular grid size, source distribution s(r), and pencil beam kernel k(r,d), used in the calculation. The discussion of the source distribution s(r) and pencil beam kernel k(r,d) are given in a later section. The values listed in table 5–1 gave good agreement with the measurement, but may not be appropriate if used under different conditions.

Expression (1) is valid only for points whose y-coordinate is in the mid-leaf region. For those points in the interleaf region, expression (1) is now modified as:

$$\phi_k^{\text{interleaf}}(x,y) = \begin{cases} 1 & \text{if } L_{j,k} \le x \le R_{j,k} \text{ for both neighboring leaves,} \\ \lambda & \text{if } x \text{ covered by both neighboring leaves,} \\ \tau & \text{if } x \text{ covered by only one leaf.} \end{cases} \tag{4}$$

Rounded Leaf-End

To minimize the variation of dose distribution near the leaf-end with off-axis leaf positions, some manufacturers employ a rounded leaf-end design for their MLC as shown in figure 5–4. As a result, the intensity distribution has a penumbra region near the leaf-end. To account for this, the leaf position can be shifted by about 0.6 to 1.0 mm to an *effective* position (LoSasso, Chui, and Ling 1998, 2001; Arnfield et al. 2000; Graves et al. 2001) or alternatively, the penumbra shape can be taken into account explicitly. To explicitly account for the rounded leaf-end effect, expressions (1) and (4) for the mid-leaf and interleaf regions are now modified respectively as expressions (5) and (6) for points in the leaf-end penumbra region:

$$\phi_k^{midleaf}(x,y) = \begin{cases} 1 & \text{if } L_{j,k} \le x \le R_{j,k}, \\ p & \text{if } x \text{ in the penumbra region at } L_{j,k} \text{ or } R_{j,k}, \\ \varepsilon & \text{if } x \text{ outside the penumbra region at } L_{j,k} \text{ and } R_{j,k}, \end{cases} \tag{5}$$

$$\phi_k^{\text{interleaf}}(x,y) = \begin{cases} 1 & \text{if } L_{j,k} \le x \le R_{j,k} \text{ for both neighboring leaves,} \\ \lambda & \text{if } x \text{ covered by both neighboring leaves,} \\ p' & \text{if } x \text{ covered by only one leaf, and in the penumbra region,} \\ \tau & \text{if } x \text{ covered by only one leaf, and outside the penumbra regio} \end{cases} \tag{6}$$

The value p in expressions (5) is calculated from the mid-leaf transmission and the geometry of the round edge as a function of the distance between x and the leaf-end, i.e., $x - R_{j,k}$ or

Table 5–1. Transmission Factors for Varian Mark II and Millennium MLCs

MLC Type	Tongue or Groove Width δ (cm)	Mid-Leaf Transmission ε (%)	Interleaf Transmission λ (%)	Tongue or Groove Transmission τ (%)
Varian Mark II	0.127	1.4	6.0	11.0
Varian Millennium	0.078	1.1	2.6	19.0

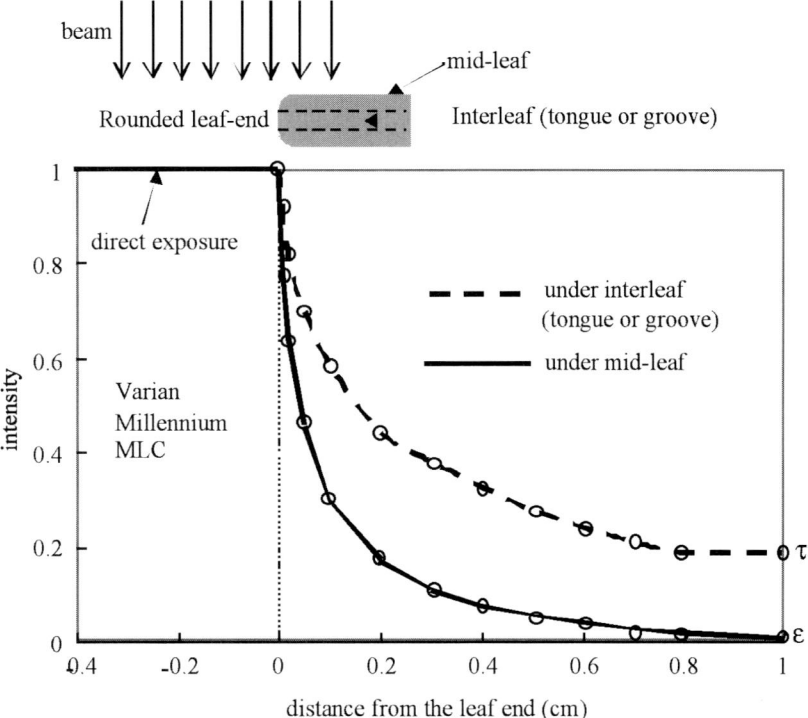

FIGURE 5–4. Intensity distribution in the penumbra region as a function of the distance from the rounded leaf end for Varian Millennium MLC.

$L_{j,k} - x$, as appropriate (LoSasso, Chui, and Ling 1998). The value ε is the mid-leaf transmission beyond the penumbra region. Similarly, p' and τ in expression (6) are the corresponding values in the tongue-and-groove, or interleaf, region. A graph of the transmission distributions for Varian Millennium MLC is shown in figure 5–4. Note that the transmission in the mid-leaf region is different from that in the interleaf region, due to the fact that the leaf thickness in the latter is approximately only one-half of that in the former. Their respective numerical values are given in table 5–2. Thus, the intensity at (x,y) for segment-k, ϕ_k, is either expression (5) or (6), depending on whether the point is in the mid-leaf region or in the interleaf region as determined by its y-coordinate.

Extra-Focal Source

In external beam radiation therapy, it is well known that the output in-air changes with the field size defined by the collimators. This is commonly referred to as the *collimator scatter* (Khan 1994) or the *head scatter* factor (Ahnesjo 1995). The main cause for this is due to the scatter in the

Table 5–2. Intensity Distribution in the Leaf-End Penumbra Region

Distance from Leaf-End (cm)	0	0.01	0.02	0.05	0.1	0.2	0.3	0.4	0.5	0.6	0.7	0.8	1.0
Mid-Leaf	1.000	0.775	0.643	0.457	0.298	0.174	0.114	0.080	0.057	0.040	0.028	0.020	0.011
Interleaf	1.000	0.925	0.805	0.703	0.587	0.448	0.385	0.331	0.281	0.251	0.222	0.19	0.19

machine head, mainly in the flattening filter, resulting in an extra-focal source distribution (Sharpe et al. 1995). Backscatter into the monitor chamber also contributes to some extent to this effect (Zhu et al. 2001). For IMRT delivered with an MLC, since the field opening is generally small and changes with the beam-on-time, the corresponding output for each segment also changes and therefore must be considered.

The first step to account for this effect is to determine the source distribution. Fix et al. had performed detailed Monte Carlo analysis for a particular machine, resulting in a 12-component source model (Fix et al. 2001a, 2001b). The source distribution, including both the focal and the extra-focal components, can also be derived from measurement. The source distribution is assumed to be radially symmetric. At large radii, the distribution was obtained from in-air output measurements using beams of different field sizes. At small radii, the distribution was empirically deduced from a comparison of calculated and measured dose profiles in the penumbra region for half-blocked fields at the depth of 5 cm. The source function was adjusted until the treatment planning calculations agreed with measurements. An example of the source distribution $s(r)$ as a function of the radial distance for the Varian 21EX machine is shown in figure 5–5.

Next, the source distribution is incorporated in the output calculation as follows. We first consider the output at the isocenter. Referring to figure 5–6a, the field opening ϕ_k for segment-k, shown as the gray area on the isocenter plane, is forward projected from the source through the MLC opening. The output at the isocenter can be calculated by backprojecting from the isocenter through the MLC opening to the source plane. For convenience, we set the source plane to be at the same distance from the MLC plane as the isocenter plane[2], so that the projected opening at the source plane is identical to that at the isocenter plane. The output-corrected intensity at the isocenter, due to segment-k, is then the amount of the source distribution enclosed by the projected opening on the source plane, that is:

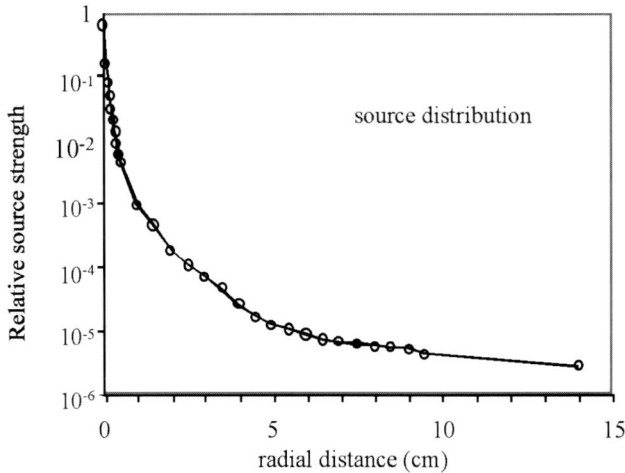

FIGURE 5–5. Source distribution as a function of the radial distance from the foclal spot for a Varian 2100EX machine.

---

[2] If the source-to-MLC-plane distance is different from the MLC-plane-to-isocenter distance, then the dimensions of the source distribution need to be properly scaled. However, the general methodology described in this section is still valid.

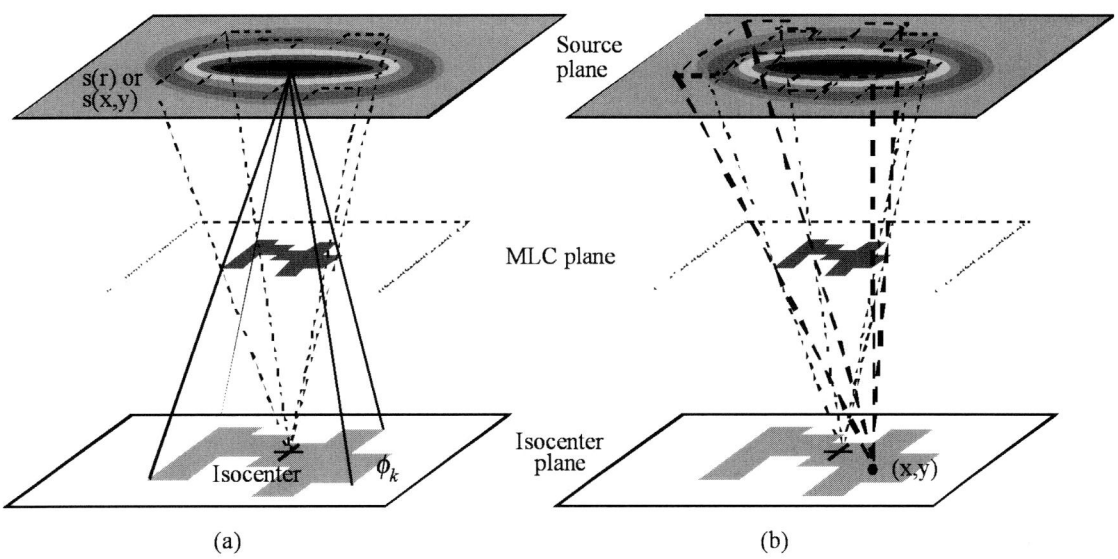

FIGURE 5–6. (a) The MLC opening is backprojected (in dotted lines) from the isocenter to the source plane. The output at the isocenter is the portion of the source distribution enclosed by the projected opening. (b) For any point (x,y) on the isocenter plane, the projected opening (in dashed lines) has the same shape and size as that in (a), except it is shifted by $(-x,-y)$.

$$\phi_k^{corrected}(0,0) = \iint\limits_{source\ plane} \phi_k(x',y')s(x',y')dx'\,dy',\qquad(7)$$

For any other point (x,y) on the isocenter plane, shown in figure 5–6b, the backprojected opening on the source plane has the same size and shape, except that it is shifted in the opposite direction by $(-x,-y)$. Thus, the output-corrected intensity at (x,y) is given by:

$$\phi_k^{corrected}(x,y) = \iint\limits_{source\ plane} \phi_k(x',y')s(x-x',y-y')dx'\,dy'\qquad(8)$$

For the entire delivery:

$$\phi^{corrected}(x,y) = \sum_{k=1}^{K}\phi_k^{corrected}(x,y) = \iint\limits_{source\ plane}\sum_{k=1}^{K}T_k\phi_k(x',y')s(x-x',y-y')dx'\,dy'$$
$$= \iint\limits_{source\ plane}\phi(x',y')s(x-x',y-y')dx'\,dy' = \phi\otimes s \qquad(9)$$

where $\phi(x',y') = \sum_{k=1}^{K}T_k\phi_k(x',y')$ is the intensity distribution from all segments, and the symbol "$\otimes$" denotes convolution.

Scatter From The MLC

When an MLC is used to shape the radiation field, in addition to the direct transmission through the leaf, ε, as discussed before, there is also scatter from the leaf. For small- and medium-sized fields, because the irradiated area on the MLC is small, the scatter from the MLC is also small, and thus can be safely ignored in calculation. For large fields, however, the contribution from MLC scatter may be significant and is no longer negligible. To compute this effect, we need to have the distribution of the MLC scatter. This can be obtained by a series of transmission measurements with the leaves fully closed and the collimators set to various openings. The amount of MLC scatter at a point in-air, say, the isocenter, increases with the irradiated area on the MLC as defined by the collimator opening. By taking the difference of transmission factors of successive collimator openings, we obtained the distribution of MLC scatter, $S^{mlc}(r)$, as a function of the radial distance, r, from the pencil beam incident on the MLC. This distribution is shown in figure 5–7. For segment-k, the blocked intensity over the area covered by the MLC, denoted as $B_k(x,y)$, is the complement of the intensity through the MLC opening, $\phi_k(x,y)$, i.e., $B_k(x,y) = 1 - \phi_k(x,y)$. For the entire delivery, the *total blocked intensity distribution* is then $B(x,y) = \sum_{k=1}^{K} B_k(x,y)$. Since blocked photons cause scatter from the MLC, the total scattered intensity from MLC is given by:

$$\phi^{mlc}(x,y) = \iint_{W \times H} B(x',y')S^{mlc}(x-x',y-y')dx'\,dy' = B \otimes S, \qquad (10)$$

where the integral is over the field opening $W \times H$, defined by the collimating jaws. The total delivered intensity distribution, combining output-corrected intensity distribution of expression (9) and scatter from the MLC of expression (10), is then:

$$\phi^{total}(x,y) = \phi^{corrected}(x,y) + \phi^{mlc}(x,y). \qquad (11)$$

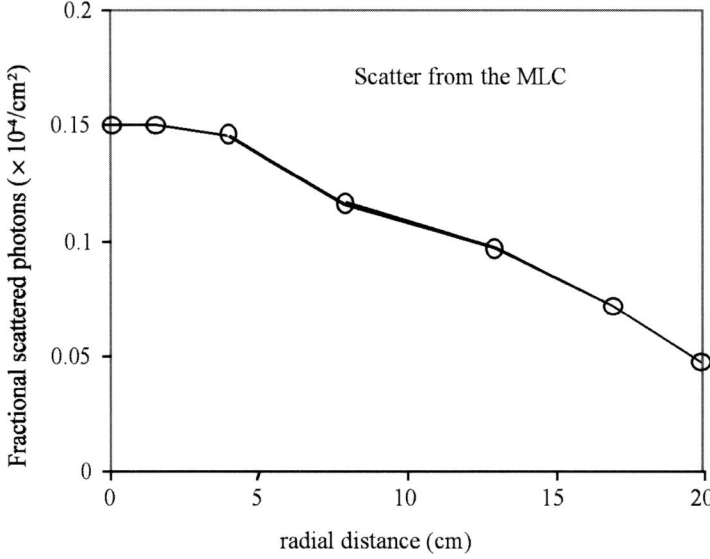

FIGURE 5–7. Distribution of scatter from the MLC as a function of the radial distance from the pencil beam incident on the leave.

Dose Calculation In A Flat Phantom

After the delivered intensity distribution in-air has been obtained, we can proceed to calculate the dose distribution in-phantom. Dose calculation for intensity-modulated field has been described elsewhere (Chui, LoSasso, and Spirou 1994). We first calculate the dose from the rectangular open field, primarily using the measured data, namely, the output factors, tissue-maximum ratios (TMRs), and off-center ratios (OCRs). The effect of intensity modulation is then accounted for by pencil beam convolution. A summary of the method is given below.

Rectangular Open Field

Assuming the machine is calibrated to deliver 1 cGy/MU at d_{max} at the isocenter for a 10×10 cm open field, then the dose to a point from a rectangular, open field is given by:

$$D_0(x,y,d) = MU \bullet OF(W \times H) \bullet \left(\frac{SAD}{s} \right)^2 \bullet TMR(d, W' \times H') \bullet OCR(x,y,d,W',H') \qquad (12)$$

where

$D_0(x,y,d)$ is the dose to a point at (x,y) from the central axis at an equivalent depth d in water;

MU is the beam-on-time;

$OF(W \times H)$ is the output factor for a collimator setting of $W \times H$;

SAD is the source-to-axis (isocenter) distance, s is the source-to-point distance projected to the central axis, and $\left(\frac{SAD}{s} \right)^2$ is the inverse-square factor;

$TMR(d,w'H')$ is the tissue-maximum-ratio at an equivalent depth d, projected to the central axis, for a field size $W' \times H'$ projected at the point of calculation; and

$OCR(x,y,d, W', H')$ is the off-center-ratio (Chui and Mohan 1986) at the lateral coordinates (x,y) and depth d for a field size $W' \times H'$ at the point of calculation.

The TMR term is interpolated from a set of measured data, which accounts for the attenuation and scatter of the open beam. Inhomogeneity correction is accounted for by the equivalent depth d. The OCR term is also interpolated from measured data. Note that the effects of horns and off-axis beam softening are already included in these measured data.

Intensity-Modulated Field

The effect of intensity modulation is accounted for as correction factors to the open field dose $D_0(x,y,d)$ by pencil beam convolution:

$$D(x,y,d) = D_0(x,y,d) \bullet \left[\frac{\iint_{W \times H} \phi^{total}(x',y')k(x-x',y-y',d)dx'\,dy'}{\iint_{W \times H} U(x',y')k(x-x',y-y',d)dx'\,dy'} \right], \qquad (13)$$

where $D(x,y,d)$ is the dose from the intensity-modulated field, $\phi^{total}(x',y')$ is the delivered intensity distribution as given by expression (11), $k(x-x',y-y',d)$ is the pencil beam kernel as a function of the relative position $(x-x',y-y')$ at depth d; and $U(x',y')$ is a uniform intensity distribution over the rectangular field $W \times H$. The pencil beam kernels $k(x,y,d)$ are typically obtained from Monte Carlo calculations at several depths (Mohan and Chui 1987). Examples of the pencil beam kernels for 6 MV and 15 MV photons are given in tables 5–3a and 5–3b, respectively. These kernels were calculated by Monte Carlo using the photon beam energy spectra from a Varian machine. For linacs from different manufacturers, the energy spectra may be different and so will be the kernel distributions. However, the general shape of the pencil beam kernels should be similar.

The integrals in expressions (9), (10), and (13) are carried out with Fast Fourier Transform and can be completed in a few seconds for a 512×512 grid.

Validation Of The Algorithms

Prior to the use of these algorithms in the clinic, their accuracy must be validated. Extensive verification has been done by comparing the calculated results to data measured with chamber and film. The test cases ranged from simple patterns to clinical intensity-modulated fields used for a variety of treatments. A few representative examples are presented here.

Interleaf Transmission, Tongue-and-Groove Effect

We first show the algorithm's ability to calculate the interleaf transmission of a Varian Millennium MLC. Figure 5–8a shows an irradiation in which all the leaves were fully closed and the beam-on-time was sufficiently long to give the dose at the interleaf region to be about 45 cGy. The horizontal isodose lines in the middle part of the field are due to interleaf transmission. The calculated dose distribution (solid lines), generally agreed well with the measured results (dotted lines). The magnitude of the interleaf transmission can be more clearly seen in figure 5–8b, which shows the dose profiles along the dashed line indicated in figure 5–8a. The peaks and troughs on the profile correspond to interleaf and mid-leaf transmission, respectively.

We next show the algorithm's ability to calculate the tongue-and-groove effect. A field was created with the jaws set to 10×10 cm and every four pairs of leaves alternately placed in and out of the field. A film was irradiated for a beam-on-time to give about 30 cGy to the open part of the field. This was followed next by reversing the previous leaf positions, and the same amount of beam-on-time given. If there was no tongue-and-groove effect, a uniform dose distribution across the field would be expected from these two complementing leaf positions. In reality, however,

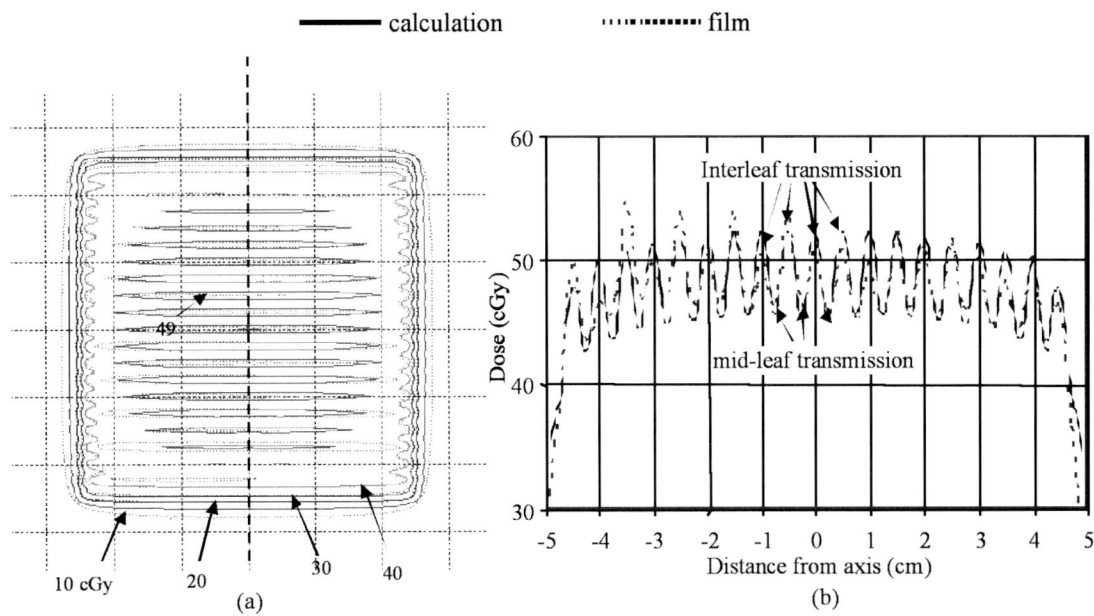

FIGURE 5–8. Interleaf transmission: comparison of calculation and film measurement. (a) Isodose distributions; (b) cross-leaf dose profiles along the dashed line in (a).

due to the tongue-and-groove effect, the isodose distribution exhibits regions of underdose as shown in figure 5–9a. The magnitude of the underdose tongue-and-groove effect is more clearly seen in figure 5–9b, which shows the cross-leaf dose profile along the dashed line indicated in figure 5–9a. The calculated results agreed well with the measured data.

Prostate

The accuracy of the algorithms for clinical fields was first tested with prostate patients. The dose at the isocenter for nearly 400 patients treated with IMRT on two machines was measured with ion chamber, and compared against the calculated results. For these calculations, we only included the effects of direct exposure, mid-leaf transmission, rounded leaf-end, and extra-focal source. The ratio of measured dose to the calculated dose is plotted in figure 5–10. For the majority of the patients, the agreement was within ±2%. The mean value of the ratios was 0.993, and the standard deviation was 0.008. Film dosimetry in flat homogeneous phantoms was also evaluated for each field for the first 20 prostate patients.

Nasopharynx

We next tested the algorithms on fields used for nasopharynx treatment. These treatment fields tend to have more intensity modulation than those used for prostate treatment. As a result, ion chamber measurements were less reliable than film for these fields. Figure 5–11 shows the measured vs. calculated isodose distribution comparison of a nasopharynx field. For the example shown, again only the effects of direct exposure, mid-leaf transmission, rounded leaf-end, and extra-focal source were included in the calculation. The capability of tongue-and-groove was intentionally turned off to show its effect. As can be seen in the figure, the agreement between the calculation and measurement is in general very good, except in areas where the tongue-and-groove underdose was seen on the film but not predicted by the calculation.

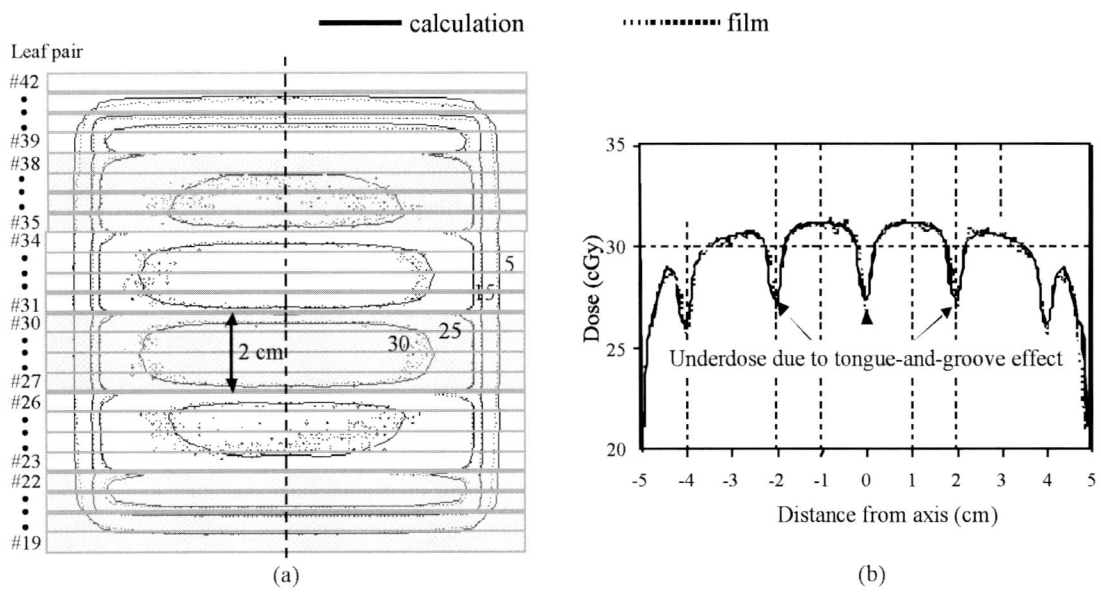

FIGURE 5–9. Tongue-and-groove effect on a Varian Millennium MLC: comparison of calculation and film measurement. (a) Isodose distributions; (b) cross-leaf dose profiles along the dashed line in (a).

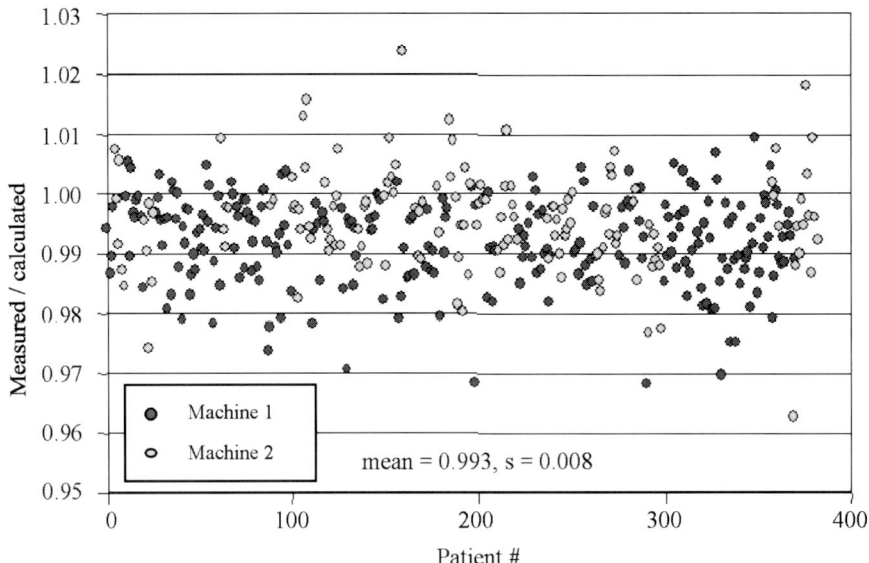

FIGURE 5–10. Comparison of calculation and chamber measurement for nearly 400 prostate patients treated on two machines. The mean value of the ratios of measured to calculated dose at the isocenter was 0.993, and the standard deviation was 0.008.

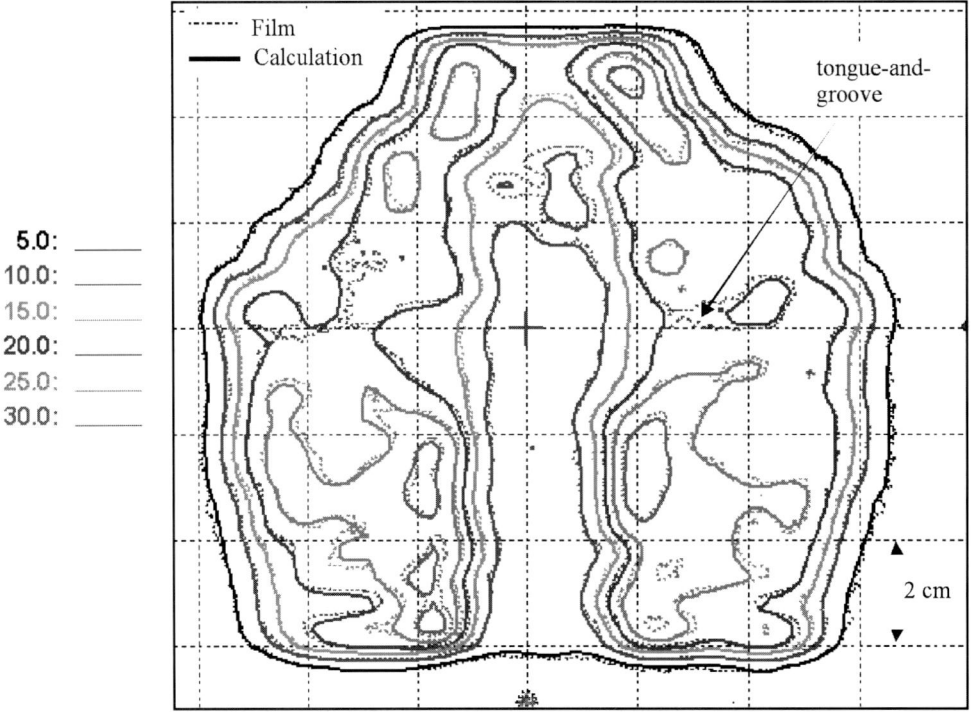

FIGURE 5–11. Comparison of calculation and film measurement for a nasopharynx field. See COLOR PLATE 7.

Lung

For lung treatment, the field size tends to be larger, the field shape more irregular, and the degrees of intensity modulation greater. In this example, we demonstrate the individual contribution from some of the effects discussed previously. Figure 5–12a shows the calculated (solid lines) and measured (dotted lines) isodose distributions of a lung field. The collimator field size was approximately 16 × 16 cm, and the maximum dose was about 60 cGy. In this figure, the calculation included only the effects of direct exposure, mid-leaf transmission, and rounded leaf-end. As seen in figure 5–12a, the overall agreement between calculation and measurement was good but there were regions of discrepancy. The dose difference (film minus calculation) is plotted in figure 5–12b. In general, the measured dose was lower than those calculated, indicated by the blue and black curves. The discrepancy in some regions was more than 10 cGy, or 16% (10 cGy out of 60 cGy). In figure 5–13, we show the effects due to individual corrections. The top row shows the dose difference between measurement and calculation, with additional corrections included in the calculation one at a time. The effects of individual corrections are shown in the bottom row. Figure 5–13a is a copy of figure 5–12b, in which the calculation included only the effects of direct exposure, mid-leaf transmission and rounded leaf-end. Figure 5–13e shows the individual effect of tongue-and-groove, which decreased the overall calculated values. When this effect is included in the calculation, the underdose discrepancy relative to measurement was reduced as shown in figure 5–13b. But regions of large discrepancy were still present, indicated by the red and black curves. Figure 5–13f shows the effect of a more accurate extra-focal source function. When this effect is included, the overall agreement was markedly improved, with

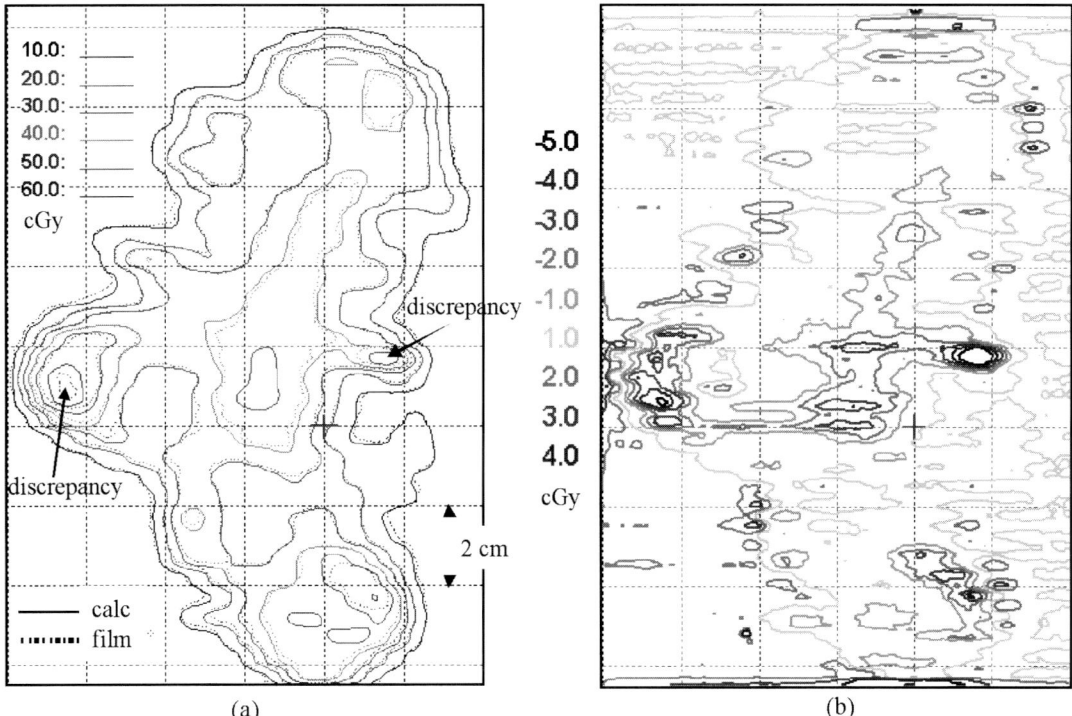

(a)   (b)

FIGURE 5–12. Comparison of calculation and film measurement for a field used in a lung treatment. (a) Overlaid isodose distributions; (b) distribution of dose differences (film-calculation). For this calculation, only the effects of direct exposure, mid-leaf transmission, and rounded leaf-end were considered. See COLOR PLATE 8.

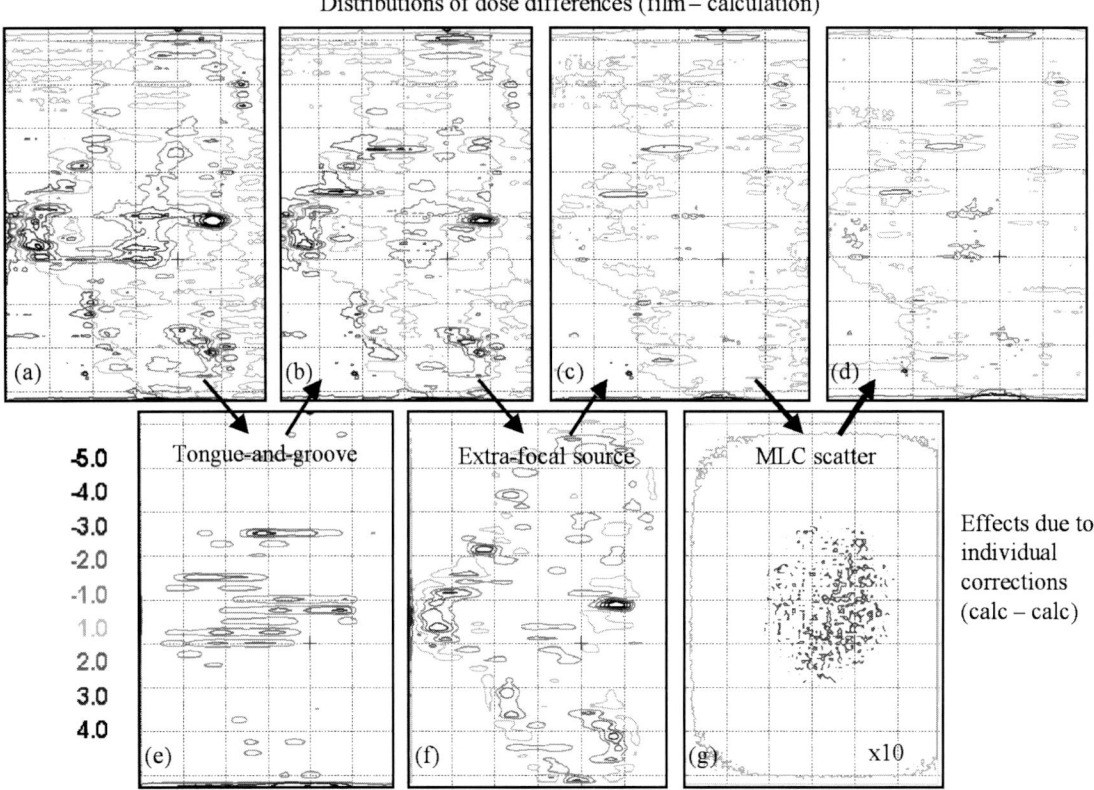

FIGURE 5–13. Top row: The distribution of dose differences (film-calculation) of the lung field in figure 12. Bottom row: The effects due to individual corrections. Descriptions of the individual panels are given in the text. See COLOR PLATE 9.

the difference reduced to about 1 cGy, or approximately 2% as shown in figure 5–13c. Finally, the effect of MLC scatter is shown in figure 5–13g. Note that this effect is relatively small, about 1%, and maximal in the center part of the field, as expected, but would increase with field size. The dose values in the figure were magnified by a factor of 10 to be visible. When this effect is included, it made additional improvement in the agreement between calculation and measurement as seen in figure 5–13d.

Summary

We have described calculation algorithms for independent verification of IMRT delivered with a conventional MLC. The calculation consists of two parts: The first part calculates the delivered intensity distribution in-air based on the leaf motion specified in the leaf sequencing file, and the second part computes absolute doses at depth in a flat phantom, either to individual points or on two-dimensional grids. The effects considered in the first part include direct exposure, interleaf transmission, rounded leaf-end, tongue-and-groove, extra-focal source, and scatter from the MLC. Dose calculation is based on measured data including the tissue-maximum ratios, off-center ratios, and output factors. The effect of intensity modulation is accounted for by pencil beam convolution.

Table 5–3a. Pencil Beam Kernels as a Function of Radial Distance and Depth in Water for 6 MV Photons

Radial Bins/ Depths (cm)	1.5	5	10	20	30	40
0–0.1	6.35E-01	5.02E-01	3.63E-01	2.09E-01	1.15E-01	7.19E-02
0.1–0.2	8.27E-02	6.69E-02	5.10E-02	3.22E-02	1.89E-02	1.25E-02
0.2–0.3	2.74E-02	2.25E-02	1.74E-02	1.14E-02	7.07E-03	4.66E-03
0.3–0.4	1.21E-02	1.03E-02	7.87E-03	5.34E-03	3.36E-03	2.24E-03
0.4–0.5	5.84E-03	5.42E-03	4.08E-03	2.95E-03	1.73E-03	1.22E-03
0.5–0.6	3.05E-03	3.11E-03	2.27E-03	1.61E-03	1.07E-03	6.92E-04
0.6–0.7	1.65E-03	1.80E-03	1.43E-03	9.58E-04	6.08E-04	4.29E-04
0.7–0.8	8.31E-04	1.04E-03	8.45E-04	6.13E-04	3.91E-04	2.89E-04
0.8–0.9	4.49E-04	6.80E-04	6.08E-04	3.97E-04	2.52E-04	1.83E-04
0.9–1	2.70E-04	5.10E-04	4.30E-04	2.89E-04	1.80E-04	1.25E-04
1–1.2	1.71E-04	2.83E-04	3.03E-04	1.88E-04	1.22E-04	8.94E-05
1.2–1.4	1.09E-04	1.84E-04	2.15E-04	1.38E-04	9.15E-05	6.50E-05
1.4–1.6	6.56E-05	1.62E-04	1.68E-04	1.11E-04	7.23E-05	4.22E-05
1.6–1.8	3.30E-05	1.03E-04	1.36E-04	9.49E-05	6.30E-05	3.93E-05
1.8–2	3.58E-05	1.02E-04	1.15E-04	7.47E-05	4.92E-05	3.68E-05
2–2.5	3.05E-05	7.42E-05	8.87E-05	6.91E-05	3.99E-05	3.04E-05
2.5–3	2.33E-05	5.16E-05	6.90E-05	5.59E-05	3.54E-05	2.46E-05
3–4	1.25E-05	3.07E-05	4.40E-05	3.78E-05	2.66E-05	1.78E-05
4–5	9.27E-06	1.83E-05	2.91E-05	2.84E-05	2.00E-05	1.49E-05
5–6	5.96E-06	1.19E-05	2.05E-05	1.97E-05	1.47E-05	1.12E-05
6–7	4.08E-06	9.10E-06	1.42E-05	1.43E-05	1.22E-05	8.49E-06
7–8	3.75E-06	6.55E-06	1.03E-05	1.19E-05	9.49E-06	7.39E-06
8–9	2.48E-06	4.95E-06	7.89E-06	8.69E-06	7.56E-06	6.04E-06
9–10	2.62E-06	4.28E-06	6.04E-06	7.74E-06	6.41E-06	5.14E-06
10–15	1.31E-06	2.37E-06	3.40E-06	4.23E-06	3.87E-06	3.11E-06
15–20	6.63E-07	1.06E-06	1.45E-06	1.90E-06	1.85E-06	1.59E-06
20–30	2.48E-07	3.85E-07	5.39E-07	7.04E-07	7.28E-07	6.69E-07
30–40	7.88E-08	1.29E-07	1.67E-07	2.23E-07	2.45E-07	2.38E-07
40–50	4.51E-08	6.80E-08	9.44E-08	1.32E-07	1.57E-07	1.53E-07

The accuracy of the algorithms has been extensively verified with chamber and film measurement for a variety of IMRT fields, ranging from simple test patterns to clinical fields. In general, the agreement between calculation and measurement is within 2% to 3% in dose or 1 mm in distance in high dose gradient regions. Typical calculation time per beam is less than 30 seconds on a 266 MHz Alpha station.

These algorithms have been incorporated into an independent verification program to provide independent checks to the results produced by the planning system. The verification program has been in routine clinical use in place of measurement since 1999. It allows us to implement IMRT in large scale in our busy clinic, which treats about 100 patients daily with IMRT.

It should be cautioned here that although the algorithms presented here have undergone extensive testing with measurement and have been in routine clinical use in place of measurement, they should not be used under new conditions without experimental verification. The new conditions include new disease sites, new treatment techniques, or new software versions. In addition, if the results from the independent verification program differ from that produced by the planning system by more than 5%, measurement is required to resolve the discrepancy.

Table 5–3b. Pencil Beam Kernels as a Function of Radial Distance and Depth in Water for 15 MV Photons

Radial Bins/ Depths (cm)	1.5	5	10	20	30	40
0–0.1	9.88E-01	8.90E-01	7.07E-01	4.77E-01	3.11E-01	2.19E-01
0.1–0.2	1.22E-01	1.17E-01	9.62E-02	6.89E-02	4.73E-02	3.47E-02
0.2–0.3	3.83E-02	3.75E-02	3.21E-02	2.34E-02	1.68E-02	1.24E-02
0.3–0.4	1.80E-02	1.77E-02	1.50E-02	1.13E-02	8.14E-03	6.10E-03
0.4–0.5	1.04E-02	9.64E-03	8.28E-03	6.19E-03	4.49E-03	3.49E-03
0.5–0.6	6.21E-03	5.86E-03	5.09E-03	3.89E-03	2.75E-03	2.21E-03
0.6–0.7	3.72E-03	3.73E-03	3.29E-03	2.50E-03	1.81E-03	1.43E-03
0.7–0.8	2.09E-03	2.49E-03	2.19E-03	1.74E-03	1.28E-03	9.51E-04
0.8–0.9	1.54E-03	1.74E-03	1.55E-03	1.19E-03	9.08E-04	6.69E-04
0.9–1	1.14E-03	1.18E-03	1.15E-03	8.82E-04	6.46E-04	5.00E-04
1–1.2	6.22E-04	7.70E-04	7.64E-04	5.87E-04	4.27E-04	3.42E-04
1.2–1.4	3.36E-04	4.41E-04	4.63E-04	3.74E-04	2.69E-04	2.09E-04
1.4–1.6	1.77E-04	2.67E-04	3.32E-04	2.74E-04	1.93E-04	1.51E-04
1.6–1.8	1.22E-04	1.96E-04	2.48E-04	2.00E-04	1.55E-04	1.10E-04
1.8–2	9.09E-05	1.46E-04	1.84E-04	1.66E-04	1.22E-04	9.74E-05
2–2.5	5.21E-05	9.62E-05	1.30E-04	1.21E-04	9.21E-05	7.13E-05
2.5–3	3.08E-05	5.05E-05	8.64E-05	9.02E-05	6.75E-05	5.06E-05
3–4	1.83E-05	3.07E-05	5.32E-05	5.70E-05	4.78E-05	3.63E-05
4–5	1.01E-05	1.78E-05	3.12E-05	3.85E-05	3.36E-05	2.65E-05
5–6	7.58E-06	1.16E-05	2.09E-05	2.68E-05	2.49E-05	2.04E-05
6–7	5.77E-06	8.54E-06	1.42E-05	2.04E-05	1.93E-05	1.59E-05
7–8	4.58E-06	6.01E-06	1.03E-05	1.44E-05	1.49E-05	1.30E-05
8–9	3.77E-06	4.81E-06	7.83E-06	1.10E-05	1.18E-05	1.00E-05
9–10	3.07E-06	3.68E-06	6.20E-06	8.93E-06	9.66E-06	8.29E-06
10–15	1.61E-06	2.18E-06	3.39E-06	4.94E-06	5.26E-06	4.80E-06
15–20	8.09E-07	1.03E-06	1.48E-06	2.16E-06	2.40E-06	2.20E-06
20–30	3.04E-07	4.00E-07	5.55E-07	7.81E-07	8.98E-07	8.34E-07
30–40	1.05E-07	1.28E-07	1.78E-07	2.49E-07	2.80E-07	2.62E-07
40–50	5.79E-08	8.09E-08	1.14E-08	1.56E-07	1.74E-07	1.61E-07

References

Ahnesjo, A. (1995). "Collimator scatter in photon therapy beams." *Med. Phys.* 22:267–278.

Arnfield, M. R., J. V. Siebers, J. O. Kim, Q. Wu, P. J. Keall, and R. Mohan. (2000). "A method for determining multileaf collimator transmission and scatter for dynamic intensity modulated radiotherapy." *Med. Phys.* 27:2231–2241.

Chang, J., G. S. Mageras, C. S. Chui, C. C. Ling, and W. Lutz. (2000). "Relative profile and dose verification of intensity-modulated radiation therapy." *Int. J. Radiat. Oncol. Biol. Phys.* 47:231–240.

Chui, C. S., and R. Mohan. (1986). "Off-center ratios for three-dimensional dose calculations." *Med. Phys.* 13:409–412.

Chui, C. S., T. LoSasso, and S. Spirou. (1994). "Dose calculation for photon beams with intensity modulation generated by dynamic jaw or multileaf collimations." *Med. Phys.* 21:1237–1244.

Deng, J., T. Pawlicki, Y. Chen, J. Li, S. B. Jiang, and C.-M. Ma. (2001). "The MLC tongue-and-groove effect on IMRT dose distributions." *Phys. Med. Biol.* 46:1039–1060.

Fix, M. K., P. Manser, E. J. Born, R. Mini, and P. Ruegsegger. (2001a). "Monte Carlo simulation of a dynamic MLC based on a multiple source model." *Phys. Med. Biol.* 46:3241–3257.

Fix, M. K., M. Stampanoni, P. Manser, E. J. Born, R. Mini, and P. Ruegsegger. (2001b). "A multiple source model for 6 MV photon beam dose calculations using Monte Carlo." *Phys. Med. Biol.* 46:1407–1427.

Graves, M. N., A. V. Thompson, M. K. Martel, D. L. McShan, and B. A. Fraass. (2001). "Calibration and quality assurance for rounded leaf-end MLC systems." *Med. Phys.* 28:2227–2233.

James, H. V., S. Atherton, G. J. Budgell, M. C. Kirby, and P. C. Williams. (2000). "Verification of dynamic multileaf collimation using an electronic portal imaging device." *Phys. Med. Biol.* 45:495–509.

Khan, F. M. *The Physics of Radiation Therapy.* 2nd ed. Baltimore, MD: Williams and Wilkins, p. 201, 1994.

Klein, E. E., and D. A. Low. (2001). "Interleaf leakage for 5 and 10 mm dynamic multileaf collimation systems incorporating patient motion." *Med. Phys.* 28:1703–1710.

Kung, J. H., G. T. Y. Chen, and F. K. Kuchnir. (2000). "A monitor unit verification calculation in intensity modulated radiotherapy as a dosimetry quality assurance." *Med. Phys.* 27:2226–2230.

Kutcher, G. J., L. Coia, M. Gillin, W. F. Hanson, S. Leibel, R. J. Morton, J. R. Palta, J. A. Purdy, L. E. Reinstein, G. K. Svensson, M. Weller, and L. Wingfield. (1994). "Comprehensive QA for radiation oncology: Report of AAPM Radiation Therapy Committee Task Group 40." *Med. Phys.* 21:581–618.

Li, J. S., A. L. Boyer, and C.-M. Ma. (2001). "Verification of IMRT dose distributions using a water beam imaging system." *Med. Phys.* 28:2466–2474.

LoSasso T., C.-S. Chui, and C. C. Ling. (1998). "Physical and dosimetric aspects of a multileaf collimation system used in the dynamic mode for implementing intensity modulated radiotherapy." *Med. Phys.* 25:1919–1927.

LoSasso, T., C. S. Chui, and C. C. Ling. (2001). "Comprehensive quality assurance for the delivery of intensity modulated radiotherapy with a multileaf collimator used in the dynamic mode." *Med. Phys.* 28:2209–2219.

Low, D.A., S. Mutic, J. F. Dempsey, R. L. Gerber, W. R. Bosch, C. A. Perez, and J. A. Purdy. (1998). "Quantitative dosimetric verification of an IMRT planning and delivery system." *Radiother. Oncol.* 49:305–316.

Mohan, R., and C. S. Chui. (1987). "Use of fast Fourier transforms in calculating dose distributions for irregularly shaped fields for three-dimensional treatment planning." *Med. Phys.* 14:70–77.

Partridge, M., P. M. Evans, A. Mosleh-Shirazi, and D. Convery. (1998). "Independent verification using portal imaging of intensity-modulated beam delivery by the dynamic MLC technique." *Med. Phys.* 25:1872–1879.

Sharpe, M. B., D. A. Jaffray, J. J. Battista, and P. Munro. (1995). "Extrafocal radiation: A unified approach to the prediction of beam penumbra and output factors for megavoltage x-ray beams." *Med. Phys.* 22:2065–2074.

Tsai, J. S., D. E. Wazer, M. N. Ling, J. K. Wu, M. Fagundes, T. DiPetrillo, B. Kramer, M. Koistinen, and M. J. Engler. (1998). "Dosimetric verification of the dynamic intensity-modulated radiation therapy of 92 patients." *Int. J. Radiat. Oncol. Biol. Phys.* 40:1213–1230.

Van Esch, A., B. Vanstraelen, J. Verstraete, G. Kutcher, and D. Huyskens. (2001). "Pre-treatment dosimetric verification by means of a liquid-filled electronic portal imaging device during dynamic delivery of intensity modulated treatment fields." *Radiother. Oncol.* 60:181–190.

Watanabe, Y. (2001). "Point dose calculations using an analytical pencil beam kernel for IMRT plan checking." *Phys. Med. Biol.* 46:1031–1038.

Webb, S., and T. Lomax. (2001). "There is no IMRT?" *Phys. Med. Biol.* 46:L7–L8.

Xing, L., and J. G. Li. (2000). "Computer verification of fluence map for intensity modulated radiation therapy." *Med. Phys.* 27:2084–2092.

Xing, L., Y. Chen, G. Luxton, J. G. Li, and A. L. Boyer. (2000). "Monitor unit calculation for an intensity modulated photon field by a simple scatter-summation algorithm." *Phys. Med. Biol.* 45:N1–N7.

Zhu, T.C., B. E. Bjarngard, Y. Xiao, and C. J. Yang. (2001). "Modeling the output ratio in air for megavoltage photon beams." *Med. Phys.* 28:925–937.

6

Treatment Planning Considerations Using IMRT

Margie A. Hunt
Chandra M. Burman

Introduction

In radiation therapy, a tumoricidal dose must be delivered while minimizing the dose to the surrounding normal tissues. Although three-dimensional conformal radiation therapy (3DCRT), with its careful delineation of target and normal tissues and volumetric evaluation of dose, has facilitated an increased target dose and/or reduced normal tissue dose in certain sites (Armstrong et al. 1993), intensity modulated radiation therapy (IMRT) could lead to even greater improvements in the therapeutic ratio.

The concept of intensity-modulated fields created through inverse planning was first proposed by Brahme in 1988 (Brahme 1988). Given a set of beams and a desired dose distribution, the optimum shape of the beams could be determined by back projection, leading to non-uniform beam intensities. Although this particular optimization technique was limited by non-physical solutions characterized by negative ray weights, the utility of the technique, particularly for concave targets surrounding an organ at risk, was obvious. It was soon realized that intensity modulation was capable of producing highly complex, tightly conforming dose distributions. Since then, many computational techniques for determining the beam intensity profiles have been proposed, including gradient techniques (Bortfeld et al. 1990), maximal entropy and maximal likelihood optimization (Llacer 1997), simulated annealing (Rosen et al. 1995), and methods relying on the genetic algorithm (Ezzell 1996). Some of these methods have been implemented in commercial planning systems including simulated annealing in CORVUS (NOMOS Corporation, Sewickley, PA) (Carol, Nash, and Campbell 1997) and gradient techniques in HELIOS (Varian Associates, Palo Alto, CA). The Memorial Sloan Kettering Cancer Center (MSKCC) optimization algorithm, developed by Spirou and Chui (1998) and discussed in detail in chapter 3, relies on a conjugate gradient minimization method and least-squares objective function.

Just as there are many optimization methods used to design IMRT plans, several methods exist for the delivery of intensity-modulated fields. Although physical compensators can be used (Purdy 1997), their fabrication is time consuming and their use is cumbersome. The relatively recent commercial availability of computer-controlled slit and multileaf collimators (MLCs) has made the delivery of intensity-modulated fields more practical. To date, extensive clinical experience exists for the following IMRT delivery methods: computer-controlled collimating slit device (Carol 1994), multiple static fields (i.e., the "step-and-shoot" method) (Galvin, Chen, and Smith 1993; Boyer et al. 1994), and dynamic multileaf collimation (DMLC) (Spirou and Chui 1994). Each method has its advantages and limitations, and although there exist some studies comparing delivery methods (Chui et al. 2001), there are really no available data that permit a true "head to head" comparison between the various IMRT techniques currently available.

Although the overall processes of IMRT and 3DCRT are quite similar, the philosophical basis of plan design differs significantly. One might say they are diametrically opposite in approach. Conventional 3DCRT treatment planning is forward based and manually optimized. That is, the treatment planner chooses all beam parameters, such as the number of beams, beam directions, shapes, weights, etc., and the computer merely calculates the resulting dose distribution. Based on the resulting dose distribution the planner intuitively iterates the various parameters in an attempt to optimize the dose distribution. A key point being, however, that the optimum, or desired dose distribution is never explicitly defined. Thus, an optimal 3DCRT plan, designed conventionally, is the result of an iterative manipulation of beam energy, weighting, and direction, and beam modifying devices such as blocks and wedges. Although computer-driven optimization of parameters such as beam direction (Vijayakumar et al. 1991; Soderstrom and Brahme 1992) and weighting (Langer and Leong 1987; Mageras and Mohan 1993) has been attempted, it has met only limited clinical success.

Conversely, with IMRT dose distributions are inversely determined, meaning that the treatment planner must specify in advance the dose distribution that is desired, and the computer then calculates a set of beam intensities that will produce, as nearly as possible, the desired dose distribution. Specification by the planner of the desired dose distribution is made by means of dose-volume constraints in which the planner defines for the computer minimum and maximum desirable doses for all structures in the plan [such as the CTV (clinical target volume), PTV (planning target volume), and radiosensitive normal tissues] plus a set of penalty weights to indicate the relative importance for meeting the specified dose constraints for each structure. For example, the planner could specify that meeting the dose constraints for the PTV are more important than for the spinal cord, or vice versa. Different penalty weights can also be applied to overdosing as opposed to underdosing certain structures. So, for IMRT the optimization parameters and structures are the primary variables used to control the dose distribution as opposed to the beam weights or shapes as in 3DCRT treatment planning. Specification of the optimization parameters and beam placement requires knowledge of how the details of the dose calculation algorithms and anatomical features of the patient such as the proximity of the normal tissues to the target affect the outcome of optimization. Hence, the combination of beams, optimization parameters, and structures needed to achieve the best plan are planning system and patient specific. Planners must develop an intuition as to how these factors affect optimized dose distributions.

This chapter will describe the general approach to treatment planning optimization that has been developed at MSKCC during the past 8 years. The techniques discussed have been found useful for the optimization algorithm and delivery methods used at our institution. Some of these techniques may not be relevant or be ineffective with other systems. A variety of site-specific information is discussed in this chapter to demonstrate general issues related to IMRT planning. Additional site-specific planning information will be found in the appropriate clinical chapters.

IMRT Or 3DCRT?

At the most basic level, IMRT could be considered in virtually all situations where there is a desire to either increase tumor dose or decrease normal tissue dose—that is, almost every site currently treated with external beam radiation therapy! However, there are increased costs associated with planning and delivering IMRT that must be considered. In addition to the necessary equipment costs, operating costs are higher (at least currently) and are proportional to the number of patients receiving IMRT treatment. Therefore, it makes sense to select sites for IMRT treatment that will reap the greatest benefit.

Multiple studies have now clearly demonstrated the value of IMRT in improving target coverage and decreasing normal tissue doses in clinical situations where concave targets surround normal tissues (Happersett et al. 1999, 2000; Burman et al. 1997; Fournier-Bidoz et al. 2001; Hunt et al. 2001; Wu et al. 2000; Chao et al. 2000; De Neve et al. 1999; Verhey 1999; Hsiung et al. 2002). Concave distributions created with IMRT have been used in prostate cancer (Zelefsky et al. 2000) to facilitate dose escalation with no increase in rectal dose and in head and neck cancer to decrease normal tissue morbidity, particularly to the parotid glands (Wu et al. 2000). As another example, consider the IMRT technique developed by Happersett et al. (2000) for the treatment of thyroid cancer. Historically, an extremely difficult site in which to achieve good dose distributions, successful treatment of thyroid carcinomas requires doses of 60 Gy or more to large target volumes including the thyroid bed and regional lymph nodes that surround the spinal cord. Happersett et al. developed an IMRT technique consisting of either five or six fields (figure 6–1) and compared it to anterior posterior/posterior anterior (AP/PA) opposed fields and a 3-D conformal plan for five patients. With IMRT the volume of the PTV receiving the prescription dose increased by 10% relative to the 3-D plan and 60% relative to AP/PA fields. Dose to the spinal cord was acceptable with all three techniques, but the volume of lung receiving in excess of 25 Gy was 10% to 15% less with IMRT.

Another situation in which IMRT may lead to significant dosimetric improvements is that in which the tumor is embedded within or surrounded by a normal tissue exhibiting a dose response with relatively strong volume dependence. An excellent example is lung tumors, the treatment of which is limited primarily by the tolerance of the spinal cord and surrounding normal lung. With conventional multi-field 3DCRT, the beam directions are carefully selected to avoid the spinal cord and to minimize the irradiated volume of normal lung. But beyond beam selection, the user has a limited armamentarium with which to manipulate target coverage and normal tissue dose. Uniform target dose and low normal tissue doses must be achieved using *uniform* or *smoothly varying* intensity patterns from each field (open or wedged fields). In contrast, IMRT combines the *non-uniform* intensity patterns from each field to obtain a *uniform* target dose. This often leads to further decreases in lung dose, particularly at the low or mid dose levels. Figure 6–2 compares 3-D and IMRT dose distributions for a lung tumor. Using identical beam arrangements, similar target coverage is achieved with both plans, but the lung dose is lower with IMRT. As discussed in greater detail in the chapter devoted to the treatment of lung cancer (chapter 12), the improvement in lung dose achieved with IMRT may facilitate dose escalation. In a retrospective review of six patients treated with 3DCRT but re-planned with IMRT by Yorke et al. (Spirou et al. 2001; Yorke 2001), the addition of IMRT would have permitted the prescription

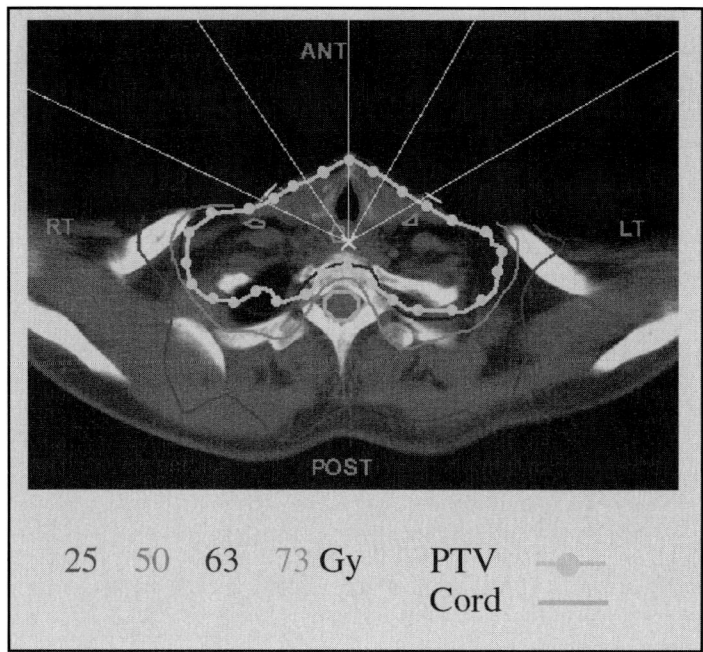

FIGURE 6–1. Five-field IMRT technique for the treatment of thyroid cancer. Note the concave shape of the dose distribution between the target and spinal cord. See COLOR PLATE 10.

dose to be escalated by 13 Gy, on average, for five of six patients. This assumes that the same biological dose limits for the lung would be required for the 3D and IMRT plans.

We have briefly described two clinical scenarios in which IMRT can produce dosimetrically superior distributions. Detailed discussions of these and other IMRT techniques can be found in the site-specific chapters. In some situations, however, the benefits of IMRT may not be as significant

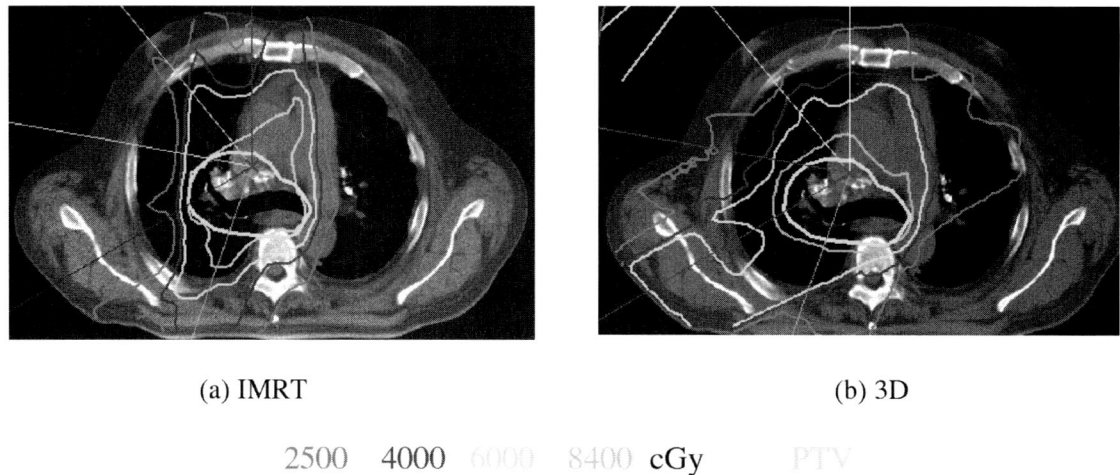

(a) IMRT   (b) 3D

2500   4000   6000   8400 **cGy**   PTV

FIGURE 6–2. A comparison of IMRT (a) and 3-D conformal (b) dose distributions for the treatment of a lung tumor. Even though the same beam arrangement was used for both plans, there is significant improvement in the lung dose with IMRT. See COLOR PLATE 11.

and may not warrant the increased cost. Therefore, it is imperative that a critical evaluation of IMRT take place prior to implementation in each clinic. At MSKCC, implementation for a new site is always preceded by a comparison of 3-D and IMRT techniques that, by itself, requires significant physics and physician time. Once the decision to use IMRT is made, a standard beam arrangement and constraint template is defined whenever possible to streamline the routine planning process. We believe that, with IMRT, communication between physicians and planners is crucial. The pre-implementation phase, during which IMRT is compared to other techniques is an ideal time to discuss the goals of treatment and to define, as precisely as possible, the desired target and normal tissues doses.

The MSKCC IMRT Planning Process

Since 1997, IMRT planning at MSKCC has relied on the optimization algorithm developed by Spirou and Chui (1998), which uses conjugate gradient minimization and a least-squares objective function (see also chapter 3). The basic planning steps are described below.

Patient Selection

The vast majority of patients receiving IMRT treatment at MSKCC are treated to sites that have been identified by the department as suitable for this type of treatment. This includes prostate, head and neck, breast, pediatric, para-spinal, and some lung tumors. For these sites, evaluations of the dosimetric benefits of IMRT have been completed and planning procedures have been written that specify the desired target and normal tissue doses and the optimization constraints used to achieve them. Other patients for whom IMRT may be beneficial, by virtue of tumor location, previous radiation treatment, etc., are discussed on a case-by-case basis prior to planning. IMRT treatment to non-routine sites may entail longer planning times (more than one week) since the target and normal tissue constraints must be determined and additional pre-treatment quality assurance (QA) including film dosimetry may be necessary.

Patient Simulation And Structure Localization

Because of the increased conformality of the dose distributions achieved with IMRT, accurate and precise patient treatment is of even more importance than with conventional treatment. All patients undergoing IMRT at MSKCC are immobilized according to tumor location and patient condition. Immobilization methods range from custom foam or thermoplastic molds for prostate, head and neck, and breast patients to stereotactic localization for patients with para-spinal lesions. After immobilization, which takes place either in a conventional simulator or CT simulator room, all patients undergo CT simulation, during which images are acquired throughout the treatment volume and an isocenter is defined. Using the CT (computed tomography) and other appropriate image sets including magnetic resonance or positron emission tomography (MR or PET), the PTV and critical organ contours are defined and transferred along with the images to the planning system.

At the planning system, additional structures used solely to control the dose distribution during optimization are often defined by the planner. Typically, these *optimization only* structures are Boolean combinations of targets and normal tissues. By defining the intersection of targets and normal tissues as separate structures, different prescription doses and constraints can easily be applied to different regions, facilitating the creation of *controlled* dose gradients between normal tissues and targets. Figure 6–3 illustrates the use of Boolean structures in prostate cancer where routinely, the PTV, rectum and PTV-rectal overlap regions are given different constraints during optimization. Special care is taken during this pre-optimization phase to ensure that an adequate number of calculation points are defined within each structure since this influences the optimization results. With the MSKCC algorithm, at least 30 points/cc, distributed quasi-randomly

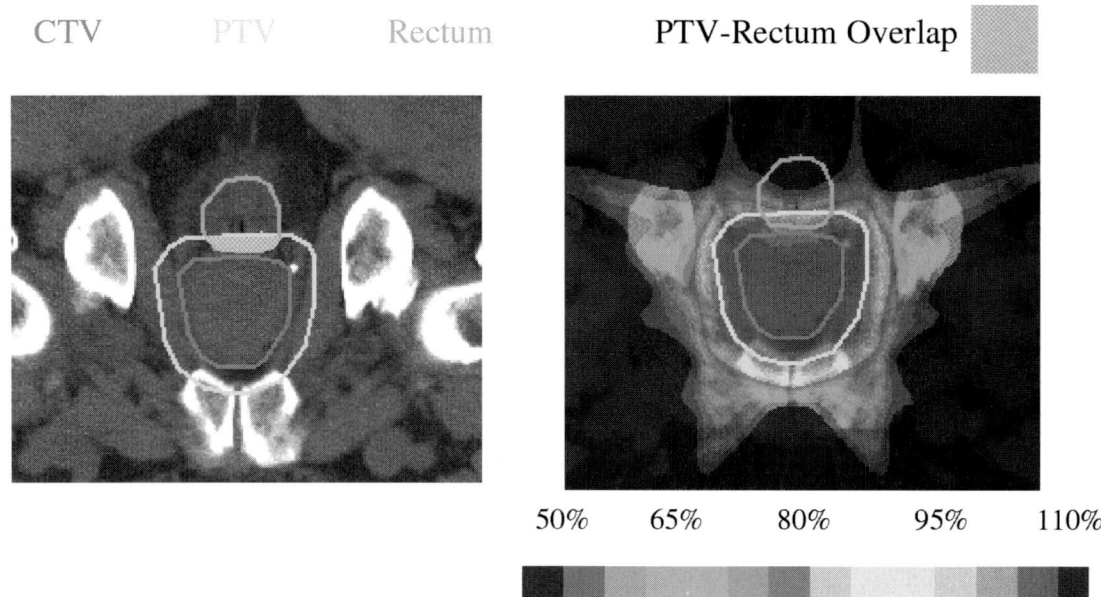

CTV   PTV   Rectum   PTV-Rectum Overlap

50%   65%   80%   95%   110%

FIGURE 6–3. Use of logical combinations of structures to control optimized dose distributions. See Color Plate 12.

throughout a structure, are needed to ensure results that are independent of point placement. Still, on occasion, portions of a structure, typically narrow outcroppings or thin superior or inferior sections, will be over- or underdosed because of a paucity of calculation points relative to the rest of the structure. Redefining these regions as separate structures for optimization usually eliminates the problem. Further details about the optimization only structures for specific sites are provided in the relevant clinical chapters.

It is important for a clinic implementing a new IMRT program to recognize that the time spent by physicians and planners contouring structures may increase relative to that with 3-D plans. One reason for this is the need to define *optimization only* structures, as discussed above. But additionally, at times, undesirably high doses are delivered to un-contoured normal tissues during optimization. The only way to lower the dose to these tissues is to contour them and to place dose constraints on them during optimization.

Selection Of Beam Energy, Number, And Direction

The selection of the optimum number and direction of beams for IMRT may be less critical than for 3DCRT as long as basic concepts are applied. The ability to modulate the beam intensity within a field can partially compensate for a relatively poor choice of beam directions. Since the complexity of treatment often increases as more fields are used, we strive for the best plan with the fewest beams and use the following guidelines:

1. In clinical situations without significant target concavities, a beam arrangement similar to that used for 3DCRT is probably sufficient, and with the addition of IMRT, a superior distribution is often achieved.
2. In the presence of significant target concavities, five to nine uniformly spaced, non-coaxial or, if beneficial, non-coplanar fields, often yield clinically acceptable dose distributions.
3. When selecting IMRT beam directions, attention should be paid to unconstrained normal tissues in the path of the beams. These tissues may receive unacceptably high doses during

optimization that may only be noticed through careful review of dose distributions and dose-volume data.

Some recent discussions have raised the concern that IMRT may increase integral dose due to the increased leakage radiation resulting from longer treatment times and the use of many treatment fields or an arc-type delivery. We would argue that the treatment time for many of the most common IMRT treatments (prostate, for example), is not significantly longer than conventional treatment and that excellent IMRT dose distributions can be achieved without increasing the number of treatment fields.In fact, IMRT, applied to a conventional beam arrangement, may actually decrease integral dose by virtue of its improved conformality. In a recent study of integral dose for prostate patients, Della Biancia evaluated various 3-D and IMRT techniques for 15 prostate patients (Della Biancia, Hunt, and Amols 2002). The integral doses for four (AP, PA, RL, LL) and six-field (RL, LL, RAO, LAO, RPO, LPO) 3-D plans were compared with those for 5, 9, and 13 non-coaxial field IMRT plans (figure 6–4). The five-field IMRT plan yielded the lowest integral dose while a six-field 3-D technique resulted in the highest. Applying intensity modulation to the 3-D four- and six-field beam arrangements decreased the integral dose by as much

3-D Plans:

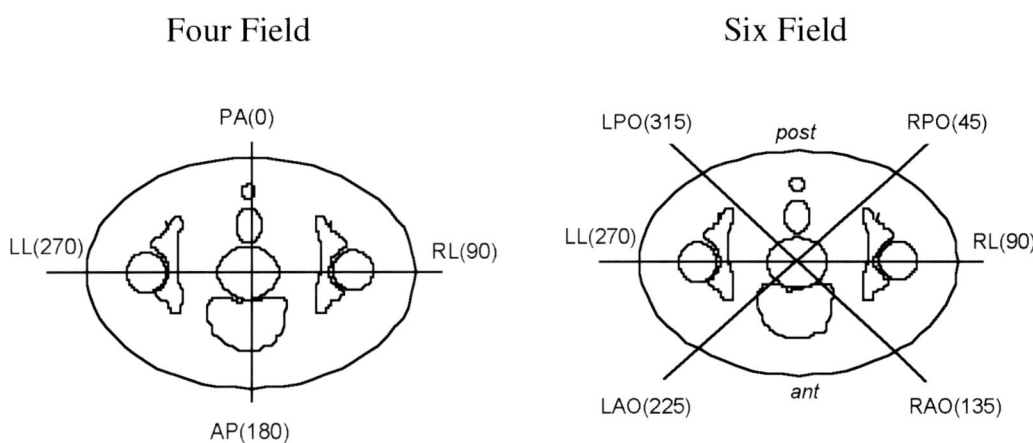

IMRT Plans:

Number of beams	Gantry Position (degree)
5	0, 75, 135, 225, 285
9	0, 40, 80, 120, 160, 200, 240, 280, 320
13	0, 40, 60, 80, 110, 130, 160, 200, 230, 250, 280, 300, 320
30	Equally distributed every 12 degrees.

FIGURE 6–4. 3-D and IMRT beam arrangements used in the study of integral dose for prostate patients.

as 10%. As a result of the lower path length and improved conformality of the five-field IMRT plan, patients receiving IMRT to 81 Gy would have received approximately the same integral dose as patients treated to 75.6 Gy with a six-field 3DCRT technique.

It has also been suggested that, with IMRT, higher energy x-ray beams ($\geq$10 MV) may not be as necessary for deep-seated targets although the recent article by Pirzkall et al. (2002) appears to cast some doubt on this. This study compared IMRT prostate plans for 6, 10, and 18 MV X-rays and found target and rectal doses for 6 MV to be comparable to those obtained with a five-field 18 MV plan, but only if at least nine fields were used. If fewer 6 MV fields were used, the doses to superficial normal tissues were significantly higher than those observed in the higher energy plans.

At MSKCC, no changes in beam energy have been made as IMRT has replaced 3DCRT treatment. Pelvis and abdominal tumors are routinely treated with 15 MV, and lung, breast, head and neck, and pediatric tumors with 6 MV. Breast patients with separations greater than 23 cm are treated using 15 MV X-rays and a beam spoiler or with a mix of 6 and 15 MV beams.

Optimization

A detailed description of the optimization algorithm can be found in chapter 3. Here, a brief description is provided with emphasis on clinical application.

For each treatment field, the collimator is adjusted to enclose the targets with a margin of 1.5 to 2 cm, which prevents the target from lying within the beam penumbra. With the MSKCC algorithm, failure to add this margin may lead to undesirable intensity peaks near the beam edges. An initial dose calculation is performed during which each beam is divided into segments 2 mm wide and 5 mm long and the dose deposition coefficients, a_{ij} representing the dose deposited to i^{th} point in a structure for a unit weight of the jth ray, are calculated. The dose deposition coefficients contain all the information needed by the optimization algorithm to determine the intensity profiles.

The desired dose distribution is described through the optimization parameters (i.e., constraints) for targets and normal tissues. Optimization is achieved by using an iterative process to minimize a quadratic objective function, shown here in a simplified form for a single target:

$$F_{obj} = \frac{1}{N}\left(\sum_j (D_j - C_p)^2 + \sum_k w_k \sum_j \Theta(D_j - C_k) \times (D_j - C_k)^2 \right)$$

where

C_p is the prescription dose for the target.
C_k is the dose of the kth target constraint.
w_k is the user-defined penalty for the kth target constraint.
D_j is the actual dose to the jth point within the target.
N is the number of target points.

$$\Theta(D_j - C_k) = \begin{cases} 1 & \begin{array}{l} \text{if } D_j - C_k < 0 \text{ and } C_k \text{ is a minimum dose constraint} \\ \text{or } D_j - C_k > 0 \text{ and } C_k \text{ is a maximum dose constraint} \end{array} \\ 0 & \text{otherwise} \end{cases}$$

For targets, a prescription dose and a *dose window* defining the maximum and minimum dose constraints are allowed. Within this window, a penalty of 1 is applied to deviations from the prescription whereas user-defined penalties are applied outside the window. For critical structures, dose and dose-volume constraints are available. Dose constraints are defined by a maximum dose and penalty while dose-volume constraints are defined by a dose-volume combination and penalty. All MSKCC optimization constraints are so-called *soft* constraints, meaning violation is possible, but at a *cost* (i.e., penalty). Hard constraints, those that may not be violated under any circumstances, are under development.

At MSKCC, IMRT templates defining the optimization parameters and clinical criteria for all targets and normal tissues are developed for each site prior to large-scale IMRT implementation. These templates are developed primarily through communication between physicians and physicists about the goals and realities of the IMRT treatment and trial-and-error planning on a significant number of test patients. The optimization parameters that define an IMRT template are *average* values, found to yield relatively good dose distributions in a majority of patients. As such, they serve as a starting point for planning individual patients. Invariably, planners must adjust the optimization parameters to *fine tune* the dose distribution for each patient. The templates for the most frequently treated IMRT sites are discussed throughout this book in the site-specific chapters. As an example, the MSKCC 81 Gy prostate template is shown in table 6–1. The optimization parameters are the actual numbers entered into the optimization software while the clinical criteria are used to evaluate the resulting dose distributions and dose-volume histograms (DVHs). It is important to recognize that the selection of optimization parameters is an empirical process and that the MSKCC templates will not necessarily apply to other optimization algorithms.

The importance of devoting significant physician and physicist effort to the development of IMRT templates cannot be stressed enough as their use helps ensure efficient and consistent results for all patients. The response of an optimization algorithm to changes in the optimization parameters is not always intuitive, and even with the use of a template, sub-optimal results can be obtained without the planner realizing it. Optimization algorithm performance studies are one way to get a better understanding of algorithm response and, although time consuming, are highly recommended. A detailed description of the methods and results of one such study that we have recently completed is given later in this chapter in **Controlling Dose Distributions Designed Using Inverse Planning**.

At times, it is useful to modify certain factors affecting the optimization process or the resulting intensity profiles. These include the form of the objective function, the number of iterations performed during optimization, and the profile smoothing method. All these affect the results of optimization and should be investigated as fully as possible prior to clinical implementation. The effect of one of these factors, intensity profile smoothing, is discussed in more detail below.

Intensity profiles may, under certain circumstances, be highly modulated, leading to delivery problems and increased treatment time. They may also deposit unacceptable dose to the critical

Table 6–1. Integral Dose Relative to That for a Five-Field (5F) IMRT Plan, Average Path Length, and Conformity Index for 3-D and IMRT Prostate Treatment Techniques. Data have been averaged over 15 prostate patients

Technique	Gantry Positions (degrees)	Relative Integral Dose (S.D.)	Average Path Length (cm) (S.D.)	Average Conformity Index (S.D.)
4F 3DS	Equally distributed every 90° Start angle = 0° deg.	1.060 (0.05)	15.2 (1.4)	1.48 (0.1)
6F 3D	Equally distributed every 45° Start angle = 45°	1.21 (0.07)	16.3 (1.5) 15.2 (1.4)	1.56 (0.1)
4F IMRT	Same as 4F 3D	0.99 (0.04)	16.3 (1.5)	1.34 (0.09)
6F IMRT	Same as 6F 3D	1.08 (0.04)	15.3 (1.5)	1.36 (0.1)
5F IMRT	0, 75, 135, 225, 285	1	15.0 (1.4)	1.20 (0.06)
9F IMRT	Equally distributed every 40° Start angle = 0°	0.998 (0.01)		1.16 (0.08)
13F IMRT	0, 40, 60, 80, 110, 130, 160, 200, 230, 250, 280, 300, 320	1.04 (0.02)	15.7 (1.4)	1.11 (0.08)
30F IMRT	Equally distributed every 12°	1.03 (0.03)	15.2 (1.3)	1.06 (0.09)

All results indicate average values obtained from 15 patients
[1] Conformity index = V_{100}(External contour)/V_{100} (PTV)

organs in the event of setup error or organ motion. Although some modulation is of course necessary, some is a result of numerical artifact, introduced during optimization. To reduce these artifacts, intensity profiles are smoothed, using one of the methods discussed in detail in chapter 3. The default method of smoothing in the MSKCC system (Spirou and Chui 1994) applies a Savitzky-Golay filter (Teukolsky, Vettering, and Flannery 1992), replacing each ray with a weighted average of itself and its neighbors. This technique significantly reduces the small intensity fluctuations arising from numerical artifact, thereby producing profiles that can be delivered easily in reasonable time. As is the case for all smoothing algorithms however, sharp dose gradients near critical structures are degraded. To minimize this degradation, a second smoothing method (*score smoothing*) was introduced (Spirou et al. 2001) that incorporates a *smoothing* factor in the objective function, thereby smoothing in regions of relatively low dose gradient while preserving the high dose gradient at target-normal tissue interfaces. Dose profiles across a target normal tissue interface for a single field resulting from the Savitzky-Golay and the score smoothing methods are compared in figures 6–5 and 6–6. Both smoothing techniques are used clinically and are chosen based on the following guidelines:

1. When extremely steep dose gradients are not required (e.g., prostate and breast plans), little difference is observed between the Savitzky-Golay and score-smoothing. Therefore, the Savitzky-Golay filter, applied at the end of each iteration, is generally applied since treatment times are minimized.

2. Steep dose gradients, such as those needed in the treatment of recurrent para-spinal or head and neck tumors, may only be achievable with more sophisticated smoothing techniques, i.e, score smoothing. Highly modulated fields delivered dynamically generally require more treatment time and are more prone to beam *hold-offs*. If step-and-shoot delivery is used, a larger number of segments may be necessary.

3. All aspects of treatment, including the effect of treatment uncertainties should be considered before selecting a smoothing technique.

Conversion Of Intensity Profiles To Leaf Motion

IMRT at MSKCC is delivered using dynamic multileaf collimation (DMLC) following the methodology developed by Spirou and Chui (1994), discussed in chapter 3. After optimization, the intensity profiles are converted into 200 segments, requiring that in each segment, at least one leaf move at the maximum allowable speed, thereby minimizing treatment time. Both transmission through the leaves and the effect of the rounded leaf edge of the Varian MLC are considered. Since the original intensity profiles cannot be converted with complete fidelity, the final leaf motion is converted back into a *deliverable* intensity profile for subsequent forward dose calculation and plan evaluation.

Forward Dose Calculation And Plan Evaluation

Once the forward dose calculation is completed using the deliverable profiles, the plan is evaluated using standard methods including planar dose distributions, DVHs, and radiobiological indices such as tumor control probability (TCP) and normal tissue complication probability (NTCP). Additionally, the *DMLC aperture*, an MLC outline defined by the initial and final leaf positions, is created for each field and can be viewed in beam's-eye view with overlaid iso-intensity contours (figure 6–7). If the dose distribution does not meet the clinical goals of treatment, the optimization parameters, optimization only structures, or beams are adjusted and the process is repeated.

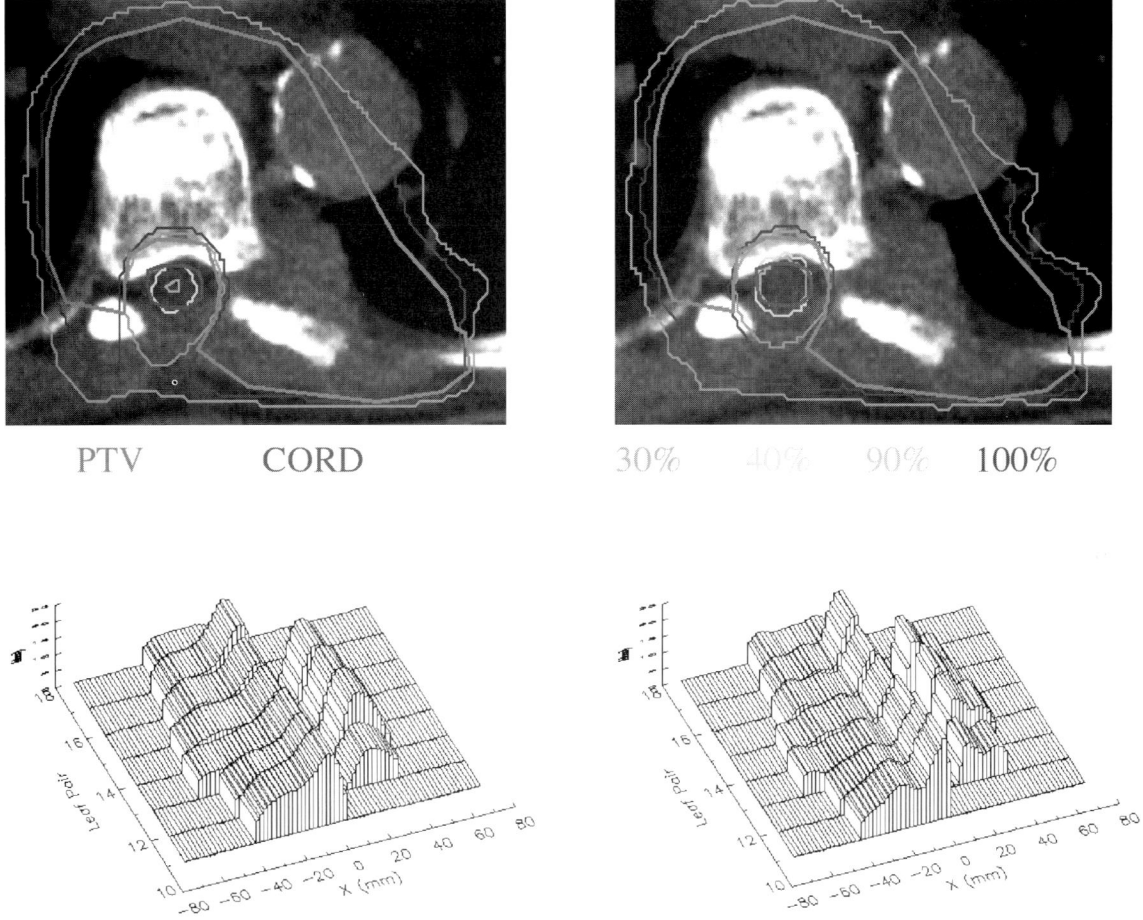

PTV   CORD   30%   40%   90%   100%

FIGURE 6–5. Effect of two different profile smoothing methods on optimized dose distributions. (a) Profile smoothing performed at end of each iteration (Savitsky-Golay). (b) Profile smoothing performed within the objective function (Score Smoothing). (c) Intensity profiles for posterior beam created by smoothing at the end of each iteration (left) and within the objective function (right). See COLOR PLATE 13.

IMRT Plan Documentation And QA

Upon completion of the plan, monitor unit (MU) settings are calculated and plan documentation is prepared. In addition to the standard documentation, the following is produced for IMRT plans.

1. Digitally reconstructed radiographs (DRRs) of each treatment field overlaid with the DMLC aperture, as described above. These images are compared with portal images, also overlaid with the DMLC aperture, obtained during the patient *setup* prior to the first treatment.

2. Independent verification of the MU setting for each field. As described in chapter 4, a stand-alone software application has been developed for this purpose that accepts, as input, the leaf motion file and MU setting for each field. From the leaf motion file, an intensity profile is generated and the dose to a user-specified point is calculated for comparison with IMRT plan output. Since 1998, this program has provided independent verification of IMRT MU settings and has supplanted the former requirement of film dosimetry for every treatment portal.

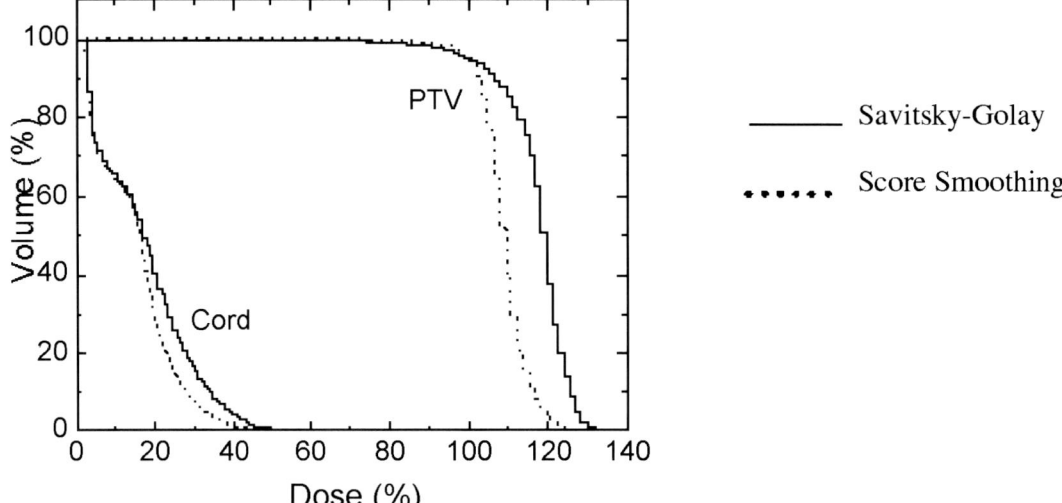

Smoothing Method	Total MU (all fields)
After each iteration	891 MU
Within Objective Function	1081 MU

FIGURE 6–6. Dose-volume histograms for the PTV and spinal cord and total MU settings for the Savitsky-Golay and score smoothing techniques for a para-spinal tumor (Figure 6–5).

3. Documentation of the *IMRT plan identification number* embedded in each leaf motion file. This number, which is indexed with each new optimization, is displayed on IMRT plan output including dose distributions and MU calculation sheets.

All IMRT plans are reviewed by a senior physicist prior to the patient's first treatment. In addition to the standard plan QA, the following is done:

1. Review of the independent MU verification for IMRT fields. Any discrepancy in excess of 2% is initially investigated by calculating doses to additional points with the independent MU software and planning system. Unresolved discrepancies are investigated with film or ion chamber dosimetry.

2. Comparison of the IMRT plan identification number displayed on plan output with that embedded in the leaf motion file for each treatment field. Discrepancies indicate that the leaf motion files do not correspond to those used to create the treatment plan. We believe that a QA check of this type is particularly crucial for institutions (such as MSKCC) using planning systems, electronic charts, and/or record and verify (R&V) systems without integrated databases. On several occasions, we have encountered situations where IMRT plans that were complete and ready for treatment were subsequently modified and planners inadvertently forgot to update the leaf motion files in the electronic chart/R&V system. Fortunately, these

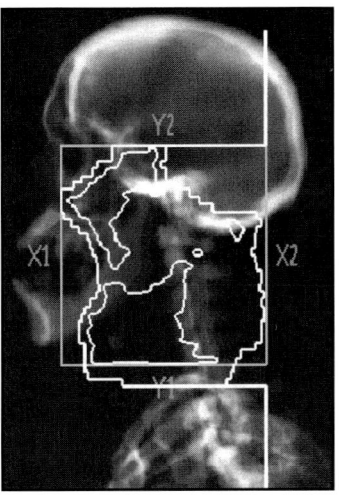

FIGURE 6–7. Digitally reconstructed radiograph demonstrating the DMLC aperture (in white) and iso-intensity line display (in gray).

potentially serious systematic errors were discovered prior to patient filming and treatment through the plan identification number QA procedure.

3. Visual review of color displays of intensity profiles for each treatment field using VARIAN Shaper or VARiS software. Profiles are examined for consistency with expected results and the presence of undesirable intensity peaks.

Controlling Dose Distributions Designed Using Inverse Planning: Optimization Algorithm Performance

Optimized dose distributions are controlled primarily through the beam arrangement, the optimization parameters, and structures. Unfortunately, the response of the optimization algorithm to changes in any of these is not always straightforward or intuitive. Optimization parameters that work well for one patient may produce only mediocre results in another. The modification of a penalty or cost for one structure may affect the dose to another structure not even in the same physical proximity. One thing is for certain: *inverse planning* requires the development of a new set of skills and intuition by the planner. In this section, we will describe some observations and studies that have helped us develop planning skills applicable to the MSKCC optimization algorithm. Although other optimization algorithms may behave quite differently, some of the observations or evaluation methods may be still be applicable.

Optimization Parameters

The goal of the optimization software is to iteratively adjust the beam intensities until the objective function is minimized, and in so doing, satisfy the criteria defined by the optimization parameters. However, often the criteria cannot be satisfied, either because they were physically unrealistic to begin with or because other factors are affecting the relationship between the objective function minimization and the optimization parameters. Our experience has indicated that, in general, optimization parameters must be more stringent than the desired clinical result and that the patient's anatomy, in particular, the proximity of the critical normal tissues to the target, must be considered when setting the optimization parameters.

To improve our understanding of the performance of the optimization software, we have undertaken a pilot study designed to evaluate optimization results for one specific target-normal tissue geometry, that of a concave target surrounding a cylindrical normal tissue (Hunt et al. 2002) (figure 6–8). The variables included the separation between the target and normal tissue and the optimization parameters applied to the normal tissue. The optimized dose distributions were evaluated using the target dose uniformity and maximum normal tissue dose. Figure 6–9 displays the relationship between target and normal tissues doses for target-normal tissue separations of 5 to 13 mm and different optimization parameters. The data points for a particular target-normal tissue separation represent the results obtained with normal tissue dose constraints of 10%, 30%, 50%, and 70% of the prescription. Clearly, the target dose uniformity and normal tissue dose depend strongly on the separation between the two structures, particularly when stringent constraints are placed on the normal tissue dose. If, for example, *acceptable* plans are those in which target dose non-uniformity is no more than 20% and the maximum normal tissue dose is less than 55%, acceptable plans cannot always be achieved. As the separation between the two structures increases, a range of constraints leads to acceptable, but different plans. The determination of the *best* plan still requires a clinical decision based on the balance between target dose uniformity and normal tissue dose.

The effect of varying the penalty placed on the normal tissue rather than the optimization dose is shown in figure 6–10. In general, an increasing penalty was associated with an increase in the PTV uniformity index and little change in the normal tissue dose. A simultaneous lowering of the normal tissue dose and increase in the PTV uniformity index was observed only for

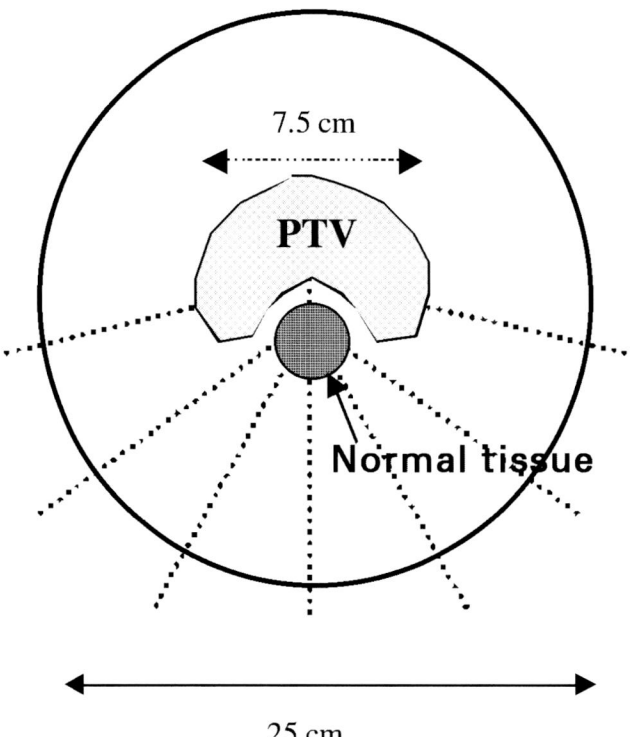

FIGURE 6–8. Phantom, target, and normal tissue geometry used in optimization performance study. A variety of beam arrangements including 5, 7, and 19 equally spaced posterio-lateral fields were studied.

small separations and stringent dose limits. Furthermore, optimization results were not as sensitive to changes in the penalty as they were to changes in constraint dose.

Beam Selection

The choice of treatment fields also affects the optimized dose distribution, particularly for concave targets. Increasing the number of fields may lead to an acceptable plan when one is not physically possible with fewer beams. Figure 6–11 shows the effect of increasing the number of posterio-lateral fields from 5 to 19 for the concave target-normal tissue geometry shown in figure 6–8. For large target-normal tissue separations and/or relatively large normal tissue doses, increasing the number of fields beyond five has only a minimal effect, consistent with several clinical studies. Nutting et al. (2001) found little change in thyroid distributions when more than five to seven fields were used, although using less than three beams led to a noticeable degradation. Happersett et al. (2000) found five to six fields yielded clinically acceptable target coverage and normal tissue sparing in thyroid IMRT.

If small target-normal tissue separations and steep dose gradients are encountered, however, increasing the number of fields does provide some advantage (figure 6–11). In the treatment of para-spinal tumors, Fournier-Bidoz et al. (2001) found at least nine fields were necessary to achieve very high dose gradients between the target and the spinal cord.

"Optimization Only" Structures

The local distribution of dose, i.e., the location of hot or cold spots or the dose gradient in one specific area, is most easily controlled using artificial structures designed for "optimization only." For example, we routinely define the intersection of the rectum and the PTV as an

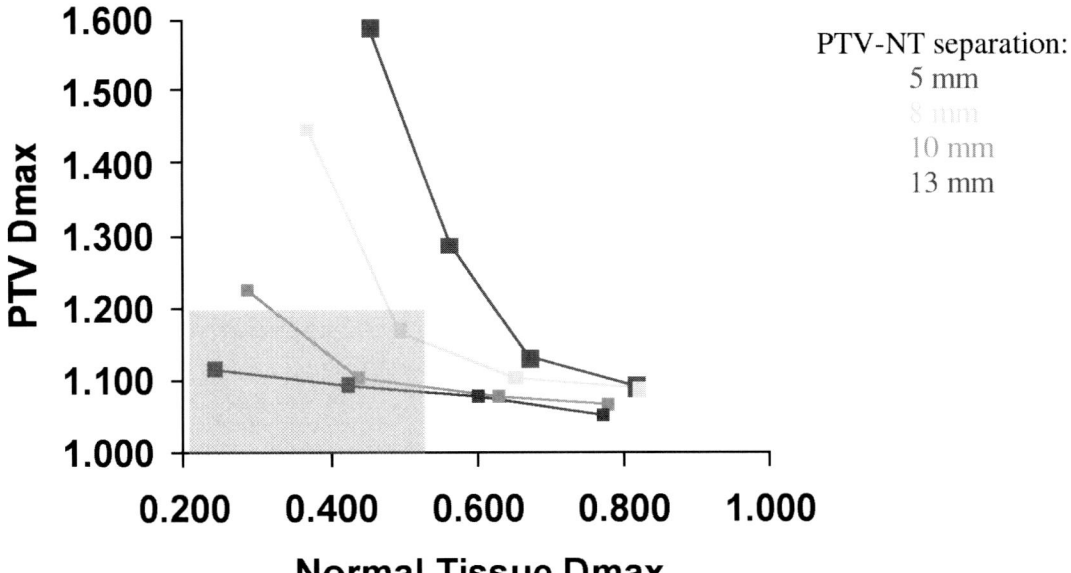

FIGURE 6–9. PTV maximum dose versus maximum normal tissue dose observed with a seven-field beam arrangement and normal tissues positioned 5, 8, 10, and 13 mm from the PTV. Each data point represents the result obtained using a normal tissue dose constraint of 10%, 30%, 50%, or 70% of the prescription and a penalty of 100. PTV constraints remained constant. The shaded region defines "clinically acceptable" plans.

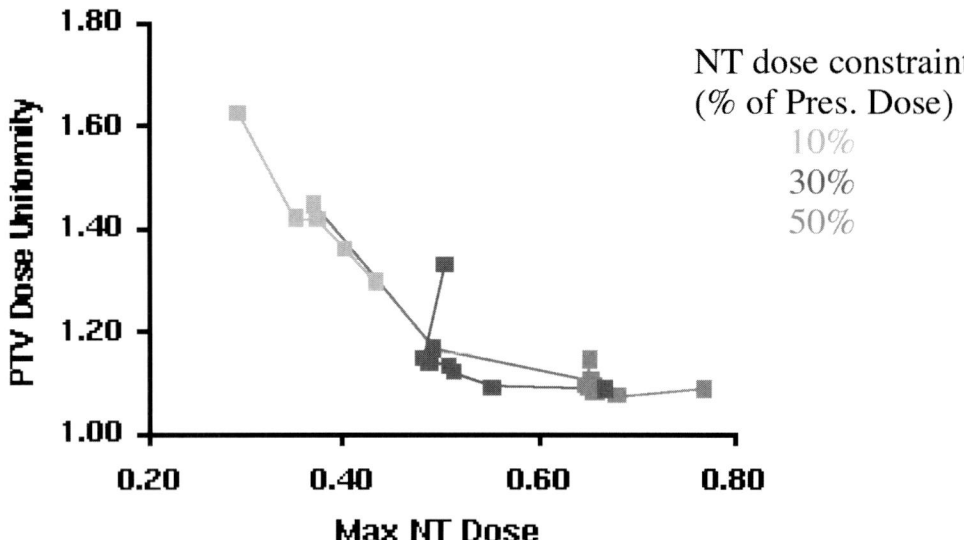

FIGURE 6–10. Effect of modifying the constraint penalty rather than dose for normal tissues positioned 8 mm from the PTV surface. Optimization results obtained with normal tissue dose constraints of 10%, 30%, and 50% of the prescription and a penalty of 100 are joined by the orange solid line. Results for a single dose constraint and penalties varying from 1 to 1000 are indicated by the different color lines.

optimization only structure for prostate patients, as shown in figure 6–3a. Different optimization parameters are prescribed to this region of the PTV (table 6–2) in order to create a well-defined dose gradient between the posterior aspect of the prostate and the anterior rectal wall (figure 6–3b).

Optimization only structures may also be useful in achieving the steepest dose gradient between two structures. In an evaluation of IMRT for para-spinal tumors, Fournier-Bidoz et al. (2001) found that optimizations using *rinds* of the PTV and adjacent spinal cord led to steeper dose gradients between the two structures than optimizations using the whole structures. For a single field, the dose gradient between the target and critical structure was improved by approximately 1.5%/mm when rinds were used.

Summary

Although many steps in the planning process for IMRT treatments are similar to those for 3-D conformal treatments, the creation of the actual treatment plan, (i.e., the optimization or inverse planning process) differs significantly and requires planners and physicians to develop new technical skills and ways of thinking about treatment planning. Initially, this method of planning may not be intuitive and, therefore, may take substantially longer than conventional 3-D planning. Eventually, though, if the planners and physicians work together to develop constraint templates and class solutions for situations where they are feasible, IMRT planning will become more "routine" and can normally be completed in approximately the same time as 3-D plans.

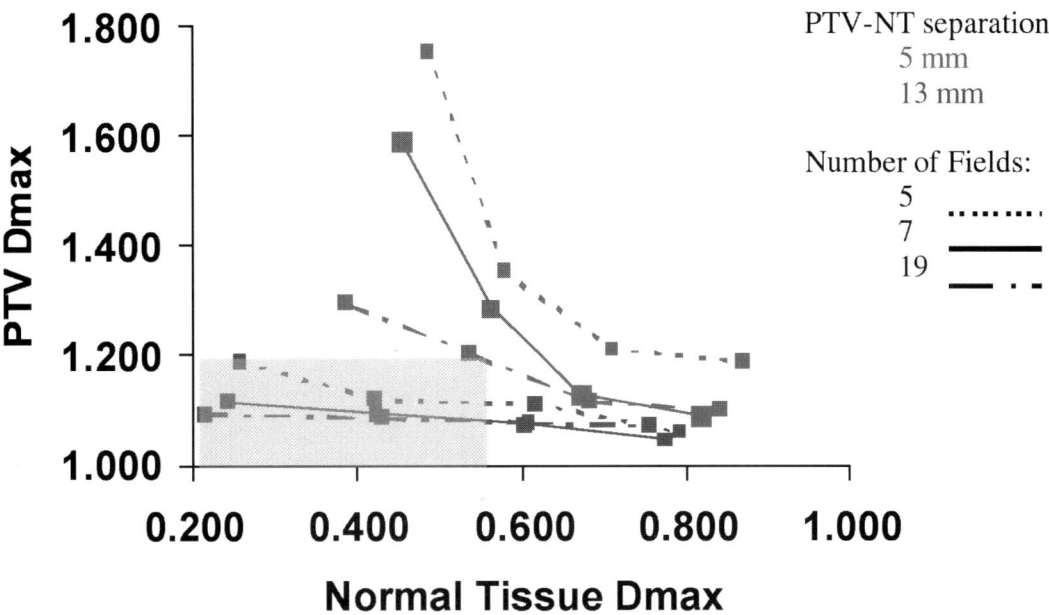

FIGURE 6–11. Effect of the number of treatment fields on optimization results for normal tissues positioned 5 and 13 mm from the PTV. Results for 5, 7, and 19 equally spaced beam arrangements directed from the posterior and posterio-lateral directions are shown. The shaded area represents "clinically acceptable" plans.

Table 6–2. MSKCC IMRT Template for 81 Gy Prostate Patients: The IMRT Template Defines the Parameters That Are Used as a Starting Point for Optimization and the Criteria Used to Evaluate the Plan

Structure	Optimization Parameters			Treatment Plan Criteria	
	Max. Dose/Penalty	Min Dose/Penalty	Volume (%)	Dose	Volume (%)
PTV (excluding Rectal Overlap)	82.6 Gy/50	79.4 Gy/50	—	90 Gy Max.	V95>90
PTV-Rectum Overlap Region	77.8 Gy/20	75.3 Gy/10	—	—	—
Rectal Wall	77 Gy/20	—	—	75.6 Gy	30
Rectal Wall	32.4 Gy/20	—	30	47 Gy	53
Bladder Wall	79.4 Gy/35	—	—	—	—
Bladder Wall	32.4 Gy/20	—	30	40 Gy	60

References

Armstrong, J. G., C. Burman, S. Leibel, D. Fontenla, G. Kutcher, M. Zelefsky, and Z. Fuks. (1993). "Three-dimensional conformal radiation therapy may improve the therapeutic ratio of high dose radiation therapy for lung cancer." *Int. J. Radiat. Oncol. Biol. Phys.* 26:685–689.

Bortfeld, T., J. Burkelbach, R. Boesecke, and W. Schlegel. (1990). "Methods of image reconstruction from projections applied to conformation radiotherapy." *Phys. Med. Biol.* 35: 1423–1434.

Boyer, A. L., T. Bortfeld, D. L. Kahler, and T. J. Waldron. "MLC Modulation of X-ray Beams in Discrete Steps" in *XIth International Conference on the Use of Computers in Radiation Therapy*. A. R. Hounsell, J. M. Wilkinson, and P. C. Williams (eds.). March 20–24, 1994. Manchester, UK. Manchester, UK: Christie Hospital NHS Trust. Madison, WI: Medical Physics Publishing, pp. 180–181, 1994.

Brahme, A. (1988). "Optimization of stationary and moving beam radiation therapy techniques." *Radiother. Oncol.* 12(2):129–140.

Burman, C., C. S. Chui, G. Kutcher, S. Leibel, M. Zelefsky, T. LoSasso, S. Spirou, Q. Wu, J. Yang, J. Stein, R. Mohan, Z. Fuks, and C. C. Ling. (1997). "Planning, delivery, and quality assurance of intensity-modulated radiotherapy using dynamic multileaf collimator: A strategy for large-scale implementation for the treatment of carcinoma of the prostate." *Int. J. Radiat. Oncol. Biol. Phys.* 39(4):863–873.

Carol, M. "Integrated 3-D Conformal Multivane Intensity Modulation Delivery System for Radiotherapy" in *XIth International Conference on the Use of Computers in Radiation Therapy*. A. R. Hounsell, J. M. Wilkinson, and P. C. Williams (eds.). March 20–24, 1994. Manchester, UK. Manchester, UK: Christie Hospital NHS Trust. Madison, WI: Medical Physics Publishing, pp. 172–173, 1994.

Carol, M. P., R. V. Nash, R. C. Campbell, R. Huber, and E. Sternick "The Development of a Clinically Intuitive Approach to Inverse Treatment Planning: Partial Volume Prescription and Area Cost Function" in *XII ICCR: XII International Conference on the Use of Computers in Radiation Therapy*. D.D. Leavitt and G. Starkschall (eds). Salt Lake City, Utah, May 27–30, 1997. Madison, WI: Medical Physics Publishing, pp. 317–319, 1997.

Chao, K. S., D. A. Low, C. A. Perez, and J. A. Purdy. (2000). "Intensity-modulated radiation therapy in head and neck cancers: The Mallinckrodt experience." *Int. J. Cancer* 90(2):92–103.

Chui, C. S., M. F. Chan, E. Yorke, S. Spirou, and C. C. Ling. (2001). "Delivery of intensity-modulated radiation therapy with a conventional multileaf collimator: Comparison of dynamic and segmental methods." *Med. Phys.* 28(12): 2441–2449.

Della Biancia, C., M. Hunt, and H. Amols. (2002). "A comparison of the integral dose from 3D conformal and IMRT techniques in the treatment of prostate cancer." (Abstract). *Med. Phys.* 29(6):1216.

De Neve, W., W. De Gersem, S. Derycke, C. De Meerleer, M. Moerman, M. T. Bate, B. Van Duyse, L. Vakaet, Y. De Deene, B. Mersseman, C. De Wagter, and C. De Waeter. (1999). "Clinical delivery of intensity modulated conformal radiotherapy for relapsed or second-primary head and neck cancer using a multileaf collimator with dynamic control." *Radiother. Oncol.* 50(3):301–314.

Ezzell, G. A. (1996). "Genetic and geometric optimization of three-dimensional radiation therapy treatment planning." *Med. Phys.* 23(3): 293–305.

Fournier-Bidoz, N., P. Giraud, S. Spirou, C. Chui, M. Lovelock, K. Yenice, and M. Hunt. (2001). "Penumbra sharpening with IMRT in Paraspinal Treatments." (Abstract). *Med. Phys.* 28(6):1260.

Galvin, J. M., X. G. Chen, and R. M. Smith. (1993). "Combining multileaf fields to modulate fluence distributions." *Int. J. Radiat. Oncol. Biol. Phys.* 27(3): 697–705.

Happersett, L., M. Hunt, C. Chui, C. Burman, C. Ling, M. Zelefsky, S. Leibel, and H. Amols. (1999). "Dose painting for prostate cancer using IMRT techniques." American Association of Physicists in Medicine (AAPM) Annual Meeting, Nashville, TN, June1999. *Med. Phys.* 26(6): No page number.

Happersett, L., M. Hunt, L. Chong, et al. (2000). "Intensity Modulated radiation therapy for the treatment of thyroid cancer." *Int. J. Radiat. Oncol. Biol. Phys.* 48(3S):351.

Hsiung, C. Y., E. D. Yorke, C. S. Chui, M. A. Hunt, C. C. Ling, E. Y. Huang, C. J. Wang, H. C. Chan, S. A. Yeh, H. C. Hsu, and H. I. Amols. (2002). "Intensity-modulated radiotherapy versus conventional three-dimensional conformal radiotherapy for boost or salvage treatment of nasopharyngeal carcinoma." *Int. J. Radiat. Oncol. Biol. Phys.* 53(3):638–647.

Hunt, M. A., M. J. Zelefsky, S. Wolden, C. S. Chui, T. LoSasso, K. Rosenzweig, L. Chong, S. V. Spirou, L. Fromme, M. Lumley, H. Amols, C. C. Ling, and S. A. Leibel. (2001). "Treatment planning and delivery of intensity-modulated radiation therapy for primary nasopharynx cancer." *Int. J. Radiat. Oncol. Biol. Phys.* 49(3):623–632.

Hunt, M., C. Y. Hsiung, S. V. Spirou, C. S. Chui, and C. C. Ling. (2002). "Evaluation of concave dose distributions created using an inverse planning system." *Int. J. Radiat. Oncol. Biol. Phys.* 54(3):953.

Langer, M., and J. Leong. (1987). "Optimization of beam weights under dose-volume restrictions." *Int. J. Radiat. Oncol. Biol. Phys.* 13(8): 1255–1260.

Llacer, J. (1997). "Inverse radiation treatment planning using the dynamically penalized likelihood method." *Med. Phys.* 24(11):1751–1764.

Mageras, G. S., and R. Mohan. (1993). "Application of fast simulated annealing to optimization of conformal radiation treatments." *Med. Phys.* 20(3): 639–647.

Nutting, C. M., D. J. Convery, V. P. Cosgrove, C. Rowbottom, L. Vini, C. Harmer, D. P. Dearnaley, and S. Webb. (2001). "Improvements in target coverage and reduced spinal cord irradiation using intensity-modulated radiotherapy (IMRT) in patients with carcinoma of the thyroid gland." *Radiother. Oncol.* 60(2):173–180.

Pirzkall, A., M. P. Carol, B. Pickett, P. Xia, M. Roach 3rd, L. J. Verhey. (2002). "The effect of beam energy and number of fields on photon-based IMRT for deep-seated targets." *Int. J. Radiat. Oncol. Biol. Phys.* 53(2):434–442.

Purdy, J. A. "The Development of Intensity Modulated Radiation Therapy" in *1st NOMOS IMRT Workshop.* Durango, Colorado. Madison, WI: Advanced Medical Publishing, 1997.

Rosen, I. I., K. S. Lam, R. G. Lane, M. Langer, and S. M. Morrill. (1995). "Comparison of simulated annealing algorithms for conformal therapy treatment planning." *Int. J. Radiat. Oncol. Biol. Phys.* 33(5):1091–1099.

Soderstrom, S., and A. Brahme. (1992). "Selection of suitable beam orientations in radiation therapy using entropy and Fourier transform measures." *Phys. Med. Biol.* 37(4): 911–924.

Spirou, S. V., and C. S. Chui. (1994). "Generation of arbitrary intensity profiles by dynamic jaws or multileaf collimators." *Med. Phys.* 21(7): 1031–1041.

Spirou, S. V., and C. S. Chui. (1998). "A gradient inverse planning algorithm with dose-volume constraints." *Med. Phys.* 25(3):321–333.

Spirou, S., E. Yorke, A. Jackson, and C. Chui. (2001). "Optimization with both dose-volume and biological constraints for lung IMRT." 43rd Annual AAPM Meeting, Salt Lake City, UT, July 22–26, 2001. *Med. Phys.* 28(6):1261.

Spirou, S. V., N. Fournier-Bidoz, J. Yang, C. S. Chui, and C. C. Ling. (2001). "Smoothing intensity-modulated beam profiles to improve the efficiency of delivery." *Med. Phys.* 28(10):2105–2112.

Teukolsky, S., W. Vettering, and B. Flannery. *Savitzky-Golay Smoothing Filters, in Numerical Recipes in C.* New York: W.H. Press, pp. 650–655, 1992.

Verhey, L. J. (1999). "Comparison of three-dimensional conformal radiation therapy and intensity-modulated radiation therapy systems." *Semin. Radiat. Oncol.* 9(1):78–98.

Vijayakumar, S., L. C. Myrianthopoulos, I. Rosenberg, H. J. Halpern, N. Low, and G. T. Chen. (1991). "Optimization of radical radiotherapy with beam's eye view techniques for non-small cell lung cancer." *Int. J. Radiat. Oncol. Biol. Phys.* 21(3):779–788.

Wu, Q., M. Manning, R. Schmidt-Ullrich, and R. Mohan. (2000). "The potential for sparing of parotids and escalation of biologically effective dose with intensity-modulated radiation treatments of head and neck cancers: A treatment design study." *Int. J. Radiat. Oncol. Biol. Phys.* 46(1):195–205.

Yorke, E. (2001). "Advantages of IMRT for dose escalation in radiation therapy of lung cancer." 43rd Annual AAPM Meeting, Salt Lake City, UT, July 22–26, 2001. *Med. Phys.* 28(6):1291–1292.

Zelefsky, M. J., Z. Fuks, L. Happersett, H. J. Lee, C. C. Ling, C. M. Burman, M. Hunt, T. Wolfe, E. S. Venktraman, A. Jackson, M. Skwarchuk, and S. A. Leibel. (2000). "Clinical experience with intensity modulated radiation therapy (IMRT) in prostate cancer." *Radiother. Oncol.* 55(3):241–249.

7

Acceptance Testing And Commissioning Of IMRT

Thomas J. LoSasso

Introduction

This chapter and the next will focus on dose delivery related issues for intensity-modulated radiation therapy (IMRT). The discussion is tailored for the dynamic multileaf collimator (DMLC) or sliding window type of delivery, but it applies to segmented or step-and-shoot (SMLC) techniques as well. We will consider two topics in this chapter: acceptance testing and commissioning.

Dose calculation and delivery are equal partners in radiation therapy; if either is inaccurate, then the treatment may be compromised. Acceptance testing with respect to IMRT, at the time the MLC is installed along with an accelerator or as a retrofit to an older linac, verifies that the MLC can deliver IMRT treatments accurately. The tests should stress the precise calibration of the gap width defined by opposing leaf positions, and the ability of the leaves to reach their target positions at the specified times.

Subsequently, the transmissions through different parts of the MLC leaves and through the rounded leaf end and beam parameters modeling the extended source distribution and the electron transport in the patient need to be carefully evaluated and applied during dose calculations. The importance of these items for IMRT is much greater than for conventional static field, three-dimensional conformal radiation therapy (3DCRT). This emphasis is justified if one considers that for static treatments these parameters only define the dose near the borders of the field; depending upon the proximity of abutting critical tissues, 1 to 2 mm uncertainty of the leaf positions is acceptable. Similarly, uncertainties in the transmissions through the leaves play a minor role for the dose outside the field, and the influence of penumbral effects is only a factor at the field edges. However, for IMRT, each of these parameters can have a significant impact on the dose delivered since the leaves modulate the dose delivered throughout the target volume. Thus the calibration of the leaves and carriages and the modeling of MLC parameters and beam parameters in the treatment planning system are critical to accurate dose delivery for IMRT.

We have refined the methods for evaluating these factors and have determined them for several MLCs at our institution. The lessons that we have learned facilitate the commissioning of future MLC at Memorial Sloan-Kettering Cancer Center (MSKCC) and may prove helpful for initiating IMRT treatment in other centers.

For most therapy facilities the MLC parameters that are necessary for commissioning the treatment planning system will be supplied by the manufacturer along with a package of IMRT-specific tools for acceptance and routine quality assurance (QA) testing. Acceptance testing the MLC for IMRT delivery requires about 2 days with the tests described in this chapter (not including verification of the treatment planning system). Periodic QA, discussed in chapter 8, *Quality Assurance of IMRT*, currently requires 1 to 2 hours per month per MLC at MSKCC. The Varian MLC, Mark I, Mark II, and Millennium(tm) models, appear very capable for the routine delivery of IMRT fields. This is quite remarkable considering that they were originally intended simply as a replacement for blocks in static fields.

Acceptance Testing

MLC Alignment with Respect to Isocenter

Misalignment of the MLC with respect to the mechanical isocenter in the direction of leaf travel shifts the intensity-modulated (IM) dose distribution for each individual field in a different direction relative to the patient's anatomy, depending upon the gantry and collimator angles. The composite dose calculation, which assumes perfect alignment, is not achieved. To place the magnitude of the problem in perspective, with IMRT dose gradients comparable to that of a 60° wedge, a 1 mm misalignment produces a dose error of less than 1% within the field, certainly not a major concern. However, the dose gradient may be much higher within IMRT fields, 10% per mm is possible, as when a critical organ is overlying the target volume in the beam's-eye view (BEV). Fortunately, the effect from one field is diluted by the dose to the target from the other IM beams. Thus, even for the combination of a 1 mm misalignment, the expected maximum dose error would be only ~2% of the prescribed dose (for a five-field plan) to that specific region. Misalignment of the MLC in the direction orthogonal to leaf motion typically shifts the dose distributions for all the fields in the same direction with respect to the patient anatomy. While alignment in the direction of leaf motion can be readjusted at any time (see next section), alignment in the direction perpendicular to leaf motion is fixed at the factory and it should be verified at installation.

Misalignment in each direction can be evaluated with the following procedure. A film is first irradiated with the leaf pattern shown in figure 7–1a followed by irradiation with the pattern in figure 7–1b. The image should appear as in figure 7–1c, a uniform field with a low-density strip from **L**eft to **R**ight, corresponding to tongue and groove underdosage, and a high-density strip from **T**op to **B**ottom, corresponding to added leakage through the rounded edges of the opposed leaves. A second image (Not Shown) is produced with the same leaf patterns, but with the collimator rotated 180° between fields. The density change of the strips in this image is proportional to the misalignment of the MLC and can be calibrated with a third image, figure 7–1d, using the same two fields as for figure 7–1c, but introducing a 1 mm shift, simulating an 0.5 mm misalignment (0.5 mm is a reasonable tolerance limit), in both the radial and transverse directions. The three images may be visually evaluated or scanned.

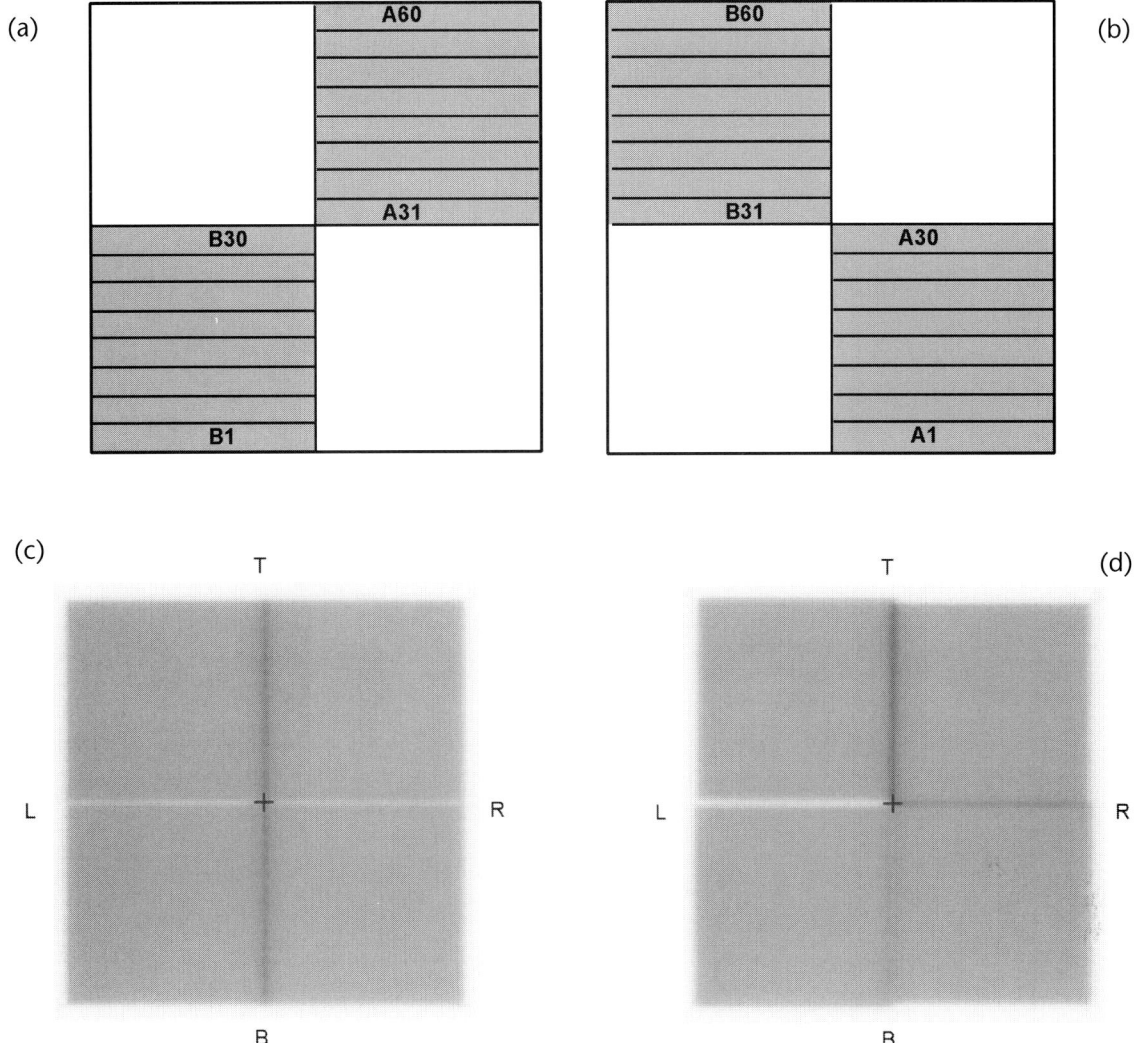

FIGURE 7–1. Procedure to quantify the misalignment of the MLC with the collimator axis of rotation. (a) and (b) Two MLC patterns used to alternately expose the film at the isocenter. (c) Image generated without collimator rotation. (d) Image generated without collimator rotation, but with a 1 mm shift in both the radial and transverse directions between fields.

Even optimal alignment of the MLC does not ensure perfect registration between fields. The instability of the mechanical isocenter due to gantry sag and roll caused by gravity, typically 1.0 mm, is unavoidable. Sag, the displacement of the field in the radial direction, is maximal at gantry angles of 0° and 180°. Depending upon the dose gradients across an adjacent pair of leaves within individual IM fields, higher or lower doses than planned may result when the fields are combined. Gantry roll, maximal at 90° and 270°, shifts these fields in the same direction, vertically downward. In this case, the shift is along the direction of the leaf motion (0° collimator angle). Similar problems exist for IM fields delivered with fixed compensators.

MLC Gap Calibration

One source of gap width error is lack of accuracy and precision in the calibration. Prior to implementation of IMRT, test procedures for leaf position consisted of simply checking the field sizes using the field light projections of the leaf ends onto graph paper at the isocenter, similar to the calibration of the jaws. The coarseness of this test reflects the fact that accuracy of static field width, whether defined by the MLC, blocks, or the jaws, only affects the borders; 1 to 2 mm errors may be tolerated. In contrast, the dose delivered with IMRT is very sensitive to the width of the gap defined by each leaf pair, as illustrated with the aid of figure 7–2. Ignoring the contribution of transmitted dose for the moment, errors in the gap in a DMLC treatment lead to proportional errors in the dose delivered. Typical average gap widths for DMLC fields are in the range of 1 to 4 cm, decreasing as the degree of modulation increases. Average gaps for prostate and head and neck fields are approximately 2.5 and 1.5 cm, respectively, based upon our experience; 2 cm will be chosen as a nominal average gap width for further discussion. The curves relate the dose error to the gap error and the average gap width. A 1 mm gap error results in a ~5% dose error for a typical DMLC treatment. Factoring in the transmitted dose (discussed later) will reduce the dose error to ~4%. Therefore, while leaf positioning accuracy of 1 mm is acceptable for static fields simply to shape their borders, we conclude that a much tighter tolerance of ~0.2 mm in

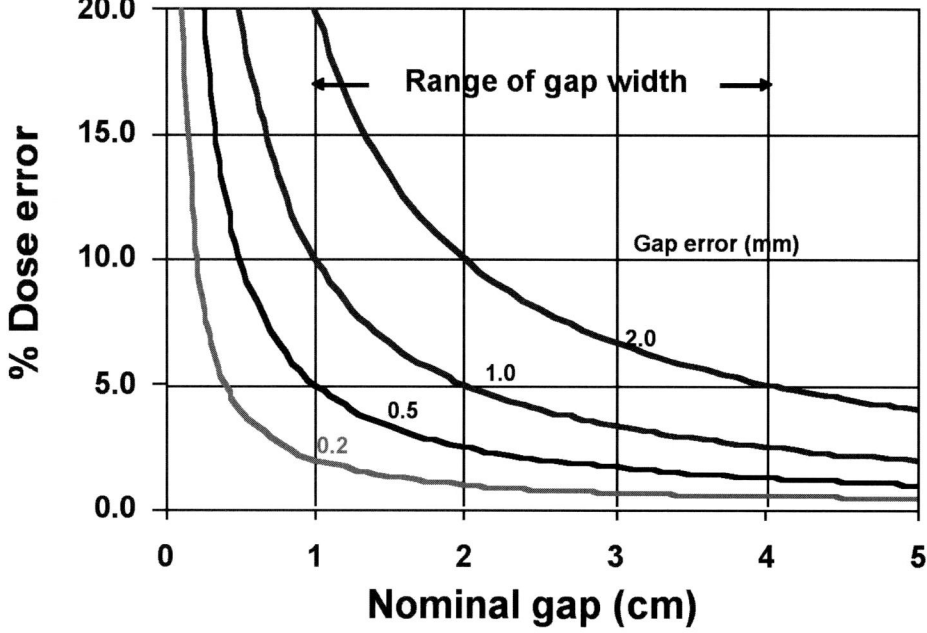

FIGURE 7–2. Relationship between dose error and gap error for DMLC fields. For the range of gap widths typical of DMLC fields, dose errors as a percent of dose delivered are shown as a function of the gap width error. Our goal is to maintain gap errors below 0.2 mm. These numbers apply to SMLC as well, although the gap and dose errors are distributed differently. These curves do not account for transmission through the leaves, which will reduce the percent dose errors somewhat.

gap width, ~1% clinical dose error, is needed for DMLC. In SMLC fields the average dose error will be similar, although it will be concentrated at the edges of the subfields.

Tools provided by the manufacturer (Varian Medical Systems, Palo Alto, CA), with parameters specified in the MLCXCAL file, are relied upon for the alignment and calibration of the leaves. The skewness of each bank of leaves and the leaf centerline offset are adjusted with respect to a reference such as the field light projections of the crosshairs, which are, in turn, aligned to the mechanical isocenter and the jaws. Alternatively, for the Millennium series the leaves can be aligned with the "Field Alignment Tool" supplied by the MLC manufacturer. The alignment in the orthogonal direction is fixed at installation of the MLC, though it should be checked (see previous section).

During installation, in preparation for acceptance, the installation engineer will calibrate the positions of the MLC leaves; however, manufacturer's specs, while adequate for the use of the MLC as a block replacement, may not be tuned to the more critical needs of IMRT. To assess the accuracy of the gap width, the leaves may be set to a nominal 1 mm gap (at isocenter). A feeler gauge with a precision of 0.001″ (0.025 mm) is inserted between the leaves' ends. If needed, the leaf gap parameter (in the MLCXCAL file) is adjusted and the MLC reinitialized so that the gauge indicates 0.51 mm (corresponding to 1.00 mm at isocenter). After this, the alignment of the two leaf banks can be further refined using the feeler gauge to ensure that the banks of leaves are parallel to each other. This assessment should also be performed at other gantry and collimator angles to assess the effects of gravity on the gap profile. Based upon our experience with 12 Varian MLCs, gap width variation vs gantry angle is generally ~0.1 mm at isocenter at time of installation, though this value will increase with use (for further discussion see chapter 7 on QA). Keep in mind that the critical parameter in MLC calibration, for accurate dose delivery with IMRT, is the accuracy of the gap width between opposing leaves, and not the absolute leaf positions. It is our experience that mechanical backlash due to gravity and frictional drag affects both banks of leaves similarly in magnitude and direction and, therefore, the effect of backlash on the gap width is less than its effect on the positions of the corresponding leaves individually.

At the time of this writing the manufacturer has been testing an additional alignment parameter. This parameter is intended to correct for "cupping" of the leaves, which can be observed subsequent to the alignment described above. Once the skewness has been minimized, the leaves form a progressively smaller gap with increasing distance from the center leaves. This characteristic, observed to vary in magnitude among MLCs, is possibly due to a slight vertical misalignment of the optical calibration system, whereby the light from the LED (light emitting diode) intercepts each leaf edge at a different point on its curved surface.

Verification of Gap Width vs. Leaf Position

Additional corrections are supplied by the manufacturer for off-axis points in the direction of leaf motion. Two sets of parameters in another file, MLCTABLE, correct for the difference between the physical leaf position and the light field edge in the plane of the isocenter. One of these, the magnification factor (~1.96), is the ratio of the isocenter distance to the midleaf distance from the source and will vary slightly with MLC model. The other set of parameters, derived solely from the geometry of the rounded leaf end, corrects for the nonlinear relationship of the field size and the leaf position. This set of parameters is independent of the MLC model and should not require adjustment. Thus, for a 0.5 mm gap defined by the projected light field in the plane of the isocenter, the physical gap-width between the leaves' ends should be 0.255 mm plus the demagnified corrections for the opposing leaves. In this regard, the off-axis leaf calibration table, MLCTABLE, from the manufacturer's pre-Millennium software provided values that are truncated in 0.01 cm increments (as measured at the isocenter), introducing gap errors of up to 0.17 mm at off-axis positions. We confirmed these discrepancies with a feeler gauge inserted between the opposing leaves at the central axis and at off-axis positions (figure 7–3). Thus, MLCTABLE should be modified for the earlier Mark I and II models as the increased precision is needed for DMLC.

Alternatively, a dynamic slit field can be used to verify the MLCTABLE parameters dosimetrically. The method compares symmetry and flatness of relative dose profiles for an open field and a 1.0 cm wide dynamic field. With the lower jaws symmetrically set to 12 cm, the DMLC field is designed such that the slit and rounded leaf ends start under one jaw and finish under the other jaw; i.e., the rounded ends are completely shielded at the initial and final stages of beam delivery.[1] To evaluate leaf positioning accuracy at extended distances from the axis, it is necessary to shift the jaws and the DMLC field asymmetrically due to the limits of leaf travel (14.5 to 15.0 cm) within a single DMLC field. Ion chamber measurements are obtained at 1 cm intervals across the width of the field at an arbitrary depth in phantom, and are normalized to the measurement at the axis. If one assumes that the MLC transmission is invariant across the field in the direction of leaf motion, then the dose profile shape for an open static field should be closely replicated for the same field size generated by a fixed-width dynamic slit moving at a constant speed. In fact, transmission constitutes a substantial portion of the total dose in this field and is sufficiently variable across the field to warrant a correction on a point-by-point basis. For this purpose leaf transmission measurements are made for each bank of leaves completely blocking the field in the same geometry as the DMLC measurements.

The gap-width accuracy at off-axis positions verified with ionization measurements for one MLC is shown as relative dose profiles (i.e., normalized to the dose at the central axis) in figures

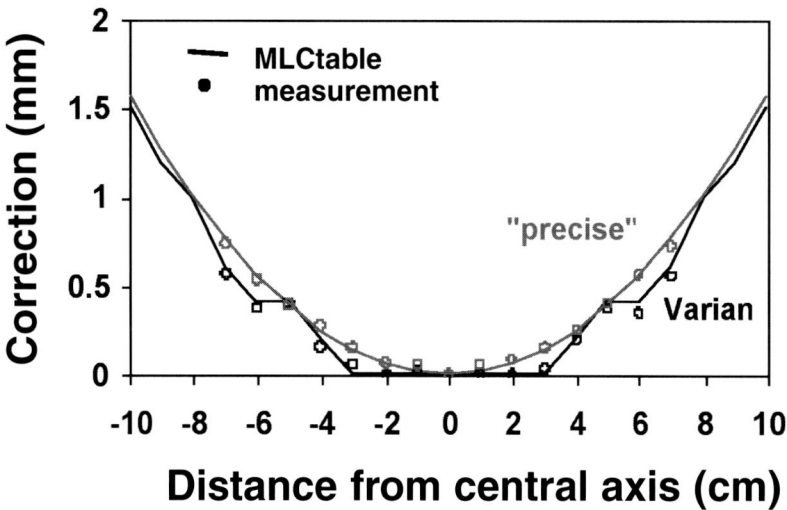

FIGURE 7–3. Comparison of leaf position corrections for pre-Millennium MLC supplied by the manufacturer in the MLCTABLE.txt file with a more precise set of values (not truncated) calculated from the geometry of the rounded edge. The solid curves are gap corrections projected to 100 cm obtained by combining leaf corrections for opposing leaves as a function of off-axis position. The open circles are gap corrections measured with a feeler gauge at the MLC, magnified to the isocenter.

---

[1]   When designing a dynamic test field to measure a beam profile, the leaf positions within the leaf sequence file should be specified at intervals of 1 cm or less. If this is not done, the off-axis corrections in MLCTABLE will not be accurately applied.

7–4a and 7–4b for 6 MV x-rays. Figure 7–4a shows the static field and DMLC field profiles and the ratios of these profiles for each of three fields, centered at the axis and at ±8 cm off-axis at a depth of 10 cm. The prominence of the horns of the beams can be observed at the centers of the off-axis open fields. Near the central axis, the open field and dynamic profiles agree within 0.5%; however, at 10 cm from the axis, the ratio of the profiles decreases by ~2%. This decrease is not due to changes in the gap-width, as the following analysis shows. For this 1 cm sliding window (15 cm of leaf travel) for the Millennium 120 MLC, the duty cycle is only 0.08, and the dose transmitted through the leaves is ~20% of the total output. Figure 7–4b shows that the measured transmission also decreases with increasing off-axis distance, to 92% of the central axis transmission at 10 cm off-axis.[2] Indeed, figure 7–4b shows that the ratios of the profiles for the DMLC

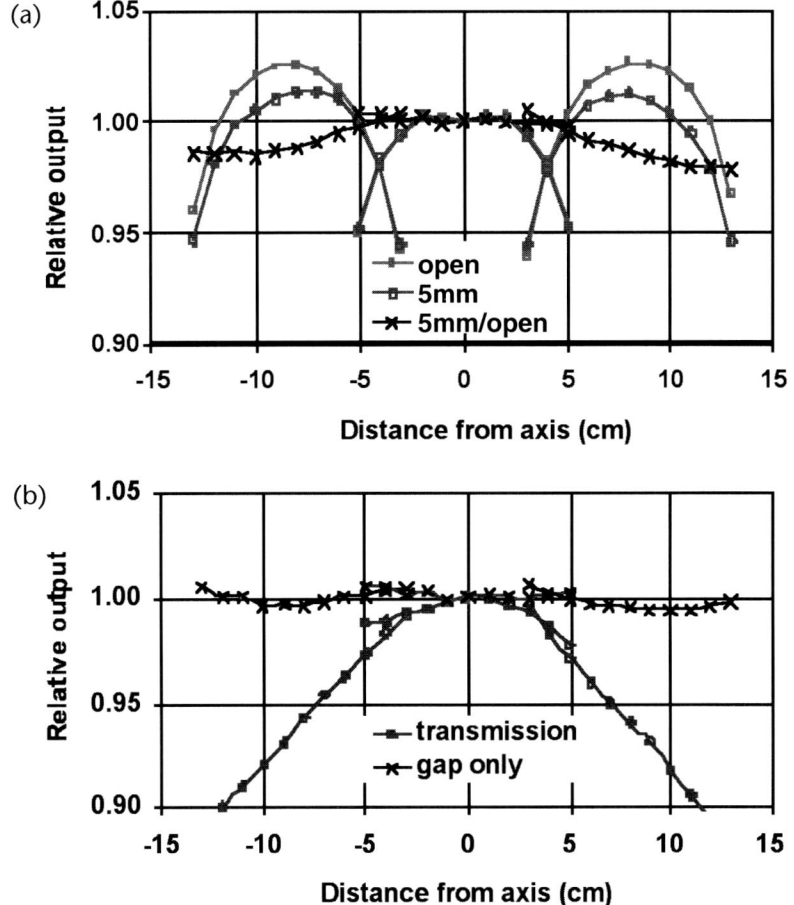

FIGURE 7–4. Measured dose profiles for a narrow, 0.5 cm, DMLC field are compared with those for an open static field. Due to the 15 cm field width limitation, fields are centered at the central axis and at ±8 cm off axis. (a) Ion chamber measurements at 1 cm intervals are normalized to the central axis. (b) The ratio of DMLC to open field is flat once the variation in off-axis transmission is measured and corrected

---

[2]   The transmitted dose is calculated from the static transmission measured at the same point and the amount of time spent under the leaf in the dynamic field.

fields and the static fields are much flatter, once the transmitted component of the dose is removed from the DMLC readings. The residual unflatness in the ratios, approximately ±0.5%, is likely due to oblique scatter mainly from the edge of the lower jaw, which "sees" a narrower DMLC window near the individual field centers than near their edges. Thus, the width of the gap is relatively constant across the field, as even an 0.5% dose change for a 1 cm sliding window would correspond to only a 0.05 mm gap-width variation. This indicates that both the leaf position corrections within the MLCTABLE file are correct and that DMLC execution is performed correctly. Having said this, it should also be noted that the dose variations due to transmission and scatter are, for the most part, artifacts of this test. Under clinical conditions, where the gaps are typically 2 to 3 cm wide and the duty cycles are 0.2 to 0.5, dose variations from these sources would be less than 0.5%.

Leaf Speed

In IMRT treatments using the Varian DMLC, leaf positions are specified as a function of fractional Monitor Units (MUs), the "index" field in the leaf sequence file. Leaf sequencing in the dynamic delivery (sliding window) takes into account the maximum leaf speed, the dose rate, and the total MUs to be delivered. Based on the leaf sequence file, the MLC control computer moves each leaf to its prescribed position corresponding to the fractional MU at that instant. During DMLC delivery, the positions of all the leaves are monitored at 55 msec intervals. Should any leaf deviate from its prescribed position, exceeding a user-specified "dynamic leaf tolerance," radiation delivery is momentarily interrupted by a "beam holdoff."[3]

The leaf tolerance can be specified in the range of 0.2 to 5 mm. It seems logical that one would want to keep this tolerance as low as possible, thereby achieving the most accurate dose delivery, since the leaves would then be closely restricted to their intended position. To examine the influence of the tolerance parameter on the delivered dose accuracy, we executed two DMLC prostate fields and measured the dose at the field center with an ion chamber for the range of tolerance. As the tolerance decreased from 5 mm to 2 mm, the dose was stable; however further decreasing the tolerance, to 0.2 mm, resulted in a dose increase, about 1% (see figure 7–5a). Similar measurements, with the ion chamber positioned far from the dynamic field in an unblocked region, yielded a similar increase in the dose. Thus we concluded that the dose variation was not due to more accurate delivery at lower tolerance; to the contrary, the increased dose was presumed to be due to instability in the beam as the holdoff was increasingly invoked, with the more accurate delivery being at the higher tolerances. We also observed that the time to deliver these same prostate fields increased dramatically, from 30 seconds to ~80 seconds as the tolerance decreased to 0.2 mm (see figure 7–5b). Both for reasons of dose accuracy and treatment time efficacy, it seems advantageous to avoid this very low tolerance region. At MSKCC, we use 2 mm as the dynamic leaf tolerance. In practice, the principal reason for this leaf position tolerance is that in the event of a catastrophic failure, e.g., a stuck leaf, the beam holdoff will terminate the beam.

The 2 mm dynamic leaf tolerance may seem inconsistent with our stated goal to maintain the gap error below 0.2 mm. However, the 0.2 mm discrepancy refers to a systematic or average gap error, and in fact, the frequency that the leaves are out of tolerance by more than 2 mm is very rare, as the following shows. Analysis of a Dynalog file[4] for a typical five-field IMRT prostate

---

[3] "Dynamic leaf tolerance" is an MLC parameter supplied by Varian specifically for IMRT treatments and operates in conjunction with the "beam holdoff" function.

[4] A log file report, generated by the DMLC control software, summarizes the deviations of the leaves from their prescribed positions. The accounting is only done for leaves that were moving during the DMLC treatment. Since the spatial information in the Dynalog file is from the primary MLC encoders, neither the accuracy of the encoders' calibration nor the backlash in the leaf and carriage mechanisms is accounted for. Thus, the log

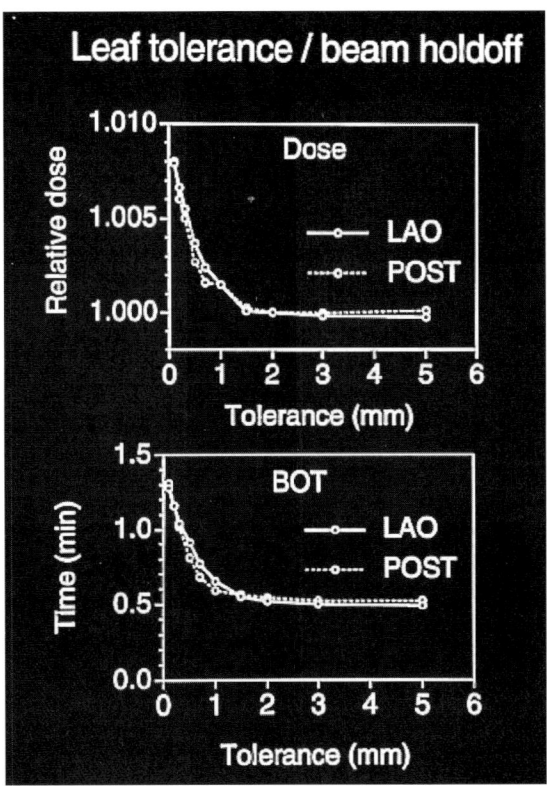

FIGURE 7–5. Dose and beam-on-time (BOT) variation vs. leaf position tolerance for two prostate fields.

treatment is shown in table 7–1. Each prostate field was delivered on two treatment machines, with the gantry at both 0° and at the planned angle. Leaf position monitoring, ~20 times per second for all moving leaves in a 30-sec delivery, yields about 12,000 leaf position checks per field. Deviations between monitored and prescribed leaf positions are tabulated by bin size. Differences greater than 1 mm are detected in ~1% of the comparisons and differences greater than 2 mm are rare, ~0.1%. Furthermore, these results show little variation with gantry angle or between machines. The log file report also provides a global rms (root mean square) shown in the right-most column. A more detailed analysis of the log file (data not shown) indicates that the average leaf position deviation is ≤0.2 mm, consistent with the global rms deviation (by definition, the true mean should be less than the rms value), which is also supplied by the control software and shown in the last column. However, for the most part, these deviations are one-sided, which would be expected from the effects of either friction or gravitation as the leaves move in one direction during DMLC delivery. This backlash in mechanical systems is normal; however, pairs of opposing leaves tend to compensate each other, resulting in an average gap error, a better indicator of DMLC performance, that is closer to 0.1 mm. Thus, deviations of individual leaves are related

---

files do not contain "actual" leaf position data; rather, from the perspective of the encoders, they document the monitored positions of the leaves and their intended positions during DMLC delivery. Nevertheless, the log file report is a useful tool for comparing leaf motions under different stresses to the MLC. Here we use it to study the impact of gantry rotations on leaf positioning. A log file analysis of the effects of repeated beam holdoffs, invoked by gating signals, on IMRT dose delivery has also been reported (LoSasso, Chui, and Ling 2001).

Table 7–1. Frequency of Deviations (Percent)

Field	Room	Angle	Logged—Prescribed Leaf Positions @ isocenter (mm)					# Checks	Global RMS
			<0.05	0.05–1.0	1.0–2.0	2.0–3.0	3.0–4–0		
LAO	245	0	49	50	0.5	0.04		12528	0.26
		225	50	50	0.5	0.04		12528	0.26
	445	0	52	48	0.5	0.02		12492	0.26
		225	48	51	0.6	0.01		12510	0.26
LPO	245	0	59	40	0.9	0.09	0.03	9918	0.26
		285	59	40	1.0	0.09	0.04	9918	0.26
	445	0	61	38	0.9	0.10		9810	0.22
		285	58	41	1.0	0.10		9810	0.26
PA	245	0	50	49	0.4	0.01	0.01	18228	0.14
		0	50	50	0.4	0.01	0.01	18198	0.14
	445	0	52	47	0.4	0.02		18180	0.14
		0	52	47	0.4	0.01		18144	0.14
RPO	245	0	62	36	1.4	0.10	0.03	10116	0.24
		75	63	36	1.4	0.11	0.03	10116	0.24
	445	0	63	35	1.4	0.12		10152	0.22
		75	63	35	1.4	0.12		10062	0.22
RAO	245	0	47	53	0.3	0.02		14490	0.14
		135	48	52	0.3	0.01		14508	0.14
	445	0	49	50	0.3	0.01		14526	0.14
		135	50	49	0.3	0.01		14436	0.14

checks = # moving leaves *BOT/55 msec

only indirectly to the quantity of interest, the accuracy of the delivered dose. Log files can be converted into "delivered" leaf sequence files, and from these files the delivered dose distributions can be calculated (within the uncertainty due to leaf calibration) and compared with the planned dose distributions (see dose verification in the next chapter).

Interleaf Transmission

Variations in transmission for a specific MLC in the direction perpendicular to leaf motion are caused by differences between the midleaf thickness and the combined thickness of tongues and grooves, plus the transmission through the minute air spaces between adjacent leaves due to mechanical tolerances. Interleaf transmission is most easily quantified with film dosimetry using a tissue-equivalent phantom and a high-resolution scanner. In general, the interleaf transmission exceeds the midleaf transmission for these MLC as shown in figures 7–6a and 7–6b. Differences between MLC models, i.e., Mark II and Millennium MLC, mainly reflect modifications to the tongue and groove shape. For the Millennium, the overlapping extrusions from the sides of the leaves are narrower in width, and thicker in the beam direction than the tongues and grooves of the Mark II; the midleaf thickness in the beam direction is also somewhat greater for the Millennium. These changes combine to reduce the average transmission of the Millennium by ~25% relative to the Mark II. As measured in phantom, the maximum interleaf transmission should not be more than twice the midleaf value for both MLC models. The average interleaf transmission is much less for the Millennium than for the Mark II, being approximately 15% and 60% of their respective midleaf values. The full-width-half-maximum (FWHM) values are ~2.5 and 2.0 mm for the Mark II and Millennium MLC, respectively.

Our own treatment planning system has the option to calculate interleaf fluence explicitly; however, most commercial systems currently allow an average MLC transmission (midleaf and interleaf combined) only. They do not separately account for the average interleaf transmission, and certainly not the variations in interleaf transmission, for a specific MLC. Thus the

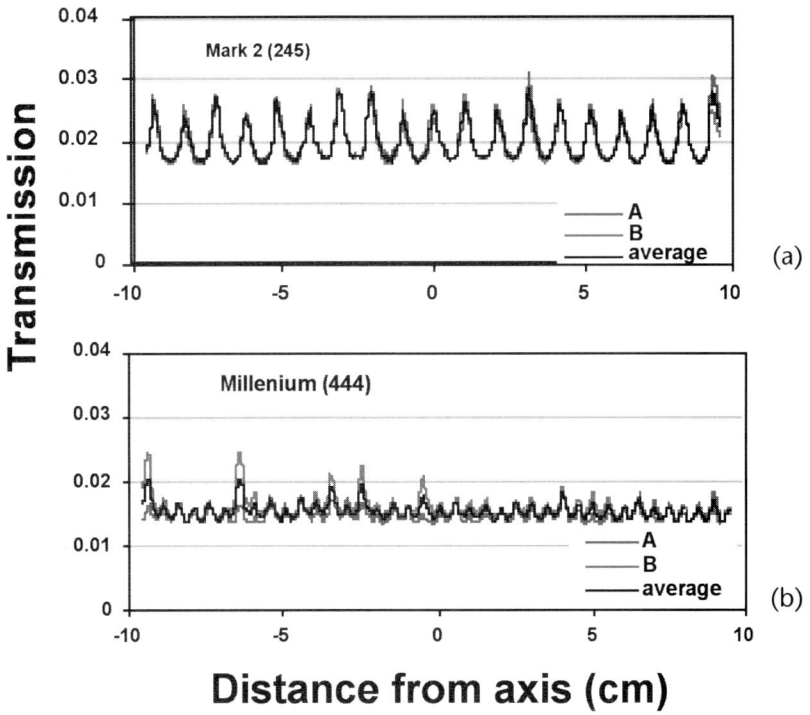

FIGURE 7–6. Interleaf and midleaf transmission profiles (A-side, B-side, and average) for the (a) Mark II and (b) Millennium 120 MLC.

difference between the actual transmission at each point under the leaves and the average MLC transmission used in the calculations is a tolerable error in the dose calculation.

Commissioning

Average MLC Transmission

Measured MLC transmission varies with MLC design; this is most obvious between the Millennium model and its predecessors, the Mark I and II models. For an individual MLC primary transmission through the leaves varies due to the structural components of the leaves, specifically the round leaf end and the tongues and grooves. The interleaf transmission, due to tongues and grooves, is typically measured with, and included in, the average MLC transmission. Less obvious are the effects of midleaf mechanical components, which alter the effective leaf thickness. For large IMRT fields scatter from the leaves is nearly equal to the primary component. Off-axis spectral effects and oblique transmission will also be addressed.

For treatments with static fields, the MLC leaves, as block replacements, define the field borders. MLC transmission in static fields principally affects the dose outside the treatment field; it is essentially a fixed fraction of the target dose, typically 1.5% to 2.0%. In such fields, the leaves block only a small portion of the beam, and scatter from the leaves is small and is often ignored in dose calculations, as is the scatter from blocks. On the other hand, MLC transmission contributes substantially to the target dose in IMRT. The leaves shield the target for significant portions of the treatment time. Furthermore, since the leaves are blocking most of the beam at each moment during irradiation, scatter from the MLC is significant, particularly for large IMRT fields.

Relative to the MLC transmission for a static field, the transmitted dose for a DMLC delivery, T_D, at any point p can be expressed as:

$$T_D = (P\,(1 - C_p) + S)/C_{av}$$

where P and S are the primary and scatter components of static transmission. The duty cycle, C_p, is the fraction of the MU for which P is under the gap, and C_{av}, the average duty cycle, is the ratio of the total MU to deliver the prescribed dose with static fields to the total MU for the DMLC fields. Thus C_p is proportional to the intensity at p, and it approaches C_{av} for the composite plan. T_D is inversely related to the duty cycle. For IMRT fields, C_{av} is typically 0.2 to 0.5, and C_p can have a larger range from 0 (outside the target) to 1 depending upon the dose uniformity. Given the sizable contribution of MLC primary and scatter transmissions to the DMLC dose, up to 10%, it is important to quantify these components during IMRT commissioning.

Static MLC transmission is determined from dose measurements in phantom using an ion chamber. The results of such measurements are shown in figures 7–7a and 7–7b for 6 and 15 MV X-rays. A cylindrical ion chamber, sensitive area ~1 cm square, was placed at the central axis in a polystyrene phantom at depths ranging from d_{max} to 30 cm, for field sizes ranging from 6 × 6 to 14 × 14. With this input the energy, field size, and depth dependency are mapped for each energy (LoSasso, Chui, and Ling 1998). For a 6 MV beam, there's actually an increase in transmission as a function of depth due to beam hardening by the MLC. This same effect is seen for wedges (Podgorsak, Kubsad, and Paliwal 1993). For 15 MV, there is no measurable increase with depth; presumably, at the high end of the spectrum, increased pair production in the MLC tends to counter the beam hardening effect. The combined effect is nearly constant transmission vs. depth. Similar profiles obtained for an 18 MV beam (not shown here), actually decrease slightly with depth. Similar observations were made for 6 and 15 MV transmission for standard wedges (Popescu et al. 1999). Transmission is also shown to increase with field size, as more photons are scattered in the MLC and reach the point of measurement.

Figures 7–7a and 7–7b also depict the clinical range of energy, field size, and depth for prostate and head and neck fields, the two sites first treated with DMLC at MSKCC. Static MLC transmission would vary by ~0.5%, which translates to 1% to 2% of the dose delivered dynamically for duty cycles in the range of 0.2 to 0.5. Having chosen an appropriate single value for MLC transmission for all prostate and head and neck treatments, then the transmitted dose error incurred for these sites should be less than 1% of the delivered dose. IMRT more recently has been extended to treat larger fields in breast and lung, and extremely large fields encompassing the entire abdominal pelvic area. These latter DMLC fields, 40×30 cm^2 or more, can be 10 times greater in area with lower duty cycles than those previously encountered. A single average value for transmission based upon the smaller fields would not be appropriate for these fields. Since typical setup parameters tend to be stable for a specific treatment site, one option was to fix the transmission for each site. Instead, a new MLC transmission model was chosen. MLC transmission is obtained by integrating the total scatter contribution to each point in the patient using a radial function to describe MLC primary and scatter components from each point in the MLC at each moment during the DMLC delivery. The radial function is derived by measuring the total transmission in air for varying blocked field sizes defined by the collimators from 3×3 cm^2 up to 40×13.5 cm^2 or 40×17 cm^2, the maximum allowed blocked field by the leaves from one carriage for the Mark II and Millennium MLCs, respectively, while shielding the rounded leaf ends under the jaw. The differential contribution of scatter for each radial increment for the measured rectangular fields is applicable to a 40×40 cm^2 area assuming radial symmetry. The primary transmission is added as a delta function at zero radius. The MLC transmission integration for clinical fields is described in the dose calculations in chapter 5. As calculated by the new transmission model, the transmitted dose is increased by 1% and 5% for typical IMRT field sizes of

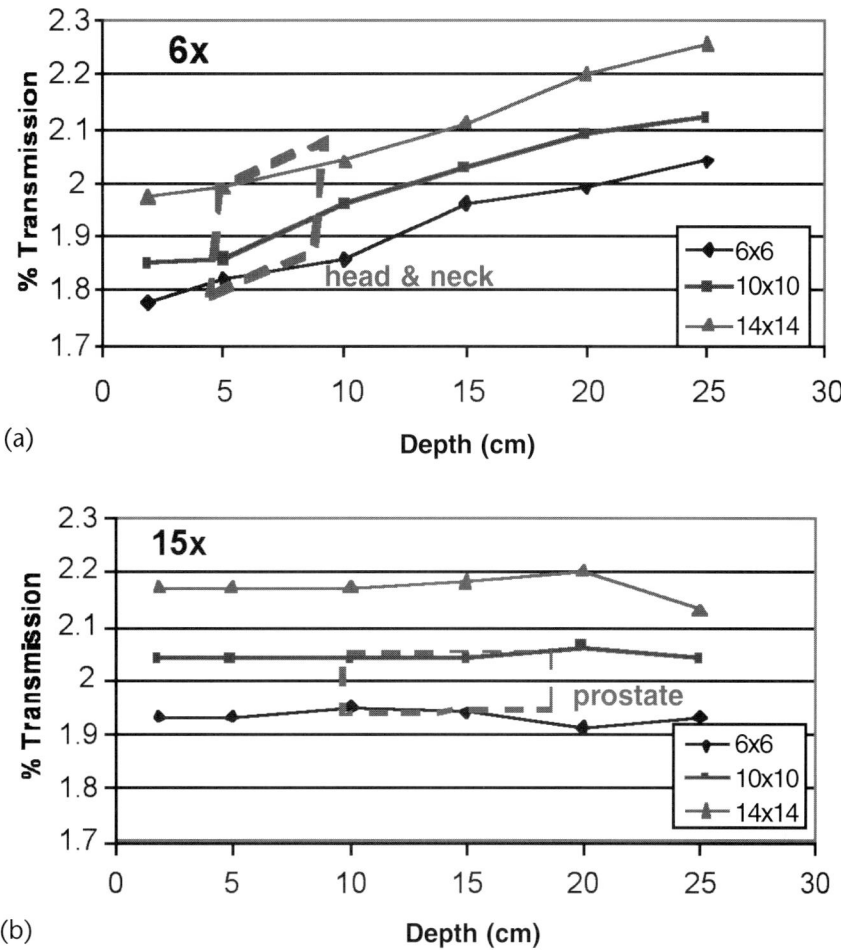

FIGURE 7–7. Static MLC transmission measured with an ion chamber vs. field size and depth for (a) 6 MV X-rays and (b) 15 MV X-rays. The ranges of field size and depth are highlighted for head and neck (6 MV) and prostate (15 MV) patients, indicating that 2% is an average transmission for these sites.

$16 \times 16 \text{ cm}^2$ and $40 \times 30 \text{ cm}^2$ (combined split fields), respectively, relative to the old model using an average value for MLC transmission.

Changes in transmission due to oblique pathlengths through the MLC and spectral effects were measured at 5 and 10 cm off-axis in the direction of leaf motion for one depth, 10 cm, for 6 MV X-rays and 15 cm for 15 MV X-rays, and for a single field size, $10 \times 10 \text{ cm}^2$. The data in figures 7–8a and 7–8b show the transmission dose as a function of off-axis distance, normalized to the dose for open $10 \times 10 \text{ cm}^2$ fields, for both 6 and 15 MV X-rays and for several MLC models (Mark I, Mark II, and Millennium 120, Varian Medical Systems, Palo Alto, CA). For comparison, the transmission through a 7 cm thick block of lead, having approximately the same attenuation as the MLC, was measured for each MLC-linear accelerator (linac) combination. Several general points can be made: (1) The MLC transmission dose is consistently slightly lower for 6 MV than for 15 MV for each dual energy machine (in rooms 445, 245, and 444), both on and off the

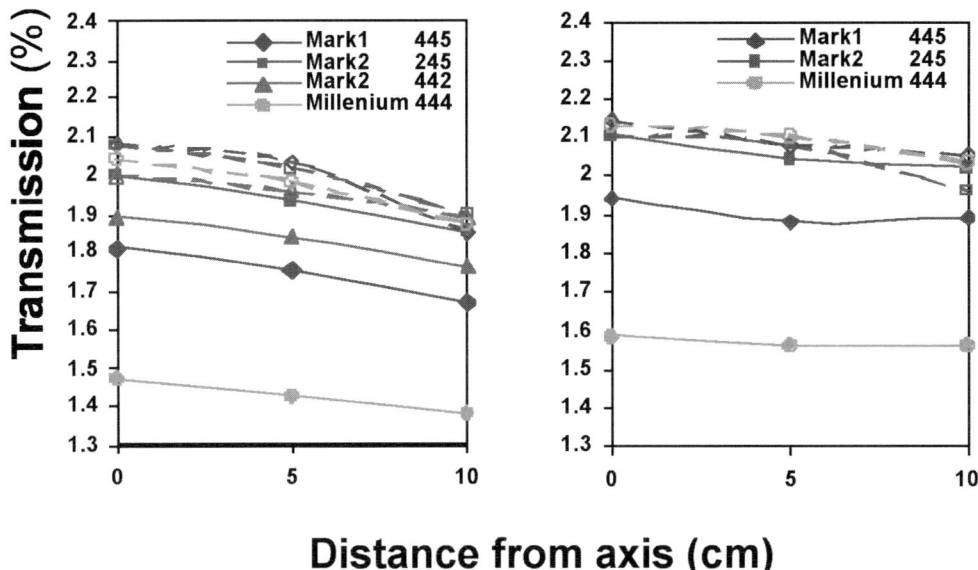

FIGURE 7–8. Variation in off-axis transmission for the Mark I, Mark II, and Millennium 120 MLCs (solid curves) and a lead block (dashed curves) due to oblique pathlength and spectral changes.

central axis. (2) The small variation in lead transmission for the four 6 MV beams is qualitatively consistent with the variation in their depth dose profiles in water. This energy shift explains the similar differences in transmission observed for the MLC and the lead block, between the two Mark II MLCs. It has been suggested that variations in MLC density are responsible for transmission variations among MLCs (Arnfield et al. 2000). While we do not disprove this, our data indicate that differences in transmission observed for our MLC are clearly related to energy variations between linacs. (3) The combined effect of off-axis spectral changes and oblique attenuation are such that at 10 cm off-axis, the transmission decreases by 0.10% to 0.15%. These off-axis variations are currently ignored by our treatment planning system. (4) The ~0.5% variations in transmission dose among the MLC models are likely due to differences in leaf design; e.g., as shown in figure 7–4, the Millennium MLC has the least midleaf and interleaf transmission.

Variation along the length of the leaves has been observed for the Mark I and Millennium models. The profiles in figures 7–9a and 7–9b indicate dose variations due to structural components of these MLC designs; the Mark II design provides a uniform profile. For the Mark I, relatively low density screws at the top and the bottom of the leaves attach the leaves to the rail and guide the leaves along their paths. The screws are positioned about every 3 cm. Point transmission measurements under the screw would exaggerate the transmission. For the Millennium MLC, the drive screw slot extends to about 3 cm from the leading edge of the leaf, allowing additional transmission. We average these components into the primary transmission value.

These ion chamber measurements only relate to the transmission through the two central pairs of leaves and their shared interleaf space. The relative variation in interleaf transmission among the other pairs of leaves is measured with film and also averaged into the primary transmission. Kodak V2 "Readypack" film is placed in a flat homogeneous phantom at depths of 10 and 15 cm for 6 MV and 15 MV, respectively, and exposed to a transmitted dose of about 30 cGy

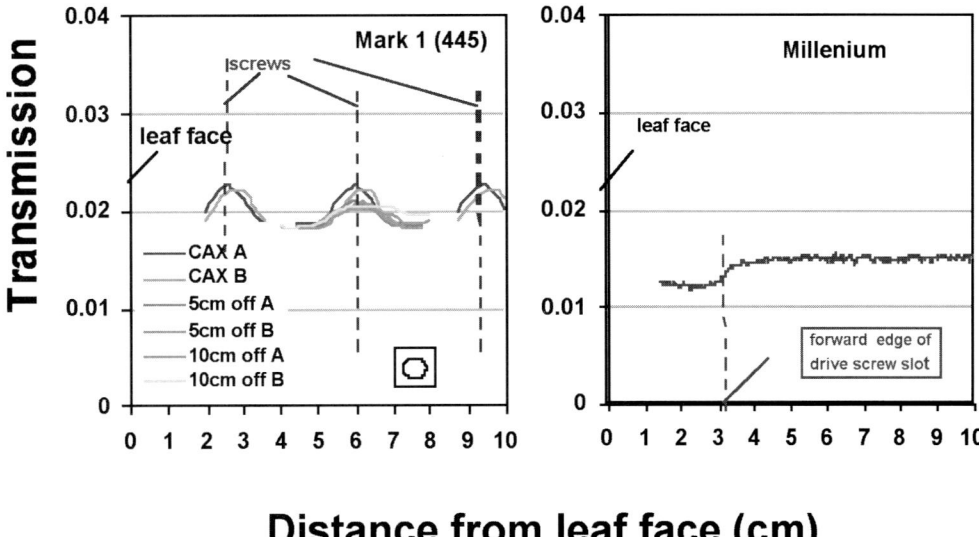

Distance from leaf face (cm)

FIGURE 7–9. Variation in off-axis transmission due to structural components of the MLC. Increased transmission is allowed by the mounting screws for the Mark I and the drive screw slot for the Millennium 120.

for the following MLC and jaw setting: (1) both banks of leaves at –7.25 cm and fully extended from their carriage (blocking the entire radiation field), with the lower jaws at –6.0 and +6.0 cm to completely shield the rounded edges of the leaves; (2) the same as (1) except that both banks of leaves at +7.25 cm; (3) the banks of leaves at 7.25 cm and –7.25 cm, respectively (i.e., an open field). The films are processed, scanned, and converted to dose profiles. The blocked profiles are normalized with the open field profile to remove the effects of beam asymmetry and unflatness. The transmission data from film dosimetry is averaged around the central axis over a 10×10 cm^2 region and over a smaller region, 1×1 cm^2 corresponding to the sensitive volume of the ion chamber. In this way, ion chamber data can be scaled to yield a spatially averaged (over a 10×10 cm^2 region) transmission factor for input to the treatment planning program.

It is obvious that the Millennium has less interleaf transmission, but also less midleaf transmission than for the Mark II, as the leaves are slightly thicker in the beam direction. The Mark I actually has less interleaf transmission than the Mark II, and its average transmission is slightly less as well. Providing MLC-specific transmission data requires unique leaf sequence files for each MLC, which could be a source of confusion in the clinic. An alternate approach is to use the same transmission data in the treatment planning system, but compensate for the variation of a particular MLC by adjusting the gap width with the MLCXCAL parameter on that machine. This method is approximate, but has the advantage of correcting for the difference at the MLC, making the solution transparent to the clinic. For our department with multiple MLC models delivering IMRT, we are using both methods. For the Mark I, the leaf gap parameter in the MLCXCAL file has been altered slightly to compensate for the small deviation in transmission relative to the Mark II MLC. For the Millennium 120, which in any case requires a separate leaf sequence file, if not a re-optimized plan, the transmission is modified in the treatment planning system.

Interleaf Effects

Interleaf effects fall into two categories, interleaf leakage and the so-called *tongue and groove* effect (T&G). They are manifested as either an over- or underdosing, respectively, of the regions between pairs of adjacent leaves. When adjacent leaves travel in synchrony across the field, the transmission through the narrow gap between the adjacent leaves is larger than the transmission through the middle of the leaf. But when the adjacent leaves are not synchronized, irradiation occurs sequentially through the tongue of one leaf and the groove of the adjacent leaf, whereby the dose in the region between the adjacent pairs of leaves is less than that of the average of the two pairs of leaves. Our early experience with IMRT prostate treatments using the Mark I and II MLC models suggested that interleaf effects were not clinically significant: the slightly higher interleaf transmission was ameliorated by interfraction "feathering," and the DMLC leaf sequences produced T&G infrequently. However, for new treatment sites and new models of MLC the significance of these interleaf effects needs to be reevaluated.

Interleaf leakage is attributed to the reduced shielding through the interlocked T&Gs, resulting in increased transmission in the region between adjacent leaves. When a radiographic film is exposed under a fully closed MLC, only transmitted dose is recorded. Figures 7–6a and 7–6b show the dose profiles obtained in the direction perpendicular to the leaf motion under both banks of leaves for the Mark II MLC (1.0 cm leaf width), and for the Millennium MLC (0.5 cm leaf width). Leakage through the narrow gap between adjacent leaves shows up as peaks above the *continuous background* of transmission dose. For the Mark II MLC, the leakage dose peaks are 30–50% higher than the continuous background; for the Millennium MLC both the midleaf and interleaf dose are smaller than for the Mark II. The signature fluctuations among the interleaf spaces for a particular MLC are likely due to minor differences in the separations of adjacent leaves due to mechanical tolerances. No significant dependence of these patterns on gantry angle was observed, but variations were observed with collimator rotation at gantry angles of 90° and 270°. Given these observed variations among the different models, it is important to characterize each new MLC prior to its implementation for IMRT. As previously indicated, the total transmission must be accounted for in treatment planning dose calculation because the MUs are typically 2 to 5 times higher for IMRT treatment than comparable static plans, and because the transmitted dose is incident on the target volume during a large portion of the treatment.

Sequential irradiation through the tongue or groove of the leading leaf of one leaf pair followed by the groove or tongue on the trailing leaf of the adjacent leaf pair occurs frequently during delivery of an IMRT field. It may result in an underdosed region; that is, the fluence in the interleaf region is less than the average of the fluences in the adjacent midleaf regions. Since the tongue and groove are each about half the midleaf thickness, ~3 half-value layers (2.4 cm and 3.2 cm thick for the Mark II MLC and the Millennium MLC, respectively), the transmitted fluence through the tongue or the groove is only ~15%. The combined irradiation through the tongue and groove would be ~30% in this situation, except that narrow air spaces between leaves also contribute to the fluence. Electron transport in tissue reduces the amplitude and broadens the width of these underdosed regions. For clinical beams, 6 to 15 MV, for both the Mark II and Millennium MLCs, this effect is best quantified as a peak measured underdose of ~15% with FWHM = 2.5 mm. The Millennium also has narrower tongues and grooves (0.8 mm vs. 1.25 mm for the Mark II). Much of the anticipated improvement afforded by the narrower T&Gs is offset by their increased thickness in the beam direction. In order to decrease interleaf leakage (apparent in figures 6a and 6b) in the design of the Millennium leaves, the thickness (in the beam direction) of the T&Gs was increased. As a result, in terms of T&G, the gain due to the reduced width of the T&Gs is offset by the increased thickness.

Potentially, T&G can be quite large and will be most noticeable on film when adjacent pairs of leaves are delivering the same dose in a portion of the field, but their gaps do not overlap at all. If the gaps partially overlap or if one gap is wholly contained within the other, then the interleaf fluence should be greater than the lower of the two adjacent midleaf doses, but still lower than the average. Under these conditions T&G effects often go unnoticed if film dosimetry is only performed for individual fields, although gradients across adjacent leaves suggest their presence in the composite dose distribution. A similar situation often occurs at the edges of irregular fields, where irradiation occurs through only the tongue or the groove.

Leaf asynchrony increases in highly modulated and irregularly shaped IMRT fields (lung and head and neck sites) increasing the severity and frequency of T&G relative to more uniform IMRT fields (prostate). T&G can also occur when the dose to adjacent regions in the target volume is primarily delivered by different subfields. In this case, T&G is present in areas shielded by the groove in one field and the tongue in the other field. Special attention to these problems is needed for specific sites such as head and neck and lung, where irregular target volumes with proximal critical organs are common. On the other hand, organ motion and random setup error will moderate these interleaf effects.

Leaf width indirectly influences T&G. The redesign of the Varian MLC from a 10 mm to a 5 mm leaf width has the obvious advantage of improved resolution; but this may actually enhance interleaf T&G, depending upon the dose gradients within the field. Narrower leaves may lead to increased modulation during inverse planning. As the leaves become thinner, the optimization can produce sharper gradients, which may produce a better plan based upon input criteria, but which may produce more T&G.

To evaluate potential T&G problems, dose distributions of IM fields encountered in clinical situations are measured with film dosimetry in flat polystyrene phantoms. The dose distribution calculations for the actual patient IMRT fields will be repeated in these phantoms and then compared with the measured distributions. One proposed technique for modeling interleaf effects in IMRT dose calculations is a ray tracing method, which incorporates interleaf effects and yields improved agreement with measurements (Chen, Boyer, and Ma 2000). Another method is well-suited to our dose calculations. Interleaf spaces have been modeled for each MLC design. Based upon the leaf pattern and measured transmission profiles for a series of reference fields developed specifically to evaluate interleaf effects (see figures 7–10a and 7–10b), appropriate values are assigned to midleaf or interleaf fluences. Interleaf fluence will be further categorized by the presence of the tongue, the groove, or the T&G combination at each point in the intensity profile, summed over the full-field irradiation. Additional corrections need to be considered for the interleaf space is irradiated through the rounded leaf edge (see next section). These dose calculations are described further in chapter 5. If T&G perturbations are unacceptably large, treatment plans can be re-optimized in a way to reduce interleaf effects, possibly by avoiding those beam angles that introduce erratic modulation, or by splitting modulated fields with bimodal dose patterns into separate, more uniform fields. It should even be possible to discriminate against the causes of T&G effects during optimization by penalizing excessive T&G.

Fortunately, interleaf leakage and T&G effects from multiple IMRT fields naturally compensate for each other to some extent in the combined dose distribution. Excess localized dose due to interleaf leakage rarely exceeds ~3% of the prescribed dose, but it occurs more frequently within fields than do T&G effects. Leaf sequencing solutions further reduce the tongue and groove effect, although at some cost to the optimized intensities and MUs. At MSKCC, leaf sequence software requires that all the leaf pairs in an IMRT field start and stop in unison (Wang et al. 1996). Others require that adjacent gaps may not partially overlap each other; instead, the smaller gap must be completely bounded by the larger adjacent gap (van Santvoort and Heijmen 1996; Webb et al. 1997). Each of these methods tends to synchronize the gaps as they move across the field, but none will completely eliminate underdosages caused by tongues and grooves.

Transmission Through The Round Leaf End

MLC leaf design is often distinguished as being either single- or double-focused (AAPM Report No. 72, 2001). In the double-focused configurations (Scanditronix, Siemens, and GE), the leaf face, as well as the leaf sides, are straight, matching the beam divergence. Single focused leaves have rounded faces (Elekta and Varian), allowing the leaves to maintain constancy of penumbra

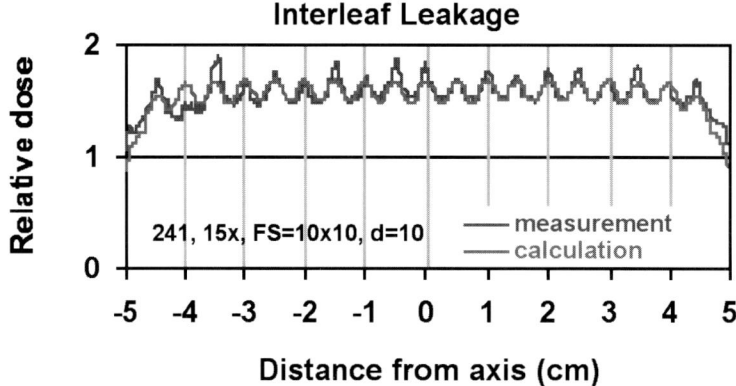

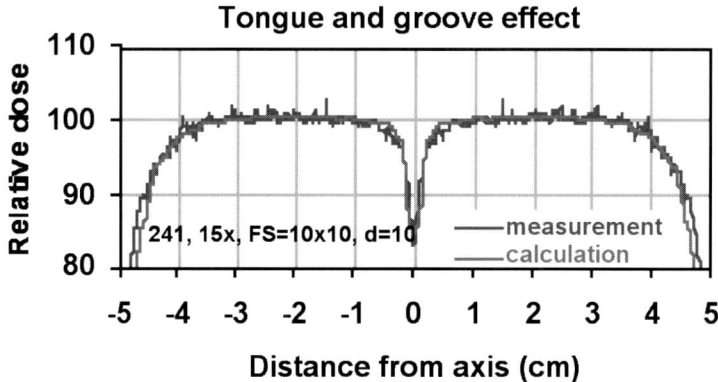

FIGURE 7–10. Comparison of calculated and measured profiles for the Mark II and Millennium 120 MLCs.

width as the non-focused leaves move away from the central axis in a simple rectilinear motion. The round end does entail more complex dosimetry issues though. The transmission through the round leaf end transmits more of the primary beam than flat leaves, and the transmission is varying with distance from the edge. For the Varian MLC, this added fluence will contribute 10% or more of the delivered dose from DMLC fields, depending upon the average gap width.

The geometry of the Varian round leaf end is shown in figure 7–11. The leaf end is curved (radius of curvature is 8 cm) in the central ~3 cm thick portion, beyond which the edges are straight, (11.3° relative to the vertical). Ray-tracing through the known geometry combined with the narrow beam linear attenuation coefficient derived from the full leaf transmission allows the leaf end transmitted fluence profile to be calculated (LoSasso, Chui, and Ling 1998). For conventional static MLC treatments the 50% isodose line defines the field edge, as for field edges defined by the jaws and blocks. For focused MLC leaves the light and radiation fields are coincident, i.e., the light field edge is presumed to represent the 50% line. For rounded leaves of the Varian MLC, the 50% line is displaced under the leaf edge. This shift between the light field edge and the 50% isodose line is about 0.3 mm, and can be ignored for static treatments.

For dynamic fields, however, the transmission through the round leaf edge is a significant portion of the delivered dose to the target volume; and it cannot be ignored. The contribution to the delivered dose is proportional to the area under the round edge transmission profile. Similarly calculated profiles are shown in figure 7–12 for leaves positioned at the axis, and at 5, 10, and

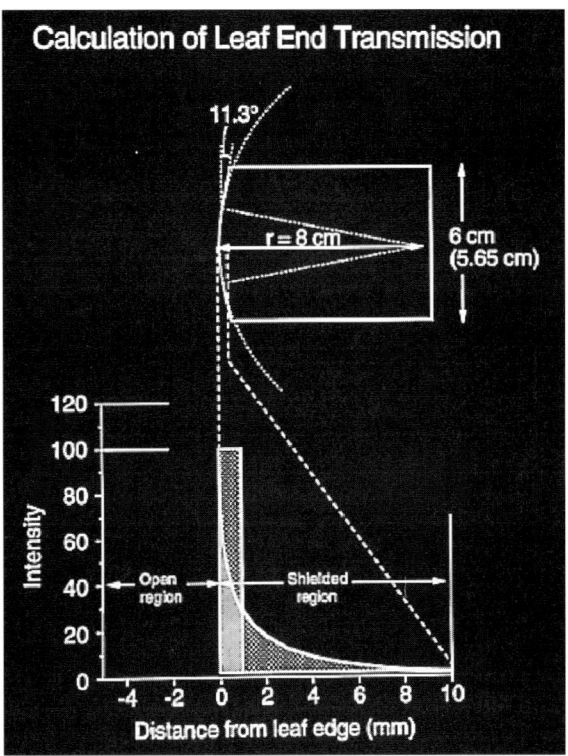

FIGURE 7–11. Calculation of the transmission through the round leaf end based upon measured MLC transmission and the geometry of the leaf face.

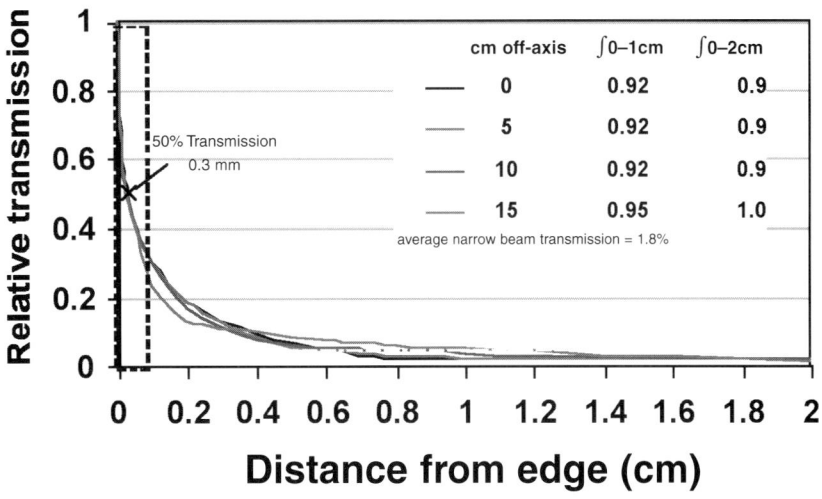

FIGURE 7–12. Comparison of calculated transmission profiles through the round leaf end at the central axis and at 5, 10, and 15 cm off-axis. The inset table gives the effective offsets.

15 cm off-axis. Off-axis the transmission drops off faster initially, and then more slowly. Although the shape of the profile changes somewhat, there is little change in the area under these curves. These small shape changes are clinically insignificant, and so a single profile is satisfactory.

Effective leaf offsets are also tabulated in figure 7–12. An effective offset for a leaf can be thought of as the amount that a leaf would need to be retracted to add the same fluence as is transmitted through the round leaf end. Two offset values are provided for each profile, corresponding to the transmission integrated over the first 1 cm and 2 cm from the projected leaf end at isocenter. For 0 to 10 cm off-axis, most of the transmission is attributed to the first centimeter. The added fluence through the rounded edge, the area under this curve, is equivalent to the added fluence obtained by retracting a focused collimator ~1 mm, as depicted by the rectangular region. And since two leaves form a gap, every gap is effectively ~2 mm larger than the nominal gap setting. Thus a 2 cm nominal gap, average for DMLC fields, is effectively 2.2 cm, yielding a 10% dose error if the rounded edge is not accounted for. The transmission profiles for all Varian MLC, the Mark I, Mark II, and the Millennium series, are the same since the geometry of the round leaf edge is unchanged. Representing the leaf end transmission as an effective offset facilitates the comparison of leaf end transmission on and off the axis, and the comparison of these calculations with measurements. However, our treatment planning calculations are based upon the central axis leaf end transmission profile itself and not the effective offset.

Transmission through the rounded leaf end can also be quantified with measurements. Film was used in our earlier work to measure total dose in phantom as a function of *nominal* static gap widths, and from these data effective gap offsets were derived for both 6 and 15 MV beams (LoSasso, Chui, and Ling 1998). Another approach to quantify this effective offset has been reported Arnfield et al. 2000; LoSasso, Chui, and Ling 2001). DMLC leaf sequence files have been created for moving gaps with fixed widths of 0.5 to 20.0 mm, with constant leaf speed to generate 12 cm wide uniform radiation fields centered at 0, 5, and 10 cm from the central axis. The jaws are set asymmetrically to shape 10×10 cm^2 fields. The MUs are the same for all the fields. An integrating ion chamber, either in a phantom or in air, is placed at the center of the DMLC field (i.e., at 0, 5, or 10 cm from the central axis) in the plane of the isocenter. For each DMLC field, the ion chamber reading is the sum of the fluence through the gap, the leaf end transmission, and the full leaf transmission. Full leaf transmission is measured using the same jaw setting and ion chamber setup (Note: measured transmission is dependent upon irradiation and chamber geometry due to scatter and interleaf transmission), but with the leaves of the MLC blocking the field completely. The full leaf transmission for each DMLC field, derived from the average of the measured transmission for the two leaf banks and adjusted for the time that the chamber is shielded by the leaves, is subtracted from the measurement. The net outputs for 6 MV and 15 MV beams, at the central axis and a depth of 10 cm, are plotted against the nominal gap width and fit with straight lines (figures 7–13a and 7–13b). The intercept at zero dose yields the effective gap correction, in agreement with twice the calculated value for individual leaves (see figure 7–12). Note that this measurement includes the effect from the uncertainty of the leaf gap calibration, the variations in attenuation due to energy and leaf design, and the limitations of the mechanical components of the MLC system.

Effective gap offsets measured for 6 MLC (one Mark I, three Mark IIs, and two Millennium 120s), are in good agreement. Figure 7–14 summarizes the measured offset values at the central axis. For the same nominal energies, variation in the offset value was within ±0.15 mm at all positions. Except for the MLC in room 442, the variation in the offset value at the central axis is within ±0.05 mm. The lower offset for the MLC in room 442, in part, is due to the slightly lower beam energy observed for this 6 MV beam (figure 7–5a). In fact, by recalculating the leaf end transmission (figure 7–12) with the narrow beam transmission reduced by 0.1%, the effective offset is reduced by 0.06 mm for an individual leaf. The effective gap offset would change by twice this amount, 0.12 mm, similar to the change in the gap offset for room 442 shown in figure 7–14. As

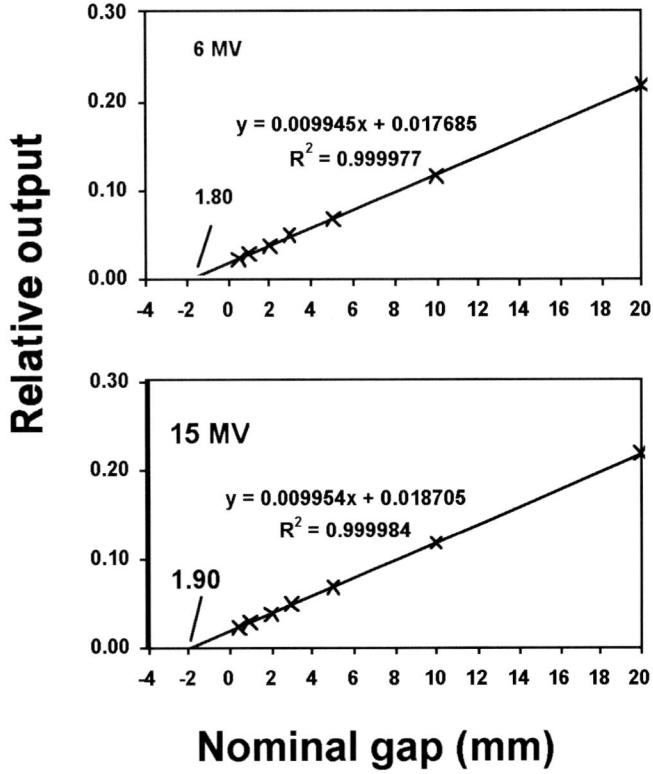

FIGURE 7–13. Output vs. DMLC gap width. T x-intercept is the effective gap offset due to the round leaf ends.

mentioned earlier, this magnitude of correction is accounted for in the MLC calibration file, MLCXCAL. Such corrections compensate for the reduced MLC leakage for both the Mark I MLC in room 445 and, after the data in figure 7–14 were acquired, for the Mark II in room 442.

Head Scatter

Dating back to our earliest 3DCRT planning, with custom blocking in the form of metal alloy blocks, head scatter has always been calculated by a backprojection to a source function. The radiation source was modeled as follows: a primary-beam component (93%) represented by a delta function, plus a scatter component (7%) represented by a conically shaped distribution with a diameter of 8 cm in the source plane. The in-air fluence at each point in a radiation field is obtained by integrating over the source distribution through the MLC delimited portal. A more realistic shape was obtained as our measurements of head scatter were extended down to very small fields, using combinations of film and ion chamber measurements (see figure 7–15),) using a special mini-phantom using a lead-film sandwich for the smallest fields.[5]

---

[5]   It has been reported that high-Z buildup materials and their areal densities affect the variation of output factor with field size for cylindrical chambers (Jursinic and Thomadsen 1999). These same data indicate that high-Z buildup actually benefits such measurements as it reduces the electron contamination reaching the detector and, therefore, responds more like a mini-columnar phantom. The lead/film sandwich used here behaves

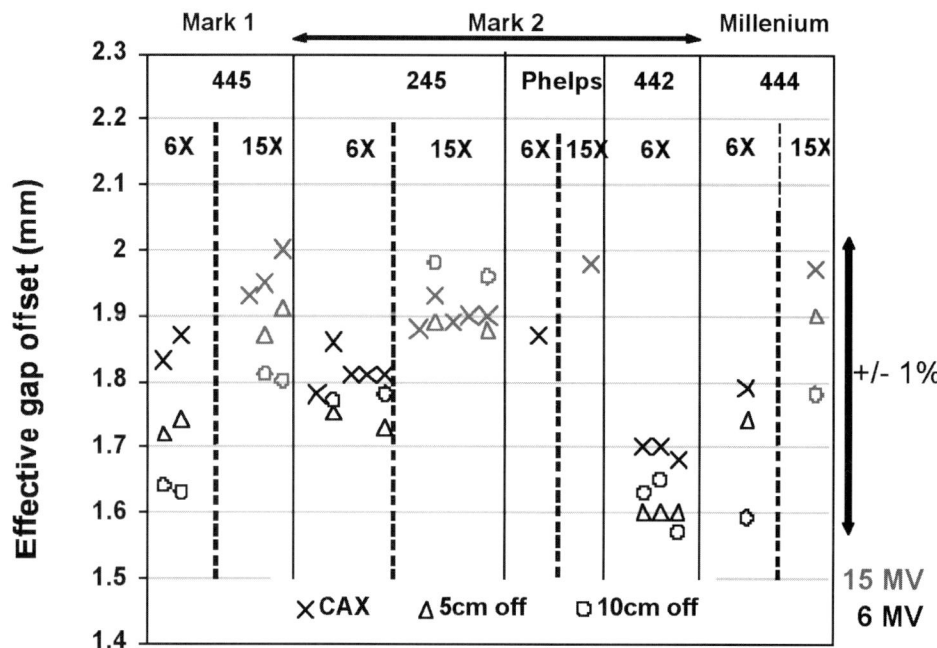

FIGURE 7–14. Comparison of measured effective offsets for five MLCs at the central axis and at 5 and 10 cm off-axis for 6 and 15 MV X-rays.

It is interesting to note from the data in figure 7–13 that the role of head scatter in determining radiation output, for uniform IMRT fields, even when generated with very narrow slits, is comparable to that for static fields with the same overall dimensions. This supports our proposition in earlier papers (LoSasso, Chui, and Ling 1998; Wang et al. 1996) that the contribution of head scatter within an IMRT field, when there is little intensity modulation within the field, is insensitive to the MLC gap width. On the other hand, as the range of modulation increases within a field, the shape of the source distribution used in the dose calculations becomes increasingly important. An accurate description of this distribution, including that of the primary source (presently approximated as a delta function), will be important for accurate dose distribution in high gradient regions of a DMLC field (e.g., at the boundary between the planning target volume (PTV) and a critical structure). This is because such high gradients are usually achieved by the use of very narrow gaps in the leaf sequence, and the dose distribution within the small window is sensitive to the source distribution. This situation may be exacerbated by the introduction of the new profile smoothing algorithm (Spirou et al. 2001) which will preserve high-gradient regions more than the previous intensity smoothing method and allow more penumbra sharpening than before (penumbra sharpening requires the use of very narrow gaps in the leaf sequence). Ongoing studies to reevaluate the source and pencil beam distributions, particularly in the first cm radius, are key to some of the remaining discrepancies between calculations and measurements (see chapter 4 on dose calculation).

---

similarly as our own measurements confirm for larger fields. Thus, we conclude that the uncertainties in our output measurements are less than 1% for 6 and 15 MV X-rays.

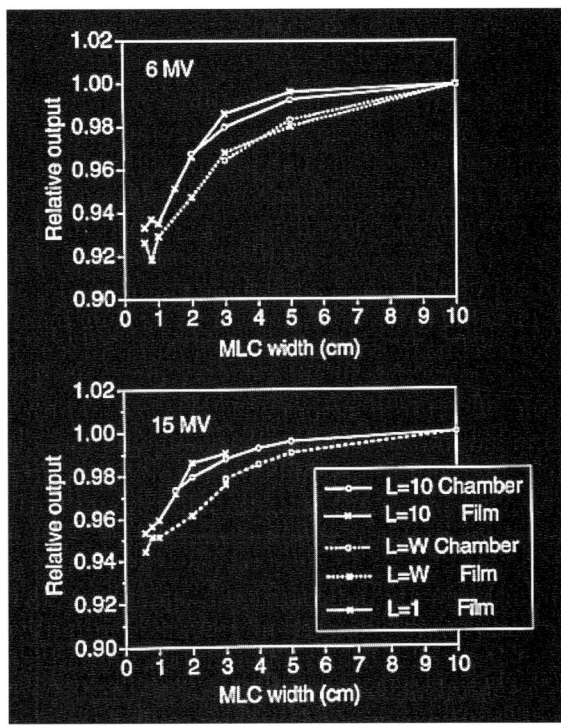

FIGURE 7–15. Head factor measurements using combination of ion chamber and film.

References

Arnfield, M. R., J. V. Siebers, J. O. Kim, Q. Wu, P. J. Keall, and R. Mohan. (2000). "A method for determining multileaf collimator transmission and scatter for dynamic intensity modulated radiotherapy." *Med. Phys.* 27:2231–2241.

American Association of Physicists in Medicine (AAPM) Report No. 72. Basic Applications of Multileaf Collimators. AAPM Radiation Therapy Committee Task Group #50. Madison, WI: Medical Physics Publishing, 2001.

Chen, Y., A. L. Boyer, and C.-M. Ma. (2000). "Calculation of x-ray transmission through a multileaf collimator." *Med. Phys.* 27:1717–1726.

Jursinic, P. A., and B. R. Thomadsen. (1999). "Measurements of head-scatter factors with cylindrical build-up caps and columnar miniphantoms." *Med. Phys.* 26:512–517.

LoSasso, T., C. S. Chui, and C. C. Ling. (1998). "Physical and dosimetric aspects of a multileaf collimation system used in the dynamic mode for implementing intensity modulated radiotherapy." *Med. Phys.* 25:1919–1927.

LoSasso, T., C. S. Chui, C. C. Ling. (2001). "Comprehensive quality assurance for the delivery of intensity modulated radiotherapy with a multileaf collimator used in the dynamic mode." *Med. Phys.* 28:2209–2219.

Podgorsak, M. B., S. S. Kubsad, and B. R. Paliwal. (1993). "Dosimetry of large wedged high-energy photon beams." *Med. Phys.* 20:369–373.

Popescu, A., K. Lai, K. Singer, and M. Phillips. (1999). "Wedge factor dependence upon depth, field size, and nominal distance—A general computational rule." *Med. Phys.* 26:541–549.

Spirou, S. V., N. Fournier-Bidoz, J. Yang, C. S., Chui, and C. C. Ling. (2001). "Smoothing intensity-modulated beam profiles to improve the efficacy of delivery." *Med. Phys.* 28:2105–2112.

van Santvoort, J. P. C., and B. J. M. Heijmen. (1996). "Dynamic multileaf collimation without 'tongue and groove' underdosage effects." *Phys. Med. Biol.* 41:2091–2105.

Wang, X., S. Spirou, T. LoSasso, J. Stein, C. S. Chui, and R. Mohan. (1996). "Dosimetric variation of intensity-modulated fields." *Med. Phys.* 23:317–327.

Webb, S., T. Bortfeld, J. Stein, and D. Convery. (1997). "The effect of stair-step leaf transmission on the 'tongue and groove problem' in radiotherapy with a multileaf collimator." *Phys. Med. Biol.* 42:595–602.

8

Quality Assurance
Of IMRT

Thomas J. LoSasso

Introduction

Certain components in the IMRT process are more obscure to the user compared to conventional 3DCRT processes. Hardware and software are still relatively new to most multileaf collimator (MLC) users, and the magnitudes and frequencies of potential dosimetric errors are not well understood. The relationship between monitor unit (MU) setting and radiation dose for intensity-modulated (IM) beams is much more complex than for non-IM fields. The leaf sequence computer files, which define the MLC leaf positions as a function of MU, are large and do not lend themselves to simple manual verification. The verification port films, usually obtained with the MLC set at the extreme leaf positions for each IMRT field, outline the entire irradiated area, but do not verify the intensity modulation patterns. Given that the present MLCs were not originally intended for used in the dynamic mode, IMRT places mechanical demands on MLC beyond their design specifications. Mechanical tolerances are significantly tighter for dynamic multileaf collimation (DMLC) than for static MLC treatments; small errors in the gap between opposing leaves may lead to a significant dose error in IMRT fields (LoSasso, Chui, and Ling 1998; Budgell et al. 2000). Nevertheless, we conclude that the present DMLC hardware and software are effective for routine clinical implementation, provided that a carefully designed routine quality assurance (QA) procedure, in addition to what is currently performed for MLC in static mode, is followed to assure the normality of operation.

This chapter describes the current status of the QA program for DMLC at Memorial Sloan-Kettering Cancer Center (MSKCC). We distinguish between the routine QA of DMLC performance (machine QA), and the verification of IM fields using DMLC (patient-specific QA). Since our initial DMLC implementation in 1995, there have been a number of incremental improvements to our overall QA program. These include methods of mechanical and dosimetric measurements, and software tools developed either by us or by the manufacturer. Dosimetric evaluation of the entire field of each IM beam for each patient, using standard verification tools (i.e., ion chambers and film), would have overtaxed our medical physics resources and was judged not to be necessary. However, the complexity of the IMRT relative to 3DCRT with static MLC fields required a reevaluation of current methodology of treatment verification. Our current approach integrates existing methods, combining periodic QA and computer verification, to provide the necessary quality assurance in a safe and efficient manner. This QA process has evolved with time; it is likely that this program will be further refined in the future.

Routine MLC QA

DMLC applications are much more sensitive to the operation of the MLC than conventional static treatments for two reasons. First, the accuracy of the mechanical settings of the MLC is critical to the accuracy of delivered dose, especially since DMLC appears to induce more wear on specific components of the MLC hardware. In particular, we have stressed the importance of the accuracy of the gap width between opposing pairs of MLC leaves in the previous chapter. Second, the MLC control system supplied by the manufacturer, designed for static MLC fields, does not respond adequately when the more restrictive clinical DMLC tolerances are exceeded. Therefore, we provide supplemental safeguards with a set of QA procedures, which detect subtle performance deterioration of the MLC that can be rectified before the mechanical imprecision becomes clinically significant.

Prior to IMRT, our routine quality assurance testing for the MLC had been to examine the position of the light field edge defined by the leaves on graph paper. This simple test, commonly used to check calibration of the jaw collimators, is consistent with the sole function of the MLC for static fields; that is, to shape the field edges with a modest degree of accuracy. Additional procedures, most of which were developed at MSKCC, extend the intent of Task Group 40 (TG 40) (Kutcher et al. 1994) to DMLC application. Our procedures for monitoring DMLC performance have been used since 1995 (Chui, Spirou, and LoSasso 1996), but modified multiple times based on our experience as to what QA is needed to provide safe and accurate dose delivery (LoSasso, Chui, and Ling 1998, 2001; Wang et al. 1996). Modifications are often based upon a better understanding of the hardware and software design of the MLC and an analysis of the frequency and severity of malfunctions.

The QA tests include semi-weekly film exposure performed by therapists and monthly measurements performed by physicists using DMLC sequences to confirm the leaf positioning/gap accuracy. These measurements can also track the long-term stability of DMLC performance. These tests relate to the physical gaps between opposing pairs of leaves. For conventional static treatments, errors in field size of 1 to 2 mm affect treatment accuracy only near the field boundaries. For DMLC, the dose delivered throughout the field is directly proportional to the leaf gap. For 1 to 2 cm average gap width, comparable errors in leaf position can translate into significant leaf gap errors, resulting in delivered dose errors of greater than 10%. Thus, calibration of the MLC is critical for IMRT treatments, and leaf (and carriage) movements need to be controlled precisely. While 1 to 2 mm accuracy is acceptable to simply shape the borders of static fields, we prescribe a much tighter tolerance of ~0.2 mm in gap width (~1% dose tolerance) at isocenter, because of its importance for accurate DMLC execution.

We have analyzed our QA data in an attempt to find patterns of MLC failure, identifying those areas that need to be monitored closely to ensure that the planned dose is delivered accurately and reproducibly. Since 1995 two of our treatment machines (Clinac-2100s in rooms 245 and 445) have been used almost exclusively for IMRT prostate patients. Since 1998 about half of all treatments on a third accelerator have been for IMRT head and neck and breast patients. Each of these three machines had been used for static MLC treatments for 3 years prior to their use for IMRT. Quality assurance records of these machines reveal interesting patterns of MLC failure that are likely related to increased wear and tear caused by IMRT operation.

Leaf Positions

One concern in DMLC delivery, error in leaf position, appears to be related to the amount of usage of individual leaf motors. Figures 8–1 and 8–2 graphically depict the record of MLC leaf motor failures on our first three DMLC machines. (Since 1995 approximately 80% of patients—~30 patients/day, five fields/patient—on two machines, rooms 245 and 445, have been IMRT prostate patients; these MLCs have been used for static MLC treatments since 1992. In figure 8–1, shading denotes motors that have been replaced at least once; the numbers indicate multiple incidences for specific motors. The solid lines in figure 8–2 graph the chronology of motor replacements. Also indicated are the approximate dates that the respective static MLC and DMLC usage were initiated. Beginning in 1995 and continuing through 1998, leaf position errors were increasingly detected during QA procedures and patient treatments. Initially, we solved these problems by increasing the frequency of MLC re-initialization during the treatment day. As we gained experience, however, we concluded that marginally performing motors needed to be replaced as part of the QA program. With the improvement of the DMLC QA and fault detection programs discussed later in this section, we have been replacing leaf drive motors at a steady rate since 1998, as shown in figure 8–2. Based on our QA records we have estimated what the

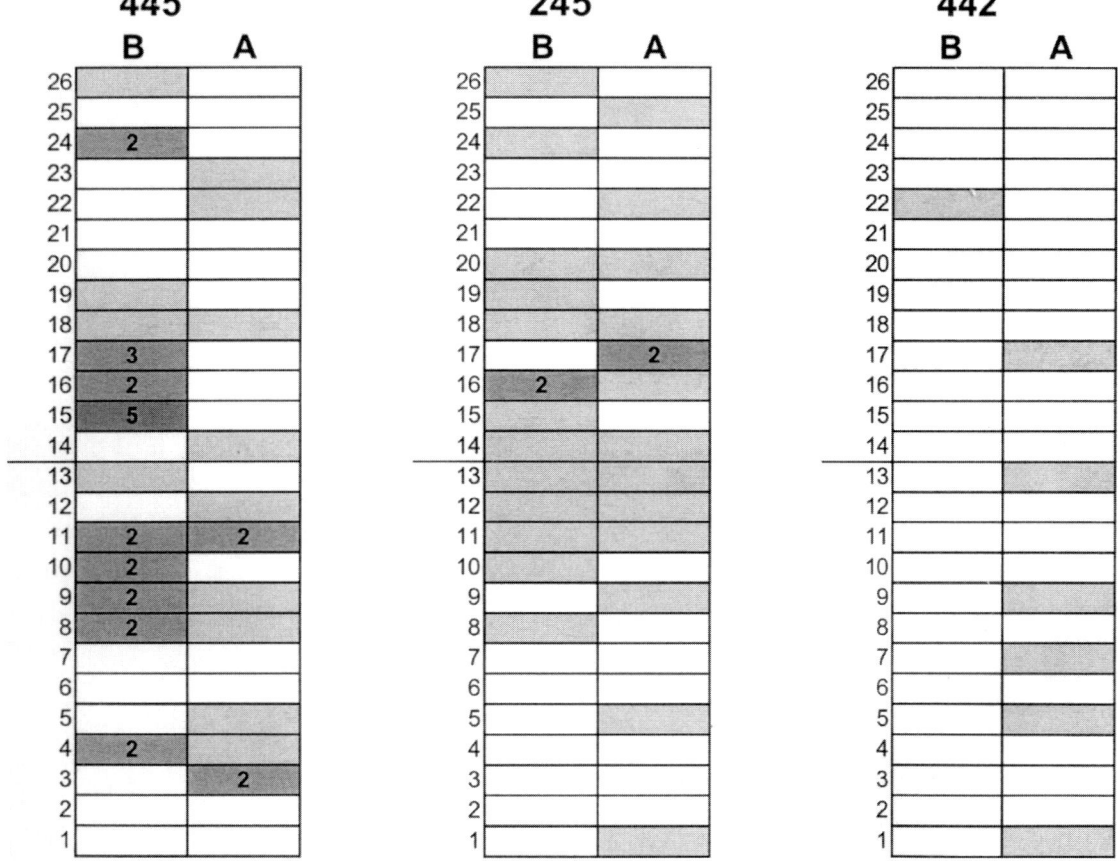

FIGURE 8–1. Leaf positions where motors have been replaced as of mid-year 2000 are shown in different shading for three MLC. Numbers indicate multiple motor replacements.

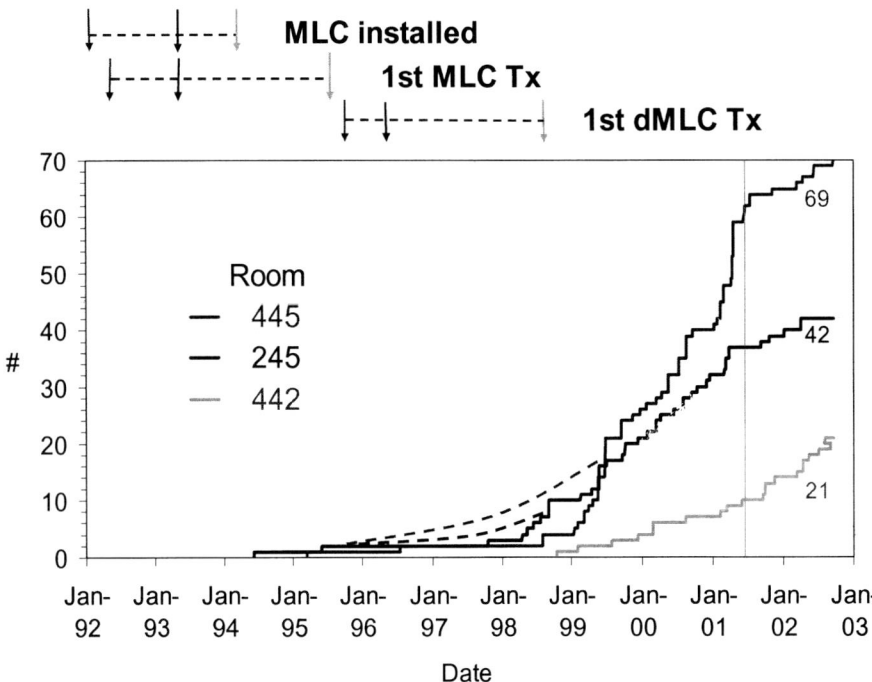

FIGURE 8–2. Cumulative motor replacements vs. time for three MLCs. The solid curves are the actual replacements. The dashed curves indicate that some motors would have been replaced earlier based upon the replacement criteria adopted in 1998. The vertical dotted line indicates the introduction to the 0.5 mm minimum gap criterion for moving leaves.

MLC motor replacement rate would have been had we initiated this prophylactic motor replacement policy in 1995. These are shown as dashed lines in figure 8–2.

Note also in figure 8–2 that for each MLC, motor failure became more frequent after the initiation of DMLC treatment. A corroborating observation is that motor failures are mostly located in the central portion of the MLC, consistent with the fact that only the central 10 leaf pairs are use for the prostate treatment, the prevalent disease site treated with IMRT in rooms 245 and 445. It is postulated that the repeated abutment of opposing leaves especially during IMRT delivery caused many of these leaf motors to fail prematurely. New software developed by the manufacturer, with a 0.5 mm minimum gap criterion, was installed in June 2001, indicated by the vertical dashed line in figure 8–2. While it is too early to be definitive, the rate of motor replacement appears to have decreased at about the same time.

One source of leaf position error stems from the count losses by the primary encoder, an integral part of the motor assembly, which become more severe during the course of the treatment day, assuming the MLC is initialized in the morning. During *initialization* of the MLC, which should be performed at least once daily (in the morning during machine warm-up as suggested by the manufacturer) (Varian 1999), the leaf positions are automatically calibrated by an optical system. However, during the day one or more of the MLC leaves may drift out of calibration, which could be due to long-term gradual performance degradation of individual motors that have been used extensively. On occasion that the count losses of a primary encoder become excessive, leaf position errors could exceed 0.5 mm at isocenter. Re-initializing the MLC will temporarily alleviate the problem, but position errors may go unnoticed since the secondary position interlock is triggered only when the discrepancy between the primary and secondary readouts reaches ~2 mm at isocenter. To safeguard against such errors, potential encoder problems are identified using a semi-weekly film test described below; the questionable motors are then replaced at the earliest convenience. This condition can exist intermittently for 2 to 3 days before detection.

We use radiographic film and a reference leaf sequencing file to produce the *picket-fence* radiation pattern depicted in figure 8–3, which, with a quick visual inspection, can discern leaf position errors of 0.2 mm and greater (LoSasso, Chui, and Ling 1998, 2001). This test is performed twice weekly, usually towards the end of the treatment day, and takes ~5 min (inclusive of film processing and a visual evaluation). Briefly, Kodak V2 Ready Pack film is placed on the treatment table and exposed to a DMLC field that produces a matrix of high intensity regions, 1 mm wide and spaced 2 cm apart. The lowest energy, typically 6 MV X-rays, is used without buildup to obtain a sharp image. Physicists evaluate these films exposed by therapists; a leaf out of position

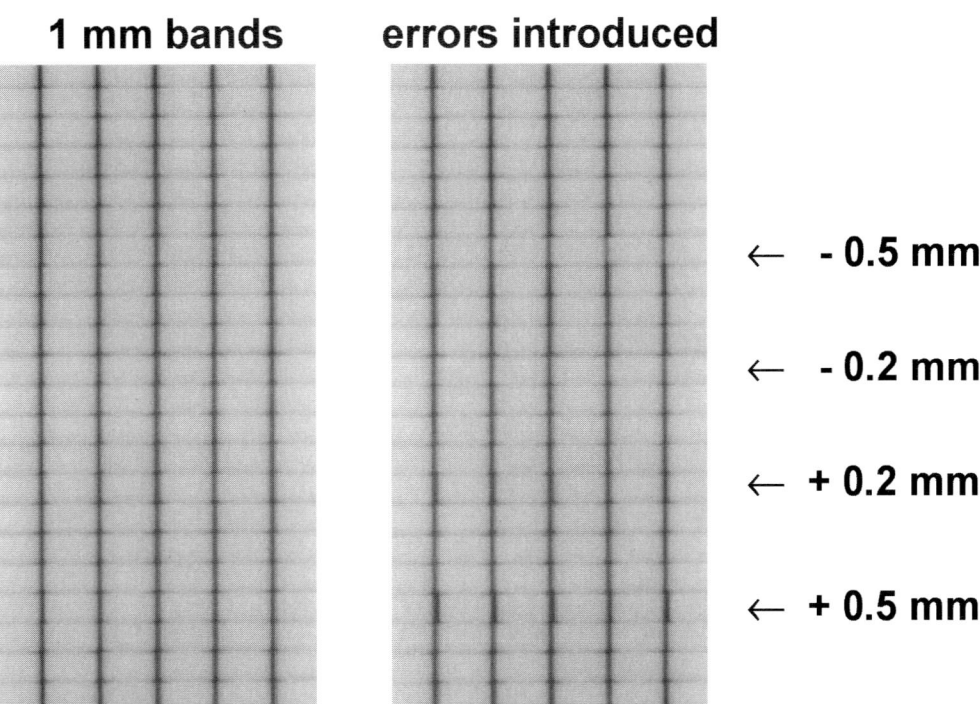

FIGURE 8–3. Semi-weekly film test. The left image showing the leaves in alignment indicates no degradation of the calibrations provided during initialization. Leaf position errors were intentionally introduced for the image on the right.

by greater than ~0.2 mm on successive films will have its motor replaced. We developed a more objective evaluation using a film digitizer, associated graphical output, and analysis, but find it to be unnecessary since variations as small as 0.2 mm in the gap width of a leaf pair can be detected visually.

The film test described above permits visual inspection of each MLC leaf individually relative to the alignment of the other leaves. However, we recognize the need for more quantitative standards and greater precision in order to track long-term changes in MLC alignment. Given that the fluence (and consequently the dose) through a narrow gap is critically dependent on gap width, dosimetric measurements can be used as a more sensitive monitor of gap width for radiation delivery with DMLC (Wang et al. 1996; LoSasso, Chui, and Ling 1998; Arnfield et al. 2000). For example, an output variation of 1%, which is easily measured, corresponds to a variation of ~0.07 mm for a 5.0 mm wide gap (with the round edge transmission factored in). Clinically, the same gap error would correspond to a clinical dose error of ~0.3% for a typical IM field with a gap width of 2.0 cm. Based upon this relationship, we developed a dosimetric procedure (figure 8–4) utilizing an ionization chamber and a 5 mm sliding window beam delivery to monitor the stability of DMLC output. Subsequent to our monthly x-ray beam output calibration, ion chamber readings for the narrow DMLC field are normalized to that measured in the static beam calibration field, using the same setup geometry to avoid uncertainties arising from changes in monitor chamber calibration, temperature, pressure, and setup accuracy. This test is performed before and after MLC re-initialization to observe changes due to encoder calibration.

The long-term results of this test for several of our treatment machines are shown in figure 8–5. Here we plot the ratio of the DMLC output (after re-initialization) to static field output versus time over a 4-year period from 1998–2002. The variation in radiation output during this period is <1% for all six machines, and is less than that corresponding to the 0.2 mm tolerance on gap width (indicated by the vertical arrow marked 0.2 mm). Given the <1% output variation for the gap width of 5 mm used for the reference DMLC field, the variation would be even less (<0.3%) for typical clinical fields with a gap width of ~2 cm. Prior to this period, larger variations were observed for the MLC in room 445; this was eventually attributed to excessive carriage movement

DMLC Output (Gap width) Check

15 MV x-rays, File: DMLC_QA2.dva, Dose = 100 MU, Dose rate = 240 MU/min
SAD: 100cm, Depth = 3.1cm, Backup = 10cm, C.S. = 10x10cm2
Tolerance: (The ratios should be within 3% (< 0.2 mm gap) of the annual calibration.)

**Measure the output for the scanned field before and after reinitializing the MLC.
Measure the output for a 10x10cm2 reference field size defined by the jaws (MLC retracted). Calculate the ratio of the scanned field to the reference field outputs.**

	Reference 10x10	DMLC - before reinitialization	DMLC - after reinitialization	Reference 10x10	
Average					Annual
Ratio					

FIGURE 8–4. Dosimetric test form to compare the variation of output at isocenter for a 5 mm wide dynamic slit and a reference 10 × 10 field. The measurements are performed monthly before and after reinitializing the MLC, and the ratios of the dynamic to the reference field are compared with the ratio determined at annual calibration.

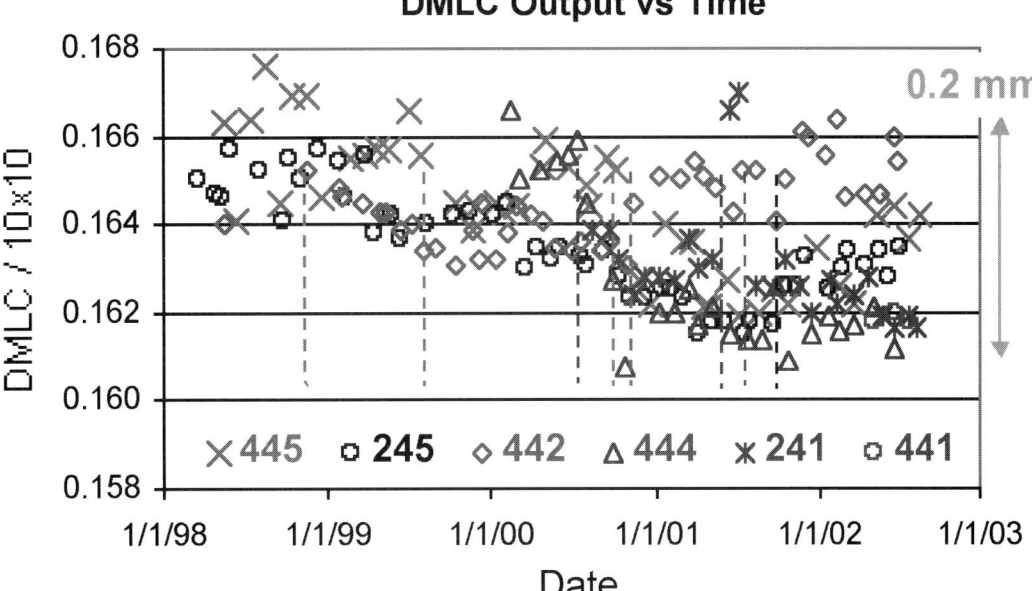

FIGURE 8–5. DMLC output stability over time. The ratios of the dynamic field to the reference field are plotted for six MLCs. The 0.2 mm range in gap width corresponds to a 3% change in the ratio. The dashed lines indicate changes to the calibration parameters in the MLCXCAL files.

and led to the replacement of the carriage bearings at the end of 1998. The dashed lines in figure 8–5 indicate adjustments to the calibration parameters in the MLCXCAL files (MLC leaf position calibration), which led to small changes in the DMLC outputs.

Another cause of leaf position error is motor fatigue. After a long period of use, a leaf may appear to move more slowly than other leaves and be unable to maintain its rated leaf speed of 3 cm/sec. Such behavior is visually obvious when the leaves are extended or retracted, and so a specific QA test is not required. Such motors must be replaced at the earliest convenience; otherwise, dose delivery errors and excessive beam holdoffs will ensue. If left uncorrected, such a motor will eventually result in a stuck leaf, perhaps during patient treatment. Motor fatigue sometimes occurs excessively at certain leaf positions, suggesting that mechanical stress for that leaf may exist in the leaf assembly (perhaps due to defective hardware). This was most apparent for motor B15 of the MLC in room 445 (figure 8–1), which had been replaced five times before the leaf assembly itself was replaced. Subsequently, the problem has not recurred at that position.

Carriage Stability

We are also concerned with the stability of the MLC carriages. IMRT beam delivery can be adversely affected by misalignments of the MLC carriages. A carriage in the accelerator head supports each bank of MLC leaves (i.e., left and right). Ideally, these two banks should be aligned parallel to each other, but alignment is affected by gravity, which causes sag and backlash in the MLC carriage supports and bearings. We have found that DMLC operation depends on gantry and collimator angles; the loads imposed by the carriages produce a mechanically and dosimetrically measurable variation in the gap width as the gantry and collimator is rotated. Our standard monthly test for this purpose (figure 8–6) is to position a cylindrical ion chamber (with appropriate buildup) at the isocenter, and to deliver a fixed dose using the 5 mm sliding window field described

Output vs Gantry and Collimator Angles

15 MV x-rays, File: DMLC_QA2.dva, Dose = 100 MU, Dose rate = 240 MU/min
SAD: 100cm, Spokas in air w/buildup, C.S. = 10x10cm2
Tolerance: (The ratios should be within 3% (< 0.2 mm gap) of each other.)

Position a cylindrical chamber in air at isocenter.
For each gantry angle, measure the open field and the scanned field outputs.
At gantry angles of 90° and 270°, also measure output for collimator angles of 90°.
Calculate the ratio of the scanned field to the reference field outputs.

Gantry	Collimator angle				
	Reference	DMLC		DMLC / Reference	
angle	0°	0°	90°	0°	90°
0°					
90°					
270°					
180°					
0°					

FIGURE 8–6. Dosimetric test form to compare the variation of output at isocenter for a 5 mm wide dynamic slit and a reference 10 × 10 field for six gantry/collimator angle combinations. The measurements are performed monthly, and the ratios of the dynamic to the reference field for a particular MLC are compared with each other.

above at different gantry and collimator angles. We graph the time trends of dose output for six combinations of gantry and collimator angles for four MLCs in figure 8–7. It is apparent that each MLC has a distinctive pattern. We surmised that the largest variation in dose output would be at the 90° and 270° gantry angles due to the influence of gravity; however, this is not always the case, as the two Millennium™ MLC in the figure and other Mark II and Millennium MLC (not shown) indicate. The horizontal dashed lines represent the output range that ±0.2 mm gap variation would impose. The vertical dashed lines are the adjustments to the MLCXCAL file as in figure 8–5. In figure 8–8, the outputs for one MLC are normalized to the output with gantry and collimator angles of 0° (the orientation used for the calibration data in figure 8–5. For this narrow test field relative to a static field, variations in output between different angles are observed to be as large as ±3% (corresponding to ±0.2 mm variation in gap width). This 0.2 mm gap variation, in turn, corresponds to ±1% dose variation for typical clinical fields with an average gap width of 2 cm. These graphs show that for any specific set of gantry and collimator angles the clinical output is stable over time to within about 1%. Furthermore, since daily treatments are generally delivered with multiple gantry angles, the deviation due to carriage instability is usually much less than 1%. Note that abrupt changes in the DMLC output seen in figure 8–8 are due to mechanical repairs: replacement of carriage bearings (indicated by *), adjustment of carriage bearings (indicated by **), and adjustment of carriage belt tension (indicated by #). These dosimetric findings are consistent with mechanical measurements of carriage positions at different gantry angles. With the exception of the Mark I MLC (room 445), the variations of output with gantry and collimator angle have been relatively stable over this 2-year period.

The relative skewness of the banks of leaves is also checked at a number of gantry and collimator angles, though on a less frequent schedule, perhaps annually. Shifting the ion chamber off-axis and repeating the DMLC output measurement can detect variation in gap width along the breadth of the carriages. Differences due to variations in interleaf leakage should be quantified

DMLC output vs gantry and collimator angle

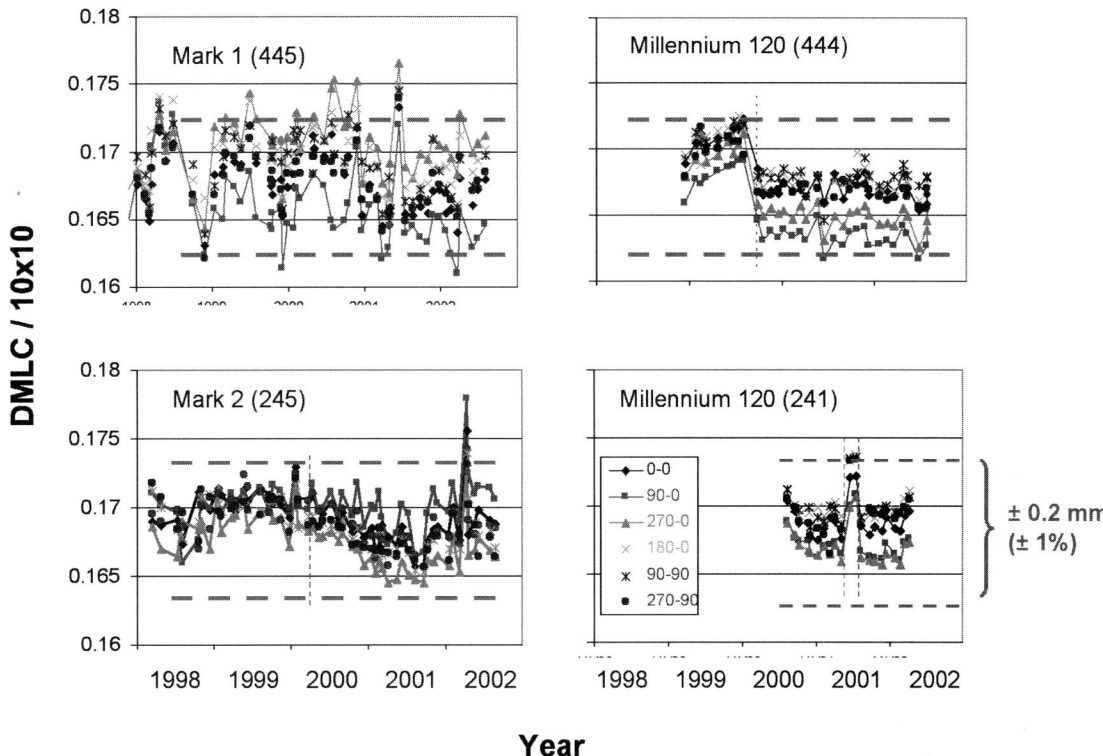

FIGURE 8–7. DMLC output stability vs. gantry and collimator angle for four MLCs over time. The DMLC outputs are normalized to the outputs for the static reference field at each angle. The dashed lines represent ±0.2 mm range in gap width (1% clinical dose variation).

with leakage measurements and removed. Alternately, a linear diode array, the Profiler (Sun Nuclear Corporation, Melbourne, FL), aligned perpendicular to the direction of leaf motion, can measure the relative DMLC output for all the leaves simultaneously. The 46 diodes are located at 0.5 cm intervals, which when mounted to the head of the machine at about 70 cm from the source, project at 0.7 cm intervals at the plane of the isocenter. Data obtained with a linear diode array for the MLC in room 245 for gantry angles of 90° and 270°, is shown in figure 8–9. The individual diode readings are normalized to the readings with both the gantry and collimator at 0°. The diodes see varying amounts of interleaf leakage, introduced by the variable interleaf transmission and the fact that the diode spacing and leaf widths are different, causing their relative response to be somewhat noisy. When the leakage is measured and removed, as in this case, residual scatter in the data points is likely due to small shifts in the detector with gantry angle and inherent variations in the diodes' responses at very low dose rates. Fitting these data points to straight lines yields the dashed lines for gantry angles of 90° and 270°. In terms of the overall trend, small variations were observed on the central axis, consistent with the ion chamber measurements, but larger variations exist at off-axis points. Apparently, the weight of the MLC induced backlash in the carriage bearings for this MLC, manifested as skewness of the leaf banks relative to the 0°-gantry position. Some backlash is unavoidable in mechanical systems. Fortunately, such

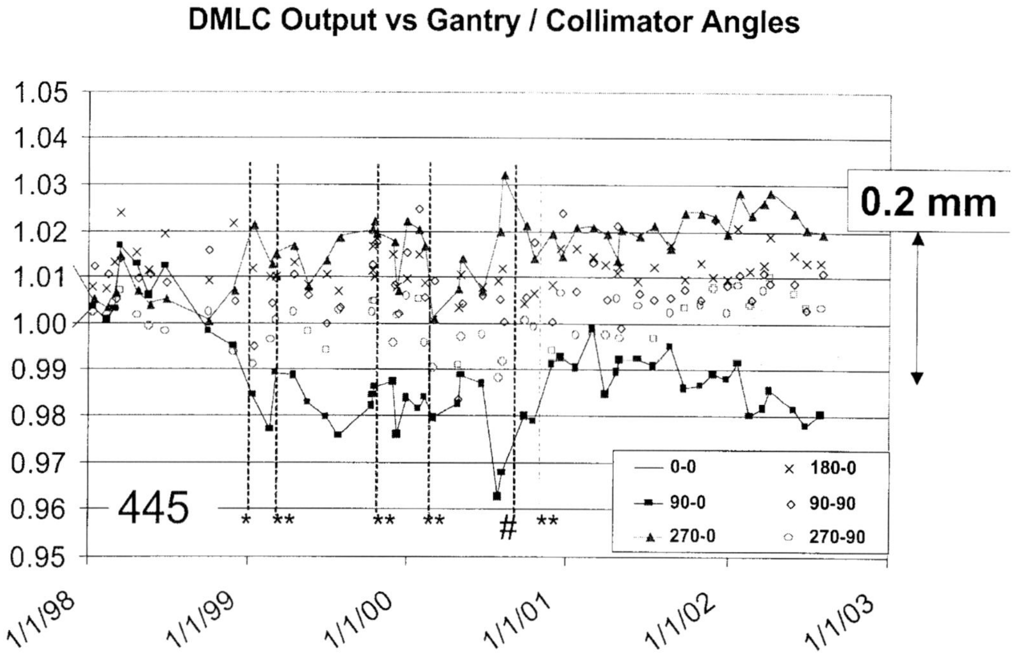

DMLC Output vs Gantry / Collimator Angles

FIGURE 8–8. DMLC output stability vs. gantry and collimator angle for one MLC. The DMLC outputs are first normalized to the static reference field at each angle and then normalized to the output with the gantry at 0°. Abrupt changes in the DMLC output correspond to mechanical repairs. *, replacement of carriage bearings; **, adjustment of carriage bearings, and # adjustment of carriage belt tension.

variations, as observed here at 90° and 270°, tend to compensate each other during treatment. It is noteworthy that the data obtained with the ion chamber and diode array have been stable during the 2-year observation period for this machine. Nevertheless, output versus gantry angle is variable among the MLC, and output changes over time may indicate mechanical problems.

Annually, we independently check the gap widths on and off the central axis by measuring the effective gap offsets at the central axis and at 5 and 10 cm off-axis and by measuring and comparing the relative dynamic dose profiles (normalized to the dose on the central axis) with open field profiles. These procedures have been described under acceptance testing in the previous chapter.

Verification Of Patient Treatment Using IMRT-DMLC

Processing of planned patient data is not failsafe, as human interventions are necessary at various stages, particularly during the transfer of the data from the planning system to the delivery system. Components of multiple versions or sequential phases of a patient's treatment plan, and even components of plans for different patients, may appear very similar and have similar leaf sequence patterns. Features are also less distinguishable since the MU and other beam parameters are calculated by the system and are not easily recognized or verified by the human operator. For these reasons a continuum of QA monitoring of patient data during the course of IMRT treatment is designed to detect human errors. Proper documentation of planned DMLC data for individual patients and pre-treatment and daily checks validate both the transfer of the planned DMLC data to the MLC control system and the delivery during every treatment. Patient-specific QA

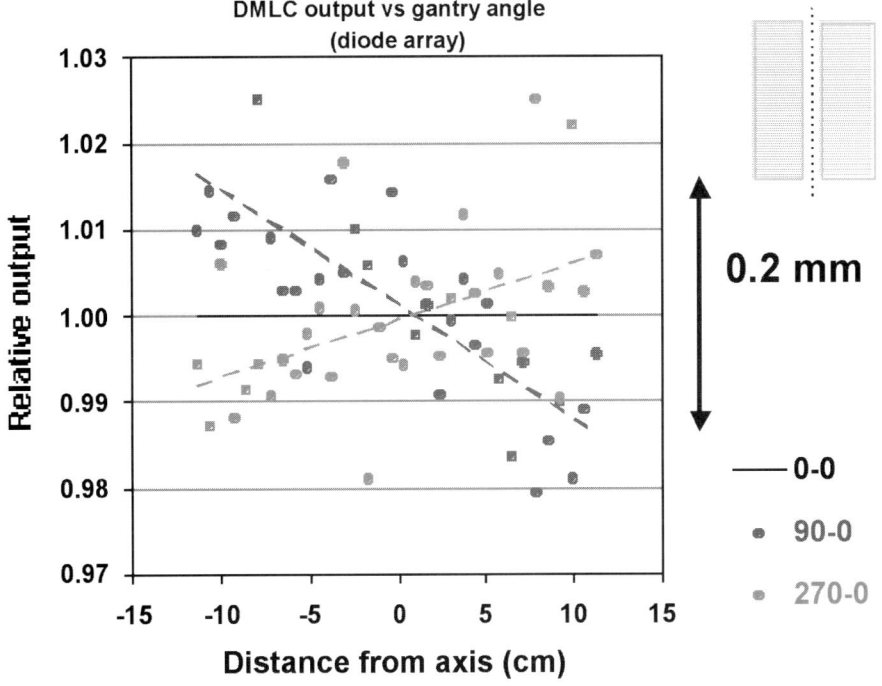

FIGURE 8–9. DMLC output stability vs. gantry angle measured with a linear diode array. The array is mounted in the blocking tray holder and aligned perpendicular to the direction of leaf motion. DMLC outputs are first normalized to the static reference field at each angle and then normalized to the output at 0°.

processes complement the machine QA described in the previous section and ensure, with a high level of confidence, the accurate delivery of the correct dose distributions for DMLC fields.

Portal Imaging

Before the first treatment session, *double exposure* portal films for all the IM fields as well as standard orthogonal images are obtained and compared visually with the corresponding digitally reconstructed radiographs (DRR), computed at the same 135 cm SSD (source-surface distance) as the portal films (Ling et al. 1996). In addition to verifying patient position alignment, portal images allow the reviewer to verify the beam aperture. For static fields the portal image defines not only the field edges, but the basic dose distribution, whether uniform or wedged, is inferred as well. Blocked regions protecting normal structures are imaged as well. For dynamic fields, portal images are MLC-shaped fields with paired leaves at the initial and final positions of the DMLC field; the superior and inferior borders are defined by the upper jaws. Thus, the portal field imaged circumscribes the target, but does not verify the protection of normal tissues, as the modulation is not imaged. All the images are displayed with a 2 cm spaced grid to facilitate spatial comparison. The development of electronic portal imaging device (EPID) technology holds significant promise for verifying dose distributions as an alternate to film dosimetry (Chang et al. 2000), but it is not clear that this will be practical with the patient anatomy superimposed.

Independent MU Check

It is a standard practice in radiotherapy to have two independent calculations of the MU for patient treatment. For static treatment fields (with either alloy blocks or MLC), independent calculations of the MU can be carried out manually; the isocenter dose is dependent on nominal beam energy, field size, and depth. For DMLC treatment, the relationship between MU and prescribed dose is more complex, with the principal variant being the duty cycle (i.e., the fraction of the time spent under the gap) to which MU is inversely related. The duty cycle typically varies from 0.2 to 0.5 for points in high-dose regions of a DMLC field, and it affects the fractional dose from radiation leakage and transmission through the round leaf ends, ~5 and 10%, respectively. Furthermore, for the consideration of output factor, the *field size* information of 200 segments per DMLC field would have to be accounted for. Consequently, manual calculation as the independent MU check is not practical.

One pre-treatment check that we initially adopted was to compare measured and calculated point doses in a flat homogeneous phantom for individual patient fields. This independent MU check was performed for the first ~400 patients treated with IMRT-DMLC. An ion chamber was placed at a reference depth (16 cm for treatment of the prostate) at isocenter and a measurement was taken for each IM field, and for a reference 10×10 cm^2 MLC field, for which the dose is known. The measured ionization is then converted to dose delivered to the isocenter in the patient [using effective pathlength and tissue-phantom ratio (TPR)], and compared to the prescribed dose. The ratios of measured and prescribed dose for approximately 380 prostate patients are shown in figure 8–10. Each point on the graph represents the ratio of the total measured to the total calculated dose for the five DMLC fields delivering the prescribed 180 cGy. The average ratio and its standard deviation for all the patients are 0.993±0.8%. When the ratio of the total dose exceeded 2%, or the ratio of an individual field exceeded 5%, further investigation including film dosimetry was conducted. Usually the discrepancy can be attributed to tongue-and-groove effects (van Sandvoort and Heijmen 1996; Webb et al. 1997) and/or setup uncertainties when the isocenter was located in a region of high dose gradients. Note that the finite size as well as the exact placement of the ion chamber may influence the verification of the isocenter dose of a DMLC

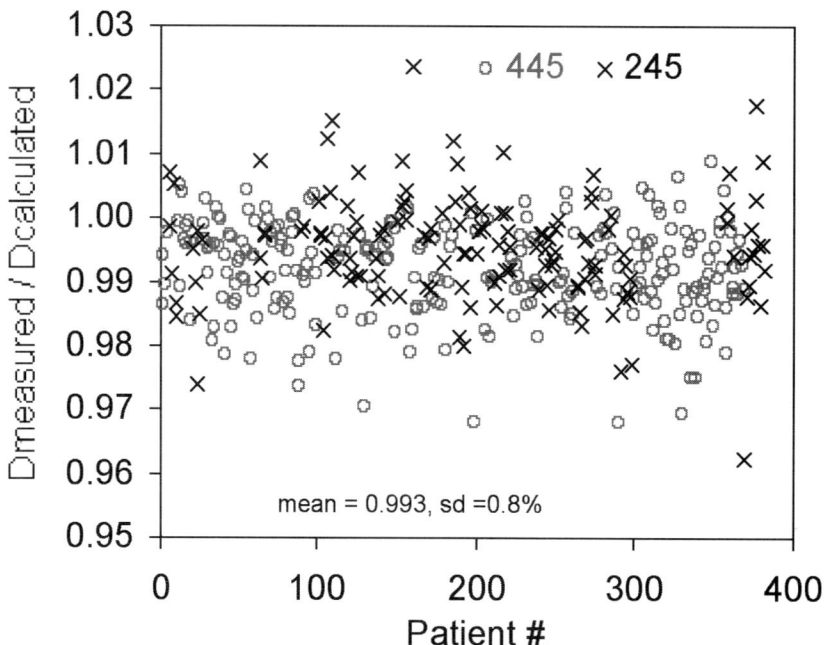

FIGURE 8–10. Ratios of measured dose to calculated dose at isocenter for ~380 prostate patients receiving IMRT treatments.

field, if it integrates over an area of non-uniform dose distribution. This problem is exacerbated when IMRT is implemented for the treatment of cancers of the head and neck, where the degree of intensity modulation is greater than that for the treatment of cancer of the prostate. In part for this reason, and because of the increasing number of IMRT patients, we developed a more efficient strategy to verify MUs.

Since January 1999 an independent MU calculation program has been in use. Persons not involved with the original treatment planning dose calculation program devised these backup calculations using the leaf sequence files, modified by a fixed leaf position offset to correct for the round edge, as input to a pencil beam algorithm. This computer program is now our principal method for independent MU calculation. Other investigators have reached similar conclusions as they developed their own independent dose calculations for MU verification (Kung, Chen, and Kuchnir 2000; Xing et al. 2000). Both these methods differentiate a fixed primary region near the point of calculation and a Clarkson summation for the more distal regions. This approach is well suited to the broad beamlets encountered with step-and-shoot delivery; however, much greater resolution is required for the sliding window delivery to accurately simulate the *primary* component. For this reason we use a pencil beam algorithm for the independent calculation. Details of this independent calculation are presented in chapter 5.

Film Dosimetry

Dosimetry at a single point verifies the dose in fixed fields with or without standard wedges, but it is insufficient for IMRT dose delivery verification. Since the intensity variation within IMRT fields is potentially different from field to field, one cannot presume to know the dose at an arbitrary point based upon the dose at any other point. Therefore, dose verification in IM fields requires the acquisition of multiple points sufficient to ensure that the correct distribution is delivered with reasonable certainty. This can be accomplished with point detectors, but only very inefficiently since each individual point would require the entire treatment exclusively. Film dosimetry is a useful tool for verifying IMRT. The high resolution of radiographic film combined with modern film density digitization provides a reasonably accurate and precise two-dimensional (2-D) distribution from a single exposure.

Our primary film technique uses flat homogeneous polystyrene phantoms to irradiate radiographic film (Kodak Ready Pack V2) perpendicular to the beam for individual fields; cylindrical phantoms polystyrene and solid water (Gammex RMI, Middleton, WI) are also available for multiple field irradiation in either the axial or longitudinal planes (see figure 8–11). With the gantry at 0°, Kodak V2 (Ready Pack) films, one for each field, are irradiated at isocenter, perpendicular to the beam axis. Film depth depends upon the average isocenter depth for a treatment site (e.g., 15 cm for prostate cases). Pinholes placed in the film locate the crosshair projections, typically at 1 cm outside the projected collimator edges for the particular field. We calibrate film at 10 MU intervals with the known dose for the same energy, depth, and approximate equivalent square field size as used for the patient verification films. Certain precautions have been added during calibration and processing. For quality control, additional films are exposed to a standard dose, and interspersed among the calibration and verification films to monitor slowly varying processor-induced changes. Each film is also flashed along its edge with a sensitometer just before processing to assess, as yet unexplained, variations in film density (up to 10%) observed for a small percent (~5%) of films within the same batch. With these precautions, and corrections if appropriate, the reproducibility of the measured dose is ~1%. Commercial hardware (Lumisys Corp, Sunnyvale, CA; Agfa Arcus II, Ridgefield Park, NJ; Vidar DosimetryPRO™, Herndon, VA) and software (Radiological Imaging Technology, RIT 113, Colorado Springs, CO) are used for film scanning and digitization. Density to dose conversion and comparison between the measured and the planned dose distributions are currently handled by in-house software. After

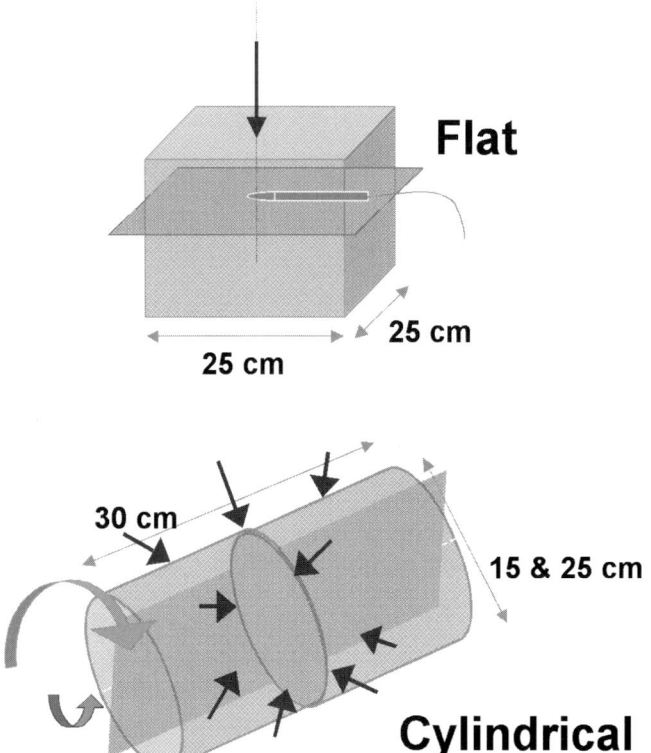

FIGURE 8–11. Homogeneous flat (single field) and cylindrical (multiple fields) phantoms for measurements with film and ion chambers.

re-computation in phantom, the calculated and the measured dose distribution are spatially registered, and 2-D overlays and dose difference displays are produced for each field.

We have used film dosimetry effectively to verify dose distributions in 2-D. With proper controls during film exposure and processing, the results obtained are generally within 2% of calculated 2-D dose distributions for small to moderate field sizes. An example of such a comparison is shown in figures 8–12a and 8–12b for a posterior field in a typical seven-field IMRT nasopharynx treatment. Figure 8–12a displays the film measurement in a flat homogeneous phantom at 10 cm depth overlaid with the dose distribution recalculated in the phantom geometry. The dose within this field varies from less than 10 cGy to the cord region up to about 30 cGy. Comparison of these dose distributions in phantom to verify the dose delivered to the patient assumes that the calculated-to-measured-dose ratios in the phantom and patient are similar in the absence of large inhomogeneities. The dose difference distribution in figure 8–12b highlights differences between the measurement and the calculation. For the most part, these differences are found in the high gradient regions at the field edges. An axial distribution for a five-field prostate and a seven-field head and neck plan in cylindrical phantoms are shown in Figure 8–13 and 8–14, respectively.

The overresponse of radiographic film to low-energy photon scatter within the phantom is a known consequence of film's high-Z sensitive layer (Burch et al. 1997; Danciu et al. 2001). When film is used for IMRT verification, energy dependency of radiographic film is uniquely problematic and is not easily predicted. While the film overresponse is proportional to the low-energy scatter component, the variation in this film response is proportional to the low-energy scatter to primary ratio. For IMRT fields there can be a large variation in the primary

Nasopharynx - PA field – coronal plane
flat phantom

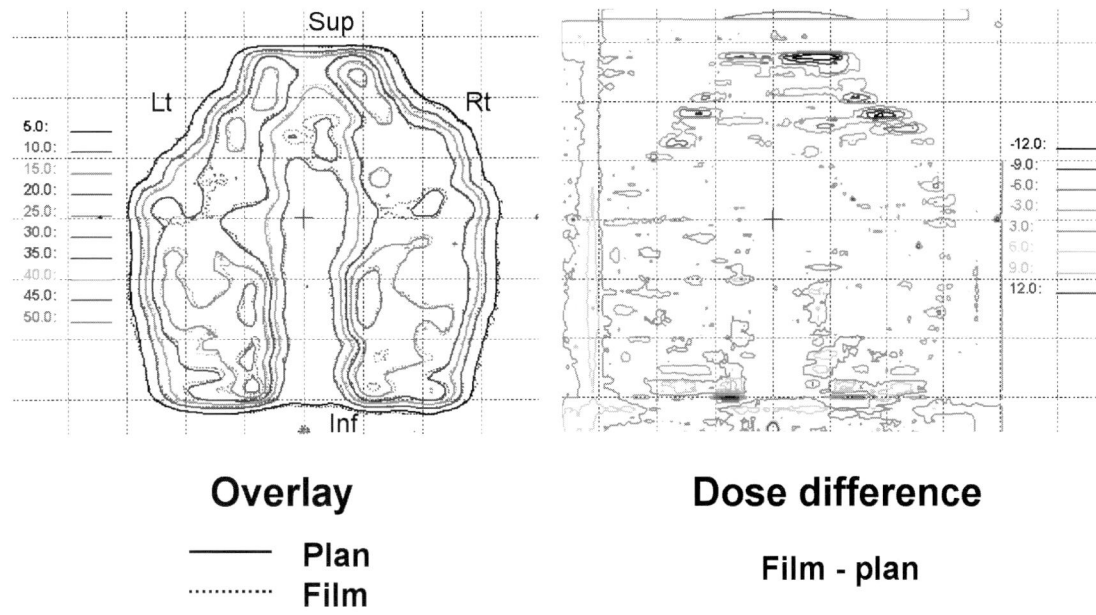

Overlay

——— Plan

············· Film

Dose difference

Film - plan

FIGURE 8–12. Film dosimetry verification in a flat phantom for a nasopharynx field. Measurements are compared with the planned dose distribution recalculated in the flat phantom. (a) overlay; (b) dose difference. See COLOR PLATE 14.

component, which varies in proportion to the degree of intensity modulation, while the low energy scatter is relatively uniform across the field. Thus, for small IMRT fields, less than 10×10 cm^2, the scatter and the overresponse are small, and so a large variation in the small overresponse is also small. For large IMRT fields, the low energy component and the overresponse are large, and so a large variation in the overresponse is also large. This complicating factor cannot be ignored under certain conditions, particularly irradiation at large field sizes and depths. Radiochromic film is manufactured to be relatively tissue equivalent avoiding the energy dependence restriction of radiographic film, but it currently suffers from insensitivity at clinical doses, nonuniformity across the film, and cost (currently ~$2 per square inch) for regular applications. For these reasons radiographic film remains popular for the verification dosimetry of IMRT. Limitations of the digitizer also add to the uncertainty of film dosimetry, particularly for steep dose gradient regions. Optical scatter artifacts associated with the scanner will be most pronounced in regions of high density (Dempsey et al. 1999).

Other Dosimetry Methods

Other verification tools have been tested with varying degrees of success. EPIDs are already found in many modern therapy facilities. Software has been developed to measure exit dose from IMRT fields and/or dose distributions with EPID, and to relate the measured data to the delivered dose in the patient (Chang et al. 2000). At present, such applications are hindered by technical difficulties in data acquisition, but these should be resolved with the new generation of detectors and software. The development of EPID holds significant promise for verifying dose

Prostate – 5-fields – axial plane
cylindrical phantom

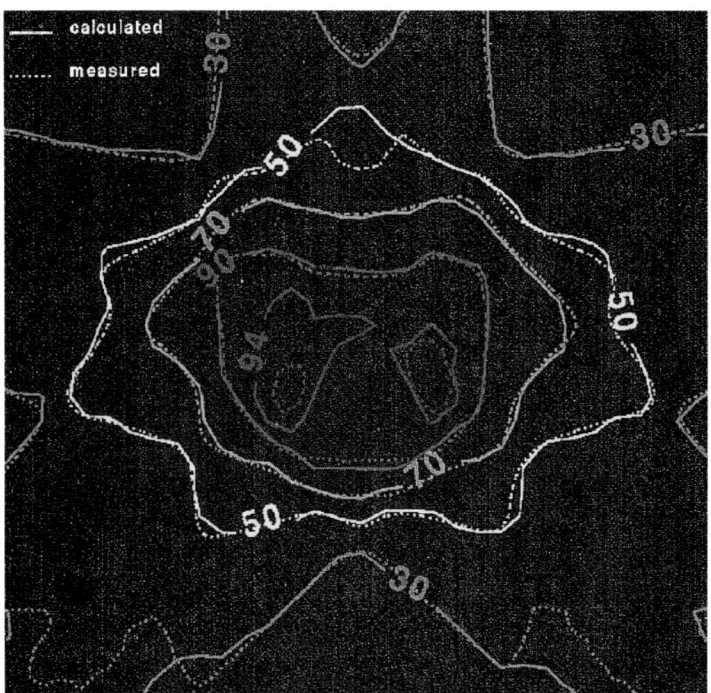

FIGURE 8–13. Film dosimetry verification in a cylindrical phantom for a five-field prostate plan. Measurements are compared with the planned dose distribution recalculated in the 25 cm diameter cylindrical phantom. See Color Plate 15.

distributions, and it may eventually provide an efficient alternative to film dosimetry for DMLC, but it is not clear that this will be practical with standard patient portal images. EPIDs have been tested for this purpose with some success, but their results are quantitatively inferior due to frame acquisition speed and/or internal scatter. Furthermore, EPIDs measure intensity patterns more similar to exit fluence, optimized for image quality, rather than patient dose.

We have found *in vivo* thermoluminescent dosimetry (TLD) to be unreliable for dose verification for DMLC fields. As is our procedure for conventional static fields, packets of three TLD chips were placed under bolus on the skin of patients, usually at the central axis. The measured doses when compared to those calculated for the same geometry occasionally deviated by as much as 10%. Similar measurements for the same DMLC fields, but with the TLD attached to the surface of a phantom, yielded better agreement of ≤3%. Thus we concluded that *in vivo* TLD was more susceptible to dosimetric error due to the combined effect of the uncertainty in the placement of the chips on the surface of a patient and the dose gradient of DMLC fields.

Linear detector arrays, (Scanditronix/Wellhöfer, Uppsala, Sweden/Schwarzenbruck, Germany; Sun Nuclear, Melbourne, FL) provide more efficient data sampling though limited to one dimension, and while beneficial for commissioning dynamic wedges (Zhu et al. 1997), do not easily accommodate IMRT fields (Kutcher et al. 1994). However, at least two companies (Sun Nuclear, Melbourne, FL and PTW, Hicksville, NY) have developed new products incorporating tissue-equivalent 2-D detector arrays specifically to address dose verification of IMRT fields. A 2-D diode array, MapCHECK (Sun Nuclear) appears to be an accurate and convenient tool for IMRT verification in recent tests at MSKCC. This array offers stability and speed for routine ver-

Nasopharynx – 7-fields – coronal plane
cylindrical phantom

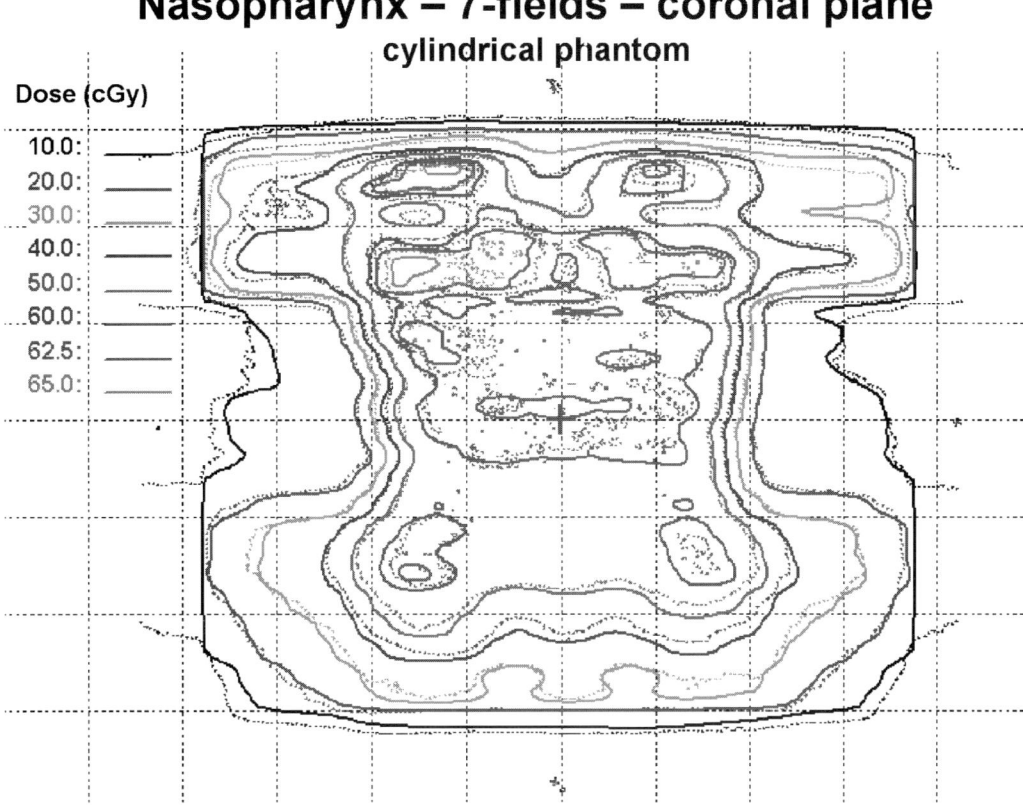

Dose (cGy)

10.0:
20.0:
30.0:
40.0:
50.0:
60.0:
62.5:
65.0:

FIGURE 8–14. Film dosimetry verification in a cylindrical phantom for a seven-field nasopharynx plan. Measurements are compared with the planned dose distribution recalculated in the cylindrical phantom. See COLOR PLATE 16.

ification; however, their resolution, dependent upon detector size and spacing, may be inadequate for the commissioning phase of highly modulated IMRT systems.

Computer-Based Verification

The routine use of dose-based verification techniques, as described above, is labor-intensive, especially if one wishes to verify the 2-D dose distributions for each IM field. Although film and ion chambers are still used periodically when commissioning new hardware and software and evaluating new anatomical sites for IMRT, we have adopted a more efficient alternate protocol for routine verification. Having established the relationship between DMLC leaf positioning and dose delivery during commissioning, combined with timely QA measurements targeting known potential mechanical problems, i.e., leaf motor failure and carriage backlash, then it should only be necessary to verify MU and leaf positions to ensure accurate dose delivery. MUs are independently calculated as described earlier. Checking that the leaf sequence files have been transferred correctly from the treatment planning system to the MLC workstation and monitoring leaf positions during delivery ensure leaf position verification. We believe that our routine protocol (figure 8–15) provides comprehensive safeguards to ensure the delivery of the correct dose distribution for DMLC fields, without resorting to routine dose-based verification.

In most cases there are several versions of an IMRT treatment plan for the same individual patient. In order to avoid erroneous or accidentally mismatched patient data, the leaf position/MU

QA – leaf positioning accuracy
- film test (semi-weekly) - leaves
- output vs gantry and collimator (monthly) - carriages

Verification – MU & leaf sequence
- independent MU check
- unique ID – plan / version
- R & V – initial leaf positions
- portal images
- leaf sequence monitor
- secondary position indicator

QA + Verification ⇒ Dose

FIGURE 8–15. Protocol for dose verification for IMRT patients.

sequences executed by the MLC are linked to the physician-approved dose distributions through a series of checks developed by MSKCC and by Varian (Varian Medical Systems, Palo Alto, CA)). To ensure that the approved version is used for treatment, the MSKCC treatment planning system automatically assigns and attaches a unique identification number to each version of the patient's data. This number is displayed on printouts and graphical plots, transferred with the leaf sequence DMLC file, and subsequently displayed with the patient name and field name at the MLC workstation. Check sums, generated within our planning system and placed within each leaf sequence file, verify the integrity of the leaf sequences at each treatment. A record and verify system developed at MSKCC *eavesdrops* on the communications between the MLC and the MLC controller, and compares initial and final leaf positions for each field with stored leaf positions, confirming that the correct data are used for treatment. Consistency between the initial values is a prerequisite for treatment. Recently we have begun to replace our record and verify system with the VARiS Treatment (Varian Medical Systems, Palo Alto, CA) software module, which provides the convenience and security in its ability to transfer treatment planning data electronically to the treatment machine control computers.

Dynamic leaf position accuracy relies on software checks provided by Varian. Since monitored leaf positions can deviate from the actual leaf positions if the primary motor encoder drifts from its calibrated state, secondary leaf position indicators provide a backup to the primary encoders, inserting an interlock when discrepancies between these readouts exceed approximately 2 mm. Primary encoders must also agree with the prescribed leaf positions in the leaf sequence file as a function of MU, also within 2 mm in our scheme (LoSasso, Chui, and Ling 1998), or the beam is held-off until the tolerance is achieved. MLC checks and interlocks are sufficiently restrictive such that undetected leaf position errors do not lead to significant dose errors when averaged over the course of treatment.

Log file → Leaf sequence file → dose distribution

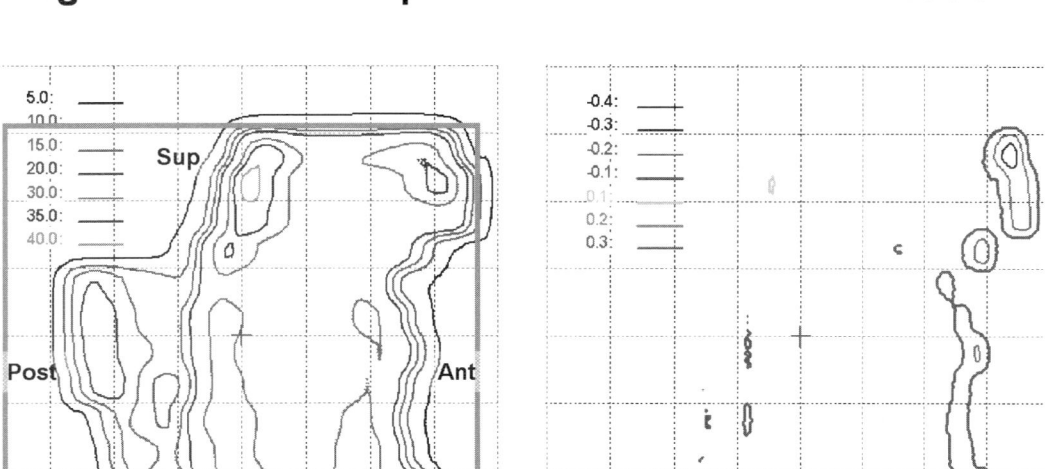

Overlay   Dose differences (log-dva)

FIGURE 8–16. Comparison of a "delivered" dose distribution derived from a Dynalog file with the planned dose distribution.

After each IM beam delivery, log files may be requested from the MLC control software. These files, which summarize the deviations of the leaves from their prescribed positions for leaves that were moving during the DMLC treatment, are useful as a verification of the proper delivery of the DMLC field. Since the spatial information in the log file is from the primary MLC encoders, neither the accuracy of the encoders' calibration nor the backlash in the leaf and carriage mechanisms is accounted for. Thus, the log files do not contain *actual* leaf position data; rather, from the perspective of the encoders, they document the deviation of the leaves from their intended positions during DMLC delivery. However, deviations of individual leaves are related only indirectly to the quantity of interest, the accuracy of the delivered dose. On average, the positional deviations of pairs of opposing leaves tend to compensate each other, resulting in an average gap error, a better indicator of DMLC performance, which is much smaller than 0.2 mm. For this reason, we convert the log files into *delivered* leaf sequence files, and from these files we calculate the delivered dose distributions (within the uncertainty due to leaf calibration and uncompensated mechanical backlash) with our treatment planning system; and we compare the delivered dose distributions with the planned dose distributions. One example is shown for a head and neck field, with the overlay of two dose distributions and their dose difference plots in figures 8–16a and 8–16b. The maximum difference ~0.3 cGy (~1% of the average dose within the field) is observed in the high gradient region. The average discrepancy is ~0.05 cGy; corresponding to average gap errors of <0.1 mm for this field. Varian has recently provided a similar feature with their Log File Viewer (Varian 2001), whereby they generate delivered intensity profiles.

Conclusions

- As a prerequisite to the clinical application of IMRT, there must be an acceptance testing and commissioning process to establish the capability for accurate IMRT, both from the treatment planning and the delivery perspectives. The combination of commissioning, routine machine (MLC) QA and patient-specific QA, are complementary to each other and provide sufficient assurance that the IMRT dose distributions are delivered as planned.
- A routine QA process must periodically check and calibrate IMRT performance, ensuring that when the treatment parameters are correctly set, the correct dose is delivered. Gap width is the critical parameter for accurate dose delivery with DMLC, and MLC position calibration and leaf motor fatigue are recognized as primary sources of gap inaccuracy. Both are systematic in nature, and can be detected by appropriately scheduled QA procedures.
- Pre-treatment checks and then continued monitoring of patient data at key points during the course of treatment have been designed to minimize the possibility of human error. This patient-specific QA process ensures that the planned DMLC data for individual patients is properly documented, and then transferred to and delivered by the MLC control system correctly.

References

Arnfield, M. R., J. V. Siebers, J. O. Kim, Q. Wu, P. J. Keall, and R. Mohan. (2000). "A method for determining multileaf collimator transmission and scatter for dynamic intensity modulated radiotherapy." *Med. Phys.* 27:2231–2241.

Budgell, G. J., J. H. L. Mott, P. C. Williams, and K. J. Brown. (2000). "Requirements for leaf position accuracy for dynamic multileaf collimation." *Phys. Med. Biol.* 45:1211–1227.

Burch, S. E., K. J. Kearfott, J. H. Trueblood, W. C. Sheils, J. L. Yeo, and C. K. C. Wang. (1997). "A new approach to film dosimetry for high energy photon beams: Lateral scatter filtering." *Med. Phys.* 24:775–783.

Chang, J., G. S. Mageras, C. S. Chui, C. C. Ling, and W. Lutz. (2000). "Relative profile and dose verification of intensity modulated radiation therapy." *Int. J. Radiat. Oncol. Biol. Phys.* 47:231–240.

Chui, C. S., S. Spirou, and T. LoSasso. (1996). "Testing of dynamic multileaf collimation." *Med. Phys.* 23:635–641.

Danciu, C., B. S. Proimos, J. C. Rosenwald, and B. J. Mijnheer. (2001). "Variation of sensitometric curves of radiographic films in high energy photon beams." *Med. Phys.* 28:966–974.

Dempsey, J. F., D. A. Low, A. Kirov, and J. F. Williamson. (1999). "Quantitative optical densitometry with scanning-laser film digitizers." *Med. Phys.* 26:1721–1731.

Kung, J. H., G. T. Y. Chen, and F. K. Kuchnir. (2000). "A monitor unit calculation in intensity modulated radiotherapy as a dosimetry quality assurance." *Med. Phys.* 27:2226–2230.

Kutcher, J. G., L. Coia, M. Gillin, W. F. Hanson, S. Leibel, R. J. Morton, J. R. Palta, J. A. Purdy, L. E. Reinstein, G. K. Svenson, et al. (1994). "Comprehensive QA for radiation oncology: Report of AAPM Radiation Therapy Committee Task Group 40." *Med. Phys.* 21:581–618

Ling, C. C., C. Burman, C. S. Chui, G. J. Kutcher, S. A. Leibel, T. LoSasso, R. M. Mohan, T. Bortfeld, L. Reinstein, S. Spirou, X. H. Wang, Q. Wu, M. Zelefsky, and Z. Fuks. (1996). "Conformal radiation treatment of prostate cancer using inversely planned intensity-modulated photon beams produced with dynamic multileaf collimation." *Int. J. Radiat. Oncol. Biol. Phys.* 35:721–730.

LoSasso, T., C. S. Chui, C. C. Ling. (1998). "Physical and dosimetric aspects of a multileaf collimation system used in the dynamic mode for implementing intensity modulated radiotherapy." *Med. Phys.* 25:1919–1927.

LoSasso, T., C. S. Chui, and C. C. Ling. (2001). "Comprehensive quality assurance for the delivery of intensity modulated radiotherapy with a multileaf collimator used in the dynamic mode." *Med. Phys.* 28:2209–2219.

van Santvoort, J. P. C., and B. J. M. Heijmen. (1996). "Dynamic multileaf collimation without 'tongue and groove' underdosage effects." *Phys. Med. Biol.* 41:2091–2105.

Varian Associates, Inc., Oncology Systems. "MLC User Guide." Palo Alto, CA: Varian, 1999.

Varian Associates, Inc., Oncology Systems. "Dynalog File Viewer, Reference Guide." Palo Alto, CA: Varian, 2001.

Wang, X., S. Spirou, T. LoSasso, J. Stein, C. S. Chui, and R. Mohan. (1996). "Dosimetric variation of intensity-modulated fields." *Med. Phys.* 23:317–327.

Webb, S., T. Bortfeld, J. Stein, and D. Convery. (1997). "The effect of stair-step leaf transmission on the 'tongue and groove problem' in radiotherapy with a multileaf collimator." *Phys. Med. Biol.* 42:595–602.

Xing, L., Y. Chen, G. Luxton, J. G. Li, and A. L. Boyer. (2000). "Monitor unit calculation for an intensity modulated photon field by a simple scatter-summation algorithm." *Phys. Med. Biol.* 45:N1–7.

Zhu, T. C., L. Ding, C. R. Liu, J. R. Palta, W. E. Simon, and J. Shi. (1997). "Performance evaluation of a diode array for enhanced dynamic wedge dosimetry." *Med. Phys.* 24:1.

9

Treatment Planning, Dose Delivery, And Outcome Of IMRT For Localized Prostate Cancer

Chandra M. Burman
Michael J. Zelefsky
Steven A. Leibel

Introduction

Dose levels in the range of 65 to 70 Gy have traditionally been used for the treatment of prostate cancer with external beam radiation therapy. However, data collected from patients treated with conventional (non-three-dimensional) techniques to ≤70 Gy demonstrated 7- to 10-year prostate-specific antigen (PSA) relapse-free survival rates of 65% in stage T1–T2 disease (Shipley et al. 1999) and 24% in more locally advanced T3 tumors (Zagars, Pollack, and Smith 1999). Higher doses have been shown to improve the outcome (Leibel et al. 1996; Hanks, Martz, and Diamond

1988; Vivini et al. 2001)). However, the ability to deliver higher dose levels with conventional techniques has been limited by a rectal and bladder complication rate of 6.9% (Leibel, Hanks, and Kramer 1984). Several clinical trials now provide evidence that the dose can be escalated with three-dimensional conformal radiation therapy (3DCRT) techniques without increasing the rate of normal tissue complications (Purdy et al. 1996; Sandler 1996; Zelefsky et al. 1995).

The prostate is surrounded by several radiosensitive critical structures, including the rectum, bladder, small and large bowel, and the femurs. The urethra is located within the prostate itself. As the treatment dose is increased, the radiation-induced damage to these tissues becomes a limiting factor. For example, Lee et al. (1996) observed that when a 1 cm margin was added to the clinical target volume (CTV) to define the planning target volume (PTV) and the dose at the target center was ≥74 Gy, the incidence of grade 2–3 rectal complications was 24%; whereas above 76 Gy, it was >30%. The patients included in that study were treated with high-energy X-rays using a four-field approach. However, it was found that by adding a lateral block to reduce the CTV to PTV margin for the final 10 Gy, the rate of rectal toxicity was significantly reduced. Thus, with the use of 3DCRT planning and dose-delivery techniques, it is possible to overcome some of these limitations and to deliver a higher dose to the PTV while keeping the dose the surrounding normal tissues within safe limits.

A dose-escalation study to determine the maximum feasible dose using 3DCRT approach was begun at the Memorial Sloan-Kettering Cancer Center (MSKCC) in October 1988 (Leibel et al. 1994; Zelefsky et al. 2001). In that study a six-field 3DCRT technique was used to treat patients to 75.6 Gy. A two-phase 3DCRT technique was used initially to administer 81 Gy. When intensity modulated radiation therapy (IMRT) became available, it was used to treat patients to 81 and, most recently, to 86.4 Gy (Ling et al. 1996; Burman et al. 1997). Here, the 3DCRT approach used to deliver 75.6 and 81 Gy is briefly reviewed, whereas IMRT planning for 81 Gy is described in more detail.

Simulation And CT Scanning

Most prostate cancer patients at MSKCC are treated in a prone position. Zelefsky et al. (1997) compared treatment plans generated for the same patients positioned both supine and prone and found the prone position to be preferable for the great majority of those undergoing 3DCRT. The data indicated that for patients treated to ≥75.6 Gy, lower rectal wall doses and displacement of the bowel out of the treatment field in the prone position may have a critical impact on reducing the risk of treatment-related morbidity. The prone position is technically reproducible and well tolerated for the majority of patients. On the other hand, the supine position appears to be more tolerable and practical for patients who are obese or those who have difficulty lying prone due to arthritis or other orthopedic problems.

A customized thermoplastic mold (Aquaplast®, Wyckoff, NJ.), designed to immobilize the pelvis, is used during simulation, CT scanning, and treatment to ensure the reproducibility of patient positioning during all procedures. The thermoplastic sheet is heated in warm water and molded to the patient. As the thermoplastic cools, it forms a hard shell. The mold extends from the mid-abdomen to the knees. As shown in figure 9–1, it attaches to a low-density wooden board. The combination of the mold and board forms a rigid shell in which the patient lies. Parts of the mold are cut away to provide ports for marking and tattooing the skin. The patient undergoes treatment simulation and CT scanning in the mold. If a CT-simulator is used, a virtual simulation is performed. The treatment isocenter is placed within the prostate gland. Figures 9–2a and 9–2b show posterior and left lateral simulation films, respectively. When a conventional simulation is performed, films are also taken for the treatment fields. For example, for a five-field IMRT plan, films are taken at the gantry angles of 0° (POST), 75° (RPO), 135° (RAO), 225° (LAO), and 285° (LPO). However, depending on the couch construction, the RAO and LAO angles may need

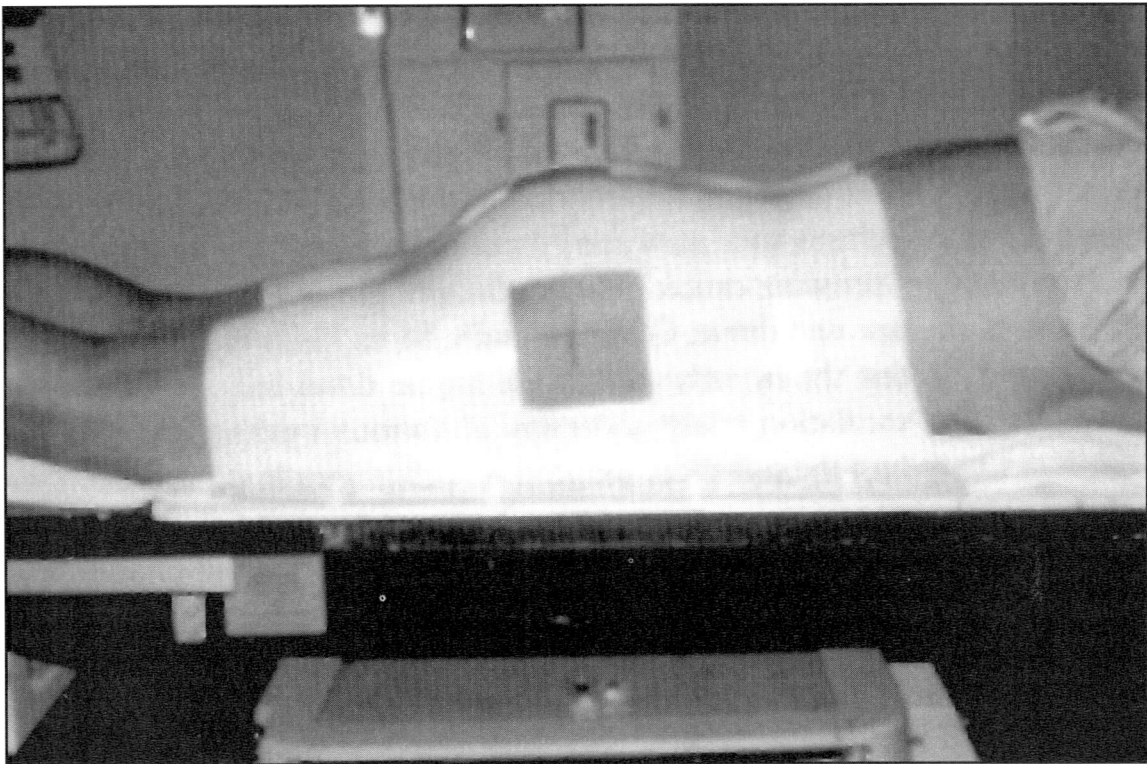

FIGURE 9–1. A patient immobilized in the prone treatment position. The thermoplastic mold attaches to the prostate board. Parts of the cast are cut away for marking the patient.

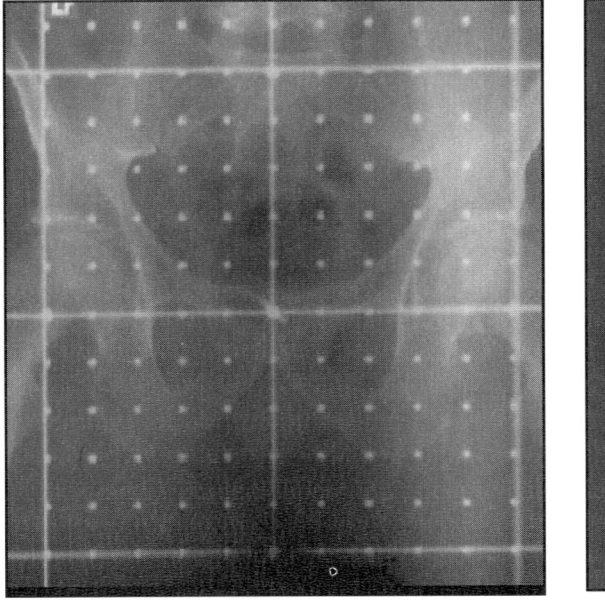 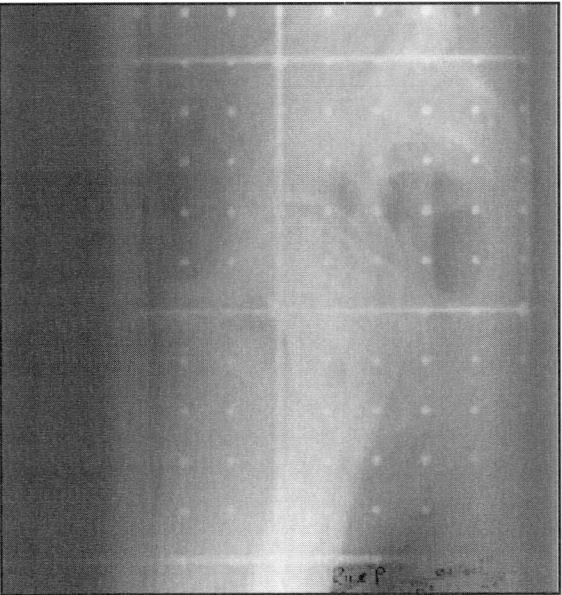

FIGURE 9–2. (a) Posterior and (b) left lateral simulation films. Cross hairs indicate the location of the isocenter.

to be adjusted by a few degrees to avoid attenuation by couch support bars. For either conventional or virtual simulation, the triangulation points for the isocenter are tattooed on the patient's skin. An additional alignment tattoo is placed, along the sagittal line, approximately 10 cm cephalad to the isocenter. To ensure reproducible leg position, tattoos are placed on the back of the legs at the level of the mid-shaft of the femurs, and the distance between the tattoos is recorded for future reference.

CT Scanning

The patient is CT scanned in the treatment position, immobilized in the thermoplastic mold. In preparation, the afternoon and evening before the imaging, the patient drinks Golytely® (Braintree Laboratories, Braintree, MA) to empty the bowel. Before scanning the patient is also asked to empty his bladder. However, some patients are CT-scanned and treated with full bladder, such as post-prostatectomy cases, or those in whom a full bladder is needed to displace the bowel away from seminal vesicles. To highlight the small bowel in the vicinity of the prostate and seminal vesicles an oral contrast agent, Readi-Cat® (barium sulfate suspension) (E-Z-EM, Inc., Westbury, NY), is administered. A rectal catheter is used to localize the rectal lumen. To obtain a volumetric data set for treatment planning, the patient is scanned in an approximately 20 cm region above and below the isocenter. A slice spacing and thickness of 3 to 5 mm is used. A few transverse images through the prostate and bladder are obtained before final imaging to ensure that the rectal lumen is clearly visible, the bladder is at the desired state of filling, and the patient is properly positioned within the scan circle. Figure 9–3a shows a posterior scout image and figure 9–3b shows a transverse image through the bladder, with the rectum and rectal catheter also visible.

If a CT-simulator is used, then the steps of the CT and the simulation are combined. A virtual simulation is carried out using the CT data set. Digitally reconstructed radiographs (DRRs) are obtained instead of simulation films.

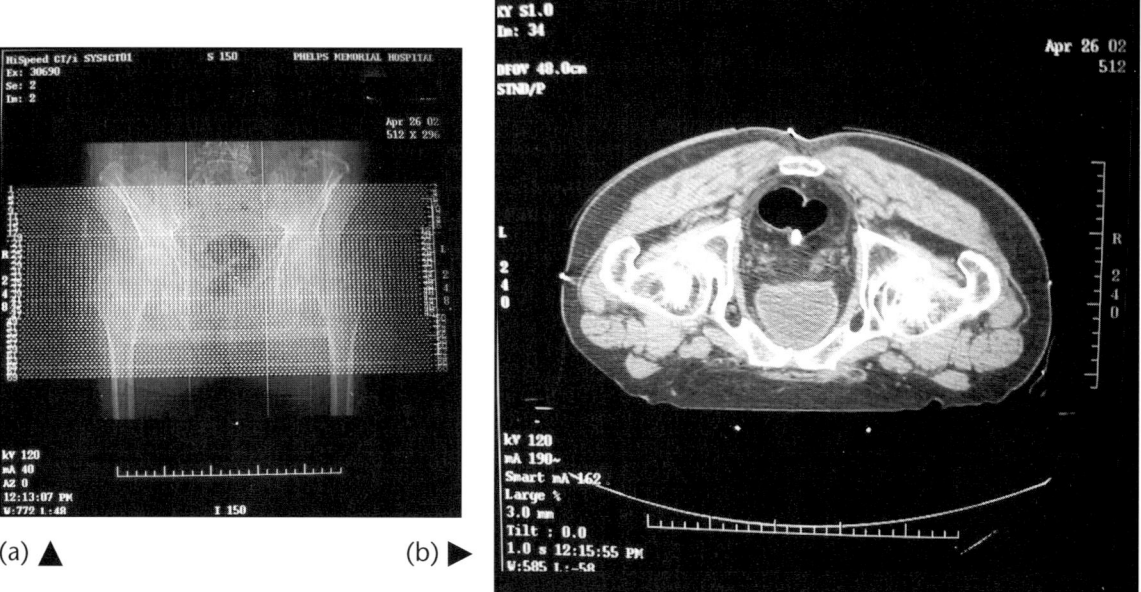

(a) ▲   (b) ▶

FIGURE 9–3. (a) A posterior scout image with the locations of the transverse images marked on it. (b) A transverse image through the bladder; the rectal catheter is also visible in the image.

Structure Contouring

The CT images are transferred to the treatment planning system. The radiation oncologist reviews the images and contours the PTV and the surrounding critical normal tissue structures.

PTV

For localized prostate cancer, the CTV includes the prostate and seminal vesicles. The PTV is defined by adding a 1 cm margin around the CTV, except posteriorly at the prostate and rectal wall interface where a 6 mm margin is used (Leibel et al. 1994). Clinically, these margins appear to be adequate in terms of the clinical response of patients treated in the phase I/II dose escalation trial (Zelefsky et al. 1995).

Normal Tissues

The normal tissue structures outlined by the physician include: the rectal wall, bladder wall, and the small and large bowel, if located in close proximity to the seminal vesicles. To define the rectal wall, the outer rectal circumference and inner rectal lumen are contoured on the CT images. A rectal catheter aids in identifying the location of the lumen. Similarly, the outer and inner surfaces of the bladder wall are outlined. Portions of the small and/or large bowel are contoured if they are located within ~1 cm of the PTV on the CT images. In addition, the planner also delineates the femoral heads and the outer surface of the skin. Figure 9–4 shows a transverse image with the outline of the PTV, rectum, bladder, and femoral heads.

Treatment Planning For 3DCRT

Six-Field 3DCRT Plan To 75.6 Gy

The volumetric CT data are used for treatment planning. A 15 MV, six-field conformal plan was used to treat patients to 75.6 Gy. As schematically shown in figure 9–5, the beams are placed at gantry angles of 45°, 90°, 135°, 225°, 270°, and 315°. The beam's-eye view display (BEV) is used to shape each beam aperture to irradiate the PTV while shielding to the extent possible the

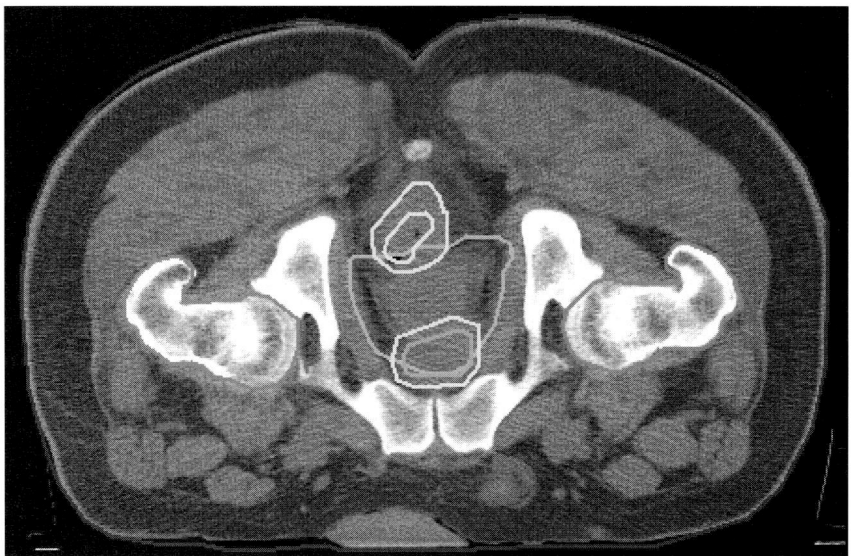

FIGURE 9–4. A transverse CT image with the PTV, rectum, bladder, femurs, and skin outlined.

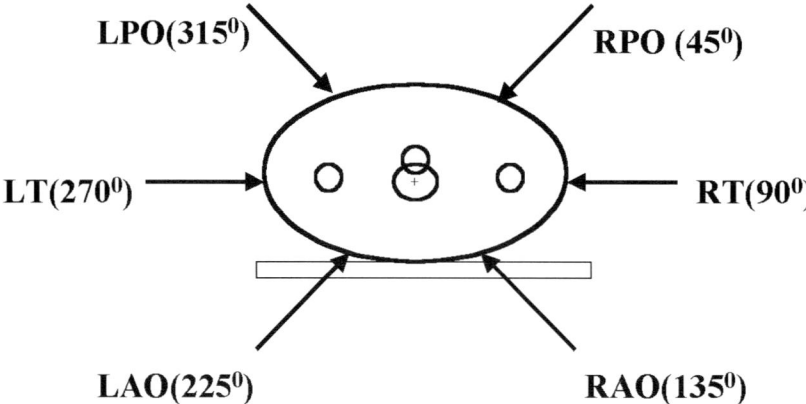

FIGURE 9–5. Schematic diagram showing the beam arrangement for the six-field 3DCRT plan.

surrounding normal tissue structures. A margin of ~5 mm is used to account for the beam penumbra in the lateral and anterior-posterior directions. Because of the overlapping beams this margin has been found to be dosimetrically adequate, whereas in the superior and inferior directions a margin of ~1 cm is needed. Either conventional blocks or multileaf collimation (MLCs) may be used for beam shaping. For MLC the cross-bound configuration, in which the area blocked by each MLC leaf is equalized with the area exposed, is used. For multiple field plans the resulting dose distributions for smooth and MLC apertures are very similar (LoSasso et al. 1993). There are many advantages to using MLC, including remote and automatic set-up of field shapes, savings in block fabrication costs, time and storage space, and the elimination of the need to handle blocks that contain lead and cadmium.

The dose distribution is calculated for representative planes, including the three mutually orthogonal planes—transverse, coronal, and sagittal—through the isocenter. A volumetric dose calculation for the PTV, rectum, bladder, and femurs is carried out. If portions of small or large bowel are located near the prostate and seminal vesicles, the dose to the bowel is also calculated. The doses from all six beams are combined to produce an acceptable dose distribution. Generally, the two lateral beams deliver half the dose to the isocenter, and the four oblique beams contribute the rest. The beam weights of the anterior-oblique and posterior-oblique beams are adjusted to deliver a uniform dose within the PTV, and to keep any hot spots away from the rectum. Figure 9–6 shows the isodose distribution in the transverse plane through the isocenter. The prescription isodose (100%) covers the PTV with a hot spot of ~6% located within the PTV. The portion of the rectal wall enclosed within the PTV receives the prescription dose, 75.6 Gy, or higher. Figure 9–7 shows the dose-volume histogram (DVH) for the PTV, bladder wall, rectal wall, and femurs. Dose constraints for normal tissues include the following: no more than 30% of the rectal wall receives ≥75.6 Gy (100%); maximum dose to large bowel is ≤60 Gy (79%); the maximum dose limit for the small bowel is <50 Gy (66%); and the maximum dose to the femurs is ≤68 Gy (90%).

Two-Phase 3DCRT Plan To 81 Gy

In October 1992, accrual to the 81 Gy level of the dose-escalation study was begun. It was found that when 81 Gy was administered using the six-field approach; a large portion of the rectum was exposed to the high dose region. It was estimated that this would lead to an unacceptably high risk of rectal complications. To keep the radiation toxicity at an acceptable level, a two-phase approach was used. The aim was to keep the dose to rectum at the same level as for the 75.6 Gy plans, while increasing the dose to the PTV not overlapping the rectum to 81 Gy. For the first phase, the six-field plan was used to deliver 72 Gy. This was followed by a boost plan to

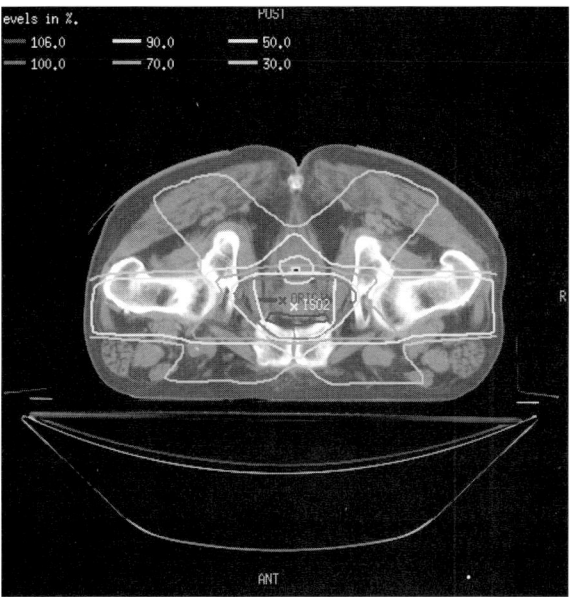

FIGURE 9–6. Isodose distribution for the six-field plan in a transverse plane through the isocenter. The outline of the PTV and the rectum are also shown. See COLOR PLATE 17.

administer the remaining 9 Gy. For the second phase, the seminal vesicles were excluded from the PTV. Six to eight posterior-lateral beams were used to treat the prostate PTV. Wedge filters were incorporated to make the dose uniform within the portion of the PTV not overlapping the rectum. For each beam the rectum was blocked in the BEV. The beam arrangement for the boost phase of the plan is schematically shown in figure 9–8, and the composite dose distribution for the 81 Gy plan on an axial plane, through the isocenter, is shown in figure 9–9. Although this technique successfully allowed the administration of 81 Gy, it was found to be very labor intensive. Each patient required two separate treatment plans. Further, the time to deliver the treatment for the second phase was longer because the therapist was required to enter the treatment room to change wedges.

Evolution Of IMRT For Prostate Cancer At MSKCC

IMRT was introduced at MSKCC in October 1995. First, only the 9 Gy boost portion of the 81 Gy 3DCRT plan was delivered with dynamic multileaf collimation (DMLC) (Ling et al. 1993). Because IMRT was used for the boost, the inconvenience of the two-phase treatment remained, as two separate plans were still required. To overcome this limitation a five-field IMRT plan was developed that treated the PTV to 81 Gy without exceeding the dose constraints for the normal tissues (Burman et al. 1997). Beginning in April 1996, the IMRT technique was used for the entire course of treatment, and in May 1997, the 86.4 Gy level of the phase I dose-escalation study was opened using the IMRT approach. The following sections describe in detail the planning and delivery of 81 Gy IMRT treatments.

Treatment Planning Of IMRT For Prostate Cancer

Five-Field IMRT Plan To 81 GY

An inverse method is used to plan IMRT (Spirou and Chui 1998). As discussed earlier, the isocenter is determined and the treatment beams are placed at specified angles as shown in

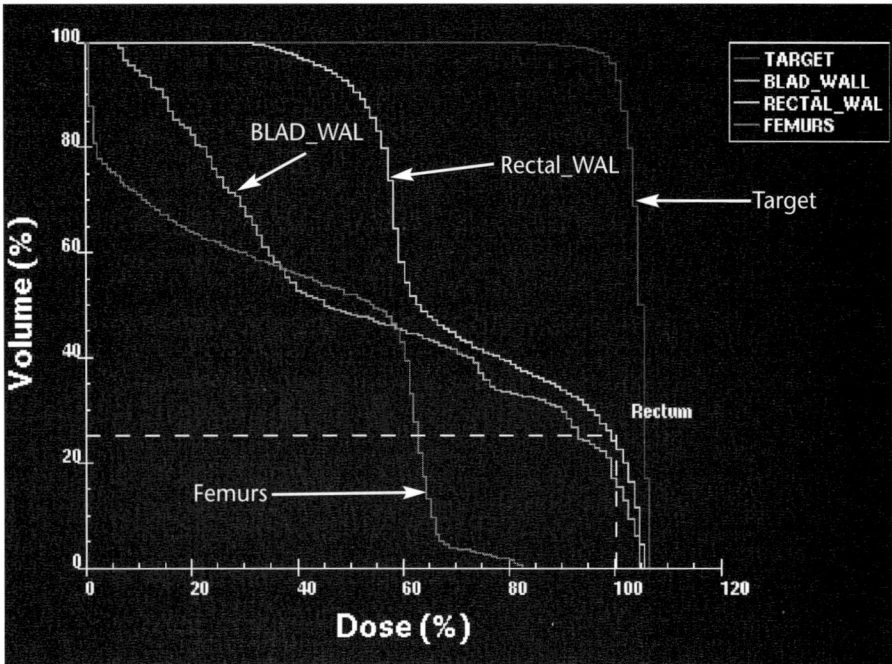

FIGURE 9–7. Dose volume histograms for a six-field 3DCRT plan. The DVHs for the PTV, rectal wall, bladder wall, and femurs are shown. The marker indicating the rectal volume receiving ≥100% dose is also displayed.

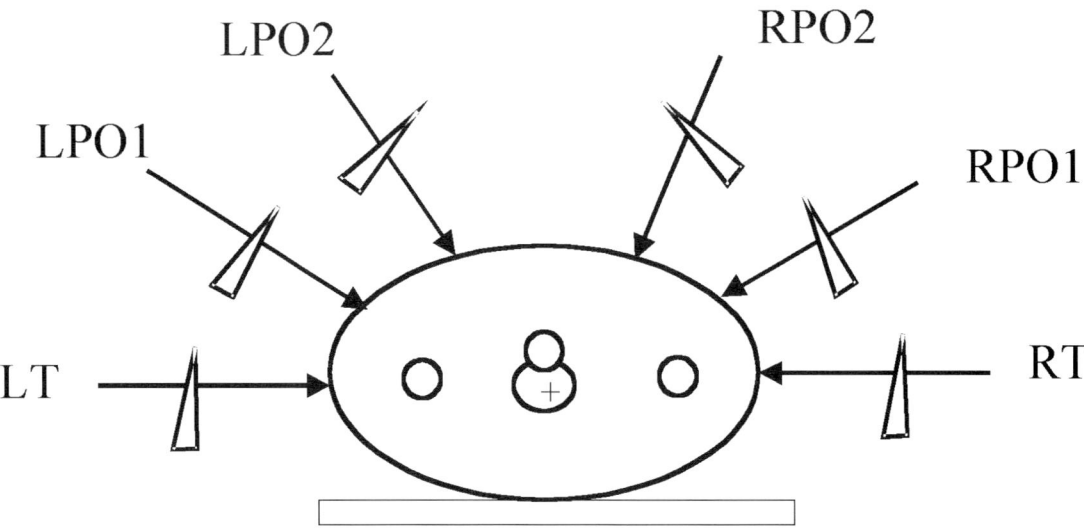

FIGURE 9–8. Schematic diagram showing the beam arrangement for the second phase of the 81 Gy 3DCRT plan. Nine Gy was delivered using the boost plan.

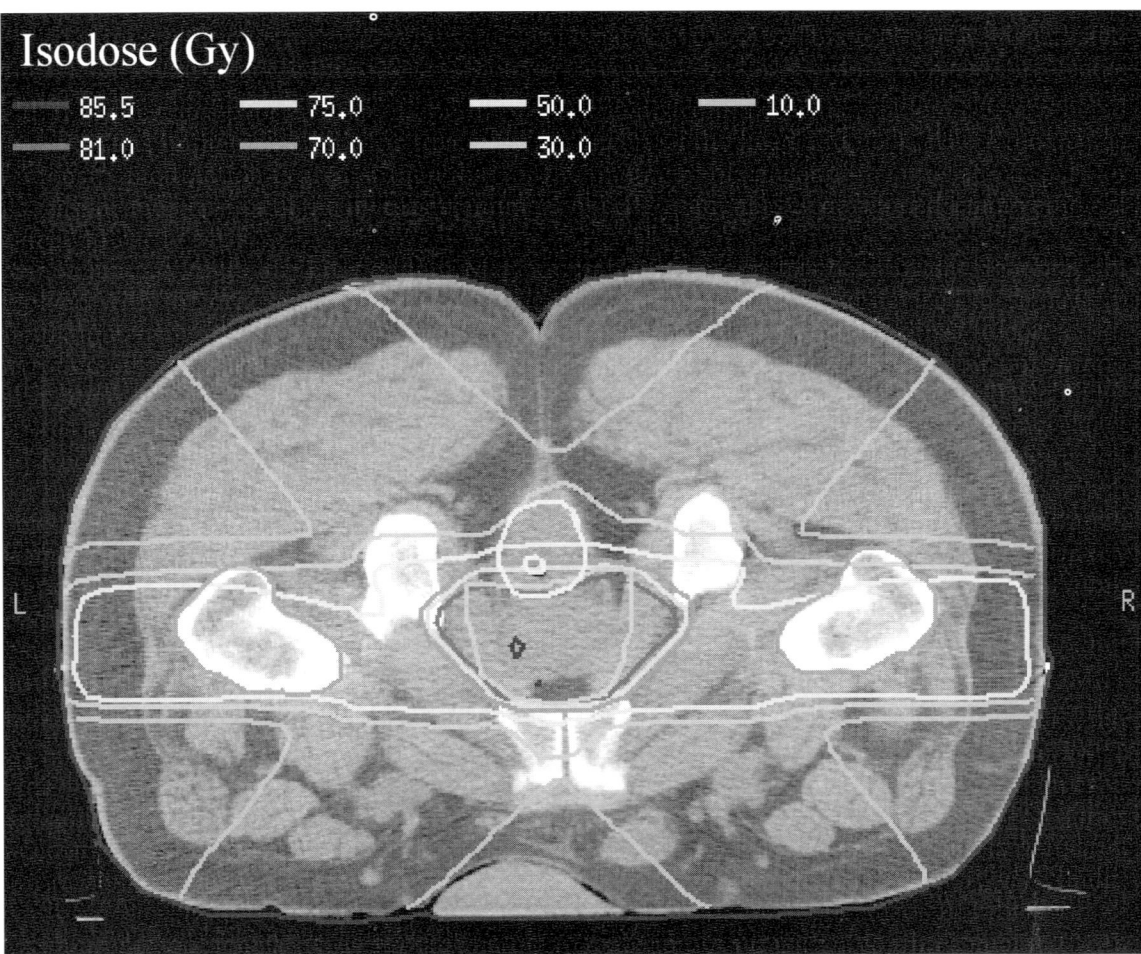

FIGURE 9–9. The composite dose distribution for the 81 Gy 3DCRT plan in a transverse plane. The isodose levels are in absolute dose (Gy). The outlines of the PTV and rectum are also shown. See COLOR PLATE 18.

figure 9–10. For optimization purposes, the following artificial structures are created by the planner: PTVE, defined by expanding the PTV by ~3 mm, except at the rectum interface; RECTOE, created by extending the outer rectal wall superiorly and inferiorly by 12 mm; BLADO, bladder outer wall. For our treatment planning system, we have found that the use of PTVE in the optimization process provides better coverage of the target near the periphery. The use of RECTOE reduces the probability of hot spots occurring within portions of the rectum that have not been contoured. A Boolean operation is used to divide the PTV into two parts: (1) PTVE_R, which is PTVE excluding the rectum overlap, and (2) OVERLAP_R, rectum and PTVE overlap region. This allows the planner to steer "hot-spots" away from the rectal wall while still obtaining satisfactory PTV coverage. As a starting point, the standard dose and dose-volume constraints, defined in table 9–1, are used. These are varied as the planner finds necessary for individual patients.

The optimization process balances the various constraints of the dose distribution. However, it should be emphasized that the parameters specifying the limits of the relative dose, the critical structures, and the overlap regions have been chosen to efficiently drive the optimization algorithm toward a desired dose distribution and are different from the evaluation criteria. Different

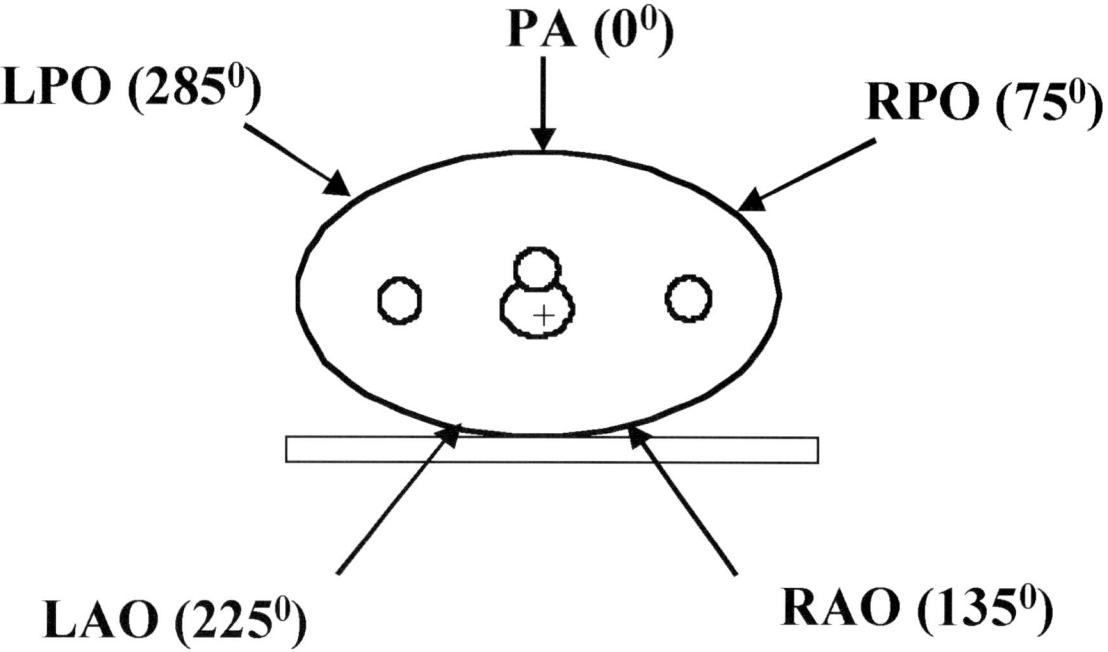

FIGURE 9–10. The schematic diagram showing the beam directions for the five-field IMRT plan.

optimization algorithms may require different constraints and penalties to reach similar dosimetric endpoints.

The optimization process produces an intensity profile for each beam. An example of the profile for the posterior-anterior (PA) field of an 81 Gy plan is shown in figure 9–11. For each intensity-modulated beam a *fluence* or *DMLC* aperture is also determined. It is defined as the entire irradiated area, with the boundaries defined by the starting and ending positions of each leaf. The corresponding DMLC aperture for the PA field is also outlined in figure 9–11. This aperture can be displayed in the BEV and can also be projected on a DRR.

The dose calculation is based on the beam intensity profiles. Combining the dose contribution from each beam with the relative weights determined by the optimization algorithm creates the treatment plan. The isodose distribution and the DVHs are evaluated according to the criteria discussed in the next section. If the dose distribution does not meet the clinical criteria, the planner adjusts the optimization constraints until the desired dose distribution is obtained.

Table 9–1. Dose and Dose Volume Constraints Used as the Starting Point for 81 Gy Prostate Plan

PTV excluding rectum overlap (PTVE_R)	Prescription dose = 100%
	Minimum dose = 98%, penalty = 50
	Maximum dose = 102%, penalty = 50
PTV and rectum overlap (OVERLAP_R)	Prescription dose = 95%
	Minimum dose = 93%, penalty = 10
	Maximum dose = 96%, penalty = 20
Rectum (RECTOE)	Maximum dose = 95%, penalty = 20
	70% of the volume receives <40% of prescription dose, penalty = 20
Bladder (BLADO)	Maximum dose = 98%, penalty = 35
	70% of the volume receives <40% of prescription dose, penalty = 20

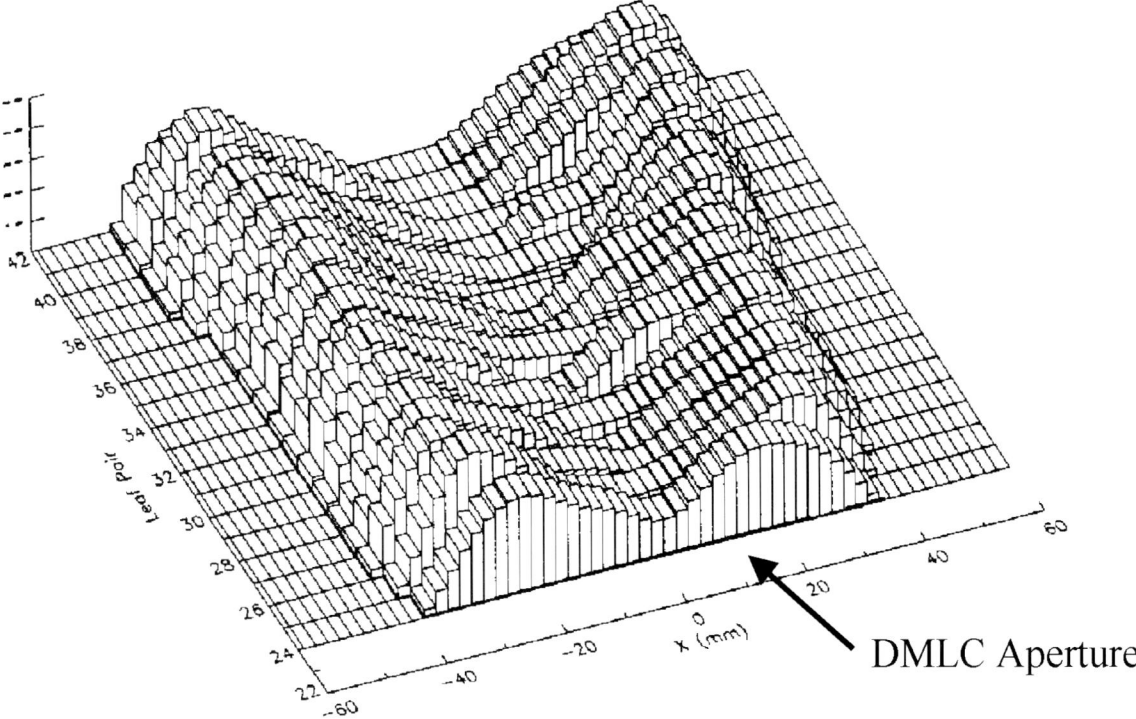

DMLC Aperture

FIGURE 9–11. The intensity profile for a posterior field. The outline of the DMLC aperture (dark line) is also shown.

Plan Evaluation

For plan evaluation, the dose distribution is displayed in a few representative planes including three mutually orthogonal planes through the isocenter: transverse, coronal, and sagittal. A volumetric dose calculation is also generated for regions of interest to determine the DVHs.

The plans are normalized so that the 100% (81 Gy) isodose surface covers the PTV adequately, except in the overlap region between PTV and rectum. Hot spots in the range of 6% to 7% typically occur within the PTV. The maximum dose limit within the PTV is 110%.

DVHs for the PTV, rectal wall, bladder wall, femurs, and the bowel are calculated. Figure 9–12 shows the DVHs for an 81 Gy IMRT plan in absolute dose. The aim for PTV coverage is to restrict the maximum dose within the PTV to <90 Gy, and ≥90% of the PTV must receive 77 Gy. There are two constraints for the rectal wall: (1) in the high-dose region no more than 30% of the rectal wall should receive ≥75.6 Gy; (2) in the intermediate-dose region no more than 53% of the rectal wall should receive ≥47 Gy. For the bladder wall there is only one dose constraint in the intermediate dose region—no more than 53% of the bladder wall should receive ≥47 Gy. The maximum dose to the femurs should be <68 Gy. The maximum dose limit for the large bowel is <60 Gy, and it is <50 Gy for the small bowel. If it is not possible to treat the initial PTV to full prescription dose without exceeding the bowel constraint, a cone down treatment is planned, using the superior jaws to shield the bowel. If this technique blocks portions of prostate, a new IMRT plan is generated for the physician-defined cone down PTV. A composite plan is designed to ensure that the maximum bowel dose constraint has not been exceeded. The total bowel dose is documented in the treatment plan.

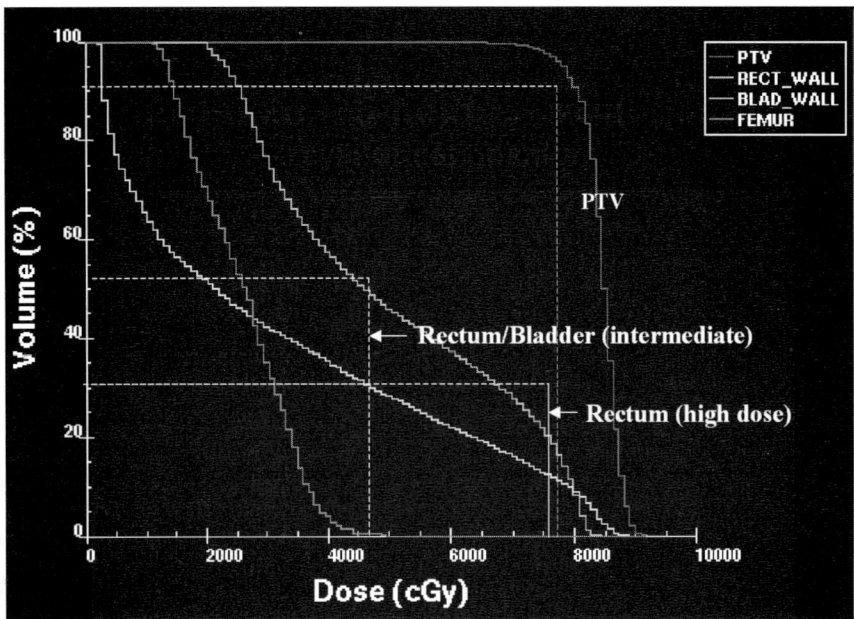

FIGURE 9–12. Dose volume histogram for the 81 Gy IMRT plan in absolute dose (Gy). The DVHs for the PTV, rectal wall, bladder wall, and femurs are shown. The markers indicate the dose volume constraints for plan evaluation.

Monitor Unit (MU) Calculation and DVA Files

The radiation oncologist reviews the plan and the prescription isodose level is selected. Based on the intensity profiles, selected treatment isodose level, dose per fraction (in the work reviewed above, 1.8 Gy), and dose rate, a computer algorithm determines the leaf motion as a function of monitor units (MUs) for each beam. The leaf-motion information is stored in a file known as leaf sequence file or the *DVA file.* The MUs are corrected for the collimator scatter factor (Sc) and phantom scatter factor (Sp). For dose delivery the DVA files are transferred to the treatment machine.

Independent MU Check

In many states, including New York, it is a requirement that the treatment monitor units be independently verified. For the first 400 patients treated with IMRT at MSKCC, independent checks were performed by ion chamber measurements and/or film dosimetry. More recently, a separate computer algorithm was written that uses the DVA file and the MU for each beam to calculate the dose delivered to a user-selected point at a given depth for the beam at normal incidence on a flat, water-equivalent phantom. For the prostate, the beam isocenter can be used as the point for independent calculation. To commission this program, comparisons were made between ion chamber measurements and the calculations. Currently, this program is used as the first-line independent MU check. The doses from the planning system and independent dose calculation are compared. No further action is required if the discrepancy is ≤3%. For a 3% to 5% difference, the planner reviews the results with a senior physicist. If the discrepancy is >5% or for unresolved discrepancies, ion chamber or film measurements are made.

Since reaching the plan evaluation objectives requires more than one optimization for the majority of patients, the optimization algorithm attaches a sequentially incrementing *run number* to each optimization. The run number also appears on the graphic printout used for plan evaluation and in the header text of the DVA file. The consistency between the run numbers appearing at these locations is confirmed as part of the pre-treatment quality assurance.

Plan Documentation

Once the plan has been finalized, the isodose distributions on three orthogonal planes through the isocenter are printed. An example is shown in figure 9–13. DVHs for the PTV, rectal wall, bladder wall, and femurs are also generated. Markers indicating the limits for $V_{77\ Gy}$ (volume receiving 77 Gy) for the PTV; $V_{75.6\ Gy}$ and $V_{47\ Gy}$ for the rectal wall; $V_{47\ Gy}$ for the bladder wall are noted on the DVHs.

If a cone down is required to limit the bowel dose, a composite plan is produced to reflect the total dose delivered. The DVHs for the PTV, femurs, rectal wall, bladder wall, and bowel are included in the patient's chart. The maximum bowel dose is also documented. The physician reviews and signs the plan.

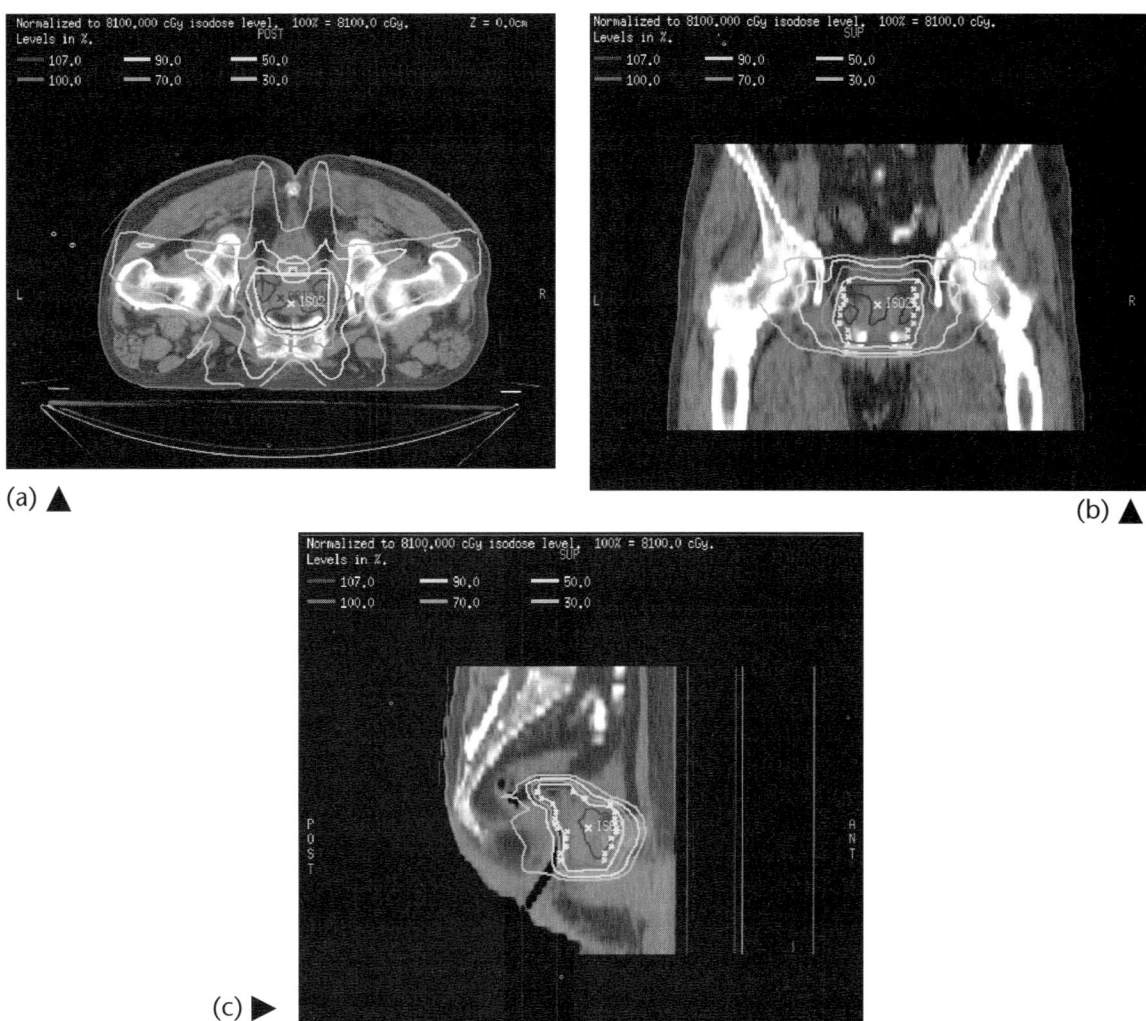

FIGURE 9–13. Isodose distribution on three planes through the isocenter: (a) transverse, (b) coronal, and (c) sagittal. The dose (81 Gy) is prescribed to the 100% isodose. The outlines of the PTV and rectum are displayed. See COLOR PLATE 19.

Treatment Aperture on Simulation Films or DRRs

The DRRs for each treatment beam are created on which the DMLC or fluence apertures, cross hairs, and fiducial grid points are superimposed. Figure 9–14 show the DRRs for a PA (a) and LAO field (b) overlaid with their respective DMLC apertures, cross hairs, and the grid points. These DRRs can be compared directly with the portal image. However, if the DRR quality is not good, the apertures can be traced onto corresponding simulation films.

Transfer of the DVA Files to the Treatment Machine

The DVA files, containing the leaf motion information, are transferred either via network or by a computer disk to the treatment machine. In addition, for port films, a computer file containing the DMLC apertures is also generated and transferred to the treatment machine. In newer machines this information can be automatically created from the DVA file.

Port Films

Before treatment, double exposure port films are taken using the DMLC aperture. These are compared with the corresponding DRRs or simulation films. The physician verifies the location of each treatment port. Thereafter, weekly portal images are taken to verify the patient setup. If the treatment machine is equipped with electronic portal imaging device (EPID), it can be used instead of the film. Work is in progress to use the EPID for *in vivo* verification of the delivered dose.

Treatment Delivery

The DVA file controls the leaf motion during the treatment delivery. The motion of each leaf pair determines the dose under that leaf pair, as shown in figure 9–15. The treatment machine computer monitors the leaf motion every 55 msec. If a leaf is lagging, it holds the beam pulses until the leaf is in the correct position. If for some reason the dose delivery is interrupted during the treatment, the leaf motion can be resumed and the remaining dose can be delivered by entering the total and remaining MUs.

Further Dose Escalation

We have completed the 86.4 Gy level of the phase I dose-escalation study. The planning dose constraints for these plans differ from those of the 81 Gy plans. However, the clinical dose constraints for the normal tissues are the same. An example of an isodose distribution for an 86.4 Gy plan is shown in figure 9–16. At 86.4 Gy the dose to the urethra is also of concern. Attempts are made to constrain the maximum urethral dose to <89 Gy. For the target dose evaluation, the prostate (CTV) without the seminal vesicles is also considered. At least 85% of the CTV should receive 86.4 Gy. In addition, the maximum dose within the PTV is constrained to ≤96 Gy. Based on the favorable toxicity outcomes observed in the phase I protocol, patients with intermediate and unfavorable risk prostate cancer are now being treated to 86.4 Gy.

We have investigated the use of IMRT to further increase the prescription dose to 91.8 Gy. We found that it is not possible to cover the target adequately while keeping the dose to the surrounding normal tissues within their tolerance limits. We are investigating the feasibility of identifying the gross disease within the prostate by magnetic resonance spectroscopic imaging and using *dose painting* to deliver higher dose levels to selected image-defined regions within the prostate (Mizowaki et al. 2002). Although several components required for this technique are still under development, our preliminary results indicate that it is feasible. We are confident that dose painting with IMRT will further improve the outcome for external beam treatment of prostate cancer.

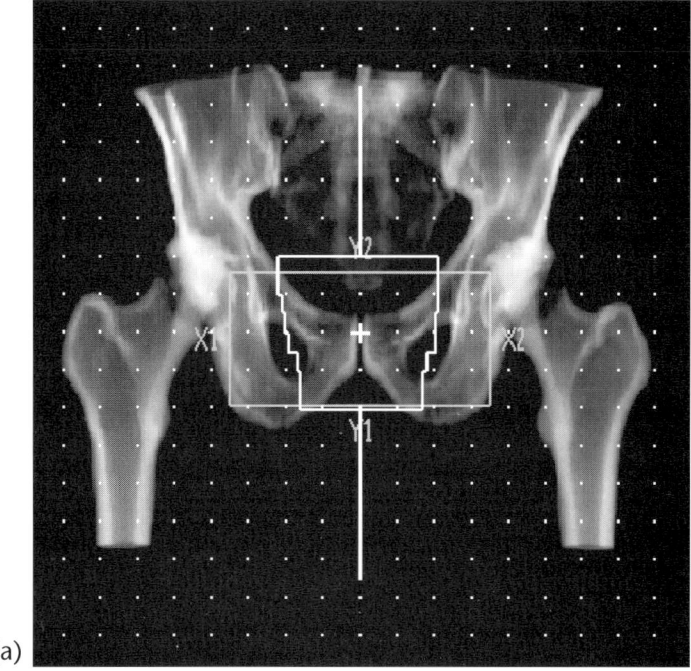

(a)

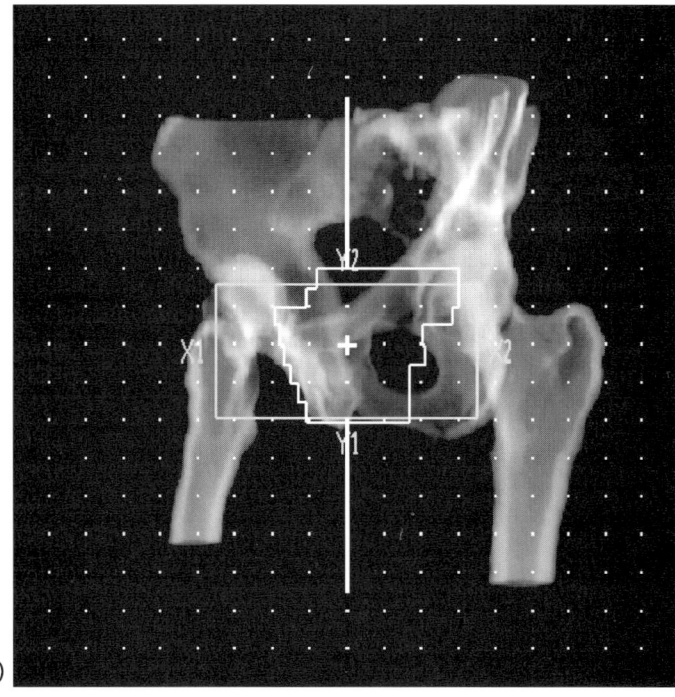

(b)

FIGURE 9–14. Examples of the DRR on which fluence aperture, field size, isocenter position, and graticule points are projected. (a) is for the PA beam and (b) is for the LAO beam.

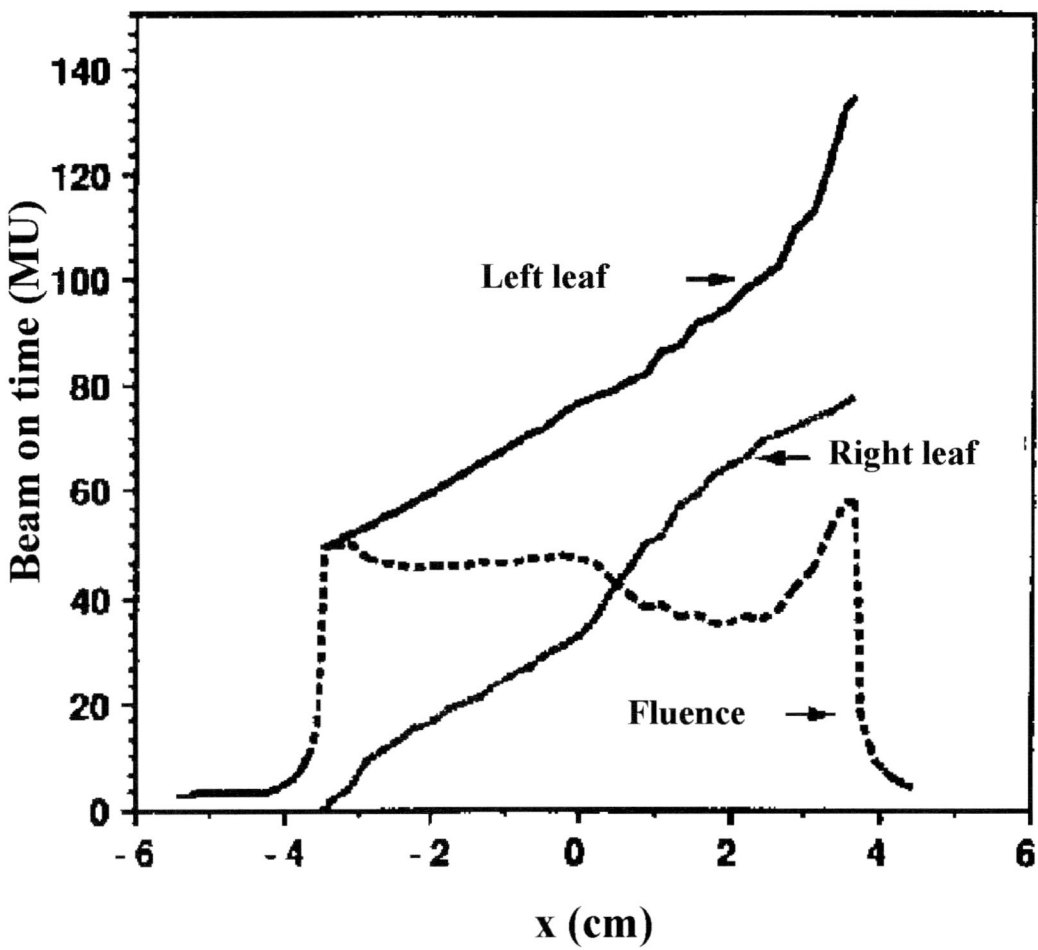

FIGURE 9–15. An example of MLC leaf-pair motion. The left and right leaf positions as function of beam on time and the resulting fluence under the leaf pair are shown.

Clinical Outcomes

The validity of the 3DCRT/IMRT concept was confirmed in a study carried out at MSKCC since October 1988. A total of 1684 patients with localized prostate cancer were treated through July 2001. The tumor dose was systematically increased from 64.8 to 86.4 Gy by increments of 5.4 Gy in consecutive groups of patients. The study was designed to assess the risk of toxicity of high-dose 3DCRT, to identify the highest feasible tumor dose, and to evaluate the impact of dose on the rates of local control.

Toxicity of 3DCRT and IMRT

An important component of the 3-D paradigm is the ability to deliver high radiation doses with a minimal risk of normal tissue toxicity. The 5-year actuarial rate of grade 3 or higher rectal toxicity for 364 patients treated with 3DCRT to 64.8 to 70.2 Gy was 3% compared to 2.2% for 507 patients treated to 75.6 to 81 Gy. In contrast, there was a dose-related increase in late grade 2 rectal bleeding. The 5-year actuarial rate of grade 2 rectal bleeding for patients receiving 64.8 to 70.2 Gy was 5% compared to 17% for those treated to 75.6 Gy ($p < 0.001$) (Zelefsky et al. 1999).

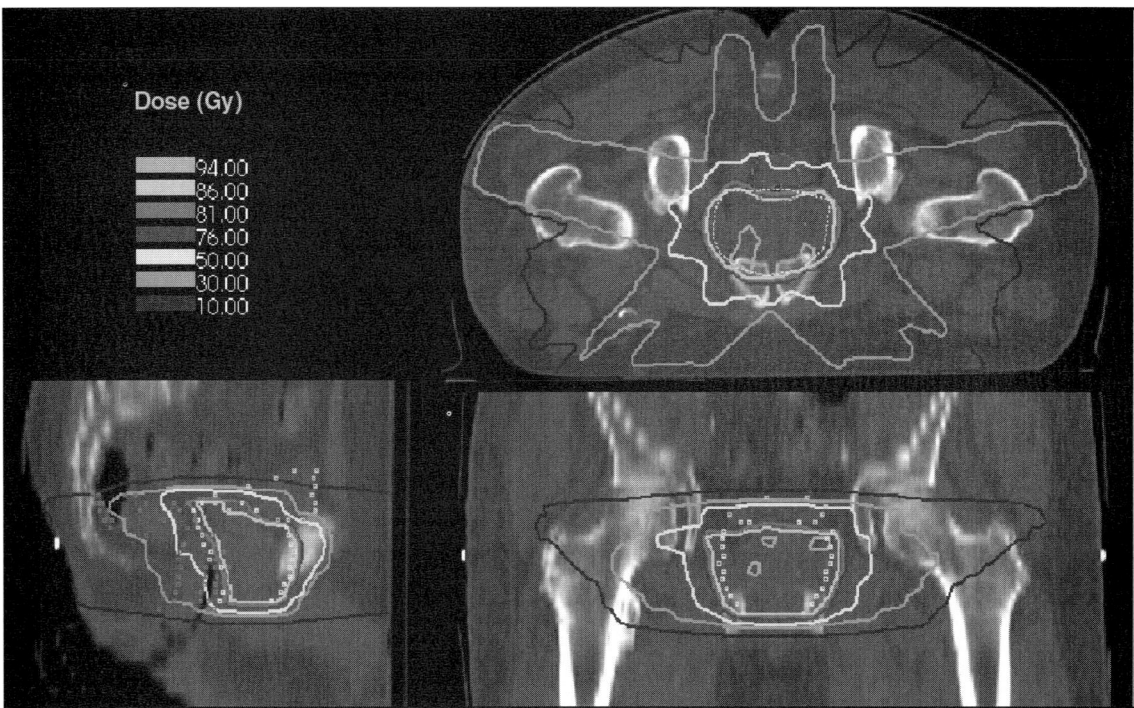

FIGURE 9–16. Isodose distribution for an 86.4 Gy plan on three orthogonal planes: transverse, sagittal, and coronal. See Color Plate 20.

When the dose was escalated to 81 Gy, a safety constraint was imposed during planning to restrict the volume of the rectal wall receiving 75.6 Gy to 30% or less. To meet this requirement, a two-phase 3DCRT technique was implemented (see above). With this approach the 5-year actuarial rate of late grade 2 rectal bleeding was 15% (Zelefsky et al. 1999, 2001). Since this technique did not substantially decrease the rate of grade 2 rectal toxicity, IMRT was introduced to improve the conformality of treatment to ≥81 Gy.

To validate the IMRT approach, the toxicity outcomes of 61 patients treated to 81 Gy with the two-phase 3DCRT approach were compared with those of 171 patients treated with IMRT to the same dose level. Acute and late urinary toxicities were not significantly different for the conventional 3DCRT and the IMRT methods. However, the combined rates of acute grade 1 and 2 rectal toxicity and the incidence of late grade 2 rectal bleeding were significantly lower ($p = 0.05$ and $p = 0.0001$, respectively) in the IMRT patients. The 3-year actuarial rate of late grade 2–3 rectal bleeding was 3% for IMRT as compared to 17% for the two-phase technique ($p < 0.001$). There was only one case of grade 3 rectal bleeding in each treatment group (Zelefsky et al. 2000).

The rates of late urinary and rectal toxicity observed thus far in 772 patients treated with IMRT to 81 to 86.4 Gy are shown in figure 9–17. A total of 698 patients were treated to 81 Gy and 74 to 86.4 Gy. The median follow-up time was 24 months with a range of 6 to 60 months. Eleven patients (1.5%) have developed grade 2 rectal bleeding and four (0.5%) have experienced grade 3 toxicity. The 3-year actuarial rate of grade 2–3 rectal toxicity was 4%. At this time, the rates of toxicity at 81 and 86.4 Gy are the same (Zelefsky et al. 2002). These findings demonstrate that the improved conformality and reduction of irradiated rectal tissue with IMRT translated into a decrease in rectal toxicity and provided an opportunity to safely increase the dose to 86.4 Gy.

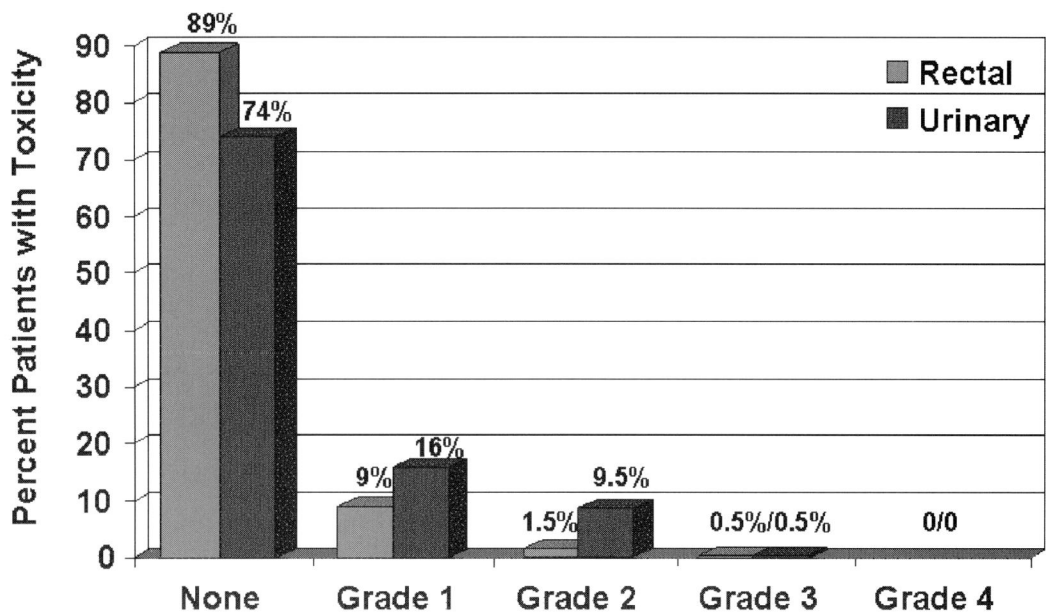

FIGURE 9–17. Late toxicity in 772 prostate cancer patients treated with IMRT to 81 to 86.4 Gy.

PSA Outcome And Local Control

The MSKCC study has generated extensive information on the efficacy of dose escalation both on long-term disease-free survival and local control. As shown in figure 9–18a, 9–18b, 9–18c, high radiation doses have significantly impacted on prostate-specific antigen (PSA) relapse-free survival. The 5-year actuarial PSA relapse-free survival rate for patients with favorable prognostic indicators (stage T1–T2, pretreatment PSA ≤10.0 ng/mL and Gleason score ≤6) receiving 64.8 to 70.2 Gy was 59% compared to 81% for those receiving 75.6 Gy and 98% for those treated to 81 Gy (64.8 to 70.2 Gy vs. 75.6 Gy, $p < 0.001$; 75.6 Gy vs. 81 Gy, $p < 0.001$) [unpublished data]. The corresponding rates for patients with intermediate prognosis (one of the prognostic indicators with a higher value) were 42% compared to 70% and 87%, respectively (64.8 to 70.2 Gy vs. 75.6 Gy, $p = 0.03$; 75.6 Gy vs. 81 Gy, $p = 0.006$) and for the group with unfavorable prognosis (two or more indicators with higher values) they were 21% compared to 42% and 70%, respectively (64.8 to 70.2 Gy vs. 75.6 Gy, $p = 0.003$; 75.6 Gy vs. 81 Gy, $p = 0.001$) (Leibel et al. 2002). The follow-up time of patients treated to 86.4 Gy is too short to assess affect of this dose on biochemical outcome.

Because radiation was confined to the prostate alone, it is reasonable to assume that the improvement in PSA relapse-free survival with increasing dose could be attributed for the most part to improved local control. To explore this notion, sextant prostate biopsies performed at ≥2.5 years (median 35 months) after 3DCRT or IMRT in 298 patients. Of the patients receiving 81 Gy, 59/67 (88%) had negative biopsies, compared with 105/137 (77%) after 75.6 Gy, 45/70 (64%) after 70.2 Gy and 11/24 (46%) after 64.8 Gy [unpublished data] (figure 9–19). When fitted to a tumor control probability (TCP) model, the mean dose to the PTV was found, within each prognostic risk group, to be directly related to the biopsy outcome. The TCD_{50} (dose required to achieve a TCP of 50%) for the favorable risk patients was 68.3 Gy compared with 72.8 Gy for the intermediate risk and 77.3 Gy for the unfavorable prognosis groups (Levegrun et al. 2002). These data indicate that the more aggressive biological phenotypes of prostate tumor cells are more radioresistant. However, the data indicate that even for more favorable phenotypes,

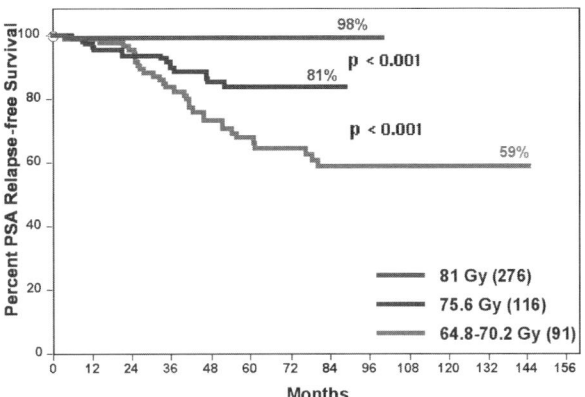

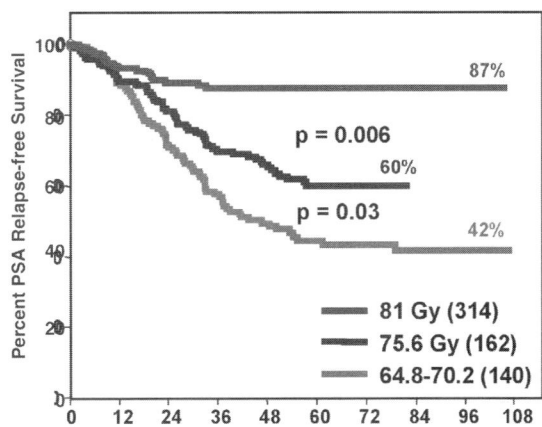

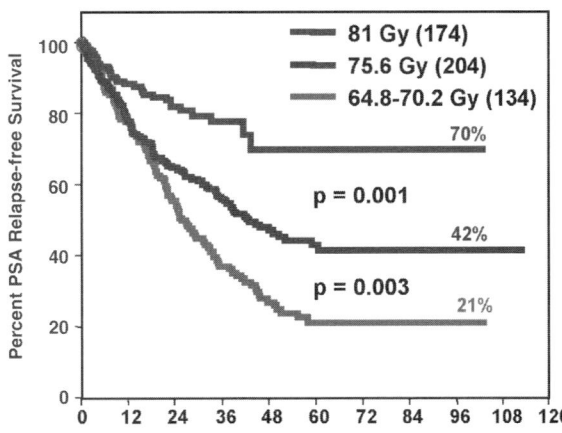

FIGURE 9–18. Actuarial PSA relapse-free survival according to dose for prostate cancer patients in the (a) favorable, (b) intermediate, and (c) unfavorable prognostic groups.

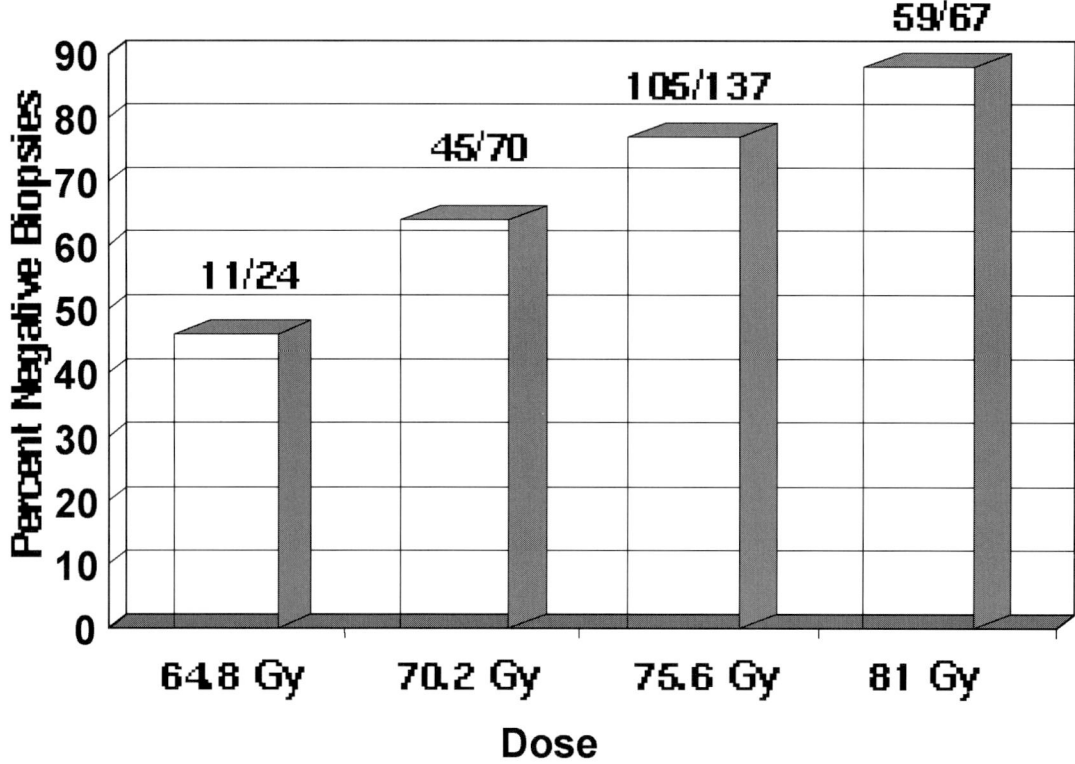

FIGURE 9–19. Effect of dose on local control as assessed in 298 patients by prostate biopsies performed at 2.5 years or longer after 3DCRT/IMRT. Above each bar is the number of patients with negative biopsy specimen/total number biopsied.

conventional doses of 65 to 70 Gy appear insufficient and that ≥81 Gy is necessary for a maximal local cure.

Conclusions

Our interim results indicate that high-dose 3DCRT is feasible and effective and represents a true advancement in the use of radiation for the cure of prostate cancer. Treatment to higher dose levels improves the disease-free survival of favorable, intermediate, and unfavorable risk patients. Further, our post-irradiation biopsy findings indicate that a dose of at least 81 Gy may be required for maximal local tumor control. Based on these data, patients at MSKCC with intermediate and unfavorable risk features are now being treated to 86.4 Gy. Whether this or higher doses will decrease the risk of local relapse will require further investigation. Nonetheless, it is clear that such high dose levels can only be delivered with advanced IMRT techniques to incrementally reduce the volume of normal tissues treated to high dose levels.

The classical approach to prostate cancer radiotherapy strives for covering the entire prostate with the prescribed dose while minimizing dose inhomogeneity within the PTV. In reality, however, some areas within the prostate may contain foci of highly resistant or hypoxic tumor clones that require locally enhanced doses. New methods of molecular- and biological-based imaging provide a potential for information on localization of tumor foci with specified phenotypic expressions of varying radiosensitivities. IMRT provides an approach for differential *dose painting* to selectively increase the dose in certain tumor-bearing areas. This departure from traditional treatment planning philosophy represents a new frontier in our IMRT prostate cancer research program.

References

Burman, C., C. S. Chui, G. J. Kutcher, S. Leibel, M. Zelefsky, T. LoSasso, S. Spirou, Q. Wu, J. Yang, J. Stein, R. Mohan, Z. Fuks, and C. C. Ling. (1997). "Planning, delivery, and quality assurance of intensity-modulated radiotherapy using dynamic multileaf collimator: A strategy for large-scale implementation for the treatment of carcinoma of the prostate." *Int. J. Radiat. Oncol. Biol. Phys.* 39:863–873.

Hanks, G., K. Martz, and J. Diamond. (1988). "The effect of dose on local control of prostate cancer." *Int. J. Radiat. Oncol. Biol. Phys.* 13:1299–1305.

Lee, W. R., G. E. Hanks, A. L. Hanlon, T. E. Schultheiss, and M. A. Hunt. (1996). "Lateral rectal shielding reduces late rectal morbidity following high dose three-dimensional conformal radiation therapy for clinically localized prostate cancer: Further evidence for a significant dose effect." *Int. J. Radiat. Oncol. Biol. Phys.* 35:251–257.

Leibel, S. A., G. E. Hanks, and S. Kramer. (1984). "Patterns of care outcomes studies: Results of the national practice in adenocarcinoma of the prostate." *Int. J. Radiat. Oncol. Biol. Phys.* 10:401–409.

Leibel, S. A., R. Heimann, G. J. Kutcher, M. J. Zelefsky, C. M. Burman, E. Melian, J. P. orazem, R. Mohan, T. J. LoSasso, Y. C. Lo, et al. (1994). "Three-dimensional conformal radiation therapy in locally advanced carcinoma of the prostate: Preliminary results of phase I dose-escalation study." *Int. J. Radiat. Oncol. Biol. Phys.* 28:55–63.

Leibel, S. A., G. J. Kutcher, M. J. Zelefsky, C. M. Burman, R. Mohan, C. C. Ling, and Z. Fuks. (1996). "3-D conformal radiotherapy for carcinoma of the prostate. Clinical experience at the Memorial Sloan-Kettering Cancer Center." *Front. Radiat. Ther. Oncol.* 29:229–237.

Leibel, S. A., Z. Fuks, M. J. Zelefsky, S. L. Wolden, K. E. Rosenzweig, K. M. Alektiar, M. A. Hunt, E. D. Yorke, L. X. Hong, H. I. Amols, C. M. Burman, A. Jackson, G. S. Mageras, T. LoSasso, L. Happersett, S. V. Spirou, C. S. Chui, and C. C. Ling. (2002). "Intensity modulated radiotherapy." *Cancer J.* 8:164–176.

Levegrun, S., A. Jackson, M. J. Zelefsky, E. S. Venkatraman, M. W. Skwarchuk, W. Schlegel, Z. Fuks, S. A. Leibel, and C. C. Ling. (2002). "Risk group dependence of dose response for biopsy outcome after three-dimensional conformal radiation therapy of prostate cancer." *Radiother. Oncol.* 63:11–26.

Ling, C. C., C. Burman, C. S. Chui, G. J. Kutcher, S. A. Leibel, T. LoSasso, R. Mohan, T. Bortfield, L. Reinstein, S. Spirou, X. H. Wang, Q. Wu, M. Zelefsky, and Z. Fuks. (1996). "Conformal radiation treatment of prostate cancer using inversely-planned intensity-modulated photon beams produced with dynamic multileaf collimation." *Int. J. Radiat. Oncol. Biol. Phys.* 35:721–730.

LoSasso, T., C. S. Chui, G. J. Kutcher, S. A. Leibel, Z. Fuks, and C. C. Ling. (1993). "The use of a multi-leaf collimator for conformal radiotherapy of carcinomas of the prostate and nasopharynx." *Int. J. Radiat. Oncol. Biol. Phys.* 25:161–170.

Mizowaki, T., M. Hunt, G. Mageras, et al. (2002). "Evaluation of external-beam radiotherapy treatment plans incorporating treatment uncertainties: A new and essential method for dose-painting IMRT plans for prostate cancer." (Abstract). *Int. J. Radiat. Oncol. Biol. Phys.* 54(supplement):185.

Purdy, J. A., W. B. Harms, J. Michalski, and J. D. Cox. (1996). "Multi-institutional clinical trials: 3-D conformal radiotherapy quality assurance. Guidelines in an NCI/RTOG study evaluating dose escalation in prostate cancer radiotherapy." *Front. Radiat. Ther. Oncol.* 29:255–263.

Sandler, H. M. (1996). "3-D Conformal radiotherapy for prostate cancer." *Front. Radiat. Ther. Oncol.* 29:238–243.

Shipley, W. U., H. D. Thames, H. M. Sandler, G. Hanks, C. A. Perez, D. A. Kuban, S. L. Hancock, and C. D. Smith. (1999). "Radiation therapy for clinically localized prostate cancer: A multi-institutional pooled analysis." *JAMA* 28:1598–1604.

Spirou, S. V., and C. S. Chui. (1998). "A gradient inverse planning algorithm with dose-volume constraints." *Med. Phys.* 25:321–333.

Vicini, F. A., A. Abner, K. L. Baglan, L. L. Kestin, and A. A. Martinez. (2001). "Defining a dose-response relationship with radiotherapy for prostate cancer: Is more really better?" *Int. J. Radiat. Oncol. Biol. Phys.* 51:1200–1208.

Zagars, G. K., A. Pollack, and L. G. Smith. (1999). "Conventional external-beam radiation therapy alone or with androgen ablation for clinical stage III (T3, Nx/N0, M0) adenocarcinoma of the prostate." *Int. J. Radiat. Oncol. Biol. Phys.* 44:809–819.

Zelefsky, M. J., S. A. Leibel, G. J. Kutcher, S. Nelson, C. C. Ling, and Z. Fuks. (1995). "The feasibility of dose escalation with three-dimensional conformal radiotherapy with patients with prostatic carcinoma." *Cancer J. Sci. Am.* 1:142–150.

Zelefsky, M. J., L. Happersett, S. A. Leibel, C. M. Burman, L. Schwartz, A. P. Dicker, G. J. Kutcher, and Z. Fuks. (1997). "The effect of treatment positioning on normal tissue dose in patients with prostate cancer treated with three-dimensional conformal radiotherapy." *Int. J. Radiat. Oncol. Biol. Phys.* 37:13–19.

Zelefsky, M. J., D. Cowen, Z. Fuks, M. Shike, C. Burman, A. Jackson. E. S. Venkatraman, and S. A. Leibel. (1999). "Long term tolerance of high dose three-dimensional conformal radiotherapy in patients with localized prostate carcinoma." *Cancer* 85:2460–2468

Zelefsky, M. J., Z. Fuks, L. Happersett, H. J. Lee, C. C. Ling, C. M. Burman, M. Hunt, T. Wolfe, E. S. Venkatraman, A. Jackson, M. Skwarchuk, and S. A. Leibel. (2000). "Clinical experience with intensity modulated radiation therapy (IMRT) in prostate cancer." *Radiother. Oncol.* 55:241–249.

Zelefsky, M. J., Z. Fuks, M. Hunt, H. J. lee, D. Lombardi, C. C. Ling, V. E. Reuter, E. S. Venkatraman, and S. A. Leibel. (2001). "High dose radiation delivered by intensity modulated conformal radiotherapy improves the outcome of localized prostate cancer." *J. Urol.* 166:876–881.

Zelefsky, M. J., Z. Fuks, M. Hunt, Y. Yamada, C. Marion, C. C. Ling, H. Amols, E. S. Venkatraman, and S. A. Leibel. (2002). "High-dose intensity modulated radiation therapy for prostate cancer: Early toxicity and biochemical outcome in 772 patients." *Int. J. Radiat. Oncol. Biol. Phys.* 53:1111–1116.

10

IMRT For Head And Neck Cancer

Lanceford M. Chong
Margie A. Hunt

Introduction

Historical Overview

Radiation therapy is a principal modality in the treatment of head and neck cancer. Its capabilities have steadily progressed with the increase in clinical knowledge and technological development. From its humble beginnings with treatment on orthovoltage units, we learned that tumors could be eradicated but that major acute and late side effects were often part of the results. Even with the availability of deeply penetrating teletherapy units (Cobalt-60) and linear accelerators (linacs), two-dimensional (2-D) treatment planning, and the cone down approach, the therapeutic ratio was still a major concern. The incorporation of a brachytherapy boost often improved the dose distribution between the tumor and the surrounding normal tissue. However, this approach is not suitable for many head and neck tumors due to anatomical, medical, or technological considerations.

Over the past two decades, there have been several major advances in the treatment of cancers of the head and neck. Effective chemotherapeutic agents have been developed for squamous cell carcinoma of the head and neck and are increasingly used sequentially or concurrently with radiation to treat unresectable cases or to promote larynx preservation (Fu 1997; Lefebvre et al. 1996; Vokes et al. 1993; Pfister et al. 1992; Bourhis and Pignon 1999; Brizel et al. 1998; Pignon et al. 2000). In response to the findings that local control was dependent on the overall duration of treatment, accelerated fractionation schemes have been devised to decrease the repopulation by tumor clonogens (Withers, Taylor, and Maciejewski 1988). Preliminary results from a recent randomized study (Fu et al. 2000) showed improved two-year local-regional control and disease-free survival using accelerated fractionation with a delayed concomitant boost compared to standard fractionation. Advances in computer and linac technology have also significantly impacted treatment of head and neck cancers by improving our ability to maximize tumor dose while minimizing the dose to adjacent normal critical structures. Image-based treatment planning and multileaf collimators have both been widely implemented, facilitating both the planning and delivery of three-dimensional conformal radiation therapy (3DCRT). More recently, the development of inverse planning systems and methods for delivering non-uniform radiation intensities have ushered in the era of intensity-modulated radiation therapy (IMRT), representing the state of the art in the treatment of many head and neck cancers (Blanco and Chao 2002).

Rationale for the Use of IMRT in Head and Neck Tumors

Based on studies comparing IMRT and other treatment approaches, IMRT appears to be clinically justifiable for cancers in the nasopharynx, sinonasal region, parotid gland, tonsil, buccal mucosa, gingiva, and thyroid as well as in tumor tracking along the cranial nerves. IMRT may also be useful in the re-treatment of previously irradiated head and neck cancers, due to its ability to spare adjacent normal tissues with acceptable target dose uniformity. Although technically superior, IMRT is costly, and its cost-effectiveness requires due consideration as the technology evolves. Only as clinical data establishing the therapeutic ratio, local-regional control, side effects, and survival with IMRT become available, will the efficacy of IMRT be established.

To date, a small number of investigators have reported on the use of IMRT for head and neck cancer. Although earlier efforts were primarily treatment planning studies, clinical studies, primarily retrospective reviews with limited patient populations and heterogenous diagnoses, have recently been reported. In one treatment planning study, Boyer et al. (1997) examined the use of IMRT in three patients with nasopharyngeal, vocal cord, and ethmoid sinus tumors. They found that IMRT was capable of producing dose distributions with invaginations, bifurcations, and internal voids, thus exhibiting significant potential for normal organ sparing.

In another treatment planning study, van Dieren et al. (2000) evaluated whether IMRT could spare parotid and submandibular glands without compromising target coverage. Thirty patients (15 with T2 tumors of the tonsillar fossa with extension into the soft palate, 15 with T3 tumors of the supraglottic larynx) were treated with lateral opposed portals. For each patient, an IMRT plan was developed retrospectively that included a parotid sparing approach. Compared to the distribution from lateral opposed portals, IMRT improved the target dose distribution. For the supraglottic larynx carcinomas, the volume receiving a biologically equivalent dose greater than 40 Gy decreased by 23% in the parotid and 7% in the submandibular gland. With tonsillar fossa cancers, the decrease in volume was 31% in the parotid and 7% in the submandibular gland.

Verellan et al. (1997) reviewed their implementation of IMRT in the treatment of nine patients with head and neck cancer using the MIMiC device (NOMOS Corporation, Sewickley, PA). Relative and absolute dosimetric measurements in anthropomorphic phantoms using a variety of detectors demonstrated excellent agreement between the measured and calculated dose distribution. For immobilization, a noninvasive system capable of achieving a setup uncertainty standard deviation of 0.3 cm (translations) and 2.0 degrees (rotations) was used in conjunction with a verification protocol capable of detecting errors as small as 0.1 cm and 1 degree. To achieve the higher degree of precision in target localization that may be necessary for IMRT treatment, the authors stated that daily on-line verification and implanted fiducial markers may be necessary.

Eisbruch et al. (1998) reported on their use of IMRT in 15 patients with stage III/IV head and neck cancer requiring bilateral neck irradiation. The minimum primary planning target volume (PTV) dose in the IMRT plans was higher than that in the standard plans (95.2% and 91% of the prescribed dose, respectively); coverage of the ipsilateral jugular nodes was also improved, but coverage of the contralateral jugular or posterior neck nodes was similar to conventional treatment. With respect to the normal critical structures, both the magnitude of dose and the volume in the high-dose regions decreased with IMRT. The mean dose to all major salivary glands, particularly the contralateral parotid gland, was much lower. It was noted that despite the normal tissue sparing, the tumor target coverage was not compromised.

Preliminary results of a retrospective study on the first 28 head and neck cancer patients treated with IMRT at Baylor College of Medicine was reported by Kuppersmith et al. (1999). The histopathologies included squamous cell carcinoma, adenoid cystic carcinoma, paraganglioma, and angiofibroma. Patients received doses from 14 to 71 Gy in daily fractions of 1.55 to 4 Gy. With respect to the normal tissue doses, the parotid gland received less than 30 Gy for midline tumors. Their incidence of acute toxicity was much lower than with conventional radiotherapy. They noted that with only a portion of an organ irradiated, the tolerance dose was likely to increase. The article highlighted the following clinical capabilities of IMRT: (1) decreased normal tissue doses during re-irradiation of previously treated patients; (2) cranial nerves could be traced to the base of skull while minimizing the dose to the parotid glands and other surrounding structures; varying doses could be administered to the primary site as opposed to the cranial nerves; (3) multiple targets could be treated simultaneously with an accelerated course and once-a-day fractionation while minimizing doses to adjacent normal structures. This technique was referred to as Simultaneous Modulated Accelerated Radiation Therapy (SMART).

The SMART technique was used between January 1996 and December 1997 on 28 patients to treat various primary head and neck sites including oropharynx, nasopharynx, larynx, oral cavity, and sphenoid sinus (Butler et al. 1999). All patients were immobilized with an invasive calvarial screw technique to yield a patient position reproducibility of better than 2 mm. The dose to the primary target was 60 Gy in 2.4 Gy fractions, while sites at risk for microscopic disease received 50 Gy in 2 Gy fractions. All targets were treated once a day, 5 days per week and were completed in 5 weeks. Sixteen of 20 patients (80%) completed the treatment in 40 days. Sixteen patients (80%) had RTOG (Radiation Therapy Oncology Group) toxicity grade III mucositis and ten patients (50%) had grade III pharyngitis. Three patients (15%) had greater than 10% weight loss. Nine patients (45%) experienced moderate acute xerostomia that significantly improved within 6 months. Nineteen patients (95%) achieved a complete response and one patient had a partial response. The mean doses to the primary and secondary targets were 64.4 Gy and 54.4 Gy, respectively. On average, 8.9% of the primary target and 11.6% of the secondary target received a dose less than that prescribed. Adjacent normal critical structure doses were as follows: 30 Gy, mandible; 17 Gy, spinal cord; 23 Gy, ipsilateral parotid; 21 Gy, contralateral parotid. The conclusion of the study was that this IMRT technique yielded encouraging initial tumor responses with acceptable morbidity.

Chao et al. (2000) implemented tomotherapy-based IMRT in patients with squamous carcinoma of the head and neck. Seven nasopharyngeal carcinoma, seven oral pharyngeal carcinoma, one supraglottic larynx carcinoma, and two patients with metastatic disease to the upper and mid cervical nodes from an unknown primary were treated with the MIMiC device. Eight patients (six nasopharyngeal carcinomas, two tonsillar carcinomas) with primary disease and one patient with recurrent nasopharyngeal carcinoma were treated with concurrent cisplatin chemotherapy. Six patients were postoperative and received radiation alone. Using IMRT, different doses were delivered to different targets simultaneously in each fraction. Acute side effects were similar to those seen with traditional radiation therapy. With IMRT, an average of 27%±8% of the parotid gland volumes received more than 30 Gy and an average of 3.3%±0.6% of the target volume received less than 95% of the prescribed dose. The authors concluded that the use of IMRT led to a high degree of target conformity and that the initial results on tumor control were promising with no severe adverse acute side effects.

Sultanem et al. (2000) reviewed the experience with IMRT in the treatment of nasopharyngeal carcinoma at the University of California, San Francisco. Thirty-five patients were treated: 4 (12%) with stage I, 6 (17%) with stage II, 11 (32%) with stage III, and 14 (40%) with stage IV disease. The target for IMRT treatment included the nasopharynx and retropharyngeal nodes but avoided the other regional lymphatics that were treated with conventional techniques. Sixty-five to 70 Gy was prescribed to the gross target volume (GTV) and positive neck nodes, 60 Gy to the clinical target volume (CTV) and 50 to 60 Gy to the clinically negative neck nodes. Eleven patients (32%) underwent a fractionated high dose rate intracavitary brachytherapy boost to the primary tumor one to two weeks following completion of external radiation therapy. Thirty-two patients (91%) were given concomitant cisplatin chemotherapy and adjuvant post-treatment cisplatin and 5FU (5-Fluorouracil) chemotherapy. With a median followup of 21.8 months, the locoregional progression free rate was 100%. At 4 years, overall survival was 94% and the distant metastasis free rate was 57%. The acute toxicity percentages were as follows: 16 patients (46%) with grade II, 18 patients (51%) with grade III, 1 patient (3%) with grade IV. Fifteen patients (43%) had grade I, 13 patients (37%) had grade II, and 5 patients (14%) had grade III late toxicity. The xerostomia evaluation at 24 months post-treatment showed 50% of the evaluated patients had grade 0, 50% had grade I, and none had grade II xerostomia. The GTV received a mean dose of 75.8 Gy while the CTV received 71.2 Gy. All normal tissue received acceptable doses including the parotid glands, which received an average dose of 43.2 Gy to 50% of the volume. The authors concluded that IMRT improved the target coverage, increased GTV dose, and improved sparing of the adjacent normal critical structures. Locoregional control for patients receiving concurrent chemotherapy was excellent.

IMRT Treatment Of Primary Head And Neck Cancer At MSKCC

The above discussion indicates that there are many situations where IMRT may improve the dose distributions for primary head and neck cancers. However, whether this improvement will prove clinically significant can only be answered on a site-by-site basis as outcome data become available. The potential improvement afforded by IMRT must also be considered in the context of its complexity and cost relative to 3DCRT or 2-D planning and treatment.

IMRT has been used routinely in the treatment of head and neck cancers at Memorial Sloan-Kettering Cancer Center (MSKCC) since May 1998. Thus far, our primary emphasis has been on the development of techniques for primary nasopharynx cancer, thyroid carcinomas, and recurrent head and neck tumors. A brief description of the technical approaches is given below, followed by a description of the planning process for one site, primary nasopharynx cancer.

Primary Nasopharyngeal Carcinoma

The MSKCC approach to the treatment of nasopharyngeal cancers with 3DCRT was described by Leibel et al. (1991). In this study, 3-D and 2-D treatment plans were compared for 10 previously untreated patients who received 3DCRT for the boost phase of treatment, and 5 others with locally recurrent disease who received 3DCRT for the entire course. 3DCRT improved the dose distribution, with a ~13% increase in tumor dose and decreased doses to the adjacent normal structures. Unfortunately, the use of a 3DCRT boost did not improve local control relative to traditional treatment (Wolden et al. 2001). It was hypothesized that this was due to the use of the 3-D plan only during the boost phase of treatment since its dose distribution was not appropriate for the entire treatment course. IMRT, on the other hand, can be used to deliver the entire treatment as shown by Hunt et al. (2001). In this study, IMRT, 3DCRT, and 2-D plans were compared for six patients, two each with negative, unilateral, and bilateral neck disease. All six patients were treated using IMRT and retrospectively planned with 3DCRT and 2-D techniques, designed to deliver 70 Gy to sites of gross disease (PTV_{gr}) and 54 Gy to the electively irradiated nodal regions (PTV_{el}). A summary of the beam arrangements and techniques employed for the three plans is given in table 10–1.

The dose distributions produced by the three techniques for a patient with N2 disease are compared in figure 10–1. The 3-D and IMRT dose distributions are similar in shape but the dose conformality, normal tissue doses and target dose uniformity are superior with IMRT. PTV coverage with the traditional parallel opposed 6 MV plan was inadequate particularly in the retropharyngeal area, base of skull, and medial aspects of bulky neck nodes.

Doses to all normal tissues improved using IMRT (table 10–2). The average maximum spinal cord dose was approximately 35, 45, and 50 Gy with the IMRT, 3-D conformal, and traditional plans, respectively. For both the mandible and temporal lobes, the volume irradiated to the higher dose levels was significantly lower with IMRT. Since no attempt was made to spare the parotid glands in this study, the dose to the parotid glands improved with IMRT but not to a level expected to preserve meaningful salivary function. The mean PTV dose increased from 68 Gy for the traditional plan to 76 Gy for IMRT, a 12% increase and *de facto* dose escalation even though the prescription dose was the same.

Thyroid Cancer

Like nasopharynx tumors, thyroid cancer is ideally suited for treatment with IMRT because of the concave shape of the target surrounding the normal critical structures, including the spinal cord and brachial plexus. Patients with unresectable thyroid cancer or those at high risk for postoperative local-regional recurrence are treated with IMRT at MSKCC. Treatment planning is image-based using fused computed tomography (CT) and FDG-PET (fluorodeoxyglucose

Table 10–1. Summary of IMRT, 3-D Conformal, and Traditional Treatment Plans

Plan Name	Field Arrangement	PTVs Included	Delivered Dose (Gy)	Cumulative Dose (Gy)
Traditional	Opposed Laterals 6 MVX	PTV_{el} PTV_{gr}	45	45
	Opposed Laterals with Cord Block Bilateral 9 MeV E⁻ Strips	PTV_{el} PTV_{gr}	9	54
	Opposed Lateral Cone Down, Involved Neck 9 MeV E⁻ Strips	PTV_{gr}	16	70
3-D Conformal	Opposed Lateral 6 MVX	PTV_{el} PTV_{gr}	36	36
	Seven Field Conformal Plan	PTV_{el} PTV_{gr}	18	54
	Seven Field Conformal Plan	PTV_{gr}	16	70
IMRT	Seven Field IMRT Plan	PTV_{el} PTV_{gr}	54	54
	Seven Field IMRT Plan	PTV_{gr}	16	70

PTV_{el} = Nasopharynx and electively irradiated nodal regions.
PTV_{gr} = Sites of gross disease in the nasopharynx and nodal regions.

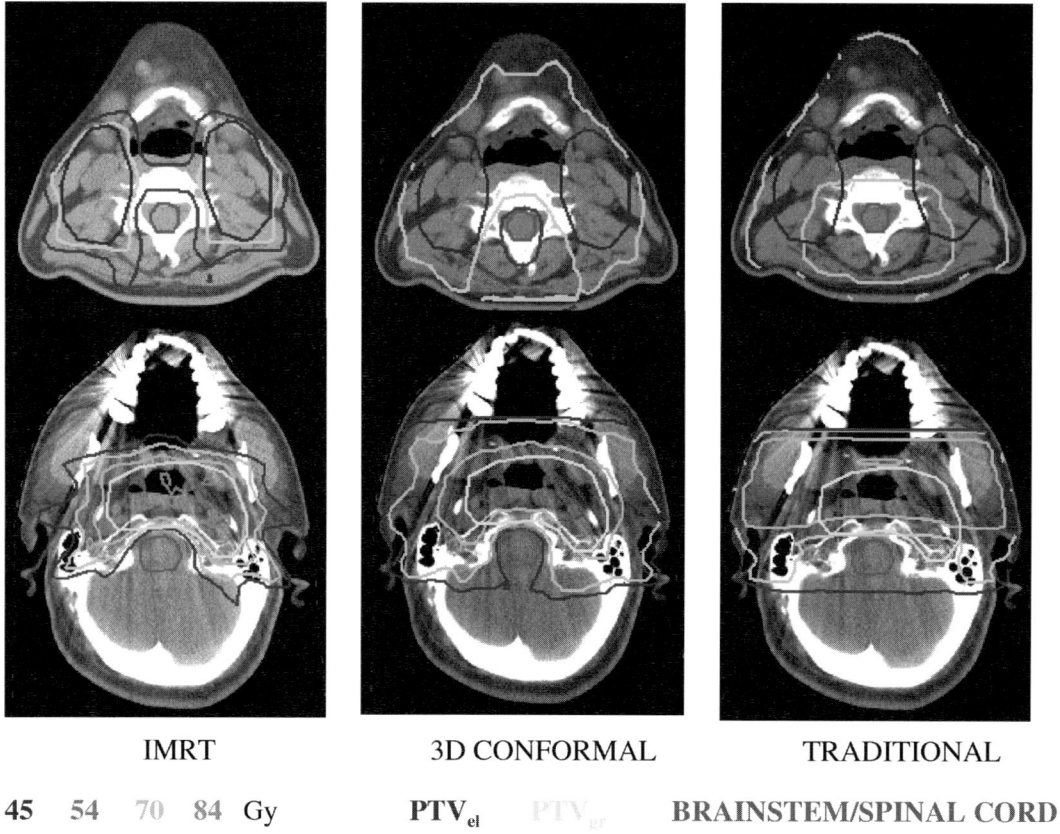

IMRT	3D CONFORMAL	TRADITIONAL

45   54   70   84   Gy   PTV$_{el}$ PTV$_{gr}$ BRAINSTEM/SPINAL CORD

FIGURE 10–1. Comparison of IMRT, 3-D conformal and traditional parallel-opposed field plans for the treatment of primary nasopharynx tumors. See COLOR PLATE 21.

radiolabeled with ^{18}F-positron emission tomography) images to localize metabolically active disease. The target volume includes the gross thyroid mass or thyroid bed, gross adenopathy, and regional lymph nodes (retropharyngeal, cervical, supraclavicular, and superior mediastinal nodes).

For papillary and follicular cancers, 50 to 54 Gy are administered to the elective nodal areas and 63 to 70 Gy to the gross disease. Anaplastic thyroid cancers are generally unresectable and

Table 10–2. Dose Volume Statistics Comparing IMRT, 3-D Conformal And Traditional Treatment Plans

Structure	Statistic	IMRT	3-D Conf.	Traditional
	Max. Dose (D$_{05}$)	81.8 Gy (3.3)	80.2 Gy (1.0)	74.2 Gy (2.5)
PTV$_{gr}$	Min. Dose (D$_{95}$)	69.4 Gy (6.2)	65.7 Gy (5.0)	54.6 Gy (1.7)
	Mean Dose	77.3 Gy (2.4)	74.6 Gy (2.2)	67.9 Gy (1.3)
Spinal Cord	Max. Dose (D$_{05}$)	34.5 Gy (5.5)	44.2 Gy (1.7)	49.1 Gy (0.9)
Brain stem	Max. Dose (D$_{05}$)	33.1 Gy (5.0)	43.3 Gy (2.7)	56.2 Gy (7.0)
Mandible	Max. Dose (D$_{05}$)	69.3 Gy (7.4)	73.9 Gy (5.3)	74.6 Gy (0.9)
	V$_{66Gy}$ (%)	9.7 % (5.9)	18.6 % (11.7)	26.8 % (13.9)
Temporal Lobes	Max. Dose (D$_{05}$)	58.7 Gy (12.5)	59.4 Gy (11.1)	67.0 Gy (3.5)
	V$_{60Gy}$ (%)	6.3% (7.1)	9.2% (13.1)	17.3% (8.8)
Parotid Gland	Mean Dose	60.5 Gy (8.9)	67.1 Gy (7.0)	67.0 Gy (4.7)
	V$_{50Gy}$ (%)	78.4% (21.2)	97.5% (2.9)	99.9% (0.1)

are administered low dose Adriamycin (10 mg/m^2) once weekly, 1.5 hours prior to the first radiotherapy fraction of the week. The radiotherapy is administered in 1.6 Gy fractions twice a day separated by 6 hours on 3 consecutive days of the week to a total dose of 57.6 Gy.

Happersett et al. (2000) compared IMRT and 3-D treatment plans for five thyroid cancer patients and determined that IMRT improved PTV dose uniformity and normal tissue doses particularly for the lung and spinal cord. As shown in figure 10–2, the IMRT technique consisted of six fields directed anteriorly, posteriorly, and obliquely. On average, the PTV dose uniformity improved by 10% and the volume of the PTV receiving at least 63 Gy increased from 37% with 3DCRT to 96% with IMRT.

Clinical Approach To IMRT Treatment For Head And Neck Cancer Patients

Over 250 patients have been treated to date, roughly half of these with primary nasopharynx cancers and the other half with thyroid carcinomas or recurrent tumors. The IMRT planning and treatment process for a typical patient with primary nasopharynx cancer is discussed below.

Consultation And Evaluation

A head and neck cancer patient is initially seen in consultation by the surgeon and radiation oncologist, and often by the medical oncologist. Consultation and evaluation for patients who will undergo treatment using IMRT is similar to that for other head and neck patients and should include the following:

1. History and physical examination of the head and neck region including indirect laryngoscopy and fiberoptic nasopharyngolaryngoscopy.
2. An illustration of the physical findings demonstrating the primary tumor extent and adenopathy.
3. Review of existing imaging studies and further workup as necessary.
4. Pretreatment dental consultation for the extraction of unsalvageable teeth in poor condition, the construction of mouth guards for patients with moderate to extensive tooth fillings, and the initiation of prophylactic fluoride therapy.
5. Pretreatment ophthalmology and audiology consultations for patients in whom the radiation may affect the orbital structures or ear.
6. Baseline thyroid function tests (T3, T4, TSH).

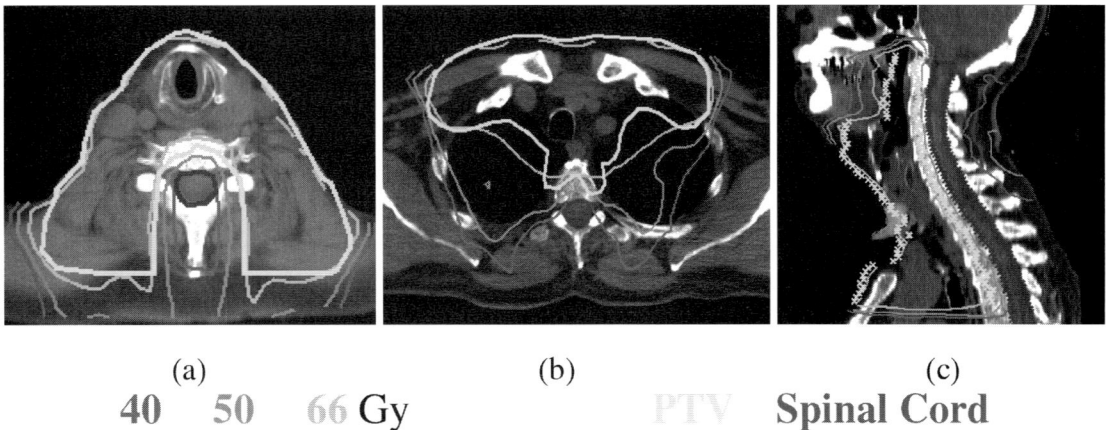

(a)   (b)   (c)

40   50   66 Gy   PTV   Spinal Cord

FIGURE 10–2. Axial and sagittal IMRT dose distributions for thyroid carcinoma designed using a six-field plan. See COLOR PLATE 22.

Simulation

Patients treated with IMRT must undergo CT-guided simulation, i.e., either conventional simulation followed by a CT in the treatment position or CT simulation. At MSKCC, all patients treated with IMRT undergo CT simulation. The patient is immobilized in the supine position, typically with the neck hyper-extended using a head rest and custom thermoplastic mold. When appropriate, a bite block is used to separate the mandible and tongue from the upper oral cavity, thereby facilitating a decrease in the irradiation of these structures and a decrease in side effects. A shoulder pull board is employed to bring the shoulders toward the feet, minimizing the amount of shoulder within the lateral or oblique fields. Palpable masses and incisional scars are outlined with radio-opaque material for later radiographic visualization.

For the CT study, intravenous contrast is used as needed to differentiate vasculature from masses or lymphadenopathy. CT images are acquired at 3 mm spacing from the vertex to a level approximately 5 cm inferior to the treatment volume. Accurate calculation of dose volume histograms (DVHs) and biological indices (e.g., normal tissue complication probability) mandate the inclusion of the entire extent of the relevant structures within the image set. The isocenter for the IMRT fields is positioned approximately in the center of the treatment volume and, if the supraclavicular nodes will be treated with separate fields, a second isocenter is placed midline at the inferior border of the IMRT fields.

Image Registration And Structure Delineation

In selected cases, other imaging studies, specifically, FDG-PET or magnetic resonance (MR), are obtained after the planning CT with the patient in the treatment position. They are registered with the planning CT and used in target and/or normal tissue delineation. The FDG-PET images can potentially improve tumor delineation over that with CT imaging alone (Jabour et al. 1993; Anzai et al. 1996; Chen et al. 1990). One limitation in the current use of PET data for treatment planning is the potentially large inaccuracy of the registration process, due to the relatively poor resolution of the PET emission and transmission images that sometimes may occur. Recently, combination PET-CT units have become available. These units provide both diagnostic quality CT and PET images without moving the patient between studies, improving the accuracy of the registration process.

MR studies are also often useful in the head and neck region, both for target and normal tissue localization. MR images may show the tumor extent much better than the CT scan alone.

It is imperative that the radiation oncologist be trained in the interpretation of all images used for structure localization. Consultation with neuroradiologists and nuclear medicine physicians may be necessary to accurately identify structures in the head and neck region or interpret PET-positive regions. An excellent reference with respect to the CT anatomy of the head and neck is the study by Nowak et al. (1999) who correlated borders of the surgical levels in the neck (I–VI) with structures seen on a CT scan, defining the six potential cervical lymph node regions and noting reproducible landmarks on the CT images. Wijers et al. (1999) developed a simplified protocol for delineating cervical target volume based on CT scans, and noted that target coverage and sparing of the major salivary glands were comparable to the more complex contouring guidelines of the above Nowak protocol. Chao et al. (2002) presented guidelines for target volume determination of head and neck lymph nodes. This was based on their analysis of nodal failure in IMRT-treated patients. The detailed and complex anatomy of the cranial nerve pathways is another important area of knowledge for the radiation oncologist. The gross anatomic information and associated axial CT images depicting these pathways are explicitly presented in the reference text by Leblanc (1995).

As mentioned previously, both MR and FDG-PET images are used in combination with CT for target and structure localization for selected head and neck cases at MSKCC including nasopharynx and thyroid cancers. The fusion of the magnetic resonance imaging (MRI) and/or

PET images with CT can aid in tumor localization for both the initial and cone down planning target volumes. Image fusion can also help ensure a minimum amount of normal tissue is treated, which is particularly important for patients undergoing high-dose irradiation with concurrent chemotherapy. Figure 10–3 depicts registered CT, MR, and PET-FDG images for a patient with advanced nasopharyngeal cancer. The treatment planning CT scan shows a very large area of abnormality that could represent either tumor or post-obstruction sinus/nasal changes depending on the area. The PET image reveals increased uptake in the bilateral retropharyngeal lymph nodes, indicating gross involvement at this level in contrast to the remaining cervical nodes. The MRI was superior to CT for evaluating the intracranial extension, due to its superior soft tissue resolution. The use of MR in this setting increases the certainty of covering the full extent of tumor, while minimizing exposure to healthy brain tissue.

Registered CT and PET images for a patient with thyroid cancer are shown in figure 10–4. This illustrates the significant extent of the thyroid carcinoma that is depicted as white areas of abnormality on the PET scan. It should be noted that normal structures may also show up as white

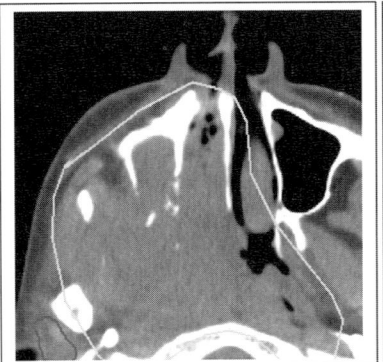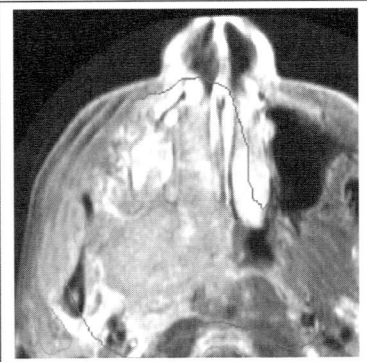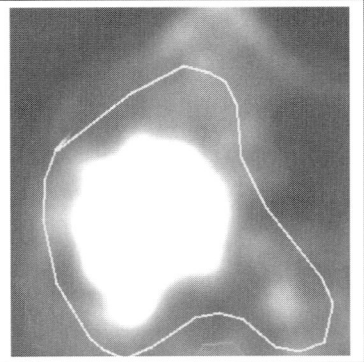

FIGURE 10–3. CT, MR, and FDG-PET images for a patient with primary nasopharynx cancer.

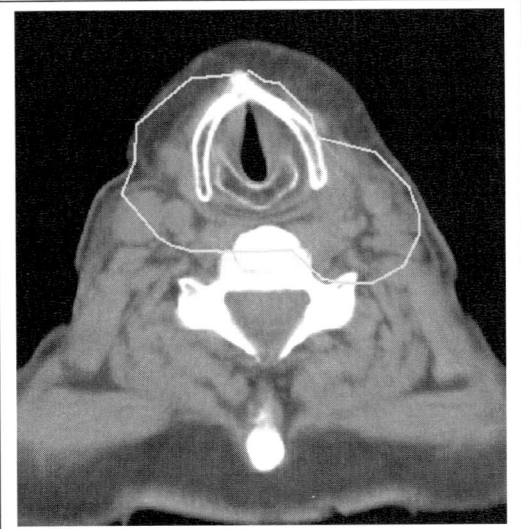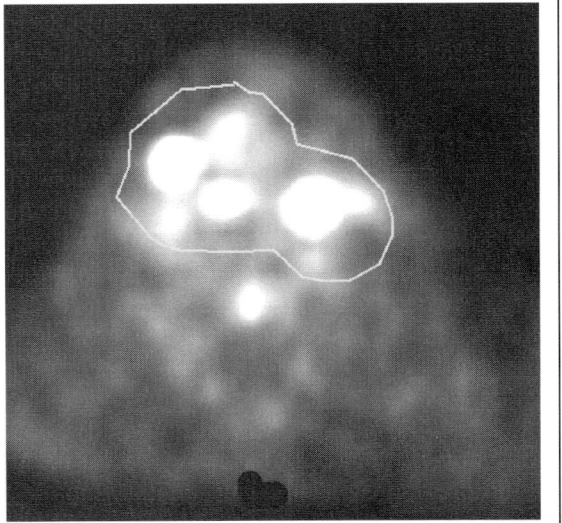

FIGURE 10–4. FDG-PET and CT images for a patient with thyroid carcinoma. The images were obtained with the patient in the treatment position, registered, and then used for localization of the target volume.

areas as is seen in the spinal cord in this slice and is not included in the PTV contours. Particularly in postoperative cases, the normal anatomy may be changed and it is quite difficult to distinguish tumor from normal structures. However, the fusion of the planning CT and PET images can allow one to more accurately contour the PTV. This approach is particularly effective in determining the cone down PTV, especially if high doses are to be administered.

Once all image sets needed for planning are acquired and registered, the target volumes (PTV) are defined by the physician. For nasopharyngeal carcinoma, two volumes are defined. PTV_{el} includes the entire nasopharynx and the elective nodal regions (CTV_{el}) with a uniform 0.5 cm margin. PTV_{gr} includes sites of gross disease in the nasopharynx and nodal regions (GTV) with a 1 cm margin everywhere except posteriorly along the skull, where a 0.5 cm margin is used. Normal tissues including the spinal cord, brainstem, bilateral parotid glands and cochlea, optical structures, and pituitary gland are also delineated as appropriate.

For target delineation using image fusion, the spatially registered image sets are displayed side by side in a split screen representation on the treatment planning computer. As the target or normal tissue is contoured in one screen, the corresponding regions are outlined automatically on the other screen. The radiation oncologist then discusses the case with the treatment planner, communicating pertinent information such as brief clinical findings, location of the primary tumor, adenopathy, high risk regions, adjacent critical structures, and the minimum dose to the tumor and maximum dose to critical structures.

Treatment Planning

The treatment planning process will be discussed in detail for one specific site, nasopharynx cancer. A similar process is used for other sites although the beam arrangements, clinical dose limits, and algorithm constraints are modified as needed.

  1. IMRT Treatment Approach

Patients with primary nasopharynx cancers are treated using seven coplanar 6 MV intensity-modulated (IM) fields, positioned every 30° from the posterior and lateral directions (figure 10–5), delivered with dynamic multileaf collimation (DMLC) (Spirou and Chui 1994).

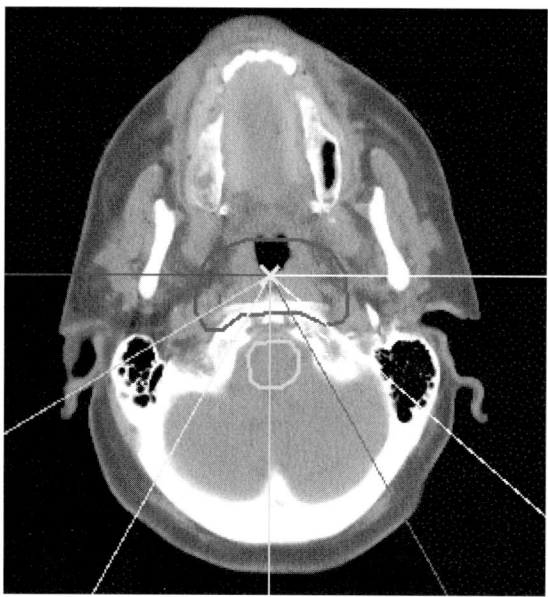

FIGURE 10–5. Beam directions for the MSKCC IMRT nasopharynx technique. Typically, ten treatment fields directed from seven gantry angles are used.

A prescription dose of 70 Gy is delivered to gross disease (PTV_{gr}) in the nasopharynx and neck, and 54 Gy to the elective nodal regions (PTV_{el}). The supraclavicular nodes are treated with a single anterior lower neck field, the superior edge of which is matched to the IMRT fields. Patients receive 1.8 Gy per fraction for the first 20 fractions (36 Gy) and thereafter, 1.8 Gy and 1.6 Gy in 2 daily fractions separated by a minimum of 6 hours for a total of 40 fractions. PTV_{el}, as defined above, is treated during the 1.8 Gy fractions, while treatment is limited to PTV_{gr} for the 1.6 Gy fractions.

2. Optimization and DMLC fields

The MSKCC inverse planning algorithm is based upon a conjugate gradient minimization method and least-squares objective function developed by Spirou and Chui (1998) that is discussed extensively in chapter 2 on optimization. During optimization, the desired dose distribution is specified in terms of optimization parameters, i.e., dose constraints for targets, dose and/or dose-volume constraints for normal tissues, and penalties that define the relative importance of each constraint. During the development of the IMRT technique for primary nasopharynx cancer, criteria for PTV dose uniformity and normal tissue doses were established by the clinicians (table 10–3). Subsequently, a set of optimization parameters were determined that produce acceptable dose distributions for most patients. This constraint template (table 10–3) serves as the starting point for planning, although, invariably, the constraints are manipulated to improve the dose distribution for individual patients.

The intensity profile derived for each IM beam is translated into a leaf-sequence file for DMLC delivery. Due to a limitation on DMLC field size on the Varian equipment, any IM fields with widths > ~14.5 cm are divided into two subfields that overlap by 1 to 2 cm. The total intensity required in the overlap is distributed between the two subfields, creating a "feathered" region. For most patients, the three most posterior fields (figure 10–5) are split, for a total of 10 DMLC fields delivered from seven gantry directions. Intensity profiles for an IM field, approximately 16 cm wide, and its "split" subfields are illustrated in figure 10–6. Note that the "split" occurs in the low-intensity region that overlies the spinal cord and brain

Table 10–3. MSKCC Clinical Dose Limits and Inverse Planning Algorithm Constraints for Primary Nasopharynx Tumors

Structure	Clinical Dose Limits	Inverse Planning Algorithm Constraint Template			
		Prescription Dose (%)	Maximum Dose (%) /Penalty	Minimum Dose (%) /Penalty	Dose (%)–%Volume Constraint/Penalty
PTV_{el}	$D_{95} \geq 50$ Gy (95% of 54 Gy) Max.Dose ≤64.8 Gy (120% of 54 Gy)	54 Gy (77%)	56.7 Gy (81%)/50	51.3 Gy (73%)/50	NA
PTV_{gr}	$D_{95} \geq 70$ Gy (100% of 70 Gy) Max.Dose ≤84 Gy (120% of 70 Gy)	70 Gy (100%)	66.5 Gy (105%)/50	73.5 Gy (95%)/50	NA
Spinal Cord	Max.Dose ≤45 Gy		28 Gy (40%)/50		NA
Brainstem	Max.Dose ≤50 Gy		35 Gy (50%)/50		NA
Parotid Gland	Mean Dose ≤26 Gy		68 Gy (98%)/50		≥21 Gy (30%) to ≤30% Volume/50
Cochlea	Max. Dose ≤60 Gy		56 Gy (80%)/50		NA

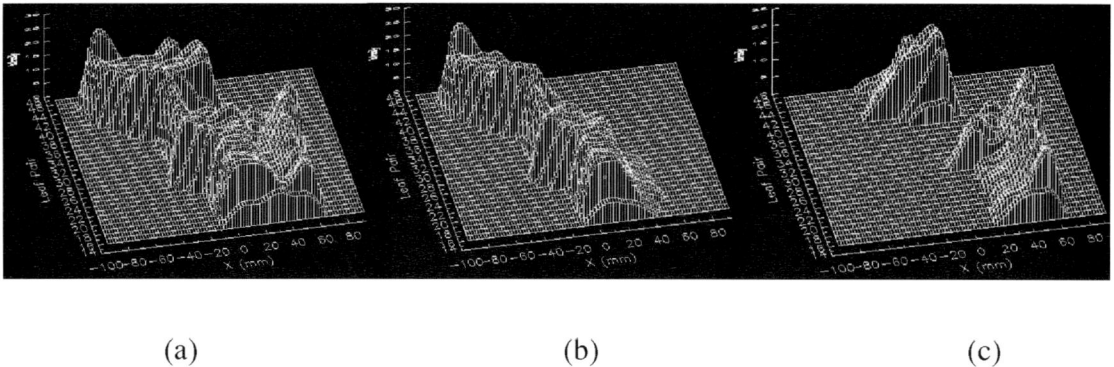

(a)   (b)   (c)

FIGURE 10–6. Intensity profiles for a left lateral nasopharynx field before (a) and after (b and c) field splitting to overcome the DMLC maximum field width limitation.

stem. This feature and the "feathering" in the overlap region help to minimize the potential dosimetric uncertainty due to field matching.

3. Plan Evaluation And Quality Assurance (QA)

The dose distributions and DVHs through the center of the nasopharynx and neck nodes for a typical patient with N0 disease are shown in figure 10–7. PTV_{gr} and PTV_{el} are well-covered by the 70 and 54 Gy prescription isodose levels while the spinal cord and brain stem receive approximately 45 Gy. The average mean dose to the parotid glands is 27 Gy and the cochleae receive an average maximum dose of 64 Gy.

For each beam of a completed IMRT plan, digitally reconstructed radiographs (DRRs) are generated displaying the so-called "DMLC aperture," corresponding to the complete irradiated area (figure 10–8). These DRRs are compared with portal images obtained with the DMLC aperture when the patient comes for treatment.

Prior to treatment, all IMRT plans undergo the following QA checks. A complete review of the plan is done by a physicist, including an evaluation of the dose distributions and DVHs and a review of all data used for patient treatment including the leaf motion files. The leaf motion files and intensity profiles are evaluated for unusual intensity peaks that might either limit treatment delivery or introduce dosimetric problems due to patient intra-fractional motion. An independent verification of the monitor unit setting for each treatment field is performed using a computer program specifically designed for this purpose. Discrepancies in excess of 2% warrant further investigation including film or ionization chamber dosimetry.

Daily Treatment with IMRT

Prior to the first treatment, the patient is positioned on the linear accelerator and images of each portal are obtained. Using the VARiS Treatment and Vision software, the DMLC aperture is automatically created, the portal images are acquired and then carefully compared against the DRRs.

IMRT for head and neck cancer patients requires precision and care on the part of the radiation therapist in every phase of the patient's daily setup and treatment procedure. Accuracy and reproducibility are vital if the tumor is to receive the proper dosage and overdosage to the normal critical structures avoided. The patient must be motivated and capable of fully cooperating with the setup and treatment procedure. Body motion must be kept to a minimum. Proper patient immobilization is absolutely essential (Chao et al. 2000). Close communication between the radiation oncologist and radiation therapist must be maintained throughout the course of treatment regarding any patient setup abnormality, problems, or difficulty. Weekly port films will

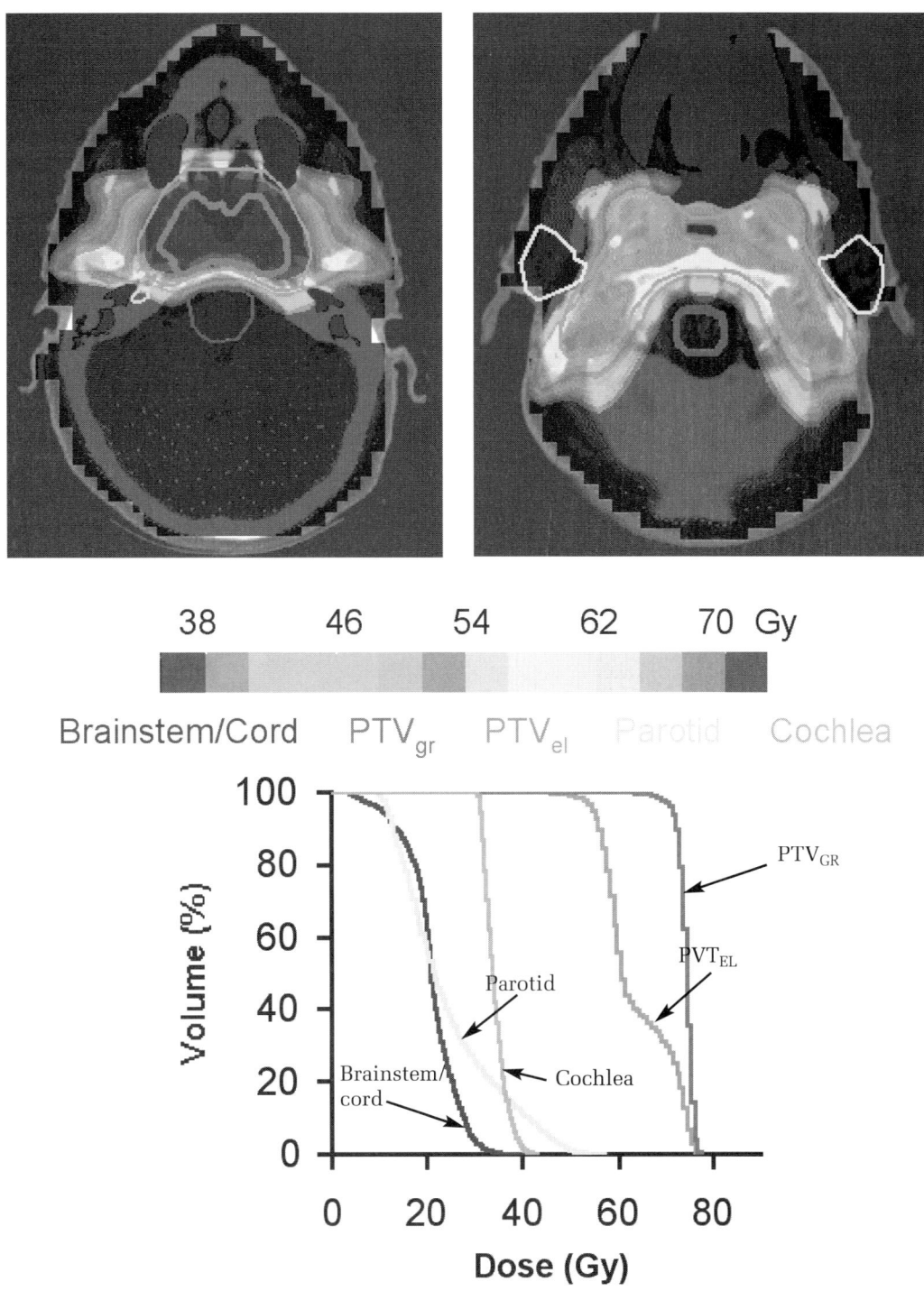

FIGURE 10–7. Axial dose distributions through the nasopharynx and neck for a seven-field IMRT plan.

be compared with the DRRs throughout the treatment course. Development of facial edema and significant weight loss may adversely affect the snug fit of the face mask and cause problems with patient immobilization and setup. Conservative modification of the mask, and on rare occasions, creation of a new mask and repeat CT simulation and treatment planning may be necessary.

Clinical Care During Radiation Therapy

The medical care of head and neck patients undergoing IMRT is the same as that required for those treated with conventional radiation therapy. For patients treated with a multi-modality approach including chemotherapy with cisplatin or carboplatin and 5FU, placement of a percutaneous endoscopic gastrostomy tube should be considered prior to initiation of treatment, particularly for elderly or frail patients, those who have lost a considerable amount of weight, or those with problems of dysphasia or odynophagia at the outset.

Weekly status evaluations are mandatory, with some patients requiring more frequent evaluations as treatment progresses. During these visits, an interval history is obtained reviewing the development of skin symptomology, a sore mouth or throat, xerostomia, decreased or abnormal taste, hoarseness, or dysphagia. A pertinent examination will note the status of the portal skin; the location and size of the primary tumor; the location, size, mobility, tenderness, and texture of lymphadenopathy; the presence of mucositis and of oral Candida. Routine measurement of the patient's weight and complete blood counts will be obtained.

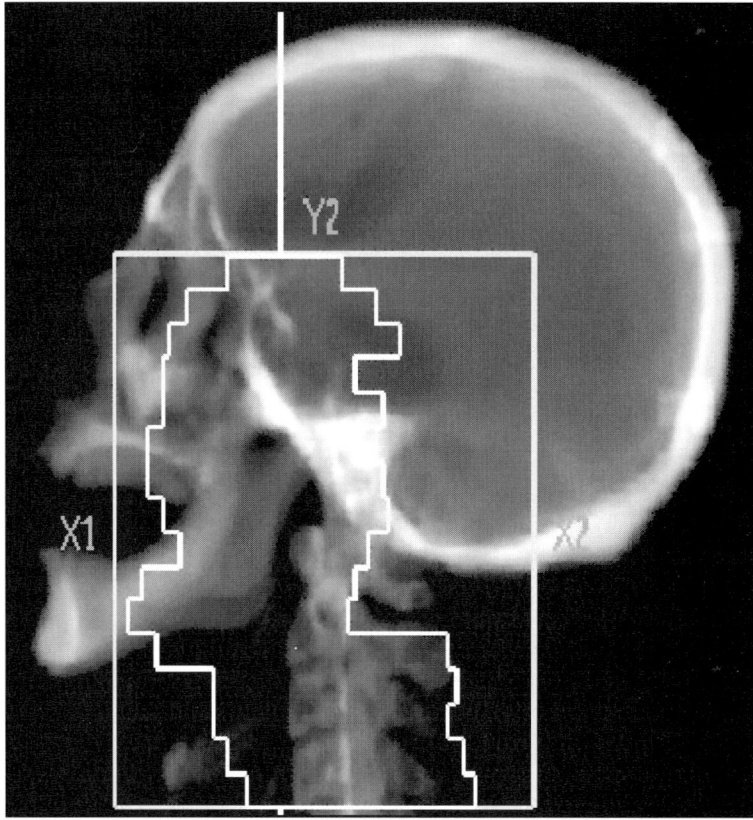

FIGURE 10–8. Digitally-reconstructed radiograph (DRR) of a left lateral IMRT field for primary nasopharynx cancer. The aperture indicates the initial and final positions of the MLC leaves used for the dynamic delivery.

Occasionally, a patient will develop acute parotitis within the first 12 hours after commencement of therapy when the treatment volume includes the parotid gland. The symptoms include swelling in the parotid regions associated with localized pain and occasionally a low-grade temperature. Although parotitis generally resolves on its own, we would prescribe a non-steroidal anti-inflammatory drug and reassure the patient should this occur.

Tumoritis can develop at ~20 Gy and is characterized by a mucosal inflammation that represents the true extent of the tumor and may necessitate subsequent modification of the portal (Wang 1997). As the primary tumor is followed during the course of treatment, those lesions that show progression or minimal regression should be reevaluated by all the physicians on the case. This may be a situation where surgery is indicated.

As patients develop the acute side effects of xerostomia, decreased or abnormal taste, and mucositis, appropriate supportive medical intervention is mandatory. Narcotic analgesics should be considered and modified as necessary to provide adequate pain relief. This may involve the use of long-acting morphine sulphate or a fentanyl patch as well as immediate-release morphine sulphate for any break-through pain. Intravenous hydration is sometimes indicated for patients who have become dehydrated due to poor oral intake or who have difficulties with their percutaneous endoscopic gastrostomy tube. A fair percentage of patients may develop oral Candida which may be asymptomatic, present with acute development or exacerbation of a sore mouth or throat or even perhaps an abnormal taste. Initiation of an antifungal medication will usually resolve the problem rapidly.

The head and neck cancer patients that we have treated with IMRT do not appear to have responses that differ from that of patients undergoing conventional treatment. Our clinical observation has been that these patients have similar acute reactions and can be managed quite adequately as presented above.

Post-Treatment Follow-up

Immediately upon completion of radiation therapy, routine follow-up evaluations should be scheduled. If the patient is elderly, frail, or having a particularly difficult time with acute mucositis, esophagitis, and weight loss, we will see this patient weekly until sufficient recovery has occurred which generally takes 3 to 4 weeks. The patients in better condition are seen monthly for 2 months and every 1 to 3 months thereafter, alternating with the other physicians on the case unless we are monitoring the response of a mass. Baseline imaging studies are considered 2 to 3 months post treatment and may include a CT or MRI of the head and neck or and/or a PET scan.

Serial endocrine screening will be important for patients who have had irradiation of these organs, including the pituitary gland and thyroid gland. Thyroid function tests, including a TSH, are obtained every 6 months post treatment for up to 5 years. Clinical hypothyroidism has been seen in ~5% of adults and a higher percentage in children whose thyroids have been irradiated. There is a 20% to 25% risk of chemical hypothyroidism overall, but this increases to 66% in patients who have also undergone a hemithyroidectomy. In patients who are found to have a significant elevation of the TSH, thyroid hormone replacement therapy is initiated irrespective of the T3 and T4 values, which oftentimes may be within normal limits. Patients who have their pituitary gland irradiated should periodically undergo irradiated screening every 1 to 2 years post irradiation. These tests should evaluate LH, FSH Serum cortisol, prolactin, TSH, free T4, and GH. For male patients, a testosterone level is also included.

Patients who have received radiation to the oral cavity or oropharynx should be seen routinely by the dental service for an indefinite period of time. Fluoride prophylaxis, initiated at the start of treatment, should be continued. These patients are advised that their dentist should be fully informed of their radiation therapy as well as the potential risk for osteoradionecrosis that may result from subsequent dental surgery.

Occasionally, a patient will develop Lhermitte's syndrome, a benign, transient myelopathy presumably due to radiation-induced demyelination in the cervical spinal cord. This can begin 1 to 3 months post therapy and last an average of 3 to 4 months and as long as 9 to 12 months. This is characterized by the development of a symmetrical, instantaneous, shooting, electrical sensation that radiates down the spine and extremities upon flexation of the neck, but it does not progress and requires no treatment.

High-dose irradiation, especially when combined with chemotherapy, can lead to late effects of the soft tissues. Particular attention should be directed towards the development of trismus as well as the decreased range of motion of the tongue, mandible, neck, and shoulders. Physical therapy should be considered as it may decrease or prevent these post treatment functional deficits resulting from fibrosis and scarring.

Some patients may develop dysphagia during treatment that could become chronic and significant. Post treatment dysphagia may be due to dysfunction of the pharyngeal muscles, the development of an esophageal stricture, or even possibly the presence of a tumor. We have observed in some patients whose pharyngeal muscles received high-dose radiation therapy, particularly in conjunction with chemotherapy, significant swallowing problems long after completion of treatment. Appropriate medical evaluation must be performed for diagnosis and appropriate therapeutic intervention.

Effect Of Setup Uncertainty

Several studies evaluating setup uncertainty specifically for head and neck patients (Rabinowitz et al. 1985; Verellen et al. 1997; Hunt et al. 1993) have measured standard deviations of systematic and random uncertainties of approximately 2 to 3 mm. Hunt et al. compared the impact of setup errors on target coverage, spinal cord, and brainstem dose for 3-D and parallel opposed dose distributions. Systematic setup errors led to target underdosage and normal tissue overdosage with both techniques, but the 3-D distributions were more susceptible to the effects of both random and systematic errors because of the increased conformality. Although, studies evaluating the effect of setup uncertainty specifically for IMRT head and neck distributions have not yet been done, the impact may be even more significant because of their exquisite conformality and the presence of steep dose gradients.

In a preliminary evaluation for nasopharynx cancer, we have modeled the effects of random and systematic setup uncertainty on IMRT dose distributions for selected patients using the following technique. Briefly, the random treatment uncertainty is modeled by convolving the planned dose distribution with a normal frequency distribution with a standard deviation of 2 mm, a technique developed by Chui, Kutcher, and LoSasso (1992). After blurring the dose distribution to show the effect of random uncertainty, systematic uncertainty is modeled using a Monte Carlo simulation. A normal frequency distribution with a standard deviation of 2 mm in each direction is sampled 500 times. For each iteration the dose distribution corrected for random uncertainty is shifted according to the sampled systematic error and the doses to the targets and normal tissues are recalculated. Confidence limit DVHs, i.e., the DVHs expected with a given statistical confidence for a population of patients, can then be calculated and analyzed. The 2 mm standard deviations of random and systematic setup uncertainty were estimated from our own analysis of setup for head and neck patients (Hunt et al. 1993). The planned, 95%, and 5% confidence limit DVHs for the PTV, brain stem, and cochlea are shown in figure 10–9 for a patient planned with the IMRT dose painting technique described below **(IMRT Dose Painting and Dose Escalation for Primary Nasopharyngeal Carcinoma)**. The tight conformality of the dose distribution and, in particular, the manner in which the high-dose region surrounds the spinal cord and brainstem is responsible for the observed increase in normal tissue dose and degradation in target coverage. Based on this preliminary analysis, we currently limit the spinal cord and

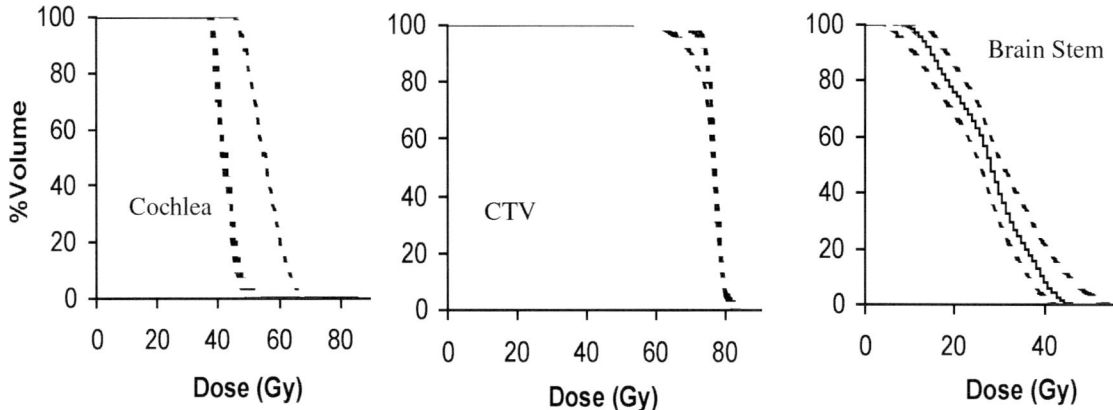

FIGURE 10–9. Dose-volume histograms for an IMRT dose painting simultaneous boost technique incorporating the effects of random and systematic setup uncertainties, each with a 2 mm standard deviation. For each structure, the planned DVH is indicated by the solid line. The 5% and 95% confidence limit DVHs are indicated by the dashed lines.

brainstem to 45 and 50 Gy, respectively, approximately 5 Gy less than what would be accepted with a less conformal distribution.

Vital Organ Sparing

Much research has been reported on the sparing of the parotid gland with IMRT. This organ is responsible for 60% to 65% of the saliva produced and xerostomia is a major acute and late side effect that can have a significant negative impact on a patient's quality of life. Salivary output may begin to decrease within 24 hours after the first fraction of 2.25 Gy (Mira et al. 1981), fall by 50% or more by the seventh day (Dreizen et al. 1977; Franzen et al. 1992; Eneroth, Herikson, and Jakobson 1972) and be barely measurable by the end of treatment (Dreizen et al. 1977; Franzen et al. 1992; Mossman 1986). Six months post treatment, the stimulated salivary flow is reduced exponentially for each parotid gland at a rate of approximately 4% per Gy of mean parotid dose (Chao et al. 2001). However, the patient's subjective evaluation of their xerostomia may not reliably correlate with objective salivary flow measurements.

Patients note a dry mouth secondary to decreased salivary output and a very thick and viscous saliva. These problems may or may not improve after completion of treatment. Recovery may be observed in some patients for the first 2 to 3 years post treatment, depending on their age, the volume of gland irradiated, the dose per fraction, and the total dose. If more than 50% of the gland was spared from radiation, the probability for some recovery is increased (Cooper et al. 1995).

The use of IMRT for head and neck cancer can reduce the parotid volume treated to high doses and result in an improved salivary status (van Dieren et al. 2000; Kuppersmith et al. 1999; Chao et al. 2001; Eisbruch et al. 1999, 2001; Wu et al. 2000). Wu et al. (2000) noted that IMRT plans were more conformal than 3DCRT plans and that the dose to the parotid glands could be reduced with an equivalent coverage of the primary tumor and regional lymph nodes. Eisbruch et al. (2001) reported that the sparing of major salivary glands by IMRT increased late salivary flow rates, and improved xerostomia. They also noted that sparing of minor salivary glands in the oral cavity was a significant independent predictor of xerostomia. An analysis of dose, volume, and function relationships in the parotid glands after IMRT suggested that a mean parotid dose of ≤26 Gy was necessary for substantial sparing of the gland (Eisbruch et al. 1999).

Butler et al. (1999) noted in a review of 20 IMRT patients with primary head and neck cancer that the mean dose to the ipsilateral parotid gland was 23 Gy and to the contralateral gland, 21 Gy. Chao et al. (2000) reported that the mean parotid dose in their series was approximately 20 Gy. They also noted that 3%±1.4% of the primary target received less than 95% of the prescribed dose due to proximity of the target volume to the critical structures such as the parotid gland. The steep dose gradient commonly noted in head and neck IMRT plans in which the tumors are in very close proximity to the parotid gland means that part of the primary target volume may be underdosed. Further research is necessary to determine whether this is of clinical importance.

At MSKCC, we attempt to limit the mean dose to at least one parotid gland to ≤26 Gy, without compromising target coverage. For patients with nasopharyngeal cancer, mean parotid doses of ≤26 Gy are usually achievable in patients with negative or unilateral neck disease for the gland on the side without neck disease. In the presence of gross adenopathy, mean parotid doses of ~35 Gy are typical. To ensure adequate target coverage in the presence of a parotid sparing technique, we require that at least 95% of the electively irradiated and gross disease PTVs receive 50 Gy and 70 Gy, respectively. Our experience indicates that, as a consequence of parotid sparing, the dose to the oral cavity and the submandibular glands may increase and therefore should be carefully evaluated. Dose distributions and DVHs derived from inverse planning with and without an attempt to spare the parotid glands are compared in figure 10–10 for a patient with a negative neck. A small section of the electively irradiated PTV receives less than 54 Gy in order to achieve a mean parotid dose of 26 Gy.

Radiation therapy may also lead to sensorineural hearing loss, particularly when the radiation is delivered in combination with chemotherapy. As discussed by Choi et al. (2000), hearing loss occurs more frequently in patients whose cochlea received ≥70 Gy. Unfortunately, the cochleae are often within or adjacent to the high dose target in the nasopharynx and could easily receive doses in excess of 70 Gy. Grau et al. (1991) demonstrated that doses of 50 to 70 Gy to the cochlea may lead to hearing loss within 18 months, but that the probability and severity of the loss was correlated with dose and the sound frequency. IMRT can be used to spare the cochlea, but similar to parotid sparing attempts, may lead to compromised target coverage (figure 10–11). A retrospective analysis of 20 of our nasopharynx patients indicated that the cochleae straddle or lie within the PTV_{gr} in approximately three-fourths of all patients and that their position greatly affects the dose they receive. At MSKCC, we currently limit the dose to the cochlea for nasopharynx patients to ≤60 Gy when possible given the target constraints outlined in table 10–3. To further guard against tumor underdosing, an additional constraint requiring ≥99% of the GTV to receive ≥70 Gy is used. IMRT dose painting and the simultaneous boost technique, as discussed in the next section, may facilitate lower cochlear doses without compromising target coverage. Figure 10–12 shows dose distributions and DVHs for a patient planned with the IMRT dose painting simultaneous boost technique according to the target criteria in table 10–3 and the additional GTV coverage criteria. The cochleae receive a maximum dose of 30 Gy.

In addition to the structures already discussed, other important normal structures to consider when planning head and neck tumors with IMRT include the orbital structures, optic nerves, optic chiasm, brain, and mandible. Generally, the dose limits for these structures can easily be achieved with IMRT except in patients with extensive superior disease or cranial extension.

IMRT Dose Painting And Dose Escalation For Primary Nasopharyngeal Carcinoma

The use of IMRT to plan non-uniform dose distributions within the target volume for head and neck patients has been described recently by Wu et al. (2000). The advantages of this technique include the delivery of a biologically higher dose to the gross disease and the simplification of

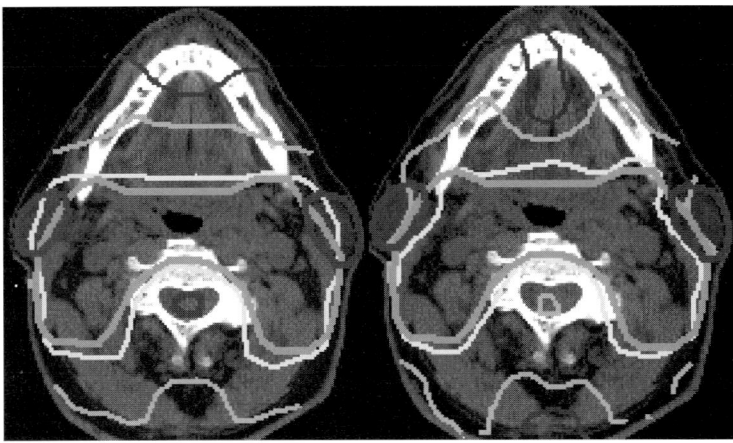

(a) Without parotid sparing (No PS) (b) With parotid sparing (PS)

20 30 54 **Gy**   **PTV**$_{el}$ **Parotid Glands**

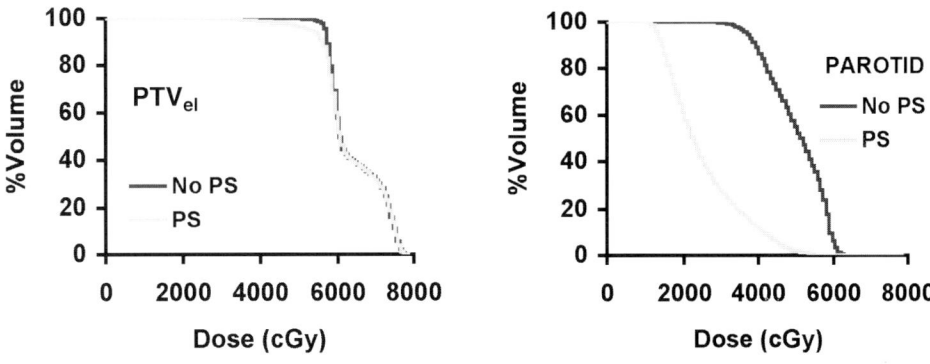

FIGURE 10–10. Axial dose distributions and DVHs for the PTV$_{el}$ and parotid glands with and without an IMRT parotid sparing technique for a patient with N0 disease. See COLOR PLATE 23.

the planning and treatment processes since only one plan is designed and used for the entire treatment course. The study by Wu examined the potential of this *concomitant* or *simultaneous integrated* boost technique for a variety of head and neck tumors and concluded that, using IMRT, they could achieve distributions similar to conventional fractionation in terms of target coverage and normal tissue doses. The technique being considered at MSKCC would deliver 70.2 Gy to the nasopharynx in 30 fractions (2.34 Gy/fraction) while concomitantly treating the neck to 54 Gy (1.8 Gy/fraction). Typical dose distributions and DVHs comparing this technique with our standard two-phase technique are shown in figure 10–13. Coverage of the PTV$_{gr}$ is very similar to that achieved with a conventional treatment strategy and IMRT, although the mean dose to PTV$_{el}$ is slightly less.

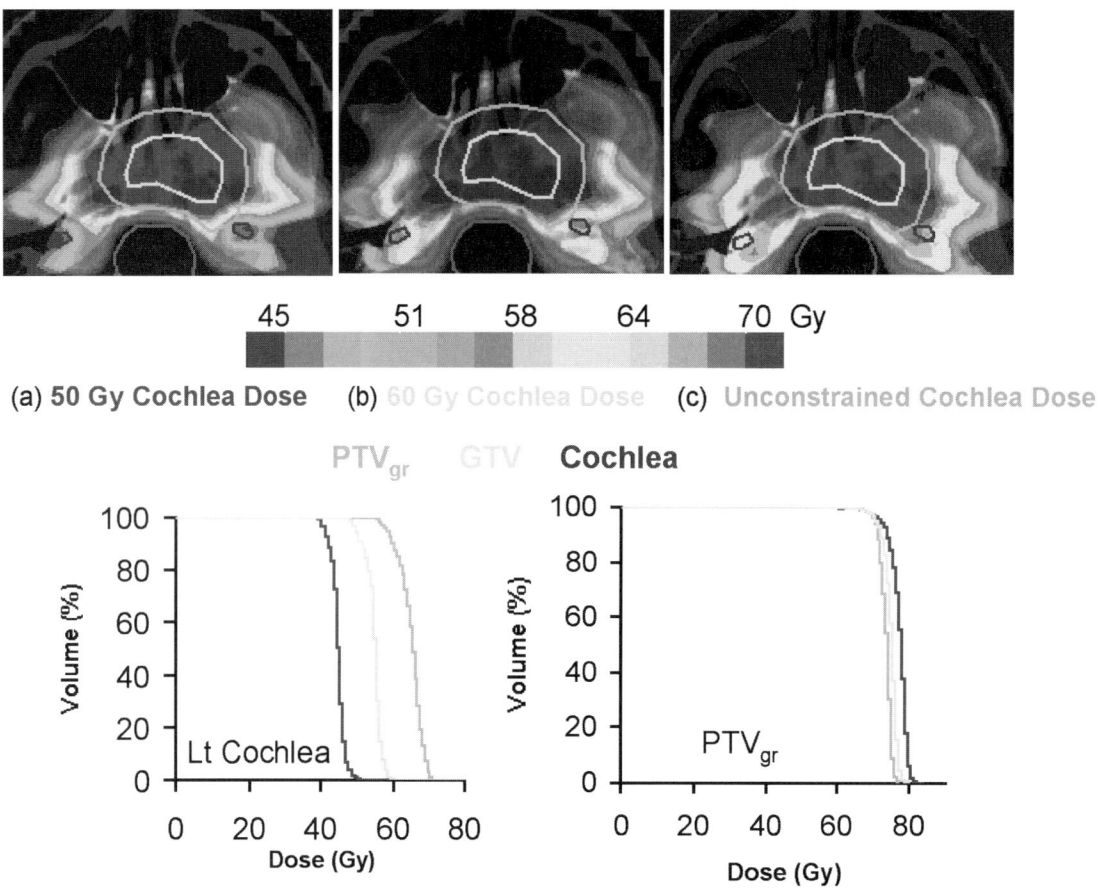

FIGURE 10–11. Dose distributions and DVHs illustrating the effect of cochlear sparing on PTV coverage when the cochleae lie within PTV_{gr}. Results are shown for three plans: unconstrained cochlear dose, 50 Gy, and 70 Gy maximum cochlear dose. See COLOR PLATE 24.

Plans designed with a simultaneous boost technique are inherently more conformal than those using a two-phase technique, leading to lower doses to critical structures in very close proximity to the 70 Gy volume such as the cochlea. Additional dose distributions in the nasopharynx for the current MSKCC two-phase treatment and the simultaneous boost technique are compared in figure 10–14. These distributions were generated with constraints on target coverage and dose uniformity, spinal cord and brainstem maximum dose, and parotid mean dose, but no constraint on the cochlea. The conformity index (Volume (70 Gy)/Volume (PTV)) is significantly improved with the simultaneous boost technique. As a result of this improved conformality, the dose to the cochlea is also less.

Treating Recurrent Head And Neck Tumors With IMRT

The management of head and neck cancer patients with recurrent disease who have previously received radical radiation therapy is a challenge. For these patients, surgery is often the treatment of choice, provided the lesion is resectable, the patient is able to tolerate the procedure, and that

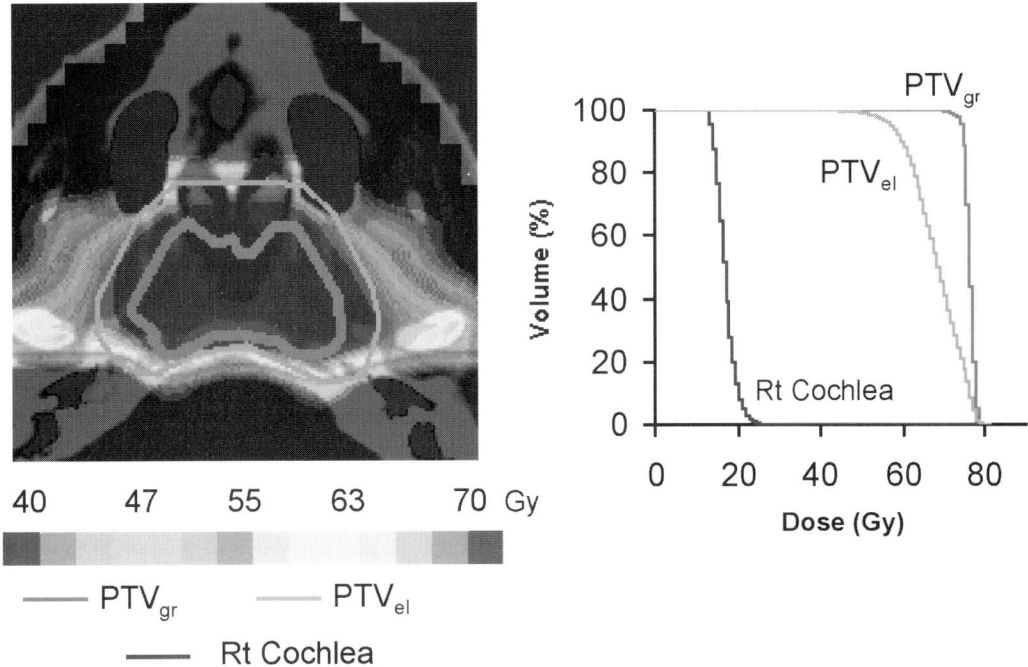

FIGURE 10–12. Dose distribution and DVH of the PTV_{gr} and cochlea for a patient planned with the 70 Gy IMRT dose painting simultaneous boost technique. The cochleae receive a maximum of 30 Gy with acceptable target coverage and dose uniformity.

recovery and rehabilitation are likely. For patients who are not surgical candidates, re-irradiation can be considered. This is a highly select group with true local-regional recurrence rather than persistence of the disease post radiation therapy. Evaluation of the following items is necessary: (1) patient condition; (2) time interval since completion of initial radiation therapy; (3) radiation dosage administered; (4) tolerance of treatment and any complications; (5) anatomic location of recurrence and adjacent normal critical structures; (6) condition of previously irradiated tissues; (7) symptoms related to the recurrence; (8) life expectancy. Relative contraindications to re-irradiation include: (1) poor condition; (2) recurrence less than 6 months from initial radiation therapy; (3) ultra-high radiation doses; (4) massive tumor recurrence equivalent to T3–T4 lesions; (5) location of recurrence in or around the central nervous system. The dose of re-irradiation will need to be in the range of 60 to 65 Gy (De Crevoisier et al. 1998; Stevens, Britsch, and Moss 1994; Wang 1994). Moderate dose re-irradiation of 45 Gy will most likely not be effective and may not even provide sufficient palliation. The more limited the disease, the better the chances for a meaningful therapeutic intervention.

Meticulous treatment planning and careful radiation technique are necessary. At MSKCC, IMRT is often used for re-irradiation cases although brachytherapy as the primary treatment or as a boost is also considered. Conservative margins around the tumor of no more than 1 cm are appropriate. The central nervous system must not be directly re-irradiated by the primary beam. Only patients who understand the high risks involved and exhibit a willingness to accept the possible complications should be considered for re-irradiation.

Re-irradiation of recurrent nasopharyngeal cancer with a stage equivalent to a T1 or T2 lesion has frequently been reported in the literature (Teo et al. 1998; Wang 1993). PET or MR image fusion can aid in localization of the tumor allowing for a limited treatment volume with a high

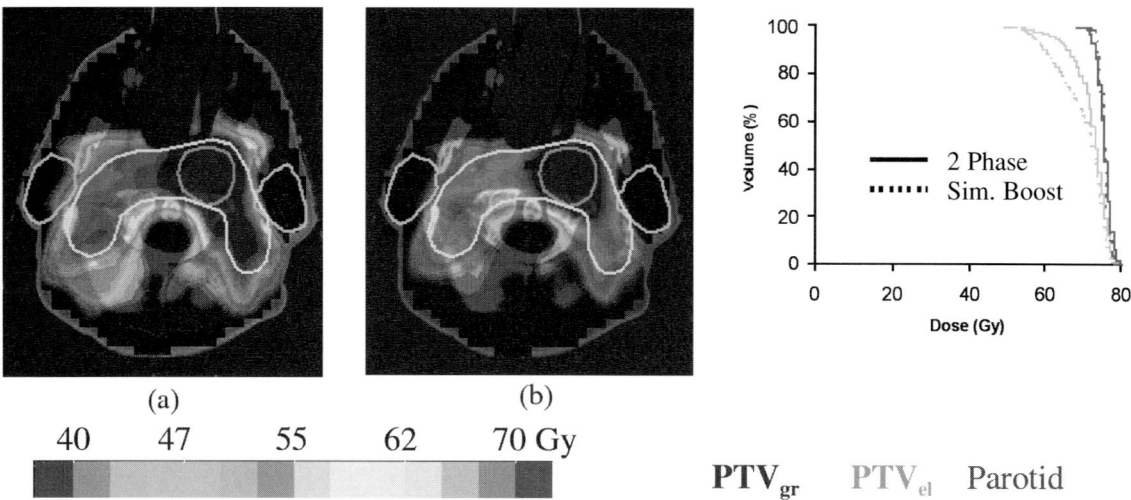

Figure 10–13. Comparison of IMRT nasopharynx dose distributions for a two-phase treatment technique (a) and a simultaneous boost treatment (b). See COLOR PLATE 25.

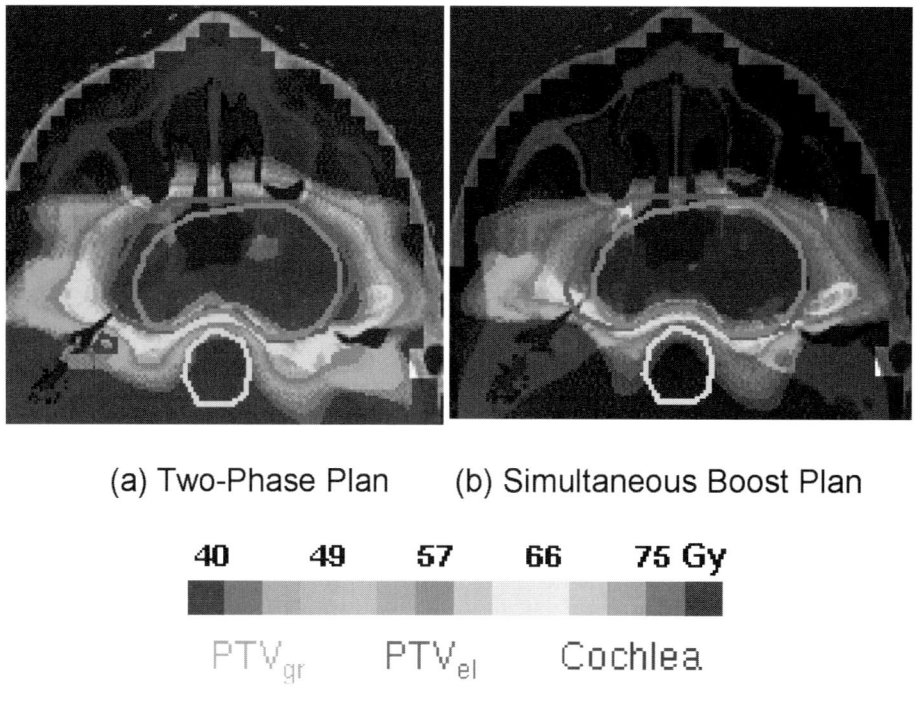

FIGURE 10–14. Comparison of the MSKCC two-phase (a) and simultaneous boost (b) techniques for the treatment of nasopharynx cancer. Using the simultaneous boost technique, the PTV$_{gr}$ receives 70.2 Gy in 30 fractions (2.34 Gy/fraction) while the electively irradiated volume, PTV$_{el}$, receives 1.8 Gy/fraction to 54 Gy. The 70 Gy conformity index is 2.4 for the two-phase plan and 1.7 for the simultaneous boost. The maximum cochlea doses with the two-phase and simultaneous boost plans are 70 and 63 Gy, respectively. See COLOR PLATE 26.

level of confidence that the disease is contained within the treatment region. Limited volume IMRT with a brachytherapy boost can potentially provide good local control although brachytherapy may not be suitable in some cases because of the size and extent of the tumor. These patients must be treated with IMRT often with concurrent chemotherapy. Special care must be exercised in analyzing the IMRT plan with respect to the central nervous system, orbit, and optic nerve and chiasm doses.

Other regions of the head and neck have been treated with re-irradiation with promising preliminary results. Studies have shown good palliation of symptoms and some have reported long term control with 20% 2-year survivals and 15% to 17% 5-year survivals (DeCrevoisier et al. 1998; Stevens, Britsch, and Moss 1994). The results appear better than those obtained with the use of chemotherapy alone.

The incidence of late toxicity is greater than that noted after primary radical radiation therapy. Several studies have suggested however that these adverse effects were still deemed acceptable (Wang 1994).

Re-irradiation with IMRT at MSKCC

At MSKCC, IMRT is used routinely in the treatment of recurrent cancers, primarily nasopharynx although additional sites including paranasal sinus have been treated. Typically, doses of ~60 Gy are prescribed with dose limits to the spinal cord and brainstem of 10 to 12 Gy. The doses delivered to other normal tissues, particularly optical pathway structures, are determined after consideration of the previous therapy.

Typically, five to nine equally spaced treatment fields are used, including non-coplanar beam arrangements when beneficial. Although PTV constraints similar to those for PTV_{gr} in table 10–3 are used, the individual needs of each patient are considered when defining the normal tissue constraints. IMRT and 3-D conformal plans for recurrent nasopharynx disease, created using seven field beam arrangements, are compared in figure 10–15. IMRT improved the target dose uniformity and led to lower doses to the optical structures. Our experience has been that it is generally not possible to achieve the extremely low normal tissue doses required for these cases with IMRT alone. Conventional cerrobend blocking is combined with dynamic multileaf IMRT when normal tissue doses must be less than approximately 30% of the prescription.

Summary

The concave shape of the target volume and close proximity of normal tissues make head and neck tumors ideal cases for IMRT. Multiple planning studies within the past 5 years have clearly demonstrated the ability of IMRT to improve target coverage and dose uniformity for many head and neck sites. More exciting, perhaps, is the opportunity to impact the significant normal tissue morbidity associated with head and neck radiotherapy and the ability to deliver different fractionation schemes using the SMART technique or IMRT "dose painting." Clinical results have already established that IMRT can be used to decrease the morbidity associated with the irradiation of the salivary glands. It remains to be seen if similar improvements in hearing loss can be achieved without sacrificing local control.

Head and neck sites have always been among the most challenging, complex and time consuming to plan. Our experience with head and neck IMRT planning has been that the complete planning process can require 10 to 12 hours of a planner's time, more if image fusion is required. Site-specific class solutions, specifying the clinical criteria for target and normal tissue doses in as much detail as possible, the beam arrangements and constraint templates to use as starting points for planning are mandatory for efficient head and neck IMRT planning. Despite the increased complexity and time required to produce them, IMRT dose distributions offer

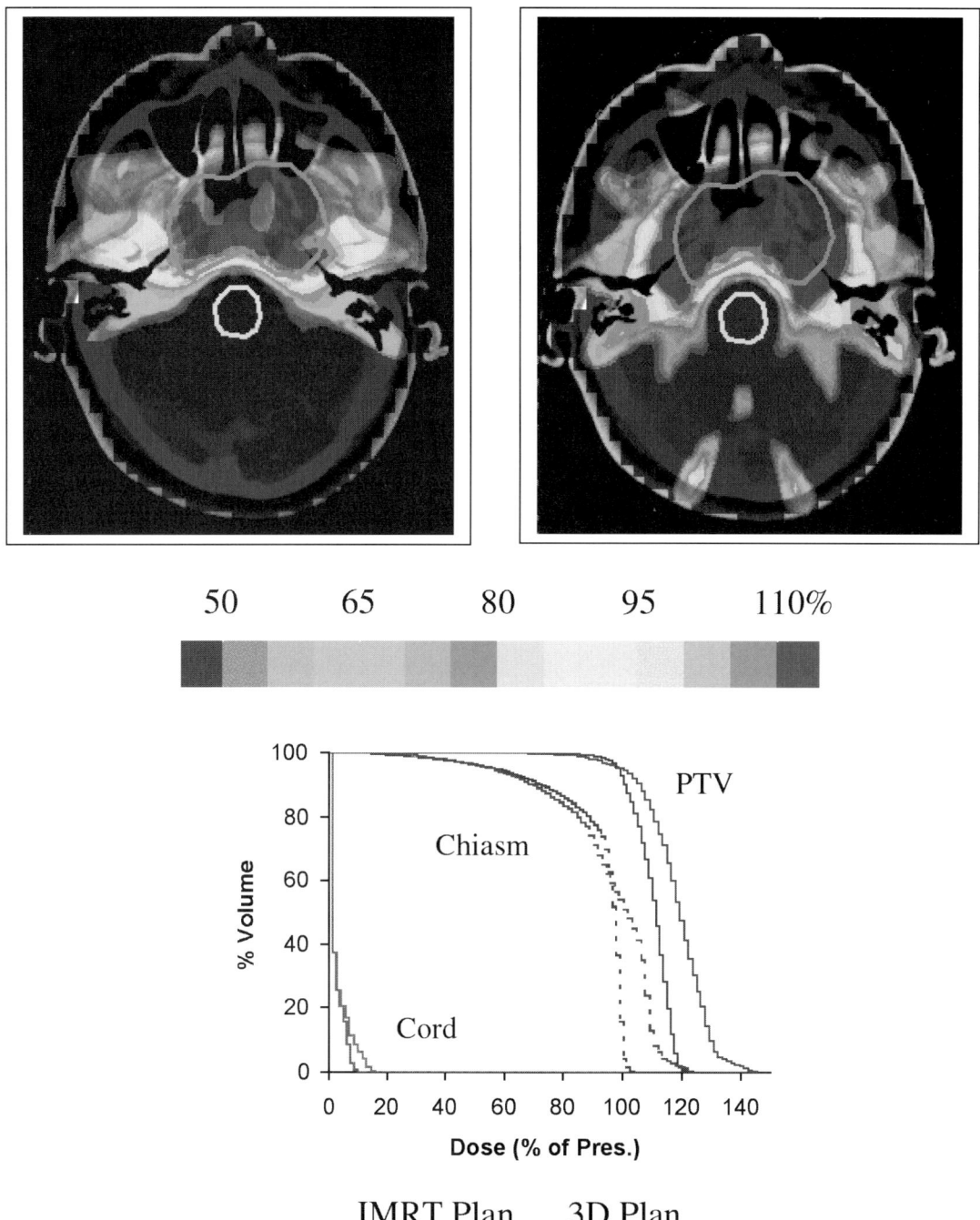

FIGURE 10–15. Dose distributions and DVHs for IMRT (left) and 3-D (right) plans for recurrent nasopharyngeal cancer. Both plans utilize a seven-field beam arrangement as shown in figure 5. For the 3-D plan, wedges and cerrobend blocks over the spinal cord and brainstem are used to create a concave dose distribution. Cerrobend blocks over the cord and brainstem are also used with the IMRT plan to achieve doses of <20% of the prescription to these structures. IMRT significantly improves target dose coverage and uniformity and normal tissue doses compared to the 3-D plan. See COLOR PLATE 27.

significant improvements over 3-D conformal plans. We believe that IMRT will become the standard method of treatment for many head and neck sites.

References

Anzai, Y., W. R. Carrol, D. J. Quint, C. R. Bradford, S. Minoshima, G. T. Wolf, and R. H. Wahl. (1996). "Recurrence of head and neck cancer after surgery or irradiation: Prospective comparison of 2-deoxy-2-[F-18]fluoro-D-glucose PET and MR imaging diagnoses." *Radiol.* 200:135–141.

Blanco, A. I., and K. S. C. Chao. "Intensity-Modulated Radiation Therapy and Protection of Normal Tissue Function in Head and Neck Cancer" in *Principles and Practice of Radiation Oncology: Updates.* Vol. 3 (No. 3). New York: Lippincott Williams & Wilkins Healthcare, 2002.

Bourhis, J., and J. P. Pignon. (1999). "Meta-analyses in head and neck squamous cell carcinoma. What is the role of chemotherapy?" *Hematol. Oncol. Clin. North Am.* 13:769–775, vii.

Boyer, A. L., P. Geis, W. Grant, and M. Carol. (1997). "Modulated beam conformal therapy for head and neck tumors." *Int. J. Radiat. Oncol. Biol. Phys.* 39:227–236.

Brizel, D. M., M. E. Albers, S. R. Fisher, R. L. Scher, W. J. Richtsmeier, V. Hors, S. L. George, A. T. Huang, and L. R. Prosnitz. (1998). "Hyperfractionated irradiation with or without concurrent chemotherapy for locally advanced head and neck cancer." *N. Engl. J. Med.* 338:1798–1804.

Butler, E. B., B. S. Teh, W. H. Grant 3rd, B. H. Uhl, R B. Kuppersmith, J. K. Chui, D. T. Donovan, and S. Y. Woo. (1999). "Smart (simultaneous modulated accelerated radiation therapy) boost: A new accelerated fractionation schedule for the treatment of head and neck cancer with intensity modulated radiotherapy." *Int. J. Radiat. Oncol. Biol. Phys.* 45:21–32.

Chao, K. S., D. A. Low, C. A. Perez, and J. A. Purdy. (2000). "Intensity-modulated radiation therapy in head and neck cancers: The Mallinckrodt experience." *Int. J. Cancer* 90:92–103.

Chao, K. S., J. O. Deasy, J. Markman, J. Haynie, C. A. Perez, J. A. Purdy, and D. A. Low. (2001). "A prospective study of salivary function sparing in patients with head-and-neck cancers receiving intensity-modulated or three-dimensional radiation therapy: Initial results." *Int. J. Radiat. Oncol. Biol. Phys.* 49:907–916.

Chao, K. S., F. J. Wippold, G. Ozyigit, B. N. Tran, and J. F. Dempsey. (2002). "Determination and delineation of nodal target volumes for head-and-neck cancer based on patterns of failure in patients receiving definitive and postoperative IMRT." *Int. J. Radiat. Oncol. Biol. Phys.* 53(5):1174–1184.

Chen, B. C., C. Hoh, B. Choi, et al. (1990). "Evaluation of primary head and neck tumor with PET-FDG." (Abstract). *Clin. Nucl. Med.* 15:758.

Choi, S., S. Wolden, D. Pfister, A. S. Budnick, S. Levegrün, A. Jackson, M. A. Hunt, M. J. Zelefsky, B. Singh, J. O. Boyle, and D. H. Kraus. (2000). "Ototoxicity following combined modality therapy for nasopharyngeal carcinoma." Proceedings of the 42nd Annual ASTRO Meeting, Boston, MA, October 22-26, 2000. *Int. J. Radiat. Oncol. Biol. Phys.* 48:261.

Chui, C. S., G. J. Kutcher, and T. Lossaso. (1992). "A convolution method for incorporating uncertainties in dose calculation." (Abstract). *Med. Phys.* 19:814.

Cooper, J. S., K. Fu, J. Marks, and S. Silverman. (1995). "Late effects of radiation therapy in the head and neck region." *Int. J. Radiat. Oncol. Biol. Phys.* 31:1141–1164.

De Crevoisier, R., J. Bourhis, C. Domenge, P. Wibault, S. Koscielny, A. Lusinchi, G. Mamelle, F. Janot, M. Julieron, A. M. Leridant, P. Marandas, J. P. Armand, G. Schwaab, B. Luboinski, and F. Eschwege. (1998). "Full-dose reirradiation for unresectable head and neck carcinoma: Experience at the Gustave-Roussy Institute in a series of 169 patients." *J. Clin. Oncol.* 16:3556–3562.

Dreizen, S., L. R. Brown, T. E. Daly, and J. B. Drane. (1977). "Prevention of xerostomia-related dental caries in irradiated cancer patients." *J. Dent. Res.* 56:99–104.

Eisbruch, A., L. H. Marsh, M. K. Martel, J. A. Ship, R. Ten Haken, A. T. Pu, B. A. Fraass, and A. S. Lichter. (1998). "Comprehensive irradiation of head and neck cancer using conformal multisegmental fields: Assessment of target coverage and noninvolved tissue sparing." *Int. J. Radiat. Oncol. Biol. Phys.* 41:559–568.

Eisbruch, A., R. K. Ten Haken, H. M. Kim, L. H. Marsh, and J. A. Ship. (1999). "Dose, volume, and function relationships in parotid salivary glands following conformal and intensity-modulated irradiation of head and neck cancer." *Int. J. Radiat. Oncol. Biol. Phys.* 45:577–587.

Eisbruch, A., H. M. Kim, J. E. Terrell, L. H. Marsh, L. A. Dawson, and J. A. Ship. (2001). "Xerostomia and its predictors following parotid-sparing irradiation of head-and-neck cancer." *Int. J. Radiat. Oncol. Biol. Phys.* 50:695–704.

Eneroth, C. M., C. O. Herikson, and P. A. Jakobson. (1972). "Effect of fractionated radiotherapy on salivary gland function." *Cancer* 30:1142–1153.

Franzen, L., U. Funegard, T. Ericson, and R. Henriksson. (1992). "Parotid gland function during and following radiotherapy of malignancies in the head and neck. A consecutive study of salivary flow and patient discomfort." *Eur. J. Cancer* 28:457–462.

Fu, K. K. (1997). "Combined-modality therapy for head and neck cancer." *Oncology* (*Huntingt*) 11:1781–1790, 1796; discussion 1796, 179.

Fu, K. K., T. F. Pajak, A. Trotti, C. U. Jones, S. A. Spencer, T. L. Phillips, A. S. Garden, J. A. Ridge, J. S. Cooper, and K. K. Ang. (2000). "A Radiation Therapy Oncology Group (RTOG) phase III randomized study to compare hyperfractionation and two variants of accelerated fractionation to standard fractionation radiotherapy for head and neck squamous cell carcinomas: First report of RTOG 9003." *Int. J. Radiat. Oncol. Biol. Phys.* 48:7–16.

Grau, C., K. Moller, M. Overgaard, J. Overgaard, and O. Elbrond. (1991). "Sensori-neural hearing loss in patients treated with irradiation for nasopharyngeal carcinoma." *Int. J. Radiat. Oncol. Biol. Phys.* 21:723–728.

Happersett, L., M. Hunt, L. Chong, et al. (2000). "Intensity modulated radiation therapy for the treatment of thyroid cancer." *Int. J. Radiat. Oncol. Biol. Phys.* 48:351.

Hunt, M. A., G. J. Kutcher, C. Burman, D. Fass, L. Harrison, S, Leibel, and Z. Fuks. (1993). "The effect of setup uncertainties on the treatment of nasopharynx cancer." *Int. J. Radiat. Oncol. Biol. Phys.* 27:437–447.

Hunt, M. A., M. J. Zelefsky, S. Wolden, C. S. Chui, T. LoSasso, K. Rosenzweig, L. M. Chong, S. V. Spirou, L. Fromme, M. Lumley, H. A. Amols, C. C. Ling, and S. A. Leibel. (2001). "Treatment planning and delivery of intensity-modulated radiation therapy for primary nasopharynx cancer." *Int. J. Radiat. Oncol. Biol. Phys.* 49:623–632.

Jabour, B. A., Y. Choi, C. K. Hoh, S. D. Rege, J. C. Soong, R. B. Lufkin, W. N. Hanafee, J. Maddahi, L. Chaiken, J. Bailet, et al. (1993). "Extracranial head and neck: PET imaging with 2-[F-18]fluoro-2-deoxy-D-glucose and MR imaging correlation." *Radiol.* 186:27–35.

Kuppersmith, R. B., S. C. Greco, B. S. Teh, D. T. Donovan, W. Grant, J. K. Chui, R. B. Cain, and E. B. Butler. (1999). "Intensity-modulated radiotherapy: First results with this new technology on neoplasms of the head and neck." *Ear Nose Throat J.* 78:238, 241–246, 248 passim.

Leblanc, A. *The Cranial Nerves.* Berlin: Springer-Verlag, 1995.

Lefebvre, J. L., D. Chevalier, B. Luboinski, A. Kirkpatrick, L. Collette, and T. Sahmoud. (1996). "Larynx preservation in pyriform sinus cancer: preliminary results of a European Organization for Research and Treatment of Cancer phase III trial. EORTC Head and Neck Cancer Cooperative Group." *J. Natl. Cancer Inst.* 88:890–899.

Leibel, S. A., G. J. Kutcher, L. B. Harrison, D. E. Fass, C. M. Burman, M. A. Hunt, R. Mohan, L. J. Brewster, C. C. Ling, and Z. Y. Fuks. (1991). "Improved dose distributions for 3D conformal boost treatments in carcinoma of the nasopharynx." *Int. J. Radiat. Oncol. Biol. Phys.* 20:823–833.

Mira, J. G., W. B. Wescott, E. N. Starcke, and I. L. Shannon. (1981). "Some factors influencing salivary function when treating with radiotherapy." *Int. J. Radiat. Oncol. Biol. Phys.* 7:535–541.

Mossman, K. L. (1986). "Gustatory tissue injury in man: radiation dose response relationships and mechanisms of taste loss." *Br. J. Cancer Suppl.* 7:9–11.

Nowak, P. J., O. B. Wijers, F. J. Lagerwaard, and P. C. Levendag. (1999). "A three-dimensional CT-based target definition for elective irradiation of the neck." *Int. J. Radiat. Oncol. Biol. Phys.* 45:33–39.

Pfister, D. G., L. B. Harrison, E. W. Strong EW, and G. J. Bosl. (1992). "Current status of larynx preservation with multimodality therapy." *Oncology* (*Huntingt*) 6:33–38, 43; discussion 44, 47.

Pignon, J. P., J. Bourhis, C. Domenge, and L. Designe. (2000). "Chemotherapy added to locoregional treatment for head and neck squamous-cell carcinoma: Three meta-analyses of updated individual data. MACH-NC Collaborative Group. Meta-Analysis of Chemotherapy on Head and Neck Cancer." *Lancet* 355:949–955.

Rabinowitz, I., J. Broomberg, M. Goitein, K. McCarthy, and J. Leong. (1993). "Accuracy of radiation field alignment in clinical practice." *Int. J. Radiat. Oncol. Biol. Phys.* 11:1857–1867.

Spirou, S. V., and C. S. Chui. (1994). "Generation of arbitrary intensity profiles by dynamic jaws or multi-leaf collimators." *Med. Phys.* 21:1031–1041.

Spirou, S. V., and C. S. Chui. (1998). "A gradient inverse planning algorithm with dose-volume constraints." *Med. Phys.* 25:321–333.

Stevens, K. R., Jr., A. Britsch, and W. T. Moss. (1994). "High-dose reirradiation of head and neck cancer with curative intent." *Int. J. Radiat. Oncol. Biol. Phys.* 29:687–698.

Sultanem, K., H. K. Shu, P. Xia, C. Akazawa, J. M. Quivey, L. J. Verhey, and K. K. Fu. (2000). "Three-dimensional intensity-modulated radiotherapy in the treatment of nasopharyngeal carcinoma: The University of California-San Francisco experience." *Int. J. Radiat. Oncol. Biol. Phys.* 48:711–722.

Teo, P. M., W. H. Kwan, A. T. Chan, W. Y. Lee, W. W. King, and C. O. Mok. (1998). "How successful is high-dose (> or = 60 Gy) reirradiation using mainly external beams in salvaging local failures of nasopharyngeal carcinoma?" *Int. J. Radiat. Oncol. Biol. Phys.* 40:897–913.

van Dieren, E. B., P. J. Nowak, O. B. Wijers, J. R. van Sornsen de Koste, H. van der Est, D. P. Binnekamp, B. J. Heijmen, and P. C. Levendag. (2000). "Beam intensity modulation using tissue compensators or dynamic multileaf collimation in three-dimensional conformal radiotherapy of primary cancers of the oropharynx and larynx, including the elective neck." *Int. J. Radiat. Oncol. Biol. Phys.* 47:1299–1309.

Verellen, D., N. Linthout, D. van den Berge, A. Bel, and G. Storme. (1997). "Initial experience with intensity-modulated conformal radiation therapy for treatment of the head and neck region." *Int. J. Radiat. Oncol. Biol. Phys.* 39:99–114.

Vokes, E. E., R. R. Weichselbaum, S. M. Lippman, and W. K. Hong. (1993). "Head and neck cancer." *N. Engl. J. Med.* 328:184–194.

Wang, C. C. (1993). "Decision making for re-irradiation of nasopharyngeal carcinoma." *Int. J. Radiat. Oncol. Biol. Phys.* 26:903.

Wang, C. C. (1994). "To re-irradiate or not to re-irradiate." *Int. J. Radiat. Oncol. Biol. Phys.* 20:913.

Wang, C. C. *Radiation Therapy for Head and Neck Neoplasms.* New York: Wiley-Liss, 1997.

Wijers, O. B., P. C. Levendag, T. Tan, E. B. van Dieren, J. van Sornsen de Koste, H. van der Est, S. Senan, and P. J. Nowak. (1999). "A simplified CT-based definition of the lymph node levels in the node negative neck." *Radiother. Oncol.* 52:35–42.

Withers, H. R., J. M. Taylor, and B. Maciejewski. (1988). "The hazard of accelerated tumor clonogen repopulation during radiotherapy." *Acta Oncol.* 27:131–146.

Wolden, S. L., M. J. Zelefsky, M. A. Hunt, K. E. Rosenzweig, L. M. Chong, D. H. Krause, D. G. Pfister, and S. A Leibel. (2001). "Failure of a 3D conformal boost to improve radiotherapy for nasopharyngeal carcinoma." *Int. J. Radiat. Oncol. Biol. Phys.* 49:1229–1234.

Wu, Q., M. Manning, R. Schmidt-Ullrich, and R. Mohan. (2000). "The potential for sparing of parotids and escalation of biologically effective dose with intensity-modulated radiation treatments of head and neck cancers: A treatment design study." *Int. J. Radiat. Oncol. Biol. Phys.* 46:195–205.

11

IMRT Of
Pediatric Cancers

Kamil Yenice
Suzanne Wolden

Introduction

Radiation therapy is an important component in the multimodality management of pediatric cancers. For many pediatric malignancies including rhabdomyosarcoma, Ewing's sarcoma, and brain tumors such as medulloblastoma, relatively high doses of radiation are required for treatment with curative intent. Most of these tumors are near multiple radiosensitive critical organs that place strict dose constraints on the radiation treatment. Because of the toxic effects of radiation on growth and development and the risk of secondary malignancies, children may especially benefit from the exquisitely conformal treatment techniques with intensity-modulated radiation therapy (IMRT). While for many adult tumors, dose escalation is an important goal of intensity modulation, for pediatric tumors, local control is often excellent with conventional techniques (Bleyer 1990). Therefore, the primary goal for pediatric patients is to decrease normal tissue doses and long-term toxicity with a secondary goal of improved target coverage.

This chapter will elucidate the specific treatment planning issues related to commonly irradiated tumors in the pediatric population. Since many of the technical aspects of treatment planning are common to those for adult population and have already been covered in the previous chapters, we will omit these details here.

Treatment Planning Of Pediatric Central Nervous System Tumors

Brain tumors are the most common solid tumors of childhood. Prognostic factors primarily depend on the tumor pathological type and grade. Treatment may consist of any combination of surgery, radiotherapy, and chemotherapy. Post-treatment 5-year survival rates vary from a bleak 5% for grade IV astrocytomas to high levels of 80% for stabdard risk medulloblastoma. In this section we will describe Memorial Sloan-Kettering Cancer Center (MSKCC) IMRT treatment planning for the two most common pediatric brain tumors, medulloblastoma and astrocytoma.

Medulloblastoma

Medulloblastoma is the most frequent malignant brain tumor in pediatric patients. Following surgery, combined radiation and chemotherapy have become the standard treatment regimen for this cancer. While this treatment is highly effective, it is associated with a number of long-term complications including neurocognitive impairment, growth retardation, and hearing impairment. The lower doses now used for treating the whole brain and spine have helped to decrease neurocognitive and growth effects. However, there remains a great concern for significant ototoxicity from cisplatin chemotherapy and high-dose radiation to the cochlea. Loss of hearing can further impair cognitive function and quality of life (McHaney et al. 1992; Schell et al. 1989). To decrease the toxic effects of radiation on hearing, 3-D conformal and IMRT techniques have been somewhat successfully applied for the treatment of medulloblastoma (Huang et al. 2002; Fukunaga-Johnson et al. 1998; Miralbell et al. 1997). In our clinic, IMRT is routinely used to deliver the posterior fossa boost treatment after craniospinal axis irradiation for medulloblastoma. Patients receive initial treatment to the whole spine and brain to either 24 or 36 Gy followed by a boost to the posterior fossa for a total dose of 55.8 Gy. The planning steps are as follows:

1. Patient Immobilization And CT Simulation

CT simulation is used to acquire a single data set that is used for planning both the craniospinal axis irradiation and the posterior fossa boost. Patients are immobilized in the prone position using a custom whole body and face mask Vac-Lock™ system. A custom Aquaplast® mold over the posterior surface of the head provides additional support and immobilization. Patients are anesthetized as necessary for simulation and treatment. CT (computed tomography) images are acquired from the vertex to the sacroiliac (SI) joints, with 3 mm spacing through the skull, regions near the spine field isocenters, and the junction planes and 5 mm spacing throughout the rest of the treatment volume. The isocenter for the cranial fields is placed just posterior to the bony orbit to minimize irradiation of the eyes and the posterior fossa boost fields are positioned by shifts from this point. The height of the patient determines whether one or two spinal fields will be needed and the position of the each spinal field isocenter. All isocenters are marked on the patient along with other alignment points used for daily setup.

2. PTV And Normal Tissues

All patients undergo MR-CT (magnetic resonance-computed tomography) image fusion to define the target volume for the posterior fossa irradiation. Rather than the entire posterior fossa, planning target volume (PTV) consists of the pre-chemotherapy volume with a 1 to 2 cm margin. Additional normal tissues of concern include the cervical spine, the brainstem, hypothalamic-pituitary axis, temporal lobes, parotid glands, and cochlear structures. For the craniospinal irradiation, the cribriform plate and eyes are contoured to aid in the design of the cranial fields.

3. Treatment Field Design

The right and left lateral cranial fields extend inferiorly to the level of C3–C4. Collimator and couch rotations are used to geometrically align the cranial and spine fields. Spinal treatment is delivered using one or two posterior 6 MV photon fields depending on patient height. The junctions between the spine fields and between the spine and brain fields are moved by 1 cm twice during the treatment course. The IMRT posterior fossa boost field arrangement is customized for each patient but typically consists of five fields: one posterior and four posterior obliques. The posterior field is frequently modified to a non-coplanar plane by 20° to 30° from the axial plane by rotating the couch 90° and adjusting the gantry angle in an attempt to minimize the exposure to the eyes.

4. Optimization And DMLC Fields

As explained in the earlier chapters, MSKCC inverse planning optimization is performed on the target and normal tissue structures defined for optimization only. Typically, these are structures generated from the clinical volumes, such as PTV, cochlea, chiasm, etc., defined by the physician. The optimization variables including dose constraints specifying the prescription dose, minimum and maximum dose for the target, and dose and dose-volume constraints for all relevant normal tissue volumes with the corresponding relative penalties are set to achieve specific dose distribution. A representative table of optimization variables for a five-field posterior fossa boost IMRT plan for a stage M2 pediatric medulloblastoma case is shown in table 11–1. These optimization parameters are specific to MSKCC inverse planning algorithm and the conversion of the beam intensities to leaf sequence. Therefore, they may not be optimal for other systems used.

5. Plan Evaluation

Forward dose calculations are performed for planar dose distributions and clinical volumes using the IMRT fields for the plan, after the calculated intensities are converted to leaf motion files. A plan is generated using all the beams computed in the optimization routine and typically normalized to the minimum peripheral dose covering the PTV. Planar dose distributions and dose-volume histogram (DVH) statistics for the PTV and each normal organ are evaluated to validate the plan. The cochlear structures are typically limited to less than 65% of the posterior fossa prescription dose, which is significantly less than conventional treatment. Doses to the oral cavity and parotid glands are also evaluated in an attempt to minimize the doses to these structures as well.

6. Comparison With 3-D Conventional And Traditional Techniques

To evaluate the relative benefit from IMRT we retrospectively compared the IMRT boost plans with conventional 3-D technique and traditional opposed laterals for four patients with medulloblastoma. Patients received 2340 to 3960 cGy craniospinal radiotherapy followed by IMRT boost to gross disease for a total of 5400 to 5580 cGy. 3-D conventional plans used shaped physical blocks, wedges, and multiple beam arrangements identical to those for the IMRT plans. All conventional 3-D plans were prescribed to the highest isodose level covering the entire PTV. Cumulative DVHs were generated for the PTV and critical organs of interest. From the calculated histograms, the maximum, minimum, and mean dose to each structure were obtained. In figure 11–1, the axial dose distributions for the traditional, conventional, and IMRT plans are illustrated for a medul-

Table 11–1. Clinical Dose Limits and Inverse Planning Algorithm Constraints for a Stage M2 Pediatric Posterior Boost IMRT Plan. The prescription for the boost phase required 1620 cGy to PTV, which had already received 3960 cGy from the cranial component of initial craniospinal axis irradiation. All parameters are relative

Optimization Structure	Clinical Structure	Inverse Planning Algorithm Constraint			
		Prescription Dose	Maximum Dose/Penalty	Minimum Dose/Penalty	Dose-Volume Constraint/Penalty
PTV_opt	PTV	100	105 /105	95 /100	N/A
Rt Cochlea_opt	Rt Cochlea	N/A	56 /60	N/A	N/A
Lt Cochlea_opt	Lt Cochlea	N/A	56 /60	N/A	N/A
Brainstem_opt	Brainstem	N/A	90 /60	N/A	N/A
Cord_opt	Cord	N/A	90 /60	N/A	N/A
Pituitary_opt	Pituitary Gland	N/A	20 /60	N/A	N/A

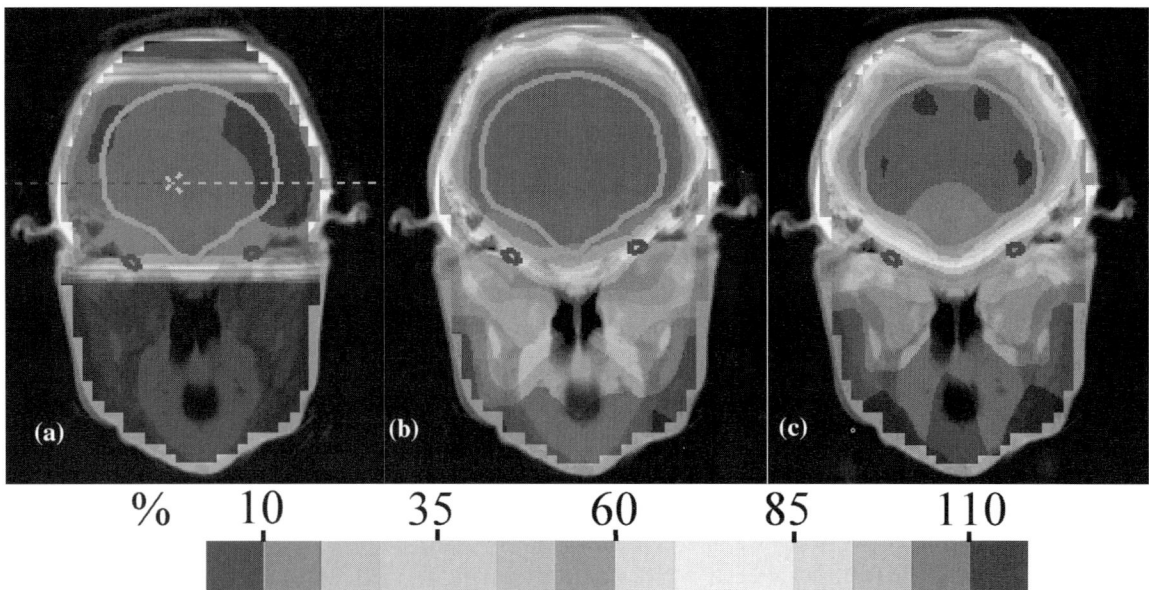

FIGURE 11–1. Axial dose distribution for posterior fossa boost treatment using (a) traditional opposed lateral fields, (b) 3-D conformal, (c) IMRT. PTV and cochlear structures are shown by green and blue contours, respectively. See COLOR PLATE 28.

loblastoma posterior fossa treatment. The corresponding DVHs for the cochlear structures are shown in figure 11–2. The IMRT plan for this case utilized the optimization parameters of table 11–1. The cochlea dose is reduced significantly with IMRT in comparison with the traditional and 3-D conventional techniques. The mean dose to cochlea from opposed laterals is 105% of the posterior fossa prescription and 68% from the 3-D technique using blocks and wedges. This is further reduced to 59% with IMRT, a dose reduction of 14% from 3-D and 44% from opposed laterals. We observed the same trend in all cases we studied. Significant dose reduction to cochlear structures as well as pituitary gland is routinely achieved with IMRT. The PTV coverage is not compromised and in general an 8% increase to the mean PTV dose is observed with IMRT.

Astrocytoma

In a recent study, Merchant et al. (2002) reported the preliminary results from a phase II trial of conformal radiation therapy for pediatric patients with localized low-grade astrocytoma and ependymoma. Based on the initial outcome for 38 patients with astrocytoma and 64 patients with ependymoma using an anatomically defined PTV, they concluded that their 3-D conformal treatment technique was safe for the pediatric population in the study. They also concluded that normal tissue sparing through the use of advanced radiation therapy treatment planning and delivery techniques should be beneficial to pediatric patients if the rate and patterns of failure are similar to conventional techniques and toxicity reduction can be objectively documented. At MSKCC, pediatric patients with such malignancies are routinely evaluated for the potential benefits from IMRT. We demonstrate a pediatric astrocytoma case in an 8-year-old patient using IMRT.

The treatment planning steps for immobilization and simulation are similar to what was described earlier for other head and neck sites. Briefly, the patient was immobilized in the supine position using a head rest and custom thermoplastic mold. For CT simulation, CT images were acquired at 3 mm spacing from the vertex to a level approximately 5 cm inferior to the treatment volume. The isocenter was marked approximately at the center of the treatment volume. The tumor for

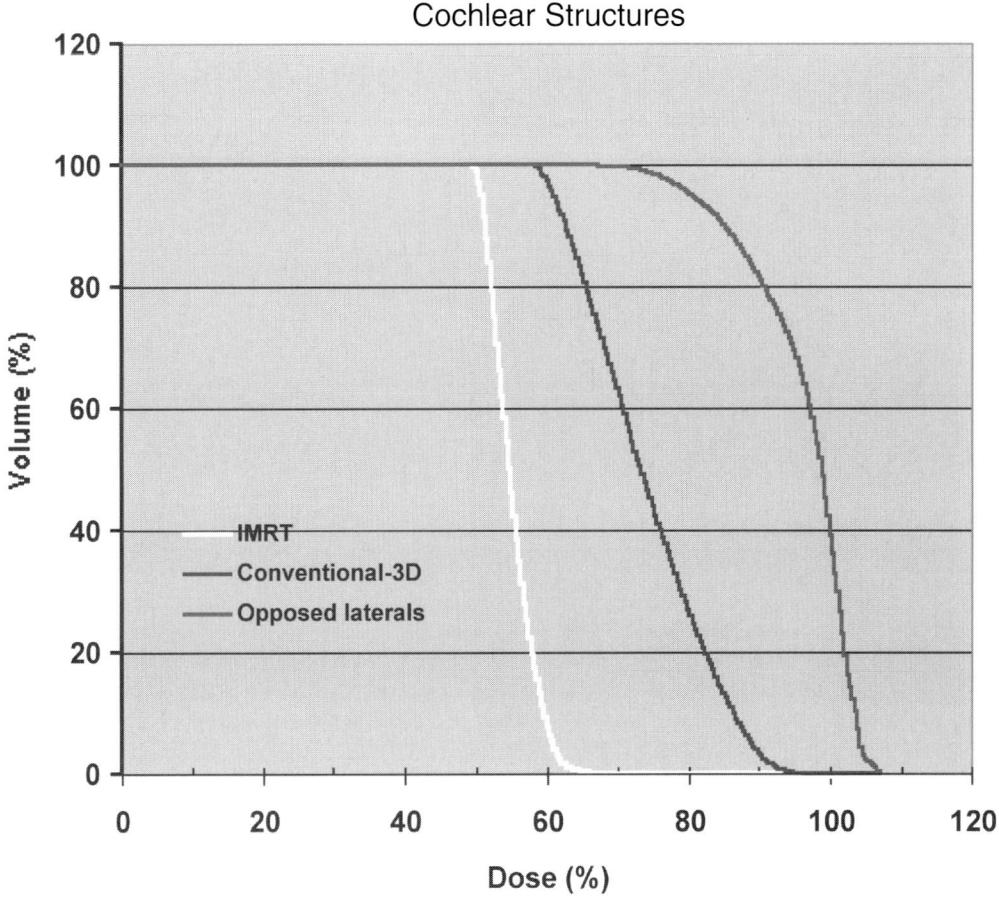

FIGURE 11–2. Dose-volume histograms for cochlear structures using traditional opposed lateral fields, 3-D conformal, and IMRT for posterior fossa boost treatment.

this patient occurred in the left thalamus region near the optical structures. The PTV partially encompassed the optic chiasm as well as the superior aspects of the brainstem neighboring the pituitary gland. All the critical organs and PTV were identified and contoured on CT images registered with a contrast enhanced MR study scanned at 3 mm slice spacing. The prescription dose was 5400 cGy in 30 fractions with the goal of keeping dose to the critical organs as low as possible. The *ideal* beam arrangement for optical structure sparing was found to be a six-field co-planar geometry mostly avoiding those structures and minimally passing through the pituitary gland and optic chiasm. This beam arrangement was used with the inverse planning algorithm parameters listed in table 11–2. The normal tissue rind that is listed in table 11–3 is a doughnut-shaped structure that was generated to aid in conformal dose shaping around the PTV, typically with 5 cm outer and 0.9 cm inner margins from the PTV. The purpose of this optimization volume is to restrict the high dose regions, *hot spots,* to within the PTV volume. The resulting dose distribution on an axial and sagittal plane was shown in figure 11–3a and 11–3b.

Although this IMRT plan satisfied all the initial clinical dose limits to the optical structures and brainstem while giving a fairly uniform dose to the PTV, because of the heavily laterally oriented beam arrangement, the shape of isodose lines (70%–90%) spread laterally and giving a higher temporal dose to the patient. In addition, the skin dose on the ipsilateral side is signifi-

Table 11–2. Clinical Dose Limits and Inverse Planning Algorithm Constraints for a Pediatric Astrocytoma IMRT Plan

Structure	Clinical Dose Limits	Inverse Planning Algorithm Constraint			
		Prescription Dose	Maximum Dose/Penalty	Minimum Dose/Penalty	Dose-Volume Constraint/Penalty
PTV	Min. Dose = 54 Gy	54 Gy (100%)	56.7 Gy (105%)/50	51.3 Gy (95%)/50	N/A
	Max. Dose = 59.4 Gy				
Optic Chiasm	Max. Dose = 54 Gy		48.6 Gy (90%)/30		N/A
Pituitary Gland	Max. Dose = 50 Gy		43.2 Gy (80%)/50		N/A
Brainstem	Max. Dose = 54 Gy		51.3 Gy (90%)/30		N/A
Eyes	Mean Dose ≤2 Gy		2 Gy (5%)/50		N/A
Normal tissue rind*	Max. Dose <54 Gy		43.2 Gy (80%)/100		N/A

Normal tissue rind is defined as a 5 cm annular ring around the PTV excluding the penumbra region of approximately 9 mm

Table 11–3. DVH Analysis for the Comparison of a Six-Field Co-Planar and an Eight-Field IMRT Plans for an Astrocytoma of the Left Thalamus

Structure	Statistics	IMRT: 6-field (%)	IMRT: 8-field (%)
PTV	Min. Dose (D_{95})	100.5	91.9
	Mean Dose	104.6	106.5
	Max. Dose (D_{05})	109.0	111.3
Pituitary	Mean Dose	73.2	71.4
	Max. Dose (D_{05})	83.9	81.9
Brainstem	Mean Dose	48.1	56.2
	Max. Dose (D_{05})	85.2	98.6
Chiasm	Mean Dose	92.0	91.3
	Max. Dose (D_{05})	99.2	98.3
Left optic nerve	Max. Dose (D_{05})	42.4	55.2
Right optic nerve	Max. Dose (D_{05})	40.7	47.3
Eyes	Max. Dose (D_{05})	6.0	6.0

Abbreviations: D_{95} = Dose received by 95% of volume, D_{05} = Dose received by 5% of volume.

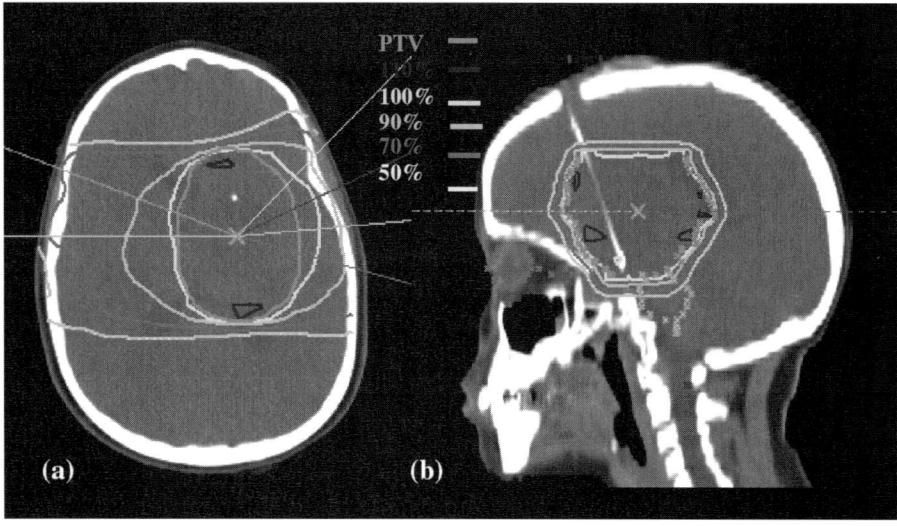

FIGURE 11–3. (a) Six-field IMRT plan with inverse planning parameters of table 11–3 in the axial plan through the isocenter. (b) Same plan in the sagittal plane of the isocenter. See COLOR PLATE 29.

cant enough to cause possible permanent alopecia in this young patient. These concerns led us to modify the plan to improve temporal lobe and skin dose.

An eight-field non-coplanar IMRT plan was generated with the same beam optimization parameters of table 11–2, except with a stricter PTV inverse planning maximum dose constraint of 102%. The number of the laterally oriented beams in the axial plane of Figure 11–3 was reduced and three additional beams were oriented at the PTV in non-coplanar oblique angles at the table positions of 25°, 50°, and –50° in an attempt to increase solid angle coverage. The beams were selected to minimize the overlap with the organs at risk, yet these beams were not optimally missing the organs at risk as in the original six-field plan. The resulting dose distributions for this eight-field non-coplanar IMRT plan are shown in the axial and sagittal plane of the isocenter in figure 11–4. The same isodose lines as in figure 11–3 are used to illustrate the differences between the two plans. In this new plan, 70% isodose line is seen to be more restricted in the lateral dimension as a result of increased solid angle coverage from the eight-field plan. However, since more of the beams are actually *seeing* the organs at risks in their projections through the isocenter, the optimization algorithm produces more inhomogeneous dose profiles for these beams. Eventually, the individual beam contributions from the IMRT plan add up to an increasingly less homogeneous PTV coverage as indicated by the larger isodose-line of 110% for the eight-field plan. The DVH statistics were summarized for the two IMRT plans in table 11–3.

The dose-volume statistics for the two plans are almost identical in delivered doses to pituitary gland, eyes, and optic chiasm. The optic nerves receive approximately 15% to 30% more doses from the 8-field IMRT plan compared with the six-field plan due to the increasing number of beams exposing the optic nerves from the eight-field plan. In addition, the dose uniformity within the PTV as determined by the ratio of D_{05} to D_{95} is 1.08 versus 1.21, an increase over 10%. However, the eight-field plan remains to be more favorable in overall dose reduction to temporal lobes and skin. This is an example of the utility of clinical evaluation of dose distributions rather than only relying on quantitative analysis of doses to identified structures.

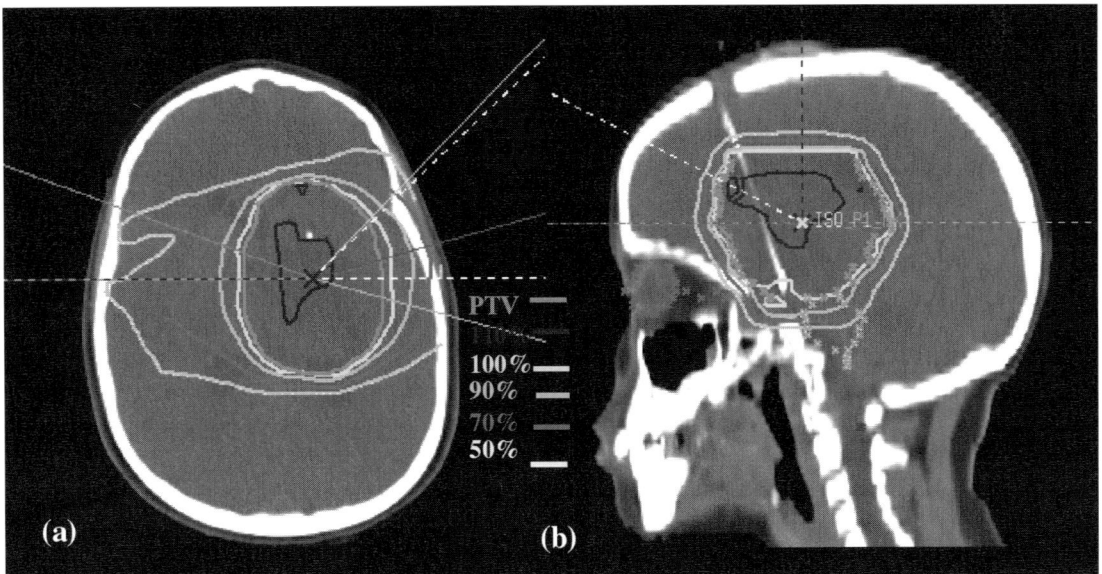

FIGURE 11–4. (a) Eight-field IMRT plan with inverse planning parameters of table 11–3 in the axial plane through the isocenter. (b) Same plan in the sagittal plane of the isocenter. See COLOR PLATE 30.

IMRT Treatment Of Pediatric Head And Neck Tumors

Rationale For Pediatric Head And Neck Tumors

The vast majority of pediatric head and neck tumors are rhabdomyosarcoma, although a number of other cancers can also occur in head and neck sites. Approximately 35% of all rhabdomyosarcomas arise in the head and neck. Of these, 25% occur in the orbit, 50% in parameningeal (skull base) sites, and 25% in other sites (oropharynx, larynx, muscles of mastication, etc.). Most children with head and neck sarcomas are younger than 10 years of age and commonly are below age 5.

All patients are treated with chemotherapy, and in a majority radiation therapy is also administered for local control. A small percentage of patients are candidates for surgery either alone or with post-operative irradiation. Tumor location is one of the strongest predictors of outcome. In the Fourth Intergroup Rhabdomyosarcoma Study (IRS IV), the 5-year relapse-free survival rates for patients without distant metastases were as follows: orbit 94%, parameningeal 71%, and other head and neck sites 81% (Crist et al. 2001). The poorer local control rate observed for parameningeal tumors may be related to the greater difficulty in defining target volumes and designing adequate radiation therapy plans in these anatomically challenging sites (Michalski et al. 1995; Wharam 1997).

Because of the young age of children treated for head and neck tumors, long-term morbidity from radiation therapy is an enormous problem. The long-term effects depend upon the location of the tumor as well as the radiotherapy fields used and dose levels administered. Patients typically develop moderate to severe cosmetic deformities due to growth arrest of facial bones (Denys et al. 1998). Reconstructive surgery is generally not an option because of tissue damage from radiotherapy. Patients may lose vision in one or both eyes because of late effects on various components of the ocular apparatus (lens, cornea, retina, and optic nerves). Chronic symptoms of a dry or painful eye may result from treatment of the conjunctiva and/or lacrimal gland (Oberlin et al. 2001).

A significant portion of the brain, including the critical frontal and temporal lobes may receive high doses of radiotherapy from treatment of skull base tumors. This may lead to impaired cognitive functioning and neuropsychological development. For skull base tumors, hearing loss may ensue if the inner ear (cochlea) is exposed to high-dose radiation. This problem can be exacerbated by the use of potentially ototoxic chemotherapy. The hypothalamus and pituitary gland may suffer radiation-induced damage that can lead to a large variety of endocrine disorders in the growing child. Exposure of the oral cavity and salivary glands to moderate doses of radiation is expected to cause xerostomia and impaired dentition. The muscles of growing children are especially prone to fibrosis. Thus, patients may develop trismus or difficulty with mastication as well as a potentially severe neck fibrosis (Wolden et al. 1999).

The possibility of a radiation-induced second malignancy is always a serious concern when treating children. The existing data on this problem seem to support the hypotheses that the risk increases with younger age at the time of radiation exposure and with increasing radiation dose. A wide variety of second cancers can develop as a result of head and neck radiation but the most common include cancers of the thyroid, salivary glands, brain, and skin (Wolden et al. 1999). Secondary sarcomas of bone or soft tissue are also relatively common (Patel et al. 1999). The long-term incidence of radiation-induced malignancies in this specific population is not well established but, based on available data, is estimated to be in the range of 5% to 10%. Limiting the exposure of high-dose radiation of all normal tissues and especially the most sensitive normal tissues such as salivary glands and thyroid should translate into a decreased risk of second malignancies.

Treatment Planning Of Pediatric Head And Neck Tumors

Tumor location, size, and shape in many pediatric head and neck patients vary considerably limiting the usefulness of standard beam arrangements, i.e., *class solutions.* Although beam arrangements may vary considerably, normal tissue dose limits have been standardized to aid in plan optimization and evaluation. In addition to the spinal cord and brainstem, other normal tissues and their maximum dose limits include: optic chiasm ($\leq$54 Gy), optic nerve ($\leq$54 Gy), retina ($\leq$45 Gy), cochlea ($\leq$50–60 Gy), hypothalamic-pituitary axis ($\leq$20–40 Gy), major salivary glands (mean dose $\leq$26 Gy), facial bones and undeveloped dentition ($\leq$20 Gy) and normal brain ($\leq$20 Gy to significant volume).

Nasopharynx

The current planning process for pediatric head and neck patients includes immobilization and CT simulation similar to what is done for adult patients, as explained in the previous chapter. Anesthesia is used on an as-needed basis. MR-CT and positron emission tomography (PET) image fusion is routinely used to aid in localization of the pre-chemotherapy tumor volumes, which include a boost gross tumor volume (GTV) and PTV for both the primary tumor and any involved lymph nodes. The PTV is typically designed by adding a 1 cm margin to the GTV. The margin is reduced to 0.5 cm along the posterior border of the nasopharynx tumor because of the proximity of the brainstem, spinal cord, and cochlea. Pediatric nasopharynx tumors are treated with a seven-field beam arrangement identical to the one described earlier for adults. All the details pertaining to inverse planning and optimization are also similar to those for adults.

We investigated the dosimetric superiority of IMRT plans in the treatment of pediatric nasopharynx lymphoepithelioma cases with respect to the traditional radiation treatment using parallel-opposed lateral fields as described in the previous chapter. This study was done retrospectively for three pediatric patients who were treated with IMRT. Parotid sparing is done with the IMRT plans either for one side, both sides, or none of the parotids for each of the three patients. In the traditional plan, parallel-opposed lateral photon fields with successive cone downs are used with cord blocks after 45 Gy and boost the gross tumor to 70 Gy. In addition, 9 MeV electron fields are used to boost bilateral posterior neck nodes to 54 Gy. Comparison of IMRT treatments with traditional plans for the PTV$_{gr}$ and parotid for each patient is summarized in table 11–4.

As can be seen from table 11–4, the coverage of PTV$_{gr}$ improves significantly with IMRT. In general, there is approximately an 8% increase in mean dose to the PTV$_{gr}$ with IMRT compared with the traditional plans regardless of sparing any or both of the parotid glands. The traditional plan has potential limitations with lateral fields because of the need to shield critical normal structures and subsequent inability to properly cover tumors that extend into the retropharynx. The currently accepted dose limit for salivary gland function is approximately a mean dose of 26 Gy (Eisbruch et al. 1999). It has been shown by Hunt et al. (2001) that a mean dose of 26 Gy or less to at least one gland may be feasible in patients with N0 or N1 disease with the current

Table 11–4. Comparison of IMRT vs. Traditional Plans for PTVgr and Parotid Gland

Patient/ PTV$_{gr}$ Vol	Parotid Sparing	PTV$_{gr}$:D$_{95}$ (Gy)		PTV$_{gr}$:D$_{mean}$ (Gy)		Parotid: D$_{mean}$		Parotid: V$_{50}$ Gy	
		IMRT	Tradi-tional	IMRT	Tradi-tional	IMRT	Tradi-tional	IMRT	Tradi-tional
Patient 1 118 cc	None	64.4	48.6	72.9	67	54	64.4	72.2	99.8
Patient 2 46 cc	Both sides	70.9	66.2	74.8	68.6	26.7	58.6	4.9	86.2
Patient 3 143 cc	One side	66.2	55.4	72.6	68.3	38.9	67.7	28.4	99.2

MSKCC IMRT technique, while it may not be possible for N2 disease due to the close proximity of the 70 Gy target to the parotid glands. Traditional plan delivers significantly higher doses to almost the entire gland, while sparing one or both of the glands with IMRT reduce the volume of parotid gland receiving 50 Gy or above (V_{50} Gy) considerably. This is achieved without sacrificing the tumor dose with the IMRT technique.

Other normal organs of interest are spinal cord, brainstem, cochlea, mandible, and temporal lobes for which the mean and maximum doses are summarized for three patients with IMRT and traditional plan in table 11–5. In almost all cases IMRT delivers lower doses to normal tissue, except to the temporal lobes where IMRT and traditional plans deliver comparable doses.

Rhabdomyosarcoma Of The Head And Neck

For most pediatric skull-based rhabdomyosarcomas, the beam arrangement is based on tumor location and typically consists of five to eight non-coplanar beams. The main treatment-planning goal is to preserve the function of many critical organs at close proximity to the PTV. The dose distribution for a 4-year-old parameningeal rhabdomyosarcoma patient achieved with eight non-coplanar IMRT beams and a conventional electron beam is shown in figure 11–5. The IMRT plan in this case delivered a mean dose of 49.1 Gy to the PTV while limiting the dose to the optical structures at risk, the chiasm getting only 50.4 Gy and both retinas getting less than 41 Gy. In addition to the IMRT photon fields, a 9 MeV electron field boosted the ethmoid area. As a result of the composite beam arrangement, dose to both optical nerves was limited at approximately 50.4 Gy.

A quantitative analysis comparing doses to targets and normal tissues for conventional 3-D treatment versus IMRT has been performed for three representative patients with rhab-domyosarcoma, including the case mentioned above. The patients in this group all had para-meningeal disease and their ages ranged from 4 to 12 years. The IMRT plans were custom designed to minimize the normal tissue toxicity and PTV was prescribed to 50.4 Gy at 100% isodose level, which covered approximately 95% of the PTV in all cases.

The 3-D plans were designed retrospectively on the same image sets and anatomical contours using coplanar beams and conventional beam shaping with beam's eye view (BEV) technology. The same clinical criteria are applied to both IMRT and conventional plans in regard to the tumor coverage and normal tissue restrictions. The DVH statistics for PTV and normal tissues are summarized in table 11–6.

The analysis in table 11–6 shows that both the IMRT and 3-D conventional plans have very similar tumor dose statistics based on the same planning goals for both plans. However, the gain in normal tissue reduction can be significant for IMRT in almost all the critical organs listed. For one patient studied, the PTV entirely encompassed the pituitary gland, optic chiasm, and the cochlear structures. The maximum dose to these structures was restricted to the prescription dose of 50.4 Gy, and they were not included in the analysis in table 11–6. In the remaining two

Table 11–5. DVH Statistics Comparing the IMRT) and Traditional Plans. Values represent average and (standard deviations) for three patients

Structure	Statistics	IMRT	Traditional
Spinal cord	Max. Dose (D_{05})	34.1 (2.2)	47.1 (0.1)
Brain stem	Max. Dose (D_{05})	43.8 (7.5)	54.7 (9.6)
Mandible	Max. Dose (D_{05})	65.2 (8.5)	72.1 (1.2)
	$V_{66\,Gy}$ (%)	5.3 (3.6)	23.0 (11.1)
Temporal lobes	Max. Dose (D_{05})	67.6 (8.1)	64.9 (1.9)
	$V_{60\,Gy}$ (%)	20.9 (14.8)	18.6 (6.5)
Cochlea	Max. Dose (D_{05})	57.0 (6.0)	64.2 (9.1)
	Mean Dose	54.8 (6.3)	61.8 (9.0)

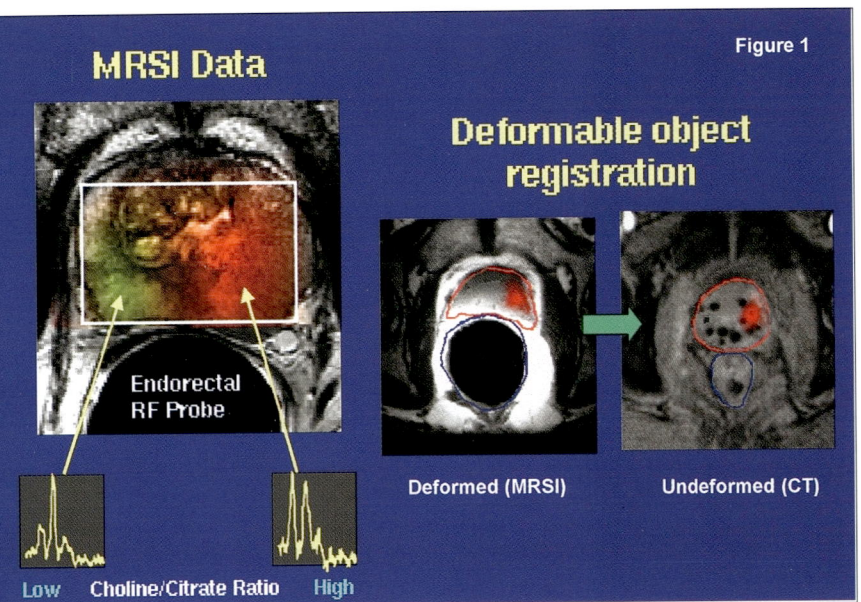

COLOR PLATE 1. FIGURE 1–1. The left panel shows magnetic resonance spectroscopy and imaging data; the color coding is based on the respective levels of choline and citrate (based on the NMR spectra below), with red indicating a high level of the former and high probability of aggressive tumor. Note the distention of the rectum due to the endorectal RF probe. The right panels illustrate that deformable object registration is needed to map the region of suspected tumor from the MRSI data to the CT scans; note the shape of the distended prostate (red contours) and of the rectum (blue contours) in the two images.

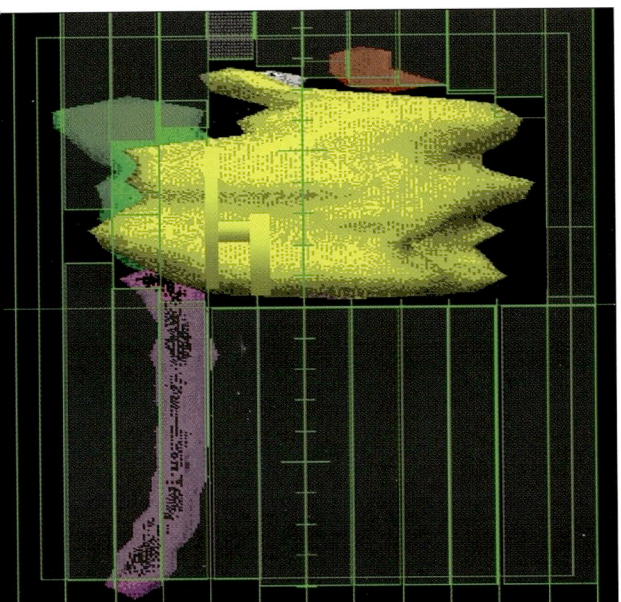

COLOR PLATE 2. FIGURE 2–2. The BEV of a right anterior oblique field for a head-neck patient in a supine position. PTV is depicted in yellow, brainstem in green, cord in purple, and pituitary gland in red. For conventional 3DCRT treatment planning the beam shape would be defined by the MLC leaves as shown. For ITP, the MLC shape would be designed automatically by the optimization algorithms.

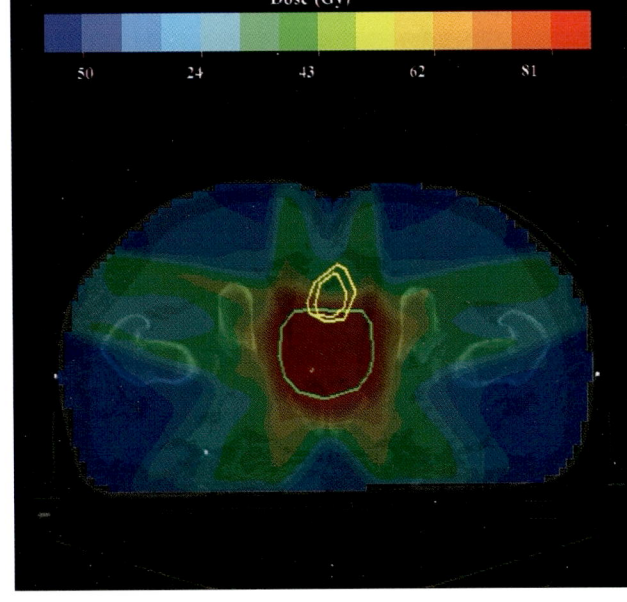

COLOR PLATE 3. FIGURE 2–20. Color wash style display isodose distribution for an 81 Gy IMRT prostate plan with beam configuration as in figure 2-19. The PTV is shown in green, and the rectal wall in yellow.

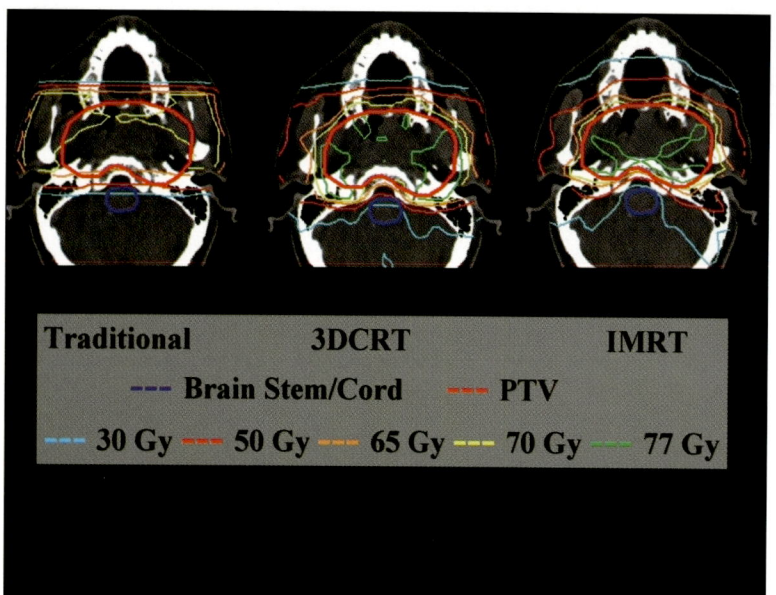

COLOR PLATE 4. FIGURE 2–22. Axial dose distributions through the nasopharynx for the IMRT, 3DCRT, and traditional treatment plans. Note the relatively poor coverage of the skull base using the Traditional plan and the improved conformality of the IMRT dose.

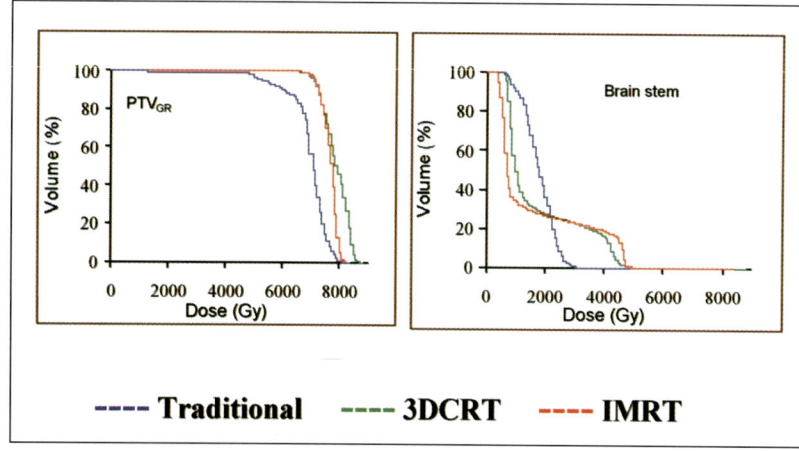

COLOR PLATE 5. FIGURE 2–23. Dose-volume histograms comparing the IMRT, 3DCRT, and traditional plans for the treatment plans in figure 2-22.

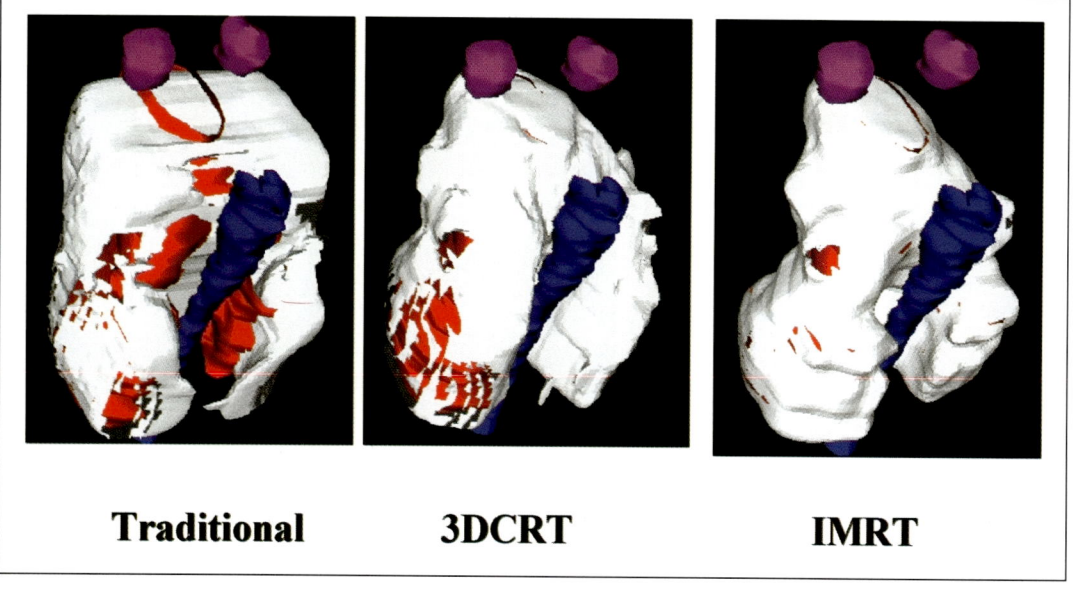

COLOR PLATE 6. FIGURE 2–24. Region of regret plot showing the 65 Gy prescription isodose surface (white), and regions of the PTV receiving less than the prescription dose (red) for traditional, 3DCRT, and IMRT nasopharyngeal treatment plans. The brainstem and cord are shown in blue, and the eyes in purple.

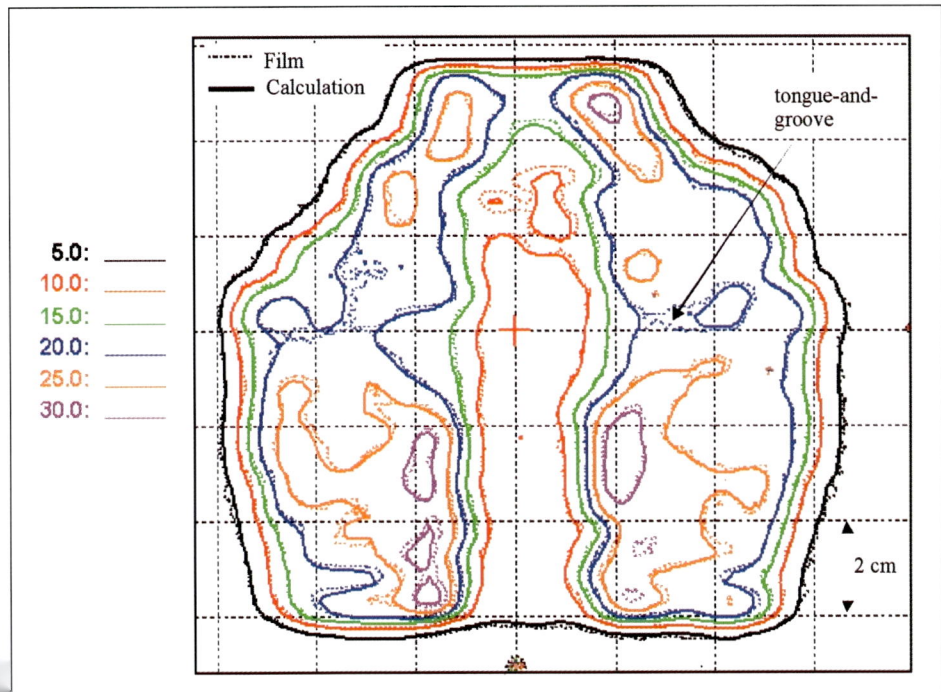

COLOR PLATE 7. FIGURE 5–11. Comparison of calculation and film measurement for a nasopharynx field.

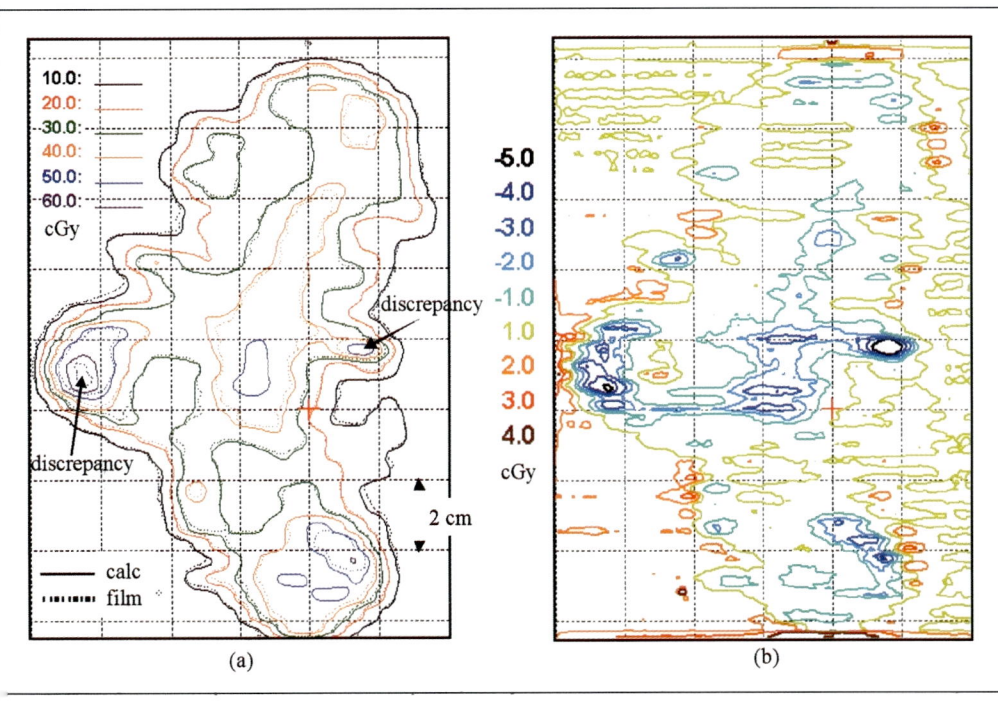

COLOR PLATE 8. FIGURE 5–12. Comparison of calculation and film measurement for a field used in a lung treatment. (a) Overlaid isodose distributions; (b) distribution of dose differences (film-calculation). For this calculation, only the effects of direct exposure, mid-leaf transmission, and rounded leaf-end were considered.

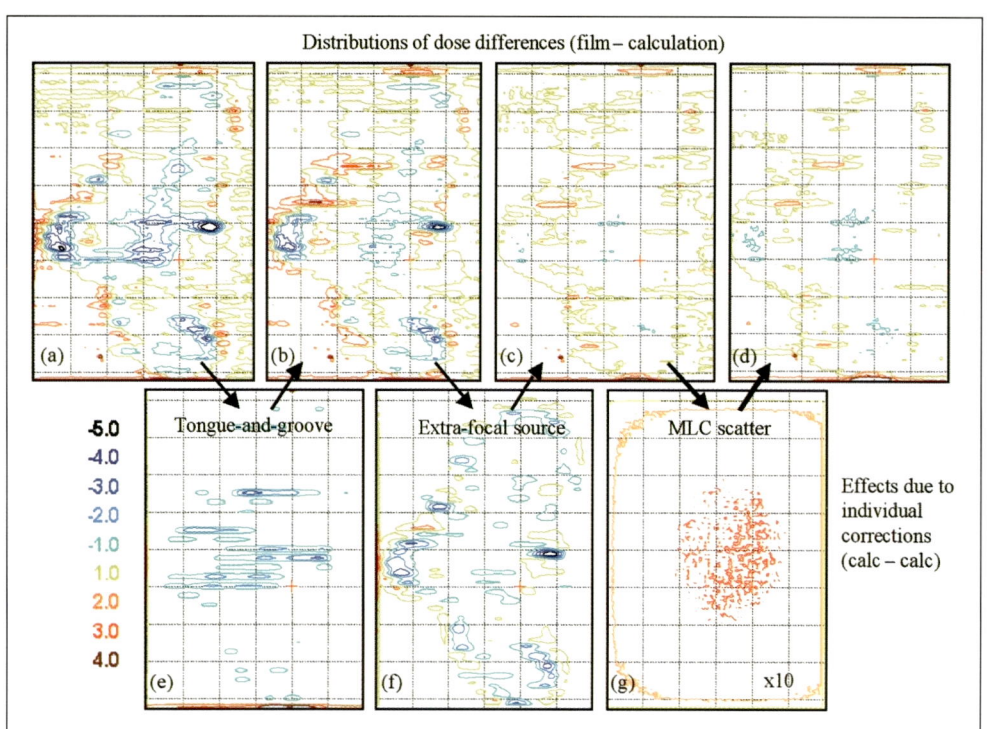

Distributions of dose differences (film – calculation)

(a) (b) (c) (d)

Tongue-and-groove   Extra-focal source   MLC scatter

Effects due to
individual
corrections
(calc – calc)

(e) (f) (g) x10

COLOR PLATE 9. FIGURE 5–13. Top row: The distribution of dose differences (film-calculation) of the lung field in figure 5-12. Bottom row: The effects due to individual corrections. Descriptions of the individual panels are given in the text.

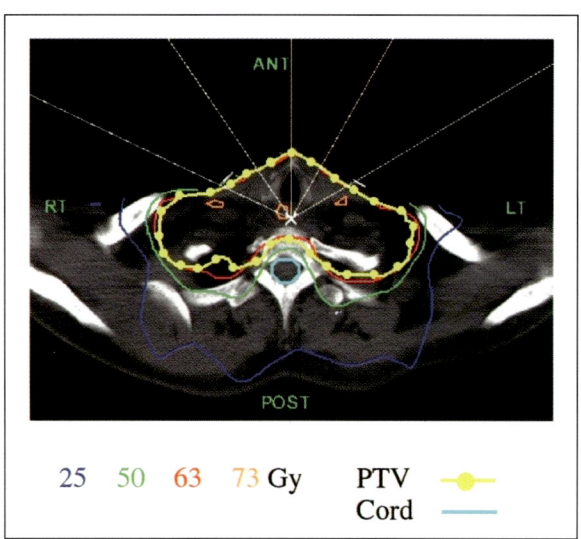

25 50 63 73 **Gy**   **PTV**   **Cord**

COLOR PLATE 10. FIGURE 6–1. Five-field IMRT technique for the treatment of thyroid cancer. Note the concave shape of the dose distribution between the target and spinal cord.

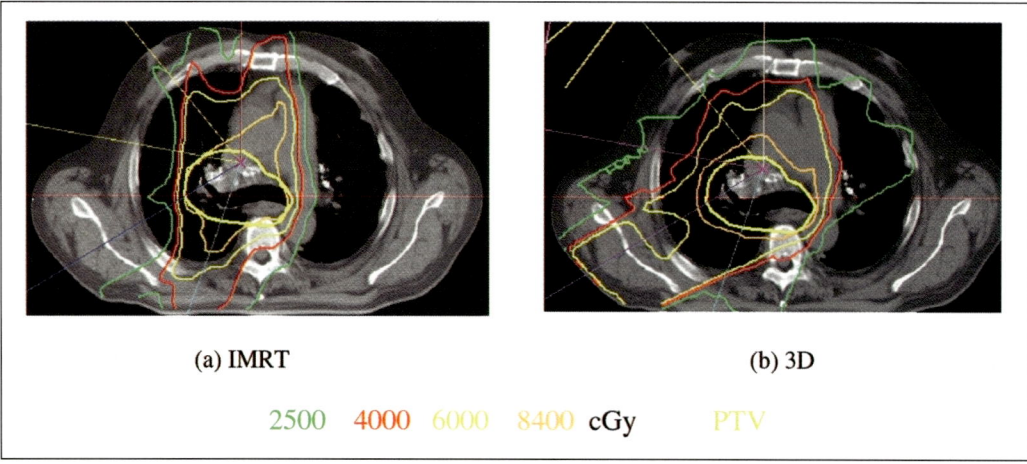

(a) IMRT   (b) 3D

2500 4000 6000 8400 **cGy**   PTV

COLOR PLATE 11. FIGURE 6–2. A comparison of IMRT (a) and 3-D conformal (b) dose distributions for the treatment of a lung tumor. Even though the same beam arrangement was used for both plans, there is significant improvement in the lung dose with IMRT.

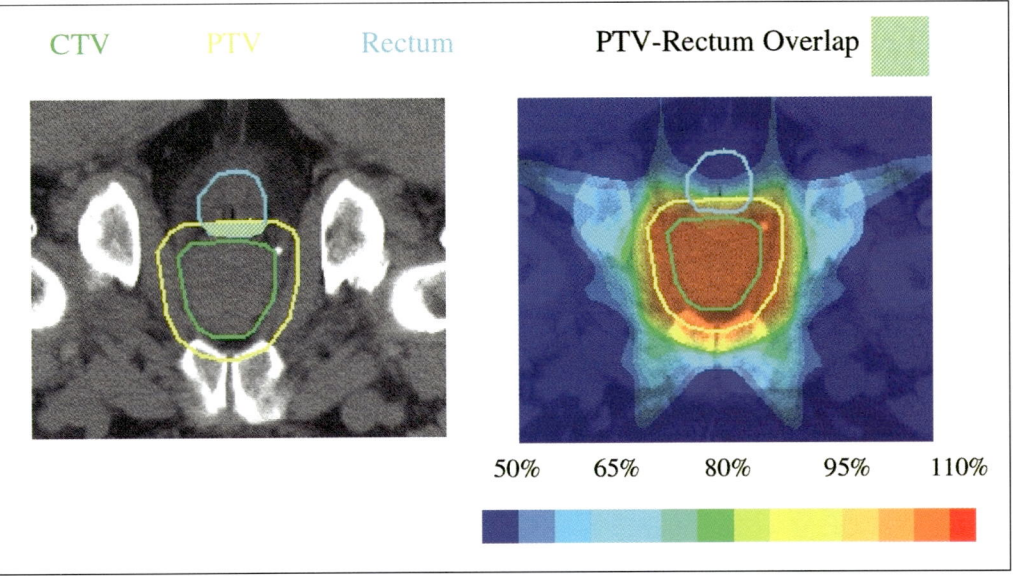

CTV   PTV   Rectum   PTV-Rectum Overlap

50%   65%   80%   95%   110%

COLOR PLATE 12. FIGURE 6–3. Use of logical combinations of structures to control optimized dose distributions.

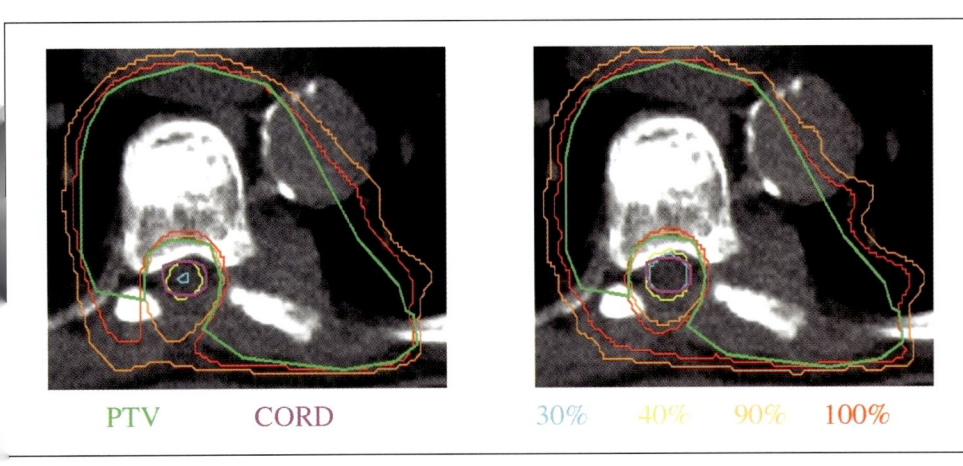

PTV   CORD   30%   40%   90%   100%

COLOR PLATE 13. FIGURE 6–5. Effect of two different profile smoothing methods on optimized dose distributions. (a) Profile smoothing performed at end of each iteration (Savitsky-Golay). (b) Profile smoothing performed within the objective function (Score Smoothing). (c) Intensity profiles for posterior beam created by smoothing at the end of each iteration (left) and within the objective function (right).

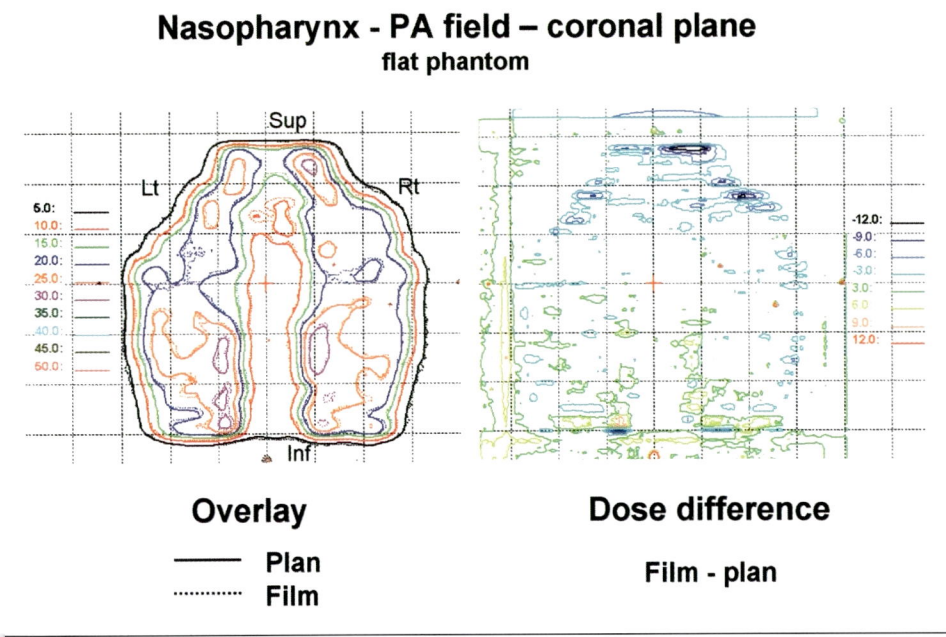

Nasopharynx - PA field – coronal plane
flat phantom

Sup   Lt   Rt

6.0:
10.0:
15.0:
20.0:
25.0:
30.0:
35.0:
40.0:
45.0:
50.0:

Inf

-12.0:
-9.0:
-6.0:
-3.0:
3.0:
6.0:
9.0:
12.0:

Overlay

——— Plan
·········· Film

Dose difference

Film - plan

COLOR PLATE 14. FIGURES 2–16 and 8–12. Film dosimetry verification in a flat phantom for a nasopharynx field. Measurements are compared with the planned dose distribution recalculated in the flat phantom. (a) overlay; (b) dose difference.

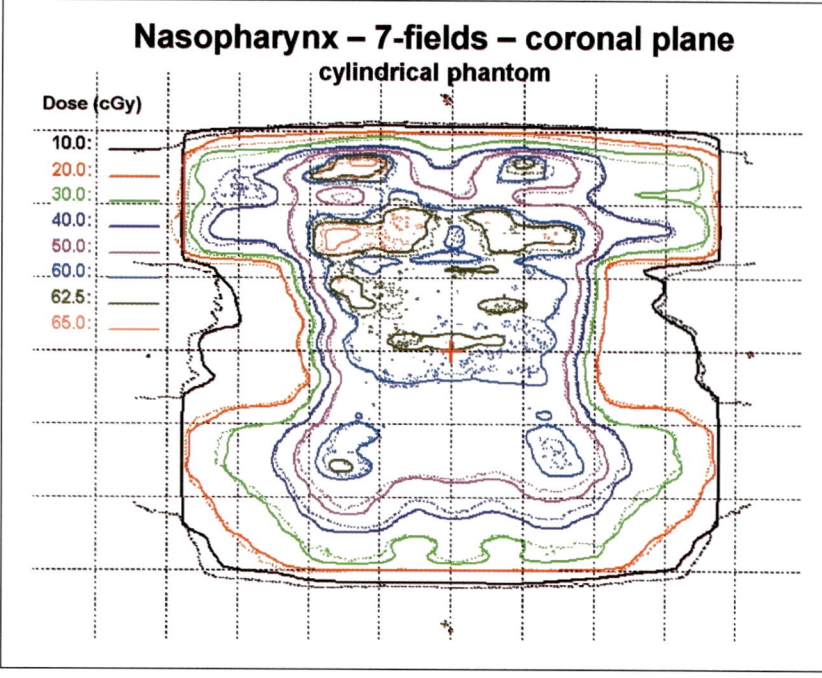

Prostate – 5-fields – axial plane
cylindrical phantom

calculated

measured

COLOR PLATE 15. FIGURE 8–13. Film dosimetry verification in a cylindrical phantom for a five-field prostate plan. Measurements are compared with the planned dose distribution recalculated in the 25 cm diameter cylindrical phantom.

COLOR PLATE 16. FIGURE 8–14. Film dosimetry verification in a cylindrical phantom for a seven-field nasopharynx plan. Measurements are compared with the planned dose distribution recalculated in the cylindrical phantom.

Nasopharynx – 7-fields – coronal plane
cylindrical phantom

Dose (cGy)	
10.0:	
20.0:	
30.0:	
40.0:	
50.0:	
60.0:	
62.5:	
65.0:	

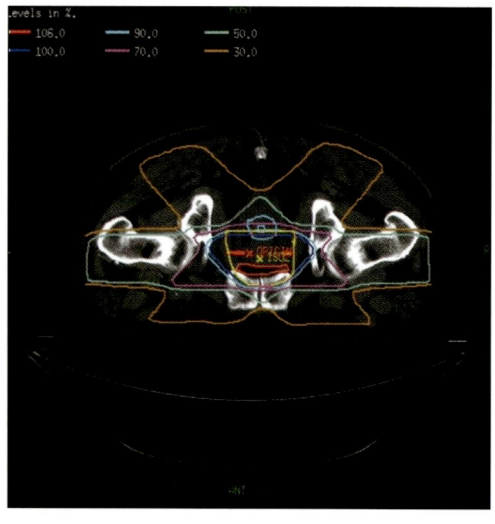

Levels in %.
106.0   90.0   50.0
100.0   70.0   30.0

COLOR PLATE 17. FIGURE 9–6. Isodose distribution for the six-field plan in a transverse plane through the isocenter. The outline of the PTV and the rectum are also shown.

Isodose (Gy)

▬ 85.5	▬ 75.0	▬ 50.0	▬ 10.0
▬ 81.0	▬ 70.0	▬ 30.0	

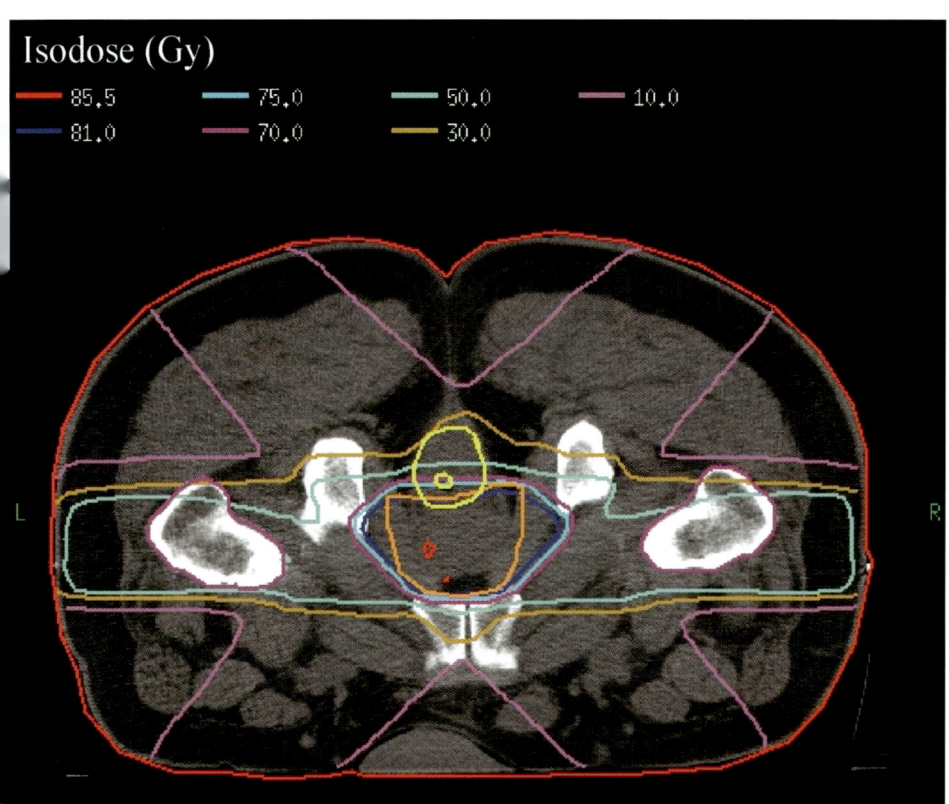

COLOR PLATE 18. FIGURE 9–9. The composite dose distribution for the 81 Gy 3DCRT plan in a transverse plane. The isodose levels are in absolute dose (Gy). The outlines of the PTV and rectum are also shown.

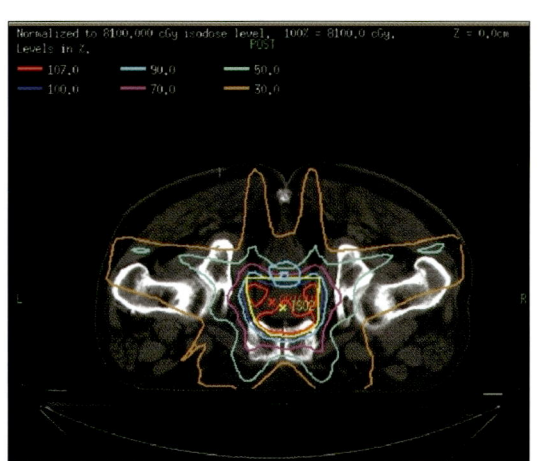

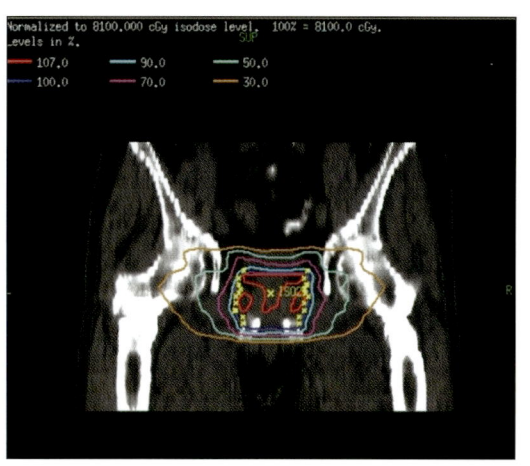

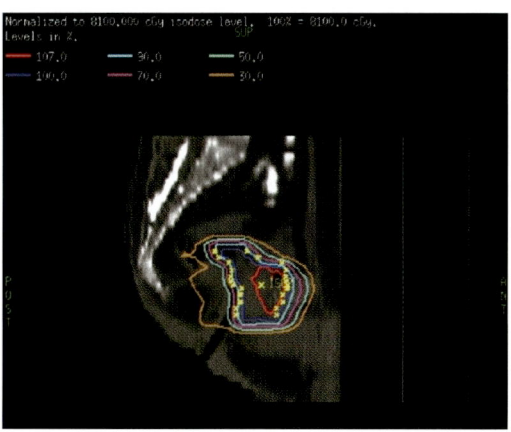

COLOR PLATE 19. FIGURE 9–13. Isodose distribution on three planes through the isocenter: (a) transverse, (b) coronal, and (c) sagittal. The dose (81 Gy) is prescribed to the 100% isodose. The outlines of the PTV and rectum are displayed.

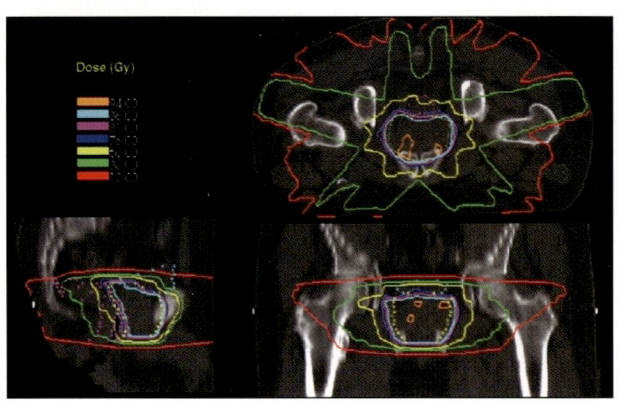

COLOR PLATE 20. FIGURE 9–16. Isodose distribution for an 86.4 Gy plan on three orthogonal planes: transverse, sagittal, and coronal.

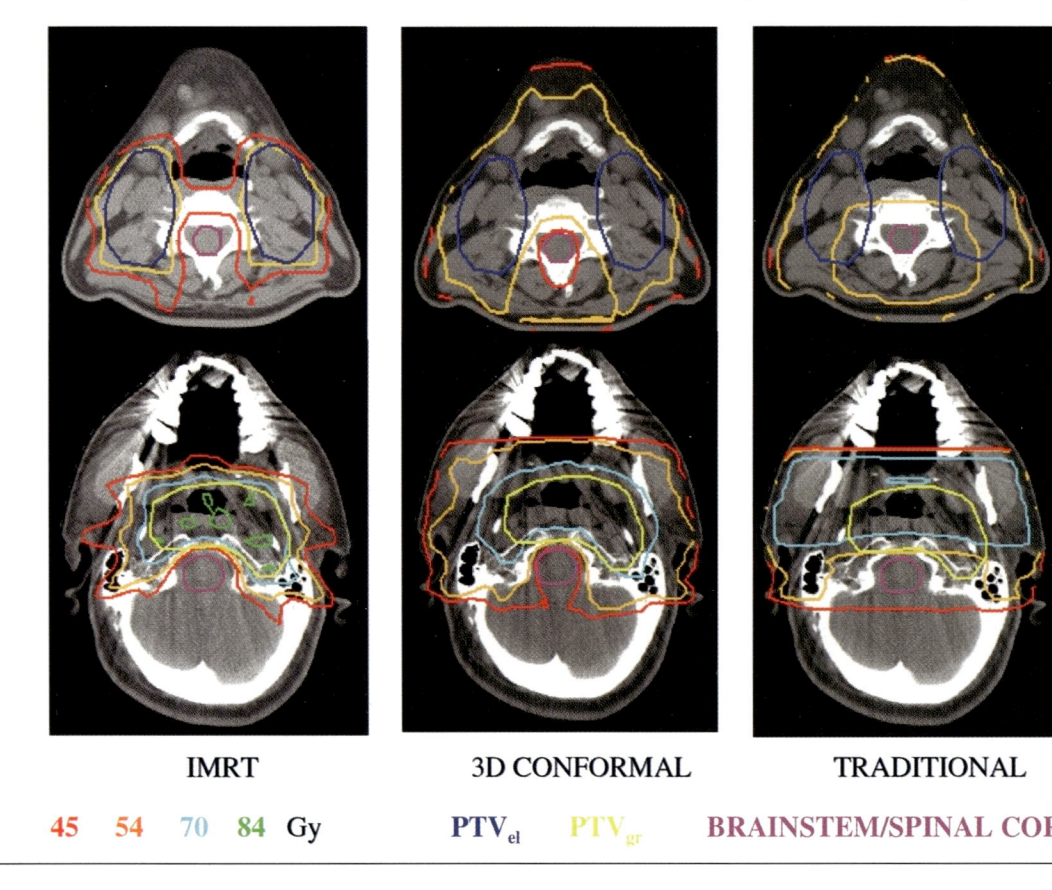

| IMRT | 3D CONFORMAL | TRADITIONAL |

45 54 70 84 Gy   **PTV**el   **PTV**gr   **BRAINSTEM/SPINAL CORD**

COLOR PLATE 21. FIGURE 10–1. Comparison of IMRT, 3-D conformal and traditional parallel-opposed field plans for the treatment of primary nasopharynx tumors.

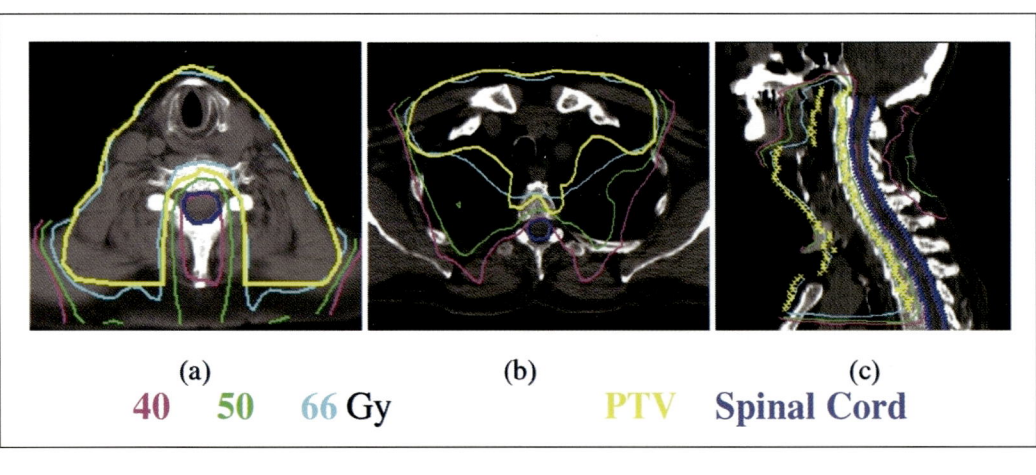

(a)   (b)   (c)

40   50   66 Gy   **PTV**   **Spinal Cord**

COLOR PLATE 22. FIGURE 10– Axial and sagittal IMRT dose distributions for thyroid carcinoma designed using a six-field plan.

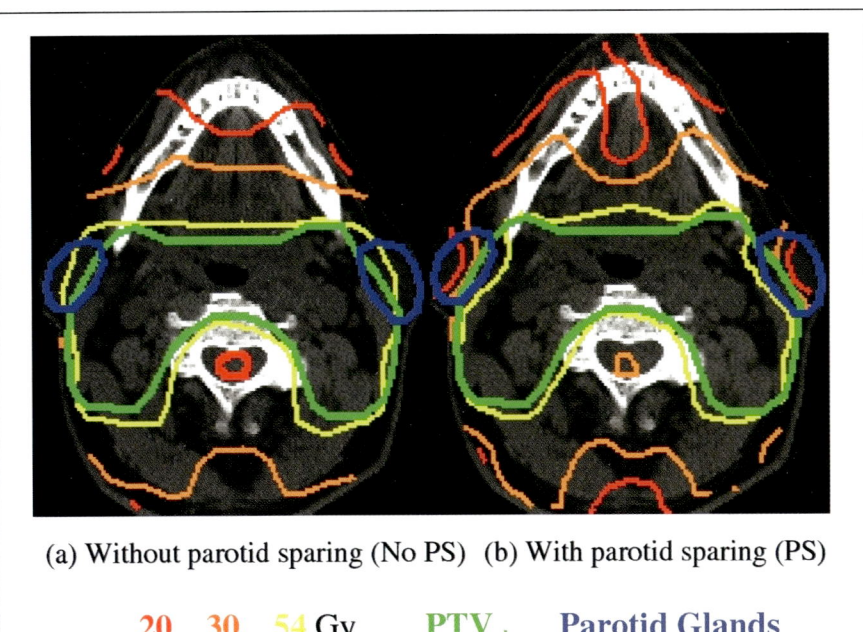

COLOR PLATE 23. FIGURE 10–10. Axial dose distributions and DVHs for the PTVel and parotid glands with and without an IMRT parotid sparing technique for a patient with N0 disease.

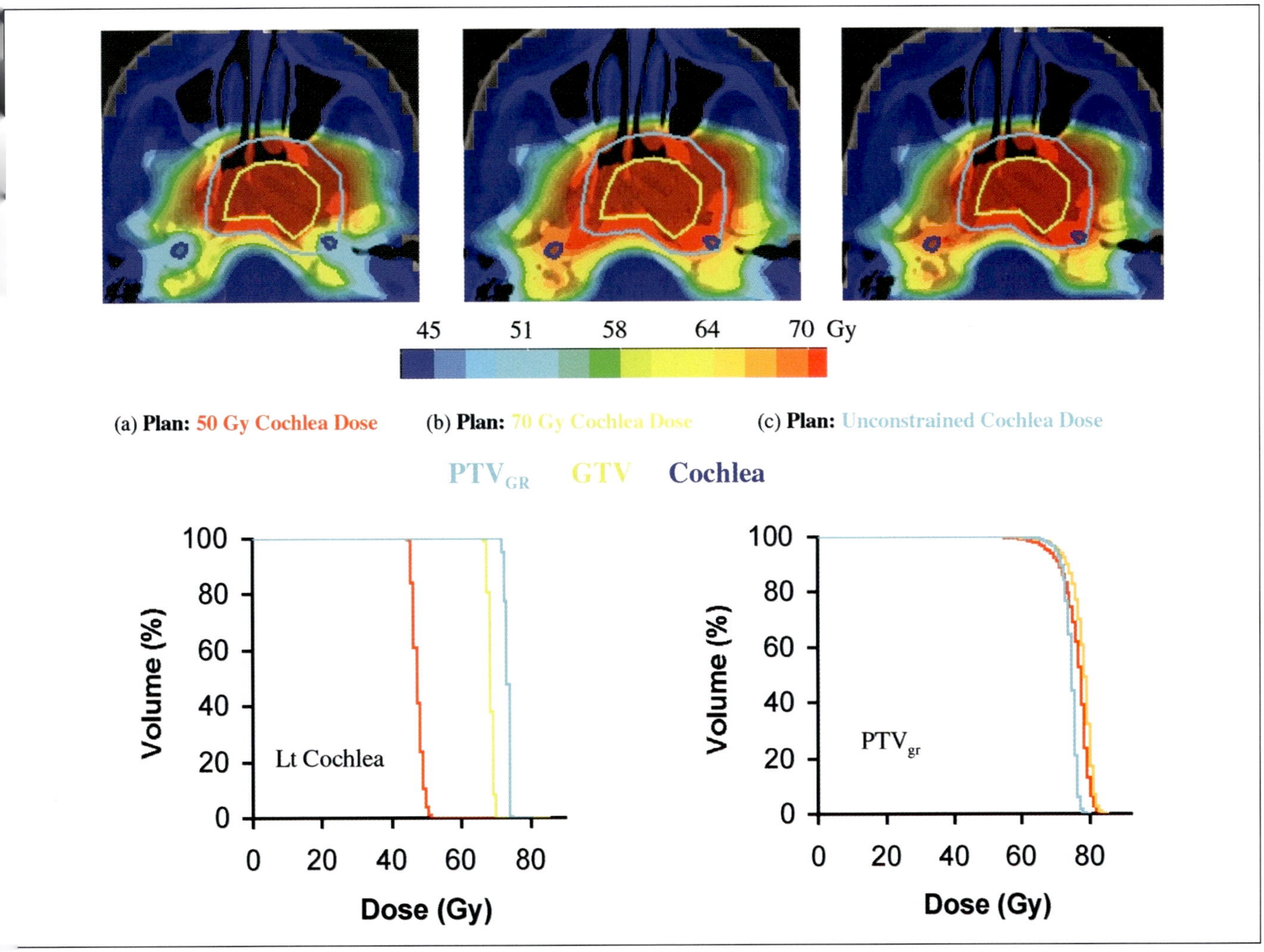

(a) Without parotid sparing (No PS)   (b) With parotid sparing (PS)

20   30   54 Gy   **PTV**₍el₎   **Parotid Glands**

45   51   58   64   70 Gy

(a) **Plan:** 50 Gy Cochlea Dose   (b) **Plan:** 70 Gy Cochlea Dose   (c) **Plan:** Unconstrained Cochlea Dose

PTV₍GR₎   **GTV**   **Cochlea**

Lt Cochlea

PTV₍gr₎

COLOR PLATE 24. FIGURE 10–11. Dose distributions and DVHs illustrating the effect of cochlear sparing on PTV coverage when the cochleae lie within PTVgr. Results are shown for three plans: unconstrained cochlear dose, 50 Gy, and 70 Gy maximum cochlear dose.

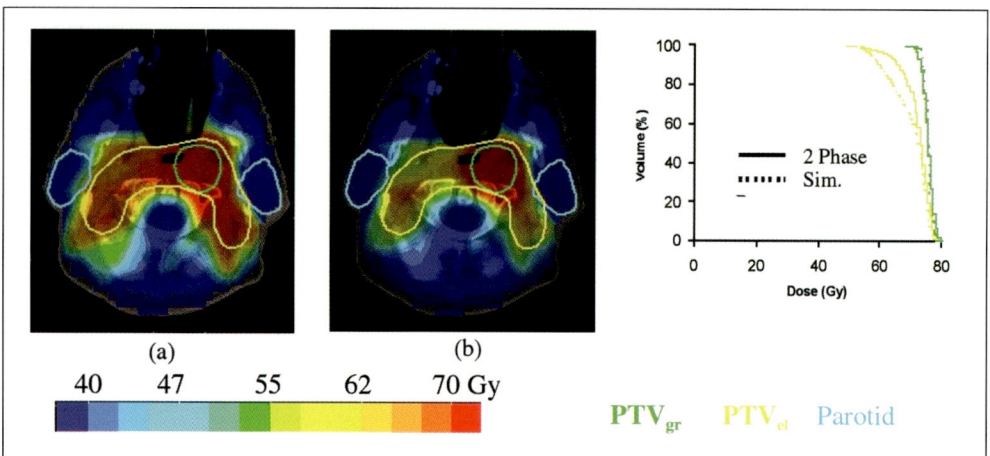

COLOR PLATE 25. FIGURE 10–13. Comparison of the MSKCC two-phase (a) and simultaneous boost (b) techniques for the treatment of nasopharynx cancer. Using the simultaneous boost technique, the PTVgr receives 70.2 Gy in 30 fractions (2.34 Gy/fraction) while the electively irradiated volume, PTVel, receives 1.8 Gy/fraction to 54 Gy. The 70 Gy conformity index is 2.4 for the two-phase plan and 1.7 for the simultaneous boost. The maximum cochlea doses with the two-phase and simultaneous boost plans are 70 and 63 Gy, respectively.

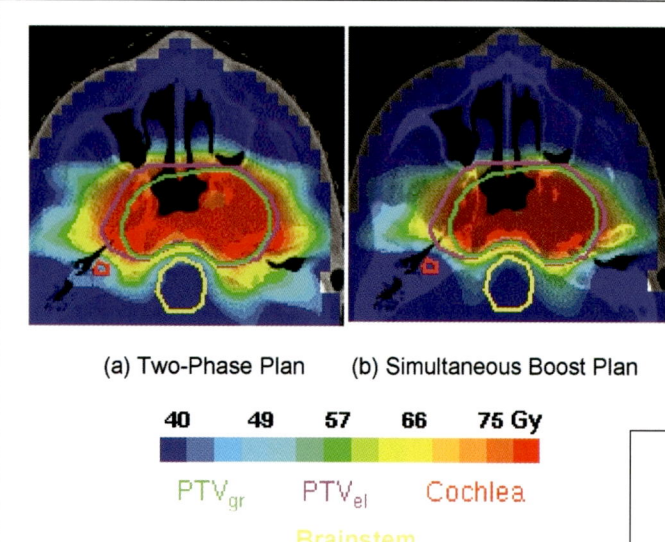

COLOR PLATE 26. FIGURE 10–14. Comparison of the MSKCC two-phase (a) and simultaneous boost (b) techniques for the treatment of nasopharynx cancer. Using the simultaneous boost technique, the PTVgr receives 70.2 Gy in 30 fractions (2.34 Gy/fraction) while the electively irradiated volume, PTVel, receives 1.8 Gy/fraction to 54 Gy. The 70 Gy conformity index is 2.4 for the two-phase plan and 2.1 for the simultaneous boost. The maximum cochlea doses with the two-phase and simultaneous boost plans are 50 and 35 Gy, respectively.

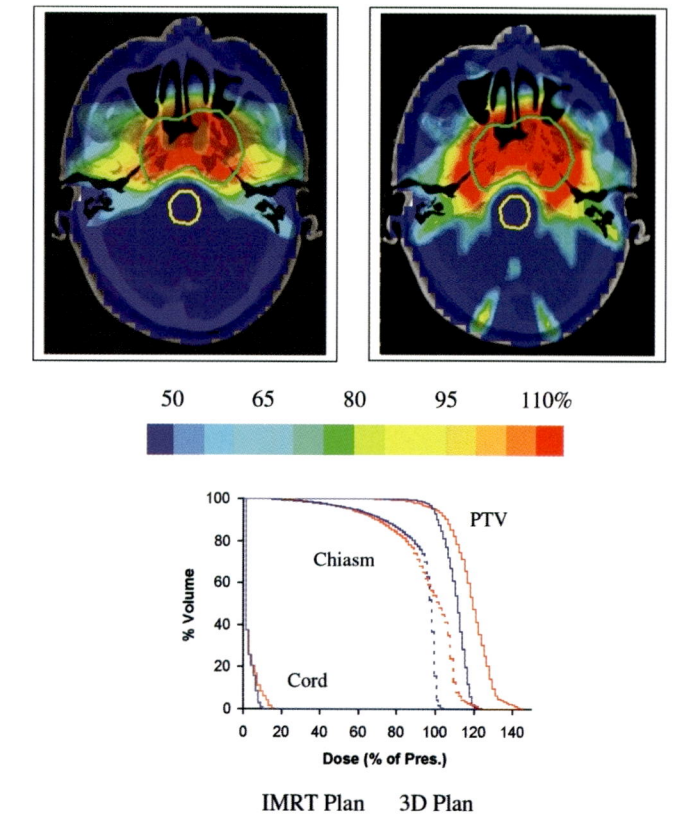

COLOR PLATE 27. FIGURE 10–15. Dose distributions and DVHs for IMRT and 3-D plans for recurrent nasopharyngeal cancer. Both plans utilize a seven-field beam arrangement as shown in figure 5. For the 3-D plan, wedges and cerrobend blocks over the spinal cord and brainstem are used to create a concave dose distribution. Cerrobend blocks over the cord and brainstem are also used with the IMRT plan to achieve doses of <20% of the prescription to these structures. IMRT significantly improves target dose coverage and uniformity and normal tissue doses compared to the 3-D plan.

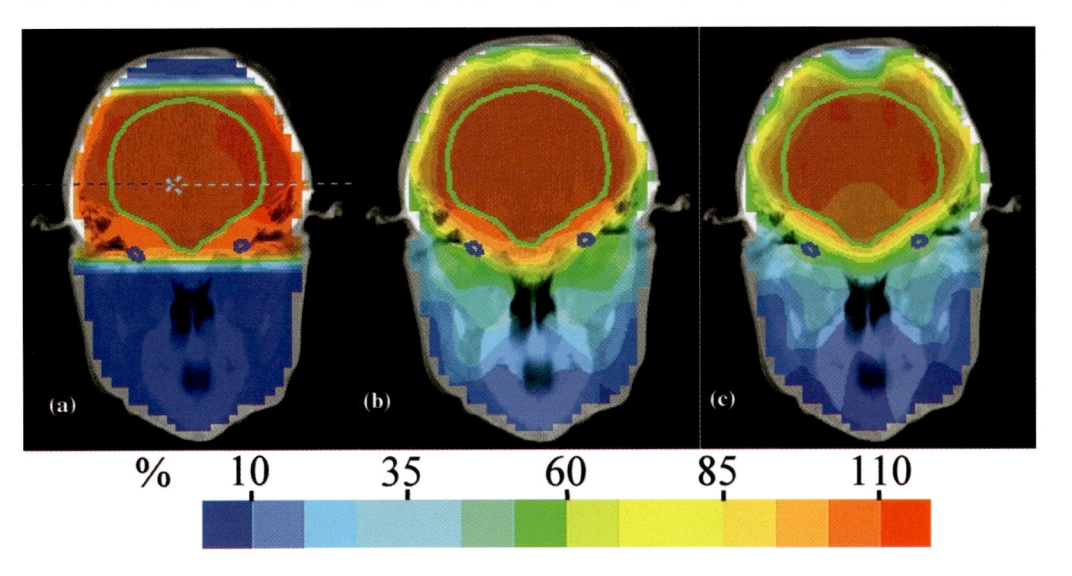

COLOR PLATE 28. FIGURE 11–1. Axial dose distribution for posterior fossa boost treatment using (a) traditional opposed lateral fields, (b) 3-D conformal, (c) IMRT. PTV and cochlear structures are shown by green and blue contours, respectively.

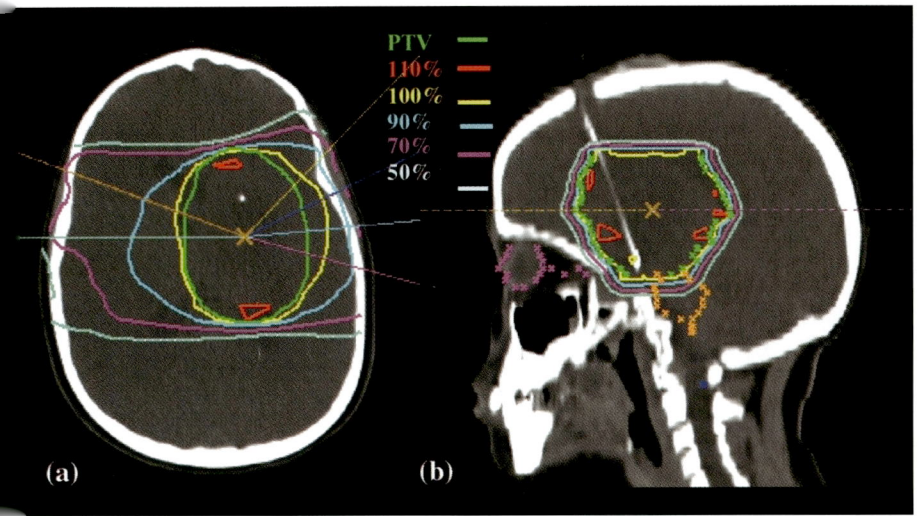

COLOR PLATE 29. FIGURE 11–3. (a) Six-field IMRT plan with inverse planning parameters of table 11.3 in the axial plan through the isocenter. (b) Same plan in the sagittal plane of the isocenter.

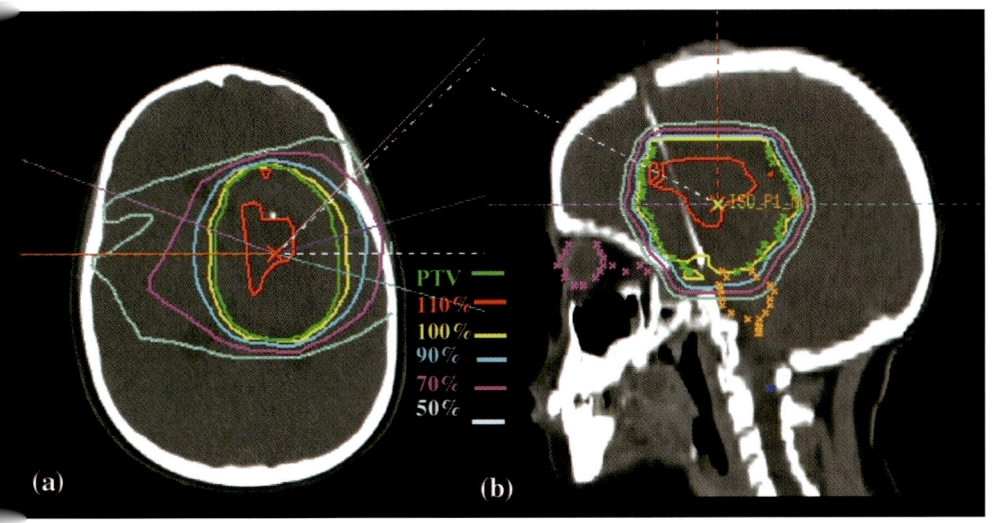

COLOR PLATE 30. FIGURE 11–4. (a) Eight-field IMRT plan with inverse planning parameters of table 11.3 in the axial plane through the isocenter. (b) Same plan in the sagittal plane of the isocenter.

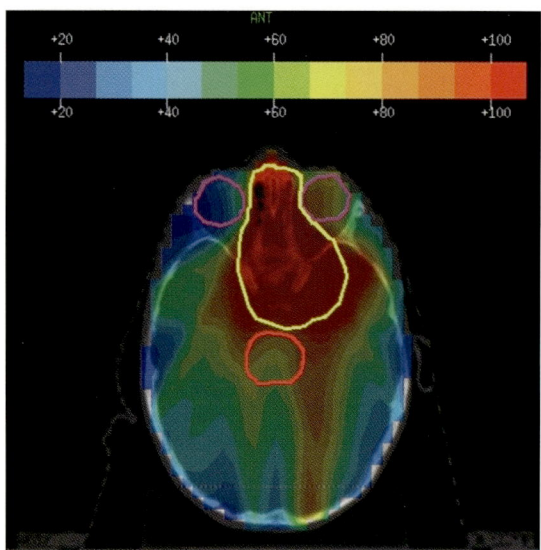

COLOR PLATE 31. FIGURE 11–5. Transverse dose distribution consisting of eight non-coplanar IMRT 6 MV photon beams and a single conventional 9 MeV electron beam for a parameningeal rhabdomyosarcoma.

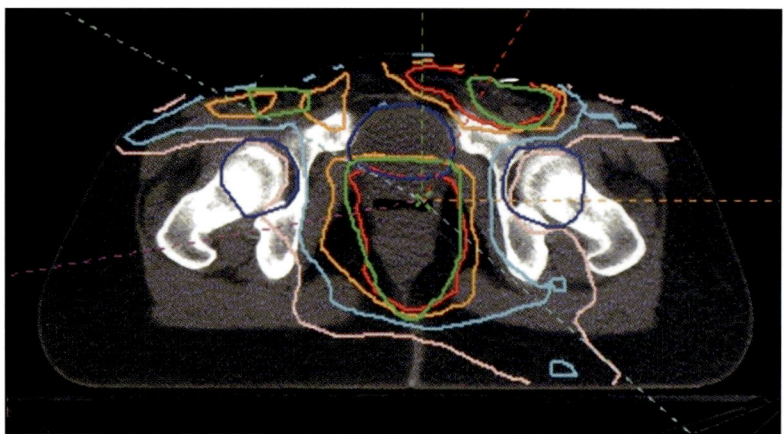

COLOR PLATE 32. FIGURE 11–6. Six-field IMRT plan for a perirectal rhabdomyosarcoma case.

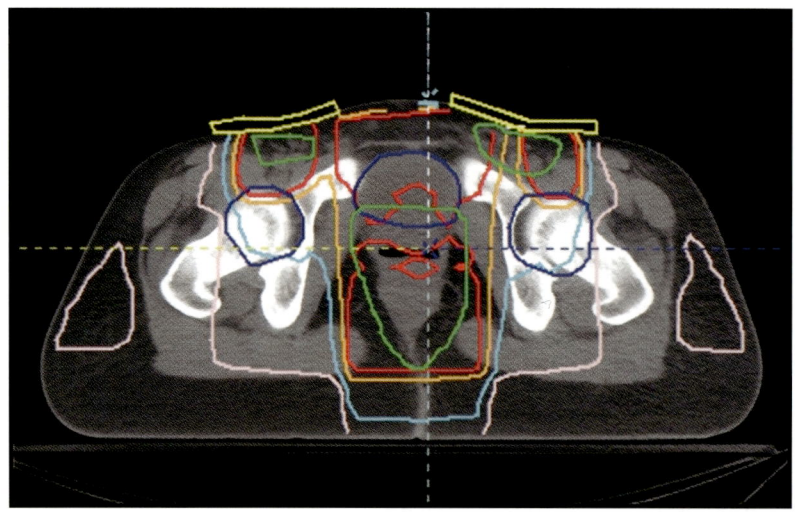

COLOR PLATE 33. FIGURE 11–7. Four-field conformal plan for a perirectal rhabdomyosarcoma case utilizing electron boost to nodal regions.

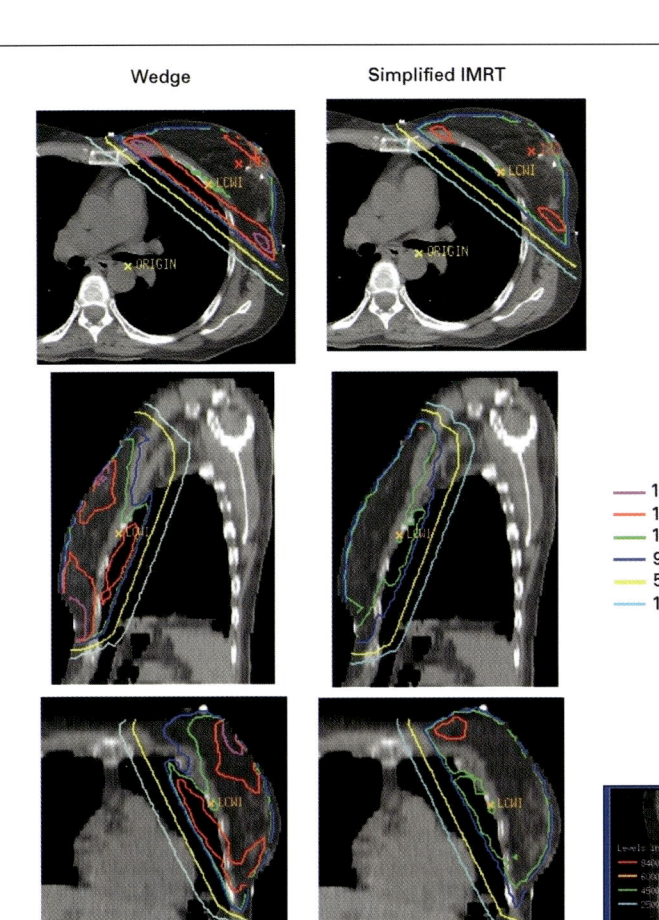

Wedge   Simplified IMRT

a)

b)

c)

— 108
— 104
— 100
— 98
— 50
— 10

COLOR PLATE 34. FIGURES 2–26 and 12–1. Dose distribution comparison. Left for standard wedged plan, right for simplified IMRT plan. Plans were normalized to the LCWI point. ISO was the location of beam isocenter. (a) Transverse plane through the LCWI point. (b) Sagittal plane through the LCWI point. (c) Coronal plane through the LCWI point.

COLOR PLATE 35. FIGURE 13–3. Colorwash display for the IMRT and 3DCRT treatment plans in figure 13-1 in the coronal (upper panels) and sagittal planes (lower panels).

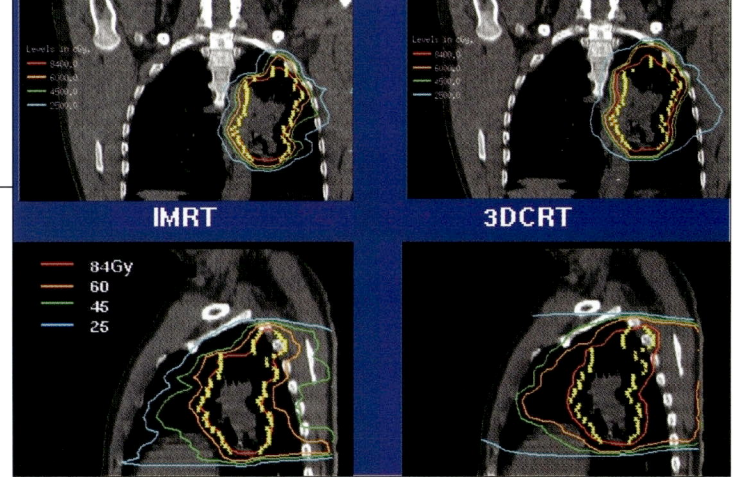

IMRT   3DCRT

— 84Gy
— 60
— 45
— 25

COLOR PLATE 36. FIGURE 13–4. Colorwash display comparing the 25 Gy isodose contour in the coronal plane (upper panels) and the 60 Gy isodose contour in the sagittal plane (lower panels) for the IMRT and 3DCRT plans.

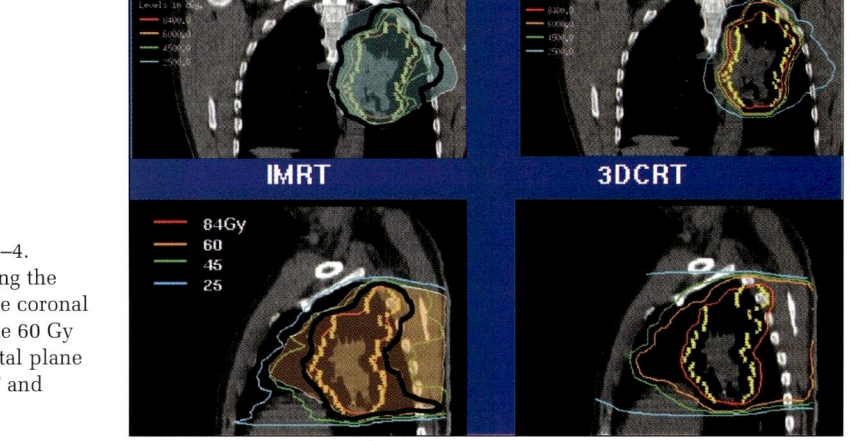

IMRT   3DCRT

— 84Gy
— 60
— 45
— 25

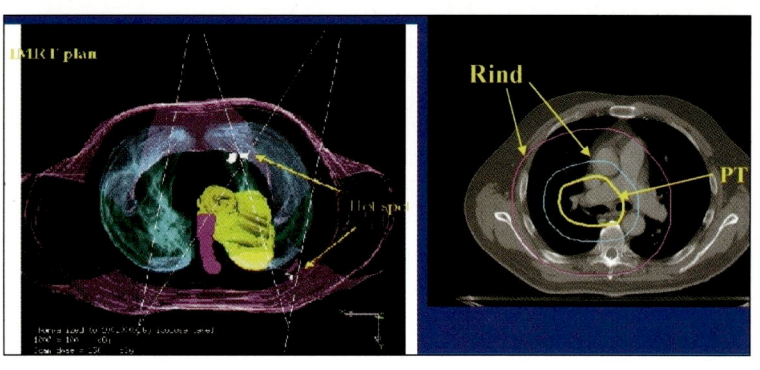

COLOR PLATE 37. FIGURES 2–7 and 13–5. Adding axillary contours to an IMRT plan (right panel, blue and pink contours, with PTV contoured in yellow) to avoid unwanted hotspots sometimes artificially created by ITP, such as shown at the intersection of two beams in the left panel.

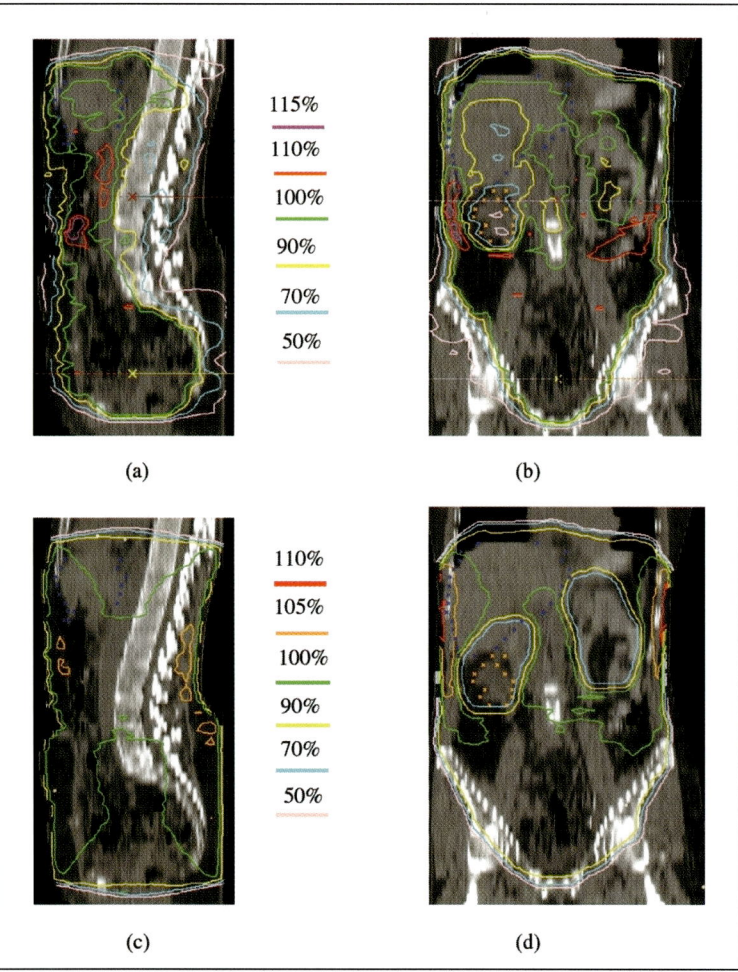

(a)

(b)

115%	—
110%	—
100%	—
90%	—
70%	—
50%	—

(c)

(d)

110%	—
105%	—
100%	—
90%	—
70%	—
50%	—

COLOR PLATE 38. FIGURE 14–3. Isodose distributions for IMRT plan with five gantry angles. Isodose levels in percent (%). The location of the two isocenters are indicated by crosses. (a) Sagittal plane. (b) Coronal plane. Also shown are isodose distributions for conventional plan with extended distance (130 SAD) AP/PA fields for the same patient. (c) Sagittal plane. (d) Coronal plane.

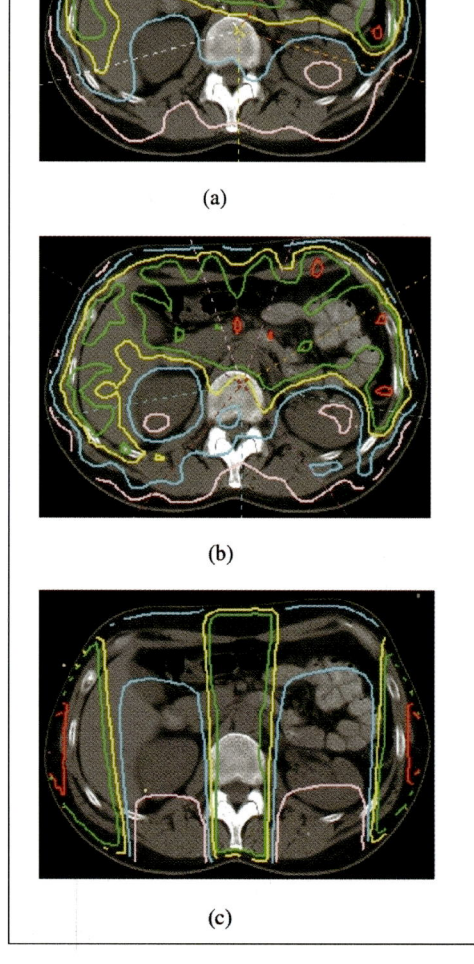

(a)

(b)

110%	—
100%	—
90%	—
70%	—
50%	—

(c)

COLOR PLATE 39. FIGURE 14–4. IMRT and conventional plan comparison. Isodose levels in percent (%). (a) IMRT plan with five gantry angles. (b) IMRT plan with nine gantry angles (from PA, with gantry angle 40° apart). (c) Conventional plan.

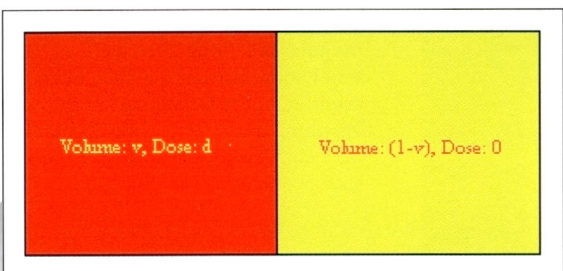

COLOR PLATE 40. FIGURE 15–1. Partial volume irradiation. Fractional volume v receives dose d (red at left), and the remaining volume (1-v) receives zero dose (yellow at right).

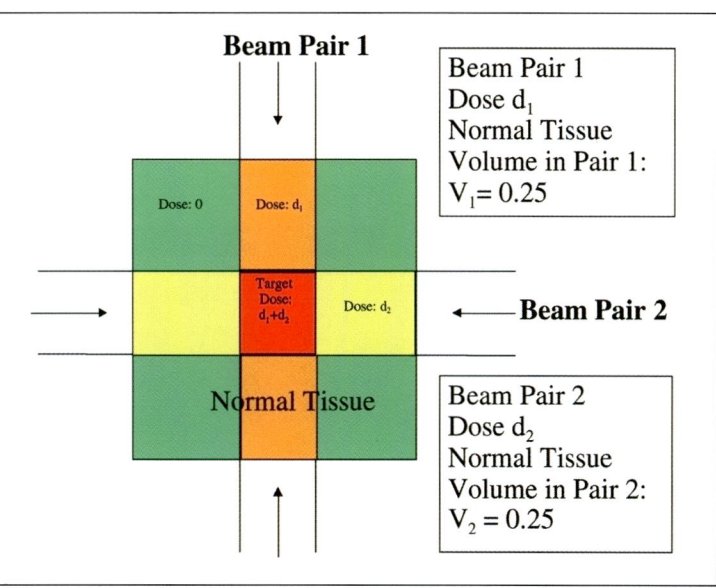

COLOR PLATE 41. FIGURE 15–7. Model target, normal tissue, and beam geometry. The square target is surrounded by a square normal tissue. The vertical beams, beam pair 1, deliver dose D_1 (orange), and the horizontal beams, beam pair 2, deliver dose D_2 (yellow). The target receives dose $(D_1 + D_2)$ (red). The un-irradiated part of the normal tissue is shown in green.

COLOR PLATE 42. FIGURE 16–2. (a) Transverse section of nasopharynx IMRT plan with dose color wash overlay. (b) Comparison of upper and lower confidence limit DVHs for CTV and left cochleum.

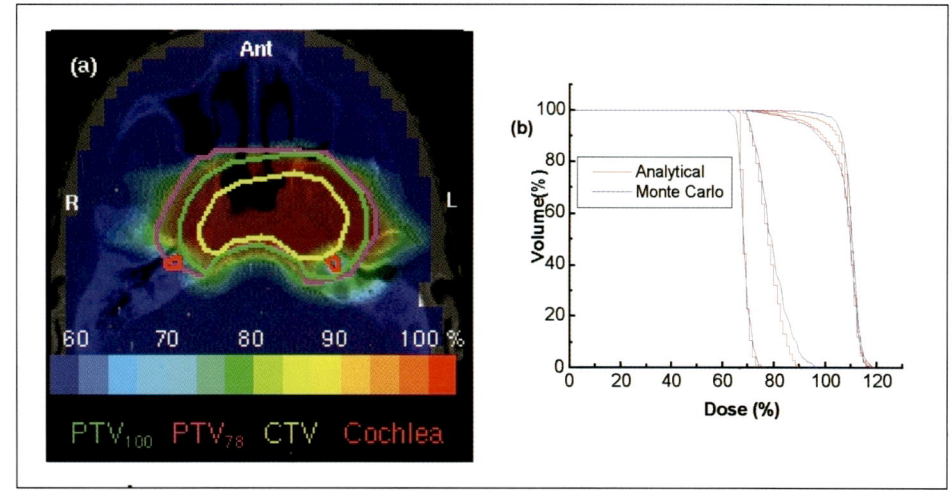

COLOR PLATE 43. FIGURE 17–4. Comparison of dose distributions for a free breathing plan (left) and deep inspiration plan (right). Both reduction of respiratory motion and lung expansion are achieved with DIBH. Note that the breathing artifacts near the diaphragm of the FB scan are absent in the DI scan and there is much less lung within the 10% (red) isodose line for the DI scan. [Reprinted with permission from Rosenzweig et al. (2000). "The deep inspiration breath-hold technique in the treatment of inoperable non-small-cell lung cancer." Int. J. Radiat. Oncol. Biol. Phys. 48(1):81-87, fig. 3. Copyright 2000 Elsevier Science Ltd., Oxford, UK.]

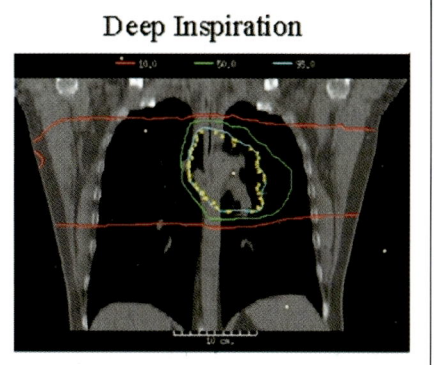

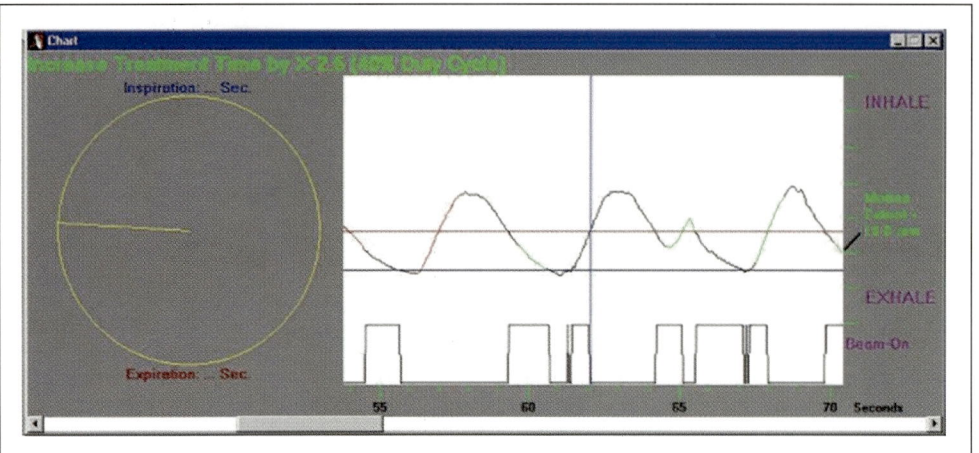

COLOR PLATE 44. FIGURE 17–11. Irregular breathing during CT acquisition can cause a slice to be acquired at the wrong part of the breathing trace. Green line segments indicate x-ray on intervals during slice acquisition.

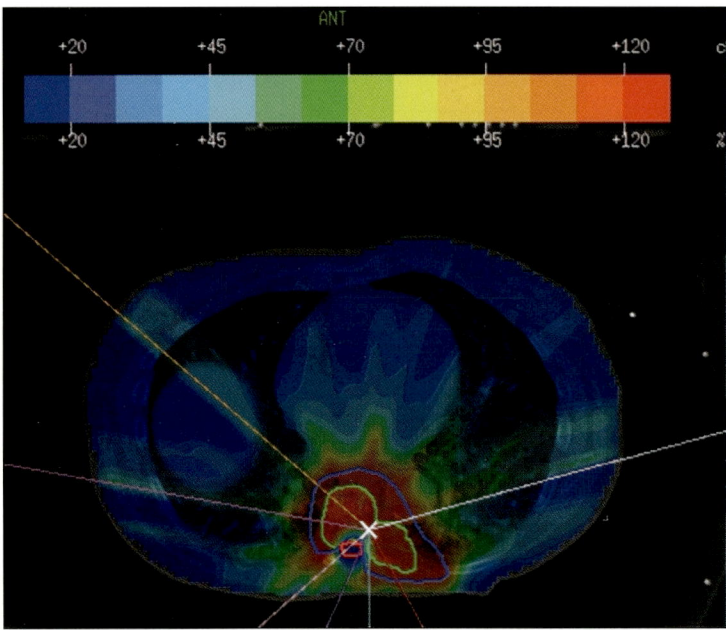

COLOR PLATE 45. FIGURE 18–4a. A 7-field IMRT plan for a T8 paraspinal case delivering 2000 cGy in 5 fractions. GTV is shown by green, PTV by blue, and cord by red colors.

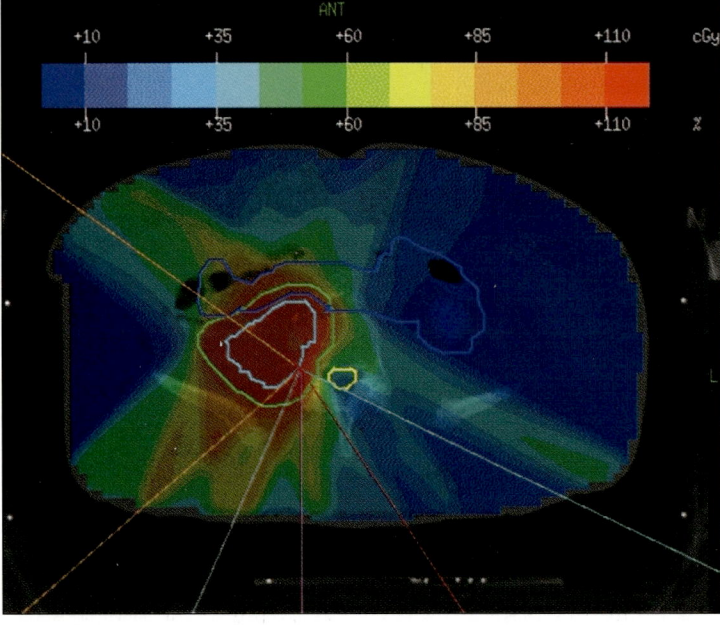

COLOR PLATE 46. FIGURE 18–4b. A 6-field IMRT plan for an L4 paraspinal lesion delivering 2000 cGy in 5 fractions. Color code for organs: green is PTV, cyan is GTV, yellow is cord, and blue is bowel. For the dose levels shown, 100% = 100 cGy.

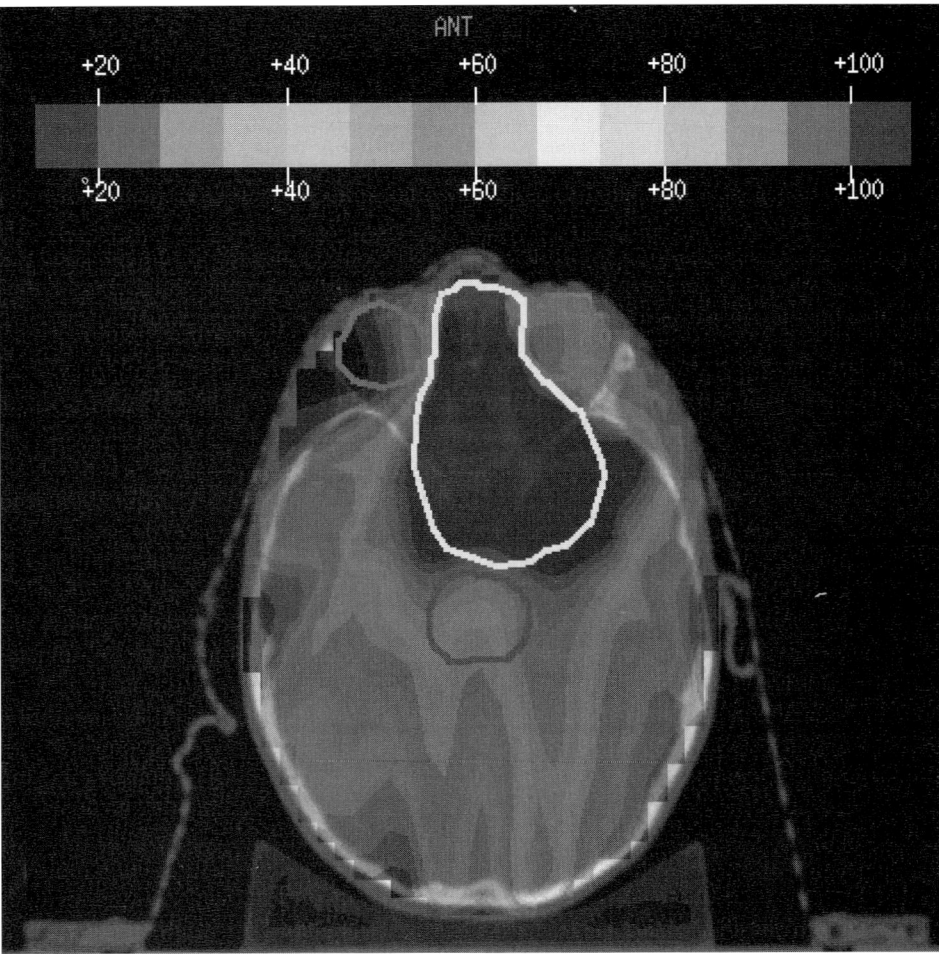

FIGURE 11–5. Transverse dose distribution consisting of eight non-coplanar IMRT 6 MV photon beams and a single conventional 9 MeV electron beam for a parameningeal rhabdomyosarcoma. See COLOR PLATE 31.

Table 11–6. DVH Analysis for the Comparison of IMRT and 3-D Conventional Treatment Planning for Three Patients with Parameningeal Rhabdomyosarcoma.

All doses are in Gy and the standard deviations are given in parentheses. The asterisk indicates the structures totally enclosed by the PTV for one patient. This patient's corresponding structures were omitted from the analysis

Structure	Statistics	IMRT	3D-conventional
PTV	Min. Dose (D_{95})	47.9 (1.4)	46.8 (1.8)
	Mean Dose	53.7 (0.6)	52.9 (0.5)
	Max. Dose (D_{05})	57.3 (2.3)	57.1 (1.3)
Pituitary*	Mean Dose	27.6 (3.5)	45.6 (2.1)
	Max. Dose (D_{05})	36.7 (6.8)	52.0 (1.3)
Brainstem	Mean Dose	21.5 (6.9)	33.5 (2.5)
	Max. Dose (D_{05})	32.2 (9.0)	41.0 (3.1)
Chiasm*	Mean Dose	31.2 (10.1)	47.0 (5.4)
	Max. Dose (D_{05})	44.0 (3.3)	51.4 (2.5)
Cochlea* (ipsilateral)	Mean Dose	23.1 (3.4)	37.0 (0.8)
Lens (ipsilateral)	Mean Dose	25.5 (7.1)	36.6 (8.0)

patients, the mean pituitary gland dose is reduced by approximately 40% with the IMRT technique in comparison to the conventional treatment planning.

Other Sites

Although, IMRT is used mostly for head and neck cancer both in adult and pediatric patients, it can significantly improve normal tissue dose in other parts of the body. Specifically, we had experience with the rhabdomyosarcoma of pelvis in a number of pediatric patients that required very conformal radiation treatment due to concerns with normal organs including, bladder, ovaries, genitals, etc. We will demonstrate an example of an IMRT plan for a 16-year-old female patient with perirectal rhabdomyosarcoma. The PTV for this patient included a perirectal volume and bilateral inguinal nodes and the treatment required 4140 cGy to be delivered in 23 fractions. A six-field IMRT plan was developed with special consideration to bladder and genital dose. The dose distribution for this plan is shown at the level of bladder and nodes in figure 11–6.

To assess the relative benefits from the IMRT plan, a retrospective analysis was done with a 3-D conformal plan. The conformal plan used a standard four-field arrangement as well as electron boost to nodal regions. The dose distribution from the standard conformal plan is shown in figure 11–7. This plan utilized standard blocks for field shaping and wedge compensators for the lateral beams.

Table 11–7 shows the comparison of the IMRT plan statistics with those of the conformal plan. The IMRT plan delivers a higher mean dose to PTV while sparing bladder, genitals, and hips by 15% to 59% with respect to conformal plan with electron boost. The treatment delivery efficiency for the IMRT plan is also improved with no need to enter treatment room between the treatment fields.

Summary

The most profound benefit of IMRT technology may be in pediatric patients. Many solid tumors of childhood have high cure rates but require relatively high doses of radiation. Long-term quality of life can be dramatically affected by late adverse sequelae of radiation exposure to normal tissues. IMRT can dramatically decrease doses to normal structures and therefore minimize potential late-effects. This chapter has illustrated a number of such examples. We are hopeful that the application

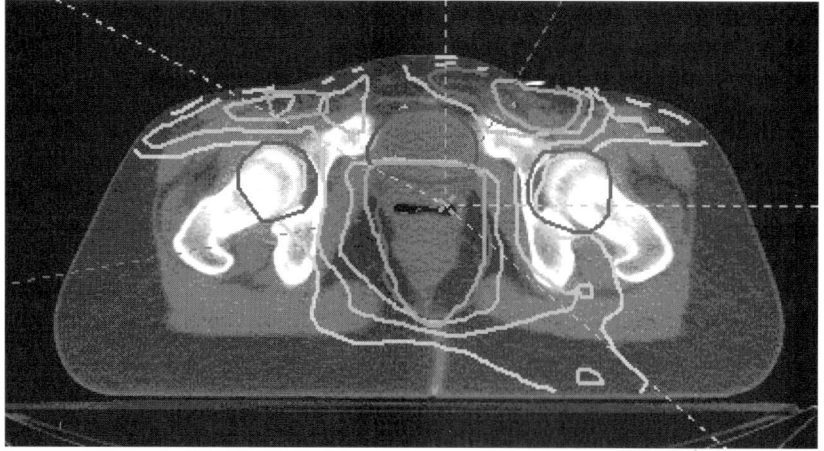

FIGURE 11–6. Six-field IMRT plan for a perirectal rhabdomyosarcoma case. See COLOR PLATE 32.

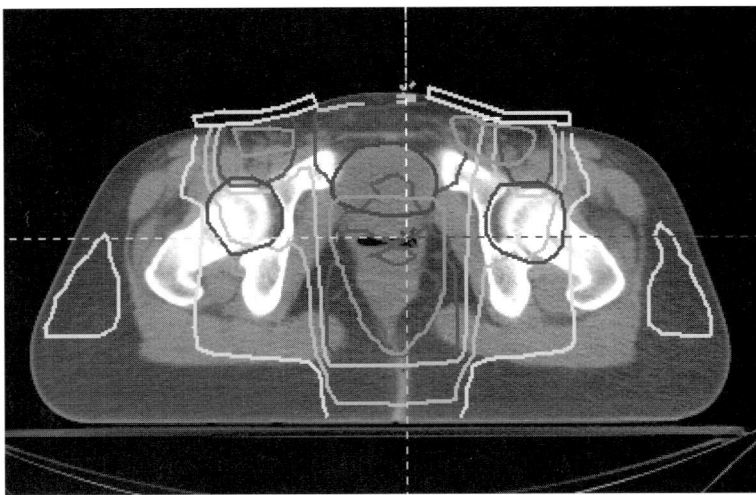

FIGURE 11–7. Four-field conformal plan for a perirectal rhabdomyosarcoma case utilizing electron boost to nodal regions. See COLOR PLATE 33.

Table 11–7. Comparison of PTV and Normal Tissue Dose from Six-Field IMRT and 3-D Conformal Plan

Structure	Statistics	IMRT	3-D Conformal
PTV	Mean dose (%)	107.0	100.4
	Min. dose (D_{95}) (%)	98.4	98.7
Bladder	Mean dose (%)	85.0	100.0
Rectum	Mean dose (%)	72.7	72.0
Genitals	Mean dose (%)	36.3	88.0
Hips	Mean dose (%)	43.0	62.0

of IMRT to the pediatric cancer population will significantly improve these children's long-term quality of life.

For planning considerations, a large number of pediatric sites with many different tumor types make finding class solutions difficult for all cases. However, inverse planning templates developed for other sites can become a very important starting point for efficient and effective IMRT plan generation. Dose distribution and dose-volume statistic comparison with more established standard conformal techniques should be used to assess for benefits in regard to normal tissue sparing and treatment delivery efficiency. Finally, increased beam-on time with IMRT can result in increased normal tissue dose in this vulnerable patient group. This effect will likely be related to treatment site and plan parameters such as number of beams and beam orientation used in planning (Verellen and Vanhavere 1999; Dorr and Herrmann 2002). This needs to be carefully assessed in risk and benefit analysis for the use of IMRT in pediatric cases.

References

Bleyer, W. A. (1990). "The impact of childhood cancer on the United States and the world." *CA Cancer Clin.* 40:355–367.

Crist, W. M., J. R. Anderson, J. L. Meza, C. Fryer, R. B. Raney, F. B. Ruymann, J. Breneman, S. J. Qualman, E. Wiener, M. Wharam, T. Loke, B. Webber, H. M. Maurer, and S. S. Donaldson. (2001). "Intergroup rhabdomyosarcoma study-IV: Results for patients with nonmetastatic disease." *J. Clin. Oncol.* 19:3091–3102.

Denys, D., S. C. Kaste, L. E. Kun, M. A. Chaudhary, L. C. Bowman, and K. T. Robbins. (1998). "The effects of radiation on craniofacial skeletal growth: A quantitative study." *Int. J. Pediatr. Otorhinolaryngol.* 45:7–13.

Dorr, W., and T. Herrmann. (2002). "Second primary tumors after radiotherapy for malignancies. Treatment-related parameters." *Strahlenther. Onkol.* 178:357–362.

Eisbruch, A., L. A. Dawson, H. M. Kim, C. R. Bradford, J. E. Terrell, D. B. Chapeha, T. N. Teknos, Y. Anzai, L. H. Marsh, M. K. Martel, R. K. Ten Haken, G. T. Wolf, and J. A. Ship. (1999). "Conformal and intensity modulated irradiation of head and neck cancer: The potential for improved target irradiation, salivary gland function, and quality of life." *Acta Otorhinolaryngol. Belg.* 53:271–275.

Fukunaga-Johnson, N., H. M. Sandler, R. Marsh, and M. K. Martel. (1998). "The use of 3D conformal radiotherapy (3D CRT) to spare the cochlea in patients with medulloblastoma." *Int. J. Radiat. Oncol. Biol. Phys.* 41:77–82.

Huang, E., B. S. Teh, D. R. Strother, Q. G. Davis, J. K. Chiu, H. H. Lu, L. S. Carpenter, W. Y. Mai, M. M. Chintagumpala, M. South, W. H. Grant 3rd, and E. B. Butler. (2002). "Intensity-modulated radiation therapy for pediatric medulloblastoma: Early report on the reduction of ototoxicity." *Int. J. Radiat. Oncol. Biol. Phys.* 52:599–605.

Hunt, M. A., M. J. Zelefsky, S. Wolden, C. S. Chui, T. LoSasso, K. Rosenzweig, L. Chong, S. V. Spirou, L. Fromme, M. Lumley, H. Amols, C. C. Ling, and S. A. Leibel. (2001). "Treatment planning and delivery of intensity-modulated radiation therapy for primary nasopharynx cancer." *Int. J. Radiat. Oncol. Biol. Phys.* 49:623–632.

McHaney, V. K., W. Meyer, W. Furman, M. Schell, and L. Kun. "The Effects of Radiation Therapy and Chemotherapy on Hearing" in *Late Effects of Treatment for Childhood Cancer.* D. M. Green and G. J. D'Angio (eds). New York: Wiley, pp. 7–10, 1992.

Merchant, T. E., Y. Zhu, S. J. Thompson, M. R. Sontag, R. L. Heiderman, and L. E. Kun. (2002). "Preliminary results from a Phase II trail of conformal radiation therapy for pediatric patients with localised low-grade astrocytoma and ependymoma." *Int. J. Radiat. Oncol. Biol. Phys.* 52:325–332.

Michalski, J. M., R. K. Sur, W. B. Harms, and J. A. Purdy. (1995). "Three dimensional conformal radiation therapy in pediatric parameningeal rhabdomyosarcomas." *Int. J. Radiat. Oncol. Biol. Phys.* 33:985–991.

Miralbell, R., A. Bleher, P. Huguenin, G. Ries, R. Kann, R. O. Mirimanoff, M. Notter, P. Nouet, S. Bieri, P. Thum, and H. Troussi. (1997). "Pediatric medulloblastoma: Radiation treatment technique and patterns of failure." *Int. J. Radiat. Oncol. Biol. Phys.* 37:523–529.

Oberlin, O., A. Rey, J. Anderson, M. Carli, R. B. Raney, J. Treuner, and M. C. Stevens. (2001). "Treatment of orbital rhabdomyosarcoma: Survival and late effects of treatment-results of an international workshop." *J. Clin. Oncol.* 19:197–204.

Patel, S. G., A. C. See, P. A. Williamson, D. J. Archer, and P. H. Evans. (1999). "Radiation induced sarcoma of the head and neck." *Head Neck* 21:346–354.

Schell, M. J., V. A. McHaney, A. A. Green, L. C. Kun, F. A. Hayes, M. Horowitz, and W. H. Meyer. (1989). "Hearing loss in children and young adults receiving cisplatin with or without prior cranial irradiation." *J. Clin. Oncol.* 7:754–760.

Verellen, D., and F. Vanhavere. (1999). "Risk assessment of radiation-induced malignancies based on whole-body equivalent dose estimates for IMRT treatment in the head and neck region." *Radiother. Oncol.* 53:199–203.

Wharam, M. D., Jr. (1997). "Rhabdomyosarcoma of parameningeal sites." *Semin. Radiat. Oncol.* 7:212–216.

Wolden, S. L., J. R. Anderson, W. M. Crist, J. C. Brenerman, M. D. Wharam Jr., E. S. Wiener, S. J. Qualman, and S. S. Donaldson. (1999). "Indications for radiotherapy and chemotherapy after complete resection in rhabdomyosarcoma: A report from the Intergroup Rhabdomyosarcoma Studies I to III." *J. Clin. Oncol.* 17:3468–3475.

12

IMRT Of Cancer Of The Breast

Linda X. Hong
Beryl McCormick
Chen-Shou Chui
Margie A. Hunt

Introduction

The possibility of late radiation sequelae after primary breast treatment remains a concern to patients and physicians. Intensity modulation with a standard tangential beam arrangement improves dose homogeneity throughout the target volume, particularly in the superior and inferior regions of the breast. Intensity-modulation radiation therapy (IMRT) can also achieve dose reduction to the heart, ipsilateral lung, and contralateral breast. Simplified IMRT (sIMRT) techniques, which do not require significant increased use of resources, are essential for routine clinical implementation in a busy department. Whether the dosimetric improvements achievable with IMRT will lead to measurable differences in clinical outcome remains to be demonstrated.

Rationale For Breast IMRT

It is widely accepted that early stage breast cancer can be managed with conservative surgery and radiation therapy as an alternative to mastectomy. The tangent techniques used to treat the entire target, the breast, have been in use for many years. In 1991, the National Cancer Institute (NCI)-sponsored Photon Treatment Planning Contract first reported on the use of 3-D treatment planning of the intact breast (Solin et al. 1991). Of 38 plans evaluated, the group was "unable to identify any beam arrangement which improved upon tangential fields." Breast conservation radiotherapy is conventionally delivered with wedged tangential fields, optimized using a single central axis isodose distribution without inhomogeneity corrections (Kutcher et al. 1996). Local control using wedged tangential beams followed by a boost to the tumor bed is excellent (Fisher et al. 1989; Solin et al. 1988; Veronesi, Zucali, and Luini 1986), and the risk of most long-term complications is low (Lingos et al. 1991; Wallgren 1992). Nonetheless, the possibility of late radiation sequelae after primary intact breast treatment remains a concern to both patients and physicians.

Several studies have demonstrated dose inhomogeneities as large as 20% due to rapid changes in the patient contour in the superior and inferior regions of the breast with standard wedged tangential beams (Buchholz et al. 1997; Cheng, Das, and Stea 1994; Chin et al. 1989; Fraass, Roberson, and Lichter 1985; Fraass et al. 1988). The unavoidable presence of lung tissue, coupled with changing patient separation near the deep border of the tangents creates additional regions of high dose in the medial and lateral aspects of the breast. These regions of increased dose may contribute to an inferior cosmetic outcome, particularly in large breasted patients (Gray et al. 1991; Moody et al. 1994; Taylor et al. 1995) as well as variability in the total dose delivered to the primary tumor bed (Buchholz et al. 1997; Taylor et al. 1995; Das et al. 1993).

Radiation-induced myocardial damage is another potential complication of breast irradiation. Cardiac morbidity associated with chest wall radiation after mastectomy has been well documented in older trials comparing post-mastectomy radiation to none (Cuzick et al. 1994; Group EBCTC 2000). The effect of *modern* radiation therapy directed to the breast only, rather than the chest wall and regional lymph nodes, has been less thoroughly studied. Although the group at Harvard concluded that there was no increased risk for cardiac-related mortality within the first 12 years of treatment (Nixon et al. 1998), other studies have documented increased mortality and cardiac toxicity related to left-sided breast radiation (Paszat et al. 1998a,b; Zambetti et al. 2001).

Still another issue in breast treatment with tangents is the possible risk of inducing contralateral breast tumors. A variety of techniques including the elimination of the medial tangent wedge and the use of local shielding (Fraass, Roberson, and Lichter 1985) have been proposed to reduce the scatter dose to the contralateral breast. All these concerns highlight the fact that improvements in the technical delivery of breast irradiation are still needed.

IMRT methods aiming to produce a uniform dose distribution in the entire target volume while protecting the critical organs such as the ipsilateral lung and the heart have been proposed (Hong et al. 1999; Landau et al. 2001). To accomplish this, the volumes of interest (target and critical organs) are normally delineated and a volume-based optimization (vIMRT) is performed, balancing the conflicting requirements of the target and the critical organs. This approach can produce superior results, when compared to conventional techniques employing tangential wedged beams. Improved dose homogeneity throughout the breast as well as reductions in dose to the heart and the lung are observed. The dose to the contralateral breast is significantly decreased, when compared to the standard wedged beams (McCormick et al. 2000). While this technique is promising, it does require the time-consuming delineation of all volumes of interest. This presents a significant increase in the planning time and thus is not suitable for large-scale implementation.

A number of simpler IMRT techniques have been reported (Evans et al. 1995; Zackrisson, Arevarn, and Karlsson 2000; van Asselen et al. 2001; Carruthers, Redpath, and Kunkler 1999; Lo et al. 2000). One of these methods (Carruthers, Redpath, and Kunkler 1999) is based on the choice of a single plane (typically a "mid-plane" on the central slice) on which a uniform dose is to be delivered by the proper design of beam intensity modulation. This method works well if the breast volume is symmetric to the chosen plane. However, since the breast volume can change shape both in the anterior-posterior and in the superior-inferior directions, symmetry may not exist. As a result, uniform dose to a single plane generally does not give the most uniform dose achievable throughout the entire breast volume. Lo et al. (2000) proposed a method that uses an additional pair of tangential wedged fields to boost extra dose to the cold region resulting from the original pair of the tangential fields. The shape and weight of these additional wedged fields were determined manually by trial and error. Similarly, Zackrisson et al. (2000) used two segments per beam to reduce the dose variation inside the planning target volume (PTV). This can be considered as a 2-level IMRT. The shape and weight of these segments were also determined manually. Both of these methods by Lo et al. (2000) and by Zackrisson et al. (2000) used manual methods to determine the shape and weight of the additional fields. This requires extra planning efforts and, for practical reasons, typically limits the number of additional fields to two, although in principle more fields can be included. Evans et al. (1995) used the portal imaging device to estimate the path length of each ray going through the treatment volume. Based on this information a pseudo-CT for the patient was constructed. The intensity of each ray can be determined automatically by several methods. In the original paper (Evans et al. 1995), the method of equalizing the average dose along each ray was suggested. Later, it was modified to equalize the maximum dose on each ray (Evans et al. 1998, 2000). The pseudo-CT used in this method assumes that the treatment volume is symmetric to the mid-plane. In general, however, the patient's breast volume is asymmetric to the mid-plane. The maximum error induced by this asymmetry was 5% for an example given in Evans et al. 1995. The magnitude of the error, however, can be reduced by considering the real contour of the patient, as suggested in Evans et al. 1998. This method indeed improved the dose uniformity in the target, but the treatment planning and preparation time was increased from 15 minutes for the standard wedged plan to 50 minutes for the intensity-modulated plan (Hansen et al. 1997). Another approach based on the path length through the treatment volume was proposed by van Asselen et al. (2001). In this approach, a path length map for each tangential beam was created using the patient's 3-D CT image set. This map is then divided into four regions, each containing approximately the same path lengths. The tangential beam is then divided into four segments, corresponding to the four regions of equal path length. Out of the four segments, the first segment covers the largest area and delivers the majority of the dose, while the other three segments cover progressively reduced areas and deliver appropriate doses so that the total dose in the treatment volume is uniform. The weightings of the first segment of the two opposing tangential fields are determined manually, and the weightings of the other three segments are based on a single point located in the center of the smallest segment. This method also produced good dose uniformity in the breast, but suffers from longer planning time in its current form.

Another multi-segment IMRT technique was described by Kestin et al. (2000). In this approach the dose distribution from the open tangential fields is first calculated. Isodose surfaces of a set of dose levels are then projected on a beam's-eye view (BEV) plane at the isocenter. The shapes of these curves on the BEV-plane are used as the shapes of the subsequent segments. The weight of each of these segments is then determined by an optimization algorithm that aims to produce a uniform dose to 100 reference points placed within the target volume. This approach also produced superior plans when compared to the conventional technique. But constructing and selecting multi-segments requires additional planning time relative to the conventional wedged technique.

We have designed a simplified IMRT (sIMRT) technique that focuses on the treatment volume only (Chui et al. 2002). A reasonable way to achieve uniform dose to the breast is for each pencil beam to deliver one half of the prescribed dose to the mid-point of the pencil beam segment that intersects the treatment volume. Using the standard two-field, opposed tangential beam arrangement, the corresponding opposed pencil beams would then together deliver the prescribed dose to the mid-point. The optimum intensity of each pencil beam can then be determined as proportional to the inverse of the mid-point dose from the open beam.

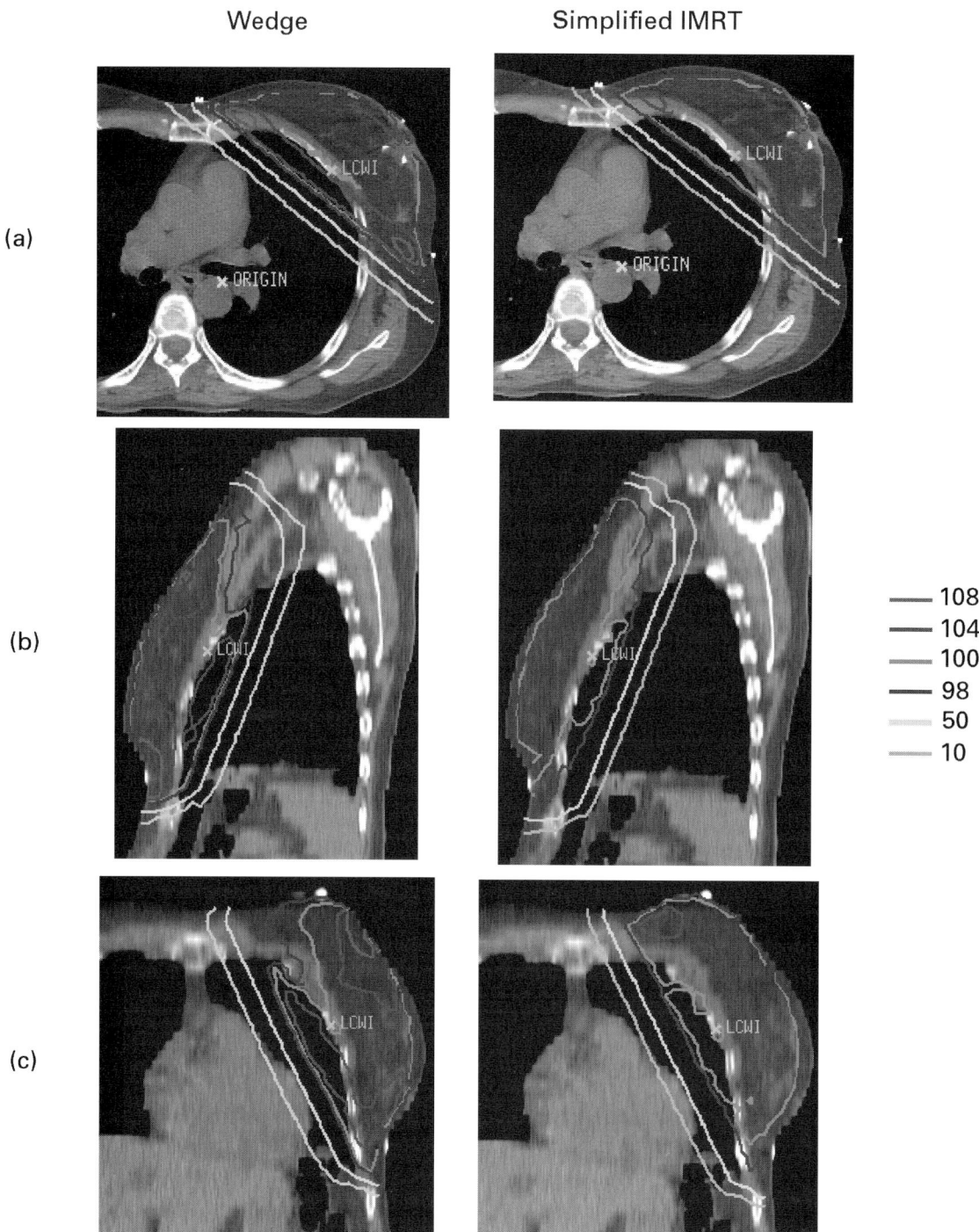

FIGURE 12–1. Dose distribution comparison. Left for standard wedged plan, right for simplified IMRT plan. Plans were normalized to the LCWI point. ISO was the location of beam isocenter. (a) Transverse plane through the LCWI point. (b) Sagittal plane through the LCWI point. (c) Coronal plane through the LCWI point. See COLOR PLATE 34.

Materials And Methods

In our initial use of this sIMRT, 15 women, ranging in age from 31 to 88 years, with stage I or early stage II breast cancer, were offered IMRT treatment. To assess the possible improvement in dose to the heart, only women with left-sided disease were offered this program. No patient required radiation to the regional lymph nodes. The patients were positioned supine in a custom Alpha-cradle (Soule Medical Company, Tampa, FL), with the ipsilateral hand and arm extended over the head. When in position, preliminary marks were placed on the skin at the palpated periphery of the breast tissue to be treated. A wire was placed over the marks, and a second wire over the site of the lumpectomy incision. The patients were scanned in this position, in a Marconi AcQSim.

Anatomic data were derived from the CT, using images at 6 mm spacing. Any surgical clips placed at the site of the lumpectomy cavity and the heart and lungs were contoured. Tangent fields were defined according to the following criteria: (1) cover the defining breast wire, or imaged breast volume, with 2 cm margin in the superior and inferior directions; (2) treat 3 cm or less of lung on the central axis (CAX), with the posterior border extending from the midline to 2 cm below the lateral wire; (3) extend at least 2 cm flash beyond the skin surface in the anterior direction. Tangent fields were finalized with information from the CT images, if the breast wires were not appropriately placed; the most common adjustments were made because of an overestimate of breast tissue in the medial direction, and an underestimate of breast tissue in the superior-lateral region. The collimator was rotated as needed to align with the contour of the lung-chest wall interface from the BEV display. If necessary, fields were further shaped in the superior and inferior corners to maintain 2 cm around the breast wire. After simulation, the oblique separation in this group of patients ranged from 14.6 cm to 26.5 cm. The amount of lung, as seen in the tangent fields at the CAX level, ranged from 1.7 cm to 3.0 cm.

Even with 3-D CT images, it was difficult to delineate accurately the breast volume in some patients, especially those with a high content of fatty tissue. Therefore, the PTV was defined as tissues included in the simulated fields less 5 mm skin and beam penumbra from the posterior field edge, excluding the lung and heart. For evaluation purpose, the contralateral breast volume was also contoured.

The following structures were defined for each patient: right and left lungs, heart, and contralateral breast.

Standard Plans

Treatment plans for the standard tangential beams were generated following the standard planning procedure for breast patients at our institution. Beam weights and wedge angles were optimized based on the dose distribution for the central axis plane calculated without inhomogeneity corrections. The criteria used for determining optimal weights and wedge angles were: (1) a symmetrical dose distribution in the central plane with equal medial and lateral high dose regions; (2) a maximum dose in the subareolar region between 102% and 105%; (3) smaller wedge angle on the medial field if the medial and lateral wedge angles were different.

Plans were normalized and prescribed to a point at the lung-chest wall interface (LCWI) along the perpendicular bisector of the posterior border of the medial and lateral fields (as shown in figure 12–1). Past studies of 3-D dose distributions performed at this institution and others demonstrated that the dose delivered to a point at the lung-chest wall interface in the central axis plane is very close to the minimum PTV dose (Solin et al. 1991; Cheng, Das, and Stea 1994; Chin et al. 1989).

Once the optimum parameters for the standard plan were determined, the dose distributions for the standard plan were recalculated using inhomogeneity corrections and the entire 3-D CT data set. Accounting for the three-dimensional information from the CT data set and using inhomogeneity corrections did not significantly alter the dose homogeneity within the PTV.

Likewise, the wedge angles and beam weights chosen during the standard plan optimization did not change with CT-based calculation. This is similar to the observations of several studies (Solin et al. 1991; Cheng, Das, and Stea 1994; Chin et al. 1989) but in contrast to the observation by the University of Michigan group (Fraass et al. 1988) that smaller wedge angles were necessary when inhomogeneity corrections were incorporated.

IMRT Techniques

Volume-Based IMRT (vIMRT)

The beam parameters (isocenter, gantry angles, collimator angles, and jaw settings) for the intensity-modulated tangential fields were the same as those used for the standard plan. For each patient, the complete 3-D CT data set was used for both the intensity optimization and the dose calculations.

Optimized intensity profiles were obtained from an inverse planning algorithm (Spirou and Chou 1998), which uses a convolution-based dose calculation (Chui, LoSaso, and Spirou 1884; Mohan and Chui 1987) and determines the intensity distributions through an iterative process. The user defines the prescribed dose and dose homogeneity constraints for the PTV, and dose and/or dose-volume constraints for the critical organs. The objective function uses summation of the squares of the differences between desired and calculated doses in the target and critical organs, and user-defined penalties applied to the PTV or critical organ points that violate their constraints. Uniformity of dose within the PTV and the maximum dose to critical structures were the constraints used for the optimization. However, tangent fields were limited by the depth dose characteristics of the opposing beams. The coverage of the target volume competed with the constraints on the heart and lung dose. For these patients, we opted to maintain adequate coverage of PTV with less strict dose constraints on the heart and lung. Skin flash was developed for the intensity profiles to extend outside the patient's skin surface by at least 2 cm as illustrated in figure 12–2.

Simplified IMRT (sIMRT) Technique

A simplified IMRT (sIMRT) technique that focuses on the treatment volume only was also developed. We surmise that a reasonable way to achieve uniform dose to the breast is for each pencil beam to deliver one half of the prescribed dose to the mid-point of the pencil beam segment that intersects the treatment volume. Using the standard opposed tangential beam arrangement, the corresponding opposed pencil beams would then together deliver the prescribed dose to the mid-point. The optimum intensity of each pencil beam is then determined as proportional to the inverse of the mid-point dose from the open beam. The advantage of this approach is that the dose is more conformal than the *planar dose compensation* method, yet faster to do in the clinic than the volume-based technique since the optimization does not depend on contours.

For each beam, a grid of pencil beams is created, typically with a spacing of 2 mm by 2 mm measured at the isocenter distance. Each pencil beam, formed by connecting a straight line from the source to the grid point, is extended to intersect the entrance and exit surface of the treatment volume. The mid-point of the intersecting segment is determined and the dose to that point from an open field is calculated. The upper half of figure 12–3 shows, schematically, two such intersecting pencil beams *a* and *b*, and the corresponding mid-points *A* and *B*. Note that since the treatment volume may not be symmetric to the plane perpendicular to the beam central axis, these mid-points may not lie on a single plane. The dose to each point from an open field is calculated as:

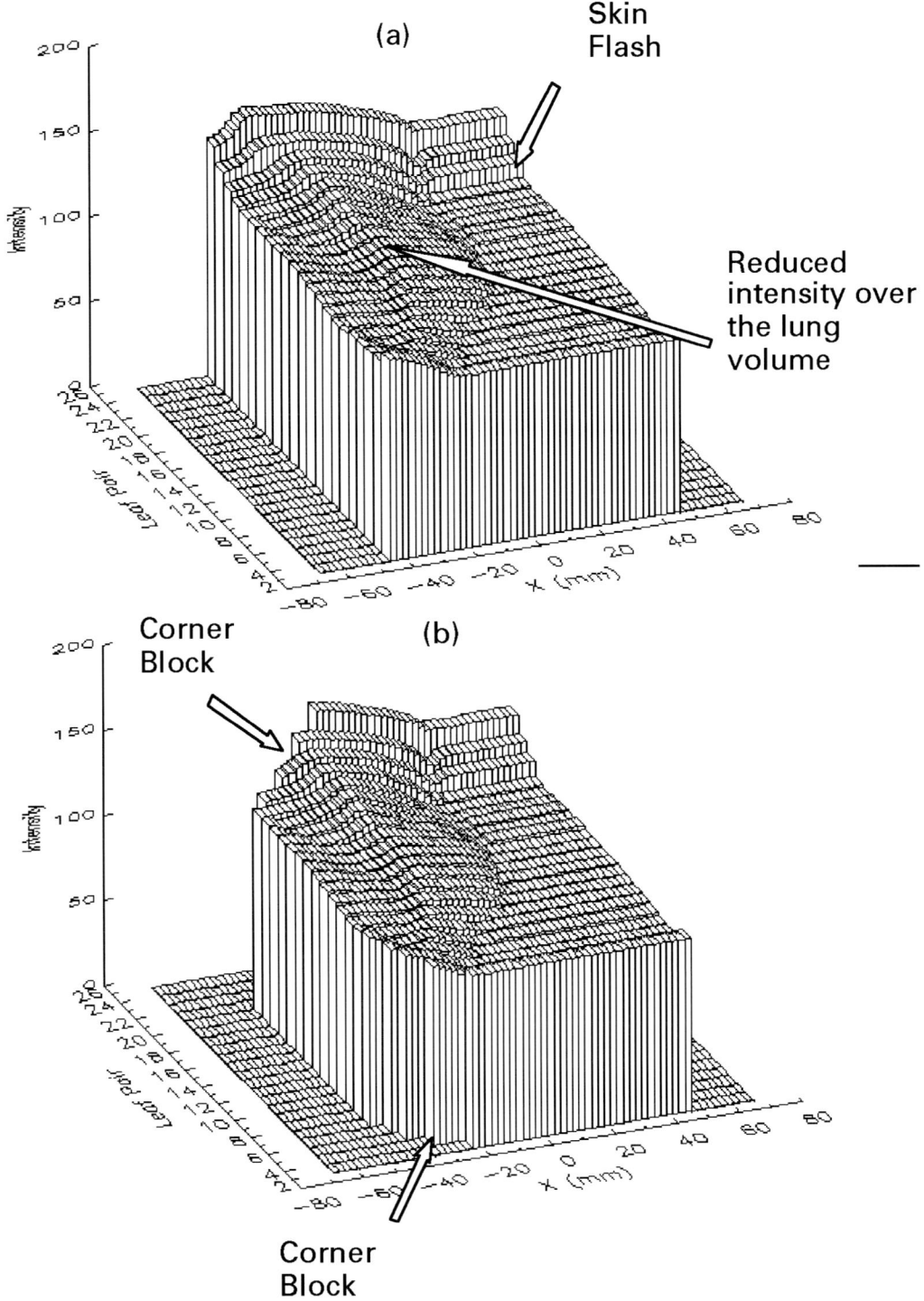

FIGURE 12–2. (a) Intensity distribution for one of the tangential fields. In the skin flash region, the intensity is flat. (b) Corner blocks were used to define the shape of the distribution.

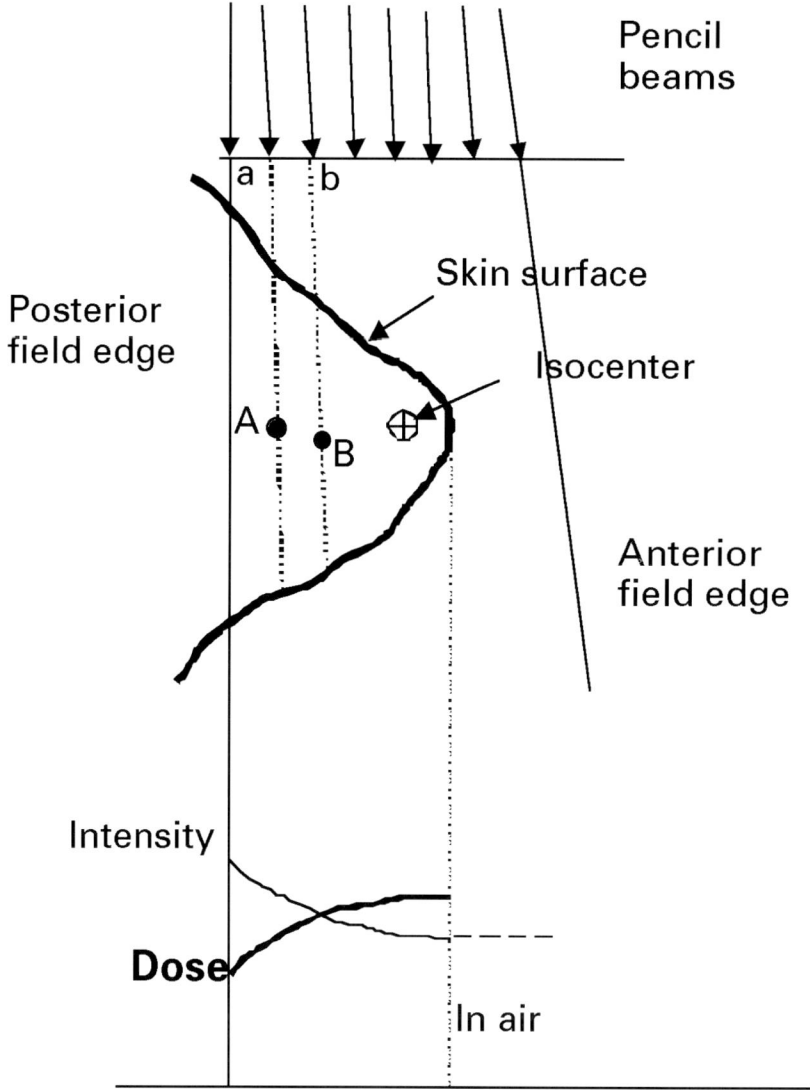

FIGURE 12–3. Schematic drawing (an axial view) of pencil beams, intensity, and dose profiles.

$$D'(x, y, d) = TMR(d, W \times H) \cdot \bullet pOCR(x, y, d) \bullet \left(\frac{SAD}{f}\right)^2, \tag{1}$$

where x,y are the BEV-coordinates of the point, d the equivalent depth, W and H the field width and height, respectively. The first term on the right-hand side of the equation is the tissue-maximum ratio (TMR). The second term is the primary "off-center ratio" (pOCR) which describes the primary beam profile, excluding the effect of field edge penumbra (Chui and Mohan 1986). This is done to avoid the sharp increase in intensity that would have resulted from the sharp decrease in dose in the penumbra region. The third term is the inverse-square correction where SAD and f are the source-axis distance and the source-to-point distance, respectively. The equivalent depth is calculated by pixel integration along the pencil beam if CT-based inhomogeneity correction is desired. Otherwise, d is simply the geometrical depth of the point. For those points in the build-up region near the apex, d is reset to d_{max}, the depth of the maximum dose. This is done for the same reason as excluding the effect of beam edge penumbra to avoid the sharp decrease in dose. A schematic drawing of the dose profile is shown as dark lines in the lower half of figure 12–3. Since the goal here is to produce a uniform dose to these mid-points, it follows that the intensity of the corresponding pencil beam is proportional to the inverse of the dose. Therefore,

$$I(x,y) \propto 1 / D'(x,y,d), \tag{2}$$

where $I(x,y)$ is the intensity of the pencil beam at the grid point. The intensity profile is shown as thin lines in the lower half of figure 12–3. For those pencil beams that do not intersect the treatment volume, the intensity is set to the minimum of those that do intersect on the same grid line. In effect, this forms the *skin flash* region.

DMLC Delivery

After the intensity profile has been determined, it needs to be delivered by some means. One commonly used method is the conventional multileaf collimator (MLC) operated in either dynamic (DMLC) or segmental mode (Convery and Webb 1998; Spirou and Chui 1994; Stein et al. 1994; Svensson, Kallman, and Brahme 1994; Xia and Verhey 1998; Bortfeld et al. 1994; Kallman et al. 1988; Ma et al. 1998; LoSasso, Chui, and Ling 1998). Since the pencil beam grid is typically calculated on a finer grid than the leaf width, intensity profiles at several grid lines need to be averaged. For example, if the calculation grid spacing is 2 mm and the leaf width is 1 cm, then intensity profiles over five consecutive grid lines need to be averaged to form a new profile for the corresponding leaf pair. In addition, if soft tissue blocking at the corners is desired, the intensity distribution can be truncated by simply closing in the leaves to the appropriate positions. This step is similar to the conventional shielding with an MLC.

Final Dose Calculation

Once the intensity distribution has been determined, the final dose to a point is calculated as (Hansen et al. 1997):

$$D(x,y,d) = TMR(d, W \times H) \bullet OCR(x,y,d) \bullet \left(\frac{SAD}{f}\right)^2 \bullet \left[\frac{\iint_{t\,arg\,et} I(x',y')k(x-x',y-y')dx'\,dy'}{\iint_{W \times H} k(x-x',y-y')dx'\,dy'}\right], \tag{3}$$

where the *TMR* and the inverse-square correction terms have the same meanings as before. The *OCR* term now also includes the penumbra effects at the field edge. The last term in the brackets is a correction factor that accounts for the effects of beam intensity distribution. The numerator is a convolution of the intensity distribution I and the pencil beam kernel k. The integral is carried out over the area that intersects the soft tissue within the treatment field, that is, excluding the pencil beams in the skin flash area. The denominator is an integration of the pencil beam kernel over the open field area defined by *WH*.

In a fashion similar to the standard tangential beam plans, the dose for the IMRT plans was normalized and prescribed to the lung-chest wall interface point (as shown in figure 12–1).

Results

Intensity Profiles

Figure 12–2 shows typical intensity profiles for the medial fields obtained by the sIMRT method. In general, the intensity profiles were fairly smooth with three distinct sections. In the section overlying the breast tissue, the intensity closely resembles that of a wedge with the slope varying in the superior and inferior directions. There is an area with slightly reduced intensity corresponding to the area overlaying the lung volume. This is because the lung density is lower than that of the soft tissue; the equivalent path length in this region is smaller and hence the reduced intensity. The area with the flat intensity provides the skin flash outside the skin surface. If corner blocks are needed to provide sparing of normal tissues, the intensity distribution can be truncated by closing in the leaves. Figure 12–2b shows such an example with the intensity distribution truncated at two corners. The intensity profiles for the lateral field are similar to those for the medial. Figures 2a and 2b also include a high-intensity ridge at the superior end of the tangential field due to increased separation and effective path length in the superior part of the field.

Dose Distribution And Dose-Volume Histogram (DVH)

Isodose distributions were examined on the transverse, coronal, and sagittal slices through the LCWI point for all 15 patients. In some cases, other transverse slices superior and inferior to the central slice were also calculated for more detailed examination. In general, the dose distribution from the sIMRT plan is similar to that from the vIMRT plan except for the heart, for which the dose from vIMRT is lower than that from sIMRT. For clarity, in the following figures only the results from the sIMRT plan and the conventional wedged-pair plan are shown for comparison. Figure 12–1 shows the dose distribution on the transverse, coronal, and sagittal slices for one of the patients. As can be seen from the figures, the dose distribution for the sIMRT technique is more uniform than the conventional method. The hottest dose encompassing 5% of the volume (D_{05}) for the sIMRT technique was 104%, while for the conventional method it was 108%. A typical DVH for the PTV was plotted in figure 12–4a. For the target, the sIMRT technique gave more uniform dose than the conventional method. This is primarily due to the fact that the intensity distribution can be optimized in the superior-inferior direction, whereas for the conventional method the intensity is flat in the non-wedged direction. For all patients, the isodose level encompassing 5% of the PTV (D_{05}) was reduced by 3% with sIMRT. Figure 12–4b shows a typical DVH for the ipsilateral lung. The sIMRT technique provided better sparing to the lung in the high-dose region than the conventional method. This is because for the sIMRT technique the intensity over the lung volume was reduced to account for the low density of the lung while the conventional method has no such flexibility as it is limited by the fixed profile of the wedge. On average, the volume of the lung receiving 100% of the prescription dose (V_{100}) was reduced by more than 35%. A summary of some of the parameters used in plan evaluation for the 15 patients is listed in table 12–1. For left-sided breast patients, dose reduction to the heart is a direct result of the decreased dose to the lung. For all patients, D_{05} of the heart was reduced by 8% and V_{95} was reduced by 34% with the sIMRT technique. With the vIMRT technique, dose to the heart was further reduced because dose constraint to the heart was explicitly specified in volume-based optimization. The mean dose to the contralateral breast was reduced by more than 50% with the sIMRT technique compared to the standard technique. The volume of the contralateral breast receiving more than 2.3 Gy (5% of prescription) was reduced by more than 80%. The main reasons for this dose reduction are the elimination of scattered dose from the wedges and reduced scattering within the patient in the region of the lung and near the posterior field edge (Hong et al. 1999; Chang et al. 1999).

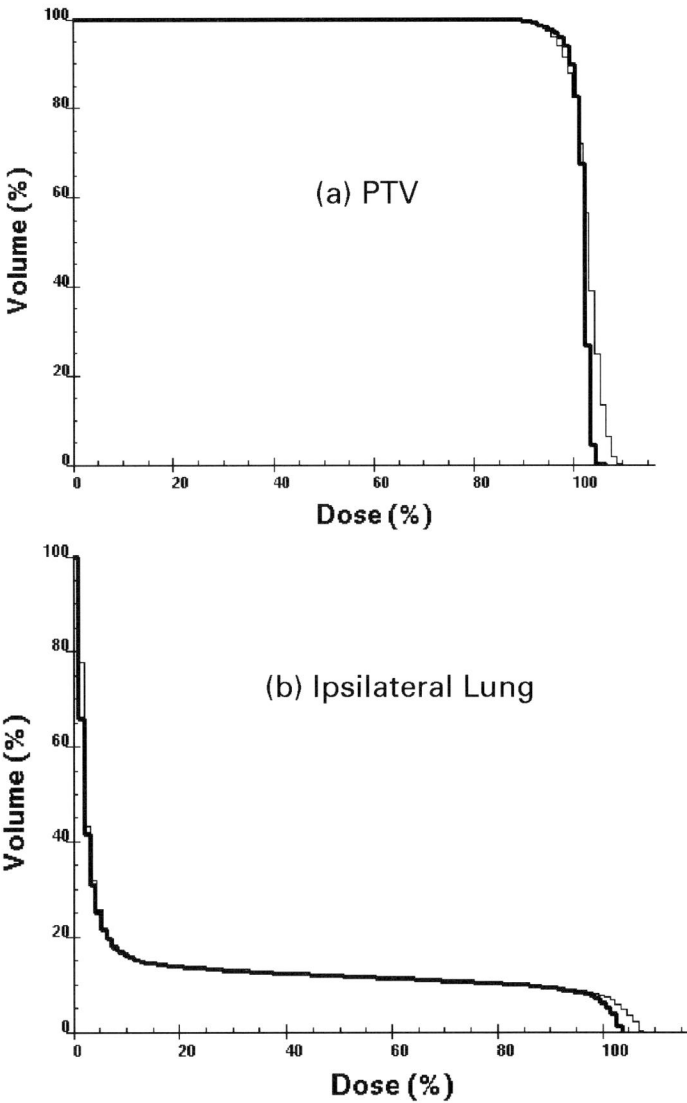

FIGURE 12–4. DVH comparison. Thick line for simplified IMRT, thin line for standard wedge plan. (a) PTV. (b) Ipsilateral lung.

Table 12–1 compares the "standard" tangent plan to the sIMRT and vIMRT tangent plans, for dose to the PTV, heart, ipsilateral lung, and contralateral breast. Comparisons were made for dose to specified percentages of each organ volume, as well as comparing the overall proportion of each organ that received a given dose level. The dose reduction in the lung and heart were mainly in the high-dose region. For example, the dose encompassing 5% of the heart volume was reduced from 20.8 Gy with "standard" tangents to 18.6 Gy with the IMRT plan, on average. The heart volume receiving 43.7 Gy (95% of prescription) or higher reduced by more than 40% with IMRT. This small volume was usually located in the left anterior part of the heart, the location of part of the right coronary and left anterior descending coronary arteries. The volume of ipsilateral lung receiving the prescription dose was 4.9% with the IMRT plan, compared to

Table 12–1. Summary of the Isodose Distributions for All Patients

		Wedge	vIMRT	sIMRT	vIMRT/wedge	sIMRT/wedge
	D_{05}	49.9 ± 0.6 Gy	48.4 ± 1.0 Gy	48.4 ± 1.0 Gy	97.0 ± 1.6%	96.9 ± 1.3%
PTV	V_{95}	96.6 ± 1.9%	98.1 ± 1.1%	97.1 ± 1.9%	101.6 ± 1.4%	100.6 ± 1.0%
	Max. Dose	52.0 ± 1.2 Gy	50.0 ± 1.3 Gy	50.3 ± 1.6 Gy	96.2 ± 1.5%	96.7 ± 1.8%
	D_{05}	46.6 ± 2.2 Gy	44.7 ± 2.4 Gy	44.7 ± 2.0 Gy	95.8 ± 2.6%	96.0 ± 0.9%
Lung	V_{100}	7.0 ± 2.8%	4.9 ± 3.0%	4.8 ± 2.8%	66.2 ± 24.4%	64.9 ± 14.9%
	Max. Dose	50.2 ± 0.9 Gy	47.7 ± 0.6 Gy	47.9 ± 0.6 Gy	95.0 ± 1.6%	95.5 ± 1.2%
	D_{05}	20.8 ±12.5Gy	18.6 ±12.7Gy	19.7 ±12.6Gy	86.4 ± 8.3%	92.1 ± 10.6%
Heart	V_{95}	1.8 ± 1.2%	1.2 ± 1.2%	1.3 ± 1.1%	58.4 ± 35.7%	66.1 ± 24.7%
	Max. Dose	47.8 ± 1.2 Gy	45.4 ± 1.3 Gy	46.0 ± 0.9 Gy	94.8 ± 1.8%	96.2 ± 1.4%
	D_{05}	2.8 ± 0.8 Gy	1.6 ± 0.6 Gy	1.7 ± 0.7 Gy	59.1 ± 15.5%	60.3 ± 15.9%
Contralateral	Mean Dose	1.7 ± 0.5 Gy	0.7 ± 0.2 Gy	0.7 ± 0.3 Gy	45.1 ± 16.9%	45.5 ± 17.0%
Breast	V_{05}	13.3 ± 9.5%	2.1 ± 2.8%	2.4 ± 3.2%	16.7 ± 15.6%	17.5 ± 15.0%

Based on a 46-Gy prescription.
D_{05}: Dose encompassing 5% of volume.
V_{05}: Percent volume receiving at least 5% of dose.
V_{95}: Percent volume receiving at least 95% of dose.
V_{100}: Percent volume receiving at least 100% of dose.

7.0% with the standard tangent plan. The high dose to 5% of lung volume was 44.7 Gy, compared with 46.6 Gy on average.

The two anatomic regions that exhibited significant improvement in dosimetry with the IMRT plan were the breast treated, when comparing homogeneity, and the contralateral breast. The mean dose to the contralateral breast was reduced by more than 50%, from 1.7 Gy with the standard plan, to 0.7 Gy with the IMRT plan. The contralateral breast volume receiving 2.3 Gy (5% of prescription) or more was reduced by more than 80%, from 13.3% to 2.1% on average. This was primarily from the elimination of the wedges with the IMRT plan and from the reduction of scattering dose in the patient from the high dose region in the lung near the posterior field edge (Hong et al. 1999; Epstein et al. 1996; Fraass, Roberson, and Lichter 1985). Homogeneity was improved across the PTV at all levels, with the maximal dose reduced from an average of 52.0 Gy to 50.0 Gy. The high dose to 5% of the PTV was reduced from 49.9 Gy to 48.4 Gy on average.

Treatment Planning Time, Monitor Units, And Treatment Delivery Time

The main timesaver on treatment planning with the sIMRT technique relative to the vIMRT technique comes from eliminating most of the contour delineation necessary for the latter. Automatic-traced external contours are the only contours needed for sIMRT. For a typical field size of 20 cm × 10 cm, the time required to compute the intensity distribution for sIMRT is less than 30 seconds on a 233-MHz Alpha workstation. The optimal plan is obtained in a single operation, without trial-and-error iterations. For the conventional technique, the planning time depends on the planner. Typically, it takes several iterations for an experienced planner to finalize the appropriate wedges and beam weights. Physician and physicist planning time was considerably longer for the vIMRT planning process because of the need to contour each organ and the actual time spent optimizing the plan. On average, the vIMRT cases took three times as long to plan, while for sIMRT plans required comparable time as that of a 3-D wedged plan.

The comparison of monitor units (MUs) for all three techniques is listed in table 12–2. The MUs for the sIMRT technique are similar to that of the conventional technique. For the conventional technique, in order to reduce the scattered dose to the contralateral breast, a smaller wedge in the medial field may be used if mixed wedges are desired. As a result, the MUs on the medial sIMRT field are often higher and those on the lateral sIMRT field often lower, compared with the wedged-pair plan. For all patients, the averaged MUs of combined medial and lateral fields

Table 12–2. Monitor Units (Medial Field/Lateral Field) Comparison For All Patients. MUs Are for 180 cGy/fraction

Patient #	Wedge (MU)	VIMRT (MU)	SIMRT (MU)	vIMRT/wedge	sIMRT/wedge
1	163/163	242/193	148/138	1.48/1.19	0.9 /0.85
2	158/158	210/206	172/161	1.33/1.30	1.09/1.02
3	161/161	193/200	155/147	1.20/1.24	0.96/0.91
4	160/160	180/180	173/139	1.13/1.12	1.08/0.87
5	133/162	185/192	166/132	1.39/1.19	1.25/0.81
6	125/156	157/158	156/153	1.26/1.01	1.26/0.98
7	134/129	200/205	167/143	1.50/1.59	1.25/1.11
8	126/129	173/175	140/122	1.37/1.36	1.10/0.95
9	126/154	165/171	167/155	1.31/1.11	1.32/1.00
10	131/158	175/184	149/141	1.33/1.17	1.14/0.89
11	150/158	146/170	140/130	0.97/1.08	0.93/0.82
12	137/160	190/183	178/158	1.39/1.14	1.30/0.99
13	128/159	179/173	152/149	1.39/1.09	1.19/0.94
14	125/160	216/200	142/130	1.73/1.25	1.14/0.81
15	128/128	154/159	157/122	1.21/1.24	1.23/0.96
Average	139/153	184/183	157/141	1.33/1.21	1.14/0.93
std. dev	15/13	26/16	12/13	0.18/0.14	0.13/0.09

from sIMRT was 298 MUs while that from the standard wedged plans was 292 MUs. The MUs for vIMRT were generally more, with an average of 367 MUs. This is because the intensity profiles of the vIMRT fields are generally less smooth than that of the sIMRT fields, thus resulting in more MUs.

The treatment delivery time is expected to be less for sIMRT than conventional treatment, provided clearance between the machine and the patient is adequate, since it is no longer necessary for the therapists to go into the treatment room between fields, and no wedge insertion is required. The actual delivery time is about one minute at the dose rate of 300 MU/min. Opening and closing the treatment room door takes about an additional minute. Relative to vIMRT, the delivery time for sIMRT is also less, due to the fewer MUs as discussed above.

Discussion

Modest reductions in dose to the heart and lung and improved homogeneity throughout the breast were observed using IMRT tangent fields. Contralateral breast dose was significantly decreased, when compared to standard wedged tangents.

Throughout the treatment volume, dose homogeneity was improved with IMRT tangents and was constrained solely by the depth dose characteristics of opposed beams. The ipsilateral lung volume receiving prescription dose in the IMRT plan was 66% of that in the standard plan. The IMRT reduced the D_{05}, the isodose level encompassing 5% of the heart volume, and an indicator of the high dose region within the heart, by 14%. The IMRT reduced the heart volume receiving 43.7 Gy (95% of prescription) or more by more than 40%. Most significantly, the IMRT plan reduced the mean dose to the contralateral breast by more than 50%, when compared to the conventional wedged-pair tangents. The volume of contralateral breast receiving 2.3 Gy or more was reduced by more than 80% with IMRT.

The clinical significance of these reductions is not well known. Unlike other types of cancer patients who receive some cardiac and lung radiation during primary treatment for their disease, such as those with lung cancer, the majority of women who receive radiation for early stage breast cancer are likely to live for many years. Increasingly, more of these women are also receiving adjuvant chemotherapy regimens with agents that are potentially damaging to the lungs and the heart. Preliminary research from the group at Duke University, prospectively comparing pre-radiation single photon emission computed tomography (SPECT) cardiac perfusions scans to 6-month post radiotherapy scans in five women, found visible perfusion defects in four patients, correlating with the tangent beam location. Further follow up exploring the role of doxorubicin and tangent field radiation on cardiac physiology was planned (Hardenbergh et al. 2001).

More prospective studies like this are needed in this group of often young and otherwise healthy women to better understand the possible long-term sequelae of high doses of radiation to a small volume of the heart and to the lung. As 3-D planning systems become widely available, for the first time information documenting not only how much of the heart but exactly which part, i.e., the wall or the coronary arteries themselves, represents the organ constraints on treatment planning will finally become known.

The increase in homogeneity in the target breast should result in a better cosmetic outcome in this group of patients. The reduction of radiation to the uninvolved side may also play some role in later events in the contralateral breast, although recent studies indicate no increased risk of contralateral cancers associated with this radiation except in patients under the age of 45 at the time of their first breast treatment (Boice et al. 1992; Obedian, Fischer, and Haffty 2000).

Challenges

The initial goal of this project was to reduce the dose delivered to the heart and ipsilateral lung in women with left-sided breast cancers. The reduction seen was modest, with the beam arrangement we used in this study. Several criteria have to be met in designing a different technique. They include:

1. Continue to maintain acceptable dose inhomogeneity throughout the target volume.
2. Avoid increasing the dose to the contralateral breast.
3. Exclude the contralateral lung from the beam.

Dose to the contralateral breast and lung are the limiting factors for utilizing multiple photon beams, an IMRT solution in other sites. And because of the location and size of the breast as a target volume, non-coplanar beams are not feasible due to collision avoidance. Therefore, it is our conclusion that it is not possible to achieve a concave distribution along the chest wall with multiple photon beams.

With available gating or breathing control technologies, it is possible to further reduce dose to lung and heart in breast radiotherapy with standard tangential beams (Sixel, Azner, and Ung 2001).

References

Boice, J. D., Jr., E. B. Harvey, M. Blettner, M. Stovall, and J. T. Flannery. (1992). "Cancer in the contralateral breast after radiotherapy for breast cancer." *N. Engl. J. Med.* 326:781–785.

Bortfeld, T. R., D. L. Kahler, T. J. Waldron, and A. L. Boyer. (1994). "X-ray field compensation with multileaf collimators." *Int. J. Radiat. Oncol. Biol. Phys.* 28:723–730.

Buchholz, T. A., E. Gurgoze, W. S. Bice, and B. R. Prestridge. (1997). "Dosimetric analysis of intact breast irradiation in off-axis planes." *Int. J. Radiat. Oncol. Biol. Phys.* 39:261–267.

Carruthers, L. J., A. T. Redpath, and I. H. Kunkler. (1999). "The use of compensators to optimise the three dimensional dose distribution in radiotherapy of the intact breast." *Radiother. Oncol.* 50:291–300.

Chang, S. X., K. M. Deschesne, T. J. Cullip, S. A. Parker, and J. Earnhart. (1999). "A comparison of different intensity modulation treatment techniques for tangential breast irradiation." *Int. J. Radiat. Oncol. Biol. Phys.* 45:1305–1314.

Cheng, C. W., I. J. Das, B. Stea. (1994). "The effect of the number of computed tomographic slices on dose distributions and evaluation of treatment planning systems for radiation therapy of intact breast." *Int. J. Radiat. Oncol. Biol. Phys.* 30:183–195.

Chin, L. M., C. W. Cheng, R. L. Siddon, R. K. Rice, B. J. Mijnheer, and J. R. Harris. (1989). "Three-dimensional photon dose distributions with and without lung corrections for tangential breast intact treatments." *Int. J. Radiat. Oncol. Biol. Phys.* 17:1327–1335.

Chui, C. S., and R. Mohan. (1986). "Off-center ratios for three-dimensional dose calculations." *Med. Phys.* 13:409–412.

Chui, C. S., T. LoSasso, and S. Spirou. (1994). "Dose calculation for photon beams with intensity modulation generated by dynamic jaw or multileaf collimations." *Med. Phys.* 21:1237–1244.

Chui, C. S., L. Hong, M. Hunt, and B. McCormick. (2002). "A simplified intensity modulated radiation therapy technique for the breast." *Med. Phys.* 29:522–529.

Convery, D. J., and S. Webb. (1998). "Generation of discrete beam-intensity modulation by dynamic multileaf collimation under minimum leaf separation constraints." *Phys. Med. Biol.* 43:2521–2538.

Cuzick, J., H. Stewart, L. Rutqvist, J. Houghton, R. Edwards, C. Redmond, R. Peto, M. Baum, B. Fisher, H. Host, et al. (1994). "Cause-specific mortality in long-term survivors of breast cancer who participated in trials of radiotherapy." *J. Clin. Oncol.* 12:447–453.

Das, I. J., C. W. Cheng, H. Fosmire, K. R. Kase, and T. J. Fitzgerald. (1993). "Tolerances in setup and dosimetric errors in the radiation treatment of breast cancer." *Int. J. Radiat. Oncol. Biol. Phys.* 26:883–890.

Epstein, R., S. Kelly, M. Cook, A. Bateman, I. Paddick, K. C. Kam, P. Dunn, I. W. Hanham, R. G. Dale, and P. M. Price. (1996). "Active minimization of radiation scatter during breast radiotherapy: Management Implications for young patients with good-prognosis primary neoplasms." *Radiother. Oncol.* 40:69–74.

Evans, P. M., V. N. Hansen, W. P. Mayles, W. Swindell, M. Torr, and J. R. Yarnold. (1995). "Design of compensators for breast radiotherapy using electronic portal imaging." *Radiother. Oncol.* 37:43–54.

Evans, P. M., E. M. Donovan, N. Fenton, V. N. Hansen, I. Moore, M. Partridge, S. Reise, B. Suter, J. R. Symonds-Taylor, and J. R. Yarnold. (1998). "Practical implementation of compensators in breast radiotherapy." *Radiother. Oncol.* 49:255–265.

Evans, P. M., E. M. Donovan, M. Partridge, P. J. Childs, D. J. Convery, S. Eagle, V. N. Hansen, B. L. Suter, and J. R. Yarnold. (2000). "The delivery of intensity modulated radiotherapy to the breast using multiple static fields." *Radiother. Oncol.* 57:79–89.

Fisher, B., C. Redmond, R. Poisson, R. Margolese, N. Wolmark, L. Wickerham, E. Fisher, M. Deutsch, R. Caplan, Y. Pilch, et al. (1989). "Eight-year results of a randomized clinical trial comparing total mastectomy and lumpectomy with or without irradiation in the treatment of breast cancer." *N. Engl. J. Med.* 320:822–828.

Fraass, B, P. Roberson, and A. Lichter. (1985). "Dose to the contralateral breast due to primary breast irradiation." *Int. J. Radiat. Oncol. Biol. Phys.* 11:485–497.

Fraass, B. A., A. S. Lichter, D. L. McShan, B. R. Yanke, R. F. Diaz, K. S. Yeakel, and J. van de Geijn. (1988). "The influence of lung density corrections on treatment planning for primary breast cancer." *Int. J. Radiat. Oncol. Biol. Phys.* 14:179–190.

Gray, J. R., B. McCormick, L. Cox, and J. Yahalom. (1991). "Primary breast irradiation in large-breasted or heavy women: Analysis of cosmetic outcome." *Int. J. Radiat. Oncol. Biol. Phys.* 347–354.

Group EBCTC (Early Breast Cancer Trialists Collaborative Group). (2000). "Favourable and unfavourable effects on long-term survival of radiotherapy for early breast cancer: an overview of the randomized trials." Lancet 355:1757–1770.

Hansen, V.N., P. M. Evans, G. S. Shentall, S. J. Helyer, J. R. Yarnold, and W. Swindell. (1997). "Dosimetric evaluation of compensation in radiotherapy of the breast: MLC intensity modulation and physical compensators." *Radiother. Oncol.* 42:249–256.

Hardenbergh, P. H., M. T. Munley, G. C. Bentel, R. Kedem, S. Borges-Neto, D. Hollis, L. R. Prosnitz, and L. B. marks. (2001). "Cardiac perfusion changes in patients treated for breast cancer with radiation therapy and doxorubicin: Preliminary results." *Int. J. Radiat. Oncol. Biol. Phys.* 49:1023–1028.

Hong, L., M. Hunt, C. Chui, S. Spirou, K. Forster, H. Lee, Y. Yahalom, G. J. Kutcher, and B. McCormick. (1999). "Intensity-modulated tangential beam irradiation of the intact breast." *Int. J. Radiat. Oncol. Biol. Phys.* 44:1155–1164.

Kallman, P., B. Lind, A. Eklof, and A. Brahme. (1988). "Shaping of arbitrary dose distributions by dynamic multileaf collimation." *Phys. Med. Biol.* 33:1291–1300.

Kestin, L. L., M. B. Sharpe, R. C. Frazier, F. A. Vicini, D. Yan, R. C. Matter, A. A. Martinez, and J. W. Wong. (2000). "Intensity modulation to improve dose uniformity with tangential breast radiotherapy: Initial clinical experience." *Int. J. Radiat. Oncol. Biol. Phys.* 48:1559–1568.

Kutcher, G. J., A. R. Smith, B. L. Fowble, J. B. Owen, A. Hanlon, M. Wallace, and G. E. Hanks. (1996). "Treatment planning for primary breast cancer: A patterns of care study." *Int. J. Radiat. Oncol. Biol. Phys.* 36:731–737.

Landau, D., E. J. Adams, S. Webb, and G. Ross. (2001). "Cardiac avoidance in breast radiotherapy: A comparison of simple shielding techniques with intensity-modulated radiotherapy." *Radiother. Oncol.* 60:247–255.

Lingos, T. I., A. Recht, F. Vicini, A. Abner, B. Silver, and J. R. Harris. (1991). "Radiation pneumonitis in breast cancer patients treated with conservative surgery and radiation therapy." *Int. J. Radiat. Oncol. Biol. Phys.* 21:355–360.

Lo, Y. C., G. Yasuda, T. J. Fitzgerald, and M. M. Urie. (2000). "Intensity modulation for breast treatment using static multi-leaf collimators." *Int. J. Radiat. Oncol. Biol. Phys.* 46:187–194.

LoSasso, T., C. S. Chui, and C. C. Ling. (1998). "Physical and dosimetric aspects of a multileaf collimation system used in the dynamic mode for implementing intensity modulated radiotherapy." *Med. Phys.* 25:1919–1927.

Ma, L., A. L. Boyer, L. Xing, and C.-M. Ma. (1998). "An optimized leaf-setting algorithm for beam intensity modulation using dynamic multileaf collimators." *Phys. Med. Biol.* 43:1629–1643.

McCormick, B., L. Hong, C. Chui, and M. Hunt. (2000). "Breast IMRT: The potential for treatment improvement with intensity modulation in left-sided disease." (Abstract). *Int. J. Radiat. Oncol. Biol. Phys.* 48:298.

Mohan, R., and C. S. Chui. (1987). "Use of fast Fourier transforms in calculating dose distributions for irregularly shaped fields for three-dimensional treatment planning." *Med. Phys.* 14:70–77.

Moody, A. M., W. P. Mayles, J. M. Bliss, R. P. A'Hern, J. R. Owen, J. Regan, B. Broad, and J. R. Yarnold. (1994). "The influence of breast size on late radiation effects and association with radiotherapy dose inhomogeneity." *Radiother. Oncol.* 33:106–112.

Nixon, A., J. Manola, R. Gelman, B. Bornstein, A. Abner, S. Hetelekidis, A. Recht, and J. R. Harris. (1998)." No long-term increase in cardiac-related mortality after breast-conserving surgery and radiation therapy using modern techniques." *J. Clin. Oncol.* 16:1374–1379.

Obedian, E., D. B. Fischer, and B, G. Haffty. (2000). "Second malignancies after treatment of early-stage breast cancer: Lumpectomy and radiation therapy versus mastectomy." *J. Clin. Oncol.* 18:2406–2412.

Paszat, L., W. Mackillop, P. Groome, C. Boyd, K. Schulze, and E. Holowaty. (1998a). "Mortality from myocardial infarction after adjuvant radiotherapy for breast cancer in the surveillance, epidemiology and end-results cancer registries. *J. Clin. Oncol.* 16:2625–2631.

Paszat, L., W. Mackillop, P. Groome, K. Schulze, and E. Holowaty. (1998b). "Mortality from myocardinal infarction following postlumpectomy radiotherapy for breast cancer: A population-based study from Ontario, Canada." *Int. J. Radiat. Oncol. Biol. Phys.* 43:755–761.

Sixel, K. E., M. C. Aznar, and Y. C. Ung. (2001). "Deep inspiration breath hold to reduce irradiated heart volume in breast cancer patients." *Int. J. Radiat. Oncol. Biol. Phys.* 49:199–204.

Solin, L. J., B. Fowble, K. L. Martz, and R. L. Goodman. (1988). "Definitive irradiation for early stage breast cancer: The University of Pennsylvania experience." *Int. J. Radiat. Oncol. Biol. Phys.* 14:235–242.

Solin, L. J., J. C. Chu, M. R. Sontag, L. Brewster, E. Cheng, K. Doppke, R. E. Drzymala, M. Hunt, R. Kuske, J. M. Manolis, et al. (1991). "Three-dimensional photon treatment planning of the intact breast." *Int. J. Radiat. Oncol. Biol. Phys.* 21:193–203.

Spirou, S. V., and C. S. Chui. (1994). "Generation of arbitrary intensity profiles by dynamic jaws or multi-leaf collimators." *Med. Phys.* 21:1031–1041.

Spirou, S. V., and C. S. Chui. (1998). "A gradient inverse planning algorithm with dose-volume constraints." *Med. Phys.* 25:321–333.

Stein, J., T. Bortfeld, B. Dorschel, and W. Schlegel. (1994). "Dynamic X-ray compensation for conformal radiotherapy by means of multi-leaf collimation." *Radiother. Oncol.* 32:163–173.

Svensson, R., P. Kallman, and A. Brahme. (1994). "An analytical solution for the dynamic control of multi-leaf collimators." *Phys. Med. Biol.* 39:37–61.

Taylor, M. E., C. A. Perez, K. J. Halverson, R. R. Kuske, G. W. Philpott, D. M. Garcia, J. E. Mortimer, R, J. Myerson, D. Radford, and C. Rush. (1995). "Factors influencing cosmetic results after conservation therapy for breast cancer." *Int. J. Radiat. Oncol. Biol. Phys.* 31:753–764.

van Asselen, B., C. P. Raaijmakers, P. Hofman, and J. J. Lagendijk. (2001). "An improved breast irradiation technique using three-dimensional geometrical information and intensity modulation." *Radiother. Oncol.* 58:341–347.

Veronesi, U., R. Zucali, and A. Luini. (1986). "Local control and survival in early breast cancer: The Milan trial." *Int. J. Radiat. Oncol. Biol. Phys.* 12:717–720.

Wallgren, A. (1992). "Late effects of radiotherapy in the treatment of breast cancer." *Acta Oncol.* 31:237–242.

Xia, P., and L. J. Verhey. (1998). "Multileaf collimator leaf sequencing algorithm for intensity modulated beams with multiple static segments." *Med. Phys.* 25:1424–1234.

Zackrisson, B., M. Arevarn, and M. Karlsson. (2000). "Optimized MLC-beam arrangements for tangential breast irradiation." *Radiother. Oncol.* 54:209–212.

Zambetti, M., A. Moliterni, C. Materazzo, M. Stefanelli, S. Cipriani, P. Valagussa, G. Bonadonna, and L. Gianni. (2001). "Long-term cardiac sequelae in operable breast cancer patients given adjuvant chemotherapy with or without doxorubicin and breast irradiation." *J. Clin. Oncol.* 19:37–43.

13

IMRT For Non-Small-Cell Lung Cancer

Howard I. Amols
Kenneth Rosenzweig

Introduction

Lung cancer remains one of the most common cancers in the United States and the most common cause of cancer death. Approximately 170,000 people are diagnosed with lung cancer annually with approximately 157,000 deaths per year (Greenlee et al. 2001). There are more female deaths from lung cancer than breast, ovarian, and cervical cancers combined. Despite some progress with the use of chemotherapy (ChT) (Arriagada et al. 1991; Furuse et al. 1999; Pisters 2000), radiation therapy (RT) remains the main curative modality for inoperable non-small cell lung cancer (NSCLC). However, the treatment of lung cancer with RT is one of the most technically challenging procedures in radiation oncology, with 5-year survival rates ranging from 5% to 10% and median survival approximately 10 months (Emami 1996; Cox et al. 1990; Perez, Pajak, and Rubin 1987). In patients receiving 65 Gy without chemotherapy, only 15% are disease free one year after treatment when assessed by bronchoscopy (Arriagada et al. 1991).

Previous studies have demonstrated the value of dose escalation. Protocol 73-01 of the Radiation Therapy Oncology Group (RTOG) reported a decrease in in-field local failure as dose increased from 40 to 60 Gy (Cox et al. 1990). Three-dimensional CT assisted radiation therapy (3DCRT) treatment of NSCLC provides improved tumor coverage while decreasing the dose to the ipsilateral and contralateral lung (Armstrong et al. 1993), and at the University of Michigan (Hayman et al. 2001) 3DCRT was used to deliver localized doses as high as 102.9 Gy to small solitary lung tumors. RTOG 93-11 is currently treating small lesions to 90.3 Gy and intermediate sized tumors to 77.4 Gy (Graham et al. 1995). There has been little investigation, however, into dose escalation for large tumors where such doses result in high probability of lung toxicity.

The focus of this chapter is to explore the feasibility of further dose escalation for NSCLC via the use of intensity-modulated radiation therapy (IMRT) and other advanced RT techniques. It has been hypothesized that doses as high as 85 to 100 Gy may be required to control NSCLC (Rosenzweig et al. 2000; Martel et al. 1999). Escalation to such high doses with conventional RT, and even with 3DCRT, however, is impossible because radiation pneumonitis becomes a serious treatment complication when mean lung doses exceed 20 Gy (Emami et al. 1991; Kwa et al. 1998). For patients receiving concurrent ChT esophagitis may also occur at organ doses as low as 50 to 60 Gy (Furuse et al. 1999). Spinal cord dose must also be limited to less than 50 Gy. A key aspect of safe dose escalation, therefore, often incorporates one or another of various biological indices for estimating lung toxicity such as *fraction of functional lung units damaged* (fdam), effective volume (V_{eff}), normal tissue complication probability (NTCP), or mean lung dose (Jackson, Kutcher, and Yorke 1993; Jackson et al. 1995). At Memorial Sloan-Kettering Cancer Center (MSKCC) we use the fdam model, which is based on the assumption that lung functions as a *parallel* tissue with radiation pneumonitis being correlated with the *fraction* of lung volume subject to radiation damage rather than the mean or integral dose (although they are obviously correlated).

If IMRT is to be successful in improving the RT of NSCLC, however, we believe that it must be used in conjunction with other emerging technologies to address not only the problems of dose escalation, but also those of geographic miss caused by poor initial identification of the gross target volume (GTV) and errors caused by respiratory motion. The integration of Positron Emission Tomography (PET) scanning using the tracer FDG (^{18}F-2-Fluoro-2-deoxy-d-glucose) and CT simulation, for example, has recently emerged as a useful tool to augment tumor detection and treatment planning. The extent of many lung tumors is not fully visible on CT (computed tomography) scans, and inadequate delineation of the GTV contributes to geographic miss, which limits the success of RT. Using surgery as the gold standard, FDG-PET imaging has been shown to have a higher sensitivity, specificity, and accuracy than CT: 76% to 92% sensitivity for PET vs. 56% to 75% for CT; 81% to 100% specificity for PET vs. 73% to 87% for CT; and 80% to 100% accuracy for PET vs. 77% to 82% for CT (Kiffer et al. 1998; Sasaki et al. 1996; Vansteenkiste et al. 1998; Wahl et al. 1994).

Respiratory motion is another serious source of error for RT of NSCLC for many patients, and the integration of new technologies such as respiratory gated radiotherapy and electronic portal imaging device (EPID) systems for verification of patient setup and IMRT beam delivery are important components of IMRT for the lung (Chang et al. 2000; Kubo, Shapiro, and Seppi 1999; Mageras et al. 2000).

To summarize, the limited success in local control of NSCLC with RT stems primarily from two factors:

1. Inability to escalate to tumoricidal doses because of limits imposed by normal tissue complications, and
2. Geographic miss of tumor caused by the limited sensitivity and accuracy of CT for initial tumor definition, plus errors in treatment delivery caused by respiratory motion.

The first factor can potentially be mitigated via IMRT, which enables the delivery of higher tumor doses, conformly shaped to the tumor geometry, with a concomitant reduction in the volume of normal tissues irradiated (particularly the lung itself, but for patients receiving RT + ChT, also esophagus). Errors of the second nature may be reduced via the incorporation of improved imaging modalities (such as FDG-PET), and control of respiratory motion during both imaging scans and radiation therapy delivery. These issues are the focus of this chapter. Specifically, we shall discuss approaches for:

1. Inverse treatment planning (ITP) and IMRT for delivery of improved dose distributions.
2. Respiratory gating (RG) for IMRT delivery, and for CT and PET imaging studies.
3. Implementation of FDG-PET and CT image registration in RT treatment planning.
4. Application of EPIDs for improved treatment verification.

IMRT Treatment Planning for Non-Small-Cell Lung Cancer (NSCLC)

Patients with newly diagnosed stage T1-4, N0-3, M0, and recurrent NSCLC are most suitable for IMRT dose escalation. Generally, there are three patient populations referred for definitive thoracic RT:

1. Early stage (T1-2 N0) lung tumors. We expect this subgroup to increase significantly due to the renewed popularity and efficacy of early lung cancer screening (Henschke et al. 1999). Although surgical resection is the standard of care for these patients, for those who are inoperable due to medical co-morbidities, RT is the main curative treatment option.
2. Locally advanced disease (T3-4 N0, T1-4 N1-3) receiving sequential ChT and RT. These patients frequently receive induction ChT in the hopes of undergoing surgical resection. However, if they remain unresectable then RT is the standard of care. RT is also used for patients unable to tolerate concurrent ChT/RT.
3. Locally advanced disease receiving concurrent ChT and RT. Recent studies (Furuse et al. 1999) show that for unresectable patients, concurrent ChT/RT has improved outcome as compared to sequential ChT/RT.

Excellent reviews on the technical aspects and clinical advantages of inverse treatment planning (ITP) and IMRT are described in other chapters of this monograph, as are the advantages and disadvantages of sliding window technique versus step-and-shoot IMRT techniques. At MSKCC we have focused on the former which permits continuous variations in beam intensity via customized, continuous motion of the individual leaf pairs of a dynamic multileaf collimator (DMLC) (Chui, LoSasso, and Spirou 1994; Spirou and Chui 1994). We present here only a brief summary with particular reference to NSCLC. Central to IMRT are the following steps:

1. Three-dimensional (3-D) CT simulation, often augmented with PET imaging images (as described later in the section on PET scanning).
2. Selection of beam angles and definition of the planning target volume (PTV), normal lung, spinal cord, and esophagus from fused CT-PET images.
3. Specification of dose-volume constraints for each relevant tissue in the treatment plan. Typically fdam less than 0.28 or mean lung dose less than 20 Gy; maximum spinal cord less than 50 Gy, and for patients receiving concurrent chemotherapy esophageal dose less than 40 Gy.
4. Definition of an objective function (OF), or mathematical expression quantifying the differences between the planners specified dose-volume constraints and the computed dose distribution, with the goal of the ITP process being to minimize the numerical value of the OF.
5. Computer-controlled linear accelerator (or linac) and MLC for delivery of IMRT, plus respiratory gating system for beam delivery.
6. Treatment verification using EPID.

Different types of NSCLC tumors may require different strategies for treatment plan optimization and beam delivery. Small stage I tumors, for example, may prove relatively simple to treat, especially for tumors in the periphery of the lung. For these tumors the field sizes may be small enough to permit meeting normal tissue dose-volume tolerances with only minimal intensity modulation. Respiratory gating for these patients may also be less beneficial than it might be for larger tumors. Stage II–III tumors present greater challenges because larger treatment volumes usually result in higher lung toxicity. Similarly, patients receiving concurrent ChT present the additional challenge of including esophageal toxicity as a dose-volume constraint, plus the added complication of both lung and esophageal motion within the respiratory cycle. We expect that RG and IMRT will be particularly beneficial to this group of patients.

Patients are simulated and treated in the supine position, hands over head, immobilized via custom alpha cradle body molds. Consecutive CT images with 3 mm slice thickness are obtained from the larynx to L2 to encompass the entire thoracic cavity. Currently, patients for whom respiratory gated RT is deemed beneficial are scanned with slightly larger slice thickness of 5 mm in order to reduce total scanning times, although improved CT gating techniques are expected to reduce the slice thickness back to 3 mm in the near future (see the next section for details on respiratory gating).

The physician identifies and outlines the GTV on the scans. The PTV is defined by a physician-specified margin of approximately 10 to 15 mm beyond the GTV to allow for microscopic tumor extension, treatment setup errors, organ motion, and other uncertainties. Tumors probably extend microscopically 6 to 8 mm beyond what is visible on imaging studies (Giraud et al. 2001), which, we believe, justifies a 10 to 15 mm margin. There is not yet enough clinical data to determine whether or not respiratory gating will permit a reduction in field margins. In theory, this should be possible, as gating reduces tumor movement, which in turn should reduce the need to expand the PTV beyond the GTV and clinical target volume (CTV); but without definitive clinical demonstration of this hypothesis we are currently using the conventional field margins described above, with or without respiratory gating assisted RT.

The carina is outlined as a landmark that is useful for comparing digitally reconstructed radiographs (DRRs) to portal images. The lungs, heart, esophagus, spinal cord, and body surface are all contoured. Complete contouring of all normal tissues is particularly important when biological dose response models are being used to predict treatment toxicity, as these models are based on dose-volume effects and require accurate information regarding tissue and organ volumes. Selection of beam directions is similar to conventional 3DCRT planning and is usually made with the aid of beam's-eye view (BEV) computer display. Typically 3 to 5 coplanar treatment beams are used, occasionally with the addition of non-coplanar beams. Treatment beams are almost exclusively 6 MV photons, and dose distributions include pixel-by-pixel inhomogeneity corrections.

Treatment planning criteria are specified via dose-volume constraints, with typical constraints being a maximum spinal cord dose constraint of 50 Gy, and fdam for lung less than of 0.28. This fdam constraint is not *explicitly* incorporated into the treatment plan optimization calculations, but is rather calculated from the ensuing dose-volume histogram (DVH) after the dose distribution has been calculated. If the fdam constraint is not met, then the planner modifies the dose-volume constraints and runs the ITP optimization again. If this fails several times, then the physician must consider decreasing the prescription dose in order to meet normal tissue tolerance limits. For patients receiving concurrent ChT an esophageal dose constraint of 40 to 50 Gy is also incorporated into the optimization process. We should note that it is often necessary to set dose constraints lower than what is actually desired, and/or to adjust penalties and dose-volume constraints during planning optimization to achieve acceptable plans. Both PTV coverage and normal tissue doses are evaluated by examining isodose distributions and DVHs. Dose distributions are usually renormalized such that the isodose contour covering the PTV is defined to be 95%.

The advantages of ITP and IMRT for NSCLC were demonstrated in a pilot study of six patients previously treated with 3DCRT who were retrospectively replanned using ITP and IMRT (Yorke, Jackson, and Chui 2001; Yorke et al. 2001). 3DCRT and IMRT plans were calculated using identical dose-volume constraints. Comparisons were made between the maximum dose achievable using the treatment planner's best 3DCRT plan (typically 3 to 5 wedged fields), and an IMRT plan using the same or similar beam orientations. For all plans the prescription dose was escalated until the biological dose constraint for lung was violated (fdam > 0.28). For the six patients, PTV ranged from 229 to 556 cm^3, and total lung volumes from 1940 to 3730 cm^3. The results are summarized in table 13–1 and figures 13–1 to 13–4. In table 13–1 we see that in five of six cases the prescription dose could be increased with IMRT, on average by 13 Gy. In figure 13–1 we show a comparison between IMRT and 3DCRT treatment plans using three coplanar beams. Note the decreased lung dose and improved PTV uniformity for the IMRT plan, which is also evident in the DVHs of figure 13–2. This reduced lung dose enables escalation of the prescription dose for this patient to 84 Gy using IMRT, as compared to only 66 Gy using 3DCRT (for the same fdam constraint of 0.28).

Table 13–1. Comparison of 3DCRT and IMRT Plans

Case	PTV (cc)	Lung (cc)	Maximum dose (Gy)		IMRT gain (Gy)
			3DCRT	IMRT	
1	462	2500	62	80	18
2	311	3500	66	76	10
3	556	1940	80	86	6
4	490	3730	88	88	0
5	312	2280	64	88	24
6	229	3110	80	98	18
average	393	2843	73	86	13

The capability of sparing normal lung irradiation using IMRT is more graphically illustrated in figures 13–3 to 13–4 where we compare the same two treatment plans in the coronal and sagittal planes. In a related study of 10 patients we found that IMRT also reduced the maximum and mean dose to the esophagus by 11% and 7% respectively, with a corresponding decrease in NTCP from 41% to 19%.

Using these advanced treatment planning techniques we have gradually escalated our prescription doses from conventional levels of 60 to 70 Gy to 81, 84, and 90 Gy for selected patients. The increased duration of radiation therapy has, in our experience, resulted in patient dissatisfaction, and we have therefore increased our daily treatment dose from 1.8 to 2.0 Gy (typically normalized to the D_{95}) to facilitate a more timely completion of treatment. We have treated eight patients to 90 Gy, but observed unacceptable toxicity, due in part to the poor

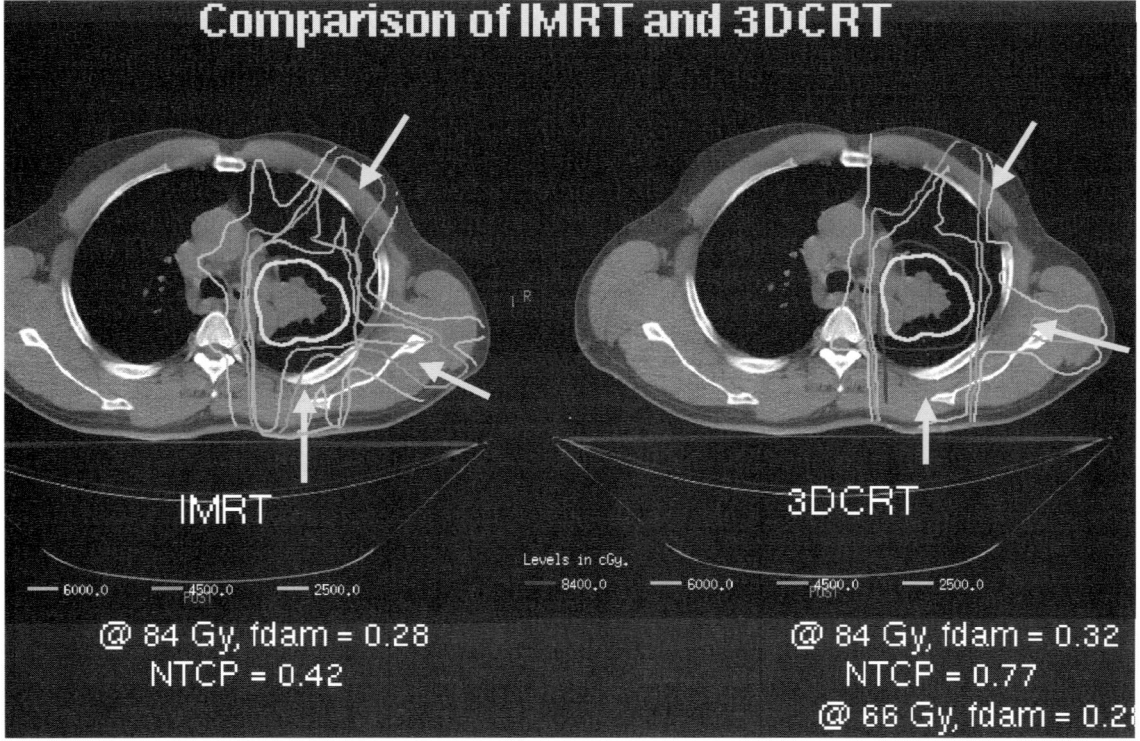

FIGURE 13–1. Comparison of IMRT (left) and 3DCRT (right) treatment plans using three coplanar beams. Note decreased lung dose for IMRT plan.

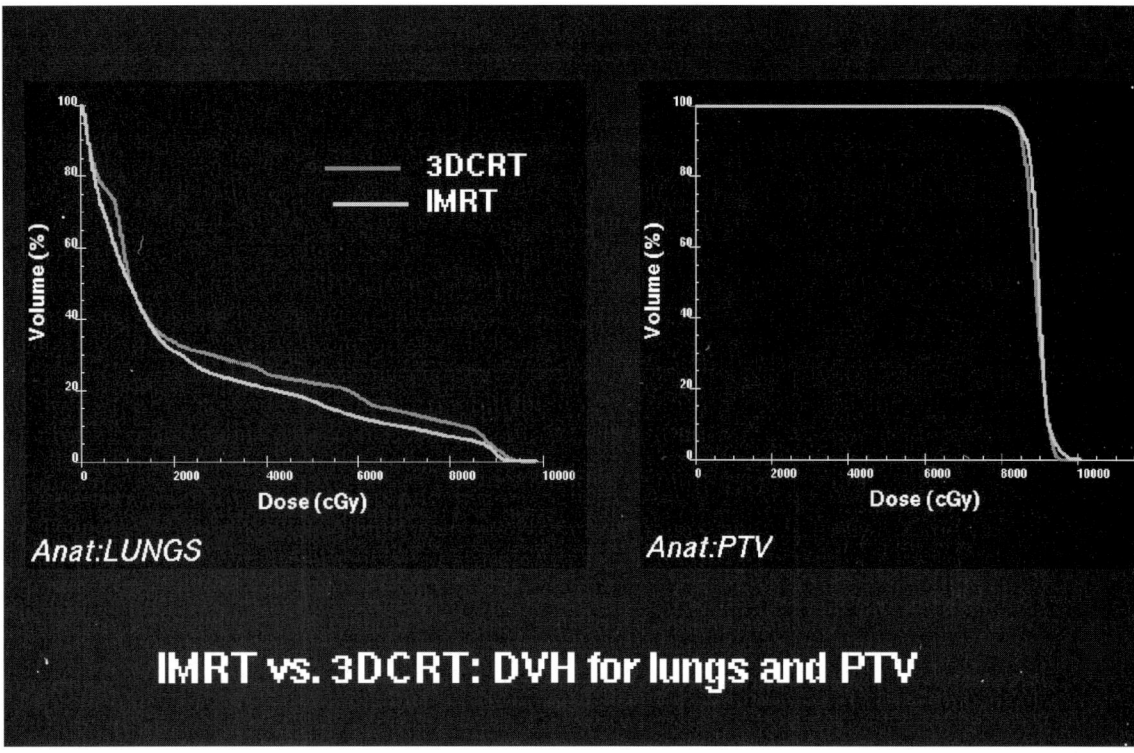

FIGURE 13–2. DVHs for lung (left) and PTV (right) showing decreased fraction of lung receiving high or intermediate dose with IMRT. IMRT for this patient enabled dose escalation to 84 Gy (using an fdam constraint of 0.28) that would not have been possible with the conventional plan.

general health of most of these patients. We have subsequently restricted maximum prescribed doses to 84 Gy.

As with other treatment sites, our experience for NSCLC has been that IMRT treatment planning requires both the physicist and the radiation oncologist to embark on something of a new learning curve, in that treatment planning strategies are sometimes different between IMRT and 3DCRT. In particular, the determination of suitable dose-volume constraints and penalty functions to be used as input to the optimization program requires some trial and error. In addition, ITP can sometimes produce unexpected results. For example, we show in figure 13–5 a situation where an ITP generated an undesirable hot spot in the isodose distribution at an unexpected location far from the intended PTV. Such anomalies can occur as the ITP algorithm did not know (since we did not inform it by means of a contour with specified dose-volume constraints) that such hot spots are unacceptable. Also in figure 13–5 we show a *work-around* to such anomalies by which an artificial contour is drawn to encompass the undesired hot spot, and the treatment plan is then redone with dose-volume constraints added for this new contour.

Respiratory Gating

Geographic miss is also a major contributing factor in the failure of RT to control NSCLC. Most obviously, geographic miss can result from incorrect initial identification of the GTV, but also from respiration induced tumor motion during RT beam delivery which can be 1 cm or more, and in some cases up to 2.5 cm. Numerous studies have demonstrated the importance of respiratory motion in patients with intrathoracic tumors (Rosenzweig et al. 2000a,b; Hanley et al. 1999; Kubo

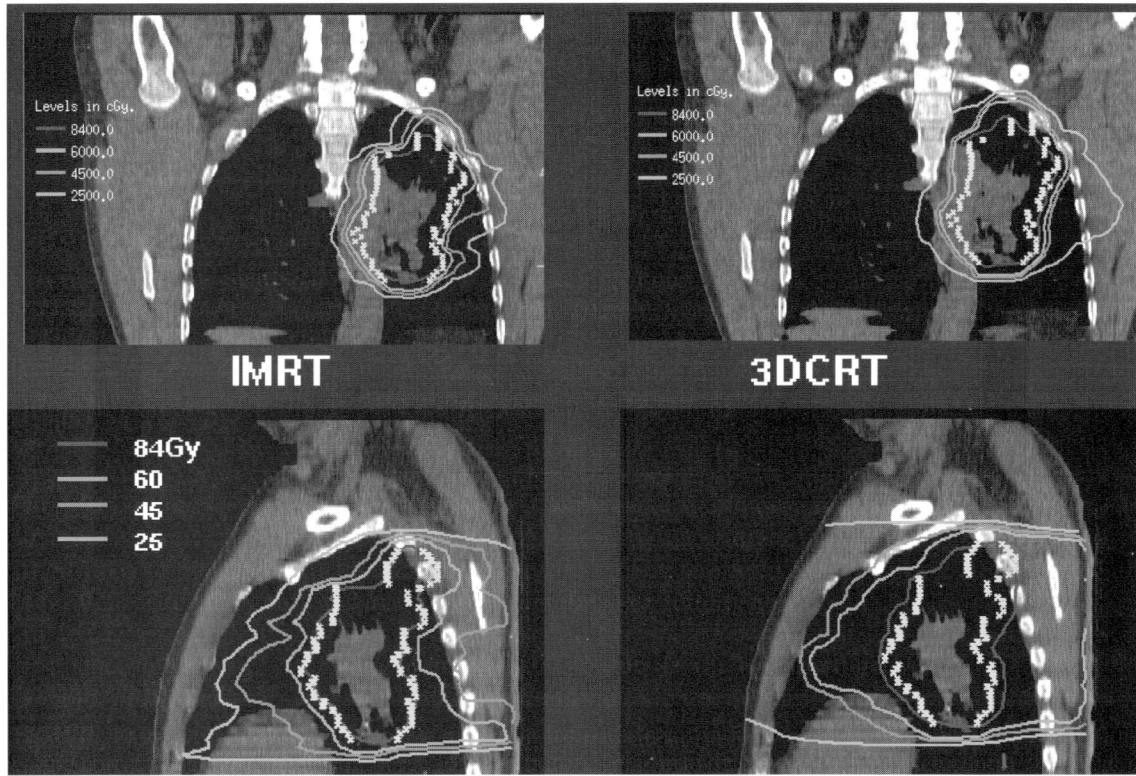

FIGURE 13–3. Colorwash display for the IMRT and 3DCRT treatment plans in figure 13–1 in the coronal (upper panels) and sagittal planes (lower panels). See COLOR PLATE 35.

and Hill 1996; Ohara et al. 1989; Paoli et al. 1999). Prescribing larger radiation fields can circumvent geographic miss due to respiratory motion, but this also increases toxicity, often to unacceptable levels. Thus, control of respiratory motion during treatment may be a key factor in improving RT results. Two different philosophies have been proposed to reduce respiratory motion. The first is coached or assisted, or force patient breath-hold techniques such as the Active Breathing Control method (Wong et al. 1999) and deep inspiration breath hold (DIBH) (Hanley et al. 1999). The second approach permits free breathing, but triggers radiation delivery to only specific phases of the respiratory cycle; so called respiratory-gated (RG) beam delivery. All these methods have been restricted to use in a few research protocols, although RG is now commercially available, and has been applied clinically at many institutions (Kubo and Hill 1996; Paoli et al. 1999).

All respiratory gating systems are designed to circumvent respiratory motion via correlating RT beam delivery to a specific phase within the breathing cycle rather than via increasing field margins to "cover" the motion. The benefits of reducing the volume of normal tissues irradiated via control of respiratory motion can be as large as much as 30% reduction in volume as compared to free breathing (Hanley et al. 1999). This in turn permits an increase of prescription doses by as much as 18 Gy for the same level of lung toxicity (Rosenzweig et al. 2000a,b).

To achieve maximum benefit, care must be given to determining the optimum phase within the respiratory cycle in which to treat, although it turns out that compromises must be made towards this end. For most patients the mean dose delivered to normal lung is best minimized by gating at end inspiration, due to displacement of normal lung tissue outside of the high dose volume. However, in many patients, tumor and organ position is more reproducible at

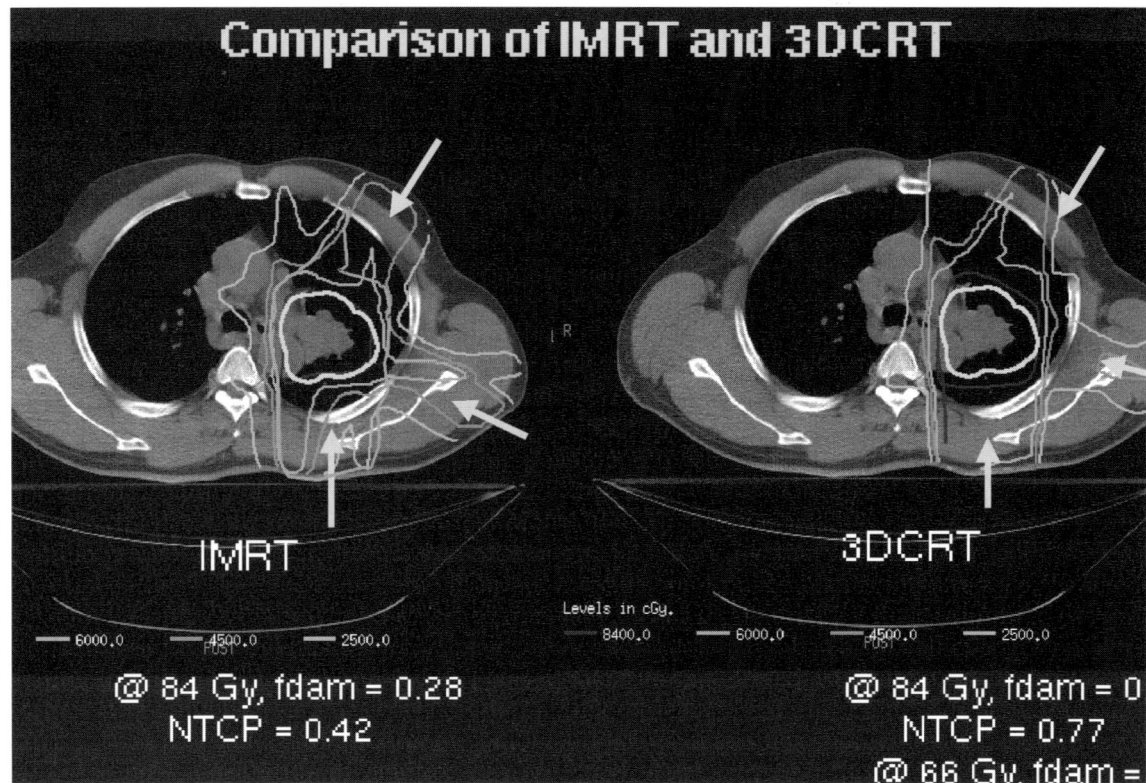

FIGURE 13–4. Colorwash display comparing the 25 Gy isodose contour in the coronal plane (upper panels) and the 60 Gy isodose contour in the sagittal plane (lower panels) for the IMRT and 3DCRT plans. See COLOR PLATE 36.

expiration rather than at inspiration. Fluoroscopic images can be useful in measuring the magnitudes and directions of diaphragm and chest wall motion, to determine the optimal point in the breathing cycle for radiation treatment.

DIBH was first implemented at MSKCC in February 1998 expressly to control the respiratory motion. DIBH permits:

1. Expansion of lung volume to reduce the amount of normal lung within the treatment field.
2. Synchronization of beam delivery to a fixed phase of the breathing cycle to reduce geographic miss caused by respiratory motion.
3. Increased separation between tumor and normal tissues (in some patients). DIBH, however, requires active participation by the patients, as they must breathe through a spirometer and hold their breath for extended periods of time (Mah et al. 2000).

Our DIBH technique also required that the treatment technologist manually gate the linac x-ray beam on and off for treatment via observation of spirometer readings (plotted on a computer monitor) indicating when the patient is in the correct breathing phase. Typically 1 to 2 breath holds of 10 to 15 sec duration are required per treatment portal with the therapist gating the linac beam off and on.

While the DIBH method was shown to be quite beneficial, it is also somewhat cumbersome to use in practice, even for conventional RT beam delivery. With the addition of IMRT, respiratory-gated beam delivery using DIBH becomes even more problematic, principally because such treatments require substantially increased treatment times as compared to 3DCRT. IMRT typically increases total beam-on time [or monitor units (MU)] by a factor of 2 to 3 depending on the degree

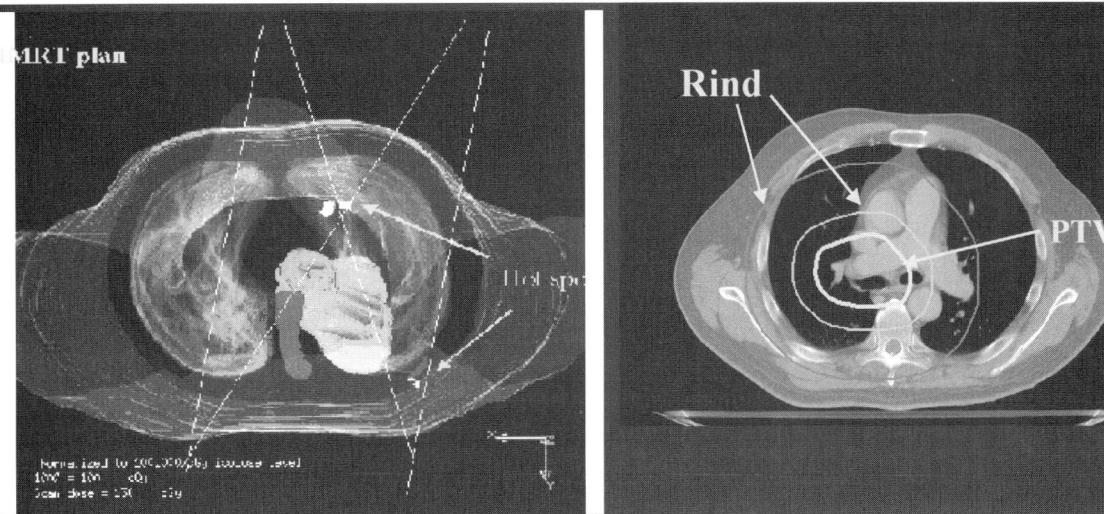

FIGURE 13–5. Adding axillary contours to an IMRT plan (right panel, blue and pink contours, with PTV contoured in yellow) to avoid unwanted hotspots sometimes artificially created by ITP, such as shown at the intersection of two beams in the left panel. See Color Plate 37.

of intensity modulation required. Treating patients with the combination of IMRT plus DIBH would therefore require as many as 5 to 6 breath holds per treatment field, or more than a dozen breath holds per treatment fraction. Such a requirement would be beyond the physical capacity of most NSCLC patients. Thus, even though we were able to demonstrate the clinical advantages of DIBH in achieving reductions in mean lung dose as compared with free-breathing treatments, DIBH is not really practical for many patients. In fact, because of their poor performance status at initial presentation, only about one-third of patients referred for definitive lung RT are capable of performing DIBH. With the addition of IMRT, even fewer patients would be able to tolerate such treatments.

Similarly, the use of DIBH for respiratory control of PET scanning would also entail prohibitively long scanning times and multiple breath holds beyond the physical capacity of most patients. Thus, a more adaptable system for control of respiratory motion is needed. In 1999 we installed a new commercially available RG system designated as *real-time position management system (RPM)* (Varian Oncology Systems, Palo Alto, CA). The RPM system, shown in figure 13–6, achieves many of the clinical advantages of DIBH, but with only passive input from the patient. The RPM system utilizes an infrared camera to track the motion of a reflective marker placed on the patient's chest. Video images of the marker position are computer analyzed to determine when the patient is in a specific user-defined phase of the breathing cycle, at which time the computer generates a gating signal to initiate linac beam delivery. The patient need only maintain a reasonably regular breathing pattern for the system to work reliably. The simplicity of this system increases patient comfort and accrual rates as compared with DIBH. The RPM system is also easily adaptable to CT and PET scanning, thus facilitating the acquisition of respiratory gated diagnostic and radiation therapy treatment simulation scans (Nehmeh et al. 2001). This capability is crucial because patient simulation, CT and PET scanning, treatment planning, and RT beam delivery must all be carried out using identical methods of synchronization to the respiratory cycle.

Studies by us, and others, have established that the RPM system functions nearly as well as DIBH for synchronization of beam delivery to patient breathing cycle. Some of the lung expansion benefits of DIBH are lost with RPM, but much of this loss can be regained (or at least compensated for) via IMRT, which as explained above, is not feasible with DIBH. We have also

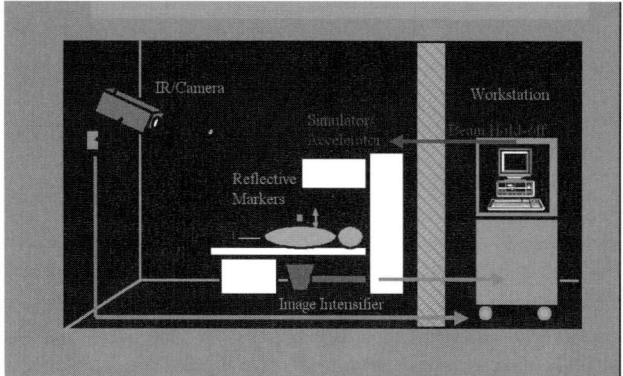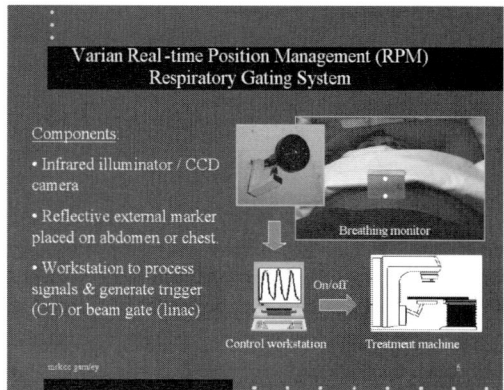

FIGURE 13–6. The RPM respiratory gating system. On the left is a schematic showing the infrared camera tracking a reflective marker placed on the patient's skin. On the right are photographs of the camera and marker, as placed on the patient's chest. The computer tracing provides visual confirmation of the respiratory cycle.

found that the reproducibility of patient breathing and the depth of inhalation can be improved via patient coaching which enhances the effectiveness of the RPM system. In particular, we have found that when patients are given verbal instructions on when to inhale and exhale, they tend to inhale more deeply and breathe more consistently as compared with normal breathing. Coaching also improves the correlation between the external marker and diaphragm positions, as assessed during simulation sessions using fluoroscopic images. Upgraded versions of the RPM system include in-room video displays that provide the patients with real-time video feedback of breathing patterns to further assist them in maintaining regular breathing.

Thus, even though the combination RPM plus IMRT increases the number of monitor units by a factor of 2 to 3 as compared to conventional RT, and the total treatment time by a factor of 6 to 12, the system is well suited to patient comfort and reproducibility (Note: Although respiratory gating increases the total time of treatment by a factor of 3 to 4 above that of IMRT alone, it does not additionally increase the total number of MU, as the x-ray beam is gated off when the patient is not in the desired phase of the breathing cycle). As an example, consider the delivery of a 2 Gy daily fraction of radiation at 300 MU/min, which requires approximately 300 total MU or 1.00 min total treatment time for conventional RT beam delivery. A similar treatment delivered with IMRT requires approximately 600–900 MU, or 2 to 3 min total treatment time. With the addition of respiratory gating, and assuming that beam on occurs during approximately one quarter of the total respiratory cycle this treatment still requires only 600 to 900 MU (i.e., the same as IMRT alone), but a total treatment time of approximately 8 to 12 min. While this is not an excessive period of time, it does raise questions about patient comfort, other types of intrafraction motion, patient immobilization techniques, and overall efficiency of resources. These issues aside, we shift the discussion now to the application of the RPM respiratory gating system to CT and PET imaging. A more detailed discussion of the special advantages of FDG-PET imaging for NSCLC is reserved for the next section, Pet Scanning.

The RPM system can be integrated with CT data acquisition in two very different modes. The first mode, which is easier both in implementation and concept, is designated as *respiration-triggered CT* (*RTCT*). In RTCT mode, the RPM system interfaces with the CT scanner in a manner almost identical to its operation on the linac. During CT simulation the physician *preselects* a phase in the breathing cycle during which imaging and treatment will occur. Typically, this selection is made after viewing fluoroscopic images of the patient's respiratory motion while in treatment position on the simulator. A reflective marker, identical to that used for linac treatments, is placed on the patient's skin, and during CT scanning the RPM records and analyzes the motion of the reflective marker, and sends a trigger signal to the CT scanner each time the patient enters the desired phase of the breathing cycle. The trigger signal initiates the acquisition of a *single* axial CT slice, which is followed by a table advance to the next couch position, at which point the CT scanner waits for the next trigger signal, images a second slice, etc. Since modern CT scanners can acquire a single axial slice in 1 to 2 sec or less, and since a typical respiratory cycle has a period of 4 to 5 sec, there is ample time during each trigger signal (approximately 1 to 2 sec) to acquire the CT data, and there is also ample time between trigger signals to translate the couch position. However, since only one slice is acquired per respiratory cycle (assuming one has access only to a single slice CT scanner), a data set of 100 slices requires 9 to 10 min. Multislice CT scanners, of course, decrease the total time required for image acquisition. Nonetheless, the relatively long acquisition times can result in artifacts from patient movement due to discomfort, or irregular breathing patterns. This can be seen in figure 13–7 where we show several different coronal CT reconstructions obtained using the RTCT technique. The free breathing scan shown (i.e., no respiratory gating) in the leftmost panel shows artifacts (arrows) near the diaphragm resulting from respiratory motion between axial slices. An RTCT scan taken at end expiration (center panel) shows virtually no artifacts, as expected. The right panel, however, also taken using RTCT but taken at end inspiration shows some small artifacts, indicating that the RTCT system is not working perfectly.

A second limitation of the RTCT method is the need to pre-select the desired point in the respiratory phase for RTCT. This restriction precludes, for example, a quantitative assessment of the preferred phase for each particular patient, or determination of the maximum width of the respiratory window to be used for beam delivery. Some patients, for example, might be best treated at end expiration owing to the increased lung inflation and separation between tumor and normal tissues; but for many patients we have found that position reproducibility is better at end expiration. Hence, assessment of the ideal portion of the respiratory cycle in which to treat may involve a compromise between ideal position and ideal reproducibility, and this cannot be easily assessed in RTCT mode. Further, the fraction of the respiratory cycle over which it is *safe* to treat (i.e., during which there is minimal respiratory motion) also cannot be easily assessed with RTCT. Another factor to be considered in using the RPM system is the fact that some patients have a tendency to lapse into irregular breathing patterns. Early version of the RPM system, which only measured respiratory amplitude, could not detect this. Improved software, however, now records both amplitude and phase, which does permit detection of irregular breathing patterns.

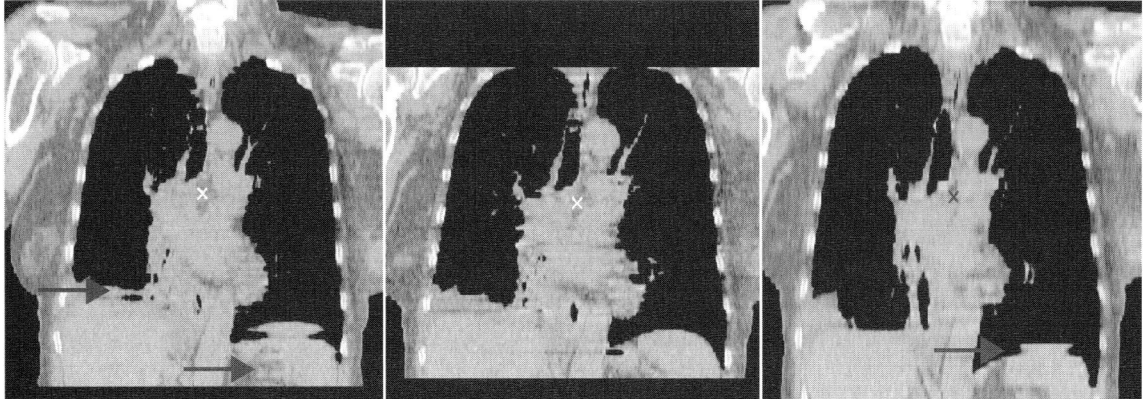

FIGURE 13–7. Coronal CT scans at isocenter showing effects of breathing and gating. Left: Unsynchronized free-breathing CT scan of a lung cancer patient. Note motion artifacts (arrows) near diaphragm. Center: Respiration triggered CT at end expiration, showing virtually no artifacts. Right: Gated scan at end inspiration showing small artifact (arrow, right figure) due to irregular breathing. Note also the change in shape of the mediastinum between expiration and inspiration scans.

For these reasons, we have developed a more sophisticated method of gating CT scans, which we designate as *respiration-correlated spiral CT* (*RCCT*). Implementation of RCCT requires a spiral (rather than an axial) CT scanner, with the CT scanner operating in a near-normal acquisition mode. The only change to normal spiral CT scanning with the addition of RCCT is that all CT images are time-labeled to designate the phase of the respiratory cycle at which the scan was acquired. This is achieved by recording a data file of RPM readings (i.e., phase and amplitude of the reflective respiratory marker) simultaneously with CT image acquisition. In this manner, each reconstructed axial CT slice can be retrospectively labeled and binned according to its appropriate phase within the respiratory cycle. RCCT scanning must be done using a very small couch pitch (i.e., slow couch speed during scanning) to ensure that there is an overlap between CT slices which ensures that any desired slice can be reconstructed at whatever respiratory phase one chooses. Thus, with RCCT it becomes possible to obtain a complete CT data set at *all* respiratory phases in a single spiral acquisition sequence, yielding a four-dimensional (4-D) data set. From such a 4-D data set the physician can choose, after CT scanning is completed, the ideal position within the respiratory cycle for each particular patient. RCCT also enables the treatment planner to select the optimum window width for beam delivery, which is a compromise between accuracy of treatment and absolute reduction of motion (i.e., narrow window width) and practicality of total treatment time (i.e., broad window width).

If PET scanning is to be used along with CT to aid in definition of the CTV, then the thoracic PET scans used for simulation must also be respiratory gated. Normal PET scans often require 45 to 60 minutes total scanning time, and patient respiratory motion can result in overestimation of lesion size, reduction in signal to noise ratio, and reduction in specific uptake value (SUV) [SUV is defined as the ratio of (decay corrected ^{18}F activity/lesion weight)/(injected activity/patient weight)]. An example of the changes in apparent PET lesion size with and without respiratory gating corrections is shown in figure 13–8. At present, the RPM system cannot directly gate PET scan data acquisition. Instead, for the scans shown in figure 13–8 the RPM system was used to generate a gating signal at the start of each patient breathing cycle (as per used with the linear accelerator and CT scanner) which was retrospectively used to bin all PET events, which are already labeled with the manufacturer's data acquisition software, into different time bins within the respiratory cycle. This is similar in concept to RCCT data acquisition described previously, except for the fact that at present the RPM system does not directly label PET events with an exact time signature, although modifications to the system to enable this are in progress.

Nonetheless, in figure 13–8 we can see that respiratory gating can have a significant impact on improving the accuracy of PET scanning, with reductions in the measured GTV of 20% to 30% as compared to standard mode. Thus, gating aids in more accurate definition of GTV and reduction in PTV size. Gating both CT and PET scans in a similar manner also facilitates more accurate fusion of image data sets.

PET Scanning

To date, radiological imaging has been largely anatomical, based on physical properties of tissue such as x-ray attenuation (CT imaging) or magnetic susceptibility (MR imaging). New imaging modalities based on the biological, metabolic, or chemical properties of tissues are beginning to provide new dimensions in tumor diagnosis (Ling et al. 2000). At present, the most promising vis-à-vis NSCLC is FDG-PET. With the approval of FDG-PET for staging lung cancer by the Food and Drug Administration, its role in cancer detection has increased dramatically (Coleman and Tesar 1997), as the tracer FDG enables PET detection of increased glucose metabolism in cancer cells. Approximately 85,000 lung cancer patients per year will benefit from FDG-PET (Gambhir et al. 1998). PET is a valuable complement to CT scanning, which has known limitations for detecting the full extent of many lung tumors. PET often identifies involved lymph nodes not identified

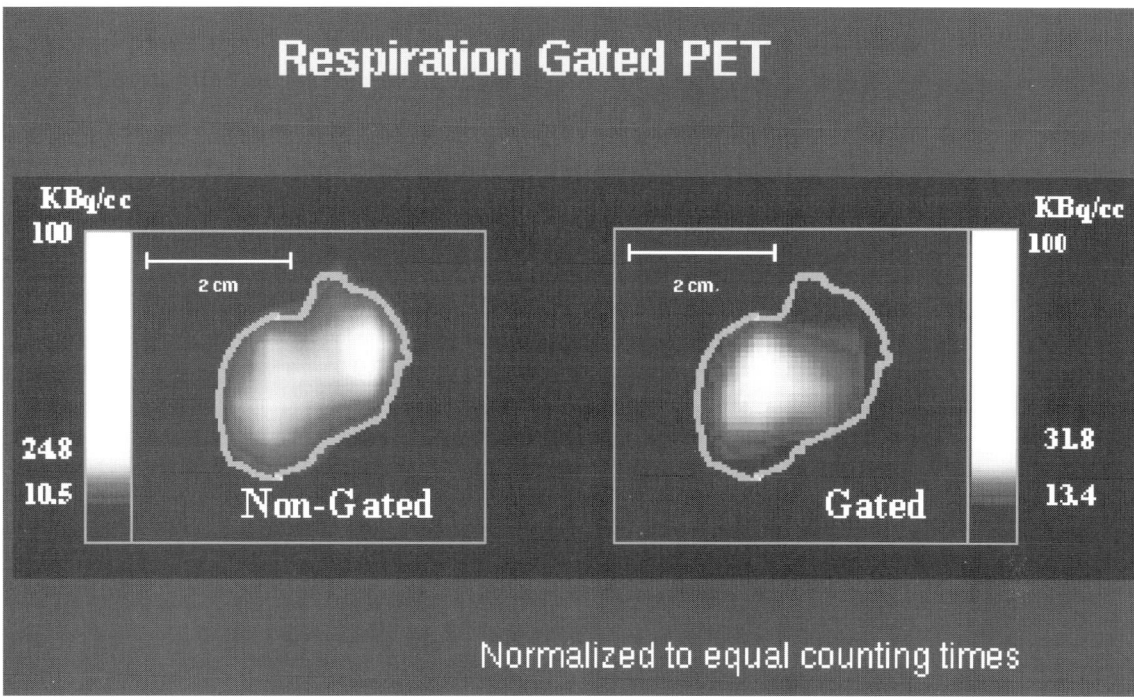

FIGURE 13–8. Apparent (but false) size of a lung lesion as seen in FDG-PET scanning without respiratory gating corrections (left panel), compared to a respiratory-gated scan (right panel). The gray contour represents the apparent lesion size in the ungated scan.

by CT, with a typical example shown in figure 13–9. Conversely, however, PET sometimes shows over-definition of GTV by CT scan alone especially in areas of atelectasis (Nestle et al. 1999). The impact of PET images on PTV definition can be profound. Prospective and retrospective studies have shown that FDG-PET images influence the design of radiation treatments in 23% to 65% of all cases (Kiffer et al. 1998; Munley et al. 1999; Hebert et al. 1996) and many patients are

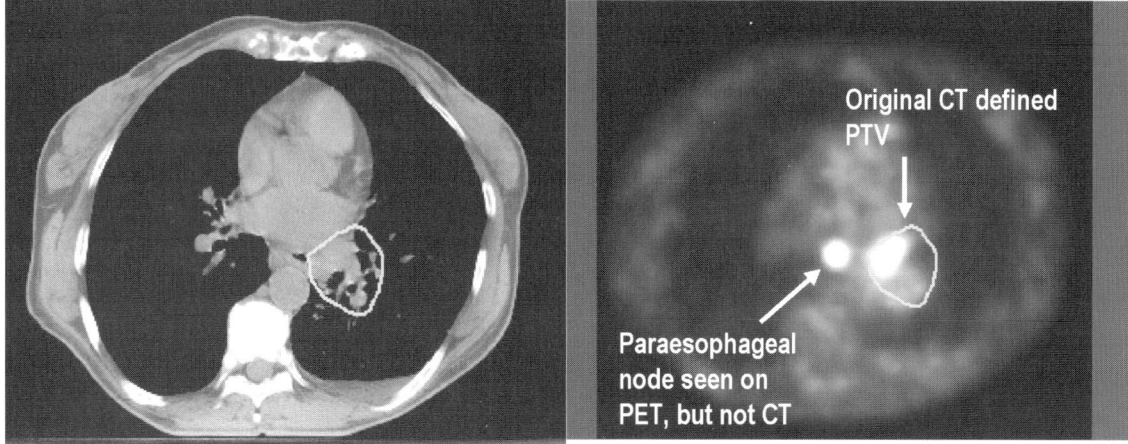

FIGURE 13–9. CT defined PTV (left) was drawn for a hilar tumor. PET image (right) shows CT defined PTV, plus paraesophageal lymph node which was not seen on CT, and which resulted in redefinition of PTV.

re-staged as a result of PET. The addition of fused PET-CT images for treatment planning will allow more accurate definition of the GTV, thereby reducing geographic miss.

However, with current technology the fusion of CT and PET images for treatment planning purposes is problematic. PET emission scans contain only limited structural anatomy with little density information, making them difficult to correlate with CT images that are entirely density based. The PET transmission scans do contain anatomical details and are currently used to register the emission scans to CT, but even the transmission scans are of poor quality, with only bony landmarks and very low-density structures being well visualized, although for lung this is often sufficient. CT-PET image registration is still in its development phases (Wahl et al. 1994; Tai et al. 1997). Many centers still rely on manual registration techniques, wherein the user must translate and/or rotate images on a computer screen to obtain the best *visual match,* although automated registration are currently being developed by several groups, most of which utilize *mutual information* algorithms (Erdi, Hu, and Chui 2000; Maes et al. 1997).

For virtually all image registration techniques the CT image is used as the reference scan. In manual method the user contours several anatomical structures that are visible in both CT and PET image sets. These contours are observed on orthogonal image planes that are reconstructed for both CT and PET, allowing the user to translate and/or rotate the PET images in three dimensions to best match the reference CT images. Once the best match has been achieved, a transformation matrix is calculated and the PET images are reformatted to best match the CT scans. One serious problem vis-à-vis image registration is that patient positioning can differ between scanning units. The recent introduction of combined PET/CT machines should greatly reduce many of these image registration difficulties.

These caveats of image registration not withstanding, our current technique entails 30 to 60 min of ungated PET scan acquisition, usually in 3 to 4 segments (due to the technical limitation on our current PET scanner of a 14 cm maximum field of view [FOV]). A *rest and stretch* follows this initial data acquisition, followed by a 16-min respiratory-gated scan sequence. The gated scan is usually focused in on the single FOV strip containing the PTV, and thus requires less scanning time. For the gated PET, each recorded PET event is placed in its appropriate time bin (based on the time lapse between the PET event and the start trigger signal from the RPM as described above), in the same manner as a multichannel analyzer operating in time scaling mode. At the completion of image acquisition one can reconstruct the individual PET emission images from each time bin, as well as the integrated image (which is the equivalent of a free breathing scan).

Another possible application of PET to the treatment of NSCLC, although still speculative, is its use for providing early information on treatment response (Erdi et al. 2000; Humm et al. 1999). Abe et al. (1990) reported on five lung cancer patients who underwent FDG scanning before and after RT. Before therapy all PET scans were positive, but after therapy only two patients remained positive. Negative scans corresponded to complete response, but patients with positive scans later showed tumor regrowth. Hebert et al. (1996) found that complete response on PET scans appears to indicate a true local remission although they did reported cases in which changes on FDG-PET scans after RT did not correlate with clinical response.

EPID-Based Treatment Verification

Recent improvements in EPIDs such as amorphous silicon (aSi) detectors now provide an attractive alternative to conventional portal films. Their speed and convenience coupled with the advantages of digital image processing make them a useful tool for on-line treatment verification and transit dosimetry measurements of IMRT treatments, in addition to being simply a replacement for conventional static portal films (Chang et al. 2000; Curtin-Savard and Podgorsak 1999; Pasma et al. 1999). The additional complications in treatment delivery introduced by the combination of IMRT plus respiratory gating render treatment verification even more important. Future advances may even make possible 3-D rather than 2-D treatment verification. In particular, real-time 3-D megavoltage cone beam imaging is currently being investigated as new generation aSi EPID technology (Jaffray et al. 1999; Midgley, Millar, and Dudson 1998; Mosleh-Shirazi 1998; Swindell et al. 1983) brings this approach within the realm of possibility.

Amorphous silicon EPIDs also have the potential for verification of IMRT and RG, wherein the EPID can be used as a movie camera, capturing as many as five images/second, which may be used to record MLC leaf positions *and* internal anatomy as a function of time, an example of which is shown in figure 13–10.

Summary and Conclusions

IMRT offers an exciting potential for improving the radiation therapy of NSCLC, a disease that generally responds poorly to conventional RT. NSCLC is in many aspects a textbook example of a treatment site for which IMRT was designed, as it is a disease for which dose escalation is clearly required, but for which improved normal tissue dose sparing is also critically important. It is also a site where significant improvements over conventional RT are clearly needed, being the most common cause of cancer death in the United States. Even with implementation of IMRT,

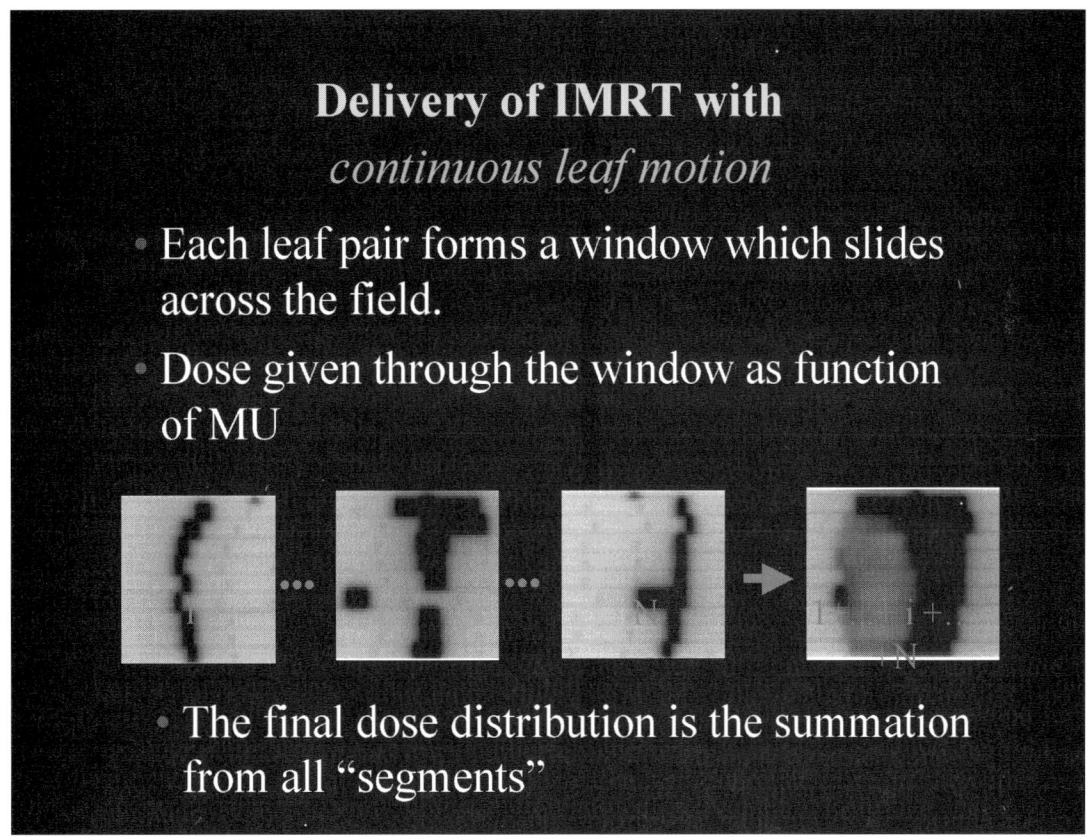

FIGURE 13–10. Using an EPID in "movie mode" to verify accuracy of IMRT treatment delivery. The three leftmost images show snapshots of the IMRT treatment at various times. The rightmost panel shows the composite image, or integration of entire treatment from this field.

however, there are still many technical details that could mitigate any possible benefits of increased radiation doses. In particular, conventional RT does poorly in NSCLC in large part because of poor initial identification of the tumor volume, and errors associated with respiratory motion during treatment. In this chapter we have therefore highlighted the concept that application of IMRT to NSCLC must be clinically tested in conjunction with other cutting edge technologies such as FDG-PET imaging, respiratory gating, and EPID-based treatment verification. We believe that such a multi-faceted approach represents the best, and perhaps the only possibility for improved outcome in the treatment of this disease with radiation.

References

Abe, Y., T. Matsuzawa, T. Fujiwara, M. Itoh, H. Fukuda, K. Yamaguchi, K. Kubota, J. Hatazawa. M. Tada, I. Ido et al. (1990). "Clinical assessment of therapeutic effects on cancer using 18F-2-fluoro-2-deoxy-D-glucose and positron emission tomography: Preliminary study of lung cancer." *Int. J. Radiat. Oncol. Biol. Phys.* 19:1005–1010.

Armstrong, J. G., C. Burman, S. A. Leibel, D. Fontenla, G. Kutcher, M. Zelefsky, and Z. Fuks. (1993). "Three-dimensional conformal radiation therapy may improve the therapeutic ratio of high dose radiation therapy for lung cancer." *Int. J. Radiat. Oncol. Biol. Phys.* 26:685–689.

Arriagada, R., T. Chevalier, E. Quoix, P. Ruffie, H. de Cremoux, J. Y. Douillard, M. Tarayre, J. P. Pignon, and A. Laplanche. (1991). "ASTRO (American Society for Therapeutic Radiology and Oncology) plenary: Effect of chemotherapy on locally advanced non-small cell lung carcinoma: A randomized study of 353 patients. GETCB (Groupe d'Etude et Traitement des Cancers Bronchiques), FNCLCC (Federation Nationale des Centres de Lutte contre le Cancer) and the CEBI trialists." *Int. J. Radiat. Oncol. Biol. Phys.* 20:1183–1190.

Chang, J., G. Mageras, C. S. Chui, C. C. Ling, and W. Lutz. (2000). "Relative profile and dose verification of intensity-modulated radiation therapy." *Int. J. Radiat. Oncol. Biol. Phys.* 47:231–240.

Chui, C. S., T. LoSasso, and S. Spirou. (1994). "Dose calculation for photon beams with intensity modulation generated by dynamic jaw or multileaf collimations." *Med. Phys.* 21:1237–1244.

Coleman, E., and R. Tesar. (1997). "Clinical PET: Are we ready?" *J. Nucl. Med.* 12:16N.

Cox, J. D., N. Azarnia, R. W. Byhardt, K. H. Shin, B. Emami, and T. F. Pajak. (1990). "A randomized phase I/II trial of hyperfractionated radiation therapy with total doses of 60.0 Gy to 79.2 Gy: Possible survival benefit with greater than or equal to 69.6 Gy in favorable patients with Radiation Therapy Oncology Group stage III non-small-cell carcinoma of the lung: A report of Radiation Therapy Oncology Group 83-11." *J. Clin. Oncol.* 8:1543–1555.

Curtin-Savard, A. J., and E. B. Podgorsak. (1999). "Verification of segmented beam delivery using a commercial electronic portal imaging device." *Med. Phys.* 26:737–742.

Emami, B. (1996). "Three-dimensional conformal radiation therapy in bronchogenic carcinoma." *Semin. Radiat. Oncol.* 6(2):92–97.

Emami, B., J. Lyman, A. Brown, L. Coia, M. Goitein, J. E. Munzenrider, B. Shank, L. J. Solin, and M. Wesson. (1991). "Tolerance of normal tissue to therapeutic irradiation." *Int. J. Radiat. Oncol. Biol. Phys.* 21:109–122.

Erdi, A. K., Y. C. Hu, and C. Chui. (2000). "Using mutual information (MI) for automated 3D registration in the pelvic and lung region for CT images. *Proc. SPIE* 3979:416–425.

Erdi, Y. E., H. Macapinlac, K. E. Rosenzweig, J. L. Humm, S. M. Larson, A. K. Erdi, and E. D. Yorke. (2000). "Use of PET to monitor the response of lung cancer to radiation treatment." *Eur. J. Nucl. Med.* 27(7):861–866.

Furuse, K., M. Fukuoka, M. Kawahara, H. Nishikawa, Y. Takada, S. Kudoh, N. Katagami, and Y. Ariyoshi. (1999). "Phase III study of concurrent versus sequential thoracic radiotherapy in combination with mitomycin, vindesine, and cisplatin in unresectable stage III non-small-cell lung cancer." *J. Clin. Oncol.* 17(9):2692–2699.

Gambhir, S. S., J. E. Shepherd, B. D. Shah, E. Hart, C. K. Hoh, P. E. Valk, T. Emi, and M. E. Phelps. (1998). "Analytical decision model for the cost-effective management of solitary pulmonary nodules." *J. Clin. Oncol.* 16(6):2113–2125.

Giraud, P., M. Antoine, A. Larrouy, B. Milleron, P. Callard, Y. De Rycke, M. F. Carette, J. C. Rosenwald, J. M. Cosset, M. Housset, and E. Touboul. (2001). "Evaluation of microscopic tumor extension in non-small-cell lung cancer for three-dimensional conformal radiotherapy planning." *Int. J. Radiat. Oncol. Biol. Phys.* 48:1015–1024.

Graham, M. V., J. A. Purdy, B. Emami, J, W. Matthews, and W. B. Harms. (1995). "Preliminary results of a trial using three dimensional radiotherapy for lung cancer." *Int. J. Radiat. Oncol. Biol. Phys.* 33:993–1000.

Greenlee, R. T., M. B. Hill-Harmon, T. Murray, and M. Thun. (2001). "Cancer statistics." *CA Cancer J. Clin. 2001* 51:15–36.

Hanley, J., M. Debois, D. Mah, G. S. Mageras, A. Raben, K. Rosenzweig, B. Michalczak, L. H. Schwartz, P. J. Gloeggler, W. Lutz, C. C. Ling, S. A. Leibel, Z. Fuks, and G. J. Kutcher. (1999). "Deep inspiration breath-hold technique for lung tumors: The potential value of target immobilization and reduced lung density in dose escalation." *Int. J. Radiat. Oncol. Biol. Phys.* 45:603–611.

Hayman, J. A., M. K. Martel, R. K. Ten Haken, D. P. Normolle, R. F. Todd 3rd, J. F. Little, M. A. Sullivan, P. W. Possert, A. T. Turrisi, and A. S. Lichter. (2001). "Dose escalation in non-small-cell lung cancer using three-dimensional conformal radiation therapy: Update of a phase I trial." *J. Clin. Oncol.* 19(1):127–136.

Hebert, M. E., V. J. Lowe, J. M. Hoffman, E. F. Patz, and M. S. Anscher. (1996). "Positron emission tomography in the pretreatment evaluation and follow-up of non-small cell lung cancer patients treated with radiotherapy: Preliminary findings." *Am. J. Clin. Oncol.* 19(4):416–421.

Henschke, C. I., D. I. McCauley, D. F. Yankelevitz, D. P. Naidich, G. McGuinness, O. S. Miettinen, D. M. Libby, M. W. Pasmantier, J. Koizumi, N. K. Altorki, and J. P. Smith. (1999). "Early lung cancer action project: Overall design and findings from baseline screening." *Lancet* 354:99–105.

Humm, J. L., J. B. Lee, J. A. O'Donoghue et al. (1999). "Changes in FDG tumor uptake during and after radiation therapy in a rodent tumor xenograft." *J. Clin. PET* 2:289–296.

Jackson, A., G. J. Kutcher, and E. D. Yorke. (1993). "Probability of radiation-induced complications for normal tissues with a parallel architecture subject to non-uniform irradiation." *Med. Phys.* 20:613–625.

Jackson, A., R. K. Ten Haken, J. M. Robertson, M. L. Kessler, G. J. Kutcher, and T. S. Lawrence. (1995). "Analysis of clinical complication data for radiation hepatitis using a parallel architecture model." *Int. J. Radiat. Oncol. Biol. Phys.* 31:883–891.

Jaffray, D. A., D. G. Drake, M. Moreau, A. A. Martinez, and J. W. Wong. (1999). "A radiographic and tomographic imaging system integrated into a medical linear accelerator for localization of bone and soft-tissue targets." *Int. J. Radiat. Oncol. Biol. Phys.* 45:773–789.

Kiffer, J. D., S. U. Berlangieri, A. M. Scott, G. Quong, M. Feigan, W. Schumer, C. P. Clarke, S. R. Knight, and F. J. Daniel. (1998). 'The contribution of F18-fluoro-2-deoxy-glucose positron emission tomographic imaging to radiotherapy planning in lung cancer." *Lung Cancer* 19:167–177.

Kubo, H. D., and B. C. Hill. (1996). "Respiration gated radiotherapy treatment: A technical study." *Phys. Med. Biol.* 41:83–91.

Kubo, H. D., E. G. Shapiro, and E. J. Seppi. (1999). "Potential and role of a prototype amorphous silicon array electronic portal imaging device in breathing synchronized radiotherapy." *Med. Phys.* 26:2410–2414.

Kwa, S. L., J. V. Lebesque, J. C. Theuws, L. B. Marks, M. T. Munley, G. Bentel, D. Oetzel, U. Spahn, M. V. Graham, R. E. Drzymala, J. A. Purdy, A. S. Lichter, M. K. Martel, and R. K. Ten Haken. (1998). "Radiation pneumonitis as a function of mean lung dose: An analysis of pooled data of 540 patients." *Int. J. Radiat. Oncol. Biol. Phys.* 42:1–9.

Ling, C. C., J. Humm, S. Larson, H. Amols, Z. Fuks, S. Leibel, and J. A. Koutcher. (2000). "Towards multidimensional radiotherapy (MD-CRT): Biological imaging and biological conformality." *Int. J. Radiat. Oncol. Biol. Phys.* 47:551–560.

Maes, F., A. Collignon, D. Vandermeulen, G. Marchal, and P. Seutens. (1997). "Multimodality image registration by maximization of mutual information." *IEEE Trans. Med. Imaging* 16(2):187–198.

Mageras, G. S., E. Yorke, K. Rosenzweig, F. Fontenla, E. Keatley, and C. Ling. (2000). "Initial clinical evaluation of a respiratory gated radiotherapy system." 42nd Annual AAPM Meeting Abstract WE-EBR-04, Chicago, IL, July 23–27, 2000. *Med. Phys.* 27(6):1419.

Mah, D., J. Hanley, K. E. Rosenzweig, E. Yorke, L. Braban, C. C. Ling, S. A. Leibel, and G. Mageras. (2000). "Technical aspects of the deep inspiration breath-hold technique in the treatment of thoracic cancer." *Int. J. Radiat. Oncol. Biol. Phys.* 48:1175–1185.

Martel, M. K., R. K.Ten Haken, M. B. Hazuka, M. L. Kessler, M. Strawderman, A. T. Turrisi, T. S. Lawrence, B. A. Fraass, and A. S. Lichter. (1999). "Estimation of tumor control probability model parameters from 3-D dose distributions of non-small cell lung cancer patients." *Lung Cancer* 24:31–37.

Midgley, S., R. M. Millar, and J. Dudson. (1998). "A feasibility study for megavoltage cone beam CT using a commercial EPID." *Phys. Med. Biol.* 43:155–169.

Mosleh-Shirazi, M. A., P. M. Evans, W. Swindell, S. Webb, and M. Partridge. (1998). "A cone beam megavoltage CT scanner for treatment verification in conformal radiotherapy." *Radiother. Oncol.* 48:319–328.

Munley, M. T., L. B. Marks, C. Scarfone, G. S. Sibley, E. F. Patz Jr., T. G. Turkington, R. J. Jaszczak, D. R. Gilland, M. S. Anscher, and R. E. Coleman. (1999). "Multimodality nuclear medicine imaging in three-dimensional radiation treatment planning for lung cancer: Challenges and prospects." *Lung Cancer* 23(2):105–114.

Nehmeh, S., Y. Erdi, K. Rosenzweig, J. Humm, E. Yorke, O. Squire. K. Sidhu, L. Braban, S. Larson, and C. Ling. (2001). "Gated positron emission tomography in lung cancer: A novel technique to reduce lung tumor motion effect for radiotherapy." 43rd Annual AAPM Meeting Abstract MO-D-BRB-01, Salt Lake City, Utah, July 22–26, 2001. *Med Phys.* 28(6):1232.

Nestle, U., K. Walter, S. Schmidt, N. Licht, C. Nieder, B. Motaref, D. Hellwig, M. Niewald, D. Ukena, C. M. Kirsch, G. W. Sybrecht, and K. Schnabel. (1999). "18F-deoxyglucose positron emission tomography (FDG-PET) for the planning of radiotherapy in lung cancer: High impact in patients with atelectasis." *Int. J. Radiat. Oncol. Biol. Phys.* 44:593–597.

Ohara, K., T. Okumura, M. Akisada, T. Inada, T. Mori, H. Yokota, and M. J. Calaguas. (1989). "Irradiation synchronized with respiration gate." *Int. J. Radiat. Oncol. Biol. Phys.* 17:853–857.

Paoli, J., K. Rosenzweig, E. Yorke et.al. (1999). "Comparison of different phases of respiration in the treatment of lung cancer: Implications for gated treatment." *Int. J. Radiat. Oncol. Biol. Phys.* 45(1):386–387.

Pasma, K. L., M. L. Dirkx, M. Kroonwijk, A. G. Visser, and B. J. Heijmen. (1999). "Dosimetric verification of intensity modulated beams produced with dynamic multileaf collimator using an electronic portal imaging device." *Med. Phys.* 26:2373–2378.

Perez, C. A., T. F. Pajak, and P. Rubin. (1987). "Long-term observations of the patterns of failure in patients with unresectable non-oat cell carcinoma of the lung treated with definitive radiotherapy." *Cancer* 59:874–881.

Pisters, K. M. W. (2000). "The role of chemotherapy in early-stage (stage I and II) resectable non-small cell lung cancer." *Semin. Radiat. Oncol.* 10(4):274–279.

Rosenzweig, K. E., J. Hanley, D. Mah, G. Mageras, M. Hunt, S. Toner, C. Burman, C. C. Ling, B. Mychalczak, Z. Fuks, and S. A. Leibel. (2000a). "The deep inspiration breath-hold technique in the treatment of inoperable non-small cell lung cancer." *Int. J. Radiat. Oncol. Biol. Phys.* 48(1):81–87.

Rosenzweig, K. E., B. Mychalczak, Z. Fuks, J. Hanley, C. Burman, C. C. Ling, J. Armstrong, R. Ginsberg, M. G. Kris, A. Raben, and S. Leibel. (2000). "Final report of the 70.2-Gy and 75.6-Gy dose levels of a phase I dose escalation study using three-dimensional conformal radiotherapy in the treatment of inoperable non-small cell lung cancer." *Cancer J.* 6(2):82–87.

Sasaki, M., Y. Ichiya, Y. Kuwabara, Y. Akashi, T. Yoshida, T. Fukumura, T. Ishida, F. Sugio, and K. Masuda. (1996). "The usefulness of FDG positron emission tomography for the detection of mediastinal lymph node metastases in patients with non-small-cell lung cancer: A comparative study with x-ray computed tomography." *Eur. J. Nucl. Med.* 23:741–747.

Spirou, S. V., and C. S. Chui. (1994). "Generation of arbitrary intensity profiles by dynamic jaws or multi-leaf collimators." *Med. Phys.* 21:1031–1041.

Swindell, W., R. G. Simpson, J. R. Oleson, C. T. Chen, and E. A. Grubbs. (1983). "Computed tomography with a linear accelerator with radiotherapy applications." *Med. Phys.* 10(4):416–420.

Tai, Y. C., K. P. Lin, C. K. Hoh, and E. J. Hoffman. (1997). "Utilization of 3D elastic transformation in the registration of chest X-ray CT and whole body PET." *IEEE Trans. Nucl. Sci.* 44(4):1606–1612.

Vansteenkiste, J. F., S. G. Stroobants, P. R. De Leyn, P. J. Dupont, J. Bogaert, A. Maes, G. J. Deneffem K. L. Nackaerts, J. A. Verschalkalen, T. E. Lerut, L. A. Mortelmans, and M. G. Demedts. (1998). "Lymph node staging in non-small-cell lung cancer with FDG-PET scan: A prospective study on 690 lymph node stations from 68 patients." *J. Clin. Oncol.* 16:2142–2149.

Wahl, R. L., L. E. Quint, R. L. Greenough, C. R. Meyer, R. I. White, and M. B. Orringer. (1994). "Staging of mediastinal non-small-cell lung cancer with FDG PET, CT, and fusion images: Preliminary prospective evaluation." *Radiol.* 191:371–377.

Wong, J. W., M. B. Sharpe, D. A. Jaffray, V. R. Kini, J. M. Robertson, J, S. Stromberg, and A. A. Martinez. (1999. "The use of active breathing control (ABC) to reduce margin for breathing motion." *Int. J. Radiat. Oncol. Biol. Phys.* 44:911–919.

Yorke, E., A. Jackson, and C. Chui. (2001). "Optimization with both dose-volume and biological constraints for lung IMRT." 43rd Annual AAPM Meeting Abstract TU-D-BRB-04, Salt Lake City, Utah, July 22–26, 2001. *Med. Phys.* 28(6):1261.

Yorke, E., A. Jackson, L. Braban et al. (2001). "Advantages of IMRT for dose escalation in radiation therapy of lung cancer." 43rd Annual AAPM Meeting Abstract WE-D-BRB-01, Salt Lake City, Utah, July 22–26, 2001. *Med. Phys.* 28(6):1291–1292.

14

IMRT Of Disease Sites Requiring Large Radiation Fields

Linda X. Hong
Kaled Alektiar
Chen-Shou Chui
Tom LoSasso
Margie A. Hunt
Spiridon Spirou
Jie Yang
Howard Amols
Clifton Ling
Zvi Fuks
Steve Leibel

Introduction

In this chapter, we present our investigation into the planning and delivery of large IMRT fields using current linear accelerators (linacs) and multileaf collimator (MLC) technology. We demonstrate techniques, such as accounting for scattered dose contributions with iterative process during optimization, beam-splitting with intensity feathering, and multiple isocenters with intensity-feathered junctions for the treatment of the whole abdomen while sparing the kidneys and bone marrows. These methods can be applied to other sites requiring large-field irradiation or involving multiple sets of large matched fields.

There have been some discussions on IMRT techniques for large treatment volumes (Roeske et al. 2000; Low and Mutic 1997; Wu et al. 2000; Chui et al. 2001), but the issues related to their actual implementation have not been adequately addressed. Multiple photon beams are usually necessary to achieve target coverage and normal tissue sparing, with isocentric setup being the preferred technique. To plan and deliver such treatment, a number of technical challenges must be met. Large IMRT fields cannot usually be treated using current linacs because of issues related to the MLC design. For example, on the Millennium™ MLC (Varian 2100EX series, Varian Associates, Palo Alto, CA), the distance between any two leaves on the same MLC carriage cannot exceed 15.0 cm (maximum leaf span) and carriage cannot be positioned more than 16.0 cm beyond isocenter in dynamic mode. This means that IMRT fields wider than 15 cm must be subdivided into subfields and fields wider than 32 cm cannot be treated using a single isocenter. For field lengths greater than 40 cm, multiple isocenters must be used. To split a dynamic multileaf collimator (DMLC) field widthwise, Wu et al. (2000) have previously developed a schema that involved *feathering* at the junction of the sub-fields. This approach has been further extended to feather fields in the cephalad-caudad direction.

In addition, large field sizes have large components of scattered dose, which are not properly accounted for in most IMRT treatment planning systems. Therefore, the impact of scattered dose on the inverse planning process and dosimetry verification was also investigated in this study.

Several prospective randomized trials have established the role of adjuvant chemotherapy in the management of ovarian cancer (Lawton et al. 1990; Whelan et al. 1992; Vergote et al. 1992; Lambert et al. 1993; Chiara et al. 1994; McGuire et al. 1996; Reisinger et al. 1996)). This had led to a significant decline in the use of radiation in the management of this disease despite its theoretical and clinically proven efficacy (Dembo 1992; Lindner, Willich, and Atzinger 1990). However, many of the randomized trials that compared whole abdomen radiation (WAR) to chemotherapy have failed to show a statistically significant difference in overall survival between the two treatments (Sell et al. 1990; Redman et al. 1993; Reisinger et al. 1996; Dent et al. 2000; Wong et al. 1999; Kojs et al. 2001). Thus, radiotherapy as an effective locoregional treatment should still play a role in the management of ovarian cancer, in addition to systemic chemotherapy (Pickel et al. 1999; Cardenes and Randall 2000). To better integrate whole abdomen radiation with chemotherapy, it is important to evaluate the toxicity profile of such radiation.

The pattern of failure in patients with ovarian cancer is primarily transperitoneal and thus the entire abdomen must be irradiated (Thomas and Dembo 1993). Typically, 30 Gy is given to the whole abdomen with an AP/PA beam arrangement with posterior shielding of the kidneys. Because these fields are usually very large (40×30 cm^2 or larger), most patients experience acute side effects, including the serious side effect of myelosuppression that can deplete bone marrow reserve and interfere with current or subsequent chemotherapy. The reported frequency of treatment interruptions due to such side effects is ~12% when radiation is used alone (Fyles et al. 1992), and ~28% when combined with chemotherapy (Ben-Josef and Court 1995). The purpose of this study is to explore the potential use of IMRT to circumvent the above side effect by minimizing the dose to the bone marrow while maintaining dose constraint to the kidneys and providing adequate coverage of the whole abdomen. Here, the goal is to treat the retroperitoneal lymph nodes and peritoneal surfaces while reducing the dose to kidneys and bone marrow.

Materials And Methods

Patient Data

The anatomical data used in this study were those of 10 patients with endometrial cancer who were treated with AP/PA postoperative WAR. All patients underwent total abdominal hysterectomy/bilateral salpingo-oophorectomy and pelvic and para-aortic lymph node sampling. Computed tomography (CT) simulation was performed with the patient supine and arms above the head. Patients were scanned from the mid-thorax to 5 cm below the ischial tuberosities. An average of 110 CT images per patient were acquired at 0.5 cm spacing.

Definition Of Target Volumes And Critical Structures

The gross target volume (GTV) included the entire peritoneal surface, defined on each CT slice as the junction between the abdominal wall and the peritoneal lining. Pelvic and para-aortic lymph node regions were also included in the GTV, and were identified by the surgical clips as well as the aorta/inferior vena cava and common, external, and internal iliac blood vessels. The planning target volume (PTV) encompassed an axial margin of 5 mm around the GTV, and extended 1 cm beyond the GTV in both the superior and the inferior directions. Due to the close proximity of the peritoneum to the liver surface (especially around the hilum) the outer 1 cm border of liver was included in PTV. The rest of the liver was considered normal tissue. Because of the potential nephrotoxicity from cisplatin chemotherapy, both kidneys were restricted to receive a maximum dose of less than 18 Gy, and therefore were excluded from PTV. The left renal hilum, where the left ovary's lymphatics drain, was included in the PTV. Normal structures relevant to planning were also outlined on the axial CT images, including the kidneys, liver, spinal cord, thoracic and lumbosacral vertebral bodies, pelvic bones, and femurs. Among these 10 patients, the ranges of PTV dimensions were: volumes 5629–12,578 cc (median 7935 cc), length 37–46 cm (median 43 cm), width 27–33 cm (median 29 cm), and depth 19–23 cm (median 20 cm).

Conventional Plans

Dose distributions for conventional AP/PA 6 MV photon beams at extended distance [130 SAD (source-axis distance)] were generated, documenting the actual treatment received by each patient. The superior edge of the field was placed 1 to 1.5 cm above the dome of the diaphragm. The inferior edge was at the bottom of the obturator foramina and the lateral edge was extended at least 1 to 2 cm beyond the skin surface. Skin flash was needed in the conventional technique because the peritoneal reflections could not be visualized accurately from simulation films (LaRouere et al. 1989). Kidneys were blocked in the PA field only with 0.5 cm margin to keep the kidney dose less than 18 Gy. Customized blocks were also utilized in the superior and inferior corners. A total dose of 30 Gy was given in 1.5 Gy fractions, with the prescription to the mid-plane point on the central axis. Graphical dose distributions were not performed at the time of treatment but were generated retrospectively for comparison with the IMRT. At the time of the retrospective review, we found that the coverage of the PTV with these AP/PA fields was quite non-uniform and varied from patient to patient. In general, the treatment portals had larger margin (>2 cm) in the pelvis and tighter margin (<1 cm) around the diaphragm domes. For the purpose of this study, the portals were modified for each patient to allow a 1 cm margin around PTV. The kidney blocks were unmodified from what was used for treatment.

Intensity-Modulated (IMRT) Plans

Beam Arrangement

For the IMRT plan, five equally spaced and non-opposing beams were found suitable for the roughly cylindrical PTVs. Beams were placed approximately 70° apart, at gantry angles of 255°, 325°, 180°, 105°, and 35°. We used 15 MV photons for the IMRT fields, instead of the 6 MV used in conventional treatment. The increased depth of maximum dose of 15 MV photons is unlikely to cause underdosing of superficial regions because of the use of multiple beam directions. The decreased integral dose due to improved depth dose of 15 MV is an advantage.

All patients but one had at least one field with length >40 cm, thus requiring separate isocenters at the abdominal and the pelvic regions. The two isocenters were placed 20 cm apart in the superior-inferior direction, with the same left-right and anterior-posterior coordinates (figure 14–1). Beams were arranged at identical gantry angles for both isocenters. For the PA field, the inferior jaw of the abdominal fields was set to 20 cm, coincident with pelvic isocenter. The superior jaw of the pelvis fields was set to at least 3 cm, resulting in an intentional overlap with the abdominal fields. For the rest of the abdominal fields, the inferior jaws were set to ±5 mm, ±1 cm relative to the PA field while maintaining the same amount of overlaps with their corresponding pelvic fields. The staggering of the overlapping regions from different field pairs would ensure smooth junctions in the treatment volumes. All patients with two isocenters required pelvic

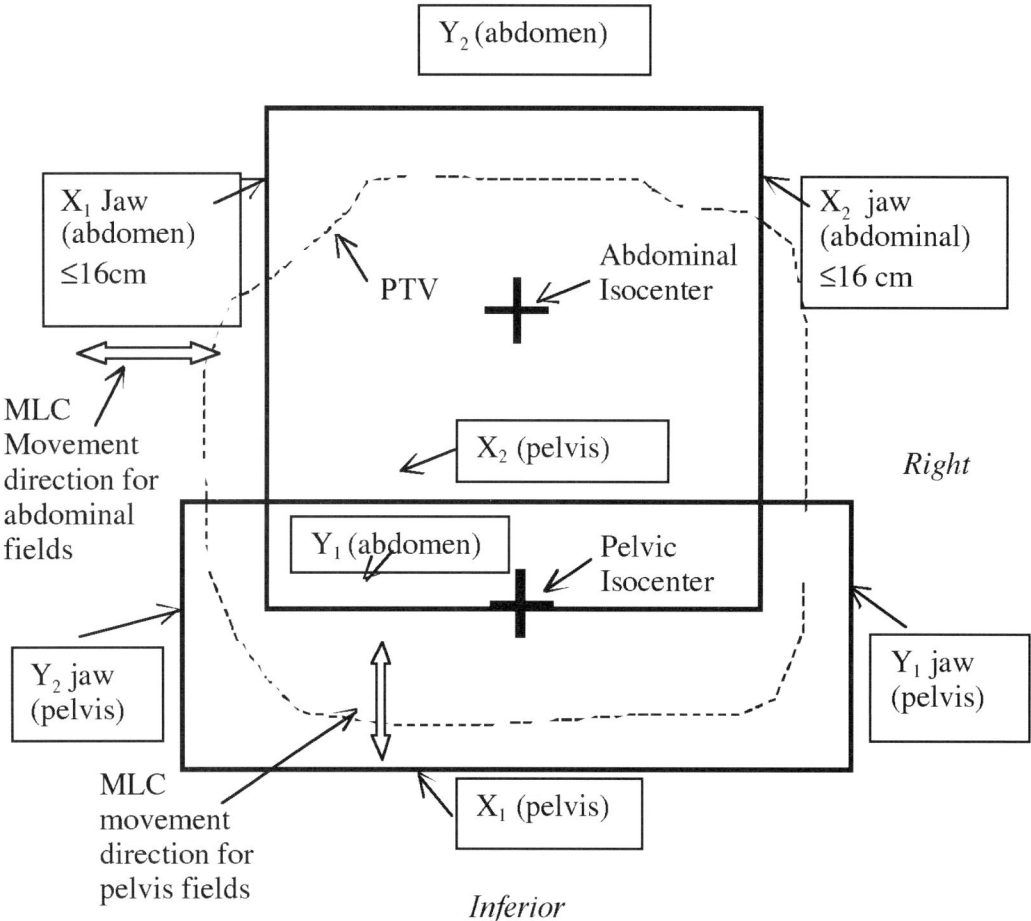

FIGURE 14–1. Schematic illustration of PA fields for two isocenter whole abdomen IMRT treatments.

field widths of ≥15 cm and lengths of ≤15 cm. This enabled us to rotate the collimator by 90° for the pelvic fields, so that the leaves could travel across the length of the pelvic fields therefore avoiding the need to split these fields (see figure 14–1 and the paragraphs on beam splitting below).

Optimization

Optimized intensity profiles were obtained from an inverse planning algorithm, which uses a convolution-based dose calculation and determines the intensity distributions through an iterative process (Spirou and Chui 1998). The user defines the prescribed dose and dose homogeneity constraints for PTV, and dose and/or dose-volume constraints for critical organs. The kidneys were constrained to a maximum dose of 18 Gy. The optimization goal was to minimize the dose to bones with reasonable PTV coverage and acceptable PTV uniformity.

To achieve a prescription dose of 30 Gy for the PTV while satisfying the dose constraint on the kidneys, a high degree of modulation was introduced into the intensity profiles in some beams.

To reduce numeric noise and artifacts in the calculated intensity profiles, it was necessary to increase the number of calculation points in the structures used for optimization to approximately 30 points per cc. Due to the large field sizes, the scattered dose contribution is significant and must be accurately calculated. However, inclusion of full scatter during the iterative inverse planning process would be computationally prohibitive. To circumvent this problem, we have devised a method briefly discussed here. During the optimization, scatter from within 2 mm radius of the pencil beam kernel was included. After the first-round optimization was completed and intensity distribution produced, the dose distribution was recalculated with full scatter. The difference between the two dose distributions (one with full scatter and the other with partial scatter) was then incorporated as a correction to be applied during another round of optimization. Three to four iterations using this corrective process led to an accurate accounting of scatter dose without incurring excessive extra computation time.

For patients with multiple isocenters, leaf motions in the abdomen and pelvic fields are perpendicular to each other because of the 90° collimator rotation for the pelvic fields (figure 14–1). In the overlap region, the intensity from the abdominal fields was modulated with left-right leaf motions with resolution of 1 cm in the superior-inferior direction, while the pelvic fields were modulated with leaf motions in superior-inferior direction with a resolution of 2 mm or better. This difference in superior-inferior resolution between fields created additional noise and numerical artifacts in the intensity files in the overlap region, usually producing sharp peaks in the intensity profiles of the pelvic fields. By including a term within the objective function that specifies the acceptable profile noise as an optimization criterion (Spirou et al. 2001), we were able to obtain intensity profiles that were feathered in the junction region after optimization.

If needed, intensity profiles could be extended beyond the skin surface (Chui, Hong, and Hunt 2002). Since we have five different beam angles and CT-based peritoneal reflection with 5 mm margin, it is decided that skin flash was not necessary for the IMRT plan.

Beam Splitting

Linear accelerators available at our institution (Varian 2100EX series) restrict the distance between any two leaves on the same carriage to no more than 15 cm and the carriage position to no more than 16.0 cm beyond isocenter in the dynamic mode. Each X-jaw is therefore limited to 16 cm and field widths for dynamically delivered fields are limited to 15 cm. As a result, targets between 15 and 32 cm wide must be treated using dynamic fields divided into multiple segments, and targets wider than 32 cm cannot be completely covered. We found that beams from other directions could compensate for the inadequate PTV coverage from any particular beam.

Also because of MLC restrictions, it is necessary to split fields with widths larger than 15 cm into subfields (Wu et al. 2000; Chui et al. 2001). To minimize field match errors, the adjacent subfields overlapped by at least 2 cm, with intensity feathering in the overlapping region. Fields wider than 27 cm had to be split into three subfields.

Dose Calculations

For conventional whole abdomen treatment with AP/PA beam arrangement, monitor unit (MU) calculations were done based only on the midplane separation. To avoid fluctuations in path length from breathing near the diaphragm, all dose calculations including IMRT plans were done without inhomogeneity correction. For 15 MV photon beams, the path length correction will be a small effect.

IMRT plans were normalized to the highest isodose line that covered the PTV excluding the region adjacent to the kidneys. The conventional plan was normalized to midplane on the central axis.

Results

Intensity Profiles And Beam Splitting

Optimization time ranged from 20 min to 80 min on a 500 MHz DEC alpha workstation. Figure 14–2a shows typical intensity profiles for the abdominal PA field for a two-isocenter patient. For this patient, the PTV width from the PA direction exceeded 32 cm and part of the PTV extended outside the field. The intensity profile for this field was characterized by a large amount of modulation, but relatively little high-frequency noise due to the use of the smoothing algorithm during the optimization. In the area overlying the kidneys, the intensity decreases significantly to reduce the dose to the kidney, as shown in figure 14–2a. Figure 14–2b shows typical intensity profiles for the pelvic PA field. Feathering of the intensity in the 6 cm overlap region between the abdominal and pelvic fields can also be observed.

The initial intensity profile for a PA abdominal field with width exceeding 27 cm is shown in figure 14–2a. Because of the excessive width, this field was divided into three subfields with intensity profiles as shown in figures 14–2c to 14–2e. The adjacent subfields overlapped by at least 2 cm to provide feathering in the junction region, as shown in figure 14–2f. For patients with two isocenters, the width of the separate pelvic fields is kept below 15.0 cm by rotating the collimator 90°, so no splitting is necessary for the pelvic fields, as shown in figure 14–2b.

After splitting, the average total number of DMLC beams was 17 for patients with two isocenters. The dose distributions from these fields closely matched the distribution of the *original fields*, i.e., those generated from optimization without considering MLC dynamic movement constraints. Delivery of 150 cGy with DMLC required, on average, 1442 MU. The need for subfields resulted in an increase in MU by 5% to 30% depending on the amount of modulation in the beam profiles. For conventional treatment, the total MU averaged 290. Beam data for all patients are listed in table 14–1.

Dose Distributions And DVH Comparison

Dose distributions with IMRT for a typical patient are shown in figure 14–3. In the vicinity of kidneys, the dose distributions were highly inhomogeneous due to the severe dose constraints to the kidneys. There were hot spots, usually between 110% to 115%, in abdominal fields. Dose homogeneity and PTV coverage in the kidney region could be improved by increasing the number of beams, as shown in figure 14–4. Outside of the kidney region, the dose distributions were fairly homogeneous with hot spots typically less than 110%, as shown in figure 14–5. The corresponding dose distributions with the conventional plan for the same patient are also shown in figures 14–3, 14–4, and 14–5.

DVH data for all patients are summarized in table 14–2. For the prescription dose of 30 Gy, a significant reduction in the dose to the bones and improved target coverage to target were achieved with IMRT. For all patients, PTV volume receiving 95% (V_{95}) of prescription dose increased on average from 71.7±4.8% with the conventional treatment to 83.9±3.9 % with IMRT ($p = 0.002$). With IMRT, loss of target coverage occurred primarily near the kidneys and was due to the strict dose constraints imposed on these structures. Dose inhomogeneity within the target increased slightly with IMRT, with the D_{05} (dose to 5% of PTV volume) averaging 32.8±0.2 Gy compared to 31.2±0.6 Gy with the conventional plan. Compared with the conventional plan, the volume of the pelvic bones receiving more than 21 Gy was reduced by nearly 60%. The mean dose to all bones (pelvic bones, spinal column, and femoral head combined) was 18.5±1.0 Gy with IMRT compared to 24.0±1.5 Gy with the conventional plan ($p = 0.002$). DVH comparisons for a typical patient are displayed in figure 14–6.

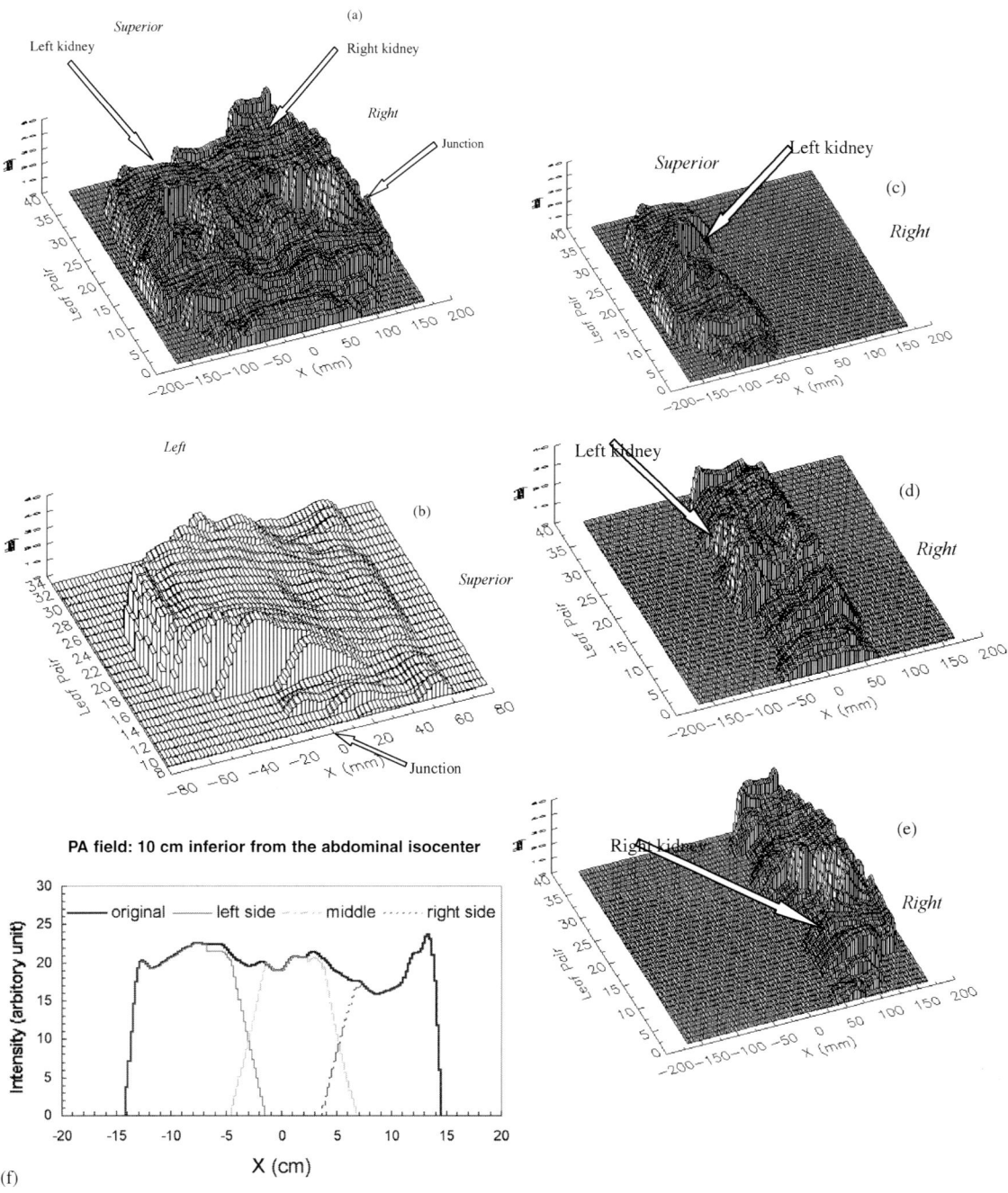

FIGURE 14–2. Intensity profiles for a PA field with 1 cm leaf width. (a) PA abdominal field before beam splitting. (b) PA pelvic field illustrating superior-inferior leaf motion created using a 90° collimator angle. (c) Left portion of PA abdominal field after beam splitting. (d) Mid portion of PA abdominal field after beam splitting. (e) Right portion of PA abdominal field after beam splitting. (f) Intensity profile for a single leaf pair before and after beam splitting.

Table 14-1. Summary Patient Beam Data

Patient No.	PTV Dimensions				No. of Isocenters	No. of Beams After Beam Splitting	Total Monitor Units
	Volume (cc)	Length (cm)	Width (cm)	Depth (cm)			
1	6990	42	27	19	2	15	1374
2	5629	42	29	23	2	17	1363
3	12,578	46	33	22	2	18	1750
4	8283	44	27	19	2	15	1369
5	8883	41	29	20	2	16	1502
6	9032	43	31	22	2	18	1580
7	5839	37	29	18	1	11	1172
8	7240	43	31	20	2	18	1491
9	7586	43	29	20	2	17	1610
10	8822	42	33	21	2	18	1211

Dosimetry Verification

Dosimetric verification of the IMRT fields was performed using film dosimetry, ionization chamber, and diode measurements. Because of the field sizes, only Kodak X-OMAT V films were practical for film dosimetry. It is known, however, that these films overrespond to scattered radiation, which is a significant component of the total dose for large fields. Separate calculations were done to estimate the scatter dose contribution and these calculations were done to correct the calibration of the film. We found about 4% overall dose over-response with film.

The contribution of scatter and leakage dose from the MLCs between adjacent subfields was also of concern. It was determined that the contribution to one subfield was proportional to the MU setting for the adjacent subfields. Film dosimetry was used to estimate that the scatter and leakage from the MLC contributed about 4% to the total dose for WAR fields.

Discussion

Given the same dose constraints to the kidneys, IMRT improved dose coverage to the PTV and reduced dose significantly to the bone marrow.

With IMRT, dose coverage to the PTV was improved. The V_{95} was 83.5% with IMRT compared to 71.7% with conventional planning. This improvement could be attributed to better coverage of the PTV around the kidneys. With IMRT, dose inhomogeneity increased slightly compared with the conventional plan, with D_{05} increasing by 5% with IMRT. To improve dose homogeneity throughout the PTV, additional beams with different gantry angles could be introduced, as shown in figures 14–4 and 14–6. It is in fact advantageous to use as many beams as clinically feasible. The need to divide most of the treatment fields into multiple segments because of excessive width, however, significantly increases the treatment time and makes more than five gantry angles impractical, particularly for patients requiring two-isocenter treatment. In the future, improved hardware and software on the accelerator and ancillary systems such as the record-and-verify system may make it feasible to increase the total number of beams without undue hardship on the clinical operation.

With further tightening of the dose constraints to kidneys, IMRT has the potential of decreasing the renal dose lower than the 18 to 20 Gy that is usually delivered with conventional AP/PA fields arrangement. This would also require increasing the number of beams to maintain reasonable dose inhomogeneity within the target. Nonetheless, this might present a significant advantage especially if WAR is to be combined with adjuvant chemotherapy. Schneider et al. (1999) analyzed the creatinine clearance over time of 56 patients treated with WAR for ovarian cancer and found a decline in renal function above and beyond what could be attributed to cis-platinum alone.

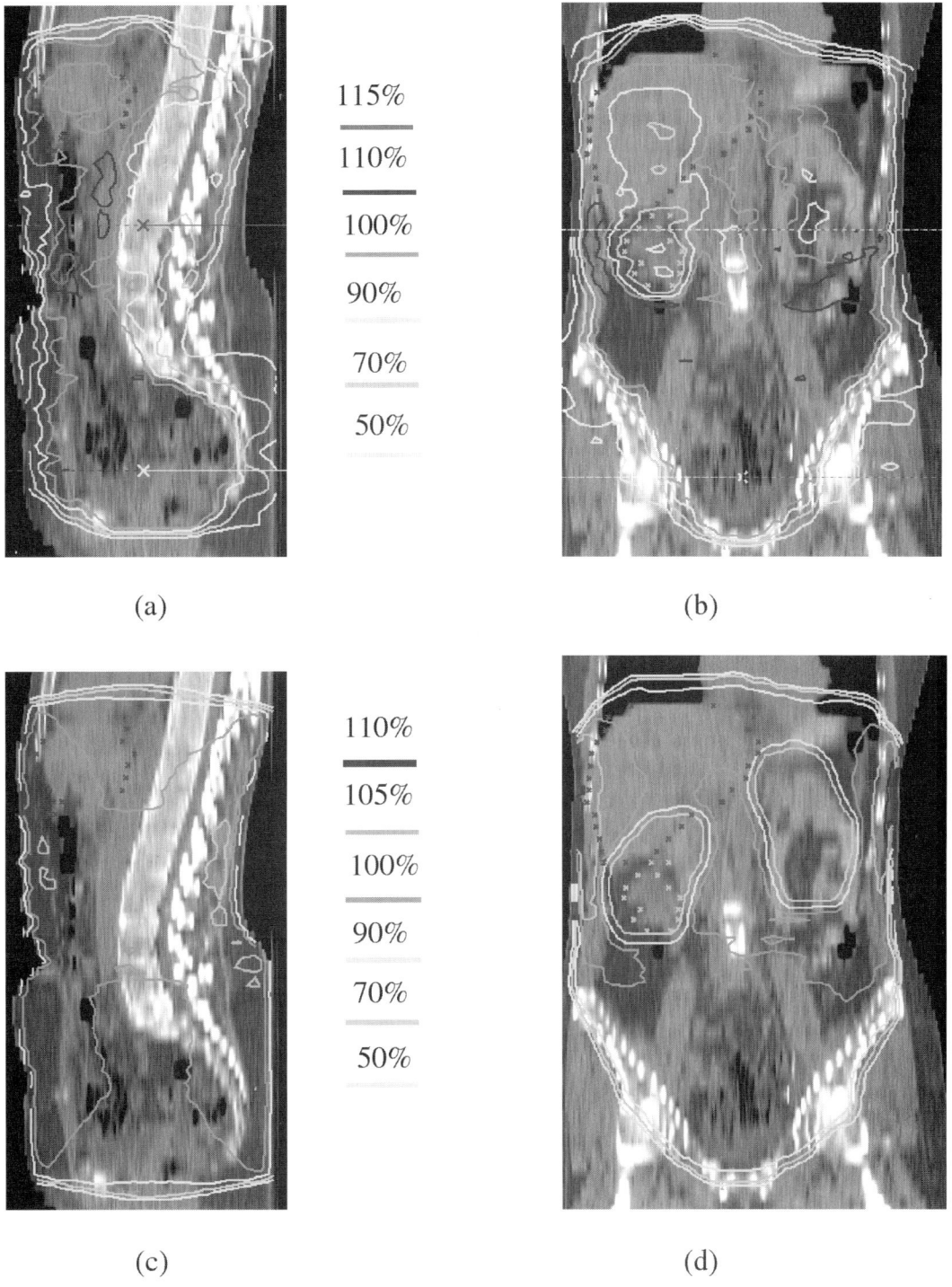

115%

110%

100%

90%

70%

50%

(a)

(b)

110%

105%

100%

90%

70%

50%

(c)

(d)

FIGURE 14–3. Isodose distributions for IMRT plan with five gantry angles. Isodose levels in percent (%). The location of the two isocenters are indicated by crosses. (a) Sagittal plane. (b) Coronal plane. Also shown are isodose distributions for conventional plan with extended distance (130 SAD) AP/PA fields for the same patient. (c) Sagittal plane. (d) Coronal plane. See COLOR PLATE 38.

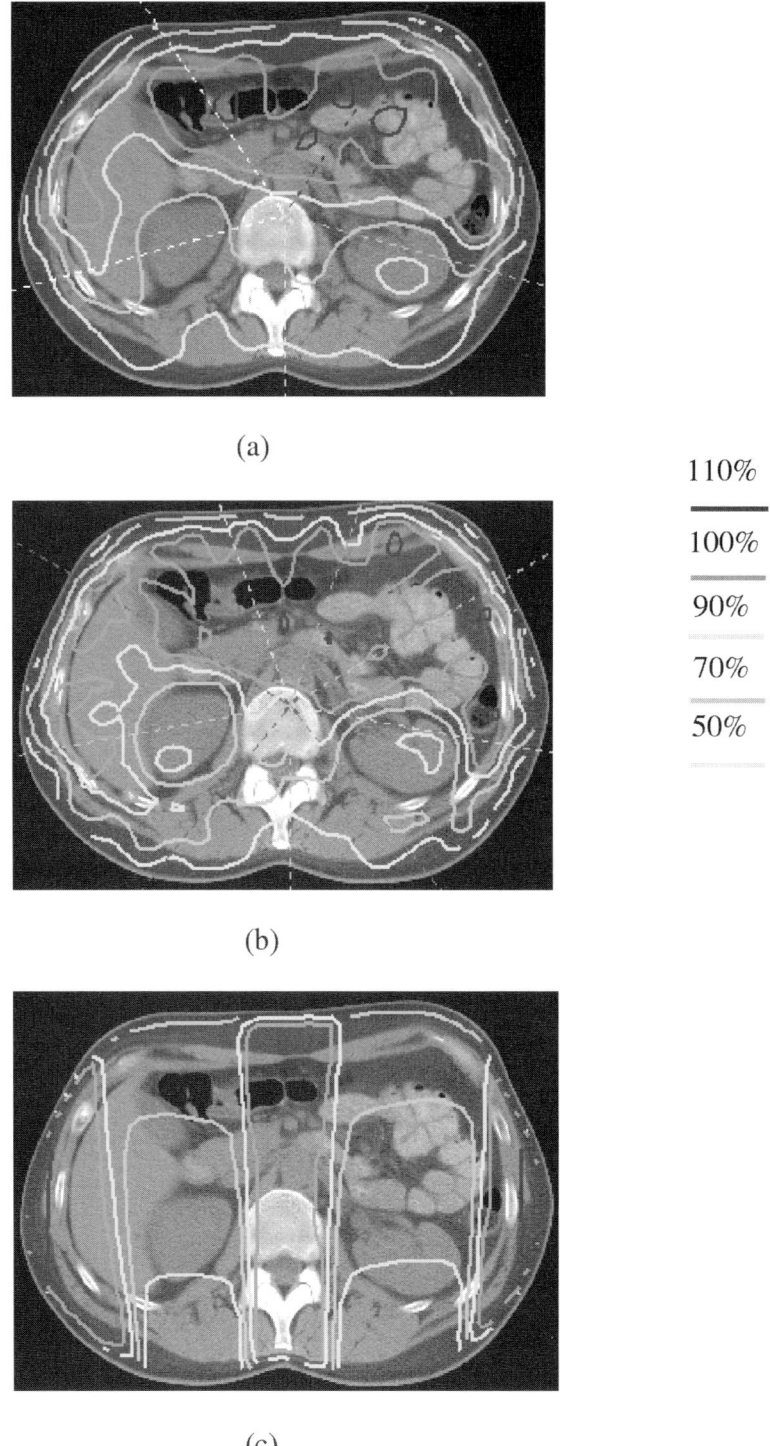

(a)

110%

100%

90%

70%

50%

(b)

(c)

FIGURE 14–4. IMRT and conventional plan comparison. Isodose levels in percent (%). (a) IMRT plan with five gantry angles. (b) IMRT plan with nine gantry angles (from PA, with gantry angle 40° apart). (c) Conventional plan. See COLOR PLATE 39.

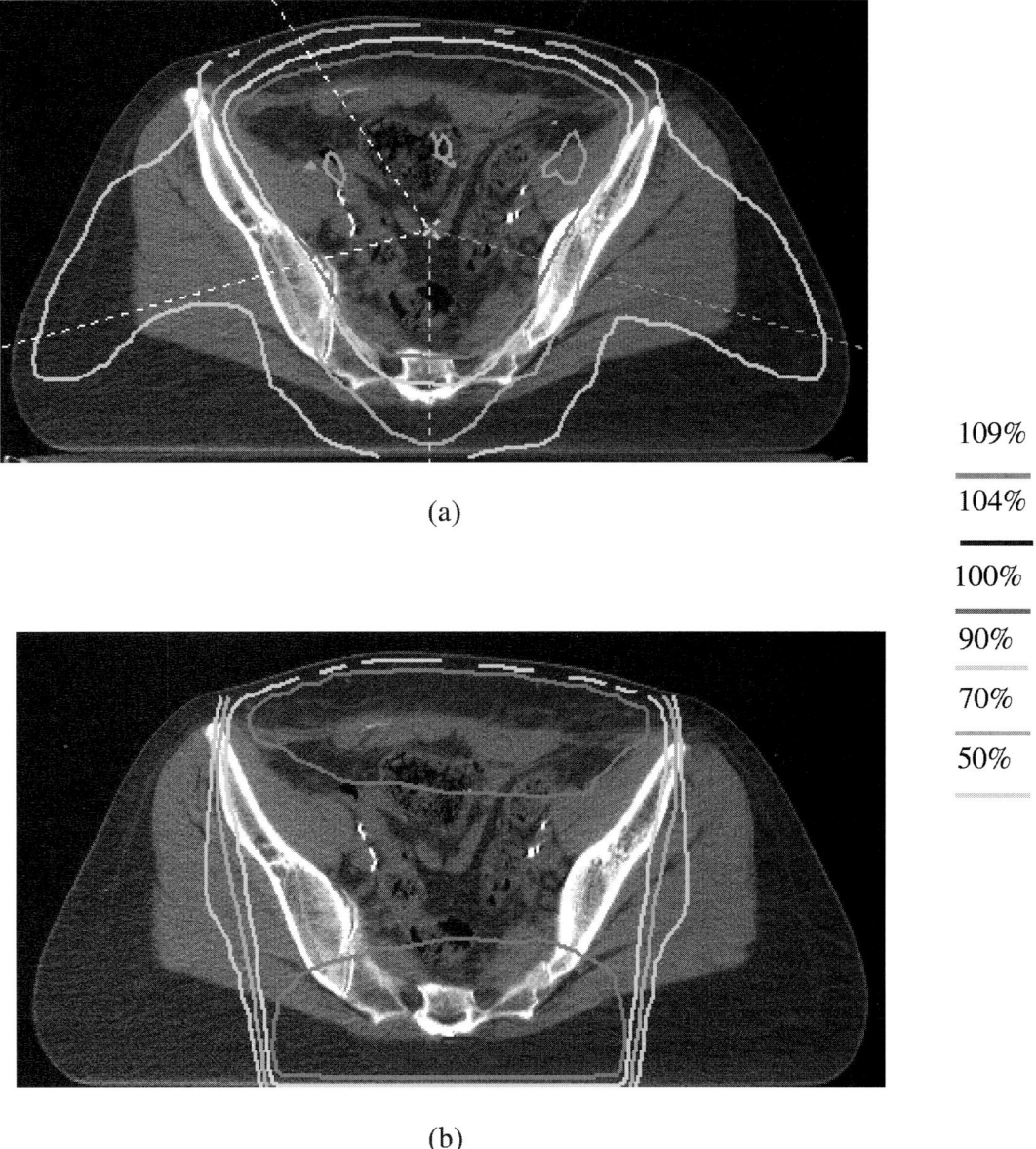

109%

104%

100%

90%

70%

50%

(a)

(b)

FIGURE 14–5. IMRT and conventional plan comparison. Isodose levels in percent (%). (a) IMRT plan with five gantry angles. (b) Conventional plan.

Dose to the bone marrow was significantly reduced with IMRT. The volume of pelvic bones receiving 21 Gy or more was 34.5% with IMRT compared to 85.5% with the conventional plan. The mean dose to all bones (including spine, pelvis, and femurs) was 18.5 Gy with IMRT compared to 24.0 Gy with the conventional plan. The anatomy of the peritoneal cavity is very challenging due to the complex arrangement of the peritoneal mesentery which supports the stomach, spleen, and liver. In addition, the boundaries between the retroperitoneum and the peritoneal

Table 14–2. Summary DVH Data Showing Averages, Standard Deviations, and p Value for 10 Patients (Based on a 30 Gy Prescription Dose).

		IMRT	CONV	Ratio (%) (IMRT/CONV)	p Value
PTV	D_{05}	32.8 ± 0.2 Gy	31.2 ± 0.6 Gy	105.1 ± 2.3	0.002
	V_{95}	83.5 ± 3.9%	71.7 ± 4.8%	117.0 ± 8.9	0.002
	D_{mean}	30.2 ± 0.3 Gy	27.8 ± 0.5 Gy	108.6 ± 2.4	0.002
Kidneys	D_{05}	17.7 ± 0.2 Gy	17.7 ± 0.9 Gy	100.3 ± 5.3	0.846
	D_{mean}	16.1 ± 0.4 Gy	15.8 ± 0.7 Gy	102.5 ± 5.2	0.275
Liver	D_{05}	31.5 ± 0.2 Gy	30.8 ± 0.6 Gy	102.4 ± 2.6	0.020
	D_{mean}	28.5 ± 0.3 Gy	27.3 ± 1.5 Gy	104.8 ± 5.7	0.027
Cord	D_{05}	28.3 ± 1.3 Gy	30.9 ± 0.4 Gy	91.5 ± 4.4	0.002
	D_{max}	30.4 ± 0.2 Gy	31.1 ± 0.4 Gy	97.7 ± 1.6	0.002
Pelvis	D_{05}	30.0 ± 0.4 Gy	30.5 ± 0.3 Gy	98.2 ± 1.0	0.014
	V_{70}	34.5 ± 2.0%	85.5 ± 4.9%	40.4 ± 3.4	0.002
	V_{95}	10.1 ± 1.6%	57.7 ± 10.8%	17.9 ± 3.3	0.002
	D_{mean}	19.1 ± 0.3 Gy	26.0 ± 1.0 Gy	73.7 ± 2.8	0.002
Bones*	D_{05}	29.8 ± 0.4 Gy	30.8 ± 0.4 Gy	96.6 ± 1.4	0.002
	V_{70}	44.8 ± 3.1%	76.6 ± 4.4%	58.5 ± 3.6	0.002
	V_{95}	9.8 ± 2.0%	62.8 ± 7.2%	15.5 ± 2.3	0.002
	D_{mean}	18.5 ± 1.0 Gy	24.0 ± 1.5 Gy	77.4 ± 2.6	0.002

* Bones: Pelvis, spinal column and femoral head combined.
D_{05}: Dose encompassing 5% of volume; D_{95}: Dose encompassing 95% of volume; V_{70}: Volume receiving 70% of dose; V_{95}: Volume receiving 95% of dose; D_{mean}: Mean dose; D_{max}: Maximum dose.

cavity are very difficult to delineate. Advances in spiral CT and magnetic resonance imaging (MRI) sequences might allow for better imaging of these structures. In this series, we did not attempt to separate the peritoneal cavity from the retroperitoneum. If this could be done, it would allow further reduction in the dose to the underlying bone and kidneys.

Reducing the myelotoxicity of WAR is of significant importance. First, it would reduce the risk of prolonged treatment interruption even when WAR is used alone. Second, and perhaps more importantly, it would allow for better integration with chemotherapy by reducing total toxicity.

We use two isocenters for patients with field sizes longer than 40 cm. The abdominal fields and pelvic fields are intentionally overlapped by at least 3 cm. By incorporating a smoothing algorithm to reduce noise and numerical artifacts in the intensity profiles, we were able to produce intensity profiles feathered in the cephalad-caudad direction in the junction region. The resulting dose distributions in the junction region between abutting IMRT fields was very homogenous. We expect that abutting IMRT fields longitudinally will not add to toxicity.

The large field sizes necessary for whole abdomen irradiation pose challenges for inverse planning, DMLC delivery, and dosimetric verification. An optimized plan for large fields must account for the scattered dose component in some manner. We have used an iterative process, where only primary radiation (or primary with minimal scatter component) was included during the optimization, and the dose distribution was recalculated with full scatter contributions at the end of each optimization cycle (Spirou et al. 2001). The difference between the actual dose and the desired dose was then used for the next optimization cycle, and the process was iterated until further improvement became minimal. This method was chosen because inclusion of full scatter during the iterative inverse planning process would require prohibitive computer memory and computation power.

The large field sizes also necessitate beam splitting for our particular MLC when using DMLC to deliver large IMRT fields. Other MLC designs may not require field splitting. Our splitting technique uses a feathering technique to reduce field-matching errors (Wu et al. 2000; Chui et al. 2001).

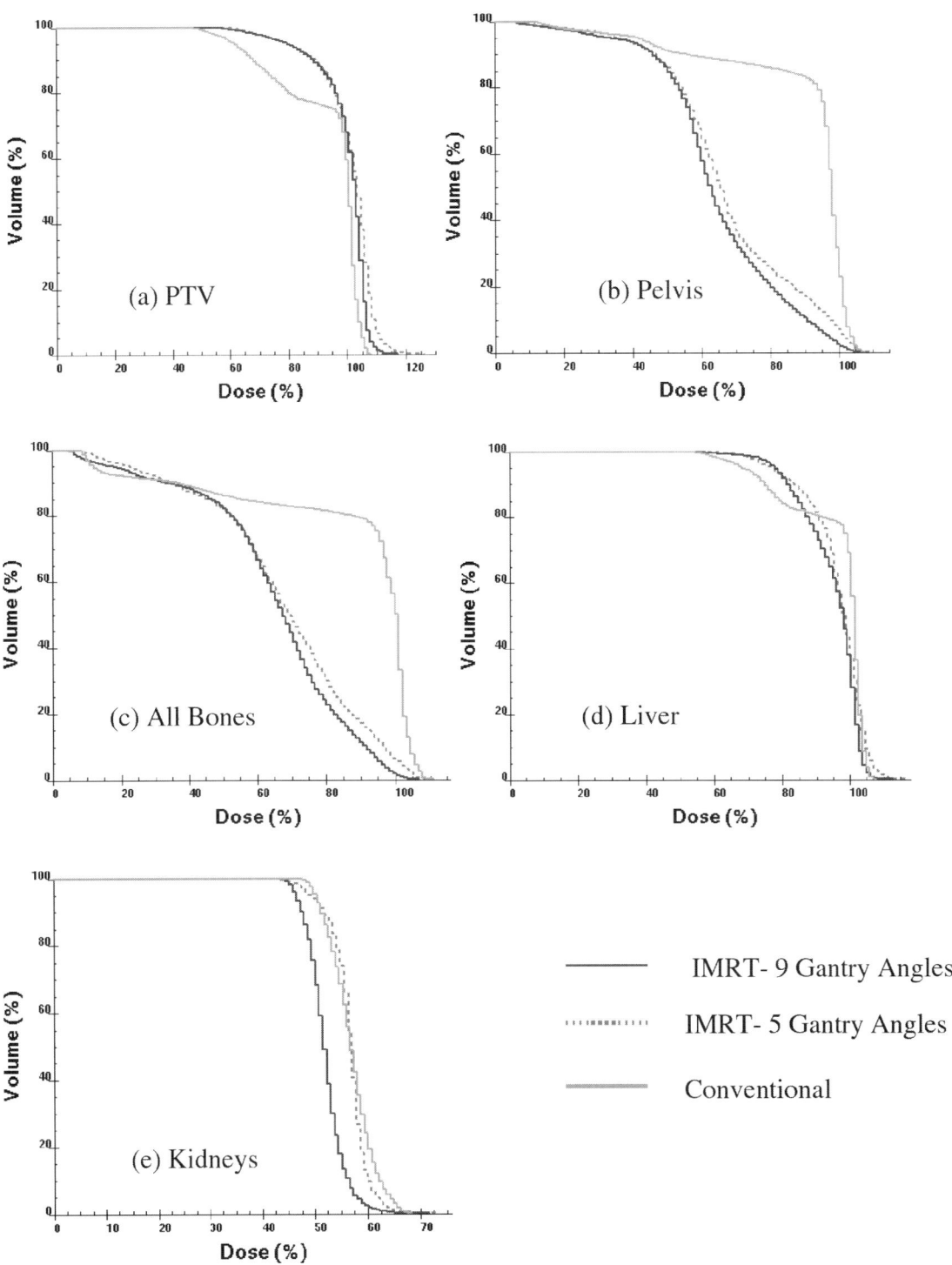

FIGURE 14–6. DVH comparison. For IMRT plan with nine gantry angles, the gantry angles were 40° apart starting from PA.

The need for subfields, however, causes some loss of efficiency during DMLC delivery, increasing MU by 5% to 30% depending on the amount of modulation in the intensity profiles. Due to overlapping of subfields, scatter and leakage dose from the MLCs of one subfield contribute to adjacent subfields, and these doses were proportional to the monitor units. To deliver 150 cGy, the average total MU was about 1442, corresponding to a duty factor of 0.10. Arnfield et al. (2000) have shown that for large fields (10×40 cm^2) the MLC scatter contribution from a Varian 120 leaf MLC is approximately 0.4% of the open field dose. With a duty factor of 0.1, the MLC scatter contributions would be about 4% of the open field dose, which agrees with the results of our film dosimetry.

IMRT treatment with a large number of beams inevitably increases treatment complexity and treatment time. Most patients will require two separate isocenters. For five different gantry angles with at least two subfields each, a total of 1500 MU, on average, will be required to deliver 150 cGy. The beam-on-time alone will be about 5 min. So it is necessary to have some kind of patient immobilization. We propose to use a thermoplastic mold to immobilize the patient's abdominal and pelvic area. Acquiring all the necessary pre-treatment portal images in one session may be impractical for both the patient and staff, thereby necessitating either multiple imaging sessions or the use of electronic portal imaging. It is also essential to have auto sequencing available to streamline the treatment process for each field as much as possible. With Varian 2100EX series accelerators, it is practical to setup and treat a patient using this technique within 20 min.

The techniques presented here can be applied to other sites requiring large-field irradiation such as soft tissue sarcoma of the thigh (Hong et al. 2002) or other treatments involving multiple sets of large matched fields.

It could be argued that by using more beam angles in a conventional treatment [such as 3-D conformal technique using 15 MV photon beams and beam's-eye view (BEV) blocking to the kidneys] one could improve over the AP/PA conventional treatment. However, for a majority of patients with field length larger than 40 cm, it is impossible to set up multiple extended distance fields. Field matching and junctions are impractical to implement with non-modulated 3-D techniques using multiple isocenters and complex beam arrangements. It is also impossible for conventional 3-D conformal techniques to achieve concave dose distributions with bone marrow sparing, especially the pelvic bones, without compromising PTV coverage or creating unacceptable dose inhomogeneity (Nutting, Dearnaley, and Webb 2000; Hunt et al. 2001).

The issues of uncertainties in CT-based radiation therapy treatment planning associated with patient breathing and setup were not specifically addressed in this chapter. Many investigators have studied the effects of breathing motion on organ definition and path length variation for conformal treatment planning (Balter et al. 1996, 1998). More recently, Bortfeld et al. (2002) investigated specifically effects of intra-fraction motion on IMRT dose delivery with statistical analysis and simulation. They concluded that the main effect of organ motion in IMRT is an averaging of the dose distribution without motion over the path of the motion, the same as for treatments with conventional beams. In this study, the movement of the diaphragm, liver and kidneys was considered using added margin as is done in conventional planning. Dose calculations were done without inhomogeneity correction to avoid fluctuations in path length from breathing near the diaphragm. With the advent of respiratory gating technology (Kubo and Hill 1996), it is possible to incorporate breathing motion into simulation, planning, and treatment process, but the time required to deliver IMRT gated WAR would currently be prohibitive.

Conclusions

We have developed methods to plan and deliver whole abdomen IMRT using available linear accelerators. IMRT can achieve dose sparing to the kidneys and bone marrow, and improve PTV coverage. These methods can be applied to other sites requiring large field irradiation. The techniques presented here, such as accounting for scattered dose contributions with iterative process during optimization, beam-splitting with intensity feathering and multiple isocenters with intensity-feathered junctions, can be applied to other sites requiring large-field irradiation or involving multiple sets of large matched fields.

References

Arnfield, M. R., J. V. Siebers, J. O. Kim, Q. Wu, P. J. Keall, and R. Mohan. (2000). "A method for determining multileaf collimator transmission and scatter for dynamic intensity modulated radiotherapy." *Med. Phys.* 27:2231–2241.

Balter, J. M., R. K. Ten Haken, T. S. Lawrence, K. L. Lam, and J. M. Robertson. (1996). "Uncertainties in CT-based radiation therapy treatment planning associated with patient breathing." *Int. J. Radiat. Oncol. Biol. Phys.* 36:167–174.

Balter, J. M., K. L. Lam, C. J. McGinn, T. S. Lawrence, and R. K. Ten Haken. (1998). "Improvement of CT-based treatment-planning models of abdominal targets using static exhale imaging." *Int. J. Radiat. Oncol. Biol. Phys.* 41:939–943.

Ben-Josef, E., and W. S. Court. (1995). "Whole abdominal radiotherapy and concomitant 5-fluorouracil as adjuvant therapy in advanced colon cancer." *Dis. Colon Rectum* 38:1088–1092.

Bortfeld, T., K. Jokivarsi, M. Goitein, J. Kung, and S. B. Jiang. (2002). "Effects of intra-fraction motion on IMRT dose delivery: statistical analysis and simulation." *Phys. Med. Biol.* 47:2203–2220.

Cardenes, H., and M. E. Randall. (2000). "Integrating radiation therapy in the curative management of ovarian cancer: Current issues and future directions." *Semin. Radiat. Oncol.* 10:61–70.

Chiara, S., P. Conte, P. Franzone, M. Orsatti, M. Bruzzone, A. Rubagotti, F. Odicino, S. Ruziati, F. Carnino, R. Rosso et al. (1994). "High-risk early-stage ovarian cancer. Randomized clinical trial comparing cisplatin plus cyclophosphamide versus whole abdominal radiotherapy." *Am. J. Clin. Oncol.* 17:72–76.

Chui, C. S., L. Hong, and M. Hunt. (2002). "A simplified intensity modulated radiation therapy technique for the breast." *Med. Phys.* 29:522–529.

Chui, C. S., L. Hong, J. Yang, and S. Spirou. (2001). "Splitting of large intensity-modulated fields." 43rd Annual AAPM Meeting Abstract TU-C-BRB-05, Salt Lake City, Utah, July 22–26, 2001. *Med. Phys.* 28(6):1252.

Dembo, A. J. (1992). "Epithelial ovarian cancer: the role of radiotherapy." *Int. J. Radiat. Oncol. Biol. Phys.* 22:835–845.

Dent, S. F., D. Klaassen, J. L. Pater, B. Zee, and M. Whitehead. (2000). "Second primary malignancies following the treatment of early stage ovarian cancer: update of a study by the National Cancer Institute of Canada—Clinical Trials Group (NCIC-CTG)." *Ann. Oncol.* 11:65–68.

Fyles, A. W., A. J. Dembo, R. S. Bush, W. Levin, L. A. Manchul, J. F. Pringle, G. A Rawlings, J. F. Sturgeon, G. M. Thomas, and J. Simm. (1992). "Analysis of complications in patients treated with abdomino-pelvic radiation therapy for ovarian carcinoma." *Int. J. Radiat. Oncol. Biol. Phys.* 22:847–851.

Hong, L., K. Alektiar, M. Hunt et al. (2002). "Intensity modulated radiotherapy for soft tissue sarcoma of thigh." *Int. J. Radiat. Oncol. Biol. Phys.* 54:139–140.

Hunt, M. A., M. J. Zelefsky, S. Wolden, C. S. Chui, T. LoSasso, K. Rosenzweig, L. Chong, S. V. Spirou, L. Fromme, M. Lumley, H. A. Amols, C. C. Ling, S. A. Leibel. (2001). "Treatment planning and delivery of intensity-modulated radiation therapy for primary nasopharynx cancer." *Int. J. Radiat. Oncol. Biol. Phys.* 49:623–632.

Kojs, Z., B. Glinski, M. Reinfuss, J. Pudelek, K. Urbanski, T. Kowalska, and J. Kulpa. (2001). [Results of a randomized prospective trial comparing postoperative abdominopelvic radiotherapy with postoperative chemotherapy in early ovarian cancer]. In French. *Cancer Radiother.* 5:5–11.

Kubo, H. D., and B. C. Hill. (1996). "Respiration gated radiotherapy treatment: A technical study." *Phys. Med. Biol.* 41:83–91.

Lambert, H. E., G. J. Rustin, W. M. Gregory, and A. E. Nelstrop. (1993). "A randomized trial comparing single-agent carboplatin with carboplatin followed by radiotherapy for advanced ovarian cancer: A North Thames Ovary Group study." *J. Clin. Oncol.* 11:440–448.

LaRouere, J., C. Perez-Tamayo, B. Fraass, R. Tesser, A. S. Lichter, J. Roberts, and M. Hopkins. (1989). "Optimal coverage of peritoneal surface in whole abdominal radiation for ovarian neoplasms." *Int. J. Radiat. Oncol. Biol. Phys.* 17:607–613.

Lawton, F., D. Luesley, G. Blackledge, C. Hilton, K. Kelly, T. Latief, J. Mould, D. Spooner, T. Rollason, T. Wade-Evans et al. (1990). "A randomized trial comparing whole abdominal radiotherapy with chemotherapy following cisplatinum cytoreduction in epithelial ovarian cancer. West Midlands Ovarian Cancer Group Trial II." *Clin. Oncol. (R. Coll. Radiol.)* 2:4–9.

Lindner, H., H. Willich, and A. Atzinger. (1990). "Primary adjuvant whole abdominal irradiation in ovarian carcinoma." *Int. J. Radiat. Oncol. Biol. Phys.* 19:1203–1206.

Low, D. A., and S. Mutic. (1997). "Abutment region dosimetry for sequential arc IMRT delivery." *Phys. Med. Biol.* 42:1465–1470.

McGuire, W. P., W. J. Hoskins, M. F. Brady, P. R. Kucera, E. E. Partridge, K. Y. Look, D. L. Clarke-Pearson, and M. Davidson. (1996). "Cyclophosphamide and cisplatin compared with paclitaxel and cisplatin in patients with stage III and stage IV ovarian cancer." *N. Engl. J. Med.* 334:1–6.

Nutting, C., D. P. Dearnaley, and S. Webb. (2000). "Intensity modulated radiation therapy: A clinical review." *Br. J. Radiol.* 73:459–469.

Pickel, H., M. Lahousen, E. Petru, H. Stettner, A. Hackl, K. Kapp, and R. Winter. (1999). "Consolidation radiotherapy after carboplatin-based chemotherapy in radically operated advanced ovarian cancer." *Gynecol. Oncol.* 72:215–219.

Redman. C. W., J. Mould, J. Warwick, T. Rollason, D. M. Luesley, J. Budden, F. G. Lawton, G. R. Blackledge, and K. K. Chan. (1993). "The West Midlands epithelial ovarian cancer adjuvant therapy trial." *Clin. Oncol.* 5:1–5.

Reisinger, S. A., R. Asbury, S. Y. Liao, and H. D. Homesley. (1996). "A phase I study of weekly cisplatin and whole abdominal radiation for the treatment of stage III and IV endometrial carcinoma: A Gynecologic Oncology Group pilot study." *Gynecol. Oncol.* 63:299–303.

Roeske, J.C., A. Lujan, J. Rotmensch, S. E. Waggoner, D. Yamada, and A. J. Mundt. (2000). "Intensity-modulated whole pelvic radiation therapy in patients with gynecologic malignancies." *Int. J. Radiat. Oncol. Biol. Phys.* 48:1613–1621.

Schneider, D. P., H. P. Marti, C. Von Briel, F. J. Frey, and R. H. Greiner. (1999). "Long-term evolution of renal function in patients with ovarian cancer after whole abdominal irradiation with or without preceding cisplatin." *Ann. Oncol.* 10:677–683.

Sell, A., K. Bertelsen, J. E. Andersen, I. Stroyer, and J. Panduro. (1990). "Randomized study of whole-abdomen irradiation versus pelvic irradiation plus cyclophosphamide in treatment of early ovarian cancer." *Gynecol. Oncol.* 37:367–373.

Spirou, S. V., and C. S. Chui. (1998). "A gradient inverse planning algorithm with dose-volume constraints." *Med. Phys.* 25:321–333.

Spirou, S. V., N. Fournier-Bidoz, J. Yang, C. S. Chui, and C. C. Ling. (2001). "Smoothing intensity-modulated beam profiles to improve the efficiency of delivery." *Med. Phys.* 28:2105–2112.

Thomas, G. M., and A. J. Dembo. (1993). "Integrating radiation therapy into the management of ovarian cancer." *Cancer* 71:1710–1718.

Vergote, I. B., L. N. Vergote-De Vos, V. M. Abeler, M. Aas, M. W. Lindegaard, K. E. Kjorstad, and C. G. Trope. (1992). "Randomized trial comparing cisplatin with radioactive phosphorus or whole-abdomen irradiation as adjuvant treatment of ovarian cancer." *Cancer* 69:741–749.

Whelan, T. J., A. J. Dembo, R. S. Bush, J. F. Sturgeon, S. Fine, J. F. Pringle, C. A. Rawlings, G. M. Thomas, and J. Simm. (1992). "Complications of whole abdominal and pelvic radiotherapy following chemotherapy for advanced ovarian cancer." *Int. J. Radiat. Oncol. Biol. Phys.* 22:853–858.

Wong, R., M. Milosevic, J. Sturgeon, M. Pintilie, A. Fyles, W. Levin, B. Rosen, D. Depetrillo, A. Oza, L. Manchul, J. Murphy, and W. Chapman. (1999). "Treatment of early epithelial ovarian cancer with chemotherapy and abdominopelvic radiotherapy: Results of a prospective treatment protocol." *Int. J. Radiat. Oncol. Biol. Phys.* 45:657–665.

Wu, Q., M. Arnfield, S. Tong, Y. Wu, and R. Mohan. (2000). "Dynamic splitting of large intensity-modulated fields." *Phys. Med. Biol.* 45:1731–1740.

15

NTCP And TCP For Treatment Planning

Andrew Jackson
Ellen Yorke

Motivation

The need to balance the probability of local control against that of severe normal tissue injury underlies all of treatment planning. With the advent of intensity-modulated radiation therapy (IMRT), the use of automated optimization software to create treatment plans has become routine. The score functions used in these optimization algorithms constitute models of the influence of dose distributions on outcome. In this context it is vital to understand the way in which models of dose/volume effects are reflected in score functions, and to improve our understanding of the increase in normal tissue complication probability (NTCP) and tumor control probability (TCP) with dose and volume.

In this chapter we will give brief reviews of models of volume effects in NTCP and TCP, and discuss the relationship of these models with score functions currently in use for the automated optimization of IMRT.

NTCP For Treatment Planning

In this section we will discuss endpoints describing clinical complications, data on normal tissue tolerances, existing models of NTCP (models of TCP will be discusses in a subsequent section), implications of volume effects for plan optimization, and score functions. We will not discuss time-dose-fractionation effects. These may be incorporated by using, for example, the linear quadratic equivalent doses in two Gray fractions (LQED2) (Fowler 1989). A brief introduction to this topic is given in the section **TCP For Treatment Planning**.

Endpoints For Clinical Complications

Complications occur on some characteristic time-scale after treatment begins. The most common method of describing this is the actuarial probability that the patient will develop complication after some time t [known as Kaplan-Meier curves (Kaplan and Meier 1958)]. Eventually, new complications cease to occur and the complication probability tends to some constant value at large times. Complications are divided into *acute* and *late,* according to time of onset (Hall 1994). Acute complications are usually defined as occurring up to 4 months after treatment and may begin during treatment. In contrast, late complications may occur many years after treatment. This division is thought to arise from the difference in time scale of the cell cycle for different tissues, with late-reacting tissues having an increased sensitivity to dose fractionation (Thames and Hendry 1987). It has practical consequences, since severe acute reactions can often be prevented by carefully monitoring symptoms during treatment, whereas, in contrast, severe late reactions may not be predictable from reactions occurring during treatment, and may limit the dose and fractionation schemes available for safe treatment.

Symptoms of complications are graded by their clinical significance (Cox, Stetz, and Pajak 1995; Pavy et al. 1995) with the various schemes usually corresponding to the divisions given in table 15–1.

Recently, other, more objective data (involving measurements of significant physiological quantities) have been used in analyzing the effect of radiation treatment on normal tissue function. For example, local ventilation and perfusion of the lungs of patients have been measured before and after treatments involving the lung (Theuws et al. 2000; Woel et al. 2002), and saliva flow has been measured before and after treatment for patients undergoing treatment for head and neck cancers (Eisbruch et al. 1999; Chao et al. 2001; Roesink et al. 2001). While offering quantitative unsubjective data for analysis, the relation of the measurements to the clinical complications may not be straightforward (Field et al. 2001).

The State Of Clinical Complication Data

A basic quantity that is widely used to provide guidance in limiting treatment is the whole volume tolerance dose, defined as the dose which, when delivered to the entire organ, gives rise to a specified complication rate by a stated time. The partial volume tolerance dose is defined in the same way for a dose given to a specified fraction of the organ. Historically, because of lack of statistics, tolerance doses (especially partial volume tolerances before the advent of routine planning with CT scans) were often guessed from clinical practice. For example, Emami et al. (1991) estimated whole volume tolerance doses for the liver based on 13 cases of radiation induced liver disease (RILD, often called *radiation hepatitis*) from a series of 40 patients treated to whole volume doses between 13 to 50 Gy (Ingold et al. 1965). The partial volume tolerances given by Emami et al. were guesses, based on data from 28 patients treated to 20 Gy whole volume (2 Gy/frac) plus 18 to 20 Gy to the left lobe (2 Gy/frac) (Poussin-Rosillo, Nisce, and D'Anglo 1976), in which

Table 15–1. Clinical Significance of Grade of Complication

Grade	Clinical Significance
0 (none)	No complication
1 ("minor")	Not requiring treatment
2 ("moderate")	Requiring Treatment
3 ("severe")	Requiring Hospitalization
4	Life Threatening
5	Death

no cases of RILD were observed. In consequence, liver treatments prior to 1990 were often very conservative, with prescription doses for tumors in the liver were restricted to ~30 Gy (~the whole volume tolerance). The advent of dose escalation trials using conformal therapy has changed this. After a series of these trials at University of Michigan, it has been shown that doses over 90 Gy can be well tolerated, provided the irradiated volume of liver is small enough—an increase in possible prescription dose by a factor of 3!

Data from conformal therapy is quantitative, but still suffers from poor statistics and distortions due to organ motion. In the case of the liver, initial data from University of Michigan contained nine cases of RILD from 93 patients (Lawrence et al. 1992; Jackson et al. 1995a). Pooling of data between institutions has begun, but has problems. The advantages and disadvantages are shown in Kwa et al. (1998), who looked at 72 cases of ≥grade 2 radiation pneumonitis in 540 lung, breast, or lymphoma patients treated at five different institutions. While the large number of cases enabled them to deduce a strong dose response using the mean lung dose, this analysis required the subtraction of a base pneumonitis rate of 10% from the data of certain institutions. Such a discrepancy might arise from differences in the health of the patient populations, either by institution or by disease site, or in the diagnosis of the endpoint.

A good summary of the normal tissue data and analysis accumulated during the era of conformal therapy is provided in a special issue of *Seminars in Radiation Oncology* (vol. 11, no. 3, July 2001). This issue contains state of the art data sets and modeling for: rectal bleeding (Jackson 2001); radiation induced heart disease (Gagliardi, Lax, and Rutqvist 2001); xerostomia (parotid gland) (Eisbruch et al. 2001); radiation induced liver disease (Dawson, Ten Haken, and Lawrence 2001); radiation pneumonitis (Seppenwoolde and Lebesque 2000); and radiation induced brain injury (Levegrun, Ton, and Debus 2001). It also includes introductory articles outlining the underlying radiobiology of normal tissue complications (Travis 2001); the various models describing the volume effects involved (Yorke 2001); and last but by no means least, the limitations of the models, and the pitfalls involved in analyzing clinical data (Schultheiss 2001).

In modeling the dependence of complication probability on dose and volume, the following limitations of clinical data should be understood.

1. Good clinical practice ensures that complications are rare, so an individual clinical data set usually contains very few events (poor statistics).
2. Clinical endpoints contain large subjective components, particularly for the lower grades.
3. Pooling data between institutions is difficult since endpoints, treatment data, and treatment techniques may be inconsistent.
4. Fixed treatment techniques introduce strong correlations among treatment variables.
5. Many confounding effects present in patient populations can disguise the true causes of complications.

Models Of NTCP

At present, most models of volume effects aim to predict binary endpoints at fixed times, e.g., the probability of grade 2 or higher rectal bleeding at 5 years. The time point should be chosen so that all complications have occurred. Note that in this approach, most of the information about the distribution of patients amongst the grades and onset times is ignored.

The first step in describing the dose and volume dependence of complication probabilities is to characterize the probability of complication after uniform irradiation of whole organs to various doses, $NTCP(D,1)$, where the "1" signifies whole volume irradiation. Clinical and animal data show that complication probabilities for whole organ irradiation can be described by sigmoidal dose responses, characterized by positions, and slopes. The position is typically the dose at which the complication probability is 50%. Defining the whole volume dose resulting in q% complication after t years to be the whole volume tolerance dose $TD_t(q,1)$, this position is $TD_t(50,1)$. The dimensionless slope of the response function at this point, γ_{50} is given by:

$$\gamma_{50} = \left(D \frac{dNTCP(D,1)}{dD} \right)\Bigg|_{D=TD_t(50,1)}$$

From now on we will assume that endpoints are defined at appropriate times, and the subscript t will be dropped from our notation.

Models of the dose-volume dependence of NTCP attempt to describe the way tolerance doses increase as the volume irradiated decreases. Dose inhomogeneities in normal tissues are large, and related to the beam and anatomic geometry of individual patients. To understand the properties of models of NTCP, it is useful to simplify this complexity by considering partial organ irradiation. As shown in figure 15–1, partial organ irradiation results in a fractional volume v receiving dose d, and the rest, $(1-v)$, getting nothing.

The partial organ irradiation dose-volume response surface $NTCP(d,v)$ measures the NTCP as a function of d and v. Lines of iso-NTCP in the dose/partial volume plane are tolerance dose curves $TD(q,v)$.

From a few simple principles, we can deduce some general properties of $NTCP(d,v)$.

1. At zero dose, $NTCP(0,v) = 0$
2. At zero volume, $NTCP(d,0) = 0$
3. For whole organ irradiation, $NTCP(d,1) \to 1$ as $d \to$ large

Models of NTCP are constrained to interpolate between these limits, which restrict the form of the response surface to a kind of *waterfall function* as shown in figure 15–2 for the case of the Lyman-Kutcher-Burman (LKB) model (Lyman 1985).

In creating models of NTCP, we may use a phenomenological or biological approach. Phenomenological models attempt to describe the form of the response surface in as flexible a manner as possible with the smallest possible number of parameters, without regard to any underlying mechanisms of injury. The biological approach seeks to use our understanding of the functional architecture of the organ to predict the general form of the response function.

Phenomenological Models

The premier example of a phenomenological model is the LKB model (Lyman 1985; Kutcher and Burman 1989; Kutcher et al. 1991). This key assumption of this model is that tolerance doses increase inversely as a power (n) of the partial volume (Lyman 1985). For example, for q% complication rate, the partial volume tolerance dose, $[TD(q,v)]$, can be related to the whole volume tolerance dose by:

$$TD(q,v) = TD(q,1)(v)^{-n} \qquad (1)$$

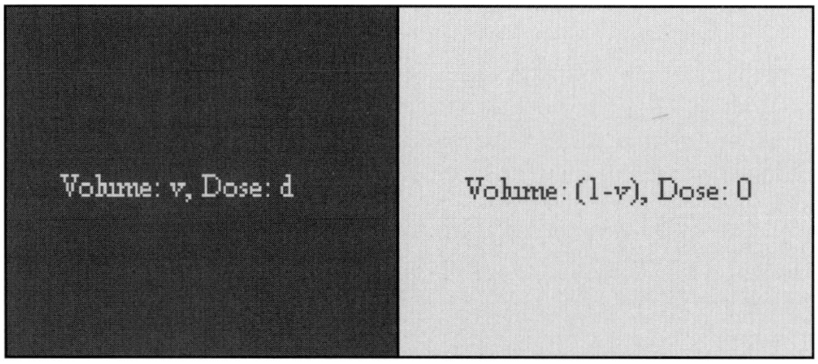

FIGURE 15–1. Partial volume irradiation. Fractional volume v receives dose d (red at left), and the remaining volume (1-v) receives zero dose (yellow at right). See Color Plate 40.

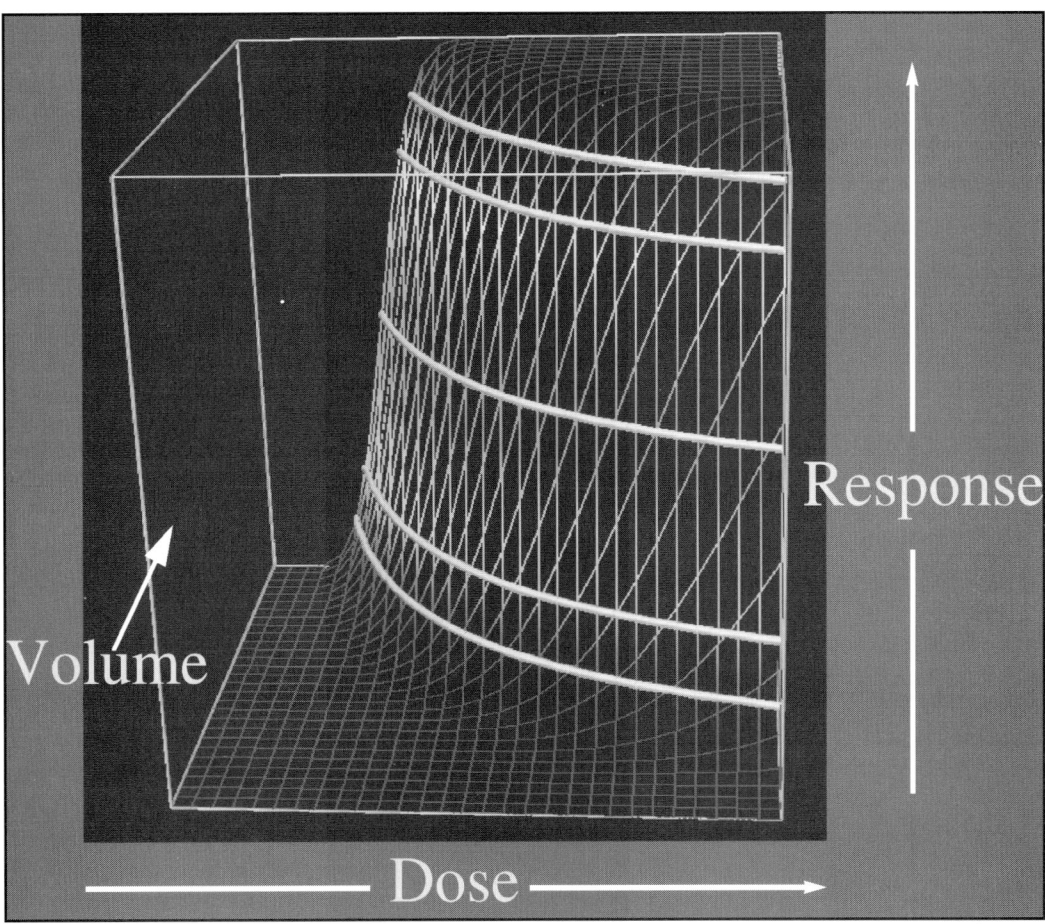

FIGURE 15–2. LKB model dose-volume response surface. (Reprinted from Ten Haken, R.K., and M. L. Kessler, "3-D RTP Plan Evaluation" in *A Practical Guide to 3-D Planning and Conformal Radiation Therapy,* J. A. Purdy and G. Starkschall (eds.), (Advanced Medical Publishing, Madison WI, 1999), fig. 16, p. 100, by permission of Advanced Medical Publishing.

Once the whole volume dose response is specified, Equation (1) fixes the entire dose-volume response surface. Repeated application of equation (1) for two different partial volumes $v1$ and $v2$ results in:

$$TD(q, v1)(v1)^n = TD(q, v2)(v2)^n. \qquad (2)$$

If $n = 1$, Equation (2) shows that complication rates are equal when mean doses are equal. Thus, the LKB model with $n = 1$ is a mean dose model.

The behavior of iso-NTCP lines as a function of n is shown in figure 15–3, for the cases $n = 1$ and $n = 0.12$, assuming (hypothetically) $TD(q,1) = 30$ Gy. The increase in tolerance dose resulting from decreasing the volume irradiated is much larger for $n = 1$ than for $n = 0.12$. For $n = 0.12$, the tolerance dose hardly changes until <20% of the organ is irradiated.

To handle inhomogeneous irradiation, the LKB model has a reduction scheme, whereby a partial volume irradiation is found that would yield the same complication rate as the inhomogeneous one. To do this, an extra assumption is required (Kutcher and Burman 1989; Kutcher et al.

1991). If Equation (1) can be taken to describe the effect of irradiating individual subvolumes within inhomogeneous organ irradiation, we can deduce the effective whole volume dose (the whole volume dose that would yield the same complication rate as the inhomogeneous dose distribution), d_{eff}.

$$d_{eff} = \left(\sum_i v_i \left(D_i \right)^{1/n} \right)^n$$

[These days, this variable is also known as the Equivalent Uniform Dose or EUD (Niemierko 1997; Wu et al. 2002)]. Another widely used variable is the effective volume, v_{eff}, defined as the partial volume treated at some dose D_{ref} (often the prescription dose), resulting in the same complication rate as the inhomogeneous dose distribution.

$$v_{eff} = \sum_i v_i \left(\frac{D_i}{D_{ref}} \right)^{1/n}$$

Tissue Architecture Models

There are two major models based on tissue architecture (Withers et al. 1988): the Serial model (Schultheiss, Orton, and Peck 1983; Niemierko and Goitein 1991), and the Parallel model (Jackson, Kutcher, and Yorke 1993; Yorke et al. 1993; Jackson et al. 1995b; Niemierko and Goitein 1993a). These models assume that complications arise from impairment of organ function, carried out by Functional Sub-Units (FSUs). These FSUs are assumed to be damaged by radiation

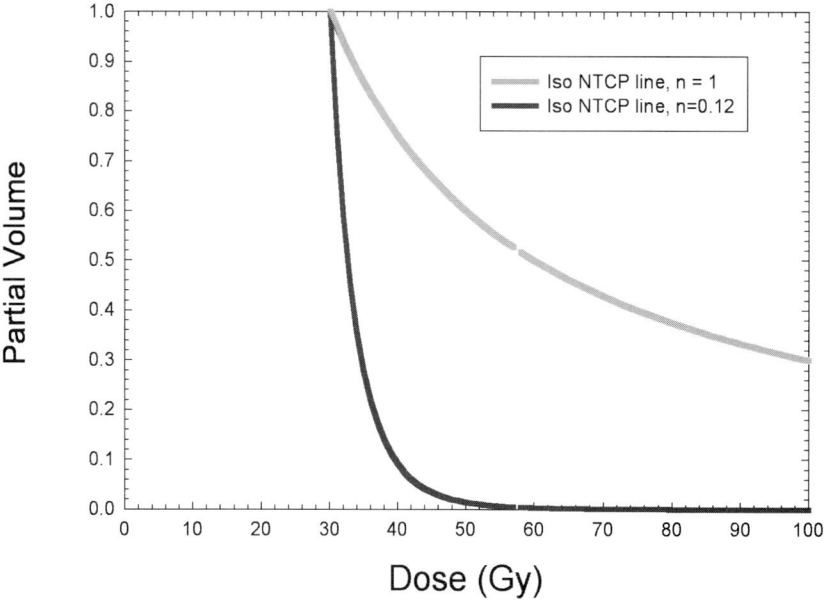

FIGURE 15–3. LKB model iso-NTCP lines for two different values of n. Both cases start from the same whole volume tolerance dose (30 Gy). The case n = 1 is shown by the dashed gray line, and the case n = 0.12 is shown by the solid black line.

independently, and small enough that they receive homogeneous irradiation. The functional organization of these FSUs (reflected in the tissue architecture) determines the dose-volume response to irradiation.

I will not discuss the relative seriality model (Kallman, Agren, and Brahme 1992) in any depth here. This model can be seen as an attempt to provide a phenomenological interpolation between serial and parallel behavior. While the serial model emerges as one limiting case, the truly parallel feature of a threshold volume is not encompassed by this model.

The Serial Model

In the Serial model, the FSUs are assumed to be organized in a linear chain; damage to any single FSU therefore impairs the function of the whole chain. Candidate serial organs include the spinal cord and optic nerve.

Given these assumptions, the probability of complication can be written as one minus the probability that no FSUs are damaged:

$$NTCP = 1 - \prod_i \left(1 - P(d_i)\right) \tag{3}$$

where d_i is the dose received by the ith FSU, and $P(d)$ is the probability that an FSU will be damaged by dose d. By rewriting Equation (3) in terms of the whole volume dose response and considering the responses of homogeneously irradiated subvolumes v_i we find:

$$NTCP = 1 - \prod_i \left(1 - \left[NTCP(d_i,1)\right]^{1/v_i}\right) \tag{4}$$

Note that once the whole volume dose response has been specified, the entire dose-volume response is determined (there is no free parameter like n in the model). At low complication probabilities, it has been shown that this model behaves like an LKB model with a (specific) low n value (Niemierko and Goitein 1991).

Due to the product over the responses of subvolumes, partial volume tolerance doses increase very little as involved volumes decrease. Therefore, the maximum dose can be used as a rough guide to the complication probability.

The Parallel Model

In the Parallel model, the FSUs are assumed to carry out the organ function independently (in parallel). The organ is assumed to have a spare capacity, or functional reserve (whose size varies from patient to patient). If the number of FSUs damaged by irradiation is greater than the patient's functional reserve, a complication will occur. In consequence, as long as irradiated volumes are small, complications do not occur. The complication probability is an increasing function of the fraction of independent functional subunits damaged (or fractional damage, f_{dam}). The condition of constant NTCP is therefore equivalent to a condition of constant f_{dam}. Candidate parallel organs are lung, liver, parotid, and kidney.

The probability of subunit damage is described by a logistic function of dose, e.g.:

$$P(d) = \frac{1}{\left[1 + \left(d_{1/2}/d\right)^k\right]} \tag{5}$$

Here $P(d)$ is the probability that an FSU will be damaged by dose d, $d_{1/2}$ is the dose which damages 50% of the FSUs, and k is a parameter describing the slope of this FSU dose response. Figure 15–4 shows the effect of varying k on the slope of the FSU dose response; for very large k, the FSU dose response becomes a step function.

The fractional damage for inhomogeneous irradiation can then be calculated as a sum over the damage done in each homogeneously irradiated subvolume:

$$f_{dam} = \sum_i P(d_i) v_i \qquad (6)$$

In the limit of very large k, the sum in Equation (6) shows that a constraint on f_{dam} (and hence NTCP) is equivalent to a dose-volume constraint. This will have important consequences below, when we consider the properties of optimization score functions.

In figure 15–5 we show the tolerance doses for a q% complication rate as a function k in the dose and partial volume plane for a parallel model. In this hypothetical example we have assumed that it requires 50% of the FSUs to be damaged to achieve a q% complication rate, and that $TD(q,1) = 30$ Gy [this requires $P(d) = 0.5$ when $d = 30$ Gy].

A couple of features of figure 15–5 deserve comment. First, there is a threshold volume: unless the volume irradiated is greater than 50%, the q% complication rate does not occur. As dose→large, the iso-NTCP curve approaches the asymptote $v = 0.5$. Second, the parameter k controls the curvature of the iso-NTCP lines in a similar fashion to n in the LKB model, (with the caveat that large k is analogous to small n and vice versa). For large k there is very little change

Parallel Model, Local Response Function

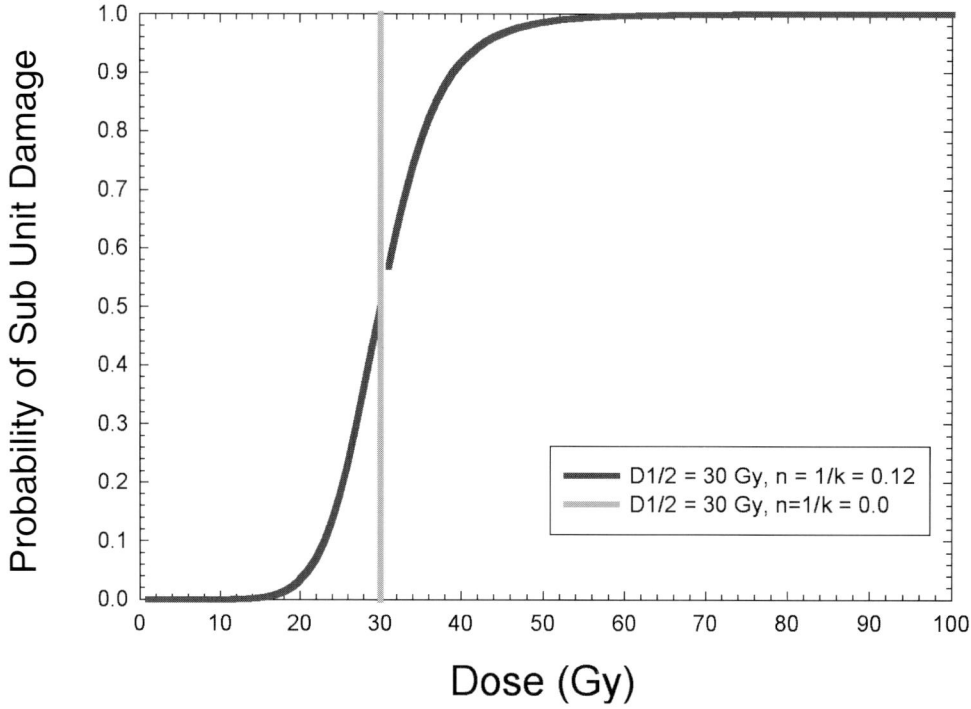

FIGURE 15–4. Examples of the dependence of the local response functions in the parallel model on the parameter k. Both examples have $d_{1/2} = 30$ Gy. The solid black line shows the case n = 1/k = 0.12; the vertical gray line shows the limit of infinite k (n = 1/ k= 0), a step function.

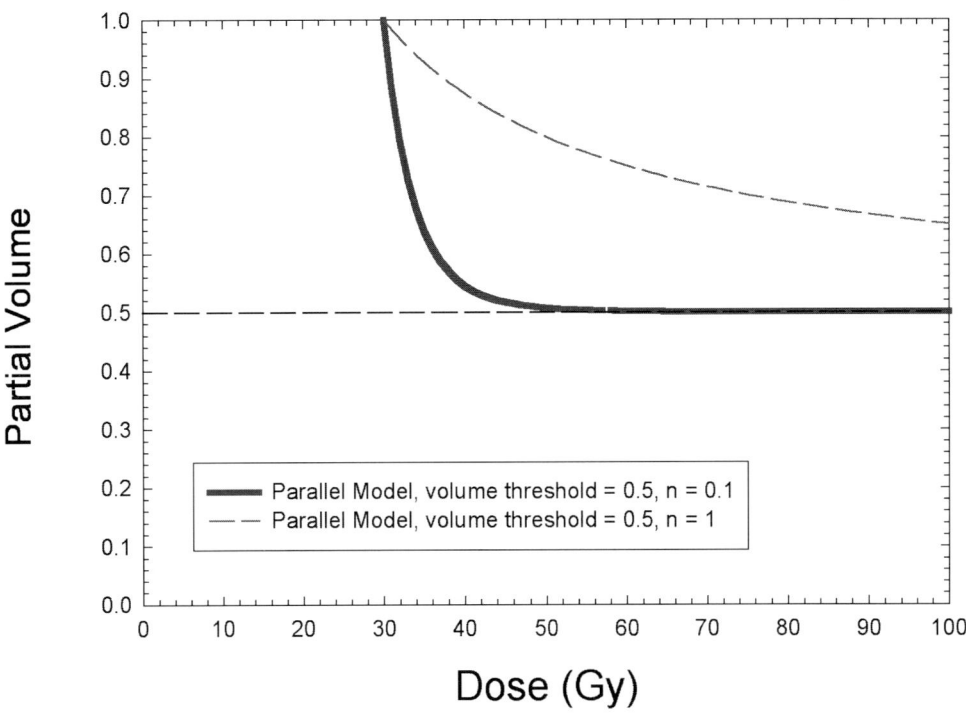

Parallel Model Iso-NTCP lines, whole volume tolerance: 30 Gy

Legend:
— Parallel Model, volume threshold = 0.5, n = 0.1
– – Parallel Model, volume threshold = 0.5, n = 1

FIGURE 15–5. Parallel model iso-NTCP lines for two different values of k. Both cases have $d_{1/2} = 30$ Gy and start from the same whole volume tolerance dose (30 Gy). The case k = 1/n = 1 is shown by the dashed gray line, and the case k = 1/n = 10 is shown by the black line. The horizontal dashed line shows the threshold $v = 0.5$.

in tolerance dose with decreasing partial volume until the threshold volume is approached. In contrast, for $k = 1$, a rapid increase in tolerance dose results as partial volume decreases from 1.

In figure 15–6 we compare the LKB and Parallel iso-NTCP curves.

The relationship between the Parallel and LKB model parameters k and n can be made rigorous. Using Equations (5) and (6), and identifying iso-NTCP curves with iso-f_{dam} curves, it can be shown that the iso-NTCP curves in the parallel model obey the relation:

$$TD(q,v) = TD(q,1)\left[(v - f_q)/(1 - f_q)\right]^{-1/k} \tag{7}$$

where f_q is the fraction of FSUs that must be damaged for a q% complication rate. Equations (1) and (7) show that by rescaling the volume axis by $v \rightarrow ((v - f_q)/(1 - f_q))$, the LKB and Parallel model iso-NTCP curves become coincident provided $n = 1/k$. This can be confirmed by inspection of figure 15–6. Further, if there is no threshold volume in the parallel model, an LKB like model results with the identification $n = 1/k$.

Limitations Of The Models

Limitations of the models are too numerous to list here, but many of them are outlined by Travis (2001), and Schultheiss (2001).

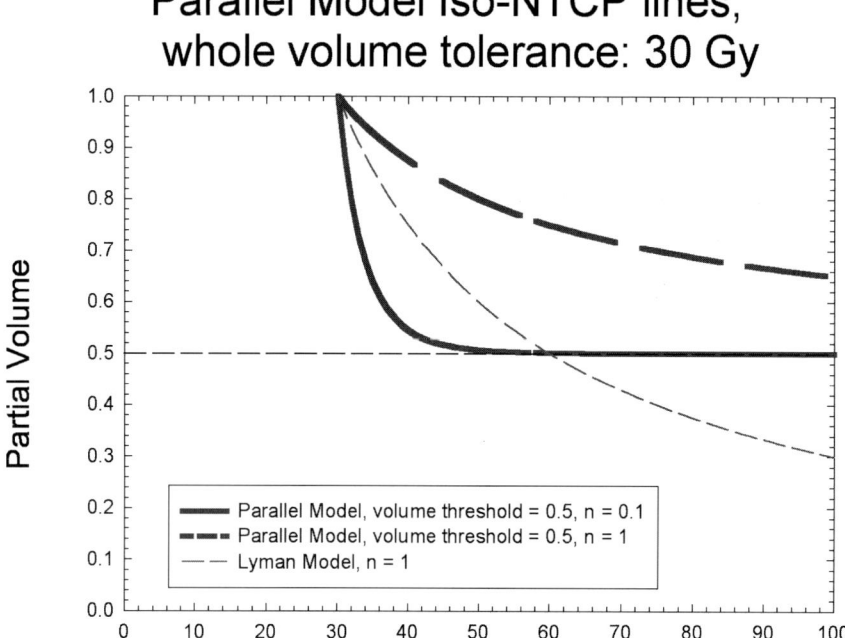

FIGURE 15-6. Comparison between Lyman and Parallel model iso-NTCP lines. All cases start from the whole volume tolerance dose 30 Gy. The thick black line shows the parallel model case $k = 1/n = 10$, and the thick long dashed line shows the parallel model case $k = 1/n = 1$ (both previously shown in figure 15–5). The thin gray dashed line shows the LKB model case $n = 1$ (previously shown in figure 15–3). The horizontal dashed line shows the threshold $v = 0.5$.

Volume Effects And Plan Optimization

Very few optimization programs use explicit models of dose-volume effects in NTCP and TCP in their score functions (as far as I am aware, no commercial systems do this). There are several reasons for this. First, *biological* optimization programs are slower than their dose-based cousins, because the non-linear functions used in evaluating TCP and NTCP take a long time to evaluate. Second, data on complication rates are inherently sparse, so many physicians and physicists are rightly skeptical of both the appropriateness of the models and the accuracy with which model parameters could be deduced. On the other hand, the dose-based score functions are themselves biological models, since they are being used in attempts to optimize the biological outcome of radiotherapy. Additionally, we do have some gross, qualitative knowledge of the volume effects present in NTCPs. For example, recent clinical data have shown that volume effects in liver and lung are large (Dawson, Ten Haken, and Lawrence 2001; Hernando et al. 2001; Yorke et al. 2001), whereas animal data indicates that those in the spinal cord are small (Schultheiss et al. 1994).

In these circumstances, we should strive to understand the volume effects implied by the score functions actually in use in planning IMRT. Our qualitative understanding of the volume effects in NTCPs can guide us in understanding how to construct clinically relevant score functions.

The major qualitative question to be answered when planning a treatment involving a critical normal tissue is simply: is it better to give a lot of dose to a small portion of the organ, or to give a little dose to a lot of the organ? In short: a lot to a little or little to a lot? (Jackson, Wang, and Mohan 1994; Soderstrom and Brahme 1995, 1996; Mohan and Ling 1995; Zaider and Amols 1998) The first strategy involves delivering most of the dose with a few beam directions, the second involves spreading the dose as evenly as possible over many beam directions. We will now examine score functions commonly in use, and ask how they resolve this question. Our qualitative understanding of volume effects in NTCPs will help us decide when these score functions are appropriate.

Limitations Of Dose-Based Quadratic Score Functions

One of the most commonly used score functions is the dose-based quadratic. This score function consists of the sum of squares of differences of doses from desired doses (from now on called *quadratics*). In the early days of IMRT, pure quadratics were widely used to design treatment plans. They are capable of producing the beautiful c-shaped dose distributions wrapping around a critical normal structure commonly found in many early test cases (Brahme 1988; Bortfeld et al. 1990). However, it was shown that these dose objectives implicitly assume that it is better to give a little to a lot than a lot to a little (Jackson, Wang, and Mohan 1994; Mohan et al. 1994; Wang et al. 1995).

To illustrate this point we will use a simplified model of radiotherapy (Jackson, Wang, and Mohan 1994; Wu and Mohan 2002). Consider an idealized target surrounded by a normal tissue, as shown in figure 15–7. We have two parallel opposed beam pairs at our disposal, denoted pair 1 and pair 2.

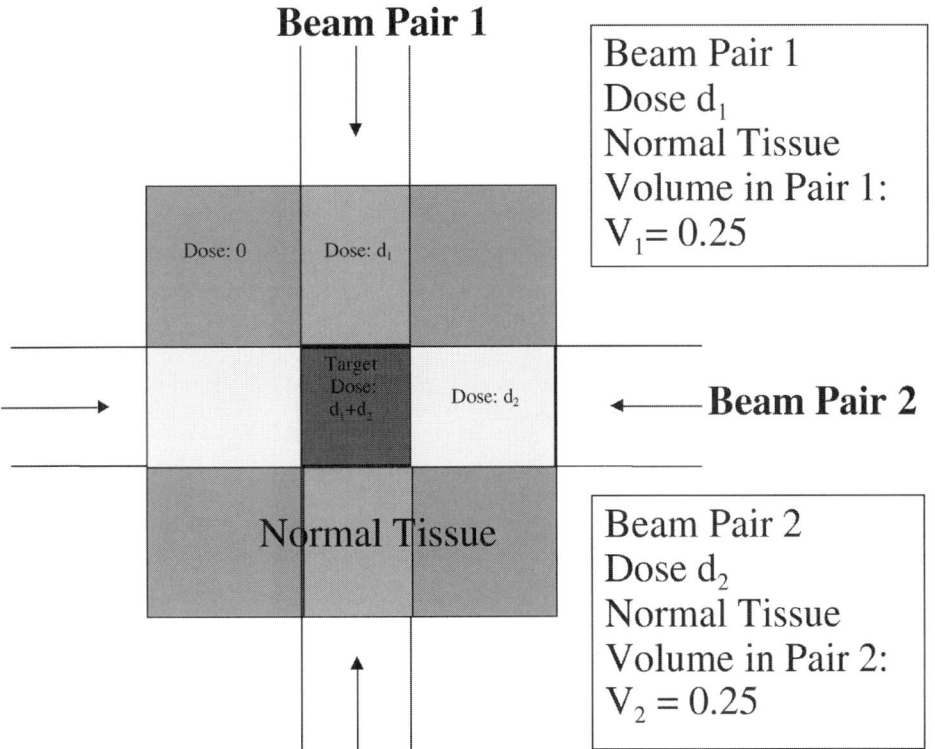

FIGURE 15–7. Model target, normal tissue, and beam geometry. The square target is surrounded by a square normal tissue. The vertical beams, beam pair 1, deliver dose D_1 (orange), and the horizontal beams, beam pair 2, deliver dose D_2 (yellow). The target receives dose $(D_1 + D_2)$ (red). The un-irradiated part of the normal tissue is shown in green. See Color Plate 41.

For simplicity, we will neglect beam divergence, the attenuation in dose delivered at depth, and also assume that the beam profile is perfectly square. Given that the prescription dose is 80 Gy, and ignoring other possible normal tissue effects such as skin reactions, we wish to know the optimal doses to be delivered to the target by each pair. The answer to this question depends on the dose-volume response of the normal tissue. Assuming the tissue is parallel (Lyman models with large n give qualitatively similar results), and that f_{dam} and the FSU response function are given by Equation (5) with $d_{1/2} = 20$ Gy and $k = 2.8$ [values used at MSKCC for the lung (Yorke et al. 2001)], a contour plot of f_{dam} as a function of the dose delivered to the target by each beam pair is shown in figure 15–8.

A cross section through this plot along the dotted line (corresponding to the constraint $D_1 + D_2 = 80$ Gy), shown in figure 15–9, demonstrates that when $D_1 = D_2 = 40$ Gy, f_{dam} is a maximum, and when either $D_1 \approx 80$ Gy, and $D_2 \approx 0$ Gy, or $D_1 \approx 0$ Gy and $D_2 \approx 80$ Gy, f_{dam} is a minimum. In other words, for this choice of Parallel model and prescription dose, the optimal doses delivered

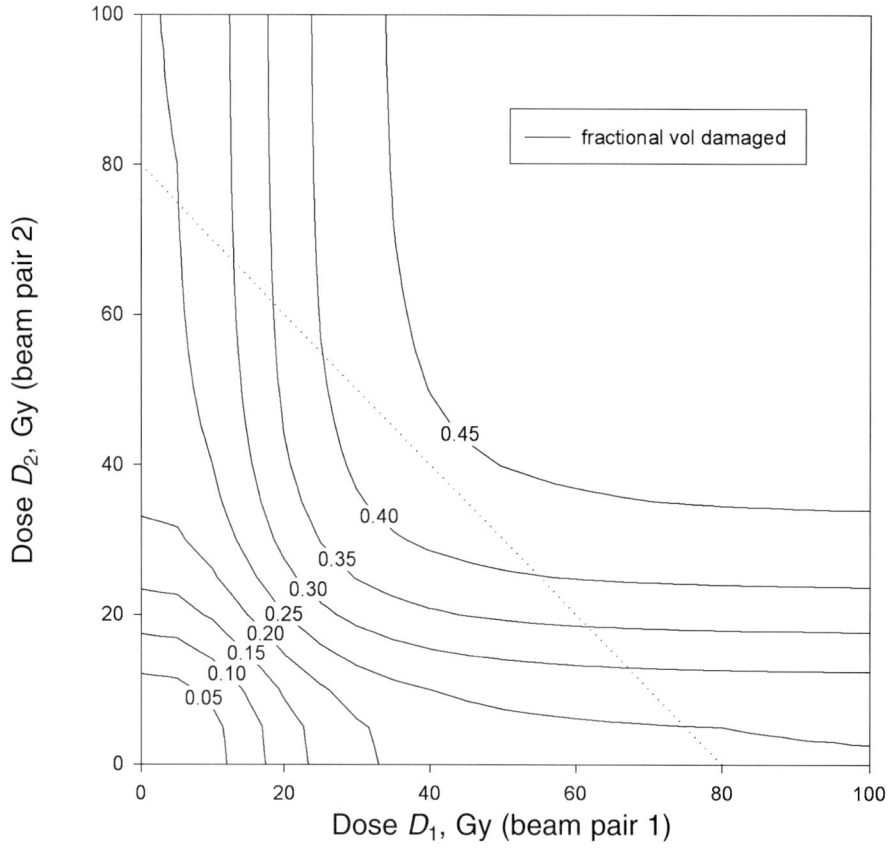

volume damaged from combined beam pairs

FIGURE 15–8. Contour plot of f_{dam} as a function of the doses from the beam pairs. Along the dotted line the target dose, $D_1 + D_2$, equals the prescription dose 80 Gy. For values of $f_{dam} > 20$ Gy ($d_{1/2}$), the contours are concave, for values <20 Gy, they are convex.

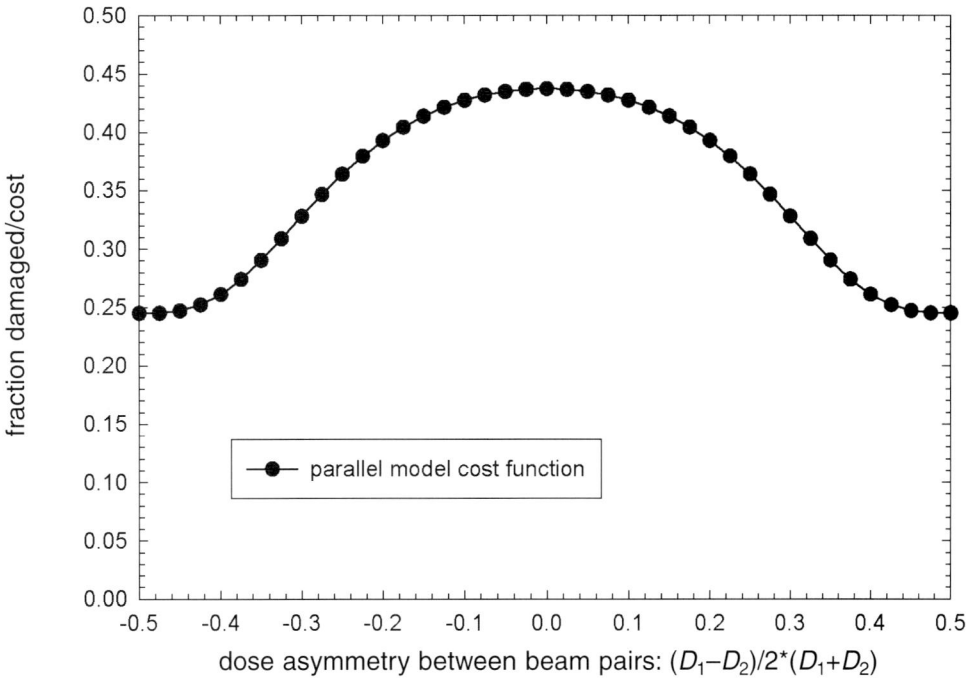

Cost function for prescription dose = 80 Gy

FIGURE 15–9. Parallel cost function vs. the asymmetry in dose from beam pairs, for $D_1 + D_2 = 80$ Gy. Note that the highest cost results when the doses from the beam pairs are equal (asymmetry = 0).

by the beam pairs occur when one pair is essentially off and all dose is delivered by the other pair (a lot to a little!).

In contrast, a quadratic dose objective with a desired normal tissue dose of 20 Gy would produce a cost function similar to that shown in figure 15–10.

For comparison, we show the section through this function at $D_1 + D_2 = 80$ Gy in figure 15–11, along with the parallel model cost function. In contrast to the parallel model, the quadratic cost function is optimized when $D_1 = D_2 = 40$ Gy (a little to a lot).

This spreading out of dose (in order to avoid high doses in normal tissues) is a feature of quadratic cost functions, and tends to distribute fluence into all available portals. (Footnote: there are ways to avoid this (Esik et al. 1997), but they require the use of unnatural weights and desired doses in the cost function, which can only be found by searching on a case-by-case basis). This behavior is appropriate when we need to limit the maximum dose to a normal tissue, as is the case when we are dealing with organs such as spinal cord or optic nerve where the volume effects are small (e.g., small n in the LKB model). There are circumstances in which the "a little to a lot" solution is appropriate for parallel tissues. For example, inspection of figure 15–8 shows that when prescription dose is such that the dose to the normal tissue is small compared with doses that damage FSUs, the "a little to a lot" solution is also preferred by parallel tissues (e.g., if the prescription dose is <40 Gy, it is possible for each beam pair to deliver <20 Gy). While it is impossible to give hard and fast rules that work in all cases, table 15–2 gives rough guidelines to the qualitative nature of the optimal solution as a function of the normal tissue

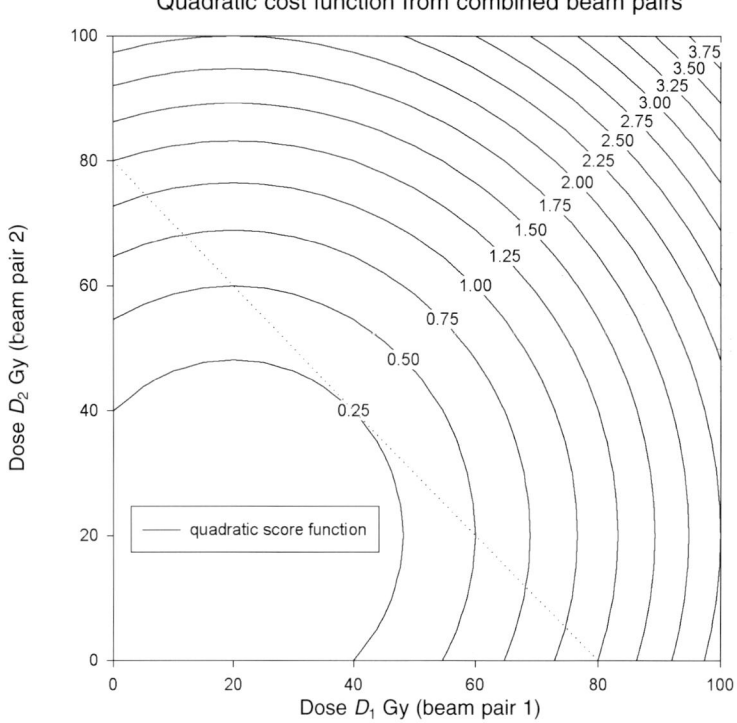

FIGURE 15–10. Contour plot of quadratic cost function as a function of dose from the beam pairs. Along the dotted line the target dose, $D_1 + D_2$, equals the prescription dose 80 Gy. Note that these contours are convex.

volume effect, the ratio of prescription dose to whole volume tolerance dose, and the relative size of tumor and normal organ. The words "large" and "small" in the table should be understood to mean large and small *enough*. At present, given our ignorance of the dose/volume dependencies of NTCPs, the boundary between these categories cannot be located for any specific complications.

While the "a little to a lot" solution works in most cases, the "a lot to a little" solution is more common than appears from the table. This is because most tissues with large volume effects have very low whole volume tolerance doses. Emami (1991) estimated TD(5%,1) to be ~17.5, 30.0, and 23.0 Gy for lung, liver, and kidney respectively.

Table 15–2. Optimal Solutions, Volume Effects, Tumor Sizes, Tolerance and Prescription Doses

	Volume Effect							
	Parallel or LKB Large n				Serial or LKB Small n			
	Prescription/Tolerance Dose				Prescription/Tolerance Dose			
	Small		Large		Small		Large	
	Tumor/Organ size		Tumor/Organ size		Tumor/Organ size		Tumor/Organ size	
Solution	Small	Large	Small	Large	Small	Large	Small	Large
A little to a lot	Y	Y	Y	N	Y	Y	Y	N
A lot to a little	N	N	N	Y	N	N	N	N

Cost functions for prescription dose = 80 Gy

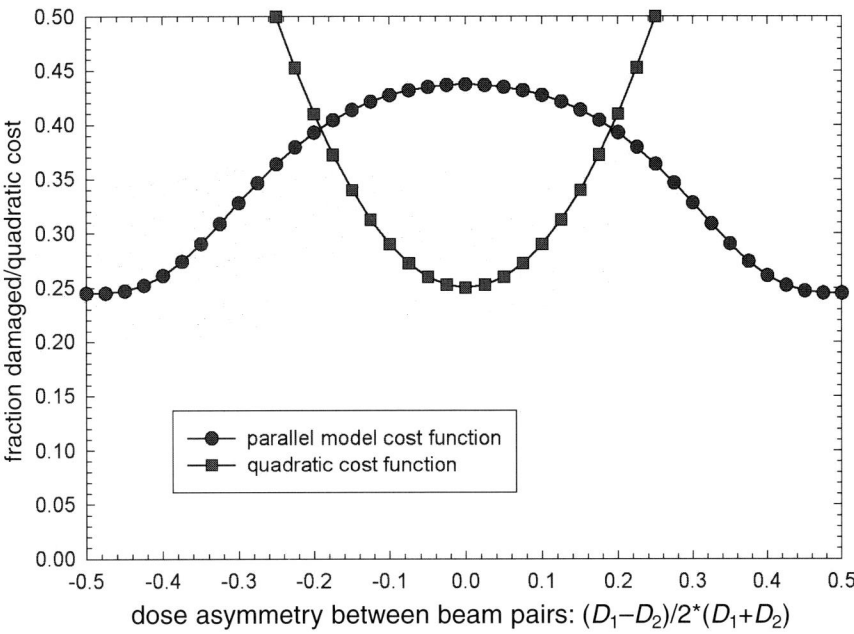

FIGURE 15–11. Parallel (circles) and quadratic (squares) cost functions vs. the asymmetry in dose from beam pairs, for $D_1 + D_2 = 80$ Gy. Note that in contrast to the parallel case, the lowest cost for the quadratic cost function occurs when the doses from the beam pairs are equal (asymmetry = 0).

Relationship Between Dose-Volume Constraints And Limits On Parallel Model NTCP

The limitations of quadratic cost functions can be alleviated by the addition of dose-volume constraints, which can mimic some of the volume effects of parallel tissues. In figure 15–4, we showed how the FSU dose response becomes a step function when k becomes large. In this case, the fraction damaged is just the volume-receiving dose greater than the threshold dose, so a limit on f_{dam} reduces to a dose-volume limit. The iso-NTCP line in the dose partial volume plane coincides with the boundary of acceptable partial volume treatments defined by the dose-volume constraint shown in figure 15–12. A dose-volume constraint that is less than 50% of the tissue receives a dose ≥ 20 Gy would exclude any partial volume treatment on or above and to the right of the red dashed line in figure 15–12.

The situation for general inhomogeneous treatments (or when k is finite) is more complicated, and recently the relative merits of using dose-volume constraints and NTCP models have been discussed in the literature (Wu et al. 2002). A range of complication probabilities is consistent with a given dose-volume constraint and vice versa and, for example, the dose and volume limits used will determine how conservative a dose-volume constraint is when compared to a constraint on NTCP in a parallel model.

Below, in the context of a parallel model, we determine: first, the range of dose-volume constraints that will guarantee that f_{dam} is $< f_q$; second, the range of f_{dam} values resulting from treatments that meet any fixed dose-volume constraint; third, the way to pick a dose-volume constraint so that the range of f_{dam} values is minimized and f_{dam} is $< f_q$.

Assuming we wish to keep the complication rate below q%, and that a parallel model gives a q% complication rate if a fraction f_q of FSUs are damaged, with $p(d) = f_q$ at $d = d_q$, then any combination d_{lim} and v_{lim} such that

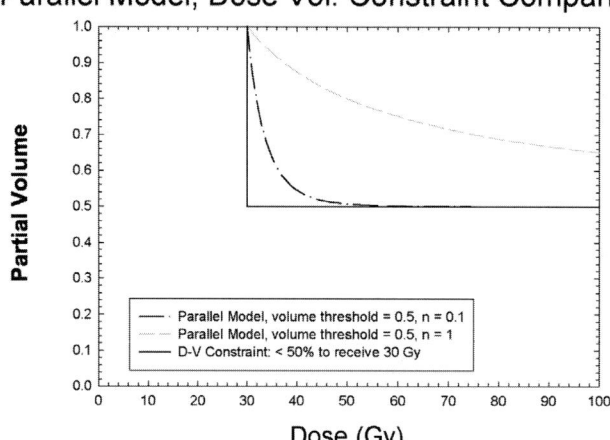

FIGURE 15–12. Comparison of parallel model iso-NTCP curves with various values of k, with that of a dose-volume constraint (<50% of the organ to receive 30 Gy). As before, the whole volume tolerance dose is 30 Gy, and the volume threshold for the parallel model is 50%. The parallel model with k = 1/n = 1 is shown with the dashed gray line, the case with k = 1/n = 10 is shown with the dash-dotted black line, the case of infinite k is coincident with the solid black line, which shows the iso-NTCP curve consistent with the dose volume constraint.

$$\left(1 - v_{\lim}\right) P\left(d_{\lim}\right) + v_{\lim} \le f_q \tag{8}$$

can be used in a dose-volume constraint to guarantee that for any inhomogeneous treatment consistent with the constraint, we will find $f_{dam} < f_q$. This follows from the fact that the *most* aggressive inhomogeneous treatment consistent with the constraint will treat a volume slightly smaller than v_{lim} to very high dose, and the rest of the volume to a dose slightly smaller than d_{lim} resulting in a fractional damage slightly less than the left hand side of the inequality (8).

Note that any treatment *that meets this constraint* will damage a minimum fraction of FSUs given by

$$\left[P\left(d_{\lim}\right) v_{\lim}\right] \le f_{dam} \tag{9}$$

This follows from the fact that the *least* aggressive treatment meeting the constraint is a partial volume treatment where volume v_{lim} receives dose d_{lim} and the rest of the volume receives nothing.

Where a given treatment lies between these limits is determined by the shape of its dose-volume histogram (DVH). The shape of the DVH is determined by the patient and beam geometries together with the field profiles.

When a treatment is driven to the dose volume constraint, our results show that f_{dam} is constrained to lie within the limits:

$$\left[P\left(d_{\lim}\right) v_{\lim}\right] \le f_{dam} < \left[\left(1 - v_{\lim}\right) P\left(d_{\lim}\right) + v_{\lim}\right] \le f_q. \tag{10}$$

Assuming the most aggressive condition is used in inequality (8), we find:

$$\left[P\left(d_{\lim}\right) + v_{\lim} - f_q\right] \le f_{dam} < f_q.$$

Eliminating $P(d_{lim})$ in favor of f_q and v_{lim}, gives:

$$f_q v_{\lim} \left[\frac{\left(1 - v_{\lim}/f_q\right)}{\left(1 - v_{\lim}\right)} \right] \le f_{dam} < f_q. \tag{11}$$

The equivalent expression in terms of f_q and $P(d_{lim})$ is:

$$f_q P\left(d_{\lim}\right) \left[\frac{\left(1 - P\left(d_{\lim}\right)/f_q\right)}{\left(1 - P\left(d_{\lim}\right)\right)} \right] \le f_{dam} < f_q.$$

The interval is a minimum when:

$$v_{\lim} = P\left(d_{\lim}\right) = 1 - \left(1 - f_q\right)^{1/2}.$$

Figure 15–13 shows the minimum size of the interval as a function of f_q. It is a maximum of 0.5 when $f_q = 0.75$. The relative size of the gap compared to f_q, shown in figure 15–14, decreases

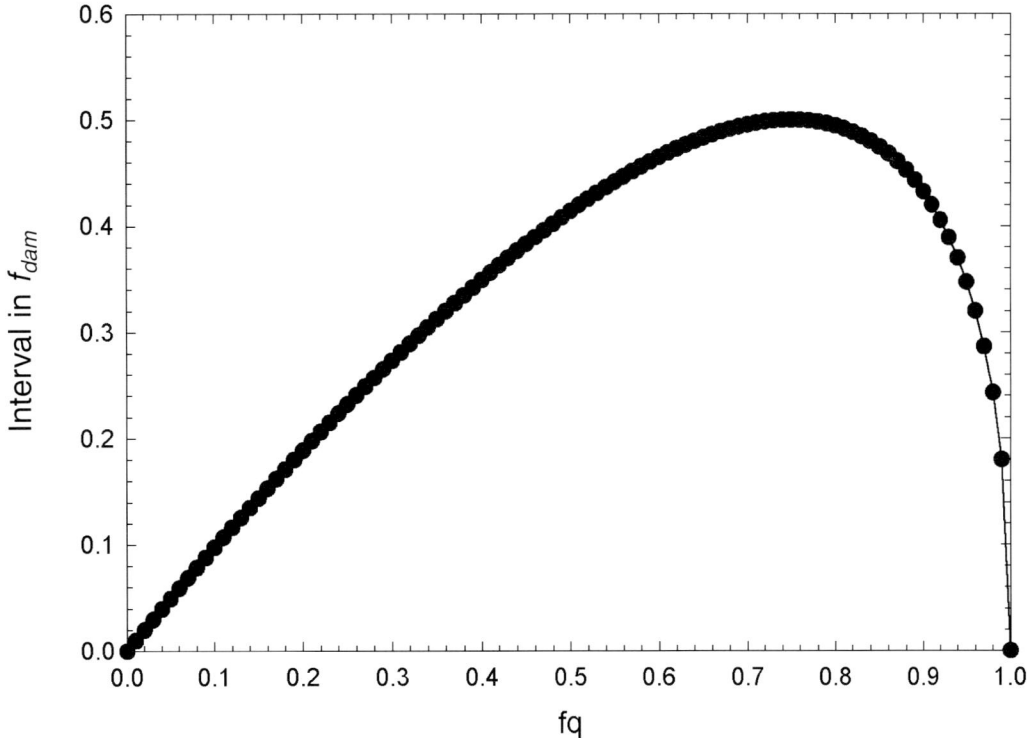

FIGURE 15–13. Minimum possible variation in f_{dam} produced by treatments obeying a dose-volume constraint to keep $f_{dam} < f_q$.

relative interval size compared to f_q-limit on f_{dam}

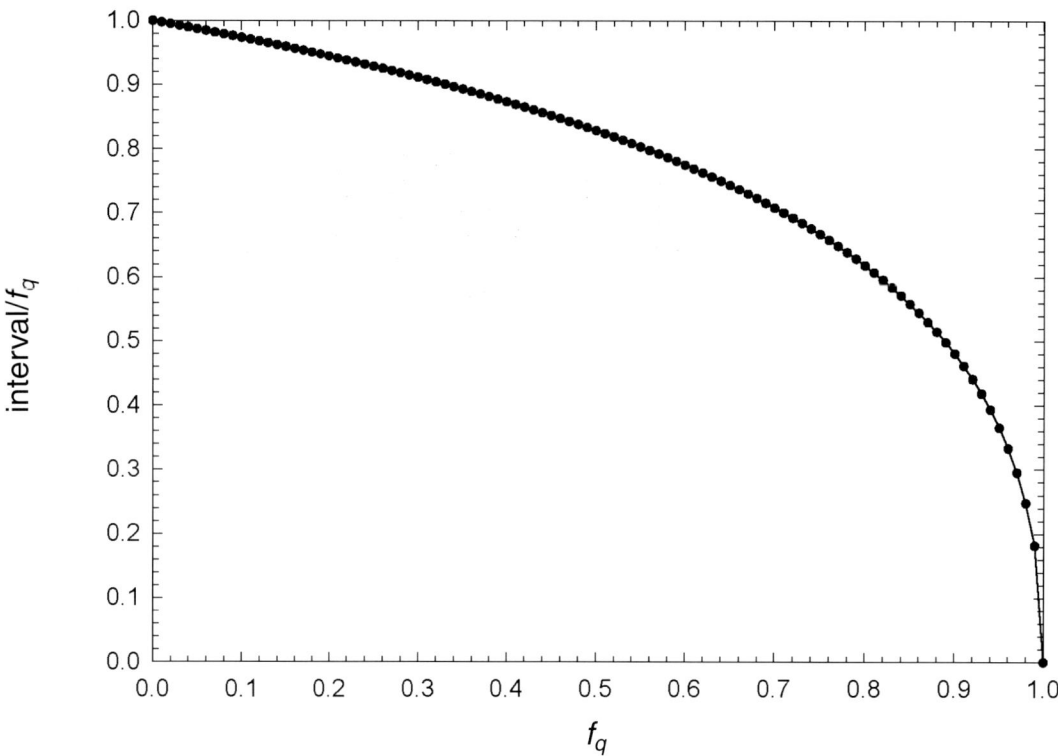

FIGURE 15–14. Size of variation in f_{dam} in figure 15–13, relative to f_q.

continuously from one to zero as f_q increases from zero to one. Thus, despite the connections between them, and our best efforts to minimize the differences between them, there is still a considerable difference between possible outcomes of optimization using a dose-volume constraint and a parallel model.

In light of the limitations of quadratic dose objectives, many commercial planning systems have added the option of using dose-volume constraints in the optimization cost function. In the absence of explicit models of the volume effects, we recommend adding dose volume constraints for parallel tissues and any tissue for which the LKB model n is large. For Serial tissues, e.g., spinal cord, and tissues with low values of n, constraints on the maximum dose should be sufficient.

Concluding Remarks

Much work remains to be done to determine the dose-volume response of normal tissue complication probabilities. It is not clear that it will ever be possible to do this, but to have a fighting chance, we certainly need:

1. Quantitative dose/volume and complication data.
   This can only come from conformal therapy.
2. Sufficient range of plan types and or dose volume combinations.
   This may require comparing data from different institutions.
3. Sufficient statistics.

It is unlikely that this will arise from any source other than conformal therapy dose escalation protocols.

It may involve pooling of data from different institutions, with consequent requirements on consistency of diagnosis and ability to account for differences in patient populations.

In addition to this formidable task, we need to understand the relationships between models on NTCP and score functions in use in the optimization of IMRT. The work presented here is but a small step in that direction.

TCP For Treatment Planning

The goal of radiation therapy is to achieve local tumor control. *In vitro* and animal studies as well as clinical experience (Bentzen 1997; Hellman 2001) demonstrate that the probability of local tumor control (TCP) increases approximately sigmoidally from near zero at low dose to near unity (or 100%) at high dose, as shown schematically in figure 15–15. The TCP vs. dose curve is roughly characterized by its location along the dose axis and its slope. These are related to two key parameters, D_{50} (or TCD_{50}) and γ_{50}. D_{50}, which determines the position of the TCP curve along the dose axis, is the uniform dose that produces 50% local control in a large sample of identically irradiated tumors. γ_{50} is a measure of the steepness of the TCP curve and is defined as the percent change in TCP per percent change in dose for a small (differential) dose interval centered at D_{50}. Clinical data suggest that D_{50} is in the range of 40 to 70 Gy for radiocurable tumors and γ_{50} in the range 1 to 3.

For small changes in dose near D_{50}, the TCP vs. dose curve can be approximated by a line through the point (dose = D_{50}, TCP = 0.5) with slope γ_{50}/D_{50}. The effect on TCP (expressed as a

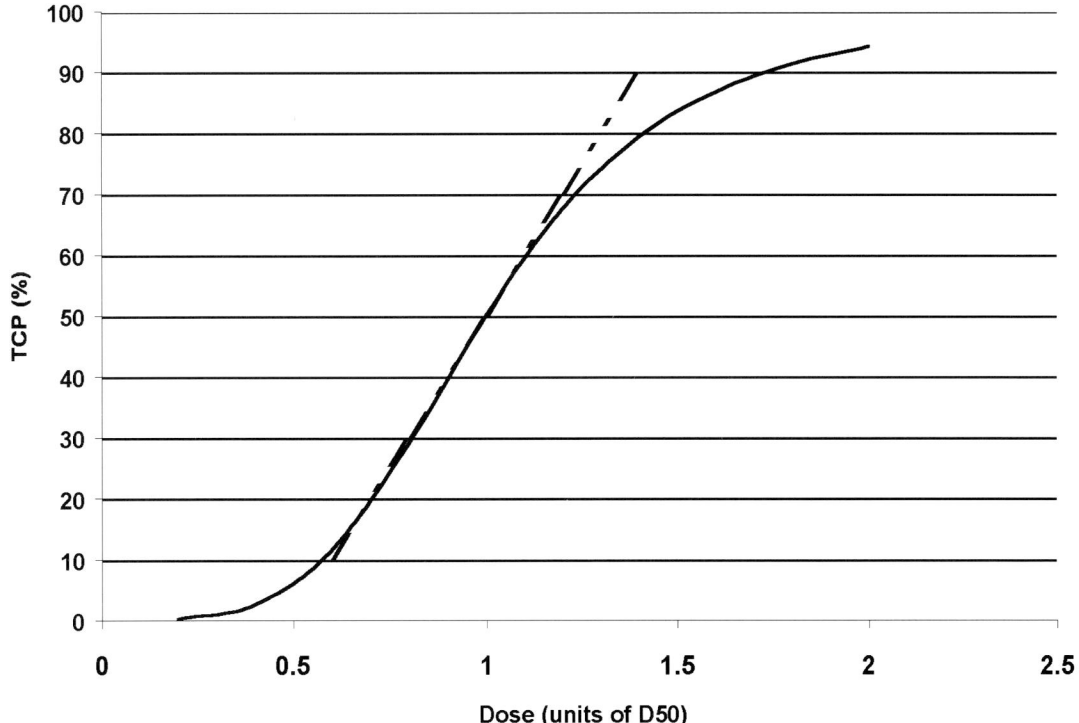

FIGURE 15–15. Schematic example of a TCP vs. dose curve.

fraction between 0 and 1.0), of a small dose change can be estimated with this linear approximation as

$$\Delta TCP \approx \gamma_{50} \, \Delta D / D_{50} \tag{12a}$$

and

$$TCP \approx 0.5 + \Delta TCP. \tag{12b}$$

For example, suppose D_{50} is 60 Gy and γ_{50} is 2.0. Then half the tumors irradiated to 60 Gy are controlled, while irradiation to 66 Gy ($\Delta D = 6$ Gy or a 10% dose increase) controls approximately a fraction of 0.7 of the tumors and irradiation to 54 Gy (10% dose decrease) controls approximately a fraction of 0.3.

Mathematically simple phenomenological expressions such as a logistic function

$$TCP(D) = 1/(1 + (D_{50}/D)^{4\gamma_{50}}) \tag{13}$$

with the prescription dose inserted for D in the denominator, are helpful for quick calculations. However, the competition between eradicating the tumor and protecting normal tissues often leads to non-uniform dose distributions in the target volume. Therefore, it is both desirable and challenging to develop TCP models based on more fundamental understanding of radiobiology, in hopes that they will guide us in devising and evaluating dose distributions. Unfortunately, this is a difficult problem for several reasons. Clinical or animal data relating local control to dose distributions in tumors are scarce and data relating dose distribution to the spatial distribution of clonogenic cells—i.e., the cells whose regrowth causes local recurrence—are even sparser. The delivered dose (including setup error, organ motion, etc.) is seldom known or even estimated. For some disease sites (e.g., prostate) it is obvious that a recurrence is truly local but for others (e.g., lung), it is difficult to distinguish an in-field from a marginal (or even same lung) recurrence. At such disease sites, the scoring of *local* failure is unclear. Finally, there are conceptual difficulties with current models, some of which are described below. While these problems are being investigated, the use of TCP in treatment plan design or evaluation should be approached with caution and skepticism.

Radiobiological Models

Most mechanistic TCP theories proceed from the hypothesis that local control is achieved if (and only if) all the clonogenic cells are destroyed by radiation. Suppose that a population of identical tumors, each initially containing a large number, N, of clonogenic cells, receives a uniform dose, D, and that, on average, a fraction SF of clonogens survive. The Munro-Gilbert hypothesis (Munro 1961) is that the TCP is equal to the probability that there are no surviving clonogens and that this probability can be calculated by Poisson statistics:

$$TCP = \exp(-N \, SF) \tag{14}$$

At zero dose, SF is unity and, for large N (≥ 100), $\exp(-N)$ is approximately zero. For large N and tumors with a slow repopulation rate relative to the dose delivery, Poisson statistics is believed to be a good approximation (Tucker, Thames, and Taylor 1990).

The surviving fraction, $SF(D,d)$, depends on the intrinsic tumor biology as well as the total dose (D), dose-per-fraction (d), and other aspects [e.g., LET (linear energy transfer)] of the radiation. It is often calculated from the linear-quadratic (LQ) theory:

$$SF(D,d) = \exp[-\alpha D(1 + d/(\alpha/\beta))] \tag{15}$$

where the radiobiological parameters α and β are estimated from laboratory data or clinical results. Then

$$TCP = \exp(-N \exp[-\alpha D(1 + d/(\alpha/\beta))]) \tag{14a}$$

Joiner et al. (1997), Hall (1994), Fowler (1989), and Thames and Hendry (1987) provide reviews and further references relating to the LQ model. The LQ model can be applied to both tumors and normal tissues and has been particularly useful in determining effective radiation therapy dose fractionation schedules.

Some authors prefer to specify the cellular radiosensitivity in terms of a more macroscopically determined quantity such as SF2, the surviving fraction of clonogens after a uniform 2 Gy dose. If dose-per-fraction effects are neglected (i.e., if α/β is large), then $SF(D)$ is related to $SF2$ by

$$SF(D) = (SF2)^{D/2} \tag{16}$$

and

$$TCP = \exp(-N \, (SF2)^{D/2}) . \tag{14b}$$

If dose-per-fraction is not neglected, the dose D in the exponent can be replaced by $BED2(D,d)$. $BED2$ is the biologically effective dose which, when delivered in 2 Gy fractions, has the same radiobiological effect as dose D delivered in fractions of size d. Using LQ expressions,

$$BED2(D,d) = D(1 + d/(\alpha/\beta))/(1 + 2/(\alpha/\beta)) \tag{17}$$

$SF2$ for human tumors is typically between 0.2 (radiosensitive) and 0.7 (radioresistant) with a value of 0.5 often being assumed for back-of-envelope estimates. If (α/β) is very large, these values of SF2 correspond to a range of α's from ≈ 0.8 Gy^{-1} to ≈ 0.18 Gy^{-1}, respectively. Although (α/β) is believed to be large for most human tumors (e.g., 10 to 20 Gy), a number of recent publications provide evidence that (α/β) is particularly low (1.5 to 3 Gy) for prostate cancer (Fowler, Chappell, and Ritter 2001). Experimentally and clinically estimated values of α and α/β are tabulated in the reviews referred to above (Fowler 1989; Hall 1994; Thames and Hendry 1987; Joiner et al. 1997). However, at present the radiosensitivity parameters for even common human tumors are only approximately known.

The parameters D_{50} and γ_{50} can be calculated from Equations (14a) or (14b). For example, starting from Equation (14b)

$$D_{50} = 2 \ln (\ln 2/N)/\ln (SF2) \tag{18a}$$

and

$$\gamma_{50} = (\ln 2/2) \ln (N/\ln 2) \tag{18b}$$

Note that γ_{50} depends *only* on the number of clonogens. The estimated cell density in typical soft tissue is in the range of 10^8–10^9/cc. For this range of Ns, Equation (18b) predicts a γ_{50} (>6.5), which is too large to match clinical observations. Equation (18b), in combination with the observed low values of γ_{50} (≈2), implies that the number of clonogenic cells is very small (≈220) compared to the total number of cells in a clinically detectable soft-tissue tumor. Further, Goitein (1987) pointed out that Equation (18a) together with the reasonable assumption that N is proportional to the volume of the tumor predicts a much stronger dependence of D_{50} on tumor size than is observed. Finally, TCP curves do not seem to be steeper for large (few hundred cc) tumors than for small (few cc) ones. To resolve these problems, one might hypothesize that the number of clonogens remains approximately constant over a wide range of tumor volumes. Despite these issues, many authors believe that the radiation response of tumors is controlled by a small number of radioresistant clonogens, while the vast majority of cells in a tumor are either not clonogenic or are radiosensitive and are killed by doses well below D50 (see, for example, Brenner 1993).

An alternative hypothesis to explain the shallow slopes of TCP curves proceeds from the observation that TCP—a statistical concept—is determined for a population of patients, not an individual (Goitein 1987; Zagars, Schultheiss, and Peters 1987; Suit et al. 1992). Even if the population is grossly similar in disease-type, tumor size, stage, etc, individual tumors are likely to differ in radiobiological characteristics (e.g.. *SF*2 or α and β, and N). The observed TCP vs. dose curve is an average over the entire population. Figure 15–16 is a simple example of a tumor population

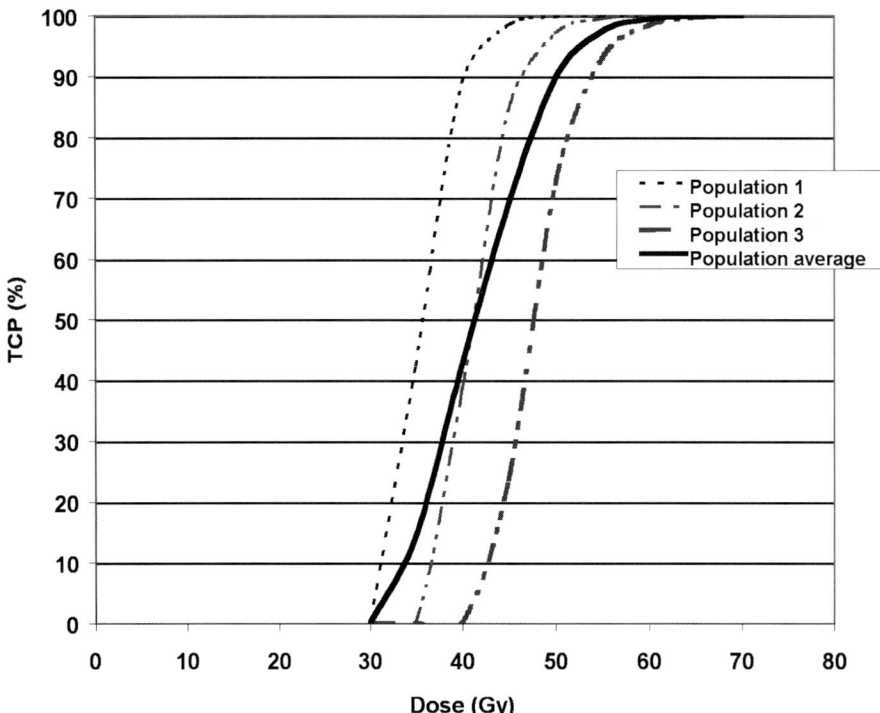

FIGURE 15–16. Simple example showing that the effect of averaging over a population with a distribution of radiosensitivities reduces the slope of the TCP vs. dose curve. In this example, each tumor in the population has 10^6 clonogens but one-third of the tumors in the population have SF2 = 0.45 [population 1], one-third have SF2 = 0.50 [population 2], and one-third have SF2 = 0.55 [population 3] and their individual TCP curves are dotted. The population average is the solid curve.

made up of three equal-sized groups, each with a different radiosensitivity ($SF2 = 0.45$, $SF2 = 0.5$, $SF2 = 0.55$). For each group, the number of clonogens is 10^6. The population average curve is notably shallower than the individual curves. Realistically, one expects *intra*-tumor as well as inter-patient variation of the radiobiological parameters ($SF2$ or α and β), and intra and inter-tumor variation of the clonogen density. Niemierko and Goitein (1993b) and others (Roberts and Hendry 1998; Webb 1994) describe methods for fitting observed TCP curves to distributions (normal or log normal) of inter-patient and intra-tumor radiobiological parameters.

Population averaging produces TCP curves with clinically realistic slopes and high clonogen densities. However, the number (or density) of clonogenic cells is not well determined by these fitting procedures, so the question of few vs. many clonogens is unresolved. In population average models, low radiosensitivity (i.e., low mean value of α or high $SF2$) can compensate for order-of-magnitude increases of clonogen density while still maintaining α within limits that agree with *in vitro* studies. The three curves in figure 15–17 show calculated TCP vs. dose for a uniformly irradiated population of 100 cc tumors using parameters from a population-average model which were derived by Webb (1994) to fit a clinical data set for control of squamous cell carcinoma of the upper respiratory and digestive tract. The assumed clonogen densities for the three curves are (left to right) 10^3 clonogens/cc (one cell in a million!), 106 clonogens/cc, and 10^9 clonogens/cc. Available clinical data did not distinguish between these very different parameterizations—including the order of magnitude of clonogen density.

Tumor repopulation and repair of sublethal damage during the course of radiation therapy lead to additional complexity in radiobiologically based theories of TCP. For the simplest approximations incorporating repopulation, the clonogen doubling time, τ, is assumed constant. If the time to deliver the entire treatment is T (e.g., 5 weeks), the surviving fraction of clonogens is approximated by the product of $SF(D,d)$ without repopulation and a growth factor, $\exp(\,(\ln 2)\,T/\tau)$ [see Hall (1994), Fowler (1989), and Thames and Hendry (1987)]. More complex expressions

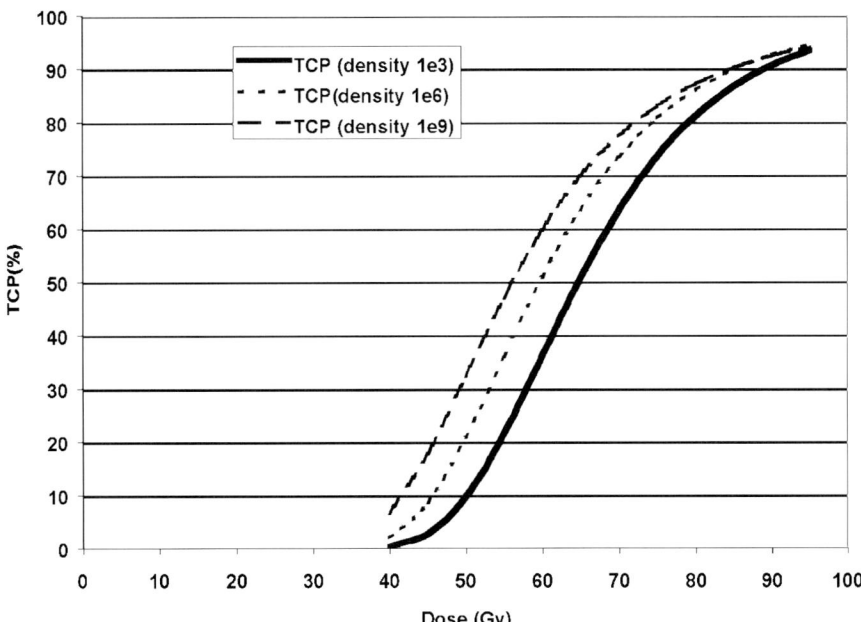

FIGURE 15–17. Calculated TCP vs. dose for three parameter sets of a population averaged model derived by Webb (1994) to fit clinical data for local control of squamous cell carcinoma of the upper respiratory and digestive tract. The minimum and maximum clonogen densities differ by 6 orders of magnitude.

appropriate to rapid repopulation, brachytherapy, or time-dependent doubling times can be found in the literature (Tucker, Thames, and Taylor 1990; Zaider and Minerbo 2000).

Non-Uniform Dose Distributions

TCP for a non-uniform dose distribution is calculated by approximating the tumor by many sub-volumes ["tumorlets" (Goitein 1987)], each small enough to receive an approximately uniform dose but large enough for Poisson statistics to hold. It is assumed that the tumorlets respond independently to radiation damage. Let v_i be the volume and n_i be the number of clonogens in the ith tumorlet and let D_i be the dose it receives (in fractions of size d_i). Here n_i is related to the clonogen density in the ith tumorlet, ρ_i, as $n_i = \rho_i \, v_i$. To control the tumor, there must be no surviving clonogenic cells in any of the tumorlets. For models without population averaging,

$$TCP = \Pi_i \, \{\exp(-n_i \, SF_i(D_i, d_i))\}. \tag{19a}$$

For population-averaged models, the product in Equation (19a) is averaged over the probability distribution of radiobiological parameters of the population of tumors,

$$TCP = [\Pi_i \{\exp(-n_i SF_i(D_i, d_i))\}]_{av}. \tag{19b}$$

For other forms of the TCP function, Equation (19a) can be generalized to

$$TCP = \Pi_i \, (TCP \, (D_i, d_i))^{vi} \tag{19c}$$

and Equation (19b) can be similarly generalized. Equations (19) can easily accommodate spatial variations of clonogen density and radiobiological parameters such as low clonogen density in regions of microscopic disease or increased density or higher $SF2$ in regions where functional imaging indicates aggressive cells. However, the associated parameters are rarely (if ever) known and the standard DVH methodology does not include spatial information.

A cold spot—a low dose in even a small volume of clonogen-bearing tumor—may seriously compromise TCP. In the limiting case of a complete geographic miss of part of the tumor, some clonogens receive zero dose and one of the factors in Equations (19) falls to zero, forcing the product to zero and guaranteeing local failure. For less extreme underdoses, the predicted decrease of TCP depends strongly on the model. Papers by Goitein and Niemierko (1996), Withers (2000), and Tome and Fowler (2002) are interesting explorations of the effects of cold spots under the assumptions of uniform clonogen density in the target volume.

The solid curves in figure 15–18 show the percent decrease in TCP caused by a 1 cc (1% of the total volume) cold spot in an otherwise uniform 60 Gy dose distribution for the three 100 cc squamous cell carcinoma models with uniform clonogen density described in relation to figure 15–17; the dotted curves are for a 10 cc cold spot of the same magnitude. No LQ correction is made for the reduced dose per fraction in the cold spot; such a correction would reduce the TCP even more. Models with a small number of clonogens are less sensitive to dosimetric inhomogeneities, as is seen in the figure 15–18.

But the predicted impact of cold spots is dramatic. If a 1 cc or a 10 cc cold volume is distributed as a shell at the periphery of a 100 cc sphere, the shell thicknesses are approximately 0.01 cm and 0.1 cm respectively! (See Withers (2000), and Tome and Fowler (2002), for emphasis of the geometric aspects of these small-volume underdoses). Small-volume cold spots have negligible effect on more "global" aspects of the DVHs, which are often used for plan evaluation. For example, an isolated 1 cc cold spot of 50% in an otherwise uniformly irradiated 100 cc tumor does not

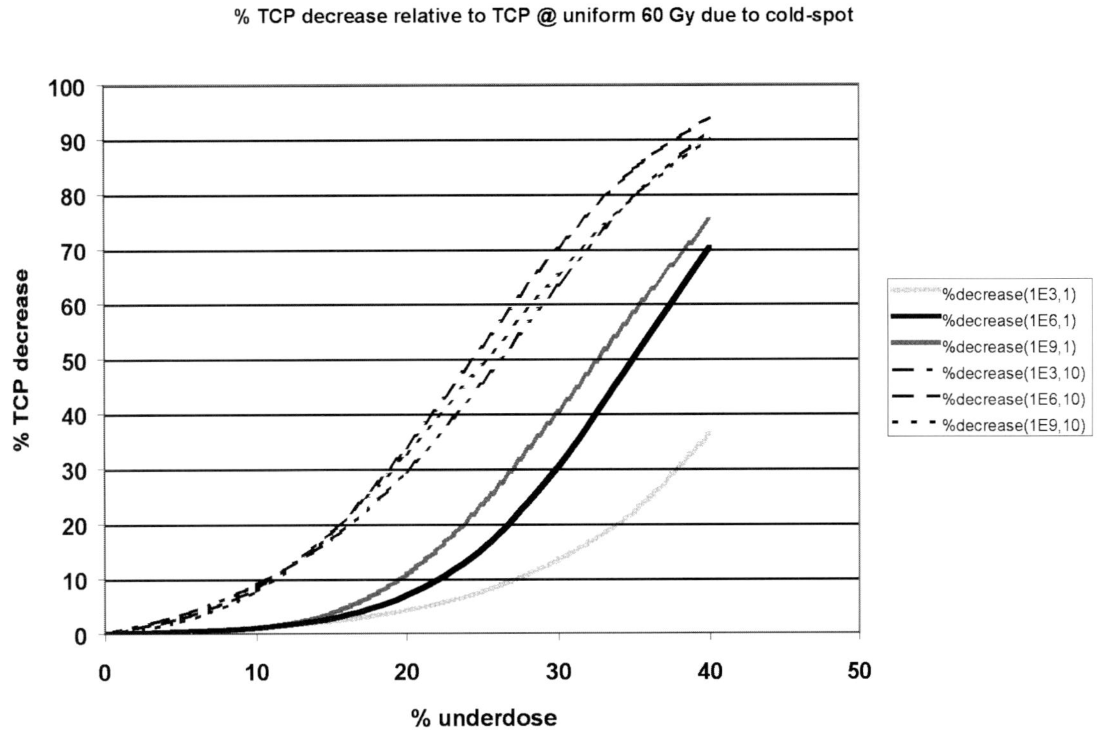

% TCP decrease relative to TCP @ uniform 60 Gy due to cold-spot

FIGURE 15–18. Percent decrease of TCP (relative to TCP for each model at a uniform 60 Gy dose) for a 1 cc cold spot (solid curves) and a 10 cc cold spot (dotted) in the same three tumor models (each tumor is 100 cc, clonogen densities are 10^3, 10^6, and 10^9) as were graphed in figure 15–17. The x-axis is the percent underdose in the cold spot.

change D_{95} (dose to 95% of the tumor). It reduces V_{95} (volume within the 95% isodose line) by 1% and reduces the mean dose by only 0.5%—yet the predicted TCP (extrapolate from figure 15–18) is reduced so much as to invalidate the treatment.

It is important to realize that in these examples the clonogen density is assumed to be uniform over the entire target region. This is almost certainly *not* the case for many cold spots that occur in radiation therapy. Dose distributions are usually designed to cover a planning target volume (PTV). The PTV is expanded beyond the gross disease (GTV) to include the physician's estimate of microscopic disease (CTV) and further to allow for setup error (PTV). Cold spots which are deemed necessary for normal tissue protection are often placed at the edges of the PTV and if the margin of expansion (CTV to PTV) is adequate, they are in a region of low or zero clonogen density. Under such circumstances, the cold spot causes minimal reduction of TCP (no reduction if there are no clonogens). For the model used to generate figure 15–18, using the radiobiological parameters of the clonogen density 10^6/cc model but assuming that the 10 cc cold spot luckily coincides with a volume where the clonogen density is only 10^3/cc, the percent decrease in TCP is less than 5% until the depth of the cold spot exceeds 40%. Levegrun et al. analyzed biopsy-proven local control of prostate cancer for a data set where the PTV was designed to extended 1 cm beyond the gland (except at the rectal interface, where the expansion was 0.6 cm) and the dose distributions were such that cold spots almost always fell outside the gland as defined on the planning CT scan. They found no significant correlation between minimum dose and local control (Levegrun et al. 2000, 2001).

However, for disease sites where the CTV is less well defined than in the prostate and for conformal radiotherapy with particularly tight field margins, our poor knowledge of clonogen distribution may lead to problems. And if a cold spot unfortunately falls in a region with high clonogen density or radioresistant cells, TCP is reduced beyond what is expected from a uniform density model. In the future, pathology-validated functional imaging may provide better localization of aggressive vs. indolent disease and allow the design of dose distributions which avoid such problems or, better yet, "dose paint" to deliver increased dose to these regions. But currently, our knowledge of clonogen distribution is weak and it may be safest to assume uniform density at least through the CTV.

While reasonable TCP models predict that hot-spots or "boost" volumes (higher doses) increase TCP, large TCP increases in models with uniform clonogen density are only achieved if the boost volume is a large fraction of the tumor. Essentially, the unboosted portion of the tumor acts as a cold spot and pulls the TCP down. Figure 15–19 is a simple example using Equation (19c) for a model without population averaging. Here a fraction of the tumor volume is irradiated to a high dose (the percent volume is plotted along the x-axis) and the remainder is irradiated to D_{50}. The dotted curve shows the maximum advantage to be expected—assuming that the dose in the high dose region is so large that TCP in that volume is unity. The solid curve is a more conservative situation where the dose in the high dose volume is 20% higher than D_{50} and TCP is calculated with the logistic function (Eq. 13) with $\gamma_{50} = 2$. A small hot spot has negligible effect on TCP. To raise TCP from 50% to 60% requires a substantial fraction of the volume (at least 25%,

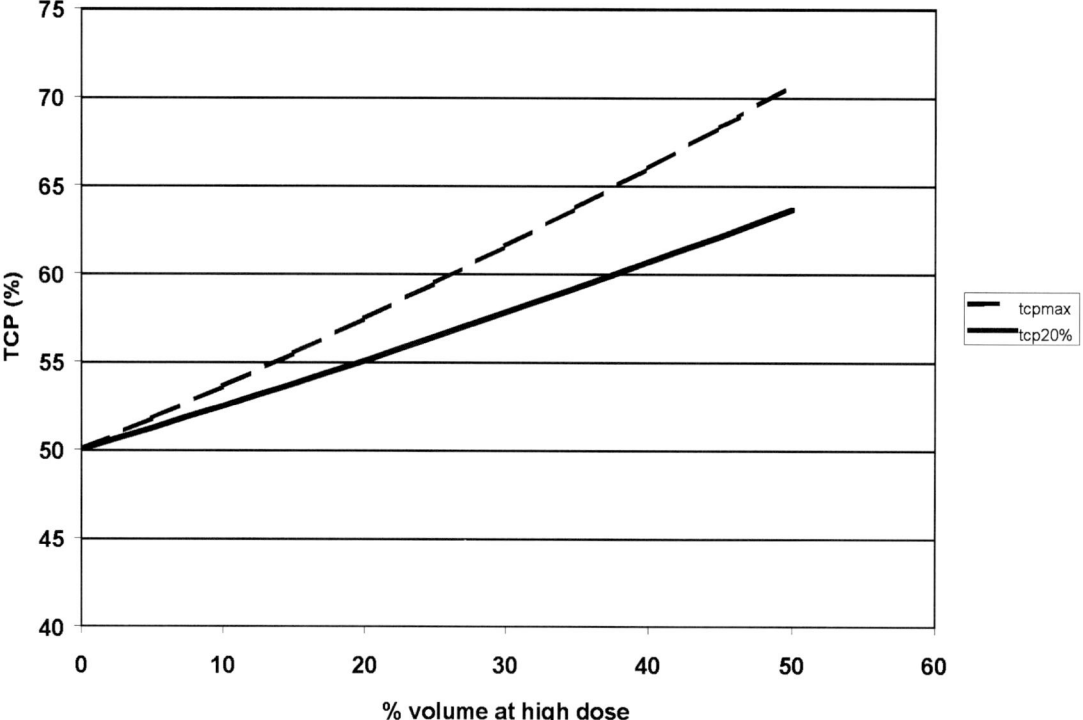

FIGURE 15–19. Simple example of the increase of TCP due to irradiating a chosen percent of the tumor volume to high-dose (x-axis) while the remainder is irradiated at D_{50}. The dotted curve is the best we can do—the high dose volume is irradiated to such a high dose that TCP within it is 100%. The solid curve is a more realistic case of a 20% "hot spot" or boost.

even going to a very high dose) be boosted. Of course physicians often select boost volumes on the basis of diagnostic or surgical/pathological studies so that these volumes do *not* have the same clonogen content as the remainder of the tumor, but rather include particularly active or aggressive disease.

It is also of interest to see how difficult it is to compensate for a small underdose with a high dose boost in another part of the target (always assuming uniform clonogen density). Figure 15–20 is an example, based on Equations (13) and (19c), showing the volume fraction needed at high dose to compensate for an underdose to 5% of the tumor—the remainder of the tumor is irradiated at D_{50}. The dose in the high dose volume, D_{hi}, is such that $TCP(D_{hi})$ is 80% and the percent of D_{50} in the cold-spot is plotted along the x-axis.

For numerous examples of TCP model calculations for simple non-uniform dose distributions see Withers (2000), and Tome and Fowler (2002).

Equivalent Uniform Dose (EUD)

TCP models are conceptually uncertain because of our inability to determine the clonogen density within at least 6 orders of magnitude. For use in plan optimization, they also suffer from the heavy computational demands of population averaging models. On the other hand, there is hope that TCP models that are consistent with clinical data can improve our quantitative theoretical understanding of non-uniform dose distributions. The equivalent uniform dose (EUD) was proposed by Niemierko (1997) as a useful intermediate quantity that could be calculated from DVHs and *SF2*

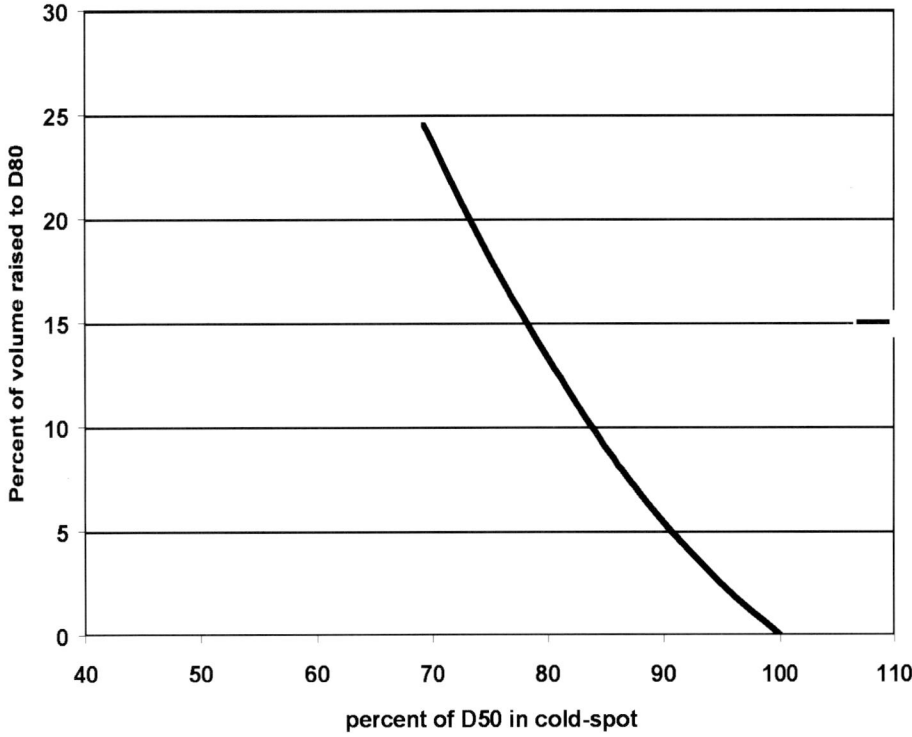

FIGURE 15–20. Percent volume which must be raised to a dose D_{80} (where $TCP(D_{80})$ = 80%) to compensate for a cold spot comprising 5% of the tumor volume where the dose in the cold spot relative to D_{50} is plotted along the x-axis. The tumor volume not included in the hot or cold volumes is irradiated to D_{50}; thus the overall TCP [from Eq. (19c)] is 50%.

or LQ model radiobiological parameters determined from clinical and animal data. The EUD is defined as the dose which, "when distributed uniformly across the target volume, causes the survival of the same number of clonogens" as the true dose distribution (Niemierko 1997). If dose-per-fraction effects are considered, the definition the EUD must be delivered in the same number of fractions as the treatment. For the simple case of uniform clonogen density and $SF2$ and negligible fractionation effects, the EUD (in Gy) of a DVH in which a fraction of the tumor volume v_i receives dose d_i is

$$EUD = 2 \ln (\Sigma v_i \, (SF2)^{D_i/2})/(\ln SF2). \qquad (20)$$

The absolute volume of the tumor, the number, and the absolute density of clonogens do not enter the expression for EUD. Only the volume fractions at various dose levels are required. Niemierko (1997) shows that an LQ-model form of Equation (20) can be generalized to include dose per fraction effects, spatial variation of clonogen density (only the relative density is required) and population averaging of radiobiological parameters. EUD was shown to be bounded above by the mean dose and below by the minimum dose in the tumor [or the structure whose DVH is used in Eq. (20)]. EUD was also shown to be fairly insensitive to β/α and to population averaging, allowing EUD to be calculated with a much less computationally expensive un-averaged expression with $SF2$ or α equal to the population mean (Niemierko 1997).

Figure 15–21 shows the percent decrease in EUD for the same situation as used for the 10 cc cold spot graphs in figure 15–18 (a 100 cc tumor with uniform clonogen density uniformly irradiated to 60 Gy except for the cold spot). The x-axis is the percent underdose in this 10 cc volume. $SF2$ was calculated for each curve as

$$SF2 = \exp(-2 \, \alpha_0) \qquad (21)$$

where α_0 is the mean value of α that was used in the population averaged TCP model described in relation to figure 15–18. The percent decreases in mean dose and in minimum dose are also plotted in figure 15–21 for comparison. EUD varies more strongly than the mean dose but less strongly than TCP. For low α (small clonogen density curve), EUD is reduced less by the cold spot than for high α (large clonogen density curve). Of course, as for the TCP calculations, if the cold spot falls in a region of low clonogen density, it would have much less of an effect on EUD.

Use Of TCP And EUD In Plan Evaluation And Optimization

The direct application of TCP or EUD to plan evaluation is at an early stage and application to plan optimization is even more so. Wang et al. (1995) and Wu et al. (2002) report two experiences with using biological indices in optimization. Given the difficulties in determining TCP models that match clinical experience and the uncertainties in determining the distribution of clonogenic cells, it is reasonable to proceed with great caution. It would be most helpful to seek qualitative characteristics of TCP or EUD which transcend differences between models and which can provide a reliable supplement to traditional dosimetric indices (e.g., mean dose, minimum dose, dose enclosing some percent of the target volume such as D_{95}) for DVH comparison.

For example, Tome and Fowler (2002) recently evaluated both TCP and EUD for a simple, four-level DVH where 1% of the tumor is deeply underdosed relative to prescription (10% to 45%), another 1% or 4% is underdosed by 10%, 80% of the target is boosted to 10% above prescription, and the remaining 15% to 18% of the target is at prescription. In this

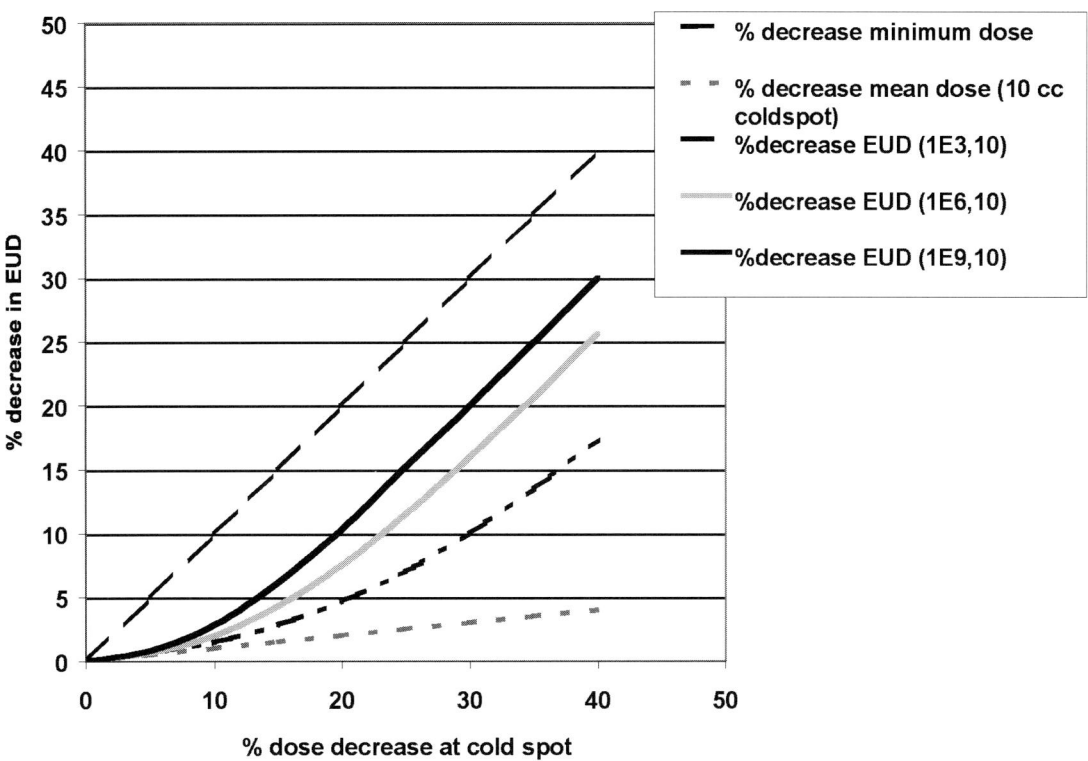

FIGURE 15–21. Percent decrease in EUD (relative to uniform 60 Gy dose) for a 10 cc cold spot in a 100 cc tumor with the same three clonogen densities (1E3, 1E6, and 1E9) as were graphed in figures 15–17 and 15–18. For each density, SF2 was calculated with the mean value of α taken from the population average model. The x-axis is the percent underdose in the cold spot.

example, D_{95} is always prescription dose and the mean dose is affected by less than 0.5% by the depth of the cold spot. They assume, further, that the target has a uniform clonogen density—thus, it is a GTV or, at worst, a CTV—and use a population averaged TCP model with a Gaussian distribution of LQ model parameters set to give clinically reasonable values of D_{50} and γ_{50}. For both TCP and EUD, they find that the boost is unable to compensate for the cold spot if the dose deficit exceeds 15% to 20% of prescription (i.e., the dose in the tiny cold spot is less than 80% to 85% of prescription). They pose this example as a strong warning against the deleterious effects of even small cold spots in regions where one knows there are clonogens. As a remedy, they suggest that an acceptable plan with dose inhomogeneity in the target should require EUD to at least equal to the prescription dose.

If TCP or EUD is to be used in optimization algorithms, the score function as well as the TCP model must be considered. There are numerous incompletely answered (or unanswered) questions for future studies to address. Can (should) TCP or EUD be combined with more "traditional" dose and dose-volume constraints or is it preferable to go biological all the way, using NTCP or related quantities for the normal tissues? How can we best prevent TCP from driving the dose higher than the physician is willing to prescribe? If we obtain similar dose distributions with dose/dose-volume and with biology-based score functions, what are the advantages of one method over another? If the distributions are qualitatively different, do we trust biological indices enough to urge adoption of the optimization results or will we be forced to tinker with the score function to get back to more familiar territory?

Summary

Normal Tissue Complication Probability

- Clinical endpoints are classified in terms of time of onset, and severity as defined by degree of medical intervention required.
- Good clinical practice ensures that severe complications occur rarely, and clinical data sets involving severe normal tissue reactions typically have poor statistics. Estimates of volume effects made before conformal therapy are not quantitative. Recent data from 3DCRT dose escalation studies has improved our understanding of volume effects.
- Whole volume doses responses are described using sigmoid functions with positions and slopes [typically TD(50,1) and γ_{50}].
- Partial volume irradiation is a useful abstraction that facilitates understanding of volume effects. Models of volume effects seek to describe the dependence of tolerance doses on the partial volume irradiated [e.g., TD(q,v)]. Histogram reduction schemes relate inhomogeneous irradiation to partial volume or whole volume treatments with the same complication probability.
- Phenomenological models (e.g., the LKB model) describe volume effects with the minimum number of parameters. Tissue architecture models (e.g., the Serial or Parallel model) use our understanding of organ function to determine the qualitative nature of volume effects.
- Score functions used in optimization of IMRT imply assumptions about the volume effects of critical normal tissues, whether these assumptions are consciously realized or not.
- Dose-volume constraints can be used in dose-based score functions to approximate the volume effects of tissues with parallel architecture.

Tumor Control Probability

- There is a well-developed formalism for evaluating TCP for a general dose distribution but the biological parameters to use in these expressions are poorly known.
- Clinically realistic TCP curves require either a small number (a few hundred) radioresistant clonogens or a population average of radiobiological parameters—or perhaps both. Current data do not discriminate between these alternatives.
- Small-volume cold spots that coincide with disease can have a drastic effect on TCP. Small volume hot spots have less dramatic effect. Cold or hot spots that do *not* coincide with disease have no effect. Unfortunately, the location of the aggressive clonogens is poorly known.
- EUD is a surrogate for TCP, which relies on fewer parameters, is less sensitive to these parameters, and is independent of clonogen number (though still dependent on their location).

References

Bentzen, S. M. "Dose Response Relationships in Radiotherapy" in *Basic Clinical Radiobiology*. G. G. Steel (ed.). New York: Oxford University Press Inc., pp. 78–86, 1997.

Bortfeld, T., J. Burkelbach, R. Boesecke, and W. Schlegel. (1990). "Methods of image reconstruction from projections applied to conformation radiotherapy." *Phys. Med. Biol.* 35:1423–1434.

Brahme, A. (1988). "Optimization of stationary and moving beam radiation therapy techniques." *Radiother. Oncol.* 12:129–140.

Brenner, D. J. (1993). "Dose, volume, and tumor-control predictions in radiotherapy." *Int. J. Radiat. Oncol. Biol. Phys.* 26(1):171–179.

Chao, K. S., J. O. Deasy, J. Markman, J. Haynie, C. A. Perez, J. A. Purdy, and D. A. Low. (2001). "A prospective study of salivary function sparing in patients with head-and-neck cancers receiving intensity-modulated or three-dimensional radiation therapy: Initial results." *Int. J. Radiat. Oncol. Biol. Phys.* 49:907–916.

Cox, J. D., J. Stetz, and T. F. Pajak. (1995). "Toxicity criteria of the Radiation Therapy Oncology Group (RTOG) and the European Organization for Research and Treatment of Cancer (EORTC)." *Int. J. Radiat. Oncol. Biol. Phys.* 31:1341–1346.

Dawson, L. A., R. K. Ten Haken, and T. S. Lawrence. (2001). "Partial irradiation of the liver." *Semin. Radiat. Oncol.* 11(3):240–246.

Eisbruch, A., R. K. Ten Haken, H. M. Kim, L. H. Marsh, and J. A. Ship. (1999). "Dose, volume, and function relationships in parotid salivary glands following conformal and intensity-modulated irradiation of head and neck cancer." *Int. J. Radiat. Oncol. Biol. Phys.* 45:577–587.

Eisbruch, A., J. A. Ship, H. M. Kim, and R. K. Ten Haken. (2001). "Partial irradiation of the parotid gland." *Semin. Radiat. Oncol.* 11(3):234–239.

Emami, B., J. Lyman, A. Brown, L. Coia, M. Goitein, J. E. Munzenrider, B. Shank, L. J. Solin, and M. Wesson. (1991). "Tolerance of normal tissue to therapeutic irradiation." *Int. J. Radiat. Oncol. Biol. Phys.* 21(1):109–122.

Esik, O., T. Bortfeld, R. Bendl, G. Nemeth, and W. Schlegel. (1997). "Inverse radiotherapy planning for a concave-convex PTV in cervical and upper mediastinal regions. Simulation of radiotherapy using an Alderson-RANDO phantom. Planning target volume." *Strahlenther. Onkol.* 173:193–200.

Field, E. A., L. P. Longman, S. Fear, S. Higham, J. Rostron, R. M. Willets, and R. S. Ireland. (2001). "Oral signs and symptoms as predictors of salivary gland hypofunction in general dental practice." *Prim. Dent. Care* 8:111–114.

Fowler, J. F. (1989). "The linear quadratic formula and progress in radiotherapy." *Br. J. Radiol.* 62:679–694.

Fowler, J., R. Chappell, and M. Ritter. (2001). "Is α/β for prostate tumors really low?" *Int. J. Radiat. Oncol. Biol. Phys.* 50:1021–1031.

Gagliardi, G., I. Lax, and L. E. Rutqvist. (2001). "Partial irradiation of the heart." *Semin. Radiat. Oncol.* 11(3):224–233.

Goitein, M. "The Probability of Controlling an Inhomogeneously Irradiated Tumor" in *Evaluation of Treatment Planning for Particle Beam Radiotherapy* (NCI contract report). S. Zink (ed.). Bethesda, MD: National Cancer Institute, 1987.

Goitein, M., and A. Niemierko. (1996). "Intensity modulated therapy and inhomogeneous dose to the tumor: A note of caution." *Int. J. Radiat. Oncol. Biol. Phys.* 36:519–522.

Hall, E. J. *Radiobiology for the Radiologist.* 4th Edition. Philadelphia: J. B. Lippincott Co., 1994.

Hellman, S. "Principles of Cancer Management: Radiation Therapy" in *Cancer: Principles and Practice of Oncology.* H. S. DeVita and S. A. Rosenberg (eds.). Philadelphia: Lippincott Williams and Wilkins, pp. 265–288, 2001.

Hernando, M. L., L. B. Marks, G. C. Bentel, S. M. Zhou, D. Hollis, S. K. Das, M. Fan, M. T. Munley, T. D. Shafman, M. S. Anscher, and P. A. Lind. (2001). "Radiation-induced pulmonary toxicity: A dose-volume histogram analysis in 201 patients with lung cancer." *Int. J. Radiat. Oncol. Biol. Phys.* 51:650–659.

Ingold, J. A., G. B. Reed, H. S. Kaplan, and M. A. Bagshaw. (1965). "Radiation hepatitis." *Am. J. Roentgenol.* 93:200–208.

Jackson, A. (2001). "Partial organ irradiation of the rectum." *Semin. Radiat. Oncol.* 11(3):215–223.

Jackson, A., G. J. Kutcher, and E. D. Yorke. (1993). "Probability of radiation-induced complications for normal tissues with parallel architecture subject to non-uniform irradiation." *Med. Phys.* 20:613–625.

Jackson, A., X. H. Wang, and R. Mohan. (1994). "Optimization of conformal treatment planning and quadratic dose objectives." Abstract. *Med. Phys.* 21:1006.

Jackson, A., R. K. Ten Haken, J. M. Robertson, M. L. Kessler, G. J. Kutcher, and T. S. Lawrence. (1995a). "Analysis of clinical complication data for radiation hepatitis using a parallel architecture model." *Int. J. Radiat. Oncol. Biol. Phys.* 31:883–891.

Jackson, A., R. K. Ten Haken, J. M. Robertson, M. L. Kessler, G. J. Kutcher, and T. S. Lawrence. (1995b). "Analysis of clinical complication data for radiation hepatitis using a parallel architecture model." *Int. J. Radiat. Oncol. Biol. Phys.* 31:883–891.

Joiner, M. C., S. M. Bentzen et al. "The Linear-Quadratic Approach to Fractionation and Calculation of Isoeffect Relationships" in *Basic Clinical Radiobiology.* G. Steel (ed.). New York: Oxford University Press Inc., pp. 106–121, 1997.

Kallman P, A. Agren, and A. Brahme. (1992). "Tumour and normal tissue responses to fractionated non-uniform dose delivery." *Int. J. Radiat. Biol.* 62:249–262.

Kaplan, E. L., and P. Meier. (1958). "Nonparametric estimation from incomplete observations." *J. Am. Stat. Assoc.* 53:447–457.

Kutcher, G. J., and C. Burman. (1989). "Calculation of complication probability factors for non-uniform normal tissue irradiation: The effective volume method." *Int. J. Radiat. Oncol. Biol. Phys.* 16:1623–1630.

Kutcher, G. J., C. Burman, L. Brewster, M. Goitein, and R. Mohan. (1991). "Histogram reduction method for calculating complication probabilities for three-dimensional treatment planning evaluations." *Int. J. Radiat. Oncol. Biol. Phys.* 21(1):137–146.

Kwa, S. L., J. V. Lebesque, J. C. Theuws, L. B. Marks, M. T. Munley, G. Bentel, D. Oetzel, U. Spahn, M. V. Graham, R. E. Drzymala, J. A. Purdy, A. S. Lichter, M. K. Martel, and R. K. Ten Haken. (1998). "Radiation pneumonitis as a function of mean lung dose: An analysis of pooled data of 540 patients." *Int. J. Radiat. Oncol. Biol. Phys.* 42(1):1–9.

Lawrence, T. S., R. K. Ten Haken, M. L. Kessler, J. M. Robertson, J. T. Lyman, M. L. Lavigne, M. B. Brown, D. J. Du Ross, J. C. Andrews, W. E. Ensminger et al. (1992). "The use of 3-D dose volume analysis to predict radiation hepatitis." *Int. J. Radiat. Oncol. Biol. Phys.* 23(4):781–788.

Levegrun, S., L. Ton, and J. Debus. (2001). "Partial irradiation of the brain." *Semin. Radiat. Oncol.* 11(3):259–267.

Levegrun, S., A. Jackson, M. J. Zelefsky, E. S. Venkatraman, M. W. Skwarchuk, W. Schlegel, Z. Fuks, S. A. Leibel, and C. C. Ling. (2000). "Analysis of biopsy outcome after three-dimensional conformal radiation therapy of prostate cancer using dose-distribution variables and tumor control probability models." *Int. J. Radiat. Oncol. Biol. Phys.* 47(5):1245–1260.

Levegrun, S., A. Jackson, M. J. Zelefsky, M. W. Skwarchuk, E. S. Venkatraman, W. Schlegel, Z. Fuks, S. A. Leibel, and C. C. Ling. (2001). "Fitting tumor control probability models to biopsy outcome after three-dimensional conformal radiation therapy of prostate cancer: Pitfalls in deducing radiobiological parameters for tumors from clinical data." *Int. J. Radiat. Oncol. Biol. Phys.* 51(4):1064–1080.

Lyman, J. T. (1985). "Complication probability as assessed from dose-volume histograms." *Radiat. Res. Suppl.* 8:S13–19.

Mohan, R., and C. C. Ling. (1995). "When becometh less more?" *nt. J. Radiat. Oncol. Biol. Phys.* 33:235–237.

Mohan, R., X. H. Wang, A. Jackson, T. Bortfeld, G. J. Kutcher, S. Leibel, Z. Fuks, and C. C. Ling. (1994). "The potential and limitations of the inverse radiotherapy technique." *Radiother. Oncol.* 32:232–248.

Munro, T. R. (1961). "The relation between tumor lethal doses and radiosensitivity of tumor cells." *Br. J. Radiol.* 34:246–251.

Niemierko, A. (1997). "Reporting and analyzing dose distributions: A concept of equivalent uniform dose." *Med. Phys.* 24(1):103–110.

Niemierko, A., and M. Goitein. (1991). "Calculation of normal tissue complication probability and dose-volume histogram reduction schemes for tissues with a critical element architecture." *Radiother. Oncol.* 20:166–176.

Niemierko, A., and M. Goitein. (1993a). "Modeling of normal tissue response to radiation: The critical volume model." *Int. J. Radiat. Oncol. Biol. Phys.* 25:135–145.

Niemierko, A., and M. Goitein. (1993b). "Implementation of a model for estimating tumor control probability for an inhomogeneously irradiated tumor." R*adiother. Oncol.* 29(2):140–147.

Pavy, J. J., J. Denekamp, J. Letschert, B. Littbrand, F. Mornex, J. Bernier, D. Gonzales-Gonzales, J. C. Horiot, M. Bolla, and H. Bartelink. (1995). "EORTC Late Effects Working Group. Late effects toxicity scoring: The SOMA scale." *Radiother. Oncol.* 35:11–15.

Poussin-Rosillo, H., L. Z. Nisce, and G. J. D'Anglo. (1976). "Hepatic radiation tolerance in Hodgkin's disease patients." *Radiol.* 121:461–464.

Roberts, S. A., and J. H. Hendry. (1998). "A realistic closed-form radiobiological model of clinical tumor-control data incorporating intertumor heterogeneity." *Int. J. Radiat. Oncol. Biol. Phys.* 41:689–699.

Roesink, J. M., M.A. Moerland, J. J. Battermann, G. J. Hordijk, and C. H. Terhaard. (2001). "Quantitative dose-volume response analysis of changes in parotid gland function after radiotherapy in the head-and-neck region." *Int. J. Radiat. Oncol. Biol. Phys.* 51:938–946.

Schultheiss, T. E. (2001). "The contoversies and pitfalls in modeling normal tissue radiation injury/damage." *Semin. Radiat. Oncol.* 11:210–214.

Schultheiss, T. E., C. G. Orton, and R. A. Peck. (1983). "Models in radiotherapy: Volume effects." *Med. Phys.* 10:410–415.

Schultheiss, T. E., L. C. Stephens, K. K. Ang, R. E. Price, and L. J. Peters. (1994). "Volume effects in rhesus monkey spinal cord." *Int. J. Radiat. Oncol. Biol. Phys.* 29(1):67–72.

Seppenwoolde, Y., and J. V. Lebesque. (2001). "Partial irradiation of the lung." *Semin. Radiat. Oncol.* 11(3):247–258.

Soderstrom, S., and A. Brahme. (1995). "Which is the most suitable number of photon beam portals in coplanar radiation therapy?" *Int. J. Radiat. Oncol. Biol. Phys.* 33:151–159.

Soderstrom, S., and A. Brahme. (1996). "Small is beautiful—and often enough: In response to Drs. Mohan and Ling, IJROBP 33(1):235–237; 1995." *Int. J. Radiat. Oncol. Biol. Phys.* 34:757–759.

Suit, H., S. Skates, A. Taghian, P. Okunieff, and J. T. Efird. (1992). "Clinical implications of heterogeneity of tumor response to radiation therapy." *Radiother. Oncol.* 25:251–260.

Thames, H. D., and J. H. Hendry. *Fractionation in Radiotherapy*. New York: Taylor and Francis, 1987.

Theuws, J. C., Y. Seppenwoolde, S. L. Kwa, I. J. Boersma, E. M. Damen, P. Baas, S. H. Muller, and J. V. Lebesque. (2000). "Changes in local pulmonary injury up to 48 months after irradiation for lymphoma and breast cancer." *Int. J. Radiat. Oncol. Biol. Phys.* 47:1201–1208.

Tome, W., and J. F. Fowler. (2002). "On cold spots in tumor subvolumes." *Med. Phys.* 29:1590–1598.

Travis, E. L. (2001). "Organizational response of normal tissues to irradiation." *Semin. Radiat. Oncol.* 11:184–196.

Tucker, S. L., H. D. Thames, and J. M. Taylor. (1990). "How well is the probability of tumor cure after fractionated irradiation described by Poisson statistics?" *Radiat. Res.* 124:273–282.

Wang, X. H., R. Mohan, A. Jackson, S. A. Leibel, Z. Fuks, Z., and C. C. Ling. (1995). "Optimization of intensity-modulated 3D conformal treatment plans based on biological indices." *Radiother. Oncol.* 37:140–152.

Webb, S. (1994). "Optimum parameters in a model for tumor control probability including interpatient heterogeneity." *Phys. Med. Biol.* 39:1895–1914.

Withers, H. (2000). "Biological aspects of conformal therapy." *Acta Oncologica* 39:569–577.

Withers, H. P., J. M. G. Taylor, and B. Maciejewski. (1988). "Treatment volume and tissue tolerance." *Int. J. Radiat. Oncol. Biol. Phys.* 15:751–759.

Woel, R. T., M. T. Munley, D. Hollis, M. Fan, G. Bentel, M. S. Anscher, T. Shafman, R. E. Coleman, R. J. Jaszczak, and L. B. Marks. (2002). "The time course of radiation therapy-induced reductions in regional perfusion: A prospective study with >5 years of follow-up." *Int. J. Radiat. Oncol. Biol. Phys.* 52:58–67.

Wu, Q., and R. Mohan. (2002). "Multiple local minima in IMRT optimization based on dose-volume criteria." *Med. Phys.* 29:1514–1427.

Wu, Q., R. Mohan, A. Niemierko, and R. Schmidt-Ullrich. (2002). "Optimization of intensity-modulated radiotherapy plans based on the equivalent uniform dose." *Int. J. Radiat. Oncol. Biol. Phys.* 52:224–235.

Yorke, E. (2001). "Modeling the effects of inhomogeneous dose distributions in normal tissues." *Semin. Radiat. Oncol.* 11:197–209.

Yorke, E. D., G. J. Kutcher, A. Jackson, and C. C. Ling. (1993). "Probability of radiation-induced complications in normal tissues with parallel architecture under conditions of uniform whole or partial organ irradiation." *Radiother. Oncol.* 26:226–237.

Yorke, E. D., et al., (2001). "Dose-volume factors contributing to the incidence of radiation pneumonitis in NSCLC patients treated with 3D-CRT." *Int. J. Radiat. Oncol. Biol. Phys.* Submitted.

Zagars, G. K., T. E. Schultheiss, and L. J. Peters. (1987). "Inter-tumor heterogeneity and radiation dose-control curves." *Radiother. Oncol.* 8:353–362.

Zaider, M., and H. I. Amols. (1998). "A little to a lot or a lot to a little: Is NTCP always minimized in multiport therapy?" *Int. J. Radiat. Oncol. Biol. Phys.* 41:945–950.

Zaider, M., and G. Minerbo. (2000). "Tumour control probability: A formulation applicable to any temporal protocol of dose delivery." *Phys. Med. Biol.* 45:279–293.

16

Geometric Uncertainties And Verification Of IMRT

Gig S. Mageras
Jenghwa Chang

Introduction

This chapter discusses two topics relevant to intensity-modulated radiation therapy (IMRT): how to account for geometric uncertainty in the treatment planning process, and how to use electronic portal imaging devices (EPIDs) to verify the relative profile and absolute dose of IMRT beam delivery. The first topic addresses patient-related factors affecting dose to target and non-target organs, while the second addresses machine-related factors affecting treatment delivery.

In the section on geometric uncertainties, the objectives are to:
- Identify sources of geometric error
- Describe ways to model the effects of geometric errors on dose to organs
- Describe ways to determine appropriate margins for planning target volumes

The section on IMRT verification describes:
- EPIDs studied: Scanning liquid ionization chamber (SLIC) and amorphous silicon (a-Si)
- EPID calibration: iterative procedure
- Dose-based verification

- EPID dosimetry software
- Comparison of SLIC and a-Si EPIDs

Geometric Uncertainties

As advances in radiotherapy technology have enabled treatment with increasing precision, they have underscored the importance of accounting for geometric uncertainties in the treatment planning process. The increasingly conformal dose distributions that are technically possible to deliver are more sensitive to treatment uncertainties that, if not dealt with properly, may result in tumor underdose or normal tissue overdose. Factors that primarily affect positional uncertainty in target and non-target organs are: (1) errors in target delineation; (2) patient positioning or setup errors; and (3) internal organ motion. Usually the patient-specific uncertainties are not known at the time the treatment is planned, since it would require substantial time and effort to measure them prior to treatment. However, they can be measured in a prior group of patients and used to estimate in a statistical fashion for a patient currently planned. We will briefly discuss each of the uncertainties listed above, and focus on methodologies for estimating their effect on a patient's plan.

In the following discussion we distinguish between systematic and random errors. A systematic error is one in which a quantity (e.g., target position) observed in the planning CT differs from its mean position over the course of treatment (e.g., as determined from measurements repeated each treatment fraction). A random error is the deviation of the quantity from its mean position on any given treatment fraction, e.g., day-to-day variation, in target position from the mean. Systematic errors in patient setup and organ position are a consequence of using a single CT for planning purposes. That is, the planning CT scan is a single measurement from a distribution of patient setup and organ positions, which in general will be different from the mean position. Systematic errors may also arise from the radiotherapy equipment itself; for example, owing to differences in the patient positioning lasers between simulation and treatment rooms, or differences in couch sag between the CT and the accelerator. In the following discussion we assume that equipment errors are small (i.e., of order 1 to 2 mm), if an adequate quality assurance (QA) program is in place.

Patient Setup Error

Variations in the position of a patient's skeletal anatomy with respect to the treatment fields are referred to as *setup errors*. Although portal films have been used in earlier studies, it is more convenient to use EPIDs, which allow precise measurements by means of image registration software. It is preferable to compare portal images to good quality digitally reconstructed radiographs (DRRs) when available, rather than simulator films, to avoid possible transfer errors between planned setup on the CT and setup on the simulator (Bel et al. 1994). In addition, by acquiring an orthogonal pair of images one can characterize the setup error in three dimensions. Table 16–1 summarizes studies of setup errors at Memorial Sloan-Kettering Cancer Center (MSKCC) for three disease sites: prostate (Hanley et al. 1997), nasopharynx (Hunt et al. 1993), and lung (Mah et al. 1998). Hurkmans et al. (2001) have reviewed setup verification practices using portal imaging.

Organ Motion

Organ motion refers to the variation of organ position and shape relative to the skeletal anatomy. During a single treatment, organ motion can be caused by respiration, heartbeat, swallowing, or peristaltic motion; between treatments it may be caused by variable filling of the bladder or gastrointestinal tract, weight change, onset or subsiding of edema, and so on. Unlike setup error, which is well approximated by a rigid body motion within a given disease site, organ motion may

Table 16–1. Summary of Patient Setup Errors in Studies Performed at MSKCC m_{sys}, SD $_{sys}$, and SD_{ran} denote mean systematic error (over patients), standard deviation systematic error, and standard deviation random error, respectively. ML, SI, and AP denote translations in the medial-lateral, superior-inferior, and anterior-posterior directions. COR, SAG, and TRANS denote in-plane rotations about the isocenter in the coronal, sagittal, and transverse planes. Positive mean values indicate right, superior, and anterior directions.

Prostate, n = 50, prone [Hanley, 1997]

Translations	m_{sys} (mm)	SD_{sys} (mm)	SD_{ran} (mm)
ML	−0.1	1.9	2.0
SI	0.4	1.4	1.7
AP	−0.3	1.3	1.9

Rotations	m_{sys} (deg)	$SD_{sys+ran}$ (deg)	
COR	9% > 1°, 0.6% > 2°	—	
SAG	−0.2	0.6	
TRANS	0.0	0.9	

Nasopharynx, n = 6 [Hunt, 1993]

Translations	m_{sys} (mm)	SD_{sys} (mm)	SD_{ran} (mm)
ML	−2.1	1.8	2.8
SI	0.6	0.8	1.6
AP	0.3	1.7	2.1

Rotations	m_{sys} (deg)	SD_{sys} (deg)	SD_{ran} (deg)
COR	0.2	0.5	1.0
SAG	−0.2	1.0	1.3
TRANS	0.1	0.4	0.9

Lung, n = 10 [Mah, 1998]

Translations	m_{sys} (mm)	SD_{sys} (mm)	SD_{ran} (mm)
ML	0.7	1.1	2.5
SI	−1.3	3.2	2.9
AP	1.1	1.6	3.1

not be fully characterized by translations and rotations owing to the deformations in organ shape that may occur. However, appropriate software tools for measuring deformations have been lacking; thus, organ motion studies have largely been limited to measurements of volume and centroid position, without explicitly attempting to quantify any deformations. Table 16–2 summarizes target motion studies at MSKCC in the prostate and lung disease sites (Melian et al. 1997; Zelefsky et al. 1999; Mah et al. 2000). It is worthwhile noting that these measurements include intra-observer random delineation uncertainties, described in the next section. Lung gross tumor volume (GTV) displacements were limited in number and varied depending on GTV location, size, and whether attached to the chest wall or mediastinum; thus, only a range of displacements are presented. Furthermore, patients in this study were stage T3 non-small-cell lung cancer, having relatively large GTV volumes. It is likely that earlier stage disease with smaller volume lung tumors would exhibit larger mobility. For a comprehensive review of organ motion and its management see Langen and Jones (2001).

Target Delineation Error

While much attention in recent years has focused on organ motion and setup error, a number of studies have shown large variations in target volume delineation between physicians and institutions. There are several contributions to uncertainty in delineation. The imaging datasets have limited resolution, particularly in the direction orthogonal to the slice planes, leading to partial volume averaging effects (Curry, Dowdey, and Murry 1990). The same observer may delineate the same object differently, referred to as *intra-observer variation*. Different observers may have different interpretations of what to include when delineating the target volume (referred to as *inter-observer error*), which can depend on the observer's training and institutional guidelines

Table 16–2. Summary of Organ Motion Studies Performed at MSKCC. See Table 16–1 caption for symbol explanations. Positive mean values indicate right, superior, and anterior directions.

Prostate, n = 13 [Melian, 1997]
4 CT scans, prone, displacement relative to initial CT

Translations	m_{sys} (mm)	$SD_{sys+ran}$ (mm)
ML	0.3	1.2
SI	0.4	3.1
AP	−0.7	4.0

Prostate, n = 50 [Zelefsky, 1999]
4 CT scans, prone, displacement relative to initial CT

Translations	m_{sys} (mm)	SD_{sys} (mm)	SD_{ran} (mm)
ML	−0.6	0.6	0.5
SI	−0.5	2.7	2.0
AP	−1.2	2.4	1.6

Lung, n = 5 [Mah, 2000]
2 breath-hold CT scans, supine, displacement from shallow expiration to shallow inspiration

Translations	Range (mm)
ML	0 to 2
SI	+5 to −10
AP	0 to 7

(Valley and Mirimanoff 1993; Ketting et al. 1997). For example, Fiorino et al. (1998) have observed small [1.8 mm standard deviation (SD)] intra-observer variations in delineation of prostate and seminal vesicles in CT; however, inter-observer variations of up to 1 cm were observed for the inferior-superior extent of the prostate plus seminal vesicles. Rasch et al. (1999) found inter-observer variations to be similar for magnetic resonance imaging (MRI) and computed tomography (CT), with variations of up to 3 mm observed at the prostate apex and base of the seminal vesicles. The availability of different image modalities and three-dimensional (3-D) image fusion software has highlighted differences in perceived target tissues. Kagawa et al. (1997) observed that prostate volumes in MRI were on average 20% smaller than in CT, with differences between the two studies usually occurring at the prostate base and apex. Similarly, Rasch et al. (1999) found that the mean ratio of CT-to-MRI prostate volumes was 1.4, with the largest differences occurring at the base of the seminal vesicles (8 mm) and prostate apex (6 mm). In head and neck, Rasch et al. (1997) found that the mean volume of the MRI-derived GTV was a factor 1.3 smaller and showed less inter-observer variation than with CT. In non-small-cell lung carcinoma, the use of PET (positron emission tomography) scanning with the tracer FDG (^{18}F-2-Fluoro-2-deoxy-d-glucose) to augment CT imaging has had significant impact on tumor definition. Prospective and retrospective studies have shown that FDG-PET images influence the design of radiation treatments in 23% to 65% of all cases and many patients are re-staged as a result of PET (Hebert et al. 1996; Kiffer et al. 1998; Nestle et al. 1999; Munley et al. 1999; Kalff et al. 2001).

Similar to setup errors and organ motion, delineation errors can be separated into systematic and random components. Intra-observer random errors may be quantified by having the observer repeatedly delineate a set of objects; however, systematic errors are more difficult to determine, owing to the need to define an object's true position and shape. Assuming that MR and PET are closer to the "truth," comparison of these image modalities to CT may provide an estimate of systematic delineation errors when using CT alone.

Accounting For Geometric Uncertainties In Treatment Plans—Monte Carlo Simulation

The standard approach (ICRU Report 50) to account for uncertainty is by adding a margin around the clinical target volume (CTV) to produce a planning target volume (PTV) (ICRU 50 1993).

The margin is chosen so as to account for all geometrical uncertainties in the CTV. Thus, delivering the prescribed dose to the PTV ensures that the CTV receives the prescribed dose as well. One then calculates a dose distribution without uncertainties that is evaluated for the PTV. The choice of margins is a trade-off: tighter margins increase the likelihood of CTV underdosage, while larger margins lead to higher volume of irradiated normal organs, which may require a reduction in the prescribed dose.

The PTV definition assumes that the prescribed dose is delivered to the CTV; however, it does not verify this. To do this, one must incorporate uncertainties directly into the dose distribution, and evaluate the dose received by the CTV. The same approach can be used to evaluate the effect of uncertainty on dose to surrounding organs at risk.

The effect of systematic and random errors on dose to the patient is qualitatively different. A systematic error in effect results in a displacement of the dose distribution from its planned position, whereas random errors result in a blurring of the dose distribution, i.e., a spreading out of isodose lines in regions of dose gradients.

Monte Carlo methods provide a powerful means of simulating physical processes, such as geometric errors and their effect on the dose to a patient. In this section we describe a method of simulating the effect of systematic setup errors, using (as input) previously measured distributions of such errors in a group of patients, the 3-D dose grid calculated for the nominal treatment plan (i.e., in the absence of uncertainties), and the dose calculation point coordinates within the CTV and relevant organs at risk. Some suppliers of commercial treatment planning systems provide access to their data files; furthermore, such an approach does not require access or detailed knowledge of the vendor's computer source code. In later sections we will describe how this method can be extended to incorporate random setup errors and organ motion.

We assume that both translational and rotational systematic errors in the three principal directions [medial lateral (ML), anterior-posterior (AP), superior-inferior (SI)] have been previously measured in a group of patients; there are no correlations between these components; and the prior distribution of errors is representative of the patient currently being planned. Further, we assume a Normal distribution for each of the six components, having standard deviations σ_x, σ_y, σ_z, $\sigma_{\theta x}$, $\sigma_{\theta y}$, and $\sigma_{\theta z}$ (x, y, z are in the right, anterior and superior directions, θ_x, θ_y, θ_z are rotations about the x, y, z axes, respectively). We use the Box-Muller method (Press et al. 1986) to generate random deviates from a Normal distribution of unit σ; and we reject deviates of greater than 3σ corresponding to probability <0.5%. In each iteration, 6 deviates $\delta_1, \delta_2, \ldots, \delta_6$ are sampled to simulate a systematic error given by a translation vector $\mathbf{T} = (\sigma_x \delta_1, \sigma_y \delta_2, \sigma_z \delta_3)$ and rotation angles $\theta_x = \sigma_{\theta x} \delta_4$, $\theta_y = \sigma_{\theta y} \delta_5$, $\theta_z = \sigma_{\theta z} \delta_6$. Next each calculation point $\mathbf{P}_i = (x_i, y_i, z_i)$ of interest (i.e., points within the CTV or an organ at risk) is transformed to location $\mathbf{P}'_i$ as caused by the simulated systematic error:

$$\mathbf{P}'_i = [R] \bullet (\mathbf{P}_i + \mathbf{T}) \tag{1}$$

where

$$[R] = [R_x] \bullet [R_y] \bullet [R_z],$$

$$[R_x] = \begin{pmatrix} 1 & 0. & 0 \\ 0 & \cos\theta_x & \sin\theta_x \\ 0 & -\sin\theta_x & \cos\theta_x \end{pmatrix}, [R_y] = \begin{pmatrix} \cos\theta_y & 0 & -\sin\theta_y \\ 0 & 1 & 0 \\ \sin\theta_y & 0 & \cos\theta_y \end{pmatrix}, [R_z] = \begin{pmatrix} \cos\theta_z & \sin\theta_z & 0 \\ -\sin\theta_z & \cos\theta_z & 0 \\ 0 & 0 & 1 \end{pmatrix} \tag{2}$$

Note that in the limit of small rotations typical for clinical situations, i.e., $\sin \theta \approx \theta$, $\cos \theta \approx 1$, the rotation matrices commute; thus, the particular order of the matrix multiplications is not important. If rotations in the prior measured group were determined about a point I other than at the origin (e.g., about the isocenter) the transformation becomes:

$$P'_i = [R] \bullet \left(P_i - I + T \right) + I \tag{3}$$

In order to reduce the calculation time, the dose at P'_i is obtained from its nearest neighbor point on the 3-D dose grid (figure 16–1a). From the set of points $\{P'_i\}$, dose-volume histograms (DVHs) are generated for the CTV and organs of interest, which represent the dose distribution resulting from this simulated systematic error. The above steps are repeated many times (typically

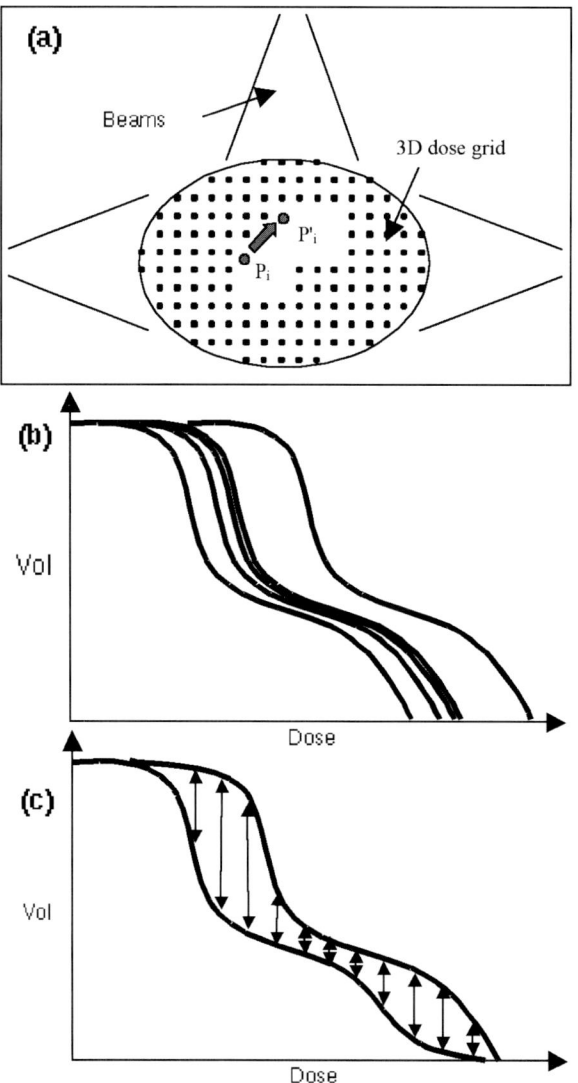

FIGURE 16–1. Schematic of Monte Carlo method to determine confidence limit "envelope" DVHs.

300 is a sufficient number) to yield an ensemble of cumulative DVHs, one for each simulated systematic error (figure 16–1b). Next, for each dose bin in the DVH, the ensemble of DVH values in that bin (i.e., the volumes) are sorted in increasing order, and two values are selected to yield a confidence limit interval in volume for that dose bin. For example, selecting the 5th and 95th percentile values give the 90% confidence limit interval. This is repeated for each dose bin to yield a confidence limit "envelope" for the DVH (figure 16–1c). Note that by its construction, the upper and lower *envelope* DVHs in general do not correspond to any single systematic displacement of the organ; rather, each dose bin may correspond to a different displacement.

The above methodology assumes that the dose distribution does not change when it is displaced with respect to the patient, i.e., it does not account for rapid changes in patient curvature or close proximity to internal inhomogeneities. In addition, the method dose not work well for organs near the patient surface, i.e., for points at a depth less than $\sim d_{max} + 3\sigma$ where σ is the standard deviation displacement in the direction of the patient surface. At shallower depths, the simulated displacements may locate the point P'_i (relative to the 3-D dose grid) in the dose buildup region or even outside the patient surface. In those instances it is better to displace the beams with respect to the patient (in a limited number of directions) and to recalculate the dose, as described in the next section. The 3-D dose grid volume must encompass the CTV and organs at risk plus a margin of $\sim 3\sigma$, to ensure that displaced points remain within the dose grid; in addition, a grid spacing of no more than 2 mm should be used to accurately determine the dose to regions of high dose gradients. We also note that in situations where the distribution of setup errors is significantly different from the Normal distribution (e.g., having long *tails*), an alternative method of sampling such a distribution is to use a numerical transformation technique, representing the distribution as a piecewise constant function (Mageras and Mohan 1993; Kalos and Whitlock 1986).

Use Of Constant Chi-Square Boundaries As Confidence Limits On Systematic Error

The methodology described in the previous section requires writing a computer program to perform, with its primary advantage being its ability to rapidly explore essentially all possible magnitudes and directions of systematic errors. A "quick and dirty" analytical scheme, which does not require a computer program, is to use boundaries of constant chi-square to define a confidence limit volume of systematic errors, then to investigate a limited number of directions within this volume that are likely to have the largest effect on dose to structures, based on inspection of the planned dose distribution. The effect of the systematic error is then simulated by moving the isocenter relative to the patient in the treatment planning system and recalculating the dose distribution.

We infer our confidence limit volume by analogy to the formalism for obtaining confidence limits on estimated model parameters, which have been determined through chi-square minimization (Press et al. 1986). In our application to systematic errors, and considering only translational errors, there are three "parameters" of interest, i.e., the components of systematic displacement, $\delta x = (\delta x, \delta y, \delta z)$, from the planned patient position. The confidence limit volume is represented by the equation for a 3-D ellipsoid:

$$\Delta\chi^2 = \delta x \bullet [C]^{-1} \bullet \delta x \qquad (4)$$

where $\Delta\chi^2$ is a function of confidence level and degrees of freedom (in this case 3) and $[C]^{-1}$ is the inverse of the covariance matrix:

$$[C] = \begin{pmatrix} \sigma_{xx}^2 & \sigma_{xy}^2 & \sigma_{xz}^2 \\ \sigma_{yx}^2 & \sigma_{yy}^2 & \sigma_{yz}^2 \\ \sigma_{zx}^2 & \sigma_{zy}^2 & \sigma_{zz}^2 \end{pmatrix} \tag{5}$$

The covariance matrix is estimated from measurements of the systematic error $\delta\mathbf{x}_i$, $i = 1,N$ in a prior group of N patients:

$$\sigma_{xx}^2 = \frac{1}{N}\sum\left(\delta x_i - \overline{\delta x}\right)^2, \sigma_{yy}^2 = \frac{1}{N}\sum\left(\delta y_i - \overline{\delta y}\right)^2, \sigma_{zz}^2 = \frac{1}{N}\sum\left(\delta z_i - \overline{\delta z}\right)^2,$$

$$\sigma_{xy}^2 = \frac{1}{N}\sum\left[\left(\delta x_i - \overline{\delta x}\right)\left(\delta y_i - \overline{\delta y}\right)\right], \sigma_{xz}^2 = \frac{1}{N}\sum\left[\left(\delta x_i - \overline{\delta x}\right)\left(\delta z_i - \overline{\delta z}\right)\right], \tag{6}$$

$$\sigma_{yz}^2 = \frac{1}{N}\sum\left[\left(\delta y_i - \overline{\delta y}\right)\left(\delta z_i - \overline{\delta z}\right)\right]$$

where $\overline{\delta x}$, $\overline{\delta y}$, $\overline{\delta z}$ are the components of the mean systematic error for the patient group (which should be close to zero). For a 90% confidence limit region (i.e., 90% of patients would have a systematic error somewhere in this volume), $\Delta\chi^2 = 6.25$. In the case of uncorrelated displacements (i.e., $\sigma_{xy} = \sigma_{xz} = \sigma_{yz} = 0$), the equation for the 90% confidence limit ellipsoid becomes

$$6.25 = \frac{(\delta x)^2}{\sigma_{xx}^2} + \frac{(\delta y)^2}{\sigma_{yy}^2} + \frac{(\delta z)^2}{\sigma_{zz}^2} \tag{7}$$

For similar magnitude errors in the three directions (i.e., $\sigma_{xx} \sim \sigma_{yy} \sim \sigma_{zz} = \sigma_{sys}$), the volume described by Equation (7) reduces to a sphere of radius $2.5\sigma_{sys}$.

We compare the prediction of this analytical model with that from the Monte Carlo simulation of the previous section, for the case of an IMRT plan for treatment of nasopharynx carcinoma. The plan goals were to deliver 100% of the prescribed dose to the smaller volume PTV (green) and 78% to the larger PTV (pink), while keeping the cochlear dose below 85% (figure 16–2). In this example, we assume translational systematic setup errors only, with $\sigma_{xx} = \sigma_{yy} = \sigma_{zz} = 2$ mm (Hunt et al. 1993), and no correlations between the components of the displacements. Equation (7) reduces to $6.25 = [(\delta x)^2 + (\delta y)^2 + (\delta z)^2]/4$, i.e., a sphere with radius $\delta r = [(\delta x)^2 + (\delta y)^2 + (\delta z)^2]^{1/2} = 5$ mm. By inspection of the dose distribution (figure 16–2a) we see that the dose falloff outside the CTV (yellow contour) is in the direction of the left cochleum (red), owing to the close proximity of that organ to the CTV. Thus, we estimate the lower confidence limit DVH for the CTV, as well as lower limit DVH for the left cochleum, by displacing the patient 5 mm along a line from the center of the CTV toward the left cochleum, or equivalently, by displacing the beams in the opposite direction. Similarly, we estimate the upper limit DVH for the left cochleum by displacing the patient 5 mm along the same line but toward the CTV. Figure 16–2b compares the resultant DVHs against the 90% confidence limit envelope DVHs from the Monte Carlo simulation. Note the good agreement between the two predictions for the left cochlear limits and for the CTV lower limit. The analytical method is less accurate at predicting an upper limit DVH for the CTV. For example, if we naively displace the patient 5 mm along the line from the cochleum toward the CTV, we obtain the DVH for the CTV shown as the rightmost red curve in figure 16–2b. The analytical prediction is below that for the Monte Carlo method, suggesting that the displacement corresponding to the true upper limit is in a direction different from our chosen one.

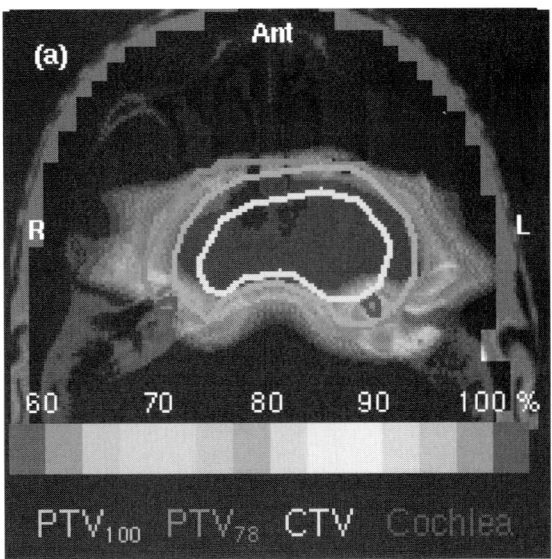

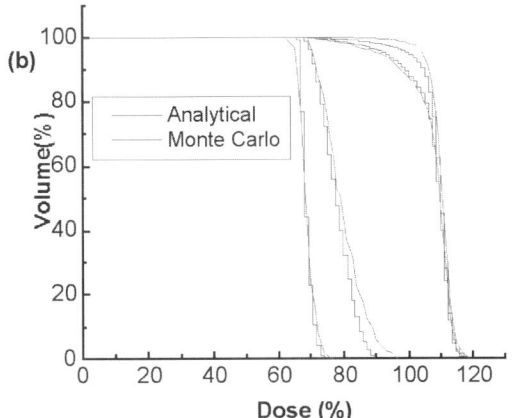

FIGURE 16–2. (a) Transverse section of nasopharynx IMRT plan with dose color wash overlay. (b) Comparison of upper and lower confidence limit DVHs for CTV and left cochleum. See COLOR PLATE 42.

Simulation Of Random Setup Errors In The Dose Calculation

Random, or day-to-day, displacements in effect blur the planned dose distribution. For treatments consisting of 30 or more fractions, the effect of random errors on dose can be reasonably approximated by a convolution:

$$D_{blur}(x,y,z) = \iiint D_{plan}(x',y',z')\, f(x-x',y-y',z-z')\, dx'dy'dz' \tag{8}$$

where $D_{blur}(x,y,z)$ is the *blurred* dose distribution, $D_{plan}(x',y',z')$ is the planned dose distribution (in the absence of random errors) and $f(x-x',y-y',z-z')$ is the distribution of random errors. Note that this formulation is limited to rigid body transformations, or to uniform expansion and contraction, i.e., f is independent of position within the patient.

A computationally faster approach, which can be used with pencil beam dose calculation algorithms, is to first convolve the pencil beam kernel with a Gaussian representing the distribution of random setup errors. In the absence of setup uncertainties, the dose at position x,y and depth d from an intensity-modulated field is computed as (Mohan and Chui 1987; Chui, LoSasso, and Spirou 1994):

$$D_{imrt}(x,y,d) = D_{open}(x,y,d) \ [D'_{imrt}(x,y,d)/D'_{open}(x,y,d)] \qquad (9)$$

where D_{open} is the dose from an open, unblocked field, and the ratio in brackets is a correction factor calculated in a flat homogeneous phantom:

$$D'_{open}(x,y,d) = \iint \varphi_{open}(x',y') \ k(x-x',y-y',d) \ dx'dy' = \varphi_{open} * k$$

$$D'_{imrt}(x,y,d) = \iint \varphi_{imrt}(x',y') \ k(x-x',y-y',d) \ dx'dy' = \varphi_{imrt} * k \qquad (10)$$

where φ_{open} and φ_{imrt} are the photon fluence distributions for the open and IMRT fields, $k(x',y',d)$ is the pencil beam kernel, and $\varphi*k$ represents the convolution operation. The convolutions can be carried out rapidly with fast Fourier transforms.

To include the effect of random setup error (Chui, Kutcher, and LoSasso 1992), the uncertainty distribution described by a function $f_r(x_r,y_r,z_r)$ in the room coordinate system is transformed into a function $f_{ran}(x,y,z)$ in the beam coordinate system, where z is the direction along the beam central axis, and x,y are the transverse coordinates. Assuming that a patient setup error is equivalent to a beam displacement in the opposite direction, and that the effect on dose of uncertainty in the z direction is negligibly small relative to transverse displacements, for mean value (over treatment fractions) dose distributions Equation (10) becomes

$$<D'_{open}> = \varphi_{open} * k * f_{ran}, <D'_{imrt}> = \varphi_{imrt} * k * f_{ran} \qquad (11)$$

If we further assume that the distribution f_{ran} is the same for all patients (a reasonable assumption, since we don't know the patient-specific f_{ran} distribution a priori), the convolution $k*f_{ran}$ may be performed in advance only once, and the results stored in computer files. The specific form of f_{ran} depends on the shape of the random error distribution, but is typically a 3-D Normal distribution, i.e., $\propto \exp[-1/2(x^2/\sigma^2_x + y^2/\sigma^2_y + z^2/\sigma^2_z)]$.

Incorporating Organ Motion In The Dose Calculation

The simplest means of incorporating organ motion in the above methodology is to add the variances for organ centroid position, σ^2_{om}, and setup error, σ^2_{se}, and do so separately for systematic and random components, i.e., $\sigma^2_{sys} = \sigma^2_{sys,se} + \sigma^2_{sys,om}$ and $\sigma^2_{ran} = \sigma^2_{sys,om} + \sigma^2_{ran,om}$. This assumes that organ displacements are uncorrelated with patient setup errors; further, it assumes no organ rotations or shape deformations. Thus, in some instances it may underestimate organ motion effects on dose. We have examined the validity of the affine transformation of modeling variability of the prostate, bladder, and rectum shape (Fontenla et al. 2000). The transformation from a point $\mathbf{P} = (x, y, z)$ inside the organ in the planning CT to the corresponding point $\mathbf{P'} = (x', y', z')$ at some later time (e.g., as observed in a CT scan during treatment) is

$$
\begin{aligned}
x' &= a_{11}x + a_{12}y + a_{13}z + b_1 \\
y' &= a_{21}x + a_{22}y + a_{23}z + b_2 \\
z' &= a_{31}x + a_{32}y + a_{33}z + b_3
\end{aligned}
\qquad (12)
$$

where $a_{11}, \ldots, a_{33}, b_1, \ldots, b_3$ are scalar coefficients. The coefficients are determined with a 3-D image registration algorithm that minimizes the distances between the surface of the delineated organ in the planning CT and that in the treatment CT (Pelizzari et al. 1989). The registration procedure uses only a subset of the possible affine transformations, i.e., translations, rotations, and expansion or contraction in the three principal directions.

We have applied this procedure to the serial CT image sets of 20 patients from our previous study of prostate motion (Zelefsky et al. 1999). The one-standard-deviation residual surface differences between the above-modeled prostate transformations and physician-drawn prostate in the "treatment" CT are 1.5 mm at the anterior, posterior and lateral prostate boundaries, increasing to 5 mm at the superior and inferior boundaries. The larger residuals in the latter direction are a result of the lower CT resolution in the superior-inferior direction (5 mm slice thickness and spacing), and the larger intra-observer delineation uncertainty arising from the difficulty in discerning the prostate-bladder boundary and prostate apex with CT. The variability in bladder and rectum is not as well modeled (1-SD residuals of 2 to 6 mm for bladder, 3 to 5 mm for rectum, depending on direction), owing to deformations of these organs that cannot be fully accounted for with the above transformation. However, the portion of rectum and bladder adjacent to the prostate can be assumed to move along with the prostate, and hence can be accurately modeled with the same transformation for prostate. It is also worth noting that this model at present is not experimentally verified for displacement of points internal to the organs.

In the Monte Carlo simulation, each point $P = (x,y,z)$ is transformed first for setup error, then for organ motion:

$$P' = T^{(om)}T^{(se)}P \tag{13}$$

where the transformations $T^{(se)}$ and $T^{(om)}$ are given by Equations (1) and (12), respectively. If there are M serial "treatment" CT scans available in the prior group of N patients, yielding a set of organ motion transformations $T_{m,n}^{(om)}$, $m = 1,M$; $n = 1,N$, a systematic error for organ motion can be simulated by randomly selecting a patient from the prior group, then averaging over the M organ motion transformations of the prior patient:

$$P'_{sys} = 1/M \sum_m T_{m,n}^{(om)} T^{(se)} P \tag{14}$$

Planning Target Volume Margin Definition

Although in the above discussion we have eschewed the PTV for plan evaluation, it is nevertheless useful for beam aperture definition or IMRT optimization. We describe two approaches for deriving PTV margins, analytical and Monte Carlo, which make use of the methodologies we have described above.[1]

In the analytical approach, we have shown earlier that we can define a 90% confidence limit volume for systematic errors given by Equation (7), where the one-standard-deviation systematic uncertainties $\sigma_{sys,x}$, $\sigma_{sys,y}$, $\sigma_{sys,z}$, are a quadrature sum of those for setup error and organ

---

[1] The more recent ICRU Report 62 advocates separation of the PTV into two separate margins: one to account for internal uncertainties (resulting in an internal target volume), and a second margin for external errors such as patient setup (ICRU 1999). However, statistics dictates that uncorrelated errors be added in quadrature, thus in conflict with the idea of separate internal and external margins, a point also made by other investigators (van Herk et al. 2000; Craig et al. 2001). Therefore, in this discussion we will consider a single PTV margin for combined uncertainties.

motion. If we displace the CTV in all possible directions allowed by this confidence limit volume, we define a resultant volume PTV_{sys} that encloses the original CTV with margins along the principal axes x,y,z given by $2.5\sigma_{sys,x}$, $2.5\sigma_{sys,y}$, and $2.5\sigma_{sys,z}$, respectively. Furthermore, we have seen that the effect of random uncertainties on dose can be represented as a convolution of pencil beam dose kernel k with the random uncertainty distribution f_{ran} [Eq. (11)], thus broadening the pencil kernel to an effective width $\sigma_{eff} = (\sigma_k^2 + \sigma_{ran}^2)^{1/2}$. We can calculate the increased spread of the isodose lines in the dose falloff region by using Equation (11) to calculate the dose at the edge of a uniform fluence distribution. Representing the fluence as a step function ($\varphi_{imrt} = 0$ for $x < 0$ and $= 1$ for $x > 0$) and the pencil beam kernel by a Normal distribution in Equation (11), we find that the 95% isodose occurs (before random uncertainties) a distance $x = 1.65\sigma_k$ inside the edge of the fluence distribution. Therefore, in the presence of random uncertainties the 95% isodose would be displaced inward by an additional amount $1.65(\sigma_{eff} - \sigma_k)$. If we now require that the CTV for 90% of the patients receive at least 95% of the prescribed dose, we need to add a margin of $1.65(\sigma_{eff} - \sigma_k)$ to PTV_{sys} to account for random uncertainties, or approximately $(1.2\sigma_{ran} - 1.4)$ mm for a 6 MV pencil beam kernel at 10 cm depth of $\sigma_k = 3.5$ mm (including the effects of finite beam source size). We note that Stroom et al. (1999) and van Herk et al. (2000) previously have derived PTV margin formulas based on similar arguments.

The Monte Carlo approach (Mageras et al. 1999) uses as input the contours of the delineated CTV, and generates an ensemble of CTV positions by repeatedly simulating systematic (random organ motion and setup errors using Equation (13). From the ensemble, the method calculates a 3-D occupancy probability distribution (OPD) for the CTV, symbolized as $\rho_{ctv,i}$. This is the likelihood that each voxel i, on a 3-D grid in the treatment room coordinate system (i.e., fixed with respect to the radiation field), is inside the CTV. The occupancy probability of each voxel is calculated by counting the number of times the voxel is inside the CTV (figure 16–3). Next, the voxels are ranked in order of decreasing ρ_{ctv}, and then summed in that order to calculate an expectation value for the fractional CTV volume within the PTV, given by $<V_{ctv}> = \Sigma^{i \in ptv} \rho_{ctv,i}$ until a pre-selected value is reached. The PTV is then defined as the volume enclosing the voxels included in the summation. In practice we determine the optimal value of $<V_{ctv}>$ by computing the PTV and the associated treatment plans for different values of $<V_{ctv}>$, then calculating confidence limit DVHs for the CTV incorporating uncertainty as described earlier. In our studies of lung and prostate, we have found that a value $<V_{ctv}> = 0.95$ results in acceptable confidence limits on dose to CTV,

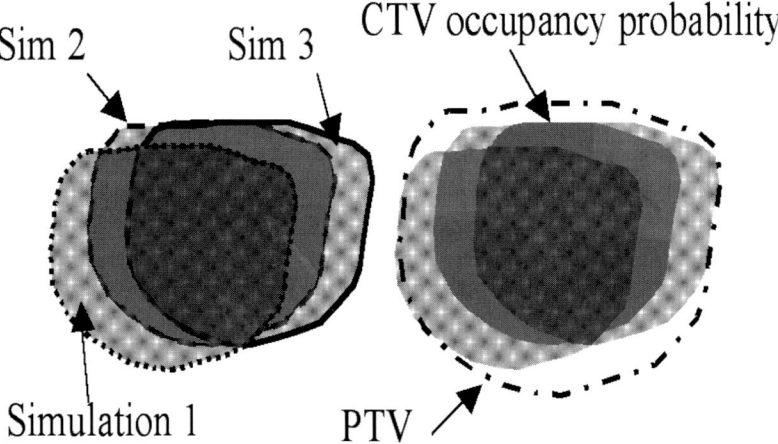

FIGURE 16–3. Schematic of Monte Carlo simulation. Simulated organ motion and setup errors produce an ensemble of possible CTV positions (left), to calculate a CTV occupancy probability (right). Darker shaded regions indicate higher probability.

i.e., at least 95% of the prescribed dose to 90% of the patients; however, the appropriate threshold should be confirmed for a given disease site and treatment technique.

The Monte Carlo approach also allows constraints to be placed on the inclusion of critical organs in the PTV, by excluding from the PTV those voxels with the large occupancy probability ρ_{crit} for the critical organ, such that the fractional volume is below a desired amount, e.g., $<V_{crit}> = \Sigma^{i \in ptv} \rho_{crit,i} = 0.10$. The critical organ constraint is implemented by first selecting voxels satisfying the $<V_{ctv}>$ requirement above, sorting them in order of increasing ρ_{crit}, and performing a running sum of $<V_{crit}>$ until the specified value (e.g., 0.10) is reached. This preferentially selects voxels in the PTV that are least likely to be in the critical organ. Note that in some circumstances this procedure may reduce the CTV coverage, i.e., $<V_{ctv}>$. In its application to prostate conformal treatments (Mageras et al. 1999), the rectal volume constraint reduced $<V_{ctv}>$ by only 1% to 2%; thus, both constraints were largely maintained. However, in general it is possible to specify a set of constraints that cannot be simultaneously satisfied. Rather than having the critical organ constraint take precedence, an alternative is to adjust the margin based on planner-specified relative importance of constraints. For example, if CTV and critical organ have equal importance, the threshold ρ_{crit} can be chosen such that both CTV and critical organ constraints are exceeded by an equal amount. An area of further development is to optimize all thresholds, for CTV and critical organs, simultaneously such that an objective function consisting of a weighted combination of volume constraints is maximized.

Following the application of volume constraints, a minimum margin around the CTV in the planning CT scan may be imposed if desired, by including in the PTV any voxels within the minimum margin to the CTV. This allows the physician to impose larger margins than those determined from the generic data. For example, based on the generic organ motion data for the prostate, left-right motion of the prostate is observed to be small, hence the resulting margin may be only a few millimeters, whereas the physician may desire a larger margin, given that are no nearby critical organs to limit the dose at that location. Finally, a binary trace algorithm (Ballard and Brown 1982) draws a contour around the PTV on each CT slice.

Since Monte Carlo simulation may be time-consuming (computation times with unoptimized code are on the order of 30 min), it is more useful for investigating "class solution" margins for a particular site and technique. Figure 16–4 shows a comparison of Monte Carlo defined PTV with the physician-drawn PTV for respiration-gated treatment of a lung tumor. The Monte Carlo

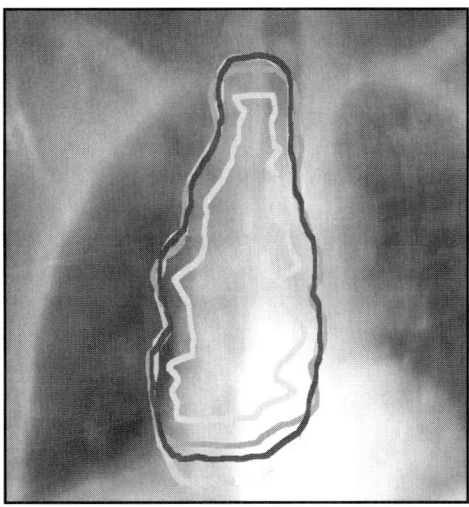

FIGURE 16–4. Comparison of Monte Carlo calculated PTV (black) with physician-drawn PTV (dark gray). Lung GTV is shown as the inner contour (light gray).

simulation assumed 5 mm systematic uncertainty (1 standard deviation) in the SI direction and 3 mm in the AP and LR directions, and 3 mm random uncertainty (1 SD) in the SI, AP, and LR directions. The larger systematic error in the SI error is to account for the additional inter-fractional variation observed in diaphragm (and by inference, GTV) position in gated treatments (Ford et al. 2002). The Monte Carlo simulation suggests that the PTV margins along the inferior and superior borders should be slightly larger than the uniform 1cm margin drawn by the physician.

Quality Assurance Of IMRT Beam Delivery Using Electronic Portal Imaging Devices (EPIDs)

Introduction

Verification of beam delivery is the end-point quality assurance (QA) for a radiation treatment; for IMRT beams, both relative profile and central-axis (CAX) dose need to be verified. Originally, we used film and ion-chamber for this purpose, and achieved better than 2% accuracy; however, both measurements are time-consuming and not suitable for routine clinical use that requires a large quantity of measurements, but does not demand a high accuracy (better than 5% is acceptable). We have studied the use of EPIDs to replace (at least a portion of) film or ionization chamber for verifying routine IMRT treatment, and have established a dose-based verification system that is accurate for better than 5% for prostate IMRT fields (Chang et al. 2000).

EPIDs Studied

We studied two EPIDs: SLIC (scanning liquid-filled ionization chamber) EPID (Mark 2, and LC250) and amorphous silicon (a-Si) EPID (aS500) of Varian Medical Systems (Palo Alto, CA), both controlled using the PortalVision™ (PV) software.

SLIC EPID – Mark 2 And LC250

A Varian Mark 2 SLIC EPID consists of image detection unit (IDU), image acquisition system (IAS), PV workstation, and cables. As shown in figure 16–5, the IDU, consisting of the SLIC panel and scanning electronics, is housed on a retractable arm (R-arm) that can move three-dimensionally and position the EPID at different source-to-detector distance (SDD) from 105 cm to 180 cm. There are collision detectors inside the cover to automatically stop the motion of gantry, couch, and R-arm. The SLIC panel is a matrix of 256×256 straight wire electrodes enclosed in a chamber filled with a special liquid Iso-Octane. The sensitive area of the matrix is 32.5×32.5 cm^2, corresponding to a pixel size of 1.27 mm. The control circuit applies high voltage (up to 450 volts) to sequentially turn on each row of the matrix for a specified period of time, during which the electrodes along the turned-on row are scanned by the electrometer amplifiers. The pixel value (ionization current) of each ionization chamber is a linear quadratic function of the square root of the portal dose rate to that pixel (van Herk 1991).

The IAS, typically found at the back of the linear accelerator (linac) contains software drivers and electronics for controlling the IDU, interfacing the linac and R-arm, and networking with the PV workstation. PV software provides user interfaces for image acquisition, treatment review, calibration, maintenance, and dosimetry. Although Mark 2 is also referred to as LC250 in Varian's reference manual, we use LC250 here to specifically mean the Mark 2 system with additional dosimetry capability, achieved by adding a dosimetry module to the IAS and a dosimetry workspace in the PV software. Image acquisition is synchronized to the linac beam pulses, or is self-triggered. Perfect timing between the linac pulses and scanning pattern is crucial to the image quality, and to the accuracy, of the EPID dosimetry.

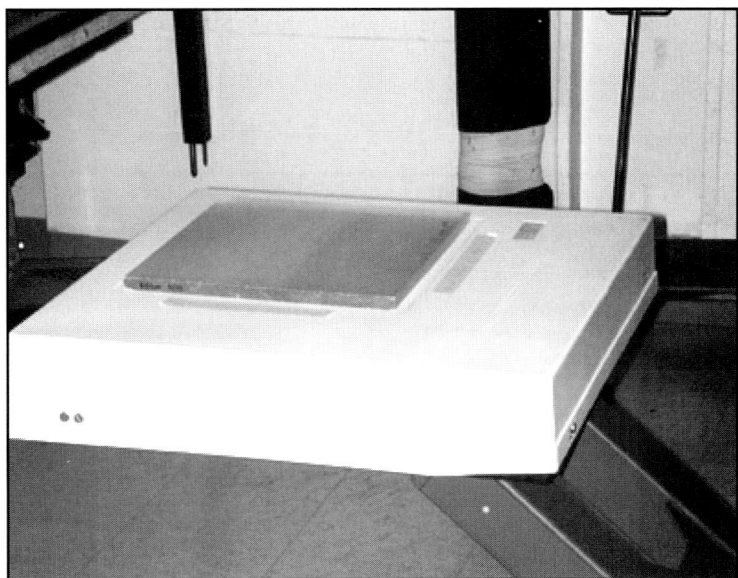

FIGURE 16–5. Photograph of the image detection unit (IDU) of a Varian Mark2 SLIC EPID. A Varian aS500 EPID has a similar appearance. The 1.5 cm polystyrene slab is added to provide a full buildup and remove electron contamination.

A-Si EPID

aS500, the latest generation of Varian's a-Si EPID, also consists of an IDU, IAS, PV workstation, and cables. The appearance of the IDU and R-arm is the same as that of Mark 2 (figure 16–5); unlike Mark 2, both IDU and IAS are mounted on the R-arm. The IDU consists of an a-Si panel, drive electronics, readout electronics, and interface electronics (to IAS). The physical buildup of the a-Si detector consists of a copper plate and a phosphor screen for transforming X-rays to visible light. The a-Si panel is a matrix (512×384) of a-Si detectors, each consisting of a light sensitive photodiode to integrate the light emitted by the phosphor screen, and a thin-film transistor (TFT) switch to the readout electronics. The sensitive area is 40×30 cm^2, corresponding to a pixel size equals 0.784 mm. Each row is sequentially scanned and read in the same manner as the Mark 2 system; the detector reading is linearly proportional to the EPID dose. Potentially high imaging speed (~10 frames/sec achievable) and excellent dose linearity make it a favorable imaging device for IMRT verification.

EPID Calibration And Iterative Procedure

In this section, we introduce the dosimetry system for our EPID verification system.

Relative Dose Unit Definitions

Instead of measuring the absolute deposited energy, we define the dose delivered to d_{max} by a 10×10 cm^2 field at the source-axis distance (SAD) as the unit dose. For example, D_{EPID} is the dose to the EPID in relative dose units (RDU), defined as the ratio of energy per unit mass deposited at the center of the EPID for the field of interest, to that of a 10×10 cm^2 field with the EPID located at SAD. $\dot{D}_{EPID}$ is the dose rate to the EPID, in units of RDU/min.

Similar definitions can be applied to other detectors (film, ionization chamber), and other geometries (full-, slab- or mini-phantom). Assuming that the detectors are stable within one experiment and the relative doses are measured for each experiment, the above definitions eliminate the absolute dose measurement, and remove the output fluctuation (±1%) between experiments. Dose conversion between different phantom geometries can be achieved using the ratio of phantom scatter factors; absolute dose can be calculated as the product of the number of relative dose units and the absolute dose of the reference field.

EPID Buildup And Scattering

Although the intrinsic buildup for both Mark 2 and aS500 EPIDs is good for the 6 MV beam, additional buildup that maximizes the detector reading is needed for high-energy beams. Both aS500 and Mark 2 EPIDs at our institution require a ~1.5 cm polystyrene slab for the 15 MV beam. We use the EPID scatter factor, S_{pe}, to incorporate the field size dependence due to scattering in the EPID. Definition of S_{pe} (Chang et al. 2000; Chang et al. 2001) is similar to that of the phantom scatter factor of water. The S_{pe} of aS500 and LC250 EPIDs are closer to the phantom scatter factor of full water phantom, S_p, than that of a slab phantom, S_{ps}.

EPID Calibration—Iterative Procedure

A typical calibration involves irradiating the EPID with different doses (dose rates) (figure 16–6a), recording the readings, and independently determining the "reference dose (rate)" (figure 16–6b). The calibration curve is a plot of detector reading *vs.* reference dose (rate), using interpolation/extrapolation or by function fitting. The dose rate is varied using either different SDDs, or with attenuator. The range of EPID positions (from 105 to 180 cm) can change the dose rate by a factor of 3. Depending on the beam energy, the beam can be further attenuated 3 to 4 orders of magnitude using lead sheets on top of the block tray. Notice that lead attenuator may significantly change (harden) the beam quality, a serious problem for energy-dependent detectors, for example, the aS500 detector for the 6 MV beam.

The reference dose (rate) is an "estimate" of the portal dose (rate) under the same irradiation conditions for acquiring the EPID readings of a given field size, and is usually converted from an ionization chamber measurement using the ratio of phantom scatter factors (Chang et al 2000), assuming that S_p, S_{ps}, and S_{pe} are known *a priori*. However, the measurement of S_{pe} requires the calibration curve to convert EPID reading to portal dose (rate). For a detector whose response is linear with dose (e.g., a-Si), S_{pe} can be obtained from the ratio of EPID readings at the given field size and reference ($10 \times 10 \text{ cm}^2$) field size, without knowing the slope of the calibration curve. For a (e.g., SLIC) detector with nonlinear dose response, we have a deadlock situation as the reference dose (rate) and S_{pe} are needed for each other, which can be solved using the iterative procedure (Chang et al. 2001).

Figure 16–7 illustrates the flow chart of this procedure. There are seven steps in this flow chart, where steps 2 through 6 are iterative. Steps 2 through 5 are used to update each variable based on the input and the results from the previous iteration, and step 6 determines the convergence of this procedure by checking the consistency among the variables. If $\left|S_{pe}^{new} - S_{pe}^{old}\right|$ is larger than a pre-determined tolerance, it goes back to update the reference dose rate and calibration curve. Otherwise, the iterative process is converged and terminated (step 7).

Dose-Based Verification

In this section, we review the theory for the dose–based verification of IMRT beam delivery (Chang et al. 2000).

Dose Integration

Two different approaches, tracking the leaf motion or measuring the delivered integral dose profile, have been developed for IMRT treatment QA using EPIDs. The first approach requires faster imaging speed, and is more susceptible to the noise effect. We adopt the second approach because

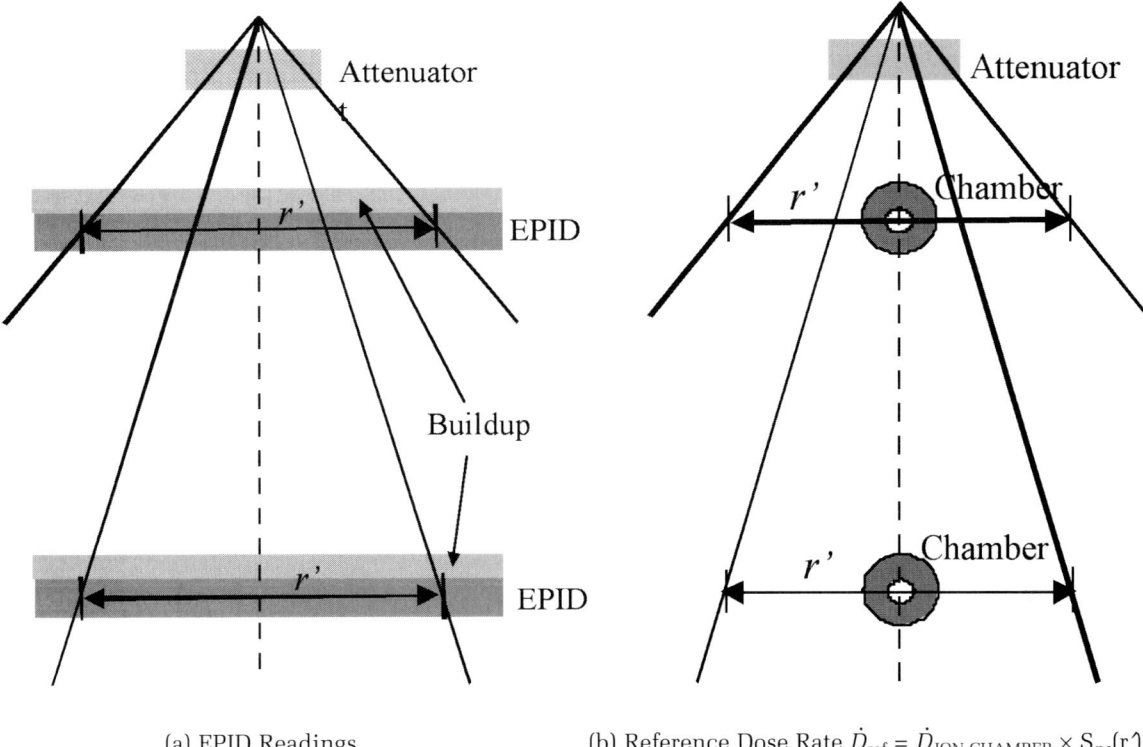

(a) EPID Readings

(b) Reference Dose Rate $\dot{D}_{\mathrm{ref}} = \dot{D}_{\mathrm{ION\text{-}CHAMBER}} \times S_{pe}(\mathrm{r}\,)$

FIGURE 16–6. Experimental setup to measure the calibration data in figure 16–5. (a) EPID readings were obtained for different radiation intensities achieved using either different source-to-detector distance (SDD, defined by the location of the SLIC detector panel in the EPID) or different attenuation. (b) An ion chamber with a 3-mm Cu buildup cap was also used to measure the dose rate for the same SDD (defined by the center of the ion-chamber,) field size and attenuation. "r" is the field size at SDD. The reference dose rate is derived from the ionization chamber measurement (see text for details). Instead of using a buildup cap, the ion chamber can also be put into a full/slab phantom. The added buildup is 1.5 cm polystyrene for 15X. (From Chang et al. 2001. Reprinted by permission of American Association of Physicists in Medicine.)

it is more robust and is a direct measure of the delivered dose. Integration over time is required if the EPID signal depends on dose rate, e.g., a SLIC matrix, or if the detector may saturate, e.g., the a-Si detector. Figure 16–8 shows eight selected dose rate maps and the integrated dose map of a treatment field from 25 images obtained during the course of irradiation. Each EPID reading, $I[i,j]$, in figure 16–8 is first converted to a dose rate, $\dot{M}[i,j]$, using the calibration curve, and then summed as

$$M[i,j] = \sum_{k=1}^{K} \dot{M}_k[i,j]\Delta t_k \tag{15}$$

where $M[i, j]$ is the integrated dose map for pixel at location $[i, j]$, K is total number of images, and Δt_k the time interval for $\dot{M}[i,j]$, (i.e., the time between images). An image acquisition rate of about 1 image per second is sufficient to reconstruct satisfactory integrated dose profiles for a typical prostate intensity-modulated (IM) field using dynamic multileaf collimation (DMLC) under normal treatment conditions—dose rate between 200 and 300 MU/min, maximum leaf speed 3 cm/s, and delivered dose between 100 to 150 MU per field. Since this approach does not track

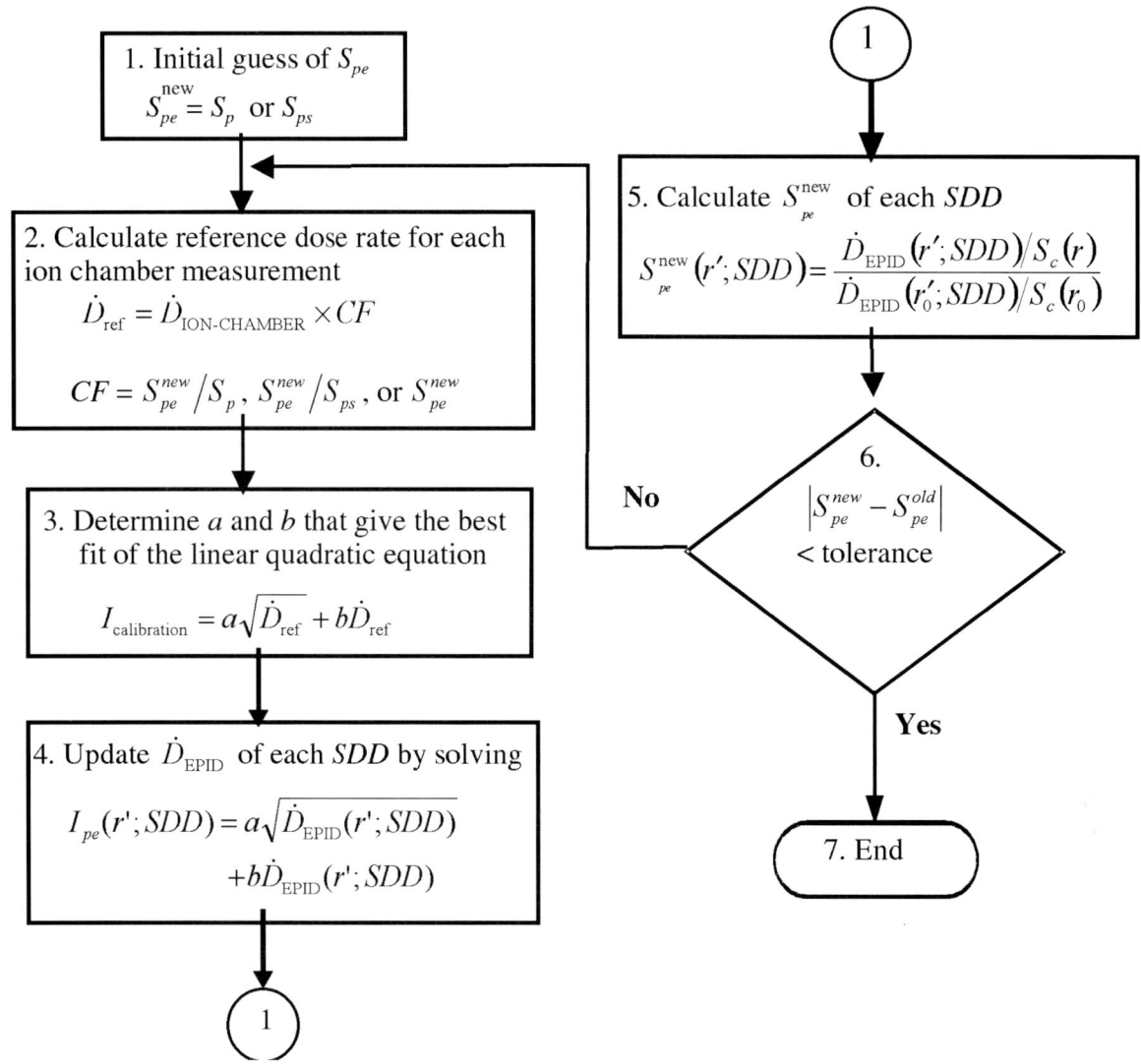

FIGURE 16–7. Flow chart of the iterative calibration procedure. r is the field at SAD and r′ is the field at the EPID plane. r_0 and r'_0 are both 10×10 cm² at the corresponding plane. $\dot{D}_{ref}$ is the reference dose rate. $\dot{D}_{ION\text{-}CHAMBER}$ is the dose rate measured using an ionization chamber for different SDD and attenuation combination. CF is the corresponding conversion factor when the ionization chamber is in a full, slab, or mini phantom. $I_{calibration}$ are the EPID readings from the same irradiation condition for the ionization chamber. I_{pe} are the EPID readings for S_{pe} measurements that are usually done for at least a subset of the SDDs used by the ionization chamber irradiation (S_{pe} for other SDD is determined by interpolation). $\dot{D}_{EPID}$ is the portal dose rate and is converted from I_{pe} using the updated calibration curve. S_{pe} of each SDD is re-calculated during each iteration. $\left| S_{pe}^{new} - S_{pe}^{old} \right|$ is the norm of the difference between the old S_{pe} and the new S_{pe} of each SDD. The iterative procedure is repeated until $\left| S_{pe}^{new} - S_{pe}^{old} \right|$ is smaller than a pre-determined tolerance. (From Chang et al. 2001. Reprinted by permission of American Association of Physicists in Medicine.)

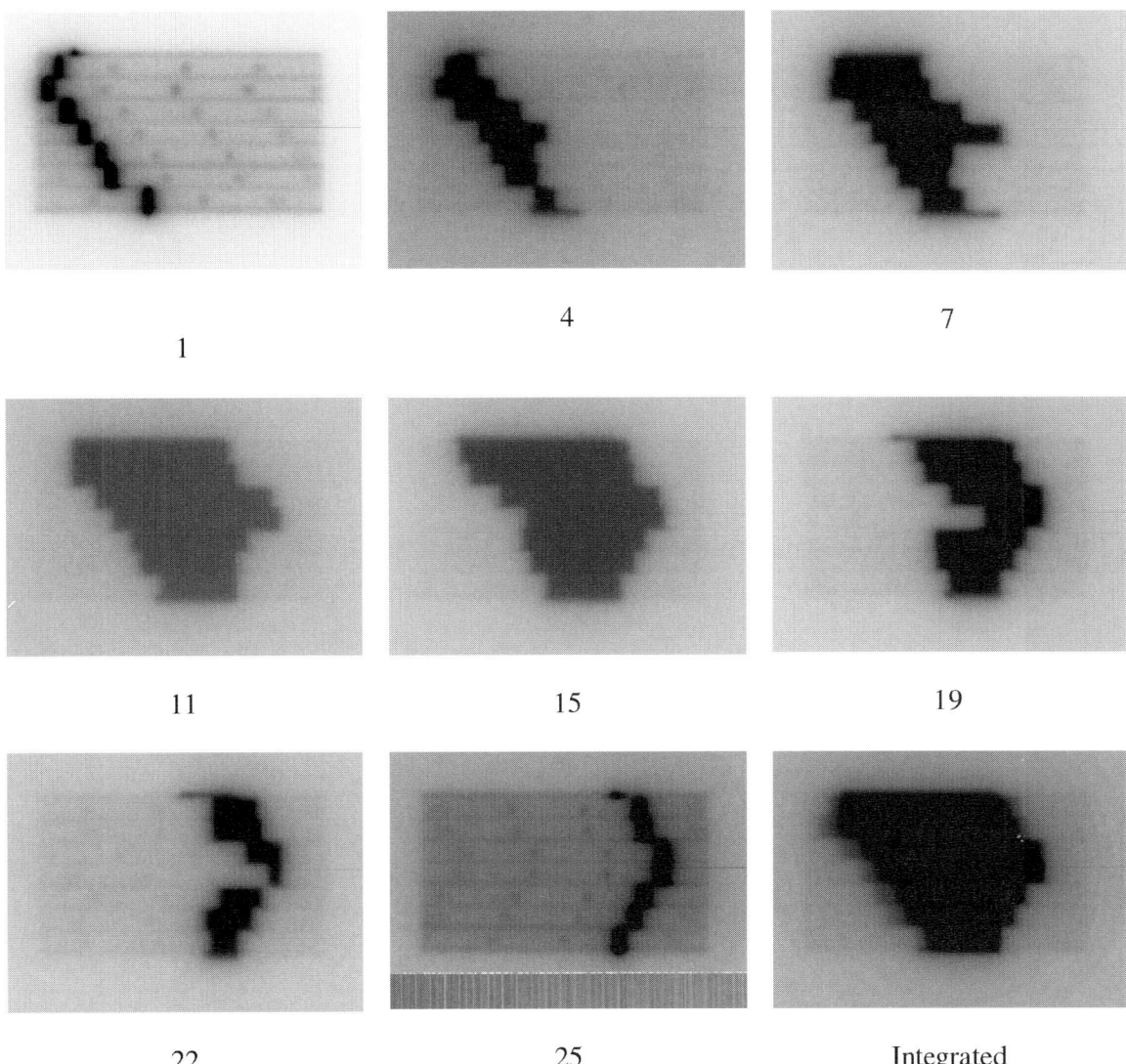

FIGURE 16–8. Sequence of images of an intensity-modulated beam for prostate treatment using sliding window technique at Memorial Sloan-Kettering Cancer Center. Twenty-five images were taken during the course of treatment and eight of them are shown here. Because the Varian Portal Vision Mark 2 and LC250 detectors measure dose rate, each image is converted to a dose rate map, then multiplied by a weighting factor corresponding to the time interval for that image. The weighted dose rate maps are then summed to obtain the integrated dose map. (From Chang et al. 2000. Reprinted by permission of American Association of Physicists in Medicine.)

the leaf motion, it can be applied to IMRT using non-sliding-window techniques, e.g., physical compensators, multiple aperture fields, or wedges.

Relative Intensity Verification

We compare the "measured" with the "planned" relative profile using a linear regression model (Chang et al. 2000) that assumes a constant scattering reaching the EPID. The comparison is made in the treatment field region at least 5 mm inside the beam border, defined to be 50% of the average intensity of the region within this border. The measured profile at SDD is projected back to the d_{max} in a semi-infinite water phantom at SAD using the inverse square law and the ratio of phantom scatter factors. We calculate the "planned" profile at d_{max} in a semi-infinite water phantom at SAD for the same field. The measured profile should reflect the intended profile if the gap between patient and imager is larger than the field size, and ray paths through the patient for a given field are approximately equal. We assume that the relation between the measured profile, $M[i, j]$, and the intended profile, $I[i, j]$, for pixel $[i, j]$ is

$$M[i, j] = k \times I[i, j] + s, \tag{16}$$

where s is the constant scattering and k accounts for normalization. A linear regression method (Mood et al. 1974; Press et al. 1992) is used to find the optimized s and k to minimize

$$\sigma^2 = \Sigma_{i,j}[M[i, j] - (k \times I[i, j] + s)]^2 / (N - 1) \tag{17}$$

where N is the total number of pixels and σ is an indicator for the goodness of match between the intended and measured profiles.

Dose Verification

The purpose is to predict dose to the prescription point from the measured portal dose

$$D_{iso} = \overline{M}[0,0] \times pcf\left(d_g, d_t, r_{p,SDD}\right) \times \left(\frac{SDD}{SAD}\right)^2 \times \frac{S_p\left(r_{p,SDD}\right)TMR\left(d, r_{p,SDD}\right)}{S_{pe}\left(r_{p,SDD}\right)} \tag{18}$$

where D_{iso} is the dose to isocenter when the patient is present; $\overline{M}[0, 0]$ is an average of integrated dose in cGy over a 1×1 cm^2 region on the central axis of the EPID; $pcf(d_g, d_t, r_p)$ is the phantom correction factor; and r_p is the equivalent square field size (figure 16–9a,b,c). The phantom correction factor pcf is an empirically measured factor to account for patient attenuation and scattering:

$$pcf(d_g, d_t, r_p) = D / D' \tag{19}$$

where d_g, d_t, r_p are the distance between patient and detector, patient thickness, and field size, respectively; D and D' are detector readings at d_{max} with and without a phantom present (figure 16–9d). A set of pcf values was measured for different phantom thicknesses, field sizes, and distances between phantom and EPID; the pcf for a specific setup is obtained by interpolation. The formalism of Equation (18) ignores the fact that intensity modulation may have changed the scatter to primary dose ratio on the central axis. This should be a very small effect unless the central axis dose is an extreme maximum or minimum in the IM profile.

Applications

Potential applications of this technique are verification of IMRT fields prior to and during treatment. Verification prior to treatment, without phantom or patient present, provides the cleanest measurement of the delivered intensity profile and absolute isocenter dose, permitting a stricter passing criterion—the measured relative profile using the EPID can be, on average, within 4% of the planned profiles, and the central axis absolute dose can be predicted within 2% of the prescribed dose. When used in patient treatment, the goal is to detect

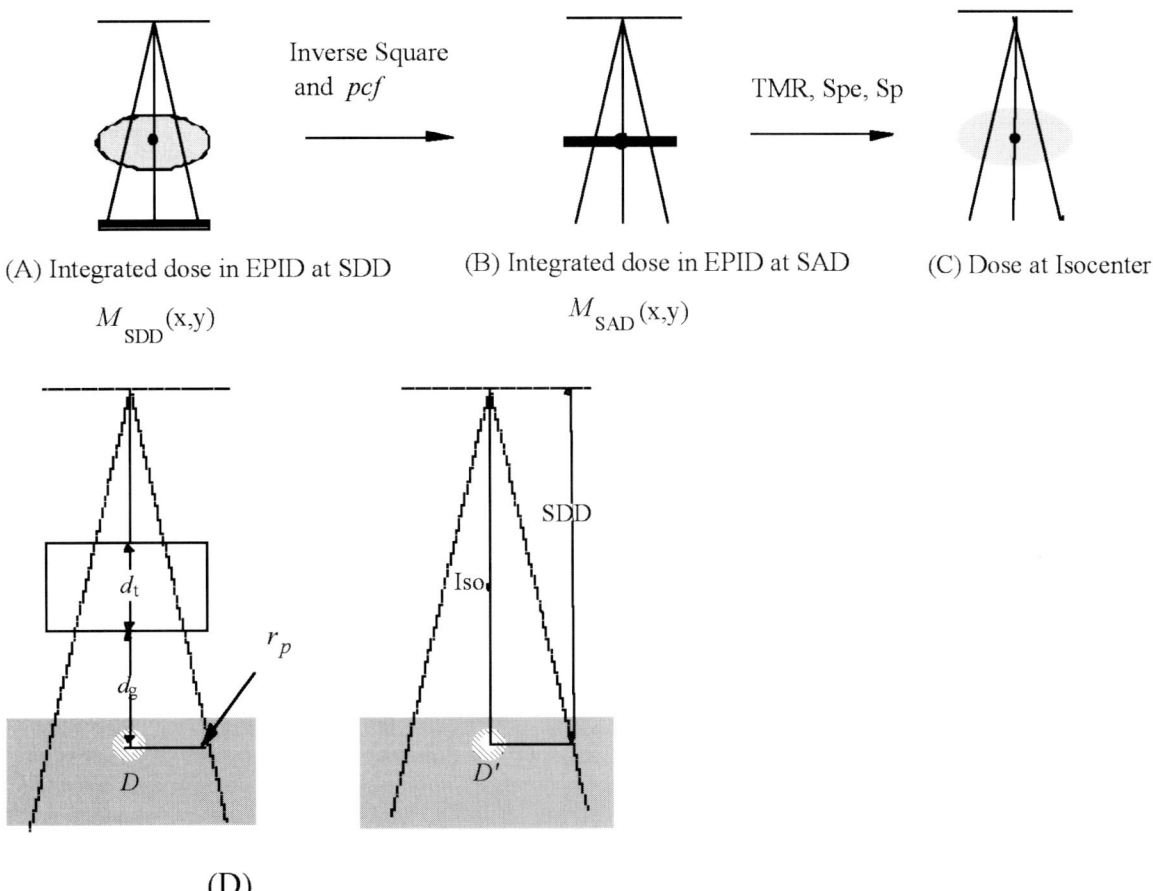

(A) Integrated dose in EPID at SDD

$M_{SDD}(x,y)$

Inverse Square and *pcf*

(B) Integrated dose in EPID at SAD

$M_{SAD}(x,y)$

TMR, Spe, Sp

(C) Dose at Isocenter

(D)

FIGURE 16–9. (a–c) Illustration of the derivation of Eq. (18) to determine dose to isocenter. (d) Measurement setup of patient correction factor *pcf*. (From Chang et al 2000. Reprinted by permission of American Association of Physicists in Medicine.)

major mistakes such as larger setup error, wrong IMRT files, and wrong jaw settings. Breathing and setup errors will likely result in less tight action levels than the ones determined from the phantom study. Also, the presence of the treatment couch will affect results at certain gantry angles. Additional correction may be needed if ray paths through the patient for a given field vary significantly, which violate the second assumption when deriving Equation (16). Uncertainties associated with the use of the phantom correction factor will therefore increase somewhat.

Dosimetry Module, Dosimetry Interface, And Dosimetry Workspace Of LC250

Off-line integration is impractical for clinical use because it takes a long processing time, demands more computing resources, and requires significant user interactions. Varian has introduced a dosimetry module and a dosimetry workspace to alleviate the problem. This section describes our experiences with, and recommendations for, using the dosimetry module.

Dosimetry Module

LC250 automates the verification process using a dosimetry module in the IAS2 that corrects the signal for flood field and dark current, converts the corrected signal into dose rate using a lookup table, and integrates the dose rate over time to produce a measured dose map. Because only one integrated profile is returned to the console, it greatly decreases the data transmission time and the required hard disk space. Because there is no off-line integration, the required computing resources [central processing unit (CPU) time and memory requirement] are significantly reduced. Once the dose map is sent to the PV computer, the verification process takes no more than 30 seconds with a few mouse clicks. In this section, we illustrate some key features for configuring the LC250.

Calibration

Most calibration procedures are similar in data acquisition scheme, but different in portal dose definition, dose conversion, and scattering correction. Significant error may be introduced if a user mixes different procedures without careful examination. To accurately obtain the calibration curve, physicists are advised to study each procedure carefully, and select a calibration procedure that they are most comfortable with. We don't recommend using the calibration tool in the PV software; instead, it is more reliable for physicists to obtain the a and b calibration coefficients using their selected procedure, and manually enter them to the file "\va_app\oncology\ vision\config\dosimetry\vaba_dol\ExternaBeam1.dat," where "\va_app" is the network directory (usually the "c:\program files\varian" directory) of the PV software. Figure 16–10 shows a sample file where the bottom three lines are the block for calibration curve. Each block consists

```
[Setup]
IDU Pos 1=-5.0
IDU Pos 2=-15.0
IDU Pos 3=-25.0
IDU Pos 4=-35.0
IDU Pos 5=-45.0
Hotspot area=10
IDU Area=20
Average frames=10
Flood field correction=1
Last Calibration=03/25/2000 14:34
Acquisition mode=Fast/Full Resolution
SW Version=6.0.60
ExternalBeam Id=2100C_1
PortImager Id=PV_2100C_1
Last Modified=superuser


[15MV_240MU]
a=1664.639
b=99.01
```

FIGURE 16–10. Contents of the file "ExternaBeam1.dat." The bottom three lines are the "a" and "b" coefficients of the calibration curve.

of the energy (15 MV), dose rate (240), and a (1664.639) and b (99.01) coefficients of the calibration curve. More blocks can be added for other energy and dose rate. A user should first enter the new coefficients to the "ExternaBeam1.dat" file using a text editor, then start the calibration facility of PortalVision, illustrated in figure 16–11, which imports the calibration coefficients "[a]old" and "[b]old" from the "ExternaBeam1.dat" file. Users should skip the "Start" but hit the "OK" button to generate the lookup table and transfer it to the IAS2.

IAS2 Service Monitor

Scanning a SLIC matrix involves very complex timings controlled by the parameters in the IAS2. Although users are generally not recommended to change these parameters but use the default values, it is inevitable for users to adjust them using the PV service monitor shown in figure 16–12, in order to find the best combinations of theses parameters for EPID verification. Before changing any parameter, users are encouraged to study the PV reference manual for detailed information of each parameter. Here, we discuss three most important parameters: "PV Sync," "Row Averages," and "Row Pattern," for controlling the timing. "PV Sync," the frequency generated by the trigger board based on LINAC pulses, is used for row synchronization. "Row Pattern," the number of rows acquired in one sync cycle, is specified by symbols: 'SM' indicates measured row after a sync pulse, and 'M' measured row without sync; there can be multiple 'M's after an 'S'. Depending on the "Row Pattern," an integer number (the number of 'M' after 'S') of rows are scanned between two sync pulses, and the imaging speed (in frames/s) can be determined as

$$\text{Imaging speed} = \text{PV Sync} \times \text{Row Pattern}/256 \,. \tag{20}$$

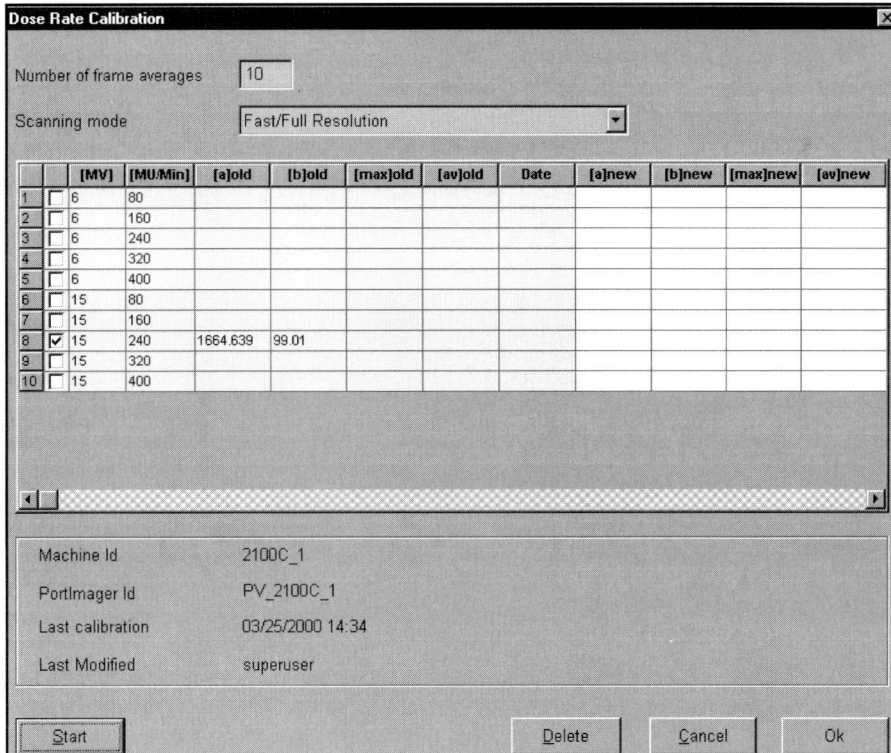

FIGURE 16–11. Window for the dose rate calibration in PortalVision™.

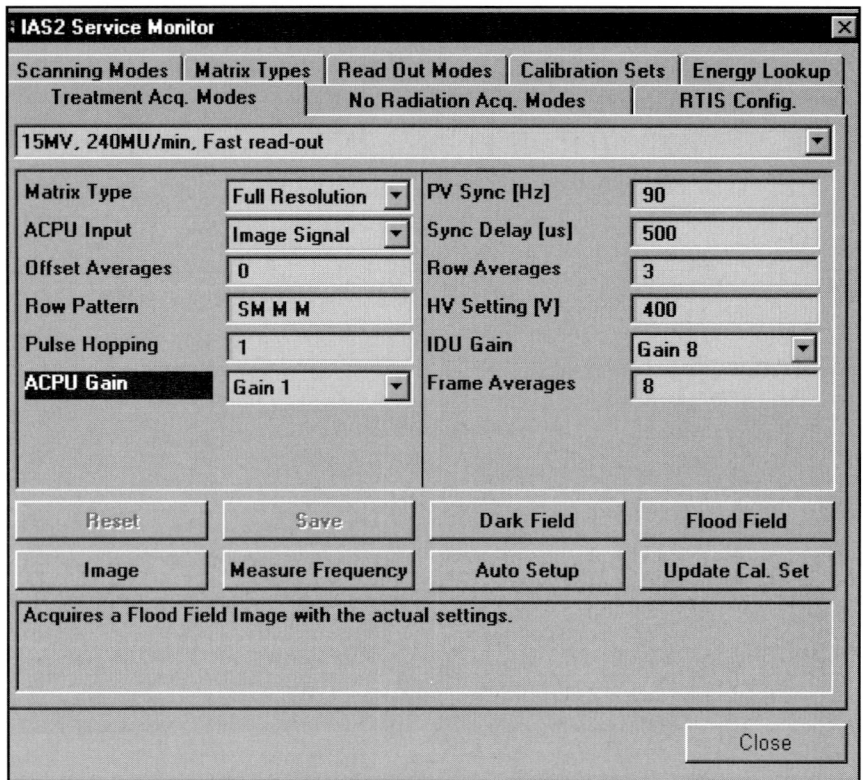

FIGURE 16–12. The window of PortalVision's service monitor that helps a user to configure the IAS2. In order for the EPID to work properly, it is vital to accurately adjust three major parameters, "PV Sync," "Row Averages," and "Row Pattern," according to the need of your Verification.

For example, the "PV Sync" in figure 16–12 is 90 and "Row Pattern" is "SM M M." Thus, three rows are scanned between two sync pulses (1/90 = 0.011 sec or 11 msec), and the imaging speed is 90 × 3/256 = 1.05 frame/sec. Although all electrometers are simultaneously turned on when a row is scanned, the current of each electrometer is sequentially scanned by an analog to digital (A/D) converter, where a pixel conversion takes ~3.5 μsec, corresponding to a "Reading Time" of 0.9 msec (= 3.5 μsec × 256) for each row. "Row Averages" specifies how many times an electrometer is scanned and averaged for each M (or scanned row). In general, a higher "Row Averages" number improves SNR and produces better image quality. However, "Row Pattern" and "PV Sync" limit "Row Averages" according to the following relation

$$\text{Row Averages} \times \text{Reading Time} + \text{Sync Delay} < 1/(\text{PV Sync} \times \text{Row Pattern}), \qquad (21)$$

where "Sync Delay" (in μsec) is the wait time after a beam pulse occurred before the row scanning starts. In figure 16–12, the value on the left of Equation (21), 3 × 0.9 + 0.5 = 3.2 msec, is smaller than that on the right, 1/(90 × 3) = 3.7 msec; the condition in Equation (21) is thus satisfied. Instead of performing the above calculation, users can use the properties of an acquired image to check the condition. As illustrated in the second line from the bottom of figure 16–13, Equation (21) is satisfied once x in the message ". . . x ms time left in . . ." is positive. Increasing "Row

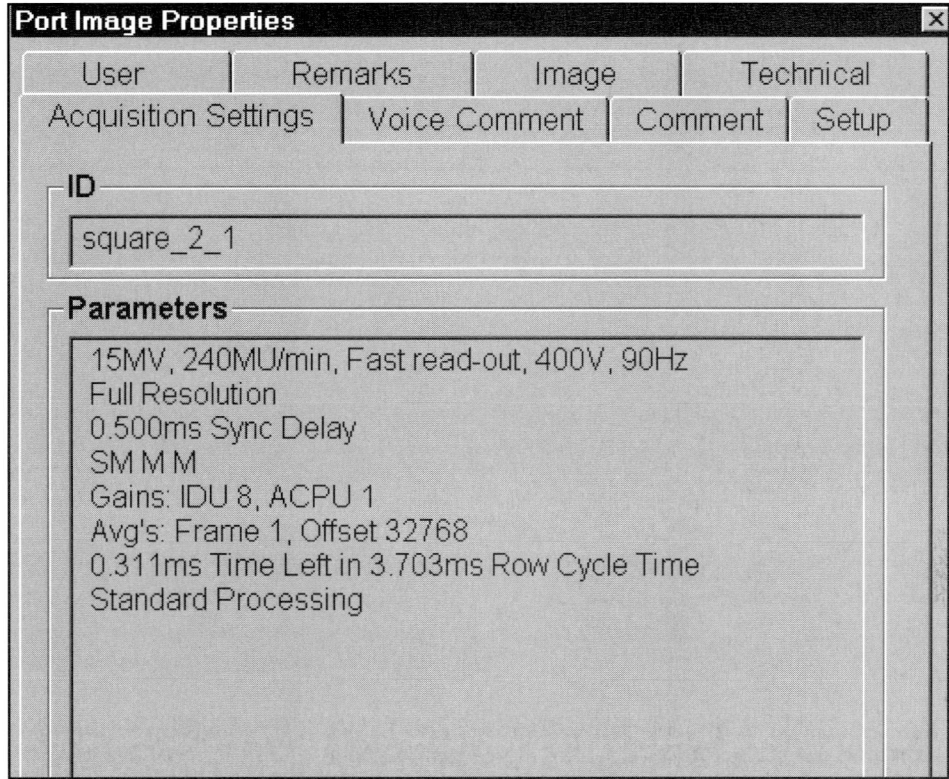

FIGURE 16–13. The window that shows the properties of a portal image taken by PortalVision.

Averages" decreases x; by trial and error (keep increasing "Row Averages" until x become negative), an optimal "Row Averages" number can be determined.

If the "Row Pattern" is set without "S," LC250 use an internal clock to control the scanning without synchronizing with the linac. This design is intended for cobalt machines that do not use pulses to generate radiation; for linac, stream artifact is observed if the acquisition is not synchronized with machine pulses. Although the severity of this artifact can be reduced using a higher "PV Sync" and by averaging multiple frames, the imaging speed of a SLIC EPID is theoretically limited to two frames/sec, corresponding to a "PV Sync" of 512, not fast enough to reduce the artifact to an acceptable level even with a frame average of 20. Since the detector memory effect also becomes serious as the imaging speed is faster than 1 frame/sec, we generally don't recommend using the non-synchronization acquisition.

Sequence Image Properties

For each image acquisition, there are "sequence" properties, as shown in figure 16–14, associated with it to control the image sequence and quality. Some of these parameters are specifically set— "During" and "Continuous acquisition" being selected, and "At Rel Position" set to 99%—to inform the IAS2 that the dosimetry module should be used for the image acquisition.

Dosimetry Interface

Currently, the EPID verification programs of PortalVision are implemented through a dosimetry interface, instead of being coded internally, an approach making it flexible to integrate user programs with PV software. Figure 16–15 illustrates the software structure of Vision applications and

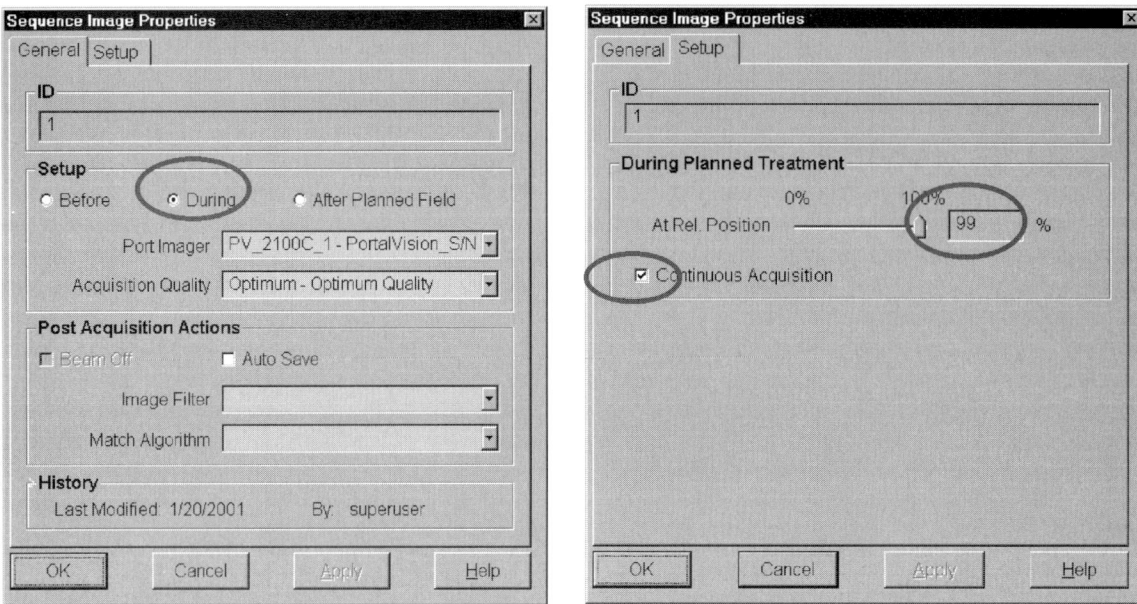

FIGURE 16–14. The sequence image property for the verification mode of the PortalVision SLIC EPID.

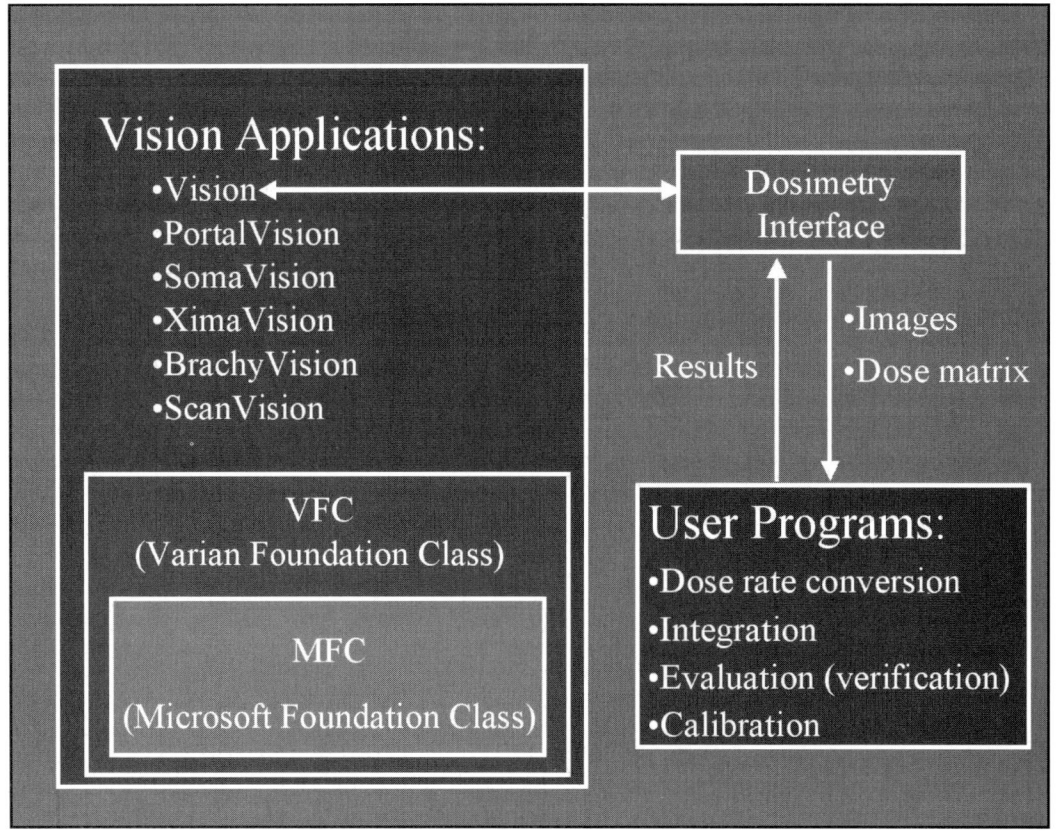

FIGURE 16–15. The software structure of Vision applications and Dosimetry interface.

dosimetry interface. On the right side is the dosimetry interface that allows users programs to directly access and quickly process the EPID data, and return the results for display. On the left side are the Vision applications developed using the Varian Foundation Class (VFC), a class library that encapsulates many classes of Microsoft Foundation Class (MFC) and is the *de facto* standard for radiation oncology software development at Varian Medical Systems. Although users do not have to know VFC to use the interface, some knowledge of VFC is essential to develop and test the user programs. The dosimetry interface provides four functions: detector reading to dose rate conversion, dose conversion, dose integration, and evaluation. Each function has a corresponding inventory object to organize dosimetry programs stored in "/va_app/oncology/vision/bin/dosimetry/customer" (for user programs), or "/va_app/oncology/vision/bin/dosimetry/customer" (for Varian programs). At the start of Vision applications, each inventory object scans both directories to find the DLL (dynamic link library) files belonging to it. When users request a function, corresponding inventory lists all the available subroutines, and dynamically links the user-selected one.

Dosimetry Workspace

Figure 16–16 illustrates the PV "Dosimetry" workspace of the "Review" task for a selected session of a simulated patient. A green icon indicates a dose (rate) matrix, and a black-and-white icon an image object; right clicking the mouse button of an object brings up a list of functions available for that object. In this example, five subroutines are available to convert image reading to dose or dose rate; clicking any of them will convert the image and return a dose rate matrix. Automatic dose (rate) conversion and integration of serial images for an irradiation field can be

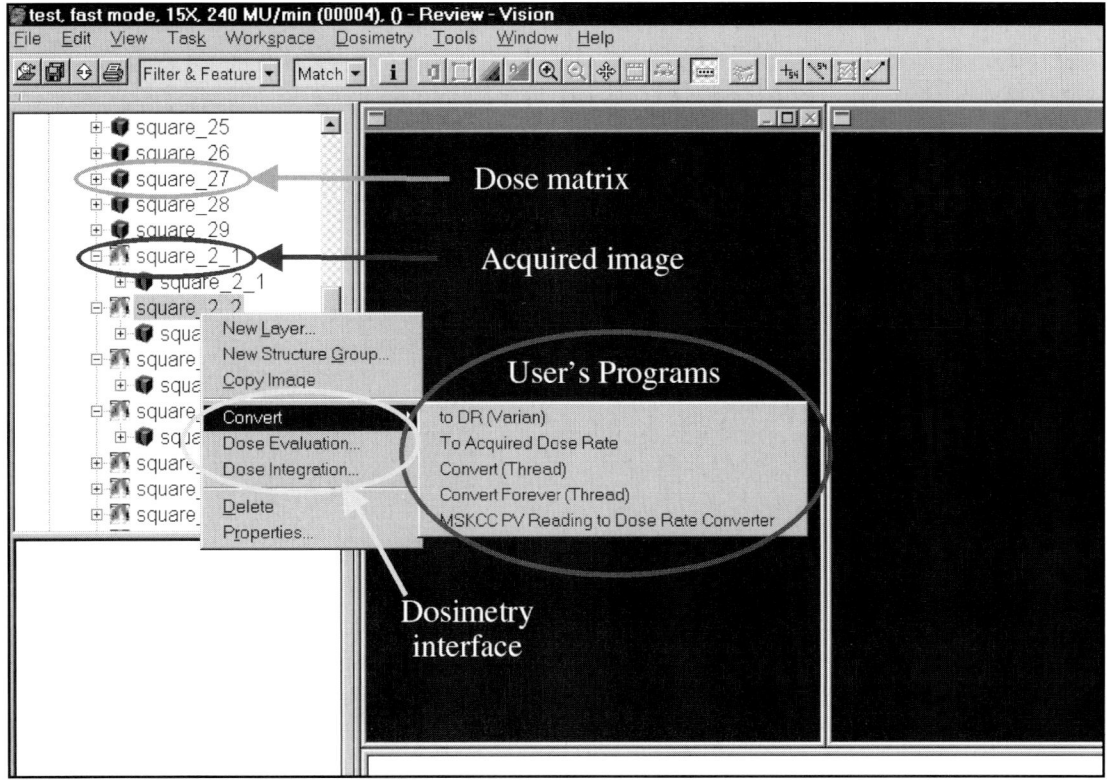

FIGURE 16–16. The window of the dosimetry workspace for a simulated patient.

achieved by clicking the right mouse button when the filed object is highlighted, as illustrated in figure 16–17. Assuming that the images are serially acquired, this function performs dose integration using the acquisition time of each image. Information in the "Position" column is used to align images; because the EPID is not expected to move during a serial acquisition, users should click the first row and enter "generic." The "Dose Rate" column indicates whether the dose rate matrix already exists. On the right side of the window are three inventories for "Image Matching," "Dose Rate Conversion," and "Dose Integration," each of which contains all the available subroutines for that function. Users should specify the subroutines in "Dose Rate Conversion" and "Dose Integration" functions. The output of "Dose Integration" is an integrated dose matrix at the lower right corner of the window, and can be saved by clicking the "Save Dose" button.

Figure 16–18 illustrates the dose evaluation, initialized by clicking the right mouse button of the highlighted field or dose (rate) matrix object. On the upper left is the planned profile, and on the lower left the measured profile; either profile can be changed by clicking and dropping another dose matrix. Clicking "Evaluation" on the upper right brings up a list of the subroutines from the inventory; the "MSKCC dose evaluation" is selected in this example. The text box on the lower right displays the evaluation results. The difference map between the two profiles is shown in the upper right box, and can be copied (click and drop) to the main display screen for further review.

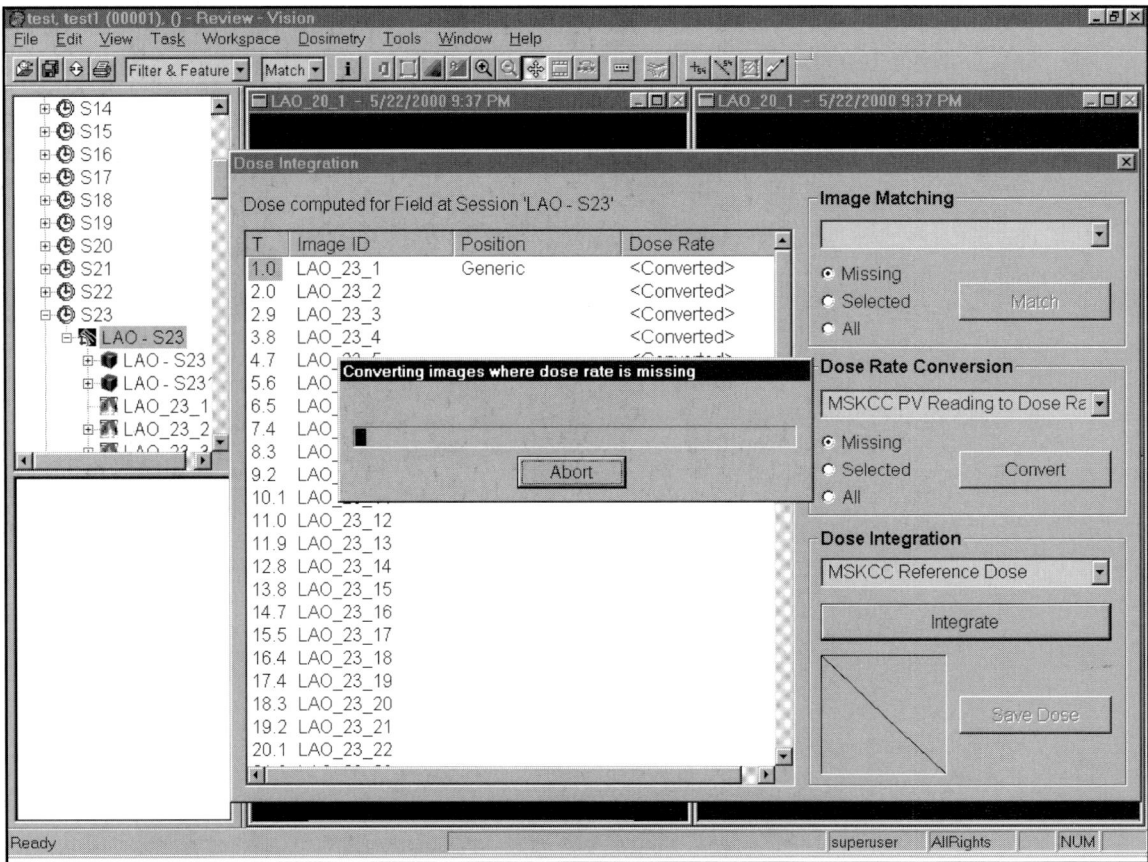

FIGURE 16–17. The window for automatic dose rate conversion and dose integration of dosimetry interface.

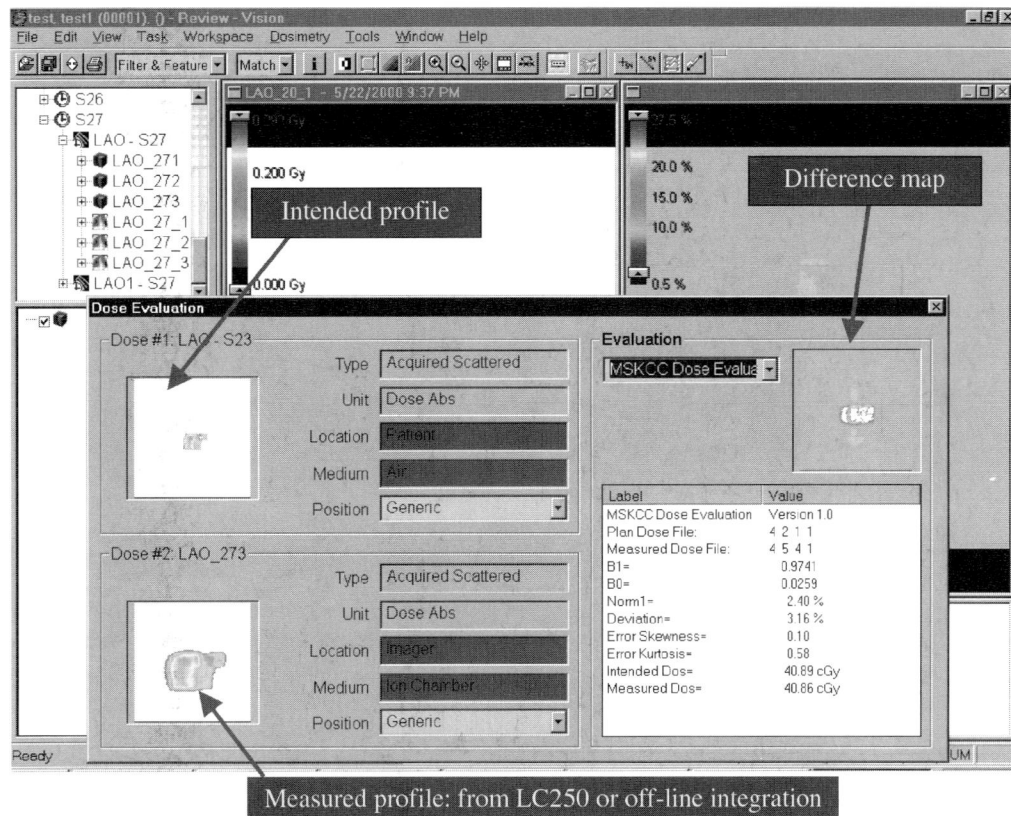

FIGURE 16–18. The window for the dose evaluation function of dosimetry interface.

Comparison Of LC250 And aS500

Detector Memory Effect

By adjusting the parameters in the Service Monitor of LC250, we can increase the imaging speed for better verification results; however, detector memory becomes significant for higher imaging speed. We have evaluated two acquisition modes for LC250: the standard mode (S-mode) operates at an imaging speed of ~2.7 sec/frame (PV sync = 90 Hz, Row Pattern = SM, and Row Average = 7), and the fast mode (F-mode) at ~1 sec/frame (PV sync = 90 Hz, Row Pattern = SM M M, and Row Average = 3), using the 15 MV beam of a Varian 2100C linac.

Figure 16–19a compares the beam profiles of a 60°-wedged field, measured using the LC250 EPID and film; we also plot the difference of these two profiles, with an average value of less than 1%. Good agreements are observed even in the penumbra and other regions where the profile tends to vary significantly within a short range. Figure 16–19b compares the measured and intended profiles of an IMRT prostate field for 130 MU and 350 MU. Although the profile of the 130 MU has a larger sampling error, both profiles agree well with the intended profiles.

Verification of the relative profile and the delivered dose was performed on 25 IM fields of five prostate patients, using both the S- and F-modes, with a 35-cm polystyrene phantom being present at the isocenter to simulate the during-treatment verification. Figure 16–20a plots the mean σ (goodness of fit between measured and intended) and its standard deviation of the 25 IM fields for three (100, 130, 350) MU settings, and figure 16–20b the mean and standard deviation of EPID/ION-CH vs. MU, where EPID/ION-CH is the ratio of the EPID-measured dose (sum of the

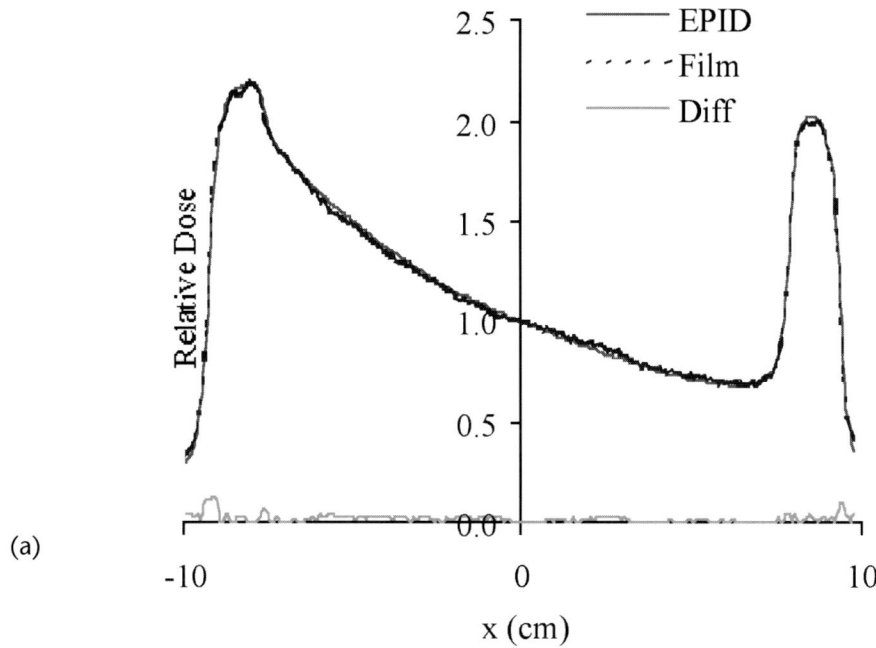

(a)

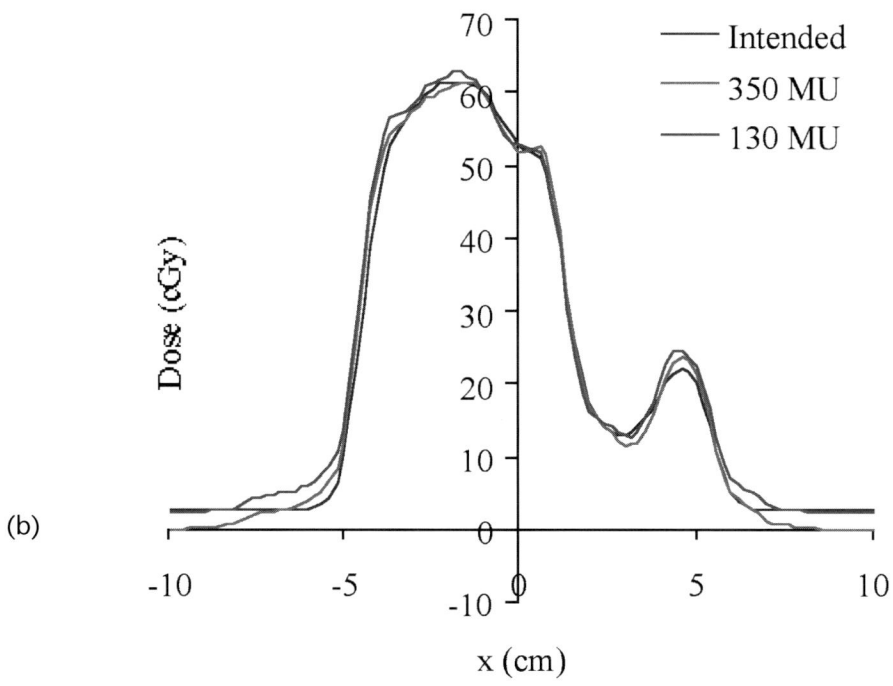

(b)

FIGURE 16–19. Comparison of (a) measured beam profiles of a 60°-wedged field using LC250 EPID and film, and (b) measured (using LC250 EPID) and intended (from treatment planning) profiles of a prostate IMRT field.

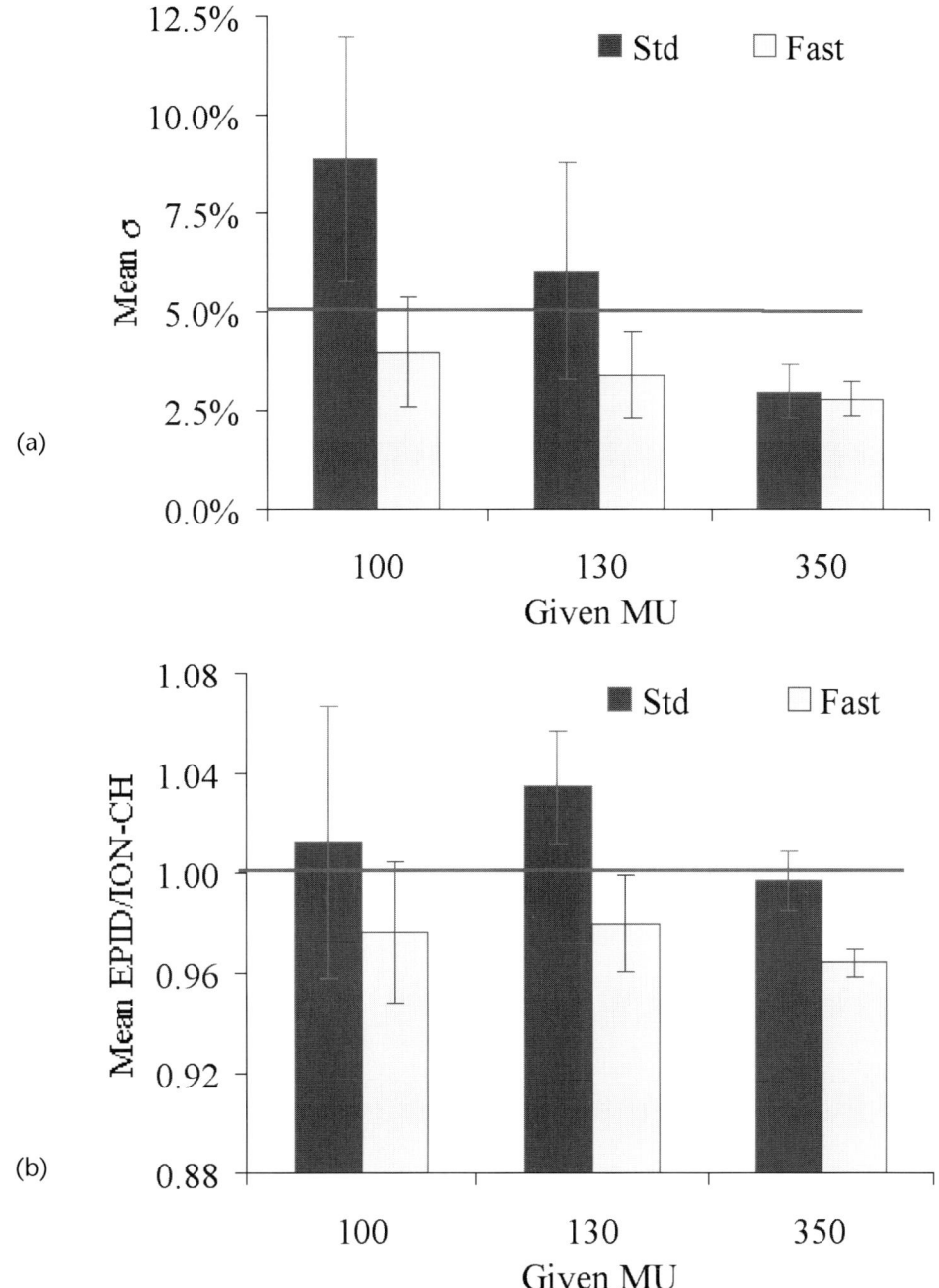

FIGURE 16–20. Results of IMRT verification using the S- or F-mode of LC250 for five patients (25 prostate fields). A 35-cm polystyrene phantom was present at the isocenter to simulate the during-treatment verification. (a) Relative profile verification: mean σ vs. MU for 25 IMRT fields where the green line indicates the acceptable 5% QA tolerance. (b) Absolute dose (sum of five prostate fields) to isocenter, mean EPID/ION-CH vs. MU for five patients, where EPID/ION-CH is ratio of the EPID-measured dose to the ion-chamber measured dose, and the green line is the ideal ratio (1.0). The error bar is one standard deviation of the averaged σ or EPID/ION-CH.

dose to isocenter for the five IM fields of each patient) to the ion-chamber measurement. Theoretically, the higher the MU, the slower the leaves move, and the more accurate the verification can be. It is observed from figure 16–20a that for the same MU, F-mode is more accurate (lower σ) than S-mode in all cases, because of a much higher imaging speed of the F-mode. This improvement, however, is not proportional to the difference in imaging speed. As a result, σ plus one standard deviation is slightly higher than 5% for 100 MU, thus not sufficiently accurate for verification during treatment.

This under-performance is due to the "detector memory" effect, that is, the loss of detector sensitivity when a SLIC detector is continuously irradiated and scanned before the equilibrium is re-established. As shown in figure 16–20b, the detector memory effect is insignificant for the S-mode, but causes the EPID measured dose to be ~3% lower than the ionization chamber measurements for the F-mode. When continuously irradiated at a dose rate of 240 MU/min, a SLIC detector loses sensitivity at a rate of 0.4% per minute. Since most clinical fields take less than a minute to irradiate, the detector memory effect should have a negligible effect on relative profile verification. Although detector memory is observed for the aS500 EPID, its effect is only a few tenths of a percent on the measured profile and is negligible.

Beam Hold-Off Effect

The beam hold-off introduces a momentary drop in dose rate, adversely affecting the signal observed in the SLIC detector and thereby increasing the standard deviation for both relative profile verification and absolute dose measurements. Correction of the beam hold-off effect involves including the fluctuation of the ion-pair concentration due to beam hold-off into the planned profile, and is yet to be developed. It is thus recommended to use the LC250 for only pre-treatment verification and to use a larger number of monitor units to avoid the beam hold-off effect. The beam hold-off has no negative effect on the aS500 EPID because it is an integration device that collects all the charges produced by the radiation.

Comparison

Table 16–3 shows and compares the results for verifying the relative profiles and beam central axis dose of the 25 IM fields, using the frame averaging method of the aS500, and the fast mode of LC250 with correction for the detector memory effect. Both EPIDs perform well at 350 MU; however, at smaller MU values, the mean σ and its standard deviation increase significantly for LC250, but remains almost a constant for aS500. Similarly, although the means of the ratio of measured/planned doses are near unity for both EPIDs, the SD increases significantly for LC250 as the MU decreases. If we require the verification procedure to be within 5% for $\sigma+SD$, and 3% for the isocenter dose, the LC250 may not be used as a QA tool in the clinical MU settings. The

Table 16–3. Comparison of IMRT Verification Results Using the Fast Mode of LC250 and the Frame Averaging of aS500 EPIDs for 25 Prostate Fields

(A) aS500	100	130	350
Mean σ	2.0%	2.0%	1.9%
SD	0.5%	0.5%	0.5%
Mean D_{iso}	0.995	0.993	1.002
SD	0.006	0.009	0.008
(B) LC250	100	130	350
Mean σ	3.98%	3.40%	2.80%
SD	1.41%	1.10%	0.45%
Mean D_{iso}	1.006	1.007	0.991
SD	0.027	0.020	0.006

aS500, on the other hand, performs much better and can satisfy the above criteria. We should point out that there is essentially no sampling error, detector memory, and beam hold-off problems for aS500 EPID at any MU setting, because the detector actually integrates all doses between two scans, instead of approximating the integration numerically (e.g., the "dose rate" method). As a result, the aS500 EPID performs significantly better (lower mean and SD of σ, and lower SD of measured/planned dose) than the LC250 EPID.

Summary

Geometric Uncertainty

- Factors that affect positional uncertainty in target and non-target organs, or geometric errors, are errors in target delineation, patient positioning or setup errors, and internal organ motion.
- Although the standard approach to accounting for uncertainty is to evaluate dose to the PTV, the PTV definition assumes that the prescribed dose is delivered to the CTV, but does not verify it. An alternative approach is to incorporate geometric uncertainties in the dose calculation, then evaluate dose to the CTV and nearby organs at risk.
- Geometric errors for the fractionated treatment of an individual patient can be separated into a systematic error and a distribution of random errors. The systematic error in effect results in a displacement of the dose distribution from its planned position, whereas the random errors result in a blurring of the dose distribution, or a spreading out of isodose lines in regions of dose gradients.
- The effect of systematic errors can be simulated by Monte Carlo methods to estimate a likely range of dose to CTV and nearby organs, using as input previously measured distributions of systematic errors and the patient-specific dose data files from the treatment planning system.
- An alternative approach is to use boundaries of constant chi-square to define a confidence limit volume of systematic errors, and investigate a limited number of directions within this volume. For uncorrelated translations only, the 90% confidence limit volume of systematic errors is an ellipsoid defined as $6.25 = \left[(\delta x)^2 / \sigma_{xx}^2 + (\delta y)^2 / \sigma_{yy}^2 + (\delta z)^2 / \sigma_{zz}^2 \right]$. For uniform errors in the three directions of magnitude σ_{sys}, the confidence limit volume reduces to a sphere of radius $2.5\,\sigma_{sys}$.
- The effect of random errors on the dose distribution, averaged over a course of fractionated treatment, can be estimated by a convolution with the distribution of random errors.
- Monte Carlo simulation of uncertainties can be useful for investigating "class solution" PTV margins for a particular disease site and technique, using as input the contour data of the CTV and organs at risk in the patient's planning CT.
- For uncorrelated translational errors, an approximate formula for the PTV margin such that 90% of patients receive 95% of the prescribed dose to the CTV, is given by $2.5\,\sigma_{sys}$ + ($1.2\,\sigma_{ran}$ – 1.4) mm, where σ_{sys} and σ_{ran} are the one-standard-deviation systematic and random errors.

IMRT Verification

- The use of EPIDs for verifying IMRT beam delivery requires a deep understanding of underlying physics of the detector, the EPID hardware and software, the calibration procedure, and the verification algorithm.
- Our approaches to use EPIDs for IMRT QA includes a linear regression model for the relative profile verification, a phantom correction factor for the absolute dosimetry, and the EPID scattering factor, S_{pe}, for dose conversion.

- The iterative procedure, incorporating the EPID scattering by iteratively updating the calibration curve and S_{pe} until convergence, is required for the SLIC EPID, but not needed for the aS500 EPID because of its linear response.
- Practical use of this technology relies on automation of the whole process, including dose rate conversion, dose integration, data transferring, and data analysis; the vendor has developed a dosimetry module and a dosimetry workspace for this purpose.
- Evaluation of LC250 indicates that it is a convenient EPID tool for pre-treatment verification of the IMRT beam delivery in real time. The detector memory effect needs to be corrected for accurate dosimetry; a larger MU should be used if the clinical MU produces significant beam hold-off effect.
- The aS500 EPID has a great potential for both pre-treatment and during-treatment verifications of IMRT beam delivery, and is a better imaging tool than the LC250 EPID.

Acknowledgments

We thank our colleagues, Yijian Cao, Agung Hertanto, Margie Hunt, Jingguo Wang, Jie Yang, and Wendall Lutz for their contributions to some of the results presented in this chapter.

References

Ballard, D. H., and C. M. Brown. Computer Vision. Englewood Cliffs, NJ: Prentice-Hall, pp. 143–145, 1982.

Bel, A., H. Bartelink, R. E. Vijlbrief, and J. V. Lebesque. (1994). "Transfer errors of planning CT to simulator: A possible source of setup inaccuracies?" *Radiother. Oncol.* 31:176–180.

Chang J., G. Mageras, and C. C. Ling. (2002). "Evaluation of rapid dose map acquisition of a scanning liquid-filled ionization chamber electronic portal imaging device." *Int. J. Radiat. Oncol. Biol. Phys.* (In press).

Chang, J., G. Mageras, C. S. Chui, C. C. Ling, and W. Lutz. (2000). "Relative profile and dose verification of intensity modulated radiation therapy." *Int. J. Radiat. Oncol. Biol. Phys.* 47:231–240.

Chang, J., G. Mageras, C. C. Ling, and W. Lutz (2001). "An iterative EPID calibration procedure for dosimetric verification that considers the EPID scattering factor." *Med. Phys.* 28:2247–2257.

Chui, C. S., G. J. Kutcher, and T. Lossaso. (1992). "A convolution method for incorporating uncertainties in dose calculation." (Abstract). *Med. Phys.* 19:814.

Chui, C. S., T. LoSasso, and S. Spirou. (1994). "Dose calculation for photon beams with intensity modulation generated by dynamic jaw or multileaf collimations." *Med. Phys.* 21:1237–1244.

Craig, T., J. Battista, V. Moiseenko, and J. Van Dyk. (2001). "Considerations for the implementation of target volume protocols in radiation therapy." *Int. J. Radiat. Oncol. Biol. Phys.* 49:241–250.

Curry, T. S., J. E. Dowdey, and R. C. Murry. *Christensen's Introduction to the Physics of Diagnostic Radiology.* 4th ed. Philadelphia: Kea & Febiger, pp. 316–317, 1990.

Fiorino, C., M. Reni, A. Bolognesi, G. M. Cattaneo, and R. Calandrino. (1998). "Intra- and inter-observer variability in contouring prostate and seminal vesicles: Implications for conformal treatment planning." *Radiother. Oncol.* 47:285–292.

Fontenla, E., G. Mageras, J. Roeske, C. Pelizzari, G. Chen, and C. Ling. (2000). "Quantifying the accuracy of affine transformations for modeling organ motion variabilities in external beam radiotherapy of prostate cancer" in "Digest of Papers of the 2000 World Congress on Medical Physics and Biomedical Engineering [CD-ROM]." G. Fullerton (Ed.). Abstract WE-EBR-10. Chicago, July 23–28, 2000.

Ford, E., G. S. Mageras, E. Yorke, K. E. Rosenzweig, R. Wagman, and C. C. Ling. (2002). "Evaluation of respiratory movement during gated radiotherapy using film and electronic portal imaging." *Int. J. Radiat. Oncol. Biol. Phys.* 52:522–531.

Hanley, J., M. A. Lumley, G. S. Mageras, J. Sun, M. J. Zelefsky, S. A. Leibel, Z. Fuks, and G. J. Kutcher. (1997). "Measurement of patient positioning errors in three-dimensional conformal radiotherapy of the prostate." *Int. J. Radiat. Oncol. Biol. Phys.* 37:435–444.

Hebert, M. E., V. J. Lowe, J. M. Hoffman, E. F. Patz, and M. S. Anscher. (1996). "Positron emission tomography in the pretreatment evaluation and follow-up of non-small cell lung cancer patients treated with radiotherapy: Preliminary findings." *Am. J. Clin. Oncol.* 19:416–421.

Hunt, M. A., G. J. Kutcher, C. Burman, D. Fass, L. Harrison, S. Leibel, and Z. Fuks. (1993). "The effect of setup uncertainties on the treatment of nasopharynx cancer." *Int. J. Radiat. Oncol. Biol. Phys.* 27:437–447.

Hurkmans, C. W., P. Remeijer, J. V. Lebesque, and B. J. Mijnheer. (2001). "Set-up verification using portal imaging; review of current clinical practice." *Radiother. Oncol.* 58:105–120.

International Commission on Radiation Units and Measurements (ICRU) Report 50. Prescribing, Recording and Reporting Photon Beam Therapy. Bethesda, MD: ICRU, pp. 3–16, 1993.

International Commission on Radiation Units and Measurements (ICRU) Publication 62. Prescribing, Recording and Reporting Photon Beam Therapy (Supplement to ICRU Report 50). Bethesda, MD: ICRU, 1999.

Kagawa, K., W. R. Lee, T. E. Schultheiss, M. A. Hunt, A. H. Shaer, and G. E. Hanks. (1997). "Initial clinical assessment of CT-MRI image fusion software in localization of the prostate for 3D conformal radiation therapy." *Int. J. Radiat. Oncol. Biol. Phys.* 38:319–325.

Kalff, V., R. J. Hicks, M. P. McManus, D. S. Binns, A. F. McKenzie, R. E. Ware, A. Hogg, and D. C. Ball (2001). "Clinical impact of (18)F fluorodeoxyglucose positron emission tomography in patients with non-small-cell lung cancer: A prospective study." *J. Clin. Oncol.* 19:111–118.

Kalos, M. H., and P. A. Whitlock. *Monte Carlo Methods: Volume I.* New York: John Wiley & Sons, pp. 48–50, 1986.

Ketting, C. H., M. Austin-Seymour, I. Kalet, J. Unger, S. Hummel, and J. Jacky. (1997). "Consistency of three-dimensional planning target volumes across physicians and institutions." *Int. J. Radiat. Oncol. Biol. Phys.* 37:445–453.

Kiffer, J. D., S. U. Berlangieri, A. M. Scott, G. Quong, M. Feigen, W. Schumer, C. P. Clarke, S. R. Knight, and F. J. Daniel. (1998). "The contribution of 18F-fluoro-2-deoxy-glucose positron emission tomographic imaging to radiotherapy planning in lung cancer." *Lung Cancer* 19:167–177.

Langen, K. M., and D. T. L. Jones. (2001). "Organ motion and its management." *Int. J. Radiat. Oncol. Biol. Phys.* 50:265–278.

Mageras, G., and R. Mohan. (1993). "Application of fast simulated annealing to optimization of conformal radiation treatments." *Med. Phys.* 20:639–647.

Mageras, G. S., Z. Fuks, S. A. Leibel, C. C. Ling, M. J. Zelefsky, H. M. Kooy, M. van Herk, and G. J. Kutcher. (1999). "Computerized design of target margins for treatment uncertainty in conformal radiotherapy." *Int. J. Radiat. Oncol. Biol. Phys.* 43:437–445.

Mah, D., J. Hanley, K. Rosenzweig et al. (1998). "Movement of anatomic landmarks and its effect on patient position determination in thoracic conformal radiation treatment." (Abstract). *Med. Phys.* 25A:127.

Mah, D., J. Hanley, K. Rosenzweig, E. Yorke, L. Braban, C. C. Ling, S. A. Leibel, and G. Mageras. (2000). "Technical aspects of the deep inspiration breath-hold technique in the treatment of thoracic cancer." *Int. J. Radiat. Oncol. Biol. Phys.* 48:1175–1185.

Melian, E., G. S. Mageras, Z. Fuks, S. A. Leibel, A. Niehaus, H. Lorant, M. Zelefsky, B. Baldwin, and G. J. Kutcher. (1997). "Variation in prostate position quantitation and implications for three-dimensional conformal treatment planning." *Int. J. Radiat. Oncol. Biol. Phys.* 38:73–81.

Mohan, R., and C. S. Chui. (1987). "Use of fast Fourier transforms in calculating dose distributions for irregularly shaped fields for three-dimensional treatment planning." *Med. Phys.* 14:70–77.

Mood, A.M., F. A. Graybill, and D.C. Boes. *Introduction to the Theory of Statistics.* Third Edition. New York: McGraw-Hill Book Company, chapter X, 1974.

Munley, M. T., L. B. Marks, C. Scarfone, G. S. Sibley, E. F. patz jr., T. G. Turkington, R. J. jaszczak, D. R. Gilland, M. S. Anscher, and R. E. Coleman. (1999). "Multimodality nuclear medicine imaging in three-dimensional radiation treatment planning for lung cancer: Challenges and prospects." *Lung Cancer* 23:105–114.

Nestle, U., K. Walter, S. Schmidt, N. Licht, C. Nieder, B. Mostaref, D. Hellwig, M. Niewald, D. Ukena, C. M. Kirsch, G. W. Sybrecht, and K. Schnabel. (1999). "18F-deoxyglucose positron emission tomography (FDG-PET) for the planning of radiotherapy in lung cancer: High impact in patients with atelectasis." *Int. J. Radiat. Oncol. Biol. Phys.* 44:593–597.

Pelizzari, C. A., G. T. Chen, D. R. Spelbring, R. R. Weichselbaum, and C. T. Chen. (1989). "Accurate three-dimensional registration of CT, PET, and/or MR images of the brain." *J. Comput. Assist. Tomogr.* 13:20–26.

Press, W. H., S. A. Teukolsky, W. T. Vetterling, and B. P. Flannery. *Numerical Recipes: The Art of Scientific Computing.* Cambridge, England: Cambridge University Press, pp. 202, 529, 1986.

Rasch, C., R. Keus, F. A. Pameijer, W. Koops, V. de Ru, S. Muller, A. Touw, H. Bartelink, M. van Herk, and J. V. Lebesque. (1997). "The potential impact of CT-MRI matching on tumor volume delineation in advanced head and neck cancer." *Int. J. Radiat. Oncol. Biol. Phys.* 39:841–848.

Rasch, C., I. Barillot, P. Remeijer, A. Touw, M. van herk, and J. V. Lebesque. (1999). "Definition of the prostate in CT and MRI: A multi-observer study." *Int. J. Radiat. Oncol. Biol. Phys.* 43:57–66.

Stroom, J. C., H. C. de Boer, H. Huizenga, and A. G. Visser. (1999). "Inclusion of geometrical uncertainties in radiotherapy treatment planning by means of coverage probability." *Int. J. Radiat. Oncol. Biol. Phys.* 43:905–919.

Valley, J. F., and R. P. Mirimanoff. (1993). "Comparison of treatment techniques for lung cancer." *Radiother. Oncol.* 28:168–173.

van Herk M. (1991). "Physical aspects of a liquid-filled ionization chamber with pulsed polarizing voltage." *Med. Phys.* 18:692–702.

van Herk, M., P. Remeijer, C. Rasch, and J. V. Lebesque. (2000). "The probability of correct target dosage: Dose-population histograms for deriving treatment margins in radiotherapy." *Int. J. Radiat. Oncol. Biol. Phys.* 47:1121–1135.

Zelefsky, M. J., D. Crean, G. S. Mageras, O. Lyass, L. Happersett, C. C. Ling, S. A. Leibel, Z. Fuks, S. Bull, H. M. Kooy, M. van Herk, and G, J. Kutcher. (1999). "Quantification and predictors of prostate position variability in 50 patients evaluated with multiple CT scans during conformal therapy." *Radiother. Oncol.* 50:225–234.

17

Advanced Treatment Techniques I

Ellen Yorke
D. Michael Lovelock

The ability to deliver a tumoricidal dose while maintaining adequate normal tissue sparing is affected by the size of the planning target volume (PTV) margin that is required to contain the mobile clinical target volume (CTV). In turn, the margin size depends in part on the daily displacements of the target from its planned position after the patient has been positioned on the treatment couch. In addition to setup errors, tumors in the thorax, abdomen, and prostate move with respect to the bony anatomy. This chapter examines techniques used to minimize target displacements in these two sites. The first section examines how respiration-gated treatment, in which both the CT scanner and the treatment machine are synchronized with the breathing cycle, can be used to improve the localization of thoracic tumors. The next section examines various strategies being used in prostate treatment to reduce the effects of setup error and prostate motion due to variations in bladder and rectal filling.

Respiratory Gating

Respiratory motion is a major source of treatment uncertainty for radiation therapy of tumors in the thorax and abdomen. During normal respiration, diaphragm excursions exceeding 2 cm are observed under fluoroscopy and motion of 1 to 3 cm has been reported for kidneys, liver, and tumors in lung (Ross et al. 1990; Davies et al. 1994; Ekberg et al. 1998; Hanley et al. 1999; Aruga et al. 2000; Shimizu et al. 1999, 2000a). In contrast, systematic setup errors of fixed bony anatomy are estimated to have standard deviations on the order of 3 to 5 mm (Ekberg et al. 1998; Mah, Hanley, and Rosenzweig 1998). To assure full dose to the CTV, the PTV must be large enough to account for respiratory motions. However, large margins include excess normal tissue in the radiation fields, compromising the dose that can be delivered to the tumor. Respiratory motion also distorts structures imaged in planning CT scans, confounding efforts to use dose-volume information based on these scans to model biological outcomes (Balter et al. 1996; Shimizu et al. 2000b).

Two strategies are used to reduce breathing motion during imaging studies and treatment: controlled breathing and respiratory gating. With *controlled breathing,* the patient is an active participant, performing a supervised breath-hold during image acquisition and/or treatment. With *automatic respiratory gating,* the patient's normal breathing is monitored by a device which turns on the treatment, CT or simulator beam only when respiratory motion is within a user-chosen gate. Ideally, respiratory gating requires no active patient participation and minimal extra staff monitoring. Below, we briefly survey the clinical implementations of these two strategies. We then describe the Memorial Sloan-Kettering Cancer Center (MSKCC) experience with controlled breathing using deep-inspiration breath hold (DIBH) and with automatic respiratory gating using the Varian Real-Time Position Management (RPM™) system (Varian Medical Systems, Palo Alto, CA).

Controlled Breathing

Pros and Cons

Controlled breathing techniques use simple equipment and are relatively inexpensive to implement. For lung patients, reproducible breath-hold at deep inspiration has been shown to shift normal lung tissue out of the treatment fields, reducing estimated lung toxicity (Hanley et al. 1999; Rosenzweig et al. 2000). However, the need for patient cooperation limits the eligible patient population, and coached and/or controlled breathing methods are labor intensive for staff. Few individuals can hold their breath longer than 15 to 20 sec and repeated breath-holds of that length, at any breathing level, are tiring. Thus controlled breathing may be incompatible with the longer beam delivery times required for IMRT.

Controlled Breathing Methods

The most straightforward controlled breathing method is an honor system. Here, the patient performs a self-monitored voluntary breath-hold at a prearranged part of the breathing cycle while the beam is activated. In a recently reported 16-patient feasibility study of self-gated radiotherapy (Kim et al. 2001), the patients used a hand-switch to control a modified accelerator interlock and the therapist could only beam-on when the patient judged that he had attained the pre-arranged breath-hold. Using the diaphragm as a surrogate for maximal tumor motion, intra-fraction reproducibility ranging from 0.13 mm to 7.65 mm and day-to-day variation (four patients/three sessions each) less than 5 mm was achieved.

Active breathing control (ABC) (Wong et al. 1999; Stromberg et al. 2000) uses a ventilator unit interfaced to a personal computer (PC) to monitor patient breathing and an occlusion valve to interrupt breathing at the chosen lung volume. Studies at normal end-inspiration or expiration and deep inspiration report that many patients can tolerate ABC breath-hold for 20 sec. Staff and patients must be specially trained and patients must be able to comply with the procedure. Imaging studies for lung and liver cancer and Hodgkins disease patients reported intra- and inter-fraction reproducibility of at worst 5 mm. A recent study of five liver patients with implanted radio-opaque microcoils, treated with ABC at normal end-expiration and monitored through the course of treatment with kilovoltage images, found no diaphragm motion during breath-hold, intra-fraction reproducibility of approximately 3 mm and long-term reproducibility of 6 mm relative to bony anatomy (Dawson et al. 2001).

Deep inspiration breath hold (DIBH), which not only reduces respiratory motion but also increases lung volume by a factor of 1.5 to 2, was developed at MSKCC (Hanley et al. 1999; Rosenzweig et al. 2000; Mah et al. 2000). The patient breathes through a PC-interfaced spirometer while being monitored and coached through a modified slow vital capacity (SVC) maneuver to a reproducible state of maximum deep inspiration (DI). Staff must be trained to coach and to interpret the PC display and patients must be able to reliably comply with coaching and sustain breath-hold for at least 10 seconds. Planning CT images are acquired and treatments are delivered during breath-hold at DI. Diaphragm positional reproducibility is 1.8±1.04 mm (intra-breath-hold) and 1.3±5.3 mm (inter-breath-hold) as determined from the study of seven treated lung cancer patients and over 92 portal images (Mah et al. 2000).

Automatic Respiratory gating

Pros And Cons

Ideally, automatic respiratory gating requires minimal effort from the patient and staff and could even be used for an unconscious patient. Since little patient effort is required, it is compatible with IMRT methods using fixed gantry angles, even if beam-on times are longer. Ideally, automatic gating eliminates the element of human judgment implicit in controlled breathing methods. However, the need to electronically interface the output signal of a motion sensor to control the accelerator beam makes automatic respiratory gating more technically complex and expensive than controlled breathing. An interface with the CT scanner is also required because planning CT images should be acquired at the treatment phase of the breathing cycle to avoid introducing systematic errors in internal organ position and shape. The fraction of the breathing cycle during which beam-on is allowed must be large enough to not overly prolong treatment. This requires some residual motion within the gate. Finally, the correlation of the signal from an external respiration sensor with internal patient anatomy cannot be assumed, but must be checked by patient-specific imaging which requires staff time. The best correlation is expected from gating on a direct image of tumor motion (Shirato et al. 2000) but this requires the invasive implantation of markers into the tumor.

Automatic Respiratory Gating Techniques

Shirato et al. (2000) describe a clinical system that gates the accelerator according to the motion of a marker (2 mm gold sphere) implanted in the tumor and tracked with orthogonal-pair images from diagnostic x-ray units fixed in the treatment room. This system had not yet been implemented for planning CT acquisition. A pre-clinical system utilizing electromagnetic tracking of implanted marker coils has also been described (Seiler et al. 2000).

Kubo and Hill (1996) investigated the dosimetric effects of sending the gating signal to the grid of the gun and using the gun-delay function on a Varian 2100C. For four patient volunteers, they also compared the output signals for three non-invasive automatic gating sensors: temperature sensors which monitor air flow in nostrils, a spirometer which monitors total respiratory airflow, and a strain gauge-band wrapped around the patient's thorax to monitor breathing motion (Respitrace). They found that the beam characteristics were not significantly altered and that the signals from all three methods were well correlated with each other. Kubo et al. (2000) further describe the clinical use of the Respitrace. Ramsey, Cordrey, and Oliver (1999) report extensive dosimetry studies of the effect of gating on static Varian 2100C beams, finding no clinically significant changes in output, flatness, and symmetry. Finally, two video-camera-based systems have been clinically used (Kubo et al. 2000; Ramsey, Cordrey, and Oliver 1999; Minohara et al. 2000; Mageras et al. 2001). For heavy-ion radiotherapy in Japan (Ramsey, Cordrey, and Oliver 1999), a charge-coupled device (CCD) camera tracks the motion of an infrared light-emitting diode attached to the patient's body for both planning CT acquisition and gated treatment. This study estimates that, with a 1-sec gate at end-expiration, it is sufficient to expand CTV to PTV by a 5 to 10 mm cranial/caudal margin, though the authors stress the need for patient-specific studies.

The Varian Real-Time Position Management (RPM) system, shown in figure 17–1, is a rather new commercial product which has been used at MSKCC since late 1999. It uses a CCD camera with attached infrared illuminator. The camera is interfaced with a PC and mounted on the wall of the accelerator or simulator room or the foot of the CT couch-insert. It monitors the motion of

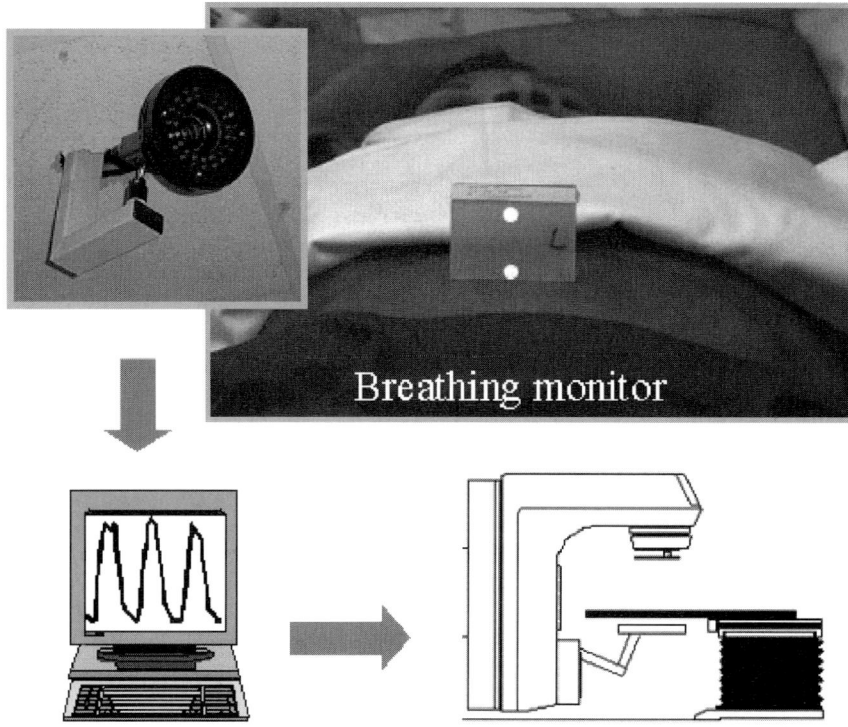

FIGURE 17–1. Components of the RPM automatic respiratory gating system. (Adapted from Wagman et al. 2002).

two infrared reflective markers that are rigidly embedded in a lightweight plastic block on the patient's chest. The system tracks the upper marker as the indicator of breathing motion and uses the 3 cm marker separation to provide a fixed distance scale. The user selects the portion of the breathing cycle for the gate. The signal from the PC is fed to the accelerator beam hold-off circuit for treatment, recorded during acquisition of breathing-synchronized fluoroscopic movies, or used to trigger the slice acquisition of a CT scanner in axial mode. Our clinical experience with the RPM system will be further described below.

MSKCC Clinical Experience

Controlled Breathing—Deep Inspiration Breath-Hold (DIBH)

DIBH has been in clinical use at MSKCC since February 1998, following imaging and treatment planning studies (Hanley et al. 1999; Rosenzweig et al. 2000; Mah et al. 2000). To date, 28 patients (all but one with lung cancer) have been treated with DIBH. Because approximately half the candidate patients cannot comply with the coaching instructions, potential patients are screened a few days before simulation. Figure 17–2 shows a typical patient setup. The patient, with nose clamped, breathes through a spirometer and is coached through a modified SVC maneuver (normal breathing followed by a maximum inspiration, a maximum expiration, and a final maximum inspiration with breath-hold). The spirometer output is connected to a laptop computer through the serial port. In-house software (Hanley et al. 1999; Mah et al. 2000) gives a graphic display of the

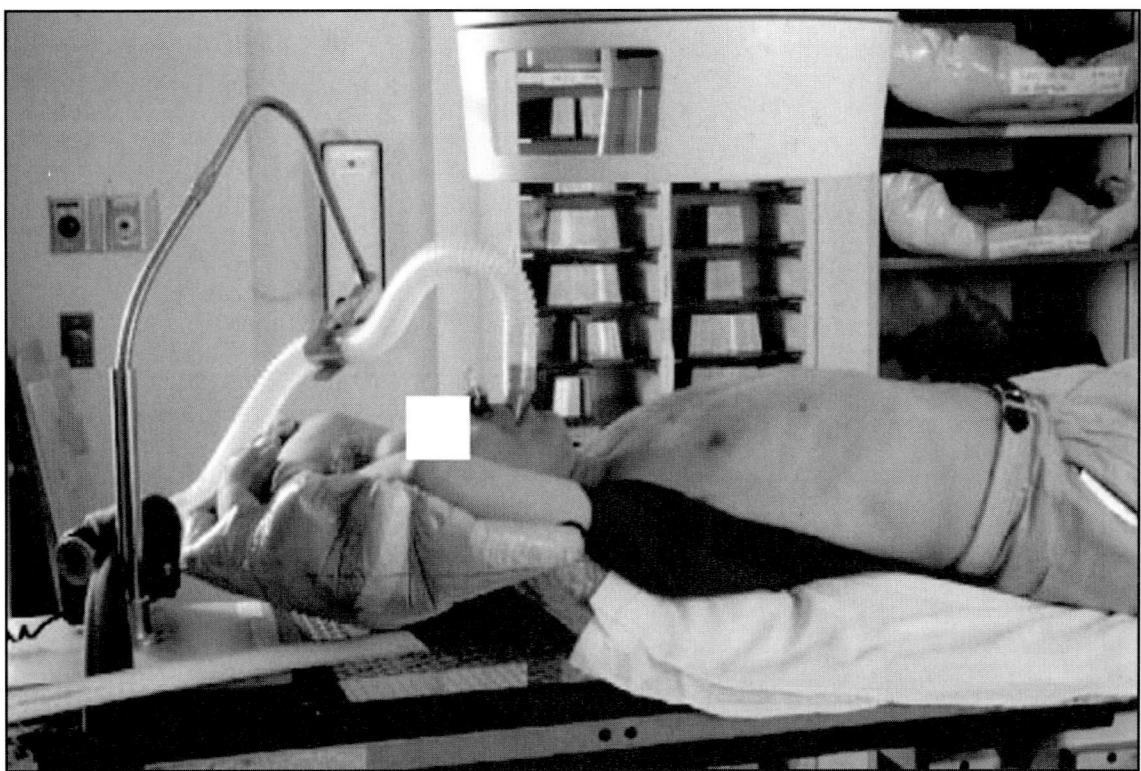

FIGURE 17–2. Patient setup for DIBH. [Reprinted with permission from Rosenzweig et al. (2000). "The deep inspiration breath-hold technique in the treatment of inoperable non-small-cell lung cancer." *Int. J. Radiat. Oncol. Biol. Phys.* 48(1):81–87, fig. 2. Copyright 2000 Elsevier Science Ltd., Oxford, UK.]

spirometer output (the *breathing trace*) and checks and saves the air flow volumes at key points in the trace to monitor that the SVC maneuver was performed as at the simulation session within preset tolerances. Figure 17–3a is a diagram of an ideal SVC breathing trace. At treatment, a bar-graph at the side of the laptop display (figure 17–3b) changes color to assist the coach in evaluating the maneuver. Breath-hold and treatment should occur when the bar turns green (Rosenzweig et al. 2000; Mah et al. 2000). Most patients can hold their breath for approximately 15 sec, which allows most single static fields to be delivered in a single breath-hold if the accelerator runs at maximum dose rate.

After immobilization, selection of isocenter in a free-breathing (FB) state, and a brief practice session, the patient receives three CT scans (from above the clavicles to well below diaphragm) in treatment position: FB, and spirometer-monitored scans at DI and shallow inspiration (SI). The DI and SI scans must be performed in 4 to 5 breath-hold segments. Because we hypothesize that the patient's state of respiration does not alter the position of her spine, the patient is set up for treatment while breathing normally. The FB scan (which requires <1 minute scanning time) is a patient-specific check of this hypothesis. The treatment plan and reference digitally reconstructed radiographs (DRRs) are based on the DI scan. The SI scan is used to set breath-hold tolerance levels by determining the motion extent of the gross target volume (GTV) for a known change in breath-hold volume (Mah et al. 2000). The simulation process, including immobilization, practice, and patient rests, takes 2 to 3 hours.

A major benefit of DIBH is the increase in lung volume, shown in figures 17–4 and 17–5. An additional benefit for some patients (figure 17–5) is the anterior shift of the GTV away from the

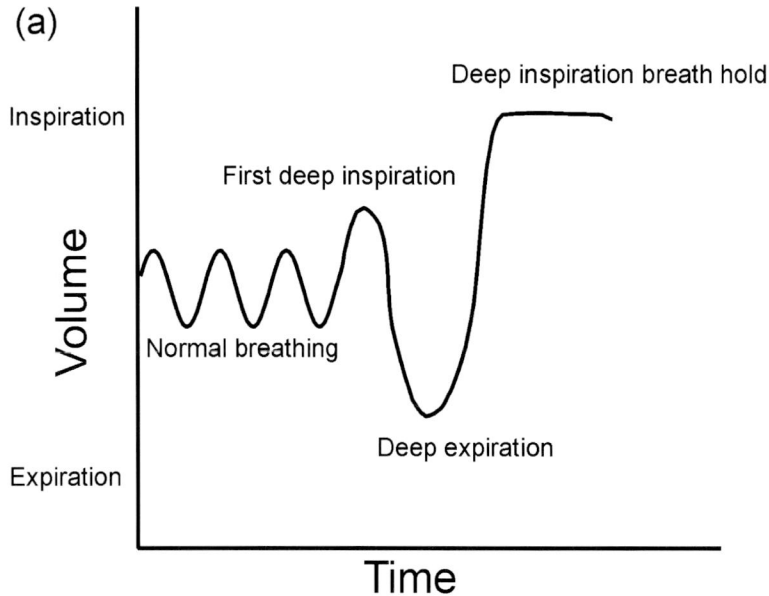

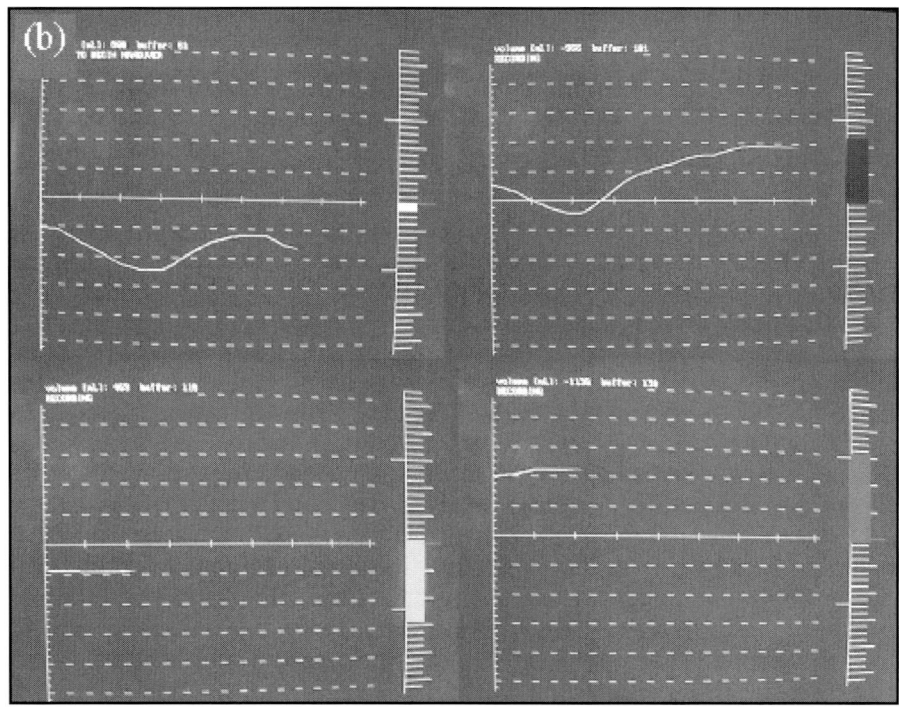

FIGURE 17–3. (a) Spirometer output signal for an ideal modified SVC maneuver, culminating in DIBH. (b) DIBH computer graphic display. Upper left: normal breathing; upper right: first deep inspiration; lower left: deep expiration; lower right: second deep inspiration and breath-hold. The bar graph at the side of the display is color-coded to signal whether the patient has performed a reproducible SVC maneuver. [Fig. 173b reprinted with permission from Mah et al. (2000). "Technical aspects of the deep inspiration breath-hold technique in the treatment of thoracic cancer." *Int. J. Radiat. Oncol. Biol. Phys.* 48(4):1175–1185, fig. 1. Copyright 2000 Elsevier Science Ltd., Oxford, UK.]

Free Breathing

Deep Inspiration

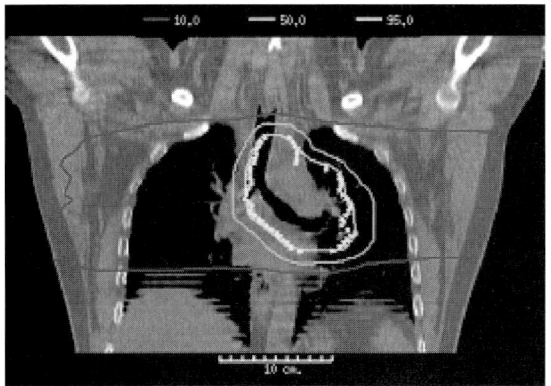 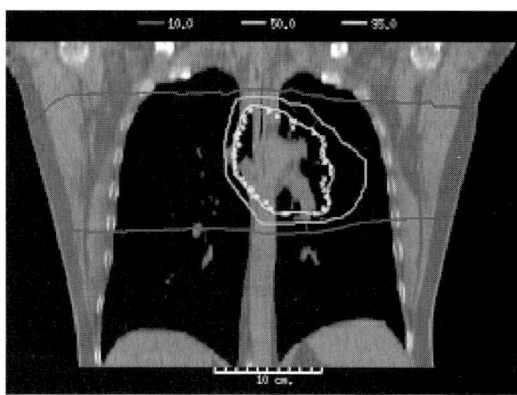

FIGURE 17–4. Comparison of dose distributions for a free breathing plan (left) and deep inspiration plan (right). Both reduction of respiratory motion and lung expansion are achieved with DIBH. Note that the breathing artifacts near the diaphragm of the FB scan are absent in the DI scan and there is much less lung within the 10% (red) isodose line for the DI scan. [Reprinted with permission from Rosenzweig et al. (2000). "The deep inspiration breath-hold technique in the treatment of inoperable non-small-cell lung cancer." *Int. J. Radiat. Oncol. Biol. Phys.* 48(1):81–87, fig. 3. Copyright 2000 Elsevier Science Ltd., Oxford, UK.] See COLOR PLATE 43.

Free Breathing

Deep Inspiration

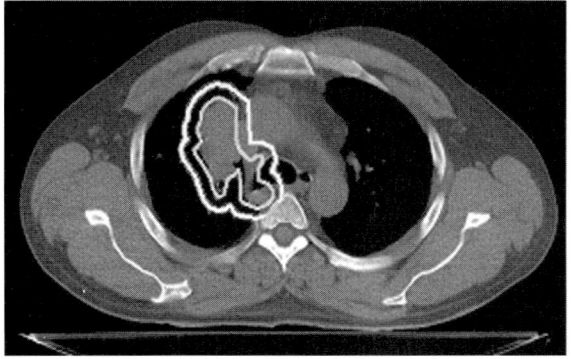 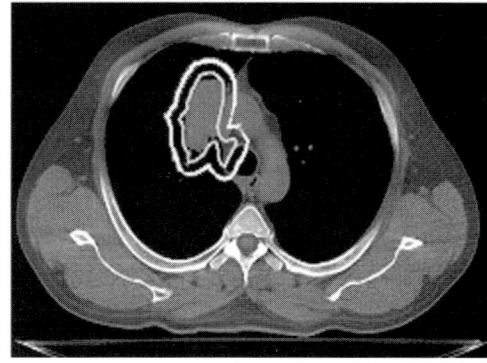

FIGURE 17–5. For some patients, the use of DIBH moves the target (PTV = outer line, GTV = inner line) away from the cord.

spinal cord. Although imaging studies using the diaphragm as a surrogate for GTV motion show that DIBH provides excellent positional reproducibility (Hanley et al.1999; Rosenzweig et al., 2000), we use the same GTV-to-PTV margin for DIBH and for FB lung cancer treatments. This is, in part, because the DI lung expansion removes enough normal lung from the irradiated region to permit treatment to higher doses without increase in estimated normal tissue complication probability (NTCP) (e.g., figure 17–4). Figure 17–6 illustrates this point with results of a treatment planning study. Also, we feel that the margins protect against possible expansion of microscopic disease outward from the GTV due to DI. Finally, Monte Carlo studies for 6 MV photons suggest that our current GTV-to-PTV margins (1 cm combined with a 0.6 cm margin from PTV to aperture edge) give adequate target coverage despite lateral disequilibrium effects, which are not handled by our treatment planning system's dose calculation algorithm (Yorke et al. 2002).

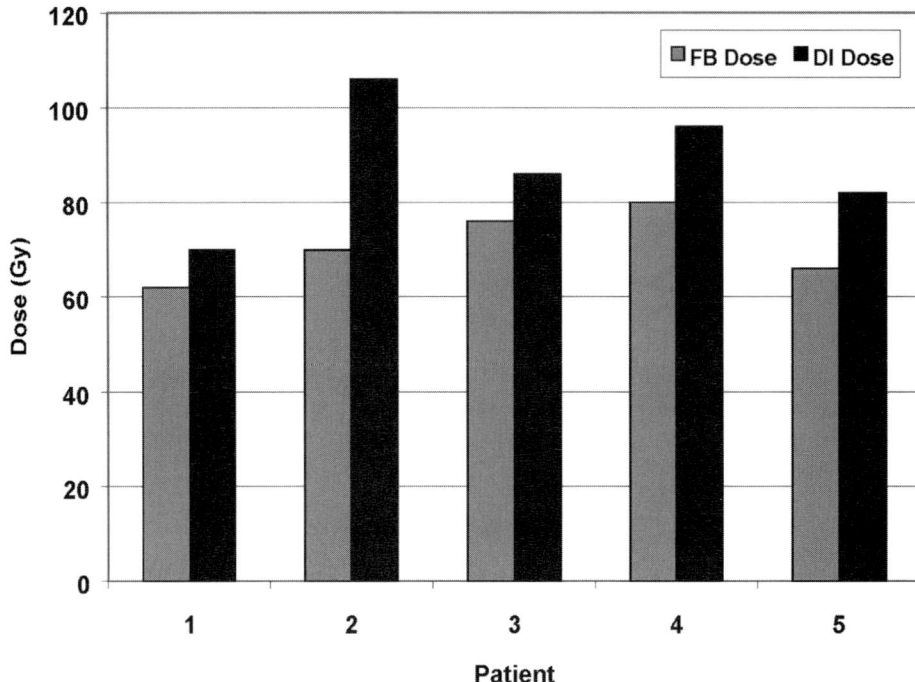

FIGURE 17–6. DIBH permits higher doses to be delivered while meeting the same normal tissue constraints on cord and lung (in this example, cord dose <50 Gy and lung NTCP calculated with the Lyman model <25%).

The therapists must be trained to use the equipment, to interpret the graphic display, and to coach the patients. It is best to have two therapists for DI treatments—one to coach and one to activate the beam, following the coach's instructions. The patient is at deep inspiration for all treatment beams and portal images. An extra anterior-posterior (A/P) port film (the "diaphragm film") showing the entire lung is taken at least weekly to confirm that the lung inflation remains constant. If the films or the spirometer traces on the computer indicate that a patient is missing or exceeding SVC maneuver levels set at simulation, the consequences and remedies are evaluated by the physicist and physician. While DIBH patients are initially scheduled for two 15-min treatment slots, an experienced patient-therapist team often require less than 20 min per four-field, wedged treatment.

Automatic Respiratory Gating—The RPM System

Since late 1999, we have treated 18 patients with gating (8 with liver cancer, 10 with lung cancer). Fifteen of these patients were treated with static three-dimensional conformal radiation therapy (3DCRT) or sliding-window IMRT based on breathing-synchronized CT scans. The liver cancer patients were gated at end-expiration, and their GTV-to-PTV expansion margin was reduced from 2 cm without gating to 1 cm (Wagman et al. 2003). Margin reduction was not performed for the lung patients and most were treated at end-expiration, although lung inflation at end-inspiration is expected to reduce estimated lung toxicity (Paoli, Rosenzweig, and Yorke 1999).

Figure 17–1 shows the components of the RPM system; note the marker block on the patient's chest and the wall-mounted camera. An in-room monitor aids in positioning the marker block, which should be securely taped about midway between the umbilicus and xiphoid. At this location, its motion amplitude under normal respiration is approximately 1 cm (the recommended minimum is 0.5 cm). Skin marks at the block corners assure its reproducible placement at each session. Since the line between the markers should be approximately vertical, we shim the bottom of the block for patients with sloping chests.

At the start of any session (simulation, CT or treatment), the system is put into *tracking mode* for a few breathing cycles. The marker block image is displayed on the screen (figure 17–7) and the system software learns the minimum and maximum vertical positions of the upper marker, which will then be used for a graphic display and to set the gating thresholds. A *periodicity filter* algorithm monitors the regularity of the marker motion and indicates the results by a series of dots at the upper left-corner of the display—the fewer the dots, the more periodic the motion.

When the periodicity-filter display is judged acceptable (≤3 or 4 dots), the system is put into *record* mode and a real-time graph of the position of the upper marker—the breathing trace—is displayed and recorded (figure 17–8). The gating thresholds are also displayed and can be interactively changed. These thresholds are always a fraction of the minimum-to-maximum distance determined during tracking. This means that, if the marker motion amplitude changes between sessions, e.g., due to altered breathing levels or marker placement variation, the absolute motion within the beam-enable gate changes by the same factor.

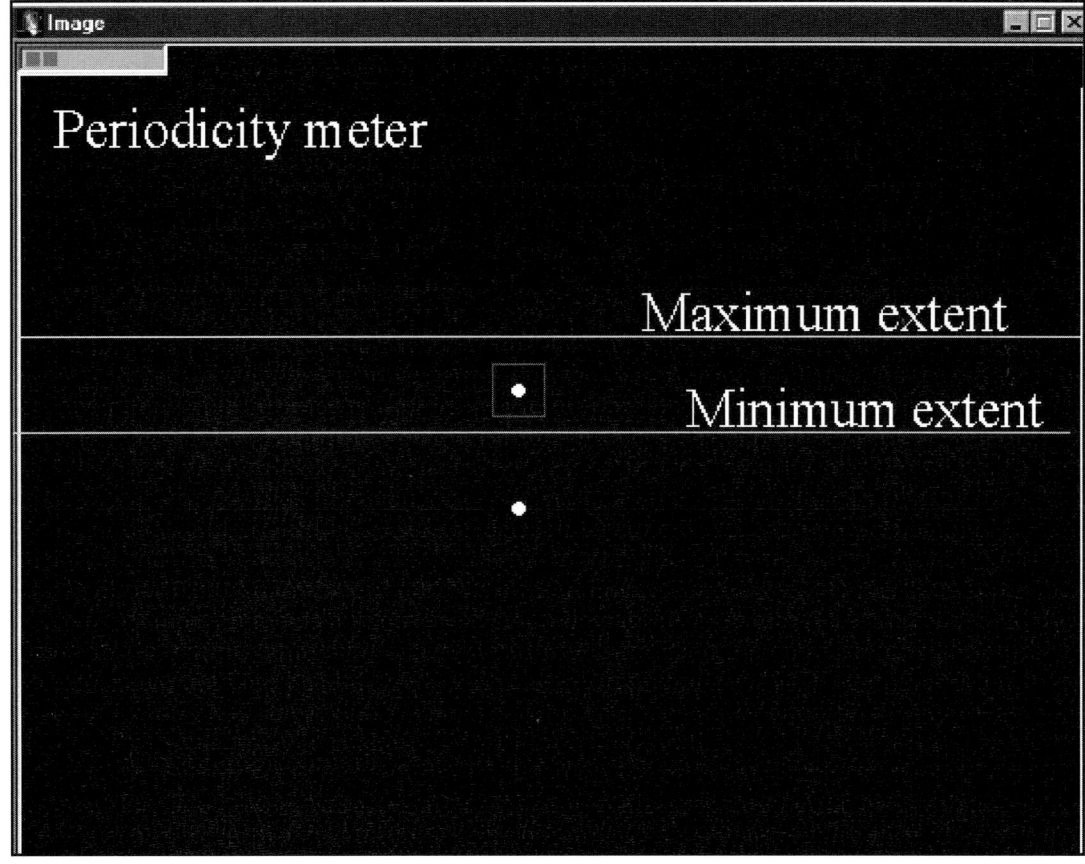

FIGURE 17–7. Appearance of the RPM system monitor screen in "track" mode. The white circles are the images of the reflective markers. The lines indicate the maximum and minimum extents detected for the motion of the upper marker. The squares in the upper left corner indicate how periodic the motion is—6 squares mean very irregular, 1 square represents perfectly periodic.

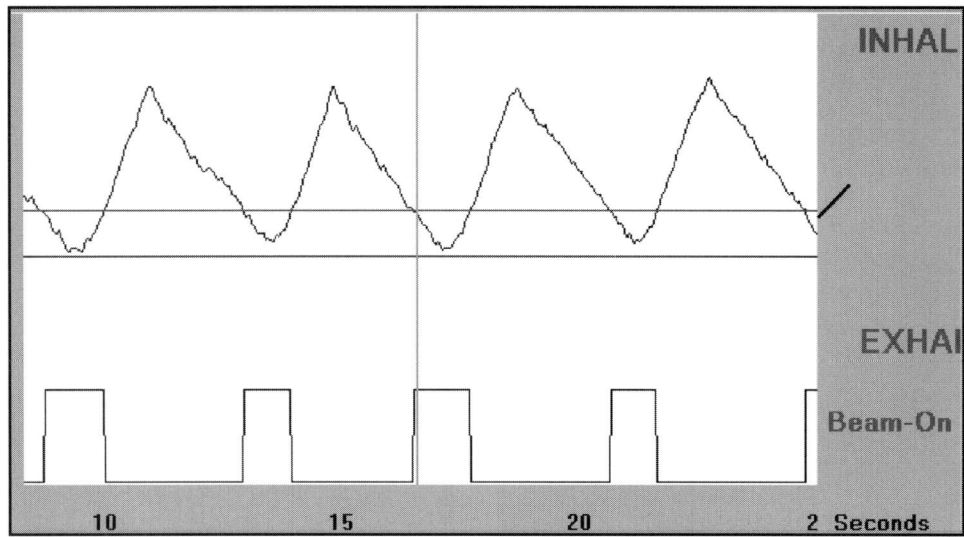

FIGURE 17–8. A patient breathing trace gated at expiration. The tick-marks on the horizontal axis are time in seconds, on the vertical axis, distance in cm. The square wave graph at the bottom of the display indicates when the treatment beam would be enabled.

The user can gate on either the amplitude or phase of the marker motion. For amplitude gating, the amplitude threshold levels are shown as horizontal lines on the breathing trace and their positions can be interactively varied. The breathing trace display in figure 17–8 shows amplitude gating at expiration. Beam-enable or CT acquisition occurs only when the trace is between those lines; these instances are shown in the display as a square-wave signal directly under the breathing trace. Amplitude gating is easy for the operator to understand and monitor. Human supervision is necessary to prevent the occurrence of a situation such as shown in figure 17–9, where irregular breathing or overall patient motion leads to beam-enable at the wrong part of the breathing cycle. This can happen even when breathing was regular at simulation.

For phase gating, the RPM system calculates the phase of the breathing trace using a running estimate of the breathing period and the user specifies the allowed range of phases. However, changes in breathing amplitude or position (either of which could produce the situation shown

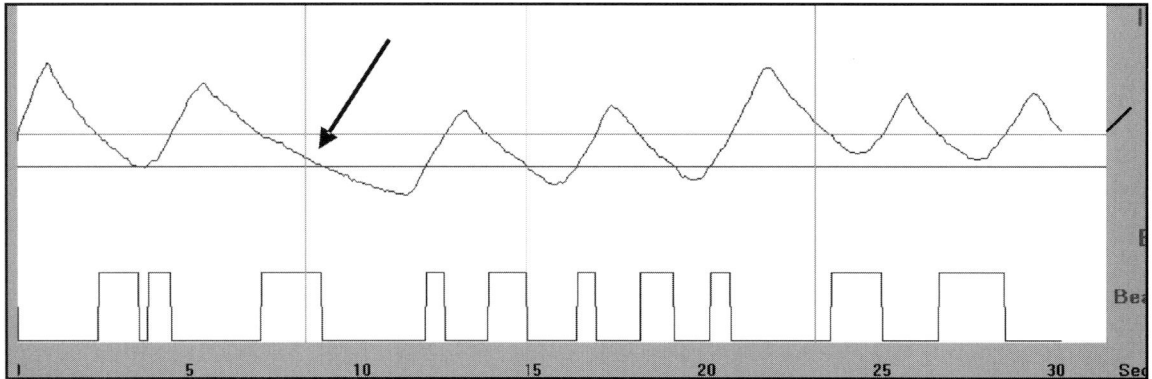

FIGURE 17–9. Irregular breathing by a patient with amplitude gating may cause beam-on at the wrong phase of the breathing cycle.

in figure 17–9) would not be easily detected with phase gating. Thus, phase gating also requires human supervision.

Before performing simulation or CT-simulation, the user must decide whether to use amplitude or phase gating, whether to gate at end-inspiration or end-expiration, and the width of the gate. Considerations for the last two decisions include the patient's disease site and breathing characteristics. An important part of this decision is the duty cycle—the ratio of the time that the trace is within the gate to the period of the entire cycle. The end-expiration part of the breathing trace is usually flatter and less variable, providing greater stability and a larger duty cycle. Thus, unless end-inspiration provides other benefits (e.g., greater lung volume at end-inspiration for lung patients), setting the gate at end-expiration is preferable. The choice of gate width is a trade-off between minimizing motion within the gate and completing treatment in a realistic time. There is little respiratory motion within a narrow (i.e., short duty cycle) gate. However, the beam delivery time is inversely proportional to the duty cycle; at a dose rate of 300 MU/min, a 100 MU beam takes 20 sec to deliver without gating and 80 sec with gating at a duty cycle of 25%. Overly long treatment times are inconvenient and may cause setup error due to motion of an uncomfortable patient.

The threshold amplitude or phase levels are set by the user at simulation and should not be changed afterward without careful consideration of the implications for the true dose distribution. Fluoroscopy is a useful tool for deciding the optimal gate width. The RPM system allows acquisition of a breathing-synchronized fluoroscopic movie. This can be displayed to determine the motion of anatomical landmarks within a chosen gate interval. The ideal performance of the system can be illustrated using a battery-driven periodic-motion phantom designed by Varian for performing quality assurance (QA) tests (figure 17–10a). We recorded a fluoroscopic movie and exposed verification films of the breathing phantom with and without gated operation. The blurred images in figure 17–10b are the average over all the fluoroscopic movie frames in one cycle (approximately 4 sec) and the verification film without gating. The sharp fluoroscopic image (figure 17–10c) is obtained by averaging image frames within the gate at end-inspiration while the corresponding verification film is taken with the accelerator beam gated at end-inspiration.

For CT imaging, the camera and illuminator are mounted on the couch insert to maintain a fixed position relative to the patient. After "tracking" and "recording" the marker motion, the user places the system in "enable" mode. The RPM system then generates a trigger when the breathing trace is within the gate. The trigger is fed to a CT simulator (Philips/Marconi PQ5000) through the contrast injector port. After a short scanner-dependent delay (we measured 330 msec) the trigger initiates acquisition of a CT slice in axial mode. At best, one slice is obtained per breathing period, leading to increased scan times. For example, 100 breathing-triggered slices take over 8 min for a patient with a 5 sec breathing period. With irregular breathing (e.g., coughing), the periodicity filter prevents image acquisition until regular breathing is restored, which further increases scan time. Careful operator supervision is required because breathing irregularities can occur shortly after a trigger signal and cause slice acquisition at the wrong breathing phase (figure 17–11). For the most part, breathing artifacts are much reduced in the breathing-triggered CT images, as shown in figure 17–12. As we ascend the learning curve for the RPM system, we acquire two respiration-triggered CT scans at the treatment breathing phase—one for treatment planning and one for quality assurance of the performance of the system. The breathing trace and gate levels, recorded at the acquisition of the planning scan, are used as the reference for treatment.

For treatment and port films, a gating-enable switch is set at the accelerator console to route the RPM system signal to the beam hold-off circuit. After setting up the patient, the therapist selects the patient from the RPM database and a movie of the reference breathing-trace is displayed on the PC monitor. The therapist next places the RPM system into *track* mode. When the periodicity meter indicates regular breathing, the therapist chooses *record* mode, observes for a

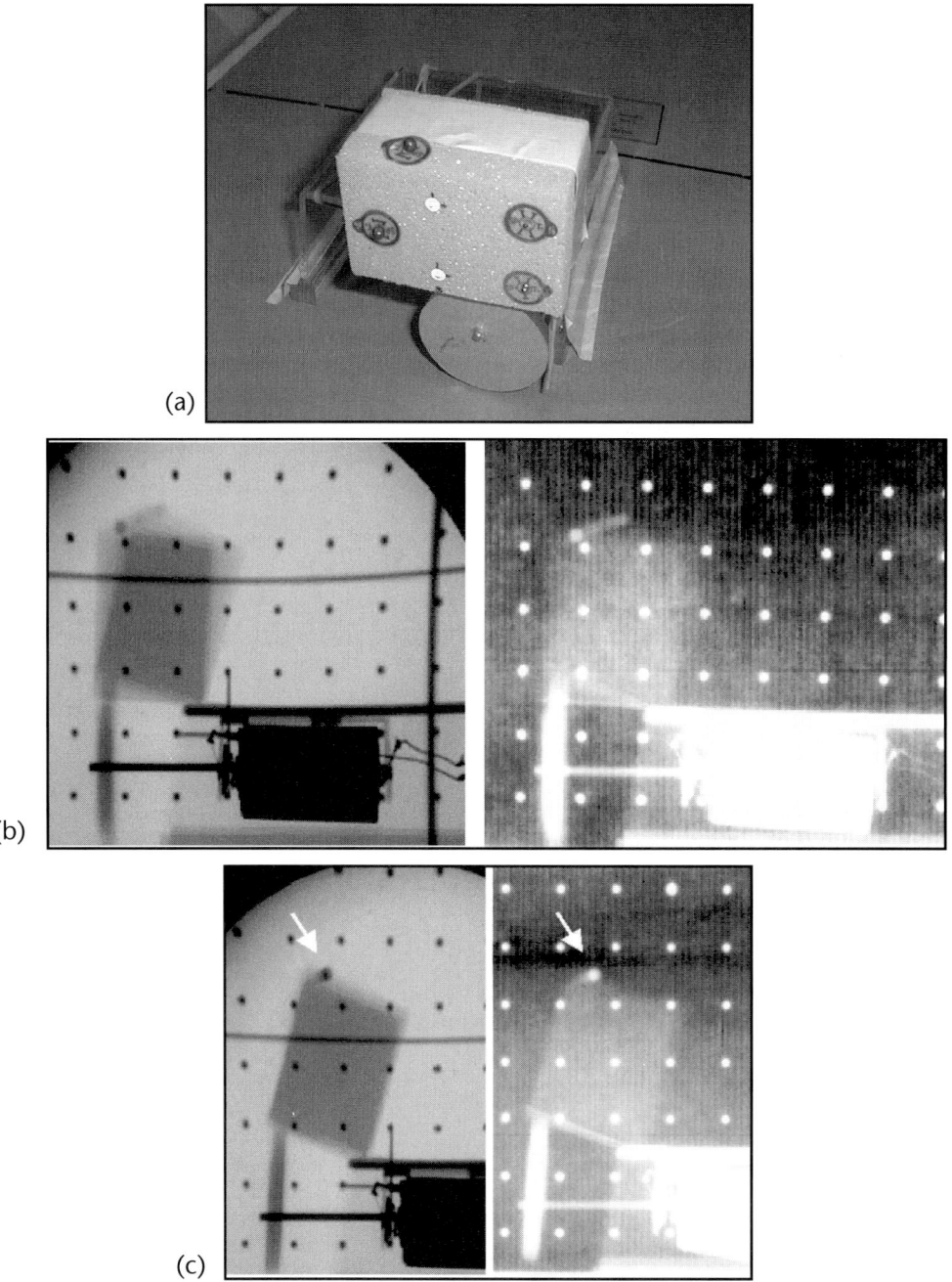

FIGURE 17–10. (a) Battery-powered breathing-motion phantom. As the eccentrically mounted wheel rotates (period approximately 4 sec), the marker block moves periodically with both vertical and longitudinal component. (b) Fluoroscopic image (left) and portal verification film (right) without gating. (c) Fluoroscopic image generated from all the frames within the gate centered at phantom end inspiration (left) and localization film acquired with accelerator beam gated at phantom end inspiration (right). [Fig. 17–10a is reprinted with permission from Mageras et al. (2001) "Fluoroscopic evaluation of diaphragmatic motion reduction with a respiratory gated radiotherapy system." *J. Appl. Clin. Med. Phys.* 2(4):191–200, fig. 1a. Copyright 2001 American College of Medical Physics.]

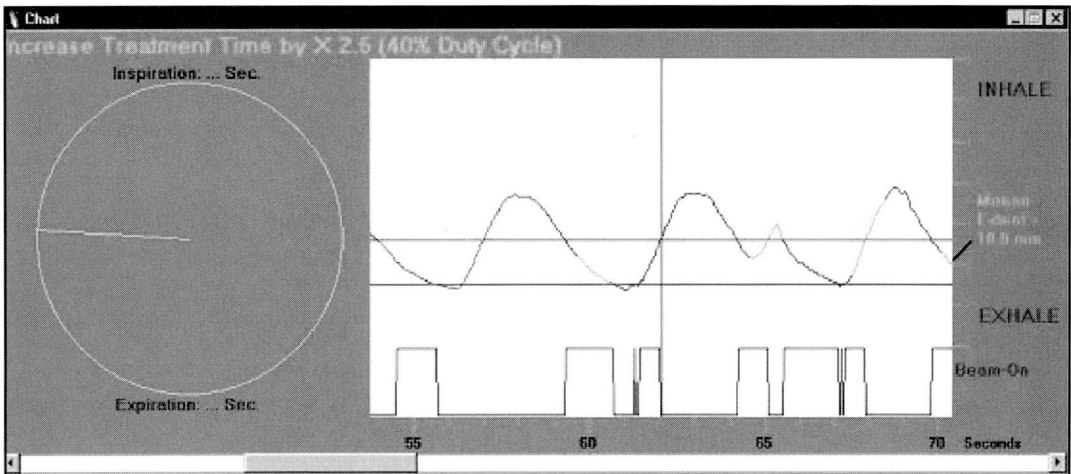

FIGURE 17–11. Irregular breathing during CT acquisition can cause a slice to be acquired at the wrong part of the breathing trace. Green line segments indicate x-ray on intervals during slice acquisition. See COLOR PLATE 44.

few breath cycles to assure that the gate levels are correct and adequate, and finally goes to *enable gating* and hits the beam-on button. Thereafter, beam is delivered only during the intervals allowed by the gate (with a measured delay of approximately 50 msec). The time (seconds) for beam delivery is inversely proportional to the duty cycle and, if the periodicity filter prevents beam delivery due to irregular breathing, the time is further increased. Thus, the therapist must remember to program adequate delivery time for each beam. For logistic reasons, we deliver gated beams at the maximum dose rate (e.g., 600 MU/min on the Varian 2100EX). To guard against systematic errors introduced by changes in respiration between simulation and treatment, extra AP (and sometimes lateral) portal images including the diaphragm are acquired—three during each of the first two weeks, two the third week with further decreases dependent on the patient.

The RPM system should be commissioned for use in treatment by measurements to assure that gating does not significantly change beam characteristics. Such changes are likely to be accelerator dependent and should be checked for each treatment unit. We found negligible changes in output, tissue-maximum ratio (TMR), flatness and symmetry for operation at 100 MU and dose rates of both 300 MU/min and 600 MU/min. Ramsey, Cordrey, and Oliver (1999) performed more extensive measurements and observed small deviations in flatness (1.9%) when <5 MU were delivered in the gating window. For use with IMRT, additional commissioning should be performed. We have delivered gated IMRT with the sliding-window DMLC method on both Varian 2100C/C and 2100 EX accelerators. We have performed ion chamber and film dosimetry for sliding window IMRT fields on both machines and found that neither RPM gating (Yorke et al. 2000) nor gated beam delivery at 600 MU/min on the 2100EX significantly changes the IMRT dose distributions. Increasing the dose rate to 600 MU/min, while not halving the gated treatment time, does decrease it by approximately 60%. Kubo and Wang (2000) also reports that RPM-type gating is consistent with sliding window DMLC.

For both simulation and treatment, the RPM system performs best if the patient breathes regularly with a pattern that is reproducible from session to session. Not only does irregular breathing lengthen treatment time, but as mentioned above (figure 17–9) we have observed situations where drift in the breathing wave form during treatment led to beam-enables at unintended points in the duty cycle. To improve breathing regularity, we have adapted a technique suggested previously and prompt the patient with simple, recorded instructions ("breathe in . . . breathe out . . .") (Kubo et al. 2000). The tempo of these instructions is customized for each patient at

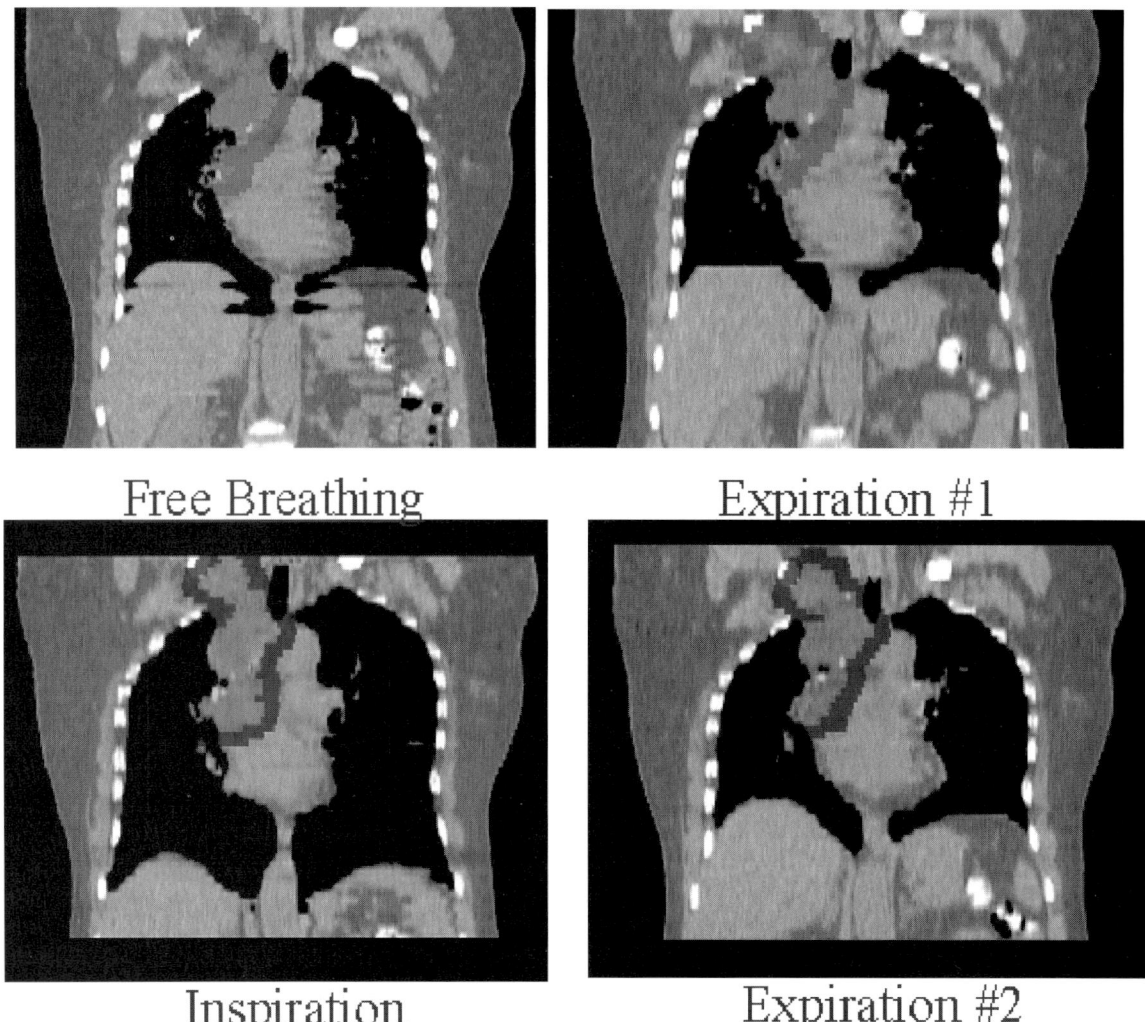

FIGURE 17–12. Respiration-triggered CT scans acquired at end inspiration or expiration show far fewer motion artifacts than do free-breathing scans.

the start of simulation and is slightly slower than the patient's natural breathing rate. The same instructions are used throughout treatment in an effort to maintain consistent respiratory conditions. For most patients, the instructed breathing trace is more regular in both period and amplitude (Mageras et al. 2001). Further, although patients are told to breathe with normal amplitude, instructed breathing is often deeper than uninstructed breathing. We have been using stand-alone commercial software (cool edit, Symantic Software Corp, Phoenix AZ) to generate breathing instructions; however, the current RPM software also provides a means of setting such audio prompts. The current software release allows visual feedback as well, which might help patients better regulate their breathing amplitude.

To what extent does the RPM system—which tracks the motion of an external marker— provide constancy of internal anatomy? We have analyzed the fluoroscopic movies, the duplicate breathing-triggered CT scans, and the gated AP portal images for our patient group. The fluoroscopy and CT studies give information about intra-fraction reproducibility while the portal images provide inter-fraction reproducibility information for the entire course of treatment. In a study of the fluoroscopic movies of the first 6 initial gating patients at MSKCC (Mageras et al. 2001), an automated algorithm was used to track the superior-inferior position of the diaphragm apex (Keatley and Mageras 2000). Average patient diaphragm excursion was reduced from 1.4 cm (range 0.7 to 2.1 cm) without gating and without breathing instruction, to 0.3 cm (range 0.2 to 0.5 cm) with instruction and with gating tolerances set for treatment at expiration for 25% of the breathing cycle.

The fluoroscopic intra-fraction results are in agreement with our analysis of respiration-triggered CT studies in terms of the center-of-mass displacements and degree of overlap of a variety of thoracic organs. For example, comparison of CT scans acquired at end-expiration (E-E) and end-inspiration (E-I) in eight patients found the average displacement, in the SI direction, of the right and left diaphragm to be 11.5 mm and 21.4 mm, respectively. In contrast, the average SI movement of the right and left diaphragm between the two (E-E) scans was 2.2 mm and 3.8 mm, respectively. Organ center-of-mass displacements were smaller between E-E than E-I scans; for all organs, average E-I displacement was 12.8 mm (range: 3.0 to 29.2 mm), and 2.0 mm for E-E (range: 0.0 to 6.4 mm).

In a study of gated portal images in eight patients (Ford et al. 2002) the patient-averaged standard deviation of the diaphragm position relative to bony anatomy was 2.8 mm, similar to that reported for other methods of reducing respiratory motion (Ekberg et al. 1998; Mah, Hanley, and Rosenzweig 1998). However, for four of these patients, the mean diaphragm position on the port films differed by more than 4 mm from the diaphragm position on the DRR (thus, during the CT session), perhaps because of changes in respiration between the CT session and treatment. For one liver-cancer patient (not among the first eight), after several weeks of treatment the diaphragm position on port films shifted approximately 2 cm superior of its simulation position due to an abdominal fluid accumulation (unnoticed by the patient!). The patient's field-apertures were adjusted to accommodate this systematic shift. These examples illustrate the need for caution in dealing with respiratory gating as it moves from phantom to real world, and emphasize the need for careful monitoring of portal images throughout the course of treatment.

Prostate Patient Setup for Conformal IMRT

Recent long-term outcome data from MSKCC (Zelefsky et al. 2001) indicates that the proportion of prostate patients with a positive biopsy taken at least 2 1/2 years after treatment decreases as the radiation dose is increased. The fractions of patients with positive biopsies after receiving 64.8, 70.2, 75.6, and 81 Gy, were 54% (13 of 24), 34% (23 of 68), 23% (27 of 119), and 10% (4 of 41) respectively (see also the chapter 9 on using IMRT for treating cancer of the prostate). This indicates that 81 Gy or higher doses are necessary for the successful treatment of prostate cancer. The delivery of such doses to the target organs (i.e., prostate and seminal vesicles), while keeping the dose to the anterior rectal wall to an acceptable level, was initially made possible by the introduction of 3-D conformal therapy (3DCRT). More recently, the introduction of IMRT has resulted in dose distributions optimized to maintain target coverage and to minimize dose to the rectal wall. The probability of late grade II rectal toxicity in patients receiving 81 Gy has been reduced from 14% with 3DCRT to 2% using IMRT. As the dose is increased beyond 81 Gy, however, it will become increasingly more difficult to maintain both low rectal complication rates and to deliver the prescribed dose to the target organs.

A limitation on the efficacy of IMRT is the daily displacement of the soft tissue structures from their planned positions. In addition to the usual problem of bony structures moving with respect to skin marks, the prostate and seminal vesicles may also be displaced relative to the bony anatomy because of daily variations in bladder and rectal filling. Shape changes of the CTV, especially in the vicinity of the seminal vesicles, may further limit efficacy. This section outlines how these target displacements are described, the magnitude of the displacements observed, and various strategies being used to reduce them.

Setup Error, Organ Motion, and Target Displacement

Let us define a 3-D coordinate system oriented such that the x direction is towards a prone patient's right, the y direction is vertically upwards, and the z direction is from the feet to the head (toward the gantry). The origin is at the patient isocenter, the point triangulated by the patient's isocenter skin marks. We can define a reference point T_{plan} to indicate the target position as the center of gravity (COG) of the target volume, as defined by the contours drawn on the planning CT scan.

For the purpose of this discussion, let us assume that CT scans are taken of the patient in the treatment position just prior to each treatment. Using the scan from day i of treatment, the position of the target COG, T_i with respect to the planned isocenter can be determined. The patient isocenter is located in the CT scan by affixing radio-opaque markers (BBs) to the patient's isocenter skin marks. The new target COG is again determined from manually drawn contours. Because of both movement of the skin with respect to underlying bone, and movement of the target with respect to bony landmarks, T_i varies from day to day. Thus on treatment day i the target is displaced by D_i from its ideal position, where $D_i = T_i - T_{plan}$ (figure 17–13a).

Setup Error And Organ Motion

The target position T can be thought of as being the sum of two pieces. If we define B as a convenient bony landmark, centered on the pubic symphysis, for example, then for the planning scan T can be written $T_{plan} = B_{plan} + T'_{plan}$, where B_{plan} is the position of the landmark with respect to the isocenter marks, and T'_{plan} is the target reference point with respect to B_{plan} (figure 17–13b). On treatment day i the skin may be displaced with respect to the underlying bone, thus the point B_i triangulated by the skin marks is also displaced. This change is the called the setup error S_i, where $S_i = B_i - B_{plan}$ (figure 17–13c). Additionally, the target reference point may also have moved with respect to the bony landmark, its relative position now being T'_i. The displacement of the soft tissue target with respect to the bony landmark is called the organ motion M_i, where $M_i = T'_i - T'_{plan}$ (figure 17–13d). The target displacement on treatment day i, D_i, is the sum of the setup error and organ motion;

$$D_i = T_i - T_{plan}$$
$$= S_i + M_i$$

Systematic And Random Errors

Clearly, acquiring CT scans of a patient in order to determine and correct the target displacement before every treatment would be very demanding of a clinic's resources. However, an important feature of the displacement components, that the means can be significantly different from zero, can be seen in the first row of histograms in figure 17–14. Histograms of the right-left (R/L), posterior-anterior (P/A), and superior-inferior (S/I) components of the displacement D for a single patient, scanned on 15 different days just prior to treatment, are shown. This situation arises partly because the plan is based on a single CT scan. This is really a snapshot of a moving object; for most patients, the position of the target with respect to the skin where the marks or tattoos will be placed is close to its mean position over the treatment course. For some patients, the target may, through various random processes, be close to its maximum displacement. As a consequence, the treatment plan is developed for the target in this unusual position. This kind of error is called a *systematic* or *preparation* error. Operationally, it is defined as the R/L, P/A, and S/I components of the displacements for an individual averaged over each treatment day. The

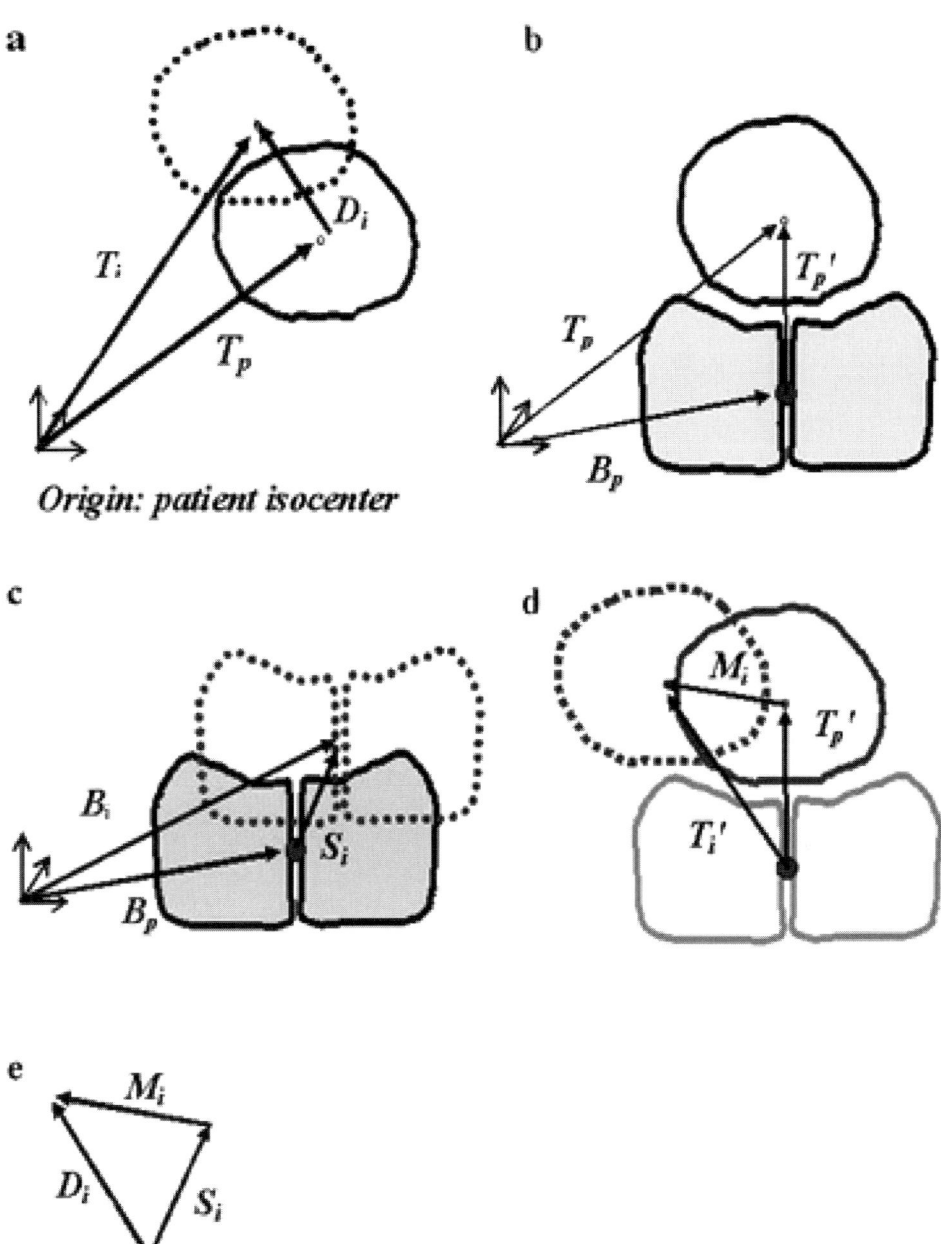

FIGURE 17–13. (a) The position of the COG of the soft tissue target in the planning scan T_p (solid), and the treatment scan T_i (dashed) with respect to the point triangulated by the skin marks, or patient isocenter. D_i is the displacement. (b) Decomposition of target position T_p in the planning scan into position of bony landmark with respect to skin marks, B_p, and position of target with respect to bony landmark T_p'. The same decomposition is applicable to the treatment scan. (c) Setup Error: the shift in the bony landmark with respect to the skin mark between the treatment scan (dashed) and the planning scan (solid). (d) Organ motion: the change in soft tissue target position from the planning scan to the treatment scan with respect to the bony landmark. (e) The displacement D_i is the vector sum of the setup error and the organ motion.

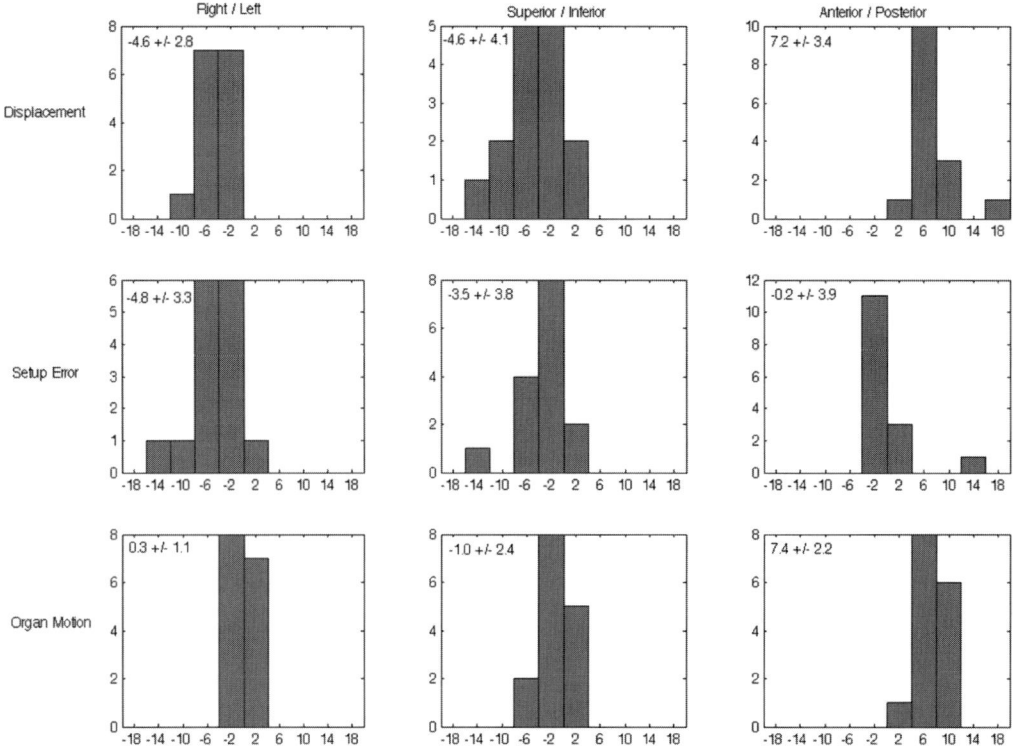

FIGURE 17–14. Distributions of the displacements D_i (first row), setup errors S_i (second row), and organ motions M_i (third row) from CT scans of a single patient made prior to treatment on 15 different days. Estimates of the systematic and random errors for each distribution are calculated as the mean and standard deviation (Shown on each plot as mean ± standard deviation). All dimensions are in millimeters.

spread of the distributions about their means arises from the day-to-day setup and organ motion variation that occurs at treatment time. This is called a *random* or *execution* error. For example, in figure 17–14 in the P/A direction, the systematic error of the target displacement is 7.2 mm in the anterior direction and the random error is 3.4 mm. A large systematic error is potentially more serious than an equally large random error because a systematic error is constant throughout the treatment, possibly leading to a cold spot in the target, or a hot spot in a nearby critical structure.

Additional Contributors To The Systematic Error

If systematic errors occurred only because the plan was designed using a single CT scan, then the distribution of systematic errors for a large number of patients would be the same as the distribution of random errors. However, other factors will increase the magnitude of the systematic errors. The conditions the patient is subject to at the time of the planning scan are different from those at the time of treatment: the time on the couch may be longer, the patient may have had an enema, a rectal catheter may be used, etc. Effects such as these can cause the mean target position in simulation to differ from the mean position under treatment conditions. Other sources of error are, for example, the uncertainty in skin mark placement, and the differences between how the CT and treatment machine's couches sag as a function of the extension from their support pedestals. These transfer errors increase the spread of the distribution of systematic errors.

Correcting For Systematic Errors

Data from a prostate motion pilot study at MSKCC are given in table 17–1. Fourteen patients had CT scans prior to treatment on 15 days. Comparison of the magnitudes of the systematic and random errors in table 17–1 reveals that the systematic errors, especially in the critical P/A direction, are larger than the random errors. This suggests a more modest improvement can be made with substantially less work: correct the systematic error alone by determining the mean displacement as early as possible into the treatment course, then adjust the patient's setup accordingly.

The properties of this kind of intervention strategy used to correct setup errors were explored by Bel et al. (1993) using a Monte Carlo computer model. The distribution of random errors was assumed to be Gaussian with a mean of zero and a standard deviation of σ. Several systematic error distributions were investigated, all Gaussian with a mean of zero but with standard deviations ranging from σ to 3σ. Their intervention protocol depends upon two parameters, N_{max}, the maximum number of consecutive imaging days, and α, the initial action level. At each treatment, the patient is imaged. After N imaging sessions, the average deviation d_N is determined and compared with a shrinking threshold $\alpha/\sqrt{N}$. A correction is made if $d_N > \alpha/\sqrt{N}$. If a correction of d_N is made, N is reset to one, and the procedure starts over. If N reaches N_{max}, the process stops.

Assuming a Gaussian distribution of systematic error with a standard deviation of 2σ (i.e., twice the standard deviation of the random error distribution), Bel et al. (1993) report several strategies capable of reducing the probability of a treatment having a setup error greater than 3σ from 18% with no intervention to less than 5%. Strategies that achieved this were those with $(\alpha/\sigma, N_{max})$ being (1,1), (2,2), and (3,4). To summarize: there is a tradeoff between α and N_{max}, and the costs of the intervention procedure, the costs being the number of measurements needed and the number of corrections needed. Reducing α increases the number of corrections, and increasing N_{max}, increases the number of measurements. Different institutions may have different precision requirements, and, depending on whether they use film or an electronic portal imager, may assess the costs differently. Bel et al. (1996) report a 3-D version of this procedure being performed at three institutions, each of which chose slightly different $(\alpha/\sigma, N_{max})$ parameters. They also augmented the procedure with a second phase designed to track any time trends (i.e., drift in patient position over several sessions) or other sudden change in setup that could occur after the initial intervention process terminated. They found that they could significantly reduce the systematic component of the setup error with an acceptably small increase in departmental workload.

MSKCC Pilot Study

Inspection of figure 17–14 reveals that the correction strategy described above, or any other strategy based on identifying bony landmarks, would not have reduced the P/A component of the displacement, as it is largely due to organ motion. Ideally, the mean target displacement should be corrected, rather than only the setup error component. The result of such an intervention

Table 17–1. The Systematic Errors Were Calculated as the Standard Deviation (SD) of the Means from the 14 Patients. A summary measure of random error was calculated as the root mean square (RMS) of the standard deviations of the individual patient distributions. All dimensions are in mm.

	Setup Error			Organ Motion			Total Displacement		
	R/L	P/A	S/I	R/L	P/A	S/I	R/L	P/A	S/I
Systematic Error: SD of Means	2.0	3.2	3.7	0.6	2.6	2.1	2.2	4.3	3.7
Random Error: RMS of SDs	2.1	1.9	2.4	0.8	2.3	2.4	2.0	3.1	3.1

strategy would be to move the means of the displacement distributions towards zero, without significantly changing their widths.

A pilot study is currently underway to examine the feasibility of correcting for the systematic error in patients treated for prostate cancer at MSKCC. The objective is to detect and correct those patients with a large systematic error in setup or organ position, thus reducing the error in the mean position of the prostate during treatment. The protocol requires the patient to undergo additional CT scans immediately prior to treatment. Patients are scanned and treated in MSKCC's Image Guided Facility, which consists of a Varian 21EX linac and a Picker IQ scanner. The scanner is positioned in the corner of the treatment room in order to facilitate a rapid and easy transfer of an immobilized patient from one couch to the other (figure 17–15).

Patients are scanned on treatment days 3 to 8. No scans are done on the first two days to allow the patient to first familiarize himself with the procedure. Following the initial six scans, scans are performed once per week to monitor any slow changes in position that may occur. Additional scans are taken immediately after treatment on three days to evaluate intra-fractional displacements. On each treatment CT scan, the planned isocenter is located by attaching BBs to the patient's isocenter skin marks. The prostate CTV and the seminal vesicles CTV from each

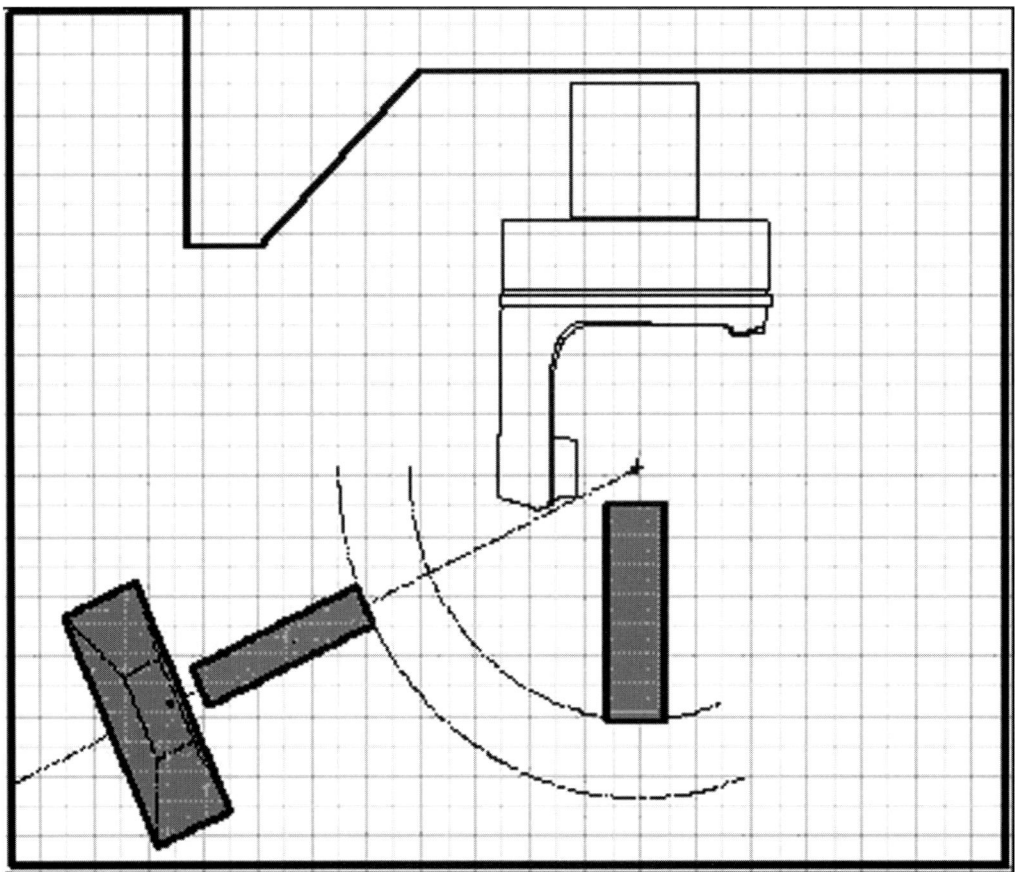

FIGURE 17–15. The Image Guided Facility at MSKCC. Patients are first set up in the treatment position on the CT couch using a second set of room lasers. After the CT scan, the treatment machine couch is rotated into alignment with the CT couch. The immobilized patient is transferred to the treatment couch using a rail system installed on both couches.

of the treatment CT scans are separately delineated by a physician, usually in the evening following the treatment. From the physician's contours, the position of the prostate COG with respect to the planned isocenter is determined. The seminal vesicles are excluded from COG calculation. The mean position of the COG is calculated as a moving average of up to five of the most recent scans. The correction to the patient's setup is the difference between the mean COG observed and the COG position seen in the planning CT. As scans are accumulated over the course of treatment, the mean position of the prostate COG may shift slightly, necessitating additional corrections to the patient's setup.

The distribution of displacements seen on scan days in the P/A direction for a single patient is shown in figure 17–16. Also shown is the distribution of displacements that would have occurred instead had the patient's setup position been monitored and corrected based on weekly portal imaging. The fraction of days on which the prostate remained inside the PTV was 93% for the CT-guided procedure, and 34% for the weekly portal image-guided procedure. Data gathered from this study will enable us to optimize the number of initial daily scans, and the frequency of monitoring scans.

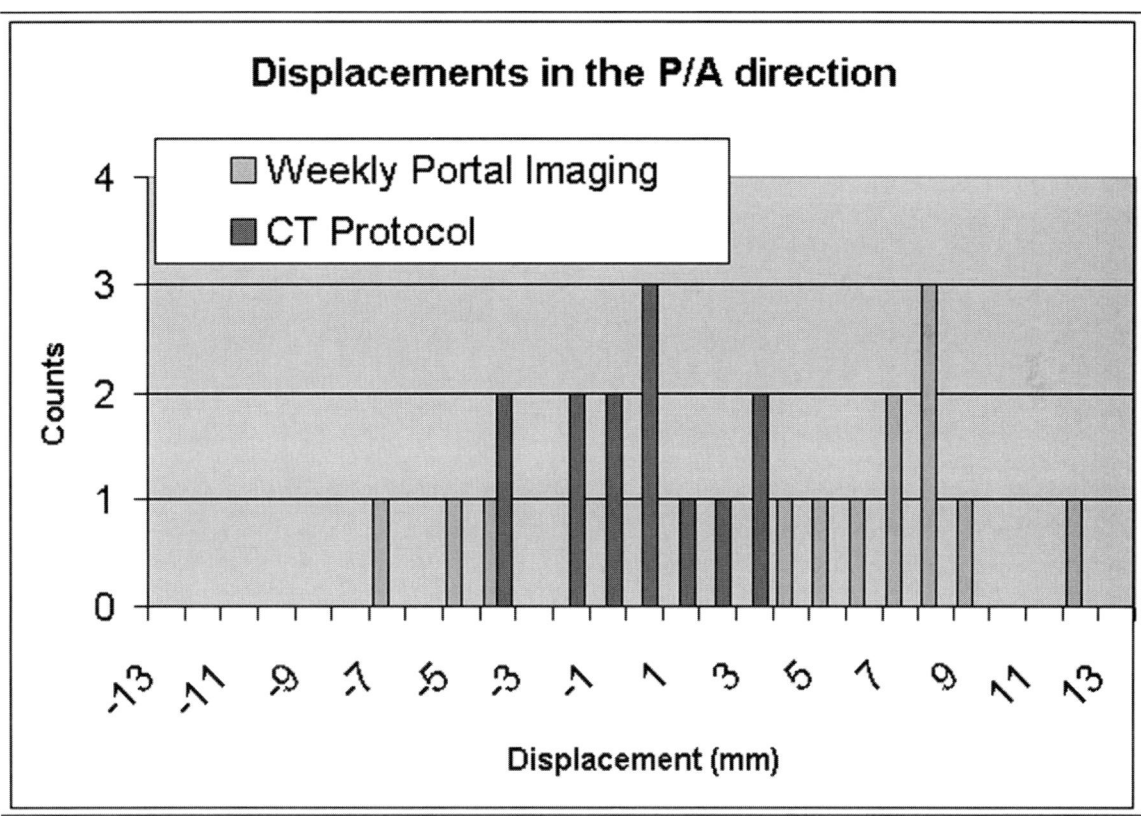

FIGURE 17–16. Comparison of the effect of the CT-guided systematic error reduction protocol and weekly portal imaging-based protocol on prostate target position in one patient. The P/A displacements of the prostate COG from the planned position on scanning days are shown in dark gray. The displacements that would have resulted from weekly portal imaging are shown in light gray. The mean ± standard deviations of the two distributions are −0.8±2.2 mm for the CT-guided protocol and 4.0±5.8 mm for weekly portal imaging.

Adaptive Radiation Therapy

A systematic approach to choosing the optimum PTV margins and dose, called *adaptive radiation therapy* (*ART*), has been described by Di Yan et al. (1997). In the ART process, an initial treatment plan is designed with PTV margins that are based on distributions of setup error and organ motion of the general population of prostate patients. During the first phase of treatment with the initial plan, multiple measurements are made of the target position. Knowing the spread in target position allows the margins to be customized for a particular patient. By constraining the rectal wall normal tissue complication probability to less than 5%, for example, the target dose that maximizes the tumor control probability (TCP) can be determined. Note that too high a dose will result in target dose non-uniformity, resulting in a lower TCP. The plan is modified and for the remainder of the therapy course, the patient receives treatment with the new beam apertures and modified prescription dose.

The ART process has been expanded to include optimization of the shape of the field apertures themselves. Multiple CT scans are taken during the first week of treatment to develop a new patient specific confidence-limited PTV (Martinez et al. 2001; Yan et al. 2000). ART is thus a comprehensive correction strategy; the cl-PTV incorporates correction modification for both systematic and random errors, and target shape change.

Correcting For Random Errors

A number of different approaches have been used to determine the prostate position and to correct for any displacement immediately before treatment. This may become increasingly important when using IMRT to deliver nonuniform dose patterns based on biological imaging. For example, magnetic resonance spectroscopic imaging (MRSI) has been proposed to identify regions of high tumor burden within the prostate, which would be treated to a higher dose than that received by the rest of the prostate. Precise target localization at the time of treatment would minimize the margin needed for the regions identified to receive higher dose, and minimize urinary or rectal complications. Since a principal aim in prostate treatment is to avoid critical structures, it is of interest to note that the first two approaches described below focus on localizing the rectal wall.

Rectal Balloon And Rectal Probe

A lucite rectal probe attached to a positioning plate has been in use since 1976 to position the rectum, and indirectly the prostate, during the boost phase (12 fractions) using a perineal 160 MeV proton beam at Massachusetts General Hospital. Patients are treated in the lithotomy position (Benk et al. 1993). An alternative method, the rectal balloon, has been used at Baylor College of Medicine (McGary et al. 2002; Teh et al. 2002) to immobilize the prostate in prone patients during IMRT. The balloon, inflated with 100 cc of air, is used daily for the entire course of treatment. Gerstner et al. (1999) have studied the effect of a rectal balloon inflated with 40 cc by taking three additional CT scans, both with and without the balloon, during the treatment course of 10 patients. They report an increase in the average distance between the prostate and the posterior rectal wall of 8 mm, and a reduction in the fraction of setups with large P/A movements (>=4mm) from 60% to 10%. In addition to better control of prostate motion, another benefit noted was the reduced dose to the posterior rectal wall for treatments in which the seminal vesicles were excluded (Wachter et al. 2002).

Radio-opaque Markers

Several studies of prostate motion with respect to pelvic bones have reported using implanted radio-opaque markers (Crook et al. 1995; Balter et al. 1995a,b; Pickett et al. 1999; Shimizu et al. 2000c; Alasti et al. 2002; Vigneault et al. 1997). Between one and four gold markers are implanted into the prostate, usually transrectally. Prior to treatment and while in the treatment position, kilovoltage (kV) or megavoltage (MV) beams are used to obtain images of the pelvic bones and the implanted makers. The efficiency of marker detection, a concern with MV portal imaging, depends on the type of imager and the size of marker. Alasti et al. (2000) report unsuccessful marker localization using a MK I liquid ionization matrix imager (Varian), but with a MK II imager had a success rate of 93% for localizing at least one cylindrical marker (1 mm diameter, 5 mm length) out of the three implanted. Success rates are higher with amorphous silicon (aSi) imagers. Nederveen, Lagendijk, and Hofman (2000, 2001) used an automatic technique to locate cylindrical markers in images acquired with an aSi imager. Success rates were 95% and 99% for markers of length 5 mm and diameters of 1.0 and 1.2 mm, respectively.

Given the observation from chest x-rays that over 20% of patients receiving ^{103}Pd seeds for prostate treatment reveal seeds in the lung (Nag et al. 1995; Older et al. 2001), the possible migration of markers from the site of implantation of even a few millimeters is a concern. If there are two or more markers, checks can be made of the inter-marker spacing. No significant migration was reported in the studies referred to here. A real-time radiotherapy tracking (RTRT) system at Hokkaido University School of Medicine using dual kV x-ray television systems mounted in the treatment room has been developed for precision setup and monitoring of patients during treatment. For prostate tracking, a single spherical marker 2 mm in diameter is implanted transurethrally. Marker migration was studied (Kitamura et al. 2002a) with follow-up CT scans in 14 patients. Marker position was compared with the COG of the organ determined from physician-drawn contours. Discrepancies between the marker and organ positions were within the accuracy of the CT measurement. The interobserver variation in prostate COG position seen among seven physicians had a standard deviation of 0.3, 0.8, and 2.0 mm in the R/L, P/A, and S/I directions.

Several investigators have reported on protocols to detect and correct prostate position prior to each treatment using implanted markers. Pickett et al. (1999) tracked prostate movement by implanting three 1.6×3 mm^2 cylindrical gold markers in the mid-transrectal wall to treat an intra-prostatic lesion to 90 Gy, and the remainder of the prostate to 73.8 Gy. Electronic portal imaging was performed daily to localize and correct target position. Marker migration was found not to be significant, based on measurements of inter-marker distances. Shimizu et al. (2000c) have corrected for prostate position during treatment with the RTRT tracking system. A single marker is implanted near the tumor. Its position relative to the isocenter is viewed during dose delivery. Should the marker deviate more than 1.5 to 2.0 mm from its planned position, the machine is gated off, the couch position corrected, and treatment re-commenced. The correction procedure has reduced the mean 3-D prostate displacement from 6.9 mm to 0.9±0.9 mm.

Urethral Catheter

Bergström, Lofroth, and Widmark (1998) have described a specially designed catheter to determine the position of the urethra at the time of treatment using portal images. The catheter is equipped with an inflatable cuff that is positioned at the floor of the bladder, and a number of radio-opaque markers positioned precisely with respect to the cuff. The catheter is inserted only during the boost phase of treatment, from 70 Gy to 74–76 Gy. They report that the technique is easy to implement and precise; displacements as small as 1 mm in the P/A direction could be detected and corrected.

Ultrasound

The NOMOS Corporation (Sewickley, PA) has developed a system that uses ultrasound to rapidly locate the position of a soft tissue structure in a patient just prior to treatment. The user first imports the isocenter and prostate, bladder and rectum contours into the BAT™ system from the CT-simulator. The device is set up adjacent to the treatment couch, and the ultrasound probe is registered to the treatment machine with a collimator-mounted docking system, thus allowing the system to determine the probe position with respect to the isocenter. After obtaining ultrasound images (prostate patients must be supine with full bladder) the operator overlays and registers the contour data to the images; the system then reports the 3-D couch correction necessary to return the prostate to its planned position. Lattanzi et al. (1999) compared measurements of prostate position from the BAT system with those from a CT scan prior to the treatment session. In 24 comparisons of 10 patients they found differences of 3.0±1.8 mm, 2.4±1.8 mm, and 4.6±2.8 mm in the A/P, lateral, and S/I directions respectively. These differences reflect an upper limit on the system's accuracy, because they include changes in position due to the elapsed time between CT and ultrasound procedures, and to possible patient displacements during the transfer between CT and treatment rooms. They found the ultrasound localization easy to perform and required only 5 min.

Serial CT Scanning

Although CT remains the predominant imaging modality for treatment planning, clinical and operational limitations have made it impractical for localizing and correcting prostate prior to each treatment. Lattanzi et al. (1998) have investigated a CT-guided correction technique limited to the boost phase. Six patients initially received 50 Gy in 2 Gy fractions with a four-field conformal technique. Prior to the second treatment phase, the patients received a second CT simulation in which the prostate was delineated and patient isocenter re-marked. For the final treatment phase from 56 Gy to 68–72 Gy, the PTV was modified to enclose the prostate with a reduced margin of 2.5 mm. The patient received a CT scan prior to each treatment session, consisting of only three slices, one through the isocenter and two at ±3 mm. Corrections to the isocenter position in the lateral and A/P directions were calculated according to the delineated prostate on the three slices. The patient was transferred from the CT to the treatment couch in the treatment position (supine with an alpha cradle support) with a specially constructed board and treated to the corrected isocenter.

Several centers have recently installed a commercial device (Siemens Primatom), consisting of a CT gantry mounted on rails along both sides of the linear accelerator (linac) couch. The CT gantry moves over the linac couch during helical CT acquisition. This device neatly resolves any difficulties in transferring an immobilized patient between CT and treatment couches. Wong et al. (2000) have reported an application of this system to CT-guided setup of prostate patients in the final five treatment fractions.

A possible alternative strategy is to move the boost phase, and hence the CT-guided correction procedure, to the initial treatment fractions. Similar to the MSKCC study, the mean displacement of the target position could then be computed and corrected, thus providing improved target positioning accuracy during the remaining non-CT-guided treatment fractions.

Discussion

Whether prone or supine setup is more suitable for IMRT depends in part on its effect on treatment plan design and target position reproducibility. In a comparison of treatment plans for patients in both positions, Zelefsky et al. (1997) found that the prone position resulted in an increase in separation between the rectal wall and target organs, especially in the region of the seminal vesicles. This resulted in a lower average dose and a lower D_{95} for the rectal wall. Prone setup also decreased the volume of bowel in the field. McLaughlin et al. (1999) have reported improved rectal sparing in patients planned lying prone on a flat surface, relative to both supine setup and an angled prone setup. Weber et al. (2000) have also reported that high dose to the rectum occurred less frequently in prone patients.

Neither setup orientation appears to have a clear advantage with respect to prostate and seminal vesicles reproducibility (McLaughlin et al. 1999; Stroom et al. 1999). In a comparison of repeat CT scans in 15 prone patients and 15 supine patients, Stroom et al. (1999) observed differences in target organ position variation between prone and supine patients along the A/P direction. The random and systematic standard deviations for prone patients were 1.5 and 3.3 mm respectively, and 2.4 mm and 2.6 mm for supine patients. Although random deviations were smaller for prone patients, systematic deviations were larger; thus, the required PTV margin was the same for both groups. Litzenberg et al. (2002) have applied a diagnostic quality radiograph system in the treatment room to daily localization of implanted markers in six prone and four supine patients. They reported a larger variation in prostate displacement (setup error and organ motion) in prone patients, particularly in the A/P direction: the errors, averaged over patients, had a standard deviation of 6.6 mm and 4.2 mm for prone and supine patients, respectively.

Prostate motion due to breathing appears to be larger for prone setup than for supine, and is further increased with the use of a thermoplastic shell. Dawson et al. (2000) observed a maximum excursion of implanted markers ranging between 2.3 and 5.1 mm in the S/I direction, and 1.0 to 3.5 mm in the A/P direction for prone patients immobilized with a themoplastic shell. Similarly, Malone et al. (2000) observed motion correlated with breathing was greatest in prone patients with immobilization shells; in 40 patients the average excursion was 2.9±1.7 mm and 1.6±1.1 mm in the S/I and A/P directions. Marker excursions for supine patients were observed to be negligible in both studies. Using a real-time tracking system, Kitamura et al. (2002b) observed increased excursions of an implanted marker in 10 patients in the prone position, compared with the same patients positioned supine. The average marker excursion was negligible for supine patients, and 1.4±0.5 mm and 1.6±0.4 mm in the S/I and A/P directions for patients positioned prone without any immobilization device. Note that the deviation from the mean position would be half of these numbers. Although the averages are small, much larger excursions were seen for some patients, and when patients breathed deeply.

In conclusion, organ motion of the prostate and seminal vesicles and the close proximity of the rectal wall make dose escalation using IMRT clinically difficult. A variety of approaches are currently under study to improve the positioning accuracy of the target organs or rectal wall, or to adapt the field apertures and dose to the target position variability of the individual patient.

References

Alasti, H., M. Petric, C. N. Catton, and P. R. Warde. (2001). "Portal imaging for evaluation of daily on-line setup errors and off-line organ motion during conformal irradiation of carcinoma of the prostate." *Int. J. Radiat. Oncol. Biol. Phys.* 49(3):869–884.

Aruga, T., J. Itami, M. Aruga, K. Nikajima, K. Shibata, T. Nojo, S. Yasuda, T. Uno, R. Hara, K. Isobe, N. Machida, and H. Ito. (2000). "Target volume definition for upper abdominal irradiation using CT scans obtained during inhale and exhale phases." *Int. J. Radiat. Oncol. Biol. Phys.* 48(2):465–469.

Balter, J. M., K. L. Lam, H. M. Sandler, J. F. Littles, R. L. Bree, and R. K. Ten Haken. (1995a). "Automated localization of the prostate at the time of treatment using implanted radiopaque markers: Technical feasibility." *Int. J. Radiat. Oncol. Biol. Phys.* 33(5):1281–1286.

Balter, J. M., H. M. Sandler, K. Lam, R. L. Bree, A. S. Lichter, and R. K. Ten Haken. (1995b). "Measurement of prostate movement over the course of routine radiotherapy using implanted markers." *Int. J. Radiat. Oncol. Biol. Phys.* 31(1):113–118.

Balter, J. M., R. K. Ten Haken, T. S. Lawrence, K. L. Lam, and J. M. Robertson. (1996). "Uncertainties in CT-based radiation therapy treatment planning associated with patient breathing." *Int. J. Radiat. Oncol. Biol. Phys.* 36(1):167–174.

Bel, A., M. van Herk, H. Bartelink, and J. V. Lebesque. (1993). "A verification procedure to improve patient set-up accuracy using portal images." *Radiother. Oncol.* 29(2):253–260.

Bel, A., P. H. Vos, P. T. Rodrigus, C. L. Creutzberg, A. G. Visser, J. C. Stroom, and J. V. Lebesque. (1996). "High-precision prostate cancer irradiation by clinical application of an offline patient setup verification procedure, using portal imaging." *Int. J. Radiat. Oncol. Biol. Phys.* 35(2):321–332.

Benk, V. A., J. A. Adams, W. U. Shipley, M. M. Urie, P. L. McManus, J. T. Efird, C. G. Willett, and M. Goitein. (1993). "Late rectal bleeding following combined X-ray and proton high dose irradiation for patients with stages T3-T4 prostate carcinoma." *Int. J. Radiat. Oncol. Biol. Phys.* 26(3):551–557.

Bergström, P., P. O. Lofroth, and A. Widmark. (1998). "High-precision conformal radiotherapy (HPCRT) of prostate cancer—a new technique for exact positioning of the prostate at the time of treatment." *Int. J. Radiat. Oncol. Biol. Phys.* 42(2):305–311.

Crook, J. M., Y. Raymond, D. Salhani, H. Yang, and B. Esche. (1995). "Prostate motion during standard radiotherapy as assessed by fiducial markers." *Radiother. Oncol.* 37(1):35–42.

Davies, S. C., A. L. Hill, R. B. Holmes, M. Halliwell, and P. C. Jackson. (1994). "Ultrasound quantitation of respiratory organ motion in the upper abdomen." *Br. J. Radiol.* 67(803):1096–1102.

Dawson, L. A., D. W. Litzenberg, K. K. Brock, M. Sanda, M. Sullivan, H. M. Sandler, and J. M. Balter. (2000). "A comparison of ventilatory prostate movement in four treatment positions." *Int. J. Radiat. Oncol. Biol. Phys.* 48(2):319–323.

Dawson, L.A., K. K. Brock, S. Kazenjian, D. Fitch, C. J. McGinn, T. S. Lawrence, R. K. Ten Haken, and J. Balter. (2001). "The reproducibility of organ position using active breathing control (ABC) during liver radiotherapy." *Int. J. Radiat. Oncol. Biol. Phys.* 51(5):1410–1421.

Ekberg, L., O. Holmberg, L. Wittgren, G. Bjelkengren, and T. Landberg. (1998). "What margins should be added to the clinical target volume in radiotherapy treatment planning for lung cancer?" *Radiother. Oncol.* 48(1): 71–77.

Ford, E. C., G. S. Mageras, E. Yorke, K. E. Rosenzweig, R. Wagman, and C. C. Ling. (2002). "Evaluation of respiratory movement during gated radiotherapy using film and electronic portal imaging." *Int. J. Radiat. Oncol. Biol. Phys.* 52(2):522–531.

Gerstner, N., S. Wachter, D. Dorner, G. Goldner, A. Colotto, and R. Potter. (1999). [Significance of a rectal balloon as internal immobilization device in conformal radiotherapy of prostatic carcinoma]. German. *Strahlenther. Onkol.* 175(5):232–238.

Hanley, J., M. M. Debois, D. Mah, G. S. Mageras, A. Raben, K. Rosenzweig, B. Mychalczak, L. H. Schwartz, P. J. Gloeggler, W. Lutz, C. C. Ling, S. A. Leibel, Z. Fuks, and G. J. Kutcher. (1999). "Deep inspiration breathhold technique for lung tumors: The potential value of target immobilization and reduced lung density in dose escalation." *Int. J. Radiat. Oncol. Biol. Phys.* 45(3):603–611.

Keatley, E., and G. Mageras. "Computer Automated Quantification of Respiratory Motion in a Fluoroscopic Movie" in *The Use of Computers in Radiation Therapy: XIIIth International Conference*. Heidelberg, Germany, May 22–25, 2000. W. Schlegel and T. Bortfeld (Eds.). Heidelberg: Springer, pp. 132–134, 2000.

Kim, D. J., B. R. Murray, R. Halperin, and W. H. Roa. (2001). "Held-breath self-gating technique for radiotherapy of non-small-cell lung cancer: A feasibility study." *Int. J. Radiat. Oncol. Biol. Phys.* 49(1):43–49.

Kitamura, K., H. Shirato, S. Shimizu, N. Shinohara, T. Harabayashi, T. Shimizu, Y. Kodama, H. Endo, R. Onimaru, S. Nishioka, H. Aoyama, K. Tsuchiya, and K. Miyasaka. (2002a). "Registration accuracy and possible migration of internal fiducial gold marker implanted in prostate and liver treated with real-time tumor-tracking radiation therapy (RTRT)." *Radiother. Oncol.* 62(3):275–281.

Kitamura, K., H. Shirato, Y. Seppenwoolde, R. Onimaru, M. Oda, K. Fujita, S. Shimizu, N. Shinohara, T. Harabayashi, and K. Miyasaka. (2002b). "Three-dimensional intrafractional movement of prostate measured during real-time tumor-tracking radiotherapy in supine and prone treatment positions." *Int. J. Radiat. Oncol. Biol. Phys.* 53(5):1117–1123.

Kubo, H. D., and B.C. Hill. (1996). "Respiration gated radiotherapy treatment: a technical study." *Phys. Med. Biol.* 41(1):83–91.

Kubo, H. D., and L. Wang. (2000). "Compatibility of Varian 2100C gated operations with enhanced dynamic wedge and IMRT dose delivery." *Med. Phys.* 27(8):1732–1738.

Kubo, H. D., P. M. Len, S. Minohara, and H. Mostafavi. (2000). "Breathing-synchronized radiotherapy program at the University of California Davis Cancer Center." *Med. Phys.* 27(2):346–353.

Lattanzi, J., S. McNeely, A. Hanlon, I. Das, T. E. Schultheiss, and G. E. Hanks. (1998). "Daily CT localization for correcting portal errors in the treatment of prostate cancer." *Int. J. Radiat. Oncol. Biol. Phys.* 41(5):1079–1086.

Lattanzi, J., S. McNeeley, W. Pinover, E. Horwitz, I. Das, T. E. Schultheiss, and G. E. Hanks. (1999). "A comparison of daily CT localization to a daily ultrasound-based system in prostate cancer." *Int. J. Radiat. Oncol. Biol. Phys.* 43(4):719–725.

Litzenberg, D., L. A. Dawson, H. Sandler, M. G. Sanda, D. L. McShan, R. K. Ten Haken, K. L. Lam, K. K. Brock, and J. M. Balter. (2002). "Daily prostate targeting using implanted radiopaque markers." *Int. J. Radiat. Oncol. Biol. Phys.* 52(3):699–703.

Mageras, G. S., E. Yorke, K. Rosenzweig, L. Braban, E. Keatley, E. Ford, S. A. Leibel, and C. C. Ling. (2001). "Fluoroscopic evaluation of diaphragmatic motion reduction with a respiratory gated radiotherapy system." *J. Appl. Clin. Med. Phys.* 2(4):191–200.

Mah, D., J. Hanley, and K. Rosenzweig. (1998). "Movement of anatomic landmarks and its effect on patient position determination in thoracic conformal radiation treatment." *Med. Phys.* 25:A80.

Mah, D., J. Hanley, K. E. Rosenzweig, E. Yorke, L. Braban, C. C. Ling, S. A. Leibel, and G. Mageras. (2000). "Technical aspects of the deep inspiration breath-hold technique in the treatment of thoracic cancer." *Int. J. Radiat. Oncol. Biol. Phys.* 48(4):1175–1185.

Malone, S., J. M. Crook, W. S. Kendal, and J. Szanto. (2000). "Respiratory-induced prostate motion: Quantification and characterization." *Int. J. Radiat. Oncol. Biol. Phys.* 48(1):105–109.

Martinez, A.A., D. Yan, D. Lockman, D. Brabbins, K. Kota, M. Sharpe, D. A Jaffray, F. Vicini, and J. Wong. (2001). "Improvement in dose escalation using the process of adaptive radiotherapy combined with three-dimensional conformal or intensity-modulated beams for prostate cancer." *Int. J. Radiat. Oncol. Biol. Phys.* 50(5):1226–1234.

McGary, J. E., B. S. Teh, E. B. Butler, and W. Grant 3rd. (2002). "Prostate immobilization using a rectal balloon." *J. Appl. Clin. Med. Phys.* 3(1):6–11.

McLaughlin, P. W., A. Wygoda, W. Sahijdak, H. M. Sandler, L. Marsh, P. Roberson, and R. K. Ten Haken. (1999). "The effect of patient position and treatment technique in conformal treatment of prostate cancer." *Int. J. Radiat. Oncol. Biol. Phys.* 45(2):407–413.

Minohara, S., T. Kanai, M. Endo, K. Noda, and M. Kanazawa. (2000). "Respiratory gated irradiation system for heavy-ion radiotherapy." *Int. J. Radiat. Oncol. Biol. Phys.* 47(4):1097–1103.

Nag, S., D. D. Scaperoth, R. Badalament, S. A. Hall, and J. Burgers. (1995). "Transperineal palladium 103 prostate brachytherapy: Analysis of morbidity and seed migration." *Urol.* 45(1):87–92.

Nederveen, A. J., J. J. Lagendijk, and P. Hofman. (2000). "Detection of fiducial gold markers for automatic on-line megavoltage position verification using a marker extraction kernel (MEK)." *Int. J. Radiat. Oncol. Biol. Phys.* 47(5):1435–1442.

Nederveen, A. J., J. J. Lagendijk, and P. Hofman. (2001). "Feasibility of automatic marker detection with an a-Si flat-panel imager." *Phys. Med. Biol.* 46(4):1219–1230.

Older, R. A., B. Synder, T. L. Krupski, D. J. Glembocki, and J. Y. Gillenwater. (2001). "Radioactive implant migration in patients treated for localized prostate cancer with interstitial brachytherapy." *J. Urol.* 165(5):1590–1592.

Paoli, J., K. Rosenzweig, and E. Yorke. (1999). "Comparison of different respiratory levels in the treatment of lung cancer: Implications for gated treatment." *Int. J. Radiat. Oncol. Biol. Phys.* 45(Suppl. 1):386–387.

Pickett, B., E. Vigneault, J. Kurhanewicz, L. Verhey, and M. Roach. (1999). "Static field intensity modulation to treat a dominant intra-prostatic lesion to 90 Gy compared to seven field 3-dimensional radiotherapy." *Int. J. Radiat. Oncol. Biol. Phys.* 44(4):921–929.

Ramsey, C. R., I. L. Cordrey, and A. L. Oliver. (1999). "A comparison of beam characteristics for gated and nongated clinical x-ray beams." *Med. Phys.* 26(10):2086–2091.

Rosenzweig, K.E., J. Hanley, D. Mah, G. Mageras, M. Hunt, S. Toner, C. Burman, C. C. Ling, B. Mychalczak, Z. Fuks, and S. A Leibel. (2000). "The deep inspiration breath-hold technique in the treatment of inoperable non-small-cell lung cancer." *Int. J. Radiat. Oncol. Biol. Phys.* 48(1):81–87.

Ross, C. S., D. H. Hussey, E. C. Pennington, W. Stanford, and J. F. Doornbos. (1990). "Analysis of movement of intrathoracic neoplasms using ultrafast computerized tomography." *Int. J. Radiat. Oncol. Biol. Phys.* 18(3): 671–677.

Seiler, P. G., H. Blattmann, S. Kirsch, R. K. Muench, and C. Schilling. (2000). "A novel tracking technique for the continuous precise measurement of tumour positions in conformal radiotherapy." *Phys. Med. Biol.* 45(9):N103–110.

Shimizu, S., H. Shirato, B. Xo, K. Kagei, T. Nishioka, S. Hashimoto, K. Tsuchiva, H. Aoyama, and K. Miyasaka. (1999). "Three-dimensional movement of a liver tumor detected by high-speed magnetic resonance imaging." *Radiother. Oncol.* 50(3):367–370.

Shimizu, S., H. Shirato, H. Aoyama, S. Hashimoto, T. Nishioka, A. Yamazaki, K. Kagei, and M. Miyasaka. (2000a). "High-speed magnetic resonance imaging for four-dimensional treatment planning of conformal radiotherapy of moving body tumors." *Int. J. Radiat. Oncol. Biol. Phys.* 48(2):471–474.

Shimizu, S., H. Shirato, K. Kagei, T. Nishioka, X. Bo, H. Dosaka-Akita, S. Hashimoto, H. Aoyama, K. Tsuchiya, and M. Miyasaka. (2000b). "Impact of respiratory movement on the computed tomographic images of small lung tumors in three-dimensional (3D) radiotherapy." *Int. J. Radiat. Oncol. Biol. Phys.* 46(5):1127–1133.

Shimizu, S., H. Shirato, K. Kitamura, N. Shinohara, T. Harabayashi, T. Tsukamoto, T. Koyanagi, and K. Miyasaka. (2000c). "Use of an implanted marker and real-time tracking of the marker for the positioning of prostate and bladder cancers." *Int. J. Radiat. Oncol. Biol. Phys.* 48(5):1591–1597.

Shirato, H., S. Shimizu, T. Kunieda, K. Kitamura, M. van herk, K. Kagei, T, Nishioka, S. Hashimoto, K. Fujita, H. Aoyama, K. Tsuchiya, K. Kudo, and K. Miyasaka. (2000). "Physical aspects of a real-time tumor-tracking system for gated radiotherapy." *Int. J. Radiat. Oncol. Biol. Phys.* 48(4):1187–1195.

Stromberg, J. S., M. B. Sharpe, L. H. Kim, V. R. Kini, D. A. Jaffray, A. A. Martinez, and J. W. Wong. (2000). "Active breathing control (ABC) for Hodgkin's disease: Reduction in normal tissue irradiation with deep inspiration and implications for treatment." *Int. J. Radiat. Oncol. Biol. Phys.* 48(3):797–806.

Stroom, J. C., P. C. Koper, G. A. Korevaar, M. van Os, M. Janssen, H. C. de Boer, P. C. Levendag, and B. J. Heijmen. (1999). "Internal organ motion in prostate cancer patients treated in prone and supine treatment position." *Radiother. Oncol.* 51(3):237–248.

Teh, B.S., S. Y. Woo, W. Y. Mai, J. E. McGary, L. S. Carpenter, H. H. Lu, J. K. Chui, M. T. Vlachaki, W. H. Grant 3rd, and E. B. Butler. (2002). "Clinical experience with intensity-modulated radiation therapy (IMRT) for prostate cancer with the use of rectal balloon for prostate immobilization." *Med. Dosim.* 27(2):105–113.

Vigneault, E., J. Pouliot, J. Laverdiere, J. Roy, and M. Dorion. (1997). "Electronic portal imaging device detection of radioopaque markers for the evaluation of prostate position during megavoltage irradiation: A clinical study." *Int. J. Radiat. Oncol. Biol. Phys.* 37(1):205–212.

Wachter, S., N. Gerstner, D. Dorner, G. Goldner, A. Colotto, A. Wambersie, and R. Potter. (2002). "The influence of a rectal balloon tube as internal immobilization device on variations of volumes and dose-volume histograms during treatment course of conformal radiotherapy for prostate cancer." *Int. J. Radiat. Oncol. Biol. Phys.* 52(1):91–100.

Wagman,, R., E. Yorke, E. Ford, P. Giraud, G. Mageras, B. Minsky, and K. Rosenzweig. (2003). "Respiratory gating for liver tumors: Use in dose escalation." *Int. J. Radiat. Oncol. Biol. Phys.* 55(3):659–668.

Weber, D. C., P. Nouet, M. Rouzaud, and R. Mirabell. (2000). "Patient positioning in prostate radiotherapy: is prone better than supine?" *Int. J. Radiat. Oncol. Biol. Phys.* 47(2):365–371.

Wong, J. W., M. B. Sharpe, D. A. Jaffray, V. R. Kini, J. M. Robertson, J. S. Stromberg, and A. A. Martinez. (1999). "The use of active breathing control (ABC) to reduce margin for breathing motion." *Int. J. Radiat. Oncol. Biol. Phys.* 44(4):911–919.

Wong, J. R., L. Grimm, M. Chow, M. Uematsu, T. Smith, and C. W. Cheng. (2001). "Precise radiation treatment of prostate cancer by correcting for the intrinsic daily movements of the prostate or rectum using a novel combination of CT scanner and linear accelerator (Primatom)." (Abstract). *Int. J. Radiat. Oncol. Biol. Phys.* 51:316.

Yan, D., F. Vicini, J. Wong, and A. Martinez. (1997). "Adaptive radiation therapy." *Phys. Med. Biol.* 42(1):123–132.

Yan, D., D. Lockman, D. Brabbins, L. Tyburski, and A. Martinez. (2000). "An off-line strategy for constructing a patient-specific planning target volume in adaptive treatment process for prostate cancer." *Int. J. Radiat. Oncol. Biol. Phys.* 48(1):289–302.

Yorke, E., G. Mageras, T. LoSasso, H. Mostafavi, and C. Ling. (2000). "Respiratory Gating of Sliding Window IMRT" in *Proceedings of the World Congress on Medical Physics and Biomedical Engineering.* July 23–27, 2000, Chicago, IL.

Yorke, E. L. Wang, K. Rosenzweig, D. Mah, J.-B. Paoli, and C. S. Chui. (2002). "Evaluation of deep inspiration breath-hold lung treatment plans with Monte Carlo dose calculation." *Int. J. Radiat. Oncol. Biol. Phys.* 53(4):1058–1070.

Zelefsky, M. J., L. Happersett, S. A. Leibel, C. M. Burman, L. Schwartz, A. P. Dicker, G. J. Kutcher, and Z. Fuks. (1997). "The effect of treatment positioning on normal tissue dose in patients with prostate cancer treated with three-dimensional conformal radiotherapy." *Int. J. Radiat. Oncol. Biol. Phys.* 37(1):13–19.

Zelefsky, M. J., Z. Fuks, M. Hunt, H. J. Lee, D. Lombardi, C. C. Ling, V. E. Reuter, E. S. Venkatraman, and S. A. Leibel. (2001). "High dose radiation delivered by intensity modulated conformal radiotherapy improves the outcome of localized prostate cancer." *J. Urol.* 166(3):876–881.

18

Advanced Treatment Techniques II

Kamil Yenice

Introduction

In radiation therapy of paraspinal tumors, highly radiosensitive spinal cord lies intimately close to the treatment volume. Radiation therapy for these lesions often requires substantial amounts of radiation dose, in excess of 50 Gy (in standard fractionation), to be delivered to the planning target volume (PTV). This is usually not an easy task to accomplish using the standard treatment techniques either due to a lack of precise treatment delivery or the limitations in the number of beam portals that can be practically used to generate isodose plans that sufficiently cover the target volume without significantly exposing the spinal cord. As it has been shown in the previous chapters, intensity-modulated radiation therapy (IMRT) can enhance the dose gradient between a target and a nearby critical organ; therefore, it can deliver higher doses to PTV while protecting the organs at risk. Paraspinal tumors without significant epidural compression (such as those of the vertebral body and chest-wall) can potentially benefit from IMRT technology. Computed tomography (CT)-guided intensity modulated radiation therapy has been used in the treatment of recurrent paraspinal tumors at Memorial Sloan-Kettering Cancer Center (MSKCC) since February 2000. Our approach to IMRT treatment for this site has focused on developing and implementing a method of generating high-gradient dose distributions near the spinal cord with a stereotactic immobilization system that is used for CT or electronic portal image device (EPID)-guided treatment setup for precise treatment delivery. Since CT-guidance requires a pre-treatment CT scan of the patient in the treatment position and subsequent treatment delivery on a treatment machine, this treatment procedure is done in a specially designed treatment room where both a CT scanner and a treatment linear accelerator are accessible. These treatments are delivered in a hypo-fractionated treatment schedule. We use the EPID-guided setup to deliver precise IMRT treatments to certain type of paraspinal tumors such as sarcomas or chordomas that may benefit from a more standard fractionation scheme. However, the setup accuracy (<3 mm) for the tumors of the abdomen or thorax cannot be guaranteed with EPID-guidance for patients without any surgical implants at this time due to the difficulty in resolving the differences between rotations and translations in 2-D EPID images. The use of CT images for daily target localization eliminates the need for implanted fiducial markers in the thoracic and lumbar sites (Murphy et al. 2001; Gilhuijs, van de Ven, and van Herk 1996).

The purpose of this chapter is to discuss the stereotactic immobilization, CT-image guided setup, and IMRT treatment techniques in use at MSKCC for paraspinal tumors. The effect of the setup uncertainty on patient dose will also be discussed.

Stereotactic Body Frame

To ensure a reliable intra-treatment patient fixation and deliver very precise conformal treatments to extracranial sites, a stereotactic body frame (SBF) has been developed and implemented at MSKCC (Yenice et al. 2000, 2003). The SBF incorporates the application of pressure to osseous points, which include the lateral and anterior pelvis, lateral ribs, and the sternum, to guarantee a reliable rigid intra-procedural immobilization of the appendicular and axial skeleton (figure 18–1). All pressure plates can be positioned mechanically inside the SBF to adjust to patient anatomy precisely. The custom body molds facilitate patient support in the SBF. The pressure plates of the SBF along with the body molds comprise a system of immobilization that prevents patient translation as well as rotation inside the SBF during the entire treatment procedure.

The frame has imbedded CT-localization fiducial rods that provide accurate anatomy localization with CT. All the fiducial rods are visible in each axial slice and therefore constitute an external fiducial system to align patient on the treatment machine based on the patient's frame coordinates. This is similar to stereotactic radiosurgery of intracranial lesions. Patient skin

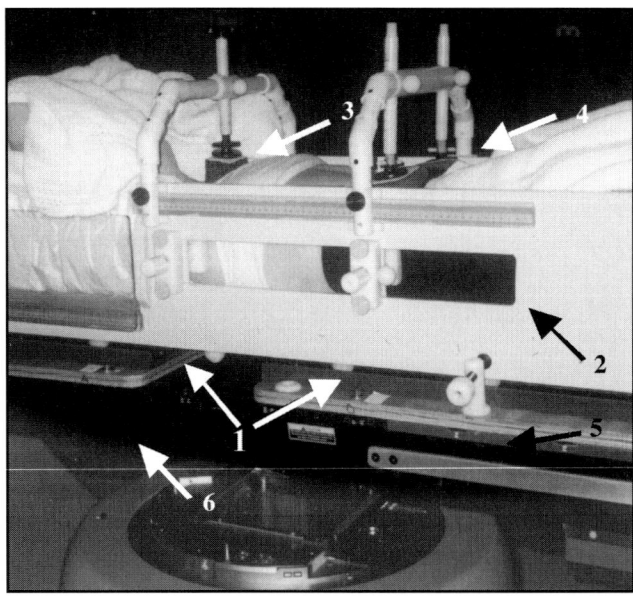

FIGURE 18–1. Memorial SBF and attachments: (1) lateral pressure plates for ribs and pelvis; (2) longitudinal frame ruler; (3) the sternal pressure arc and pressure bar; (4) pelvic bracelet with pressure plate system; (5) treatment table to SBF locking mechanism; (6) rail system.

marks are generally used with this procedure only for quick determination of gross setup errors. The SBF has rolling balls attached underneath to facilitate smooth patient transfer from a tilt table to CT couch and from CT couch to the treatment table on a hard plastic rail system (see figure 18–2).

Patient Simulation And Localization

On the day of simulation, patient molds are made and patients are fit in the SBF. Unlike other commercial systems, the Memorial SBF system is designed to fit on a vertical-to-horizontal tilt table. Patients are placed in the SBF in the standing position and are subsequently brought into the horizontal position, with patients lying in supine orientation. This vertical-to-horizontal setup is believed to improve daily patient positioning and facilitates patient setup inside the SBF. Once the immobilization procedure is complete, the patient is then transferred smoothly from the tilt table to CT couch. Axial images at 3 mm slice spacing are acquired, and the gross tumor volume (GTV) is delineated on the axial CT images and the corresponding isocenter position is marked. Frequently, it may be necessary to manually shift the isocenter to an easily recognized location such as the edge of a vertebral body to facilitate the identification of the isocenter in daily setup CT studies.

IMRT Treatment Planning

Planning Volume Segmentation

Treatment planning is based on axial CT studies fused with MR scans. Contrast-enhanced magnetic resonance (MR) scans are obtained in patient position as close as possible to the treatment position using the body molds from the SBF setup. The clinical target volume (CTV) is defined according to the definitions of International Commission on Radiation Units and Measurements (ICRU) 50 (Chavaudra and Bridier 2001). The planning target volume (PTV) is then generated from the expansion of CTV to include possible uncertainty in anatomy delineation and setup. The CTV to PTV expansion parameter ranges between 5 mm and 10 mm in 3-D–space except at the cord

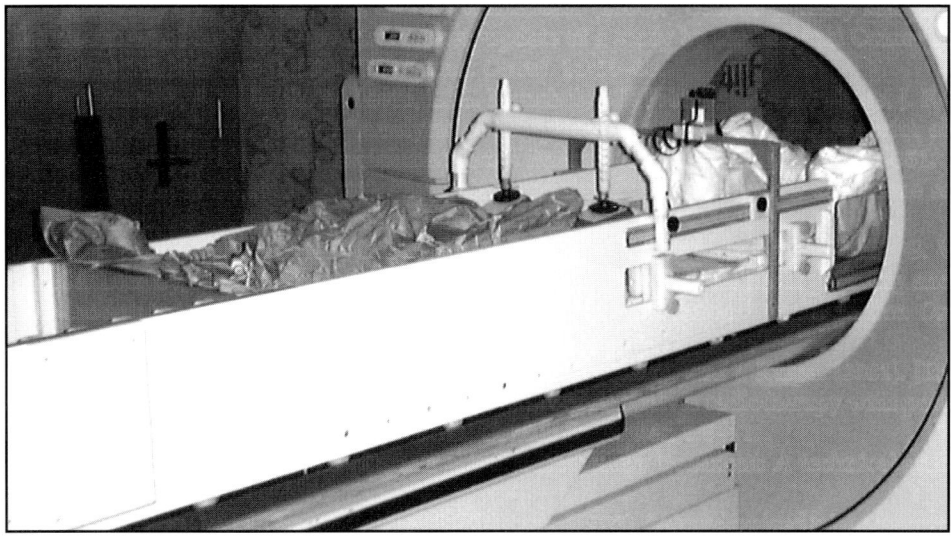

FIGURE 18–2. Memorial SBF compatibility with CT scanning and features.

border, where both CTV and PTV are placed 3 to 5 mm from the cord, but PTV is never made smaller than CTV. For cases where the tumor expands into the cord, it is expected to compromise some PTV/CTV coverage near the cord in order to restrict the cord dose.

In MSKCC IMRT planning for a given PTV and relevant risk organs optimization structures are generated from the expansion of the corresponding volumes to allow the entry of dose volume criteria in the inverse planning process, as explained in chapter 6. For paraspinal tumors surrounding the spinal cord, we generate specific optimization volumes for PTV and the cord. Planning volume generation for a concave paraspinal PTV surrounding the cord is schematically illustrated in figure 18–3 when the cord and PTV are physically separated by at least 5 mm. In such a case, the MSKCC inverse planning algorithm typically yields a compromise solution between the PTV and cord depending on the set penalties for dose-volume constraints. Most of the time, some PTV coverage is sacrificed in order to respect the cord tolerance. However, some improvement in the dose gradient in this region can be gained by defining an optimization boost region in PTV and a separate rind structure of cord as described by Fournier-Bidoz et al. (2001). For this purpose a 2-mm strip of PTV (PTV-strip) at the cord border is separately outlined for optimization. For the remainder of PTV an expanded planning PTV volume (PTV-exp) is generated by adding a margin of 2 to 3 mm to PTV excluding the region of PTV-strip (see figure 18–3c). An expanded planning cord volume is also obtained by adding a margin of 2 to 3 mm to the cord volume. The rind structure for cord dose constrain can be constructed from a Boolean combination of cord-exp and cord volumes. In addition to these structures, a 4 to 5 cm thick PTV rind structure built 6 to 9 mm away from the PTV is used for eliminating hot spots outside the PTV volume. For most expanded planning volumes a minimum of 30 points/cc is used for optimization calculations whereas the PTV-strip and cord-rind are typically assigned 50 to 100 points/cc. For cases where the cord and PTV abuts or nearly abuts, PTV-strip and PTV-exp are defined excluding a 5 mm-rind region adjacent to cord to ensure proper dose falloff between the PTV and cord.

Beam Selection And Optimization

Although tumor shape, size, and location show some variability, the close proximity of target to spinal cord and some degree of concavity seem to be common for all paraspinal lesions. To achieve very high dose gradients between the target and spinal cord, Fournier-Bidoz et al. (2001) showed that at least nine fields were necessary. Typically seven to nine clinically useable co-planar beams, spaced at 20° to 30° apart, are positioned at posterior and posterior-oblique angles.

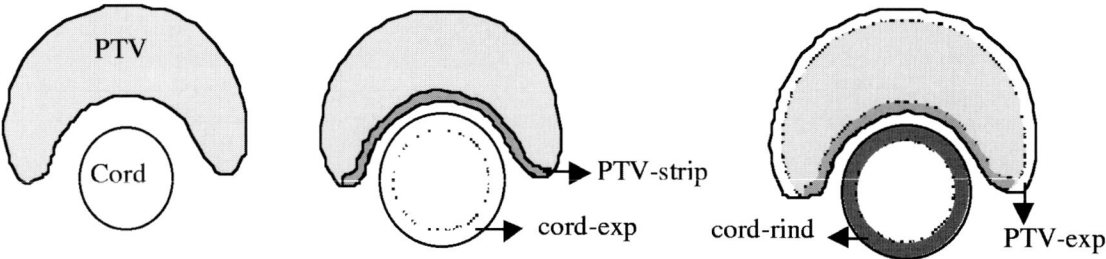

FIGURE 18–3. Optimization volume generation for a paraspinal tumor (a) schematic PTV and cord volumes in an axial plane; (b) an expanded cord volume is generated by a 2 to 3 mm radial expansion of the cord volume, and a PTV-strip is generated by drawing a 2 mm margin inward from the section of the PTV-outline bordering the cord; (c) PTV-expanded is generated by a 2 to 3 mm expansion of PTV in 3-D except at the cord border and cord-rind is obtained from cord-exp and cord volumes.

The clinically desired dose distribution is specified in terms of optimization variables including dose-volume constraints and relative penalties for each target/organ and corresponding constraint. A representative table of such variables is shown in table 18–1.

Plan Evaluation

After the computation of the optimized beam intensities, deliverable profiles are generated using the MSKCC leaf sequencer program, each consisting of 200 segments. Using these converted beam profiles forward dose calculations are run and dose-volume histograms (DVHs) are calculated for plan evaluation. The plans are re-normalized to meet the prescribed PTV dose and cord dose maximum. To limit maximum cord dose to less than 30% of prescription dose, Cerrobend™ blocks with the block edge 3 mm from the cord are typically used. The use of physical Cerrobend blocks reduces cord doses beyond that achievable by IMRT alone. Finally, digitally reconstructed radiographs (DRRs) for each dynamic field are calculated for verification of treatment setup.

The axial dose distributions for a 72-year-old female patient with a metastatic disease at T8 thoracic vertebra and a 65-year-old female patient at L4-S1 level are shown in figures 18–4a and 18–4b, respectively. The first patient had a history of leiomyosarcoma and had previously received 3000 cGy in 10 fractions to the same site. The second patient had also previous radiation of 5000 cGy to her primary endometrial cancer, partially overlapping with her current treatment site. Both patients received additional 2000 cGy in five fractions with stereotactic CT-guided setup. The IMRT plan for the T8 thoracic lesion delivered less than 30% of the prescription to the spinal cord. The seven-beam IMRT plan utilizing cord blocks for the first case resulted in a dose gradient of 10%/mm between the PTV and cord volumes. For paraspinal tumors with some epidural involvement such as this case, the final PTV/GTV coverage is dictated largely by the cord dose. As can be seen from the DVHs in figure 18–5, significant volume is receiving less than 100% of the prescription when the cord dose is restricted to less than 30% of the prescription. Approximately, 90% of PTV still receives the full prescription dose from this plan. For the L4-S1 patient, the minimum cord-PTV distance was approximately 8 mm; also the cord dose restriction was less stringent due to the fact that the overlap with the previous treatment was largely in the sacrum where the cord/cauda radiation tolerance is higher. Because of these less strict constraints on the dose distribution, it is expected to have a better target coverage, as can be seen from the histograms in figure 18–6. The entire GTV received the full prescription dose while all critical organ doses were kept below the planned dose limits for cord as well as kidneys and the small bowel. As pointed out in chapter 6 on IMRT treatment planning, the distance between the critical organ and target dictates the prescription dose and PTV dose uniformity (Hunt et al. 2002). This is clearly seen in the two examples we illustrate here, the six-field IMRT plan delivers a

Table 18–1. Optimization Variables Including Dose-Volume Constraints and Relative Penalties for Each Target/Organ and Corresponding Constraint. All variables listed are relative, although absolute dose and volume variables can also be defined.

Optimization Structure	Clinical Structure	Inverse Planning Algorithm Constraint			
		Prescription Dose	Maximum Dose/Penalty	Minimum Dose/Penalty	Dose-Volume Constraint/Penalty
PTV_strip	PTV	100	100/10	100/10	N/A
PTV_exp	PTV	100	100/10	100/10	N/A
Cord_exp	Cord	N/A	5/100	N/A	N/A
Cord_rind	Cord	N/A	5/100	N/A	N/A
PTV_rind	Normal tissue	N/A	80/50	N/A	N/A

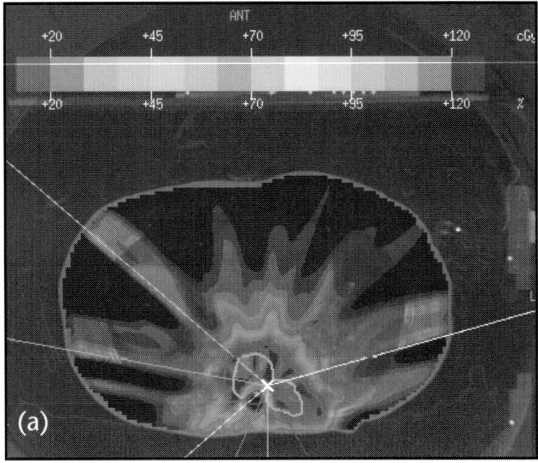 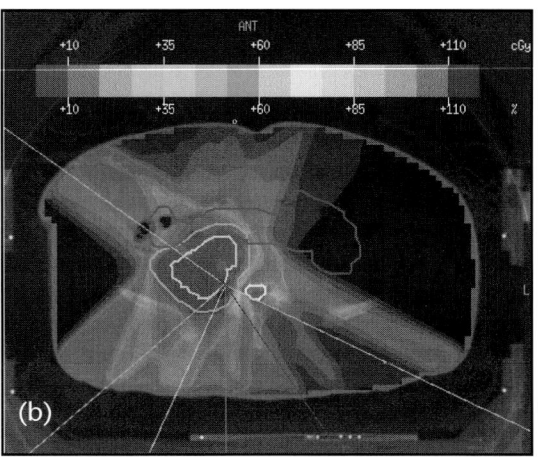

FIGURE 18–4a. A 7-field IMRT plan for a T8 paraspinal case delivering 2000 cGy in 5 fractions. GTV is shown by green, PTV by blue, and cord by red colors. See COLOR PLATE 45.

FIGURE 18–4b. A 6-field IMRT plan for an L4 paraspinal lesion delivering 2000 cGy in 5 fractions. Color code for organs: green is PTV, cyan is GTV, yellow is cord, and blue is bowel. For the dose levels shown, 100% = 100 cGy. See COLOR PLATE 46.

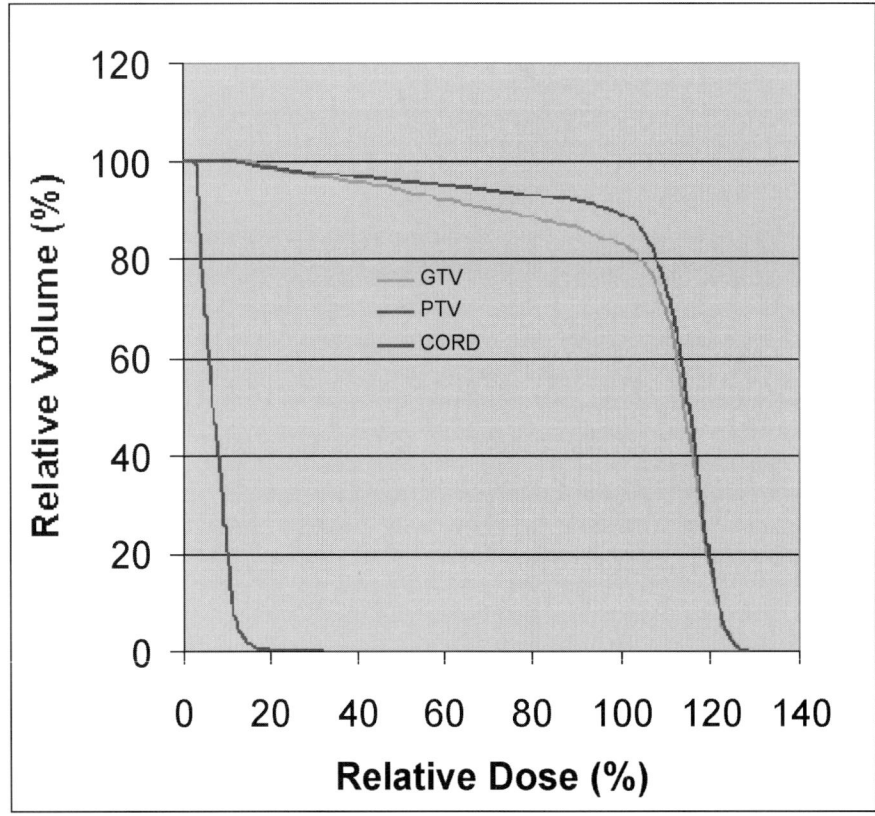

FIGURE 18–5. DVHs for a 7-field IMRT plan for a T8 thoracic paraspinal case. To reduce the cord dose beyond 30% of prescription, physical cord blocks were used for each IMRT field.

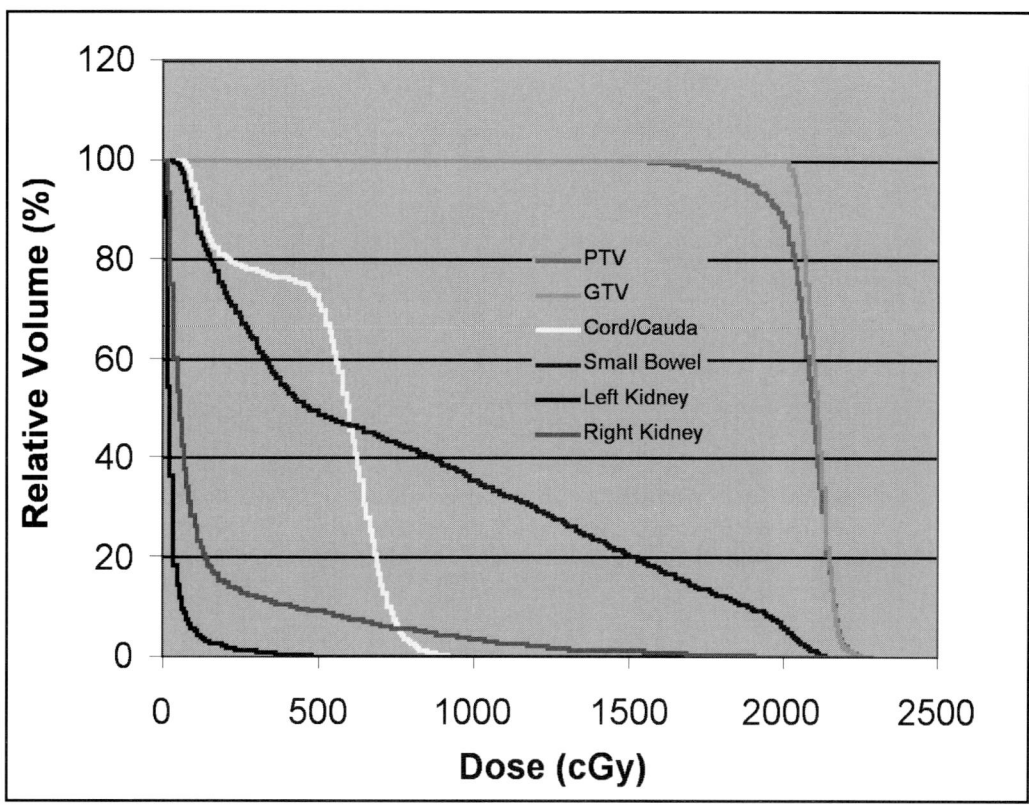

FIGURE 18–6. DVHs for a 6-field IMRT plan for an L4 lumbar paraspinal case.

more uniform dose to a larger volume of PTV for the L4 lesion than the seven-field plan for the T8 lesion where the PTV abuts the cord.

Treatment Setup And QA

For stereotactic setup using the SBF, patient setup is based on the coordinates of the isocenter in the SBF coordinate system. Currently, we use a manual method to register the patient setup coordinates from the CT study. Using the manual technique, the isocenter and other bony land-mark positions are measured on axial CT slices with respect to fixed frame fiducial markers. The coordinates of each landmark are then computed in the SBF ruler system (Yenice et al. 2003). These coordinates define the initial position of the patient with respect to the SBF on the plan-ning CT. On the treatment day, patients are immobilized in the SBF and scanned in the treat-ment room CT scanner to assess their daily treatment position immediately prior to their treatment. This is a relatively short CT scan of approximately 40 slices and takes about 5 to 6 minutes. The manual registration procedure is repeated on this CT study and patient treatment setup coordi-nates are computed. If the daily patient setup were perfect, these coordinates would be exactly the same as those determined from the planning CT. Any shift with respect to the initial coordi-nates from the planning CT gives the daily positioning error. Patients are subsequently transferred from the CT table to linear accelerator table by aligning the two tabletops and smoothly sliding the SBF on the rail system on both couches. The SBF is aligned on the treatment table using the room lasers to the calculated setup coordinates. A final verification of treatment position is done by comparing DRRs from the planning CT with portal images before the treatment is delivered.

Typically, all the beam portals are filmed on the first treatment fraction. On subsequent treatment fractions, only the orthogonal fields are filmed for final verification. On average, daily treatment verification and treatment with CT-image guidance takes 1.25 hours on the first day and 50 minutes for the subsequent fractions.

Stereotactic Setup Accuracy

The overall accuracy of the SBF setup has been measured using a number of fixed radio-opaque targets inside the SBF. These targets were first scanned at the simulator CT scanner and the SBF coordinates of each target were computed from the position of the targets and SBF fiducial localizers on the axial CT slices. Using the SBF rulers, the targets were set up on the treatment machine with room lasers and orthogonal verification images were taken. The deviation between the center of each target and the orthogonal radiation field was measured quantifying the overall error in 3-D space. For 10 randomly placed radio-opaque objects inside the SBF, scanned with 2 mm slice thickness, the standard deviations of error for system accuracy were 0.6 mm in lateral (Lat), 0.8 mm in anterior-posterior (A/P), and 1.0 mm in superior-inferior (SI) dimensions.

For patient setup using the CT-guided SBF localization, we use the edge of bony landmarks, which are not as well imaged as radio-opaque targets. Therefore, we can expect two standard deviations of the phantom measurements as a reasonable estimate of what is achievable for a patient setup. We observed that this was consistent with our final treatment position verification with beam portals that were always within 2 mm of the desired position in all directions.

Patient Setup Errors

Patient setup errors were measured from the uncorrected patient setups based on daily fraction CT studies for seven patients. The mean systematic errors and standard deviations, averaged over all seven patients, were 2.3±1.6mm in Lat, −0.1±2.0 mm in AP, and −0.2±2.4 mm in SI directions. Random setup errors representing day-to-day variation of patient setup averaged over the patient population were approximately 1.5 mm for all directions. We also determined the intra-fractional setup variation for two patients, T8 and L4 lesions, by taking before and after treatment CT scans on each treatment day. This analysis showed that the maximum standard deviation of patient setup using the SBF before and after treatment was 1.4 mm, in the A/P direction. These results indicated that patients were adequately immobilized by the SBF for these hypo-fractionated treatments.

Effect Of Setup Uncertainty On Patient Dose

We included the random and measured systematic errors in the seven-field IMRT plan for a patient with an L4 paraspinal lesion treated with this technique. The random treatment delivery uncertainty was incorporated into the dose calculation by adding a gaussian error function, with a 2 mm standard deviation as an upper limit in each direction, to the pencil beam model as described by Chui, Kutcher, and LoSasso (1992). This resulted in an increase of 10% to 15% in the cord dose with a marginal change in the PTV coverage, when no systematic errors were included. Next, we assumed the daily measured systematic errors ranging in 0.8 to 2.9 mm in Lat, 1.5 to 2.8 mm in A/P, and −0.8 to 3.6 mm in SI directions for this patient were not corrected for. We included these errors in the dose calculations, along with a maximum random error of 2 mm, by moving the patient plan isocenter by amounts equal to the measured errors. This resulted in a dramatic increase in the cord dose by up to 90% of the prescribed cord dose. The directions of systematic errors for this patient were such that when applied, the high dose gradient was moved toward the cord. When the signs of systematic errors were inverted, the high dose gradient moved away from the cord, reducing the cord dose but also degrading the target coverage from the same side. This analysis indicated that precise daily treatment setup was important for highly conformal treatment plans.

Summary

Highly conformal IMRT plans can be delivered to paraspinal tumors with high levels of confidence when patient setup accuracy is guaranteed during the entire treatment procedure. This can be achieved with a stereotactic setup where a stereotactic body frame is used to immobilize the patient and setup on the treatment table. The random and systematic errors with any such a device should be studied carefully to assess the daily delivered dose to the cord. For very conformal plans, such as the IMRT plans generated for concave targets, a 2 mm random error and no systematic error can increase the cord dose by 10% to 15%. This may be of clinical significance in the treatment of recurrent sites, such as many paraspinal patients. If neither systematic nor random errors are corrected, maximum dose received by the spinal cord can exceed 100% of planned dose in some patients. For hypo-fractionated treatment regimes daily image-guidance is feasible for precise treatment delivery. For treatments employing more standard fractionation schemes, periodic correction schemes should be considered to minimize the impact of setup errors.

References

Chavaudra, J., and A. Bridier. (2001). [Definition of volumes in external radiotherapy: ICRU reports 50 and 62]. French. *Cancer Radiother.* 5:472–478.

Chui, C. S., G. J. Kutcher, and T. LoSasso. (1992). "A convolution method for incorporating uncertainties in dose calculations." *Med. Phys.* 19:814.

Fournier-Bidoz, N., P. Giraud, S. Spirou, C. Chui, M. Lovelock, K. Yenice, and M. Hunt. (2001). "Penumbra sharpening with IMRT in paraspinal treatments." AAPM Annual Meeting , Salt Lake City, Utah, July 22–26, 2001, Abstract TU-D-BRB-02. *Med. Phys.* 28(6):1260.

Gilhuijs, K. G., P. J. van de Ven, and M. van Herk. (1996). "Automatic three-dimensional inspection of patient setup in radiation therapy using portal images, simulator images, and computed tomography data." *Med. Phys.* 23:389–399.

Hunt, M. A., C. Y. Hsiung, S. V. Spirou, C. S. Chui, H. I. Amols, and C. C. Ling. (2002). "Evaluation of concave dose distributions created using an inverse planning system." *Int. J. Radiat. Oncol. Biol. Phys.* 54:953–962.

Murphy, M., S. Chang, I. Gibbs et al. (2001). "Image-guided radiosurgery in the treatment of spinal metastases." *Neurosurg. Focus* 11:1–7.

Yenice, K. M., D. M. Lovelock, W. R. Lutz et al. (2000). "Patient immobilization and 3D-conformal therapy for paraspinal tumors." *Radiother. Oncol.* 56(Suppl 1):23–23.

Yenice, K. M., D. M. Lovelock, M. A. Hunt et al. (2003). "CT-image guided intensity modulated therapy of paraspinal tumors using stereotactic immobilization." *Int. J. Radiat. Oncol. Biol. Phys.* 55(3):583–593.

19

Radiation Protection Aspects Of IMRT

James Mechalakos
Jean St. Germain

Introduction

A commercial multileaf collimator (MLC) has been available since 1992 and MLC use continues to expand (Klein et al. 1999). The growth of intensity-modulated radiation therapy (IMRT) has been a consequence of this development. The method of dose delivery employing certain IMRT techniques can directly affect certain radiation safety parameters associated with the treatment accelerator. The question naturally arises as to whether the National Council on Radiation Protection and Measurements (NCRP) recommendations made in the mid-1970's need to be modified, and what other radiation safety issues might be affected by this technique that is rapidly growing in popularity.

The term "IMRT" can be applied to any technique in which the beam intensity is varied across the field, from the simple use of a wedge compensator to tomotherapy. This chapter will primarily focus on the use of IMRT techniques employing a dynamic multileaf collimator (DMLC) in which field modulation is accomplished by the moving leaves of a multileaf collimator. We shall also briefly mention two other types of IMRT using multileaf collimation that are discussed in radiation safety literature on the subject:

- Segmented multileaf collimation (SMLC), in which the field modulation is achieved by summing a number of static MLC fields, and
- Tomotherapy, in which consecutive "slices" of the patient's body are treated using an arc rotation technique.

Facility Shielding

If a medical facility is considering adding a treatment room that will include IMRT treatments, or if an existing treatment room is being upgraded to provide IMRT, the radiation safety/medical physicist must have an idea of the contribution that IMRT will make to the overall output of the machine, as this will affect shielding and patient monitoring. The effect of IMRT on shielding calculations as originally outlined by NCRP Reports 49 and 51 will be discussed in the following paragraphs (NCRP 49, NCRP 51).

Recommendations Of The NCRP

NCRP Reports 49 and 51 were published in 1976 and 1977 respectively. These reports recommend methodology for calculation of structural shielding and radiation protection guidelines for medical use of X-rays and gamma rays. These reports were supplemented with NCRP Report 102; and the methodology for shielding calculations presented in NCRP Report 49 and NCRP Report 51 has been used consistently over the last 25 years (NCRP 102). Both NCRP Report 49 and NCRP Report 51 are currently under revision. The pertinent relationships from these reports are summarized below.

The transmission factor, B, for a primary barrier (NCRP 49), i.e., a barrier sufficient to attenuate the useful beam to the required degree, is given by

$$B = \frac{P(d_{pri})^2}{WUT} \qquad (1)$$

where:

P = permissible weekly exposure rate on the other side of the barrier in roentgens (rad or cGy are used for clarity)

d_{pri} = distance from radiation source to person to be protected in meters

W = workload of machine at 1 meter from the "source" (rad or cGy/wk at 1 m)

U = use factor, fraction of time that the primary beam points at the barrier

T = occupancy factor, fraction of time that the area to be shielded is occupied.

The transmission factor for a secondary barrier, i.e., a barrier sufficient to attenuate the leakage to the required degree, is given in NCRP Report 49 by

$$B = \frac{1000P(d_{sec})^2}{WT} \qquad (2)$$

where d_{sec} is the distance from the radiation source to the person to be protected in meters. The factor of 1000 accounts for the minimum required attenuation of the head shielding. The other factors are defined as above.

Workload

It is obvious from even a cursory examination of NCRP Reports 49 and 51 that the values calculated by the equations above will be driven by the workload assumptions. The workload is directly related to the degree of use of either an X-ray or gamma ray source. The *standard assumptions* state that above 4 MV, workload is expressed as the weekly exposure of the useful beam at 1 meter from the source, in units of R m^2 wk^{-1} or rad or cGy m^2 wk^{-1} for X-rays. NCRP Report 49 recommends a workload of 100,000 rad or cGy m^2 wk^{-1} for machines up to 10 MV and NCRP Report 51 recommends 50,000 rad or cGy m^2 wk^{-1} for machines operating above 10 MV.

If one monitor unit (MU) is defined as the dose to water at a depth of d$_{max}$ for a 10 × 10 cm^2 field at 100 cm SAD (source-axis distance), the number of monitor units per week, MU/wk, can be used as an approximation of the workload for conventional treatments [as an example, see Kleck, Elsalim, and Kase (1994)]. For IMRT treatments, the number of MUs per field can increase substantially as compared to conventional treatment while exposure of the useful beam at 1 m from the source does not. This is due to the fact that only a small dynamic aperture is open at any particular time to allow for intensity modulation; therefore, a longer *beam-on* time is typically required to deliver the same prescribed dose to the entire area than would be required for a typical conventional treatment.

The factor by which the number of MUs is increased for IMRT fields depends on the type of IMRT delivery and the complexity of the treatment plan. A 1993 study by Kleck, Elsalim, and Kase (1994) reported total workloads of approximately 20,000 to 50,000 MU/wk. Using treatment machine schedules and patient charts, Mechalakos and St. Germain (2002) compiled a summary of workloads of five linear accelerators for a typical work week and found that the machines treating predominantly IMRT had a typical weekly workload close to 100,000 MUs, approximately a factor of 2 greater than machines using primarily conventional treatments. In a more comprehensive study covering one year's worth of treatments for three linear accelerators, the same authors had comparable results: non-IMRT machines treated approximately 50,000 MU/wk while the predominantly IMRT machine treated close to 100,000 MU/wk (Mechalakos, St. Germain, and Burman 2003). In these two studies, the prostate was the predominant site being treated with IMRT. In the larger study, an average of 578 MU was used per IMRT treatment, almost a factor of 2 higher than non-IMRT treatments, both wedged and unwedged. Other treatment sites may give rise to higher or lower numbers of MUs per plan, depending on the complexity of the case and the size of the treatment field.

Mutic et al. (2001) reported an increase of a factor of approximately 2.5 in MU for the use of DMLC treatments compared to conventional treatments. Followill, Geis, and Boyer (1997) estimated an increase of a factor of 2.8 for SMLC treatments over conventional unwedged treatments. The smaller factor reported by Mechalakos and St. Germain is partially due to the inclusion of wedged treatments in the non-IMRT group. A number of studies have reported MU increases in tomotherapy delivery as well (Followill, Geis, and Boyer 1997; Grant et al. 1994; Mutic and Low 1998; Verellen and Vanhavere 1999; Meeks et al. 2002). The Intensity-Modulated Radiation Therapy Collaborative Working Group (IMRT CWG) suggested monitor unit increases of factors between 2 and 5 depending on which intensity modulation technique is employed (IMRT CRG 2001).

Besides increases in the number of MUs per plan, increases in patient throughput need to be considered. IMRT treatments do not generally require therapists to enter the room once the patient has been set up, as would be required using wedges. With the advent of timesaving technologies such as automatic field sequencing and electronic portal imaging, treatments should become more and more streamlined, possibly allowing for shorter appointments. In short, the duty cycle of the treatment machine can be expected to increase.

Dose rate increases need to be considered as well. For example, respiratory gating techniques employed at some institutions are currently treating patients at a dose rate of 600 MU/min. As more of such technologies become available, dose rate improvement should be expected to continue. Finally, as also observed by the IMRT CWG, increases in dose rate can also be expected when calibrating machines at a depth of 10 cm rather than at d_{max}, a technique that results in 1.3 to 1.4 times the dose to d_{max} per MU (IMRT CWG 2001).

Mechalakos and St. Germain concluded that based on current results at their institution using DMLC, the NCRP estimate of 100,000 rad or cGy/wk is sufficiently conservative; however the energy dependence of the recommendations is dated. It is interesting to note that the 100,000 rad or cGy/wk NCRP recommendation was based on a load of 50 patients/day. In the studies by Mechalakos et al. (2002, 2003), the treatment machines reviewed rarely reached 40 patients per day.

Effect Of Increase In Workload On Shielding

The increase in workload of the machine does not necessarily lead to an increase in shielding. The combination of higher workload without a significantly higher integral dose delivery has to be considered when making shielding recommendations. When considering the primary barrier, if approximately the same integral dose is delivered to the patient using IMRT as with conventional treatment, the primary barrier will see a comparable amount of radiation on a per patient basis. Previously published reports are in agreement (Mutic et al. 2001; IMRT CWG 2001). One must, however, be cognizant of changes in throughput and dose rate that **will** increase the dose to the primary barrier since more patients are being treated more quickly. Mutic et al. (2001) proposed a shielding calculation formalism in which the workload in equation (1) was estimated by the weekly tumor dose for the purposes of primary barrier calculation. This proposal had the advantage of allowing for more MUs per patient yet did not affect the primary barrier thickness, all other things remaining equal.

The amount of leakage radiation and neutron contamination can be expected to increase. Since both are produced in the head before the beam strikes the MLC, leakage radiation and neutron contamination will increase by the same factor as the number of monitor units. In this case, the workload in equation (2) and in neutron contamination calculations (see NCRP 79) will be proportional to the number of monitor units delivered. Since the amount of radiation reaching the patient does not change significantly for the same prescription, the amount of radiation scattered from a patient will not change significantly. Again, higher throughput and dose rate will increase the scatter as it does radiation reaching the primary barrier because of the increase in the number of patients.

In summary, a higher number of monitor units per plan does not significantly affect the radiation reaching the primary barrier and the scattered radiation per patient as long as the prescription remains the same. However the amount of leakage radiation will be increased depending on the increase in monitor units, as will neutron contamination for higher energy beams.

Use Factors

The NCRP recommended use factors of 1 for the floor, 1/4 for the walls, and not more than 1/4 for the ceiling. In some cases, IMRT treatment plans for a particular site will use a standard set of gantry angles. In the second study by Mechalakos and St. Germain, the IMRT treatment room was using IMRT only for prostate cases and employed a prescribed set of five gantry angles in routine plans (Mechalakos, St. Germain, and Burman 2003). This resulted in those five angles dominating the angular distribution of radiation in the room, rather than the "standard" 0°, 90°, 180°, and 270°. Although facility design cannot rely on only certain types of treatments, one should be aware that this type of situation may arise.

New Machine vs. Upgrade

If one is planning on acquiring a new treatment machine that will treat using IMRT, it is important to consider a conservative design that will allow for increases in output and duty cycle. In the one-year survey of linear accelerator workloads reported by Mechalakos et al., the IMRT machine workload for IMRT techniques increased from 50% to 90%, and the total workload increased from approximately 70,000 MU/wk to almost 100,000 MU/wk (Mechalakos, St. Germain, and Burman 2003). If one is upgrading an existing machine to treat with IMRT, it is recommended that the radiation safety/medical physicist review the room surveys to see if the measured scattered radiation levels will support an increase in workload. If there is any doubt, it is recommended that the room be resurveyed. One can also monitor the workload using the record and verify (R&V) system for a specified period of time after IMRT treatments begin to determine machine workloads and use factors. If the R&V system is capable of formatted output, one can easily import data into a standard spreadsheet program and summarize the data as desired.

Whole Body Dose To Patients

With the increase in head leakage and neutron contamination for IMRT described above, one must also expect an increase in dose to the patient outside the treatment field. Most estimates of whole body dose from IMRT have been for tomotherapy, however Followill also estimated a whole body dose equivalent for SMLC of 190 mSv at 6 MV and 911 mSv for 18 MV assuming a prescription of 70 Gy to the isocenter (Followill, Geis, and Boyer 1997; Mutic and Low 1998; Verellen and Vanhavere 1999; Meeks et al. 2002). Using the lifetime risk for fatal secondary cancer of 5.0×10^{-2}/Sv found in NCRP Report 116, they estimated a secondary cancer risk of 1.0% and 4.5% for 6 MV and 18 MV respectively, slightly higher than that of conventional treatment (NCRP 116). A summary of this and other whole body dose estimates can be found in the IMRT Collaborative Working Group report (IMRT CWG 2001).

Personnel Monitoring

Personnel monitoring has a dual role: first, monitoring the dose received by occupational workers and second, monitoring the radiation environment in which staff are functioning. The dose to radiation workers in radiation oncology would not be expected to approach the occupational dose limits, and for external beam radiotherapy, this dose is typically defined by the structural shielding placed around the treatment unit. NCRP Reports 49 and 51 allowed structural shielding to be designed to reflect the occupational dose limits. This criterion was based on the assumption that the overall shielding design would be sufficiently conservative so that the actual measured doses would be below the occupational dose limits. As discussed previously, a significant change in workload could require a re-evaluation of this assumption. Both reports have been under revision for some time, and the new reports may reflect more conservative dose assumptions. Many, if not most, installations place the control desk on the other side of a secondary barrier. In these cases, the leakage radiation limits would be expected to be the limiting design factor as discussed previously.

Since many installations have several years' worth of occupational dose monitoring records, it should be possible to compare staff doses before and after the introduction of IMRT techniques. It should also be possible to use personnel monitors as area monitors to assess the maximum dose that could be received by occupational staff if there is a large staff that rotates among various treatment units.

References

Followill, D., P. Geis, and A. Boyer. (1997). "Estimates of whole-body dose equivalent produced by beam intensity modulation conformal therapy." *Int. J. Radiat. Oncol. Biol. Phys.* 38:667–672.

Grant III, W., A. Bleier, C. Campbell et al. (1994). "Leakage considerations with a multileaf collimator designed for intensity-modulated conformal radiotherapy." Abstract. *Med. Phys.* 21:921.

IMRT CWG. Intensity-Modulated Radiotherapy Collaborative Working Group. "Intensity modulated radiotherapy: Current status and issues of interest." *Int. J. Radiat. Oncol. Biol. Phys.* 51:880–914.

Kleck, J., M. Elsalim, and K. Kase. (1994). "Clinical workloads and use factors for medical linear accelerators." *Med. Phys.* 21:952.

Klein, E. E., J. Tepper, M. Sontag, M. Franklin, C. Ling, and D. Hubo. (1999). "Technology assessment of multileaf collimation: A North American users survey." Int. J. Radiat. Oncol. Biol. Phys. 44:705–710.

Mechalakos, J. G., and J. St. Germain. (2002). "Estimation of shielding factors for linear accelerators." *Oper. Radiat. Safety*, November 2002.

Mechalakos, J. G., J. St. Germain, and C. M. Burman. (2003). "One year survey of output for IMRT and non-IMRT linear accelerators." Submitted for publication, 2003.

Meeks, S. L., A. C. Paulino, E. C. Pennington, J. H. Simon, M. W. Skwarchuk, and J. M. Buatti. (2002). "In vivo determination of extra-target doses received from serial tomotherapy." *Radiother. Oncol.* 63:217–222.

Mutic, S., and D. A. Low. (1998). "Whole-body dose from tomotherapy delivery." *Int. J. Radiat. Oncol. Biol. Phys.* 42:229–232.

Mutic, S., D. A. Low, E. E. Klein, J. F. Dempsey, and J. A. Purdy. (2001). "Room shielding for intensity-modulated radiation therapy treatment facilities." *Int. J. Radiat. Oncol. Biol. Phys.* 50:239–246.

NCRP 49. National Council on Radiation Protection and Measurements Report 49. "Structural Shielding Design and Evaluation for Medical Use of X-rays and Gamma Rays of Energies Up to 10 MeV." Bethesda, MD: NCRP, 1976.

NCRP 51. National Council on Radiation Protection and Measurements Report 51. "Radiation Protection Design Guidelines for 0.1-100 MeV Particle Accelerator Facilities." Bethesda, MD: NCRP, 1977.

NCRP 79. National Council on Radiation Protection and Measurements Report 79. "Neutron Contamination from Medical Electron Accelerators." Bethesda, MD: NCRP, 1984.

NCRP 102. National Council on Radiation Protection and Measurements Report 102. "Medical X-Ray, Electron Beam and Gamma Ray Protection for Energies up to 50 MeV (Equipment Design, Performance and Use)." Bethesda, MD: NCRP, 1989.

NCRP 116. National Council on Radiation Protection and Measurements, Report 116. "Limitation of Exposure to Ionizing Radiation." Bethesda, MD: NCRP, 1993.

Verellen, D., and F. Vanhavere. (1999). "Risk assessment of radiation induced malignancies based on whole-body equivalent dose estimates for IMRT treatment in the head and neck region." *Radiother. Oncol.* 53:199–203.

List Of Terms And Abbreviations

A/D	Analog-to-digital
a-Si	Amorphous silicon
AAPM	American Association of Physicists in Medicine
ABC	Active breathing control
AP/PA	Anterior-posterior/posterior-anterior
ART	Adaptive radiation therapy
BED	Biologically effective dose
BEV	Beam's eye view
BOLD	Blood oxygenation level-dependent
BOT	Beam-on-time
BTV	Biological target volume
CAX	Central axis
cc	Cubic centimeter
CCD	Charge-coupled device
cGy	Centigray
cm	Centimeter
CNS	Central nervous system
COG	Center of gravity
CSF	Collimator scatter factor
CT	Computed tomography
CTV	Clinical target volume
DI	Deep inspiration
DIBH	Deep inspiration breath hold
DICOM	Digital Imaging and Communication in Medicine
DLL	Dynamic link library
d_{max}	Depth of maximum dose
DMLC	Dynamic multileaf collimator
DNA	Deoxyribonucleic acid
DRR	Digitally-reconstructed radiograph
DVA	Leaf sequence file
DVH	Dose-volume histogram
E-E	End-expiration
E-I	End-inspiration
EBCTC	Early Breast Cancer Trialists Collaboration (Group)
EF5	2-(2-nitro-1H-imidazol-1-yl)-N-(2,2,3,3,3-pentafluoropropyl acetamide

EORTC	European Organization for Research and Treatment of Cancer
EPI	Echo-planar dynamic imaging
EPID	Electronic portal imaging device
EUD	Equivalent uniform dose
4-D	Four-dimensional
FAZA	2-nitronidazole fluoro-azomycin arabinoside
FB	Free-breathing
fdam	Fraction of functional lung units damaged
FDG	Fluorodeoxyglucose radiolabeled with ^{18}F
FETNIM	F18-fluoro-erythronitroimidazole
FMISO	Fluorine-18-labeled fluoro-misonidazole
FOV	Field-of-view
FSU	Functional sub-unit
FWHM	Full-width-half-maximum
GABA	Gamma aminobutyric acid
Gd-DTPA	Gadolinium diethylenetriaminepenta-acetic acid,
gm	Gram
GTV	Gross tumor volume
Gy	Gray
HF	Hypoxic fraction
HIF	Hypoxia-inducible factor
IADZ	Iodo-azomycin-galactoside
IAS	Image acquisition system
IAZA	Iodo-azomycin arabinoside
ICRU	International Commission on Radiation Units and Measurements
IDU	Image detection unit
IM	Intensity-modulated
IMRT	Intensity-modulated radiation therapy
IMRT CWG	IMRT Collaborative Working Group
IRS IV	Fourth Intergroup Rhabdomyosarcoma Study
ITP	Inverse treatment plan
kV	Kilovolt
LAO	Left anterior oblique
Lat	Lateral
LCWI	Lung-chest wall interface
LED	Light emitting diode
LET	Linear energy transfer
linac	Linear accelerator
LKB	Lyman-Kutcher-Burman (model)
LL	Left lateral
LPO	Left posterior oblique
LQ	Linear quadratic
LR	Left rear
MACH-NC	Meta-analysis of Chemotherapy on Head and Neck Cancer Collaboration Group
MFC	Microsoft Foundation Class
MGH	Massachusetts General Hospital
MHz	Megahertz
ML	Medial lateral

MLC	Multileaf collimator
mm	Millimeter
MR	Magnetic resonance
MRI	Magnetic resonance imaging
MRS	Magnetic resonance spectroscopy
MRSI	Magnetic resonance spectroscopy/imaging
MSKCC	Memorial Sloan-Kettering Cancer Center
MU	Monitor unit
MV	Megavolt
NCI	National Cancer Institute
NCRP	National Council on Radiation Protection and Measurements
NMR	Nuclear magnetic resonance
NSCLC	Non small cell lung cancer
NTCP	Normal tissue complication probability
NTP	Nucleoside triphosphates
1-D	One-dimensional
OAR	Organ-at-risk
OCR	Off-center ratio
OF	Objective Functions
OPD	Occupancy probability distribution
PA	Posterior-anterior
PC	Personal computer
PC	Phosphocholine
pcf	Phantom correction phantom
PCR	Phosphocreatine
PE	Phosphoethanolamine
PET	Positron emission tomography
P_i	Inorganic phosphates
pOCR	Primary off-center ratio
PSA	Prostate-specific antigen
PSF	Phantom scatter factor
PTV	Planning target volume
PV	PortalVision™
QA	Quality assurance
R&V	Record and verify (system)
RAO	Right anterior oblique
RCCT	Respiration-correlated spiral computed tomography
RDU	Relative dose units
RF	Radiofrequency
RG	Respiratory gating
RILD	Radiation induced liver disease
RL	Right lateral
rms	Root mean square
RNA	Ribonucleic acid
RPM	Real-time position management (system)
RPO	Right posterior oblique
RT	Radiation therapy
RTCT	Respiration-triggered computed tomography

RTOG	Radiation Therapy Oncology Group
RTRT	Real-time radiotherapy tracking (system)
SAD	Source-axis distance
SBF	Stereotactic body frames
Sc	Collimator scatter factor
SD	Standard deviation
SDD	Source-to-detector distance
SF	Surviving fraction
SI	Sacroiliac
SI	Superior-inferior
sIMRT	Simplified intensity-modulated radiation therapy
SLIC	Scanning liquid ionization chamber
SMART	Simultaneous Modulated Accelerated Radiation Therapy
SMLC	Segmented or step-and-shoot multileaf collimator technique
Sp	Phantom scatter factor
SPECT	Single photon emission computed tomography
SSD	Source-surface distance
SUV	Specific uptake value
SVC	Slow vital capacity
2-D	Two-dimensional
3-D	Three-dimensional
3DCRT	Three-dimensional conformal radiation therapy
T&G	Tongue and groove
TCD_{50}	Dose required to achieve a TCP at 50%
TCP	Tumor control probability
TFT	Thin-film transistor
TG	Task Group
TLD	Thermoluminescent dosimeter
TLD	Thermoluminescent dosimetry
TMR	Tissue-maximum ratio
TPR	Tissue-phantom ratio
UCSF	University of California at San Francisco
VFC	Varian Foundation Class
Veff	Effective volume
vIMRT	Volume-based intensity-modulated radiation therapy
WAR	Whole abdomen radiation

Index